SS521-AG-PRO-010

0910-LP-106-0957

REVISION 6

U.S. Navy Diving Manual

Volume 1:	Diving Principles and Policies
Volume 2:	Air Diving Operations
Volume 3:	Mixed Gas Surface Supplied Diving Operations
Volume 4:	Closed-Circuit and Semiclosed Circuit Diving Operations
Volume 5:	Diving Medicine and Recompression Chamber Operations

SUPERSEDES SS521-AG-PRO-010, REVISION 5, Dated 15 August 2005.

PUBLISHED BY DIRECTION OF COMMANDER, NAVAL SEA SYSTEMS COMMAND

15 APRIL 2008

Published by Books Express Publishing

ISBN 978-1-907521-20-1
To purchase copies at discounted prices please contact
info@books-express.com

LIST OF EFFECTIVE PAGES

Date of issue for original is:

Original . . . . 15 April 2008

TOTAL NUMBER OF PAGES IN THIS PUBLICATION IS 992, CONSISTING OF THE FOLLOWING:

Page No.	*Change No.
Title Page	0
Title Page-2 blank	0
A through B	0
Certification Sheet	0
Certification Sheet-2 blank	0
Record of Changes	0
Record of Changes-2 blank	0
i (Foreword)	0
ii blank	0
iii through vii	0
viii blank	0
ix through lvii	0
lviii blank	0
lix through lxii	0
Vol. 1 Title Page	0
Vol. 1 Title Page-2 blank	0
1-i through 1-xiii	0
1-xiv blank	0
1-1 through 1-31	0
1-32 blank	0
2-1 through 2-36	0
3-1 through 3-63	0
3-64 blank	0
4-1 through 4-13	0
4-14 blank	0
5-1 through 5-13	0
5-14 blank	0
1A-1 through 1A-16	0
1B-1 through 1B-3	0
1B-4 blank	0
1C-1	0
1C-2 blank	0
1D-1 through 1D-9	0
1D-10 blank	0
Vol. 2 Title Page	0
Vol. 2 Title Page-2 blank	0
2-i through 2-xv	0
2-xvi blank	0
2-xvii	0
2-xviii blank	0
6-1 through 6-58	0
7-1 through 7-40	0
8-1 through 8-36	0
9-1 through 9-84	0
10-1 through 10-13	0
10-14 blank	0
11-1 through 11-14	0
2A-1 through 2A-3	0
2A-4 blank	0
Vol. 3 Title Page	0
Vol. 3 Title Page-2 blank	0
3-i through 3-vii	0
3-viii blank	0
3-ix	0
3-x blank	0
3-xi	0
3-xii blank	0
12-1 through 12-14	0

* Zero in this column indicates an original page.

Page No.	*Change No.	Page No.	*Change No.
13-1 through 13-13	0		
13-14 blank	0		
14-1 through 14-35	0		
14-36 blank	0		
15-1 through 15-39	0		
15-40 blank	0		
16-1 through 16-9	0		
16-10 blank	0		
Vol. 4 Title Page	0		
Vol. 4 Title Page-2 blank	0		
4-i through 4-vii	0		
4-viii blank	0		
4-ix	0		
4-x blank	0		
4-xi	0		
4-xii blank	0		
17-1 through 17-61	0		
17-62 blank	0		
18-1 through 18-57	0		
18-58 blank	0		
19-1 through 19-25	0		
19-26 blank	0		
Vol. 5 Title Page	0		
Vol. 5 Title Page-2 blank	0		
5-i through 5-xi	0		
5-xii blank	0		
20-1 through 20-49	0		
20-50 blank	0		
21-1 through 21-31	0		
21-32 blank	0		
5A-1 through 5A-13	0		
5A-14 blank	0		
5B-1 through 5B-7	0		
5B-8 blank	0		
5C-1 through 5C-23	0		
5C-24 blank	0		
Index-1 through Index-12	0		

* Zero in this column indicates an original page.

NAVSEA TECHNICAL MANUAL CERTIFICATION SHEET

Certification Applies to: New Manual ☐ Revision ☒ Change ☐

Applicable TMINS/Pub. No. SS521-AG-PRO-010/0910-LP-106-0957

Publication Date (Da, Mo, Yr) 15 April 2008

Title: U.S. NAVY DIVING MANUAL, Revision 6

TMCR/TMSR/Specification No.:

CHANGES AND REVISIONS:

Purpose: This revision provides new procedures for decompression using air and/or oxygen.

Equipment Alteration Numbers Incorporated:

TMDER/ACN Numbers Incorporated:

Continue on reverse side or add pages as needed.

CERTIFICATION STATEMENT

This is to certify that responsible NAVSEA activities have reviewed the above identified document for acquisition compliance, technical coverage, and printing quality. This form is for internal NAVSEA management use only, and does not imply contractual approval or acceptance of the technical manual by the Government, nor relieve the contractor of any responsibility for delivering the technical manual in accordance with the contract requirement.

Authority	Name	Signature	Organization	Code	Date
Acquisition	R. Whaley	Robert C Whaley	NAVSEA	00C3	1 Apr 2008
Technical	CAPT J. Gray	Gray	NAVSEA	00C3B	4/1/08
Printing Release					

DERIVED FROM NAVSEA 4160/8 (5 - 89)

PAGE LEFT BLANK INTENTIONALLY

RECORD OF CHANGES

CHANGE NO.	DATE OF CHANGE	TITLE AND/OR BRIEF DESCRIPTION	ENTERED BY

PAGE LEFT BLANK INTENTIONALLY

Foreword

Department of the Navy
Naval Sea Systems Command
14 April 2008

Revision Six of the U.S. Navy Dive Manual represents the largest change to diving procedures since 1956. The manual must be reviewed in its entirety, but please pay particular attention to Chapters 9, 14, 17, 18, 20, and 21.

We would like to thank the Fleet. We received thoughtful and comprehensive input from multiple Commands that were ready to embrace change. The Fleet deserves a great deal of credit. Here at NAVSEA, the counsel of my teachers, MDV Pratschner, MDV Boyd, and MDV Orns, representing over 75 years of experience, was vital to our transition. My teammates Don Fegley and John Gillispie, who make technical problems disappear, kept us on schedule. My friend, LT Jim Pearson, Royal Navy, gave us the perspective necessary to achieve our goals. The sheer intelligence of my mentors, Dr. Bob Whaley and CAPT John Murray, guaranteed academic rigor.

Away from home, both NEDU and the Dive School played pivotal roles in drafting and fielding this revision. I would like to thank my friends, CAPT Steve Reimers, CAPT Randy Getman and CDR Chris Moore, for their support.

More than anyone else, this revision is the triumph of 40 years of loyal dedication to our country by Dr. Ed Flynn.

The work on Revision Seven starts today. Revision Seven will completely change the format of the Dive Manual, eliminating duplications and inconsistencies. Most of all, it will be organized so that all requirements are found in the first chapter.

J. G. GRAY
Supervisor of Diving

PAGE LEFT BLANK INTENTIONALLY

Prologue

Department of the Navy
Naval Sea Systems Command
15 June 2008

Since the late 1950's, the U.S. Navy Diving Manual served as the internationally recognized standard for allowable exposure while breathing compressed air at varying depths. The 1956/1957 USN Diving Manual "Air Tables" also provided the prescribed decompression schedules for dive profiles that exceeded allowable exposure limits.

Recently, three events caused the Navy to revisit the 1956/1957 Air Tables and led to this revision of the diving manual.

1. Unacceptable rates of decompression sickness experienced during the salvage of TWA800 in 1996 while using the 1956/1957 USN Diving Manual "Air Tables" and Sur "D" O_2 Tables at selected depth ranges above 100 feet.
2. Commercial diving practices which migrated away from the1956/1957 USN Diving Manual "Air Tables" due to unacceptable rates of decompression sickness at long endurance/shallow depth dives.
3. Most importantly, the seminal research into hyperbaric medicine led by Dr. Edward D. Thalmann, Captain, MC, USN (Ret.) that developed a mathematical algorithm that modeled gas exchange in human tissues offered the promise of less risky guidelines for air diving.

In 1998, the Navy Experimental Diving Unit began the development process to replace the 1956/1957 Air Tables and led to the selection of the Thalmann Algorithm as the basis of a new Air Table. A variation of the Thalmann Algorithm was eventually adopted as a result of empirical testing and operational Fleet feedback.

Revision 6 of the U.S. Navy Diving Manual represents the culmination of research and empirical validation of the air diving tables contained herein. Aside from Thalmann, the efforts of two individuals stand above all others in analyzing and meticulously documenting the end product for this most comprehensive revision to the U. S. Navy Diving Manual in fifty years. Those individuals are Dr. Edward T. Flynn, Captain, MC, USN, (Ret.) of the Office of the Supervisor of Salvage & Diving and Dr. Wayne A. Gerth, Ph.D. of the Navy Experimental Diving Unit. Future Navy and civilian divers as yet unborn owe these men a debt of gratitude for the inherently safer diving that will result from their untiring labor and razor sharp intellect.

Richard Hooper
Director of Ocean Engineering
Supervisor of Salvage and Diving

PAGE LEFT BLANK INTENTIONALLY

Safety Summary

STANDARD NAVY SYNTAX

Since this manual will form the technical basis of many subsequent instructions or directives, it utilizes the standard Navy syntax as pertains to permissive, advisory, and mandatory language. This is done to facilitate the use of the information provided herein as a reference for issuing Fleet Directives. The concept of word usage and intended meaning that has been adhered to in preparing this manual is as follows:

"Shall" has been used only when application of a procedure is mandatory.

"Should" has been used only when application of a procedure is recommended.

"May" and "need not" have been used only when application of a procedure is discretionary.

"Will" has been used only to indicate futurity; never to indicate any decree of requirement for application of a procedure.

The usage of other words has been checked against other standard nautical and naval terminology references.

GENERAL SAFETY

This Safety Summary contains all specific WARNINGS and CAUTIONS appearing elsewhere in this manual and are referenced by page number. Should situations arise that are not covered by the general and specific safety precautions, the Commanding Officer or other authority will issue orders, as deemed necessary, to cover the situation.

SAFETY GUIDELINES

Extensive guidance for safety can be found in the OPNAV 5100 series instruction manual, Navy Safety Precautions.

SAFETY PRECAUTIONS

The WARNINGS, CAUTIONS, and NOTES contained in this manual are defined as follows:

WARNING **Identifies an operating or maintenance procedure, practice, condition, or statement, which, if not strictly observed, could result in injury to or death of personnel.**

CAUTION **Identifies an operating or maintenance procedure, practice, condition, or statement, which, if not strictly observed, could result in damage to or destruction of equipment or loss of mission effectiveness, or long-term health hazard to personnel.**

NOTE **An essential operating or maintenance procedure, condition, or statement, which must be highlighted.**

WARNING Voluntary hyperventilation is dangerous and can lead to unconsciousness and death during breathhold dives. (Page 3-20)

WARNING Never do a forceful Valsalva maneuver during descent. A forceful Valsalva maneuver can result in alternobaric vertigo or barotrauma to the inner ear. (Page 3-25)

WARNING If decongestants must be used, check with medical personnel trained in diving medicine to obtain medication that will not cause drowsiness and possibly add to symptoms caused by the narcotic effect of nitrogen. (Page 3-25)

WARNING Reducing the oxygen partial pressure does not instantaneously reverse the biochemical changes in the central nervous system caused by high oxygen partial pressures. If one of the early symptoms of oxygen toxicity occurs, the diver may still convulse up to a minute or two after being removed from the high oxygen breathing gas. One should not assume that an oxygen convulsion will not occur unless the diver has been off oxygen for 2 or 3 minutes. (Page 3-44)

WARNING CPR should not be initiated on a severely hypothermic diver unless it can be determined that the heart has stopped or is in ventricular fibrillation. CPR should not be initiated in a patient that is breathing. (Page 3-55)

WARNING Do not use a malfunctioning compressor to pump diver's breathing air or charge diver's air storage flasks as this may result in contamination of the diver's air supply. (Page 4-11)

WARNING Welding or cutting torches may cause an explosion on penetration of gas-filled compartments, resulting in serious injury or death. (Page 6-22)

WARNING SCUBA equipment is not authorized for use in enclosed space diving. (Page 6-27)

WARNING These are the minimum personnel levels required. ORM may require these personnel levels be increased so the diving operations can be conducted safely. (Page 6-31)

WARNING Skip-breathing may lead to hypercapnia and is prohibited. (Page 7-30)

WARNING During ascent, the diver without the mouthpiece must exhale to offset the effect of decreasing pressure on the lungs which could cause an air embolism. (Page 7-36)

WARNING During enclosed space diving, all divers shall be outfitted with a MK 21 MOD 1, KM-37, MK 20 MOD 0, or EXO BR MS that includes a diver-to-diver and diver-to-topside communications system and an EGS for the diver inside the space. (Page 8-29)

WARNING For submarine ballast tanks, the divers shall not remove their diving equipment until the atmosphere has been flushed twice with air from a compressed air source meeting the requirements of **Chapter 4**, or the submarine L.P. blower, and tests confirm that the atmosphere is safe for breathing. Tests of the air in the enclosed space shall be conducted hourly. Testing shall be done in accordance with NSTM 074, Volume 3, Gas Free Engineering (S9086-CH-STM-030/CH-074) for forces afloat, and NAVSEA-S-6470-AA-SAF-010 for shore-based facilities. If the divers smell any unusal odors they shall immediately don their EGS. (Page 8-29)

WARNING If the diving equipment should fail, the diver shall immediately switch to the EGS and abort the dive. (Page 8-29)

WARNING If job conditions call for using a steel cable or a chain as a descent line, the Diving Officer must approve such use. (Page 8-32)

WARNING The interval from leaving 40 fsw in the water to arriving at 50 fsw in the chamber cannot exceed 5 minutes without incurring a penalty. (See **paragraph 9-12.6**.) (Page 9-16)

WARNING These procedures cannot be used to make repetitive dives on air following MK 16 helium-oxygen dives. (Page 9-29)

WARNING **Table 9-4** cannot be used when diving with equipment that maintains a constant partial pressure of oxygen such as the MK 16 MOD 0 and the MK 16 MOD 1. Consult NAVSEA 00C for specific guidance when diving the MK 16 at altitudes greater than 1000 feet. (Page 9-47)

WARNING Altitudes above 10,000 feet can impose serious stress on the body resulting in significant medical problems while the acclimatization process takes place. Ascents to these altitudes must be slow to allow acclimatization to occur and prophylactic drugs may be required to prevent the oocurrence of altitude sickness. These exposures should always be planned in consultation with a Diving Medical Officer. Commands conducting diving operations above 10,000 feet may obtain the appropriate decompression procedures from NAVSEA 00C. (Page 9-50)

WARNING Mixing contaminated or non-oil free air with 100% oxygen can result in a catastrophic fire and explosion. (Page 10-10)

WARNING The interval from leaving 40-fsw in the water to arriving at 50-fsw in the chamber cannot exceed 5 minutes without incurring a penalty. (See **paragraph 14-4.14**.) (Page 14-6)

WARNING The MK 16 MOD 0 UBA provides no visual warning of excess CO_2 problems. The diver should be aware of CO_2 toxicity symptoms. (Page 17-5)

WARNING Failure to adhere to these guidelines could result in serious injury or death. (Page 17-15)

WARNING	No repetitive dives are authorized after an emergency procedure requiring a shift to the EBS. (Page 17-19)
WARNING	Hypoxia and hypercapnia may give the diver little or no warning prior to onset of unconsciousness. (Page 17-30)
WARNING	Failure to adhere to these guidelines could result in serious injury or death. (Page 18-14)
WARNING	Hypoxia and hypercapnia may give the diver little or no warning prior to onset of unconsciousness. (Page 18-26)
WARNING	The MK 25 does not have a carbon dioxide-monitoring capability. Failure to adhere to canister duration operations planning could lead to unconsciousness and/or death. (Page 19-19)
WARNING	Drug therapy shall be administered only after consultation with a Diving Medical Officer by qualified inside tenders adequately trained and capable of administering prescribed medications. (Page 20-30)
WARNING	The gag valve must remain open at all times. Close only if relief valve fails. (Page 21-20)
WARNING	This procedure is to be performed with an unmanned chamber to avoid exposing occupants to unnecessary risks. (Page 21-21)
WARNING	Do not exceed maximum pressure rating for the pressure vessel. (Page 21-26)
WARNING	Fire/Explosion Hazard. No matches, lighters, electrical appliances, or flammable materials permitted in chamber. (Page 21-30)
CAUTION	When in doubt, always recompress. (Page 3-29)
CAUTION	Do not institute active rewarming with severe cases of hypothermia. (Page 3-55)
CAUTION	GFIs require an established reference ground in order to function properly. Cascading GFIs could result in loss of reference ground; therefore, GFIs or equipment containing built-in GFIs should not be plugged into an existing GFI circuit. (Page 6-21)
CAUTION	This checklist is an overview intended for use with the detailed Operating Procedures (OPs) from the appropriate equipment O&M technical manual. (Page 6-50)
CAUTION	Prior to use of VVDS as a buoyancy compensator, divers must be thoroughly familiar with its use. (Page 7-9)

CAUTION When diving with a Variable Volume Dry Suit, avoid overinflation and be aware of the possibility of blowup when breaking loose from mud. It is better to call for aid from the standby diver than to risk blowup. (Page 8-28)

CAUTION In very cold water, the wet suit is only a marginally effective thermal protective measure, and its use exposes the diver to hypothermia and restricts available bottom time. The use of alternative thermal protective equipment should be considered in these circumstances. (Page 11-6)

CAUTION Prior to the use of variable volume dry suits and hot water suits in cold and ice-covered waters, divers must be trained in their use and be thoroughly familiar with the operation of these suits. (Page 11-6)

CAUTION There is an increased risk of CNS oxygen toxicity when diving the MK 16 MOD 1 compared to diving the MK 16 MOD 0, especially during the descent phase of the dive. Diving supervisors and divers should be aware that oxygen partial pressures of 1.6 ata or higher may be temporarily experienced during descent on N_2O_2 dives deeper than 120 fsw (21% oxygen diluent) and on HeO_2 dives deeper than 200 fsw (12% oxygen diluent). Refer to paragraph 18-10.1.1 for information on recognizing and preventing CNS oxygen toxicity. (Page 18-14)

CAUTION Defibrillation is not currently authorized at depth. (Page 20-4)

CAUTION If the tender is outside of no-decompression limits, he should not be brought directly to the surface. Either take the decompression stops appropriate to the tender or lock in a new tender and decompress the patient and new tender to the surface in the outerlock, while maintaining the original tender at depth. (Page 20-4)

CAUTION Inserting an airway device or bite block is not recommended while the patient is convulsing; it is not only difficult, but may cause harm if attempted. (Page 20-24)

CAUTION AED's are not currently approved for use under pressure (hyperbaric environment) due to electrical safety concerns. (Page 20-36)

CAUTION Acrylic view-ports should not be lubricated or come in contact with any lubricant. Acrylic view-ports should not come in contact with any volatile detergent or leak detector (non-ionic detergent is to be used for leak test). When reinstalling view-port, take up retaining ring bolts until the gasket just compresses evenly about the view-port. Do not overcompress the gasket. (Page 21-26)

PAGE LEFT BLANK INTENTIONALLY

Table of Contents

Chap/Para			Page

Chap/Para Page

Chap/Para Page

Chap/Para Page

Chap/Para		Page

Chap/Para Page

Chap/Para		Page

Chap/Para | Page

Chap/Para		Page

Chap/Para **Page**

Chap/Para **Page**

Chap/Para Page

Chap/Para **Page**

Chap/Para		Page

Chap/Para		Page

18 MK 16 MOD 1 CLOSED-CIRCUIT MIXED-GAS UBA DIVING

Chap/Para | Page

Chap/Para Page

Chap/Para		Page

Chap/Para Page

Chap/Para		Page

List of Illustrations

Figure		Page

PAGE LEFT BLANK INTENTIONALLY

List of Tables

VOLUME 1

Diving Principles and Policy

U.S. NAVY DIVING MANUAL

PAGE LEFT BLANK INTENTIONALLY

Volume 1 - Table of Contents

Chap/Para Page

Chap/Para | Page

Chap/Para		Page

Volume 1 - List of Illustrations

Volume 1 - List of Tables

PAGE LEFT BLANK INTENTIONALLY

CHAPTER 1

History of Diving

1-1 INTRODUCTION

1-1.1 **Purpose.** This chapter provides a general history of the development of military diving operations.

1-1.2 **Scope.** This chapter outlines the hard work and dedication of a number of individuals who were pioneers in the development of diving technology. As with any endeavor, it is important to build on the discoveries of our predecessors and not repeat mistakes of the past.

1-1.3 **Role of the U.S. Navy.** The U.S. Navy is a leader in the development of modern diving and underwater operations. The general requirements of national defense and the specific requirements of underwater reconnaissance, demolition, ordnance disposal, construction, ship maintenance, search, rescue and salvage operations repeatedly give impetus to training and development. Navy diving is no longer limited to tactical combat operations, wartime salvage, and submarine sinkings. Fleet diving has become increasingly important and diversified since World War II. A major part of the diving mission is inspecting and repairing naval vessels to minimize downtime and the need for dry-docking. Other aspects of fleet diving include recovering practice and research torpedoes, installing and repairing underwater electronic arrays, underwater construction, and locating and recovering downed aircraft.

1-2 SURFACE-SUPPLIED AIR DIVING

The origins of diving are firmly rooted in man's need and desire to engage in maritime commerce, to conduct salvage and military operations, and to expand the frontiers of knowledge through exploration, research, and development.

Diving, as a profession, can be traced back more than 5,000 years. Early divers confined their efforts to waters less than 100 feet deep, performing salvage work and harvesting food, sponges, coral, and mother-of-pearl. A Greek historian, Herodotus, recorded the story of a diver named Scyllis, who was employed by the Persian King Xerxes to recover sunken treasure in the fifth century B.C.

From the earliest times, divers were active in military operations. Their missions included cutting anchor cables to set enemy ships adrift, boring or punching holes in the bottoms of ships, and building harbor defenses at home while attempting to destroy those of the enemy abroad. Alexander the Great sent divers down to remove obstacles in the harbor of the city of Tyre, in what is now Lebanon, which he had taken under siege in 332 B.C.

Other early divers developed an active salvage industry centered around the major shipping ports of the eastern Mediterranean. By the first century B.C., operations in one area had become so well organized that a payment scale for salvage work was established by law, acknowledging the fact that effort and risk increased with depth. In 24 feet of water, the divers could claim a one-half share of all goods recovered. In 12 feet of water, they were allowed a one-third share, and in 3 feet, only a one-tenth share.

1-2.1 **Breathing Tubes.** The most obvious and crucial step to broadening a diver's capabilities was providing an air supply that would permit him to stay underwater. Hollow reeds or tubes extending to the surface allowed a diver to remain submerged for an extended period, but he could accomplish little in the way of useful work. Breathing tubes were employed in military operations, permitting an undetected approach to an enemy stronghold (Figure 1-1).

At first glance, it seemed logical that a longer breathing tube was the only requirement for extending a diver's range. In fact, a number of early designs used leather hoods with long flexible tubes supported at the surface by floats. There is no record, however, that any of these devices were actually constructed or tested. The result may well have been the drowning of the diver. At a depth of 3 feet, it is nearly impossible to breathe through a tube using only the body's natural respiratory ability, as the weight of the water exerts a total force of almost 200 pounds on the diver's chest. This force increases steadily with depth and is one of the most important factors in diving. Successful diving operations require that the pressure be overcome or eliminated. Throughout history, imaginative devices were designed to overcome this problem, many by some of the greatest minds of the time. At first, the problem of pressure underwater was not fully understood and the designs were impractical.

Figure 1-1. Early Impractical Breathing Device. This 1511 design shows the diver's head encased in a leather bag with a breathing tube extending to the surface.

Figure 1-2. Assyrian Frieze (900 B.C.).

1-2.2 **Breathing Bags.** An entire series of designs was based on the idea of a breathing bag carried by the diver. An Assyrian frieze of the ninth century B.C. shows what appear to be divers using inflated animal skins as air tanks. However, these men were probably swimmers using skins for flotation. It would be impossible to submerge while holding such an accessory (Figure 1-2).

A workable diving system may have made a brief appearance in the later Middle Ages. In 1240, Roger Bacon made reference to "instruments whereby men can walk on sea or river beds without danger to themselves."

1-2.3 **Diving Bells.** Between 1500 and 1800 the diving bell was developed, enabling divers to remain underwater for hours rather than minutes. The diving bell is a bell-shaped apparatus with the bottom open to the sea.

The first diving bells were large, strong tubs weighted to sink in a vertical position, trapping enough air to permit a diver to breathe for several hours. Later diving bells were suspended by a cable from the surface. They had no significant underwater maneuverability beyond that provided by moving the support ship. The diver could remain in the bell if positioned directly over his work, or could venture outside for short periods of time by holding his breath.

The first reference to an actual practical diving bell was made in 1531. For several hundred years thereafter, rudimentary but effective bells were used with regularity. In the 1680s, a Massachusetts-born adventurer named William Phipps modified the diving bell technique by supplying his divers with air from a series of weighted, inverted buckets as they attempted to recover treasure valued at $200,000.

In 1690, the English astronomer Edmund Halley developed a diving bell in which the atmosphere was replenished by sending weighted barrels of air down from the surface (Figure 1-3). In an early demonstration of his system, he and four companions remained at 60 feet in the Thames River for almost 1½ hours. Nearly 26 years later, Halley spent more than 4 hours at 66 feet using an improved version of his bell.

1-2.4 **Diving Dress Designs.** With an increasing number of military and civilian wrecks littering the shores of Great Britain each year, there was strong incentive to develop a diving dress that would increase the efficiency of salvage operations.

1-2.4.1 **Lethbridge's Diving Dress.** In 1715, Englishman John Lethbridge developed a one-man, completely enclosed diving dress (Figure 1-4). The Lethbridge equipment was a reinforced, leather-covered barrel of air, equipped with a glass porthole for viewing and two arm holes with watertight sleeves. Wearing this gear, the occupant could accomplish useful work. This apparatus was lowered from a ship and maneuvered in the same manner as a diving bell.

Lethbridge was quite successful with his invention and participated in salvaging a number of European wrecks. In a letter to the editor of a popular magazine in 1749, the inventor noted that his normal operating depth was 10 fathoms (60 feet),

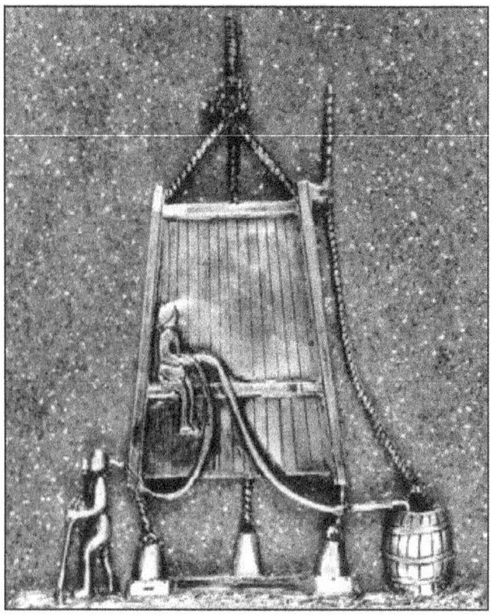

Figure 1-3. Engraving of Halley's Diving Bell.

Figure 1-4. Lethbridge's Diving Suit.

with about 12 fathoms the maximum, and that he could remain underwater for 34 minutes.

Several designs similar to Lethbridge's were used in succeeding years. However, all had the same basic limitation as the diving bell—the diver had little freedom because there was no practical way to continually supply him with air. A true technological breakthrough occurred at the turn of the 19th century when a hand-operated pump capable of delivering air under pressure was developed.

1-2.4.2 **Deane's Patented Diving Dress.** Several men produced a successful apparatus at the same time. In 1823, two salvage operators, John and Charles Deane, patented the basic design for a smoke apparatus that permitted firemen to move about in burning buildings. By 1828, the apparatus evolved into Deane's Patent Diving Dress, consisting of a heavy suit for protection from the cold, a helmet with viewing ports, and hose connections for delivering surface-supplied air. The helmet rested on the diver's shoulders, held in place by its own weight and straps to a waist belt. Exhausted or surplus air passed out from under the edge of the helmet and posed no problem as long as the diver was upright. If he fell, however, the helmet could quickly fill with water. In 1836, the Deanes issued a diver's manual, perhaps the first ever produced.

1-2.4.3 **Siebe's Improved Diving Dress.** Credit for developing the first practical diving dress has been given to Augustus Siebe. Siebe's initial contribution to diving was a modification of the Deane outfit. Siebe sealed the helmet to the dress at the collar by using a short, waist-length waterproof suit and added an exhaust valve to the system (Figure 1-5). Known as Siebe's Improved Diving Dress, this apparatus is the direct ancestor of the MK V standard deep-sea diving dress.

1-2.4.4 **Salvage of the HMS *Royal George*.** By 1840, several types of diving dress were being used in actual diving operations. At that time, a unit of the British Royal Engineers was engaged in removing the remains of the sunken warship, HMS *Royal George*. The warship was fouling a major fleet anchorage just outside Portsmouth, England. Colonel William Pasley, the officer in charge, decided that his operation was an ideal opportunity to formally test and evaluate the various types of apparatus. Wary of the Deane apparatus because of the possibility of helmet flooding, he formally recommended that the Siebe dress be adopted for future operations.

Figure 1-5. Siebe's First Enclosed Diving Dress and Helmet.

When Pasley's project was completed, an official government historian noted that "of the seasoned divers, not a man escaped the repeated attacks of rheumatism and cold." The divers had been working for 6 or 7 hours a day, much of it spent at depths of 60 to 70 feet. Pasley and his men did not realize the implications of the observation. What appeared to be rheumatism was instead a symptom of a far more serious physiological problem that, within a few years, was to become of great importance to the diving profession.

1-2.5 **Caissons.** At the same time that a practical diving dress was being perfected, inventors were working to improve the diving bell by increasing its size and adding high-capacity air pumps that could deliver enough pressure to keep water entirely out of the bell's interior. The improved pumps soon led to the construction of chambers large enough to permit several men to engage in dry work on the bottom. This was particularly advantageous for projects such as excavating bridge footings or constructing tunnel sections where long periods of work were required. These dry chambers were known as *caissons*, a French word meaning "big boxes" (Figure 1-6).

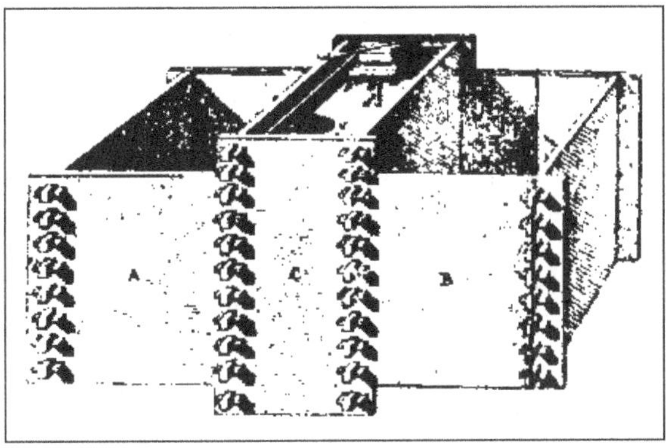

Figure 1-6. French Caisson. This caisson could be floated over the work site and lowered to the bottom by flooding the side tanks.

Caissons were designed to provide ready access from the surface. By using an air lock, the pressure inside could be maintained while men or materials could be passed in and out. The caisson was a major step in engineering technology and its use grew quickly.

1-2.6 Physiological Discoveries.

1-2.6.1 **Caisson Disease (Decompression Sickness).** With the increasing use of caissons, a new and unexplained malady began to affect the caisson workers. Upon returning to the surface at the end of a shift, the divers frequently would be struck by dizzy spells, breathing difficulties, or sharp pains in the joints or abdomen. The sufferer usually recovered, but might never be completely free of some of the symptoms. Caisson workers often noted that they felt better working on the job, but wrongly attributed this to being more rested at the beginning of a shift.

As caisson work extended to larger projects and to greater operating pressures, the physiological problems increased in number and severity. Fatalities occurred with alarming frequency. The malady was called, logically enough, caisson disease. However, workers on the Brooklyn Bridge project in New York gave the sickness a more descriptive name that has remained—the "bends."

Today the bends is the most well-known danger of diving. Although men had been diving for thousands of years, few men had spent much time working under great atmospheric pressure until the time of the caisson. Individuals such as Pasley, who had experienced some aspect of the disease, were simply not prepared to look for anything more involved than indigestion, rheumatism, or arthritis.

1-2.6.1.1 ***Cause of Decompression Sickness.*** The actual cause of caisson disease was first clinically described in 1878 by a French physiologist, Paul Bert. In studying the effect of pressure on human physiology, Bert determined that breathing air under pressure forced quantities of nitrogen into solution in the blood and tissues of the body. As long as the pressure remained, the gas was held in solution. When the pressure was quickly released, as it was when a worker left the caisson, the nitrogen returned to a gaseous state too rapidly to pass out of the body in a natural manner. Gas bubbles formed throughout the body, causing the wide range of symptoms associated with the disease. Paralysis or death could occur if the flow of blood to a vital organ was blocked by the bubbles.

1-2.6.1.2 ***Prevention and Treatment of Decompression Sickness.*** Bert recommended that caisson workers gradually decompress and divers return to the surface slowly. His studies led to an immediate improvement for the caisson workers when they discovered their pain could be relieved by returning to the pressure of the caisson as soon as the symptom appeared.

Within a few years, specially designed recompression chambers were being placed at job sites to provide a more controlled situation for handling the bends. The pressure in the chambers could be increased or decreased as needed for an individual worker. One of the first successful uses of a recompression chamber was in 1879 during the construction of a subway tunnel under the Hudson River between New York and New Jersey. The recompression chamber markedly reduced the number of serious cases and fatalities caused by the bends.

Bert's recommendation that divers ascend gradually and steadily was not a complete success, however; some divers continued to suffer from the bends. The general thought at the time was that divers had reached the practical limits of the art and that 120 feet was about as deep as anyone could work. This was because of the repeated incidence of the bends and diver inefficiency beyond that depth. Occasionally, divers would lose consciousness while working at 120 feet.

1-2.6.2 **Inadequate Ventilation.** J.S. Haldane, an English physiologist, conducted experiments with Royal Navy divers from 1905 to 1907. He determined that part of the problem was due to the divers not adequately ventilating their helmets, causing high levels of carbon dioxide to accumulate. To solve the problem, he established a standard supply rate of flow (1.5 cubic feet of air per minute, measured at the pressure of the diver). Pumps capable of maintaining the flow and ventilating the helmet on a continuous basis were used.

Haldane also composed a set of diving tables that established a method of decompression in stages. Though restudied and improved over the years, these tables remain the basis of the accepted method for bringing a diver to the surface.

As a result of Haldane's studies, the practical operating depth for air divers was extended to slightly more than 200 feet. The limit was not imposed by physiological factors, but by the capabilities of the hand-pumps available to provide the air supply.

1-2.6.3 **Nitrogen Narcosis.** Divers soon were moving into deeper water and another unexplained malady began to appear. The diver would appear intoxicated, sometimes feeling euphoric and frequently losing judgment to the point of forgetting the dive's purpose. In the 1930s this "rapture of the deep" was linked to nitrogen in the air breathed under higher pressures. Known as nitrogen narcosis, this condition occurred because nitrogen has anesthetic properties that become progressively more severe with increasing air pressure. To avoid the problem, special breathing mixtures such as helium-oxygen were developed for deep diving (see section 1-4, Mixed-Gas Diving).

1-2.7 **Armored Diving Suits.** Numerous inventors, many with little or no underwater experience, worked to create an armored diving suit that would free the diver from pressure problems (Figure 1-7). In an armored suit, the diver could breathe air at normal atmospheric pressure and descend to great depths

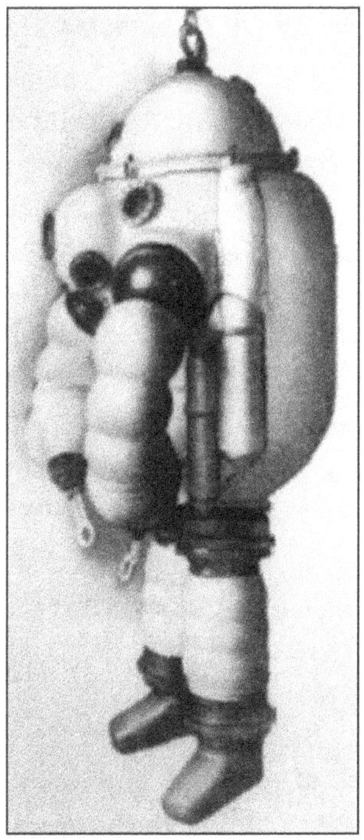

Figure 1-7. Armored Diving Suit.

without any ill effects. The barrel diving suit, designed by John Lethbridge in 1715, had been an armored suit in essence, but one with a limited operating depth.

The utility of most armored suits was questionable. They were too clumsy for the diver to be able to accomplish much work and too complicated to provide protection from extreme pressure. The maximum anticipated depth of the various suits developed in the 1930s was 700 feet, but was never reached in actual diving. More recent pursuits in the area of armored suits, now called one-atmosphere diving suits, have demonstrated their capability for specialized underwater tasks to 2,000 feet of saltwater (fsw).

1-2.8 **MK V Deep-Sea Diving Dress.** By 1905, the Bureau of Construction and Repair had designed the MK V Diving Helmet which seemed to address many of the problems encountered in diving. This deep-sea outfit was designed for extensive, rugged diving work and provided the diver maximum physical protection and some maneuverability.

The 1905 MK V Diving Helmet had an elbow inlet with a safety valve that allowed air to enter the helmet, but not to escape back up the umbilical if the air supply were interrupted. Air was expelled from the helmet through an exhaust valve on the right side, below the port. The exhaust valve was vented toward the rear of the helmet to prevent escaping bubbles from interfering with the diver's field of vision.

By 1916, several improvements had been made to the helmet, including a rudimentary communications system via a telephone cable and a regulating valve operated by an interior push button. The regulating valve allowed some control of the atmospheric pressure. A supplementary relief valve, known as the spitcock, was added to the left side of the helmet. A safety catch was also incorporated to keep the helmet attached to the breast plate. The exhaust valve and the communications system were improved by 1927, and the weight of the helmet was decreased to be more comfortable for the diver.

After 1927, the MK V changed very little. It remained basically the same helmet used in salvage operations of the USS S-51 and USS S-4 in the mid-1920s. With its associated deep-sea dress and umbilical, the MK V was used for all submarine rescue and salvage work undertaken in peacetime and practically all salvage work undertaken during World War II. The MK V Diving Helmet was the standard U.S. Navy diving equipment until succeeded by the MK 12 Surface-Supplied Diving System (SSDS) in February 1980 (see Figure 1-8). The MK 12 was replaced by the MK 21 in December 1993.

1-3 SCUBA DIVING

The diving equipment developed by Charles and John Deane, Augustus Siebe, and other inventors gave man the ability to remain and work underwater for extended periods, but movement was greatly limited by the requirement for surface-supplied air. Inventors searched for methods to increase the diver's movement without increasing the hazards. The best solution was to provide the diver with a portable,

Figure 1-8. MK 12 and MK V.

self-contained air supply. For many years the self-contained underwater breathing apparatus (SCUBA) was only a theoretical possibility. Early attempts to supply self-contained compressed air to divers were not successful due to the limitations of air pumps and containers to compress and store air at sufficiently high pressure. SCUBA development took place gradually, however, evolving into three basic types:

- Open-circuit SCUBA (where the exhaust is vented directly to the surrounding water),
- Closed-circuit SCUBA (where the oxygen is filtered and recirculated), and
- Semiclosed-circuit SCUBA (which combines features of the open- and closed-circuit types).

1-3.1 **Open-Circuit SCUBA.** In the open-circuit apparatus, air is inhaled from a supply cylinder and the exhaust is vented directly to the surrounding water.

1-3.1.1 **Rouquayrol's Demand Regulator.** The first and highly necessary component of an open-circuit apparatus was a demand regulator. Designed early in 1866 and patented by Benoist Rouquayrol, the regulator adjusted the flow of air from the tank to meet the diver's breathing and pressure requirements. However, because cylinders strong enough to contain air at high pressure could not be built at the time, Rouquayrol adapted his regulator to surface-supplied diving equipment and the technology turned toward closed-circuit designs. The application of Rouquayrol's concept of a demand regulator to a successful open-circuit SCUBA was to wait more than 60 years.

1-3.1.2 **LePrieur's Open-Circuit SCUBA Design.** The thread of open-circuit development was picked up in 1933. Commander LePrieur, a French naval officer, constructed an open-circuit SCUBA using a tank of compressed air. However, LePrieur did not include a demand regulator in his design and, the diver's main effort was diverted to the constant manual control of his air supply. The lack of a demand regulator, coupled with extremely short endurance, severely limited the practical use of LePrieur's apparatus.

1-3.1.3 **Cousteau and Gagnan's Aqua-Lung.** At the same time that actual combat operations were being carried out with closed-circuit apparatus, two Frenchmen achieved a significant breakthrough in open-circuit SCUBA design. Working in a small Mediterranean village, under the difficult and restrictive conditions of German-occupied France, Jacques-Yves Cousteau and Emile Gagnan combined an improved demand regulator with high-pressure air tanks to create the first truly efficient and safe open-circuit SCUBA, known as the Aqua-Lung. Cousteau and his companions brought the Aqua-Lung to a high state of development as they explored and photographed wrecks, developing new diving techniques and testing their equipment.

The Aqua-Lung was the culmination of hundreds of years of progress, blending the work of Rouquayol, LePrieur, and Fleuss, a pioneer in closed-circuit SCUBA development. Cousteau used his gear successfully to 180 fsw without significant difficulty and with the end of the war the Aqua-Lung quickly became a commercial success. Today the Aqua-Lung is the most widely used diving equipment, opening the underwater world to anyone with suitable training and the fundamental physical abilities.

1-3.1.4 **Impact of SCUBA on Diving.** The underwater freedom brought about by the development of SCUBA led to a rapid growth of interest in diving. Sport diving has become very popular, but science and commerce have also benefited. Biologists, geologists and archaeologists have all gone underwater, seeking new clues to the origins and behavior of the earth, man and civilization as a whole. An entire industry has grown around commercial diving, with the major portion of activity in offshore petroleum production.

After World War II, the art and science of diving progressed rapidly, with emphasis placed on improving existing diving techniques, creating new methods, and developing the equipment required to serve these methods. A complete generation of new and sophisticated equipment took form, with substantial improvements being made in both open and closed-circuit apparatus. However, the most significant aspect of this technological expansion has been the closely linked development of saturation diving techniques and deep diving systems.

1-3.2 **Closed-Circuit SCUBA.** The basic closed-circuit system, or oxygen rebreather, uses a cylinder of 100 percent oxygen that supplies a breathing bag. The oxygen used by the diver is recirculated in the apparatus, passing through a chemical filter that removes carbon dioxide. Oxygen is added from the tank to replace that consumed in breathing. For special warfare operations, the closed-circuit system has a major advantage over the open-circuit type: it does not produce a telltale trail of bubbles on the surface.

1-3.2.1 **Fleuss' Closed-Circuit SCUBA.** Henry A. Fleuss developed the first commercially practical closed-circuit SCUBA between 1876 and 1878 (Figure 1-9). The Fleuss device consisted of a watertight rubber face mask and a breathing bag connected to a copper tank of 100 percent oxygen charged to 450 psi. By using oxygen instead of compressed air as the breathing medium, Fleuss eliminated the need for high-strength tanks. In early models of this apparatus, the diver controlled the makeup feed of fresh oxygen with a hand valve.

Fleuss successfully tested his apparatus in 1879. In the first test, he remained in a tank of water for about an hour. In the second test, he walked along a creek bed at a depth of 18 feet. During the second test, Fleuss turned off his oxygen feed to see what would happen. He was soon unconscious, and suffered gas embolism as his tenders pulled him to the surface. A few weeks after his recovery, Fleuss made arrangements to put his recirculating design into commercial production.

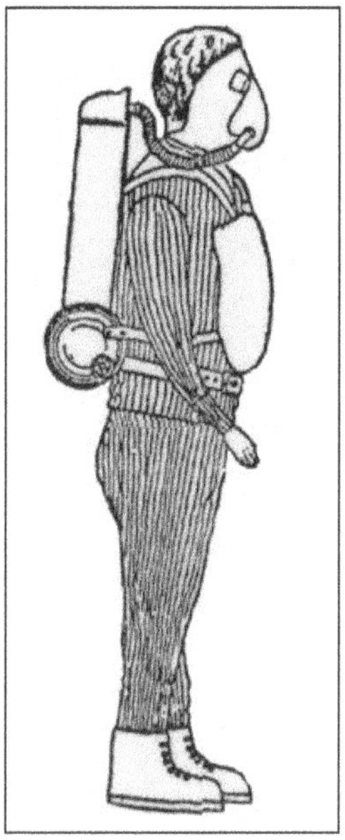

Figure 1-9. Fleuss Apparatus.

In 1880, the Fleuss SCUBA figured prominently in a highly publicized achievement by an English diver, Alexander Lambert. A tunnel under the Severn River flooded and Lambert, wearing a Fleuss apparatus, walked 1,000 feet along the tunnel, in complete darkness, to close several crucial valves.

1-3.2.2 **Modern Closed-Circuit Systems.** As development of the closed-circuit design continued, the Fleuss equipment was improved by adding a demand regulator and tanks capable of holding oxygen at more than 2,000 psi. By World War I, the Fleuss SCUBA (with modifications) was the basis for submarine escape equipment used in the Royal Navy. In World War II, closed-circuit units were widely used for combat diving operations (see paragraph 1-3.5.2).

Some modern closed-circuit systems employ a mixed gas for breathing and electronically senses and controls oxygen concentration. This type of apparatus retains the bubble-free characteristics of 100-percent oxygen recirculators while significantly improving depth capability.

1-3.3 **Hazards of Using Oxygen in SCUBA.** Fleuss had been unaware of the serious problem of oxygen toxicity caused by breathing 100 percent oxygen under pressure. Oxygen toxicity apparently was not encountered when he used his apparatus in early shallow water experiments. The danger of oxygen poisoning had actually been discovered prior to 1878 by Paul Bert, the physiologist who first proposed controlled decompression as a way to avoid the bends. In laboratory experiments with animals, Bert demonstrated that breathing oxygen under pressure could lead to convulsions and death (central nervous system oxygen toxicity).

In 1899, J. Lorrain Smith found that breathing oxygen over prolonged periods of time, even at pressures not sufficient to cause convulsions, could lead to pulmonary oxygen toxicity, a serious lung irritation. The results of these experiments, however, were not widely publicized. For many years, working divers were unaware of the dangers of oxygen poisoning.

The true seriousness of the problem was not apparent until large numbers of combat swimmers were being trained in the early years of World War II. After a number of oxygen toxicity accidents, the British established an operational depth limit of 33 fsw. Additional research on oxygen toxicity continued in the U.S. Navy after the war and resulted in the setting of a normal working limit of 25 fsw for 75 minutes for the Emerson oxygen rebreather. A maximum emergency depth/time limit of 40 fsw for 10 minutes was also allowed.

These limits eventually proved operationally restrictive, and prompted the Navy Experimental Diving Unit to reexamine the entire problem of oxygen toxicity in the mid-1980s. As a result of this work, more liberal and flexible limits were adopted for U.S. Navy use.

1-3.4 **Semiclosed-Circuit SCUBA.** The semiclosed-circuit SCUBA combines features of the open and closed-circuit systems. Using a mixture of gases for breathing, the apparatus recycles the gas through a carbon dioxide removal canister and continually adds a small amount of oxygen-rich mixed gas to the system from a supply cylinder. The supply gas flow is preset to satisfy the body's oxygen demand; an equal amount of the recirculating mixed-gas stream is continually exhausted to the water. Because the quantity of makeup gas is constant regardless of depth, the semiclosed-circuit SCUBA provides significantly greater endurance than open-circuit systems in deep diving.

1-3.4.1 **Lambertsen's Mixed-Gas Rebreather.** In the late 1940s, Dr. C.J. Lambertsen proposed that mixtures of nitrogen or helium with an elevated oxygen content be used in SCUBA to expand the depth range beyond that allowed by 100-percent oxygen rebreathers, while simultaneously minimizing the requirement for decompression.

In the early 1950s, Lambertsen introduced the FLATUS I, a semiclosed-circuit SCUBA that continually added a small volume of mixed gas, rather than pure oxygen, to a rebreathing circuit. The small volume of new gas provided the oxygen necessary for metabolic consumption while exhaled carbon dioxide was absorbed in an absorbent canister. Because inert gas, as well as oxygen, was added to the rig, and because the inert gas was not consumed by the diver, a small amount of gas mixture was continuously exhausted from the rig.

1-3.4.2 **MK 6 UBA.** In 1964, after significant development work, the Navy adopted a semiclosed-circuit, mixed-gas rebreather, the MK 6 UBA, for combat swimming and EOD operations. Decompression procedures for both nitrogen-oxygen and helium-oxygen mixtures were developed at the Navy Experimental Diving Unit. The apparatus had a maximum depth capability of 200 fsw and a maximum endurance of 3 hours depending on water temperature and diver activity. Because the apparatus was based on a constant mass flow of mixed gas, the endurance was independent of the diver's depth.

In the late 1960s, work began on a new type of mixed-gas rebreather technology, which was later used in the MK 15 and MK 16 UBAs. In this UBA, the oxygen partial pressure was controlled at a constant value by an oxygen sensing and addition system. As the diver consumed oxygen, an oxygen sensor detected the fall in oxygen partial pressure and signaled an oxygen valve to open, allowing a small amount of pure oxygen to be admitted to the breathing circuit from a cylinder. Oxygen addition was thus exactly matched to metabolic consumption. Exhaled carbon dioxide was absorbed in an absorption canister. The system had the endurance and completely closed-circuit characteristics of an oxygen rebreather without the concerns and limitations associated with oxygen toxicity.

Beginning in 1979, the MK 6 semiclosed-circuit underwater breathing apparatus (UBA) was phased out by the MK 15 closed-circuit, constant oxygen partial pressure UBA. The Navy Experimental Diving Unit developed decompression procedures for the MK 15 with nitrogen and helium in the early 1980s. In 1985, an improved low magnetic signature version of the MK 15, the MK 16, was approved for Explosive Ordnance Disposal (EOD) team use.

1-3.5 **SCUBA Use During World War II.** Although closed-circuit equipment was restricted to shallow-water use and carried with it the potential danger of oxygen toxicity, its design had reached a suitably high level of efficiency by World War II. During the war, combat swimmer breathing units were widely used by navies on both sides of the conflict. The swimmers used various modes of underwater attack. Many notable successes were achieved including the sinking of several battleships, cruisers, and merchant ships.

1-3.5.1 **Diver-Guided Torpedoes.** Italian divers, using closed-circuit gear, rode chariot torpedoes fitted with seats and manual controls in repeated attacks against British ships. In 1936, the Italian Navy tested a chariot torpedo system in which the divers used a descendant of the Fleuss SCUBA. This was the Davis Lung (Figure 1-10). It was originally designed as a submarine escape device and was later manufactured in Italy under a license from the English patent holders.

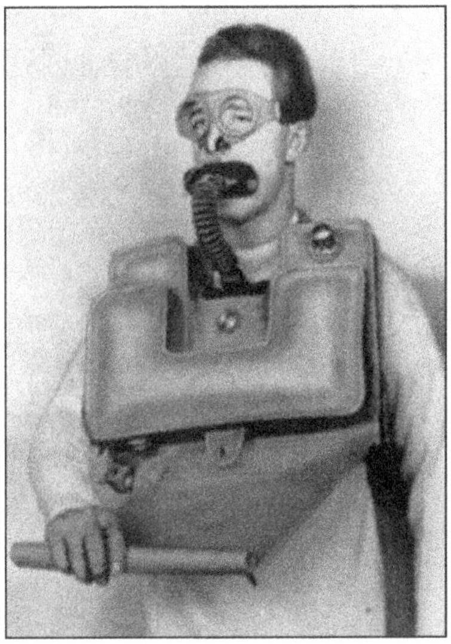

Figure 1-10. Original Davis Submerged Escape Apparatus.

British divers, carried to the scene of action in midget submarines, aided in placing explosive charges under the keel of the German battleship *Tirpitz*. The British began their chariot program in 1942 using the Davis Lung and exposure suits. Swimmers using the MK 1 chariot dress quickly discovered that the steel oxygen bottles adversely affected the compass of the chariot torpedo. Aluminum oxygen cylinders were not readily available in England, but German aircraft used aluminum oxygen cylinders that were almost the same size as the steel cylinders aboard the chariot torpedo. Enough aluminum cylinders were salvaged from downed enemy bombers to supply the British forces.

Changes introduced in the MK 2 and MK 3 diving dress involved improvements in valving, faceplate design, and arrangement of components. After the war, the MK 3 became the standard Royal Navy shallow water diving dress. The MK 4 dress was used near the end of the war. Unlike the MK 3, the MK 4 could be supplied with oxygen from a self-contained bottle or from a larger cylinder carried in the chariot. This gave the swimmer greater endurance, yet preserved freedom of movement independent of the chariot torpedo.

In the final stages of the war, the Japanese employed an underwater equivalent of their kamikaze aerial attack—the kaiten diver-guided torpedo.

1-3.5.2 **U.S. Combat Swimming.** There were two groups of U.S. combat swimmers during World War II: Naval beach reconnaissance swimmers and U.S. operational swimmers. Naval beach reconnaissance units did not normally use any breathing devices, although several models existed.

U.S. operational swimmers, however, under the Office of Strategic Services, developed and applied advanced methods for true self-contained diver-submersible operations. They employed the Lambertsen Amphibious Respiratory Unit (LARU), a rebreather invented by Dr. C.J. Lambertsen (see Figure 1-11). The LARU was a closed-circuit oxygen UBA used in special warfare operations where a complete absence of exhaust bubbles was required. Following World War II, the Emerson-Lambertsen Oxygen Rebreather replaced the LARU (Figure 1-12). The Emerson Unit was used extensively by Navy special warfare divers until 1982, when it was replaced by the Draeger Lung Automatic Regenerator (LAR) V. The LAR V is the standard unit now used by U.S. Navy combat swimmers (see Figure 1-13).

Figure 1-11. Lambertsen Amphibious Respiratory Unit (LARU).

Today Navy combat swimmers are organized into two separate groups, each with specialized training and missions. The Explosive Ordnance Disposal (EOD) team handles, defuses, and disposes of munitions and other explosives. The Sea, Air and Land (SEAL) special warfare teams make up the second group of Navy combat swimmers. SEAL team members are trained to operate in all of these environments. They qualify as parachutists, learn to handle a range of weapons, receive intensive training in hand-to-hand combat, and are expert in SCUBA and other swimming and diving techniques. In Vietnam, SEALs were deployed in special counter-insurgency and guerrilla warfare operations. The SEALs also participated

Figure 1-12. Emerson-Lambertsen Oxygen Rebreather.

Figure 1-13. Draeger LAR V UBA.

in the space program by securing flotation collars to returned space capsules and assisting astronauts during the helicopter pickup.

1-3.5.3 **Underwater Demolition.** The Navy's Underwater Demolition Teams (UDTs) were created when bomb disposal experts and Seabees (combat engineers) teamed together in 1943 to devise methods for removing obstacles that the Germans were placing off the beaches of France. The first UDT combat mission was a daylight reconnaissance and demolition project off the beaches of Saipan in June 1944. In March of 1945, preparing for the invasion of Okinawa, one underwater demolition team achieved the exceptional record of removing 1,200 underwater obstacles in 2 days, under heavy fire, without a single casualty.

Because suitable equipment was not readily available, diving apparatus was not extensively used by the UDT during the war. UDT experimented with a modified Momsen lung and other types of breathing apparatus, but not until 1947 did the Navy's acquisition of Aqua-Lung equipment give impetus to the diving aspect of UDT operations. The trail of bubbles from the open-circuit apparatus limited the type of mission in which it could be employed, but a special SCUBA platoon of UDT members was formed to test the equipment and determine appropriate uses for it.

Through the years since, the mission and importance of the UDT has grown. In the Korean Conflict, during the period of strategic withdrawal, the UDT destroyed an entire port complex to keep it from the enemy. The UDTs have since been incorporated into the Navy Seal Teams.

1-4 MIXED-GAS DIVING

Mixed-gas diving operations are conducted using a breathing medium other than air. This medium may consist of:

- Nitrogen and oxygen in proportions other than those found in the atmosphere
- A mixture of other inert gases, such as helium, with oxygen.

The breathing medium can also be 100 percent oxygen, which is not a mixed gas, but which requires training for safe use. Air may be used in some phases of a mixed-gas dive.

Mixed-gas diving is a complex undertaking. A mixed-gas diving operation requires extensive special training, detailed planning, specialized and advanced equipment and, in many applications, requires extensive surface-support personnel and facilities. Because mixed-gas operations are often conducted at great depth or for extended periods of time, hazards to personnel increase greatly. Divers studying mixed-gas diving must first be qualified in air diving operations.

In recent years, to match basic operational requirements and capabilities, the U.S. Navy has divided mixed-gas diving into two categories:

- Nonsaturation diving without a pressurized bell to a maximum depth of 300 fsw, and
- Saturation diving for dives of 150 fsw and greater depth or for extended bottom time missions.

The 300-foot limit is based primarily on the increased risk of decompression sickness when nonsaturation diving techniques are used deeper than 300 fsw.

1-4.1 Nonsaturation Diving.

1-4.1.1 **Helium-Oxygen (HeO_2) Diving.** An inventor named Elihu Thomson theorized that helium might be an appropriate substitute for the nitrogen in a diver's breathing supply. He estimated that at least a 50-percent gain in working depth could be achieved by substituting helium for nitrogen. In 1919, he suggested that the U.S. Bureau of Mines investigate this possibility. Thomson directed his suggestion to the Bureau of Mines rather than the Navy Department, since the Bureau of Mines held a virtual world monopoly on helium marketing and distribution.

1-4.1.1.1 ***Experiments with Helium-Oxygen Mixtures.*** In 1924, the Navy and the Bureau of Mines jointly sponsored a series of experiments using helium-oxygen mixtures. The preliminary work was conducted at the Bureau of Mines Experimental Station in Pittsburgh, Pennsylvania. Figure 1-14 is a picture of an early Navy helium-oxygen diving manifold.

Figure 1-14. Helium-Oxygen Diving Manifold.

The first experiments showed no detrimental effects on test animals or humans from breathing a helium-oxygen mixture, and decompression time was shortened. The principal physiological effects noted by divers using helium-oxygen were:

- Increased sensation of cold caused by the high thermal conductivity of helium
- The high-pitched distortion or "Donald Duck" effect on human speech that resulted from the acoustic properties and reduced density of the gas

These experiments clearly showed that helium-oxygen mixtures offered great advantages over air for deep dives. They laid the foundation for developing the reliable decompression tables and specialized apparatus, which are the cornerstones of modern deep diving technology.

In 1937, at the Experimental Diving Unit research facility, a diver wearing a deep-sea diving dress with a helium-oxygen breathing supply was compressed in a chamber to a simulated depth of 500 feet. The diver was not told the depth and when asked to make an estimate of the depth, the diver reported that it felt as if he were at 100 feet. During decompression at the 300-foot mark, the breathing mixture was switched to air and the diver was troubled immediately by nitrogen narcosis.

The first practical test of helium-oxygen came in 1939, when the submarine USS *Squalus* was salvaged from a depth of 243 fsw. In that year, the Navy issued decompression tables for surface-supplied helium-oxygen diving.

1-4.1.1.2 ***MK V MOD 1 Helmet.*** Because helium was expensive and shipboard supplies were limited, the standard MK V MOD 0 open-circuit helmet was not economical for surface-supplied helium-oxygen diving. After experimenting with several different designs, the U.S. Navy adopted the semiclosed-circuit MK V MOD 1 (Figure 1-15).

Figure 1-15. MK V MOD 1 Helmet.

The MK V MOD 1 helmet was equipped with a carbon dioxide absorption canister and venturi-powered recirculator assembly. Gas in the helmet was continuously recirculated through the carbon dioxide scrubber assembly by the venturi. By removing carbon dioxide by scrubbing rather than ventilating the helmet, the fresh gas flow into the helmet was reduced to the amount required to replenish oxygen. The gas consumption of the semiclosed-circuit MK V MOD 1 was approximately 10 percent of that of the open-circuit MK V MOD 0.

The MK V MOD 1, with breastplate and recirculating gas canister, weighed approximately 103 pounds compared to 56 pounds for the standard air helmet and breastplate. It was fitted with a lifting ring at the top of the helmet to aid in hatting the diver and to keep the weight off his shoulders until he was lowered into the water. The diver was lowered into and raised out of the water by a diving stage connected to an onboard boom.

1-4.1.1.3 ***Civilian Designers.*** U.S. Navy divers were not alone in working with mixed gases or helium. In 1937, civilian engineer Max Gene Nohl reached 420 feet in Lake Michigan while breathing helium-oxygen and using a suit of his own design. In 1946, civilian diver Jack Browne, designer of the lightweight diving mask that bears his name, made a simulated helium-oxygen dive of 550 feet. In 1948, a British Navy diver set an open-sea record of 540 fsw while using war-surplus helium provided by the U.S.

1-4.1.2 **Hydrogen-Oxygen Diving.** In countries where the availability of helium was more restricted, divers experimented with mixtures of other gases. The most notable example is that of the Swedish engineer Arne Zetterstrom, who worked with hydrogen-oxygen mixtures. The explosive nature of such mixtures was well known, but it was also known that hydrogen would not explode when used in a mixture of less than 4 percent oxygen. At the surface, this percentage of oxygen would not be sufficient to sustain life; at 100 feet, however, the oxygen partial pressure would be the equivalent of 16 percent oxygen at the surface.

Zetterstrom devised a simple method for making the transition from air to hydrogen-oxygen without exceeding the 4-percent oxygen limit. At the 100-foot level, he replaced his breathing air with a mixture of 96 percent nitrogen and 4 percent oxygen. He then replaced that mixture with hydrogen-oxygen in the same proportions. In 1945, after some successful test dives to 363 feet, Zetterstrom reached 528 feet. Unfortunately, as a result of a misunderstanding on the part of his topside support personnel, he was brought to the surface too rapidly. Zetterstrom did not have time to enrich his breathing mixture or to adequately decompress and died as a result of the effects of his ascent.

1-4.1.3 **Modern Surface-Supplied Mixed-Gas Diving.** The U.S. Navy and the Royal Navy continued to develop procedures and equipment for surface-supplied helium-oxygen diving in the years following World War II. In 1946, the Admiralty Experimental Diving Unit was established and, in 1956, during open-sea tests of helium-oxygen diving, a Royal Navy diver reached a depth of 600 fsw. Both navies conducted helium-oxygen decompression trials in an attempt to develop better procedures.

In the early 1960s, a young diving enthusiast from Switzerland, Hannes Keller, proposed techniques to attain great depths while minimizing decompression requirements. Using a series of gas mixtures containing varying concentrations of oxygen, helium, nitrogen, and argon, Keller demonstrated the value of elevated oxygen pressures and gas sequencing in a series of successful dives in mountain lakes. In 1962, with partial support from the U.S. Navy, he reached an open-sea depth of more than 1,000 fsw off the California coast. Unfortunately, this dive was marred by tragedy. Through a mishap unrelated to the technique itself, Keller lost consciousness on the bottom and, in the subsequent emergency decompression, Keller's companion died of decompression sickness.

By the late 1960s, it was clear that surface-supplied diving deeper than 300 fsw was better carried out using a deep diving (bell) system where the gas sequencing techniques pioneered by Hannes Keller could be exploited to full advantage, while maintaining the diver in a state of comfort and security. The U.S. Navy developed decompression procedures for bell diving systems in the late 1960s and early 1970s. For surface-supplied diving in the 0-300 fsw range, attention was turned to developing new equipment to replace the cumbersome MK V MOD 1 helmet.

1-4.1.4 **MK 1 MOD 0 Diving Outfit.** The new equipment development proceeded along two parallel paths, developing open-circuit demand breathing systems suitable for deep helium-oxygen diving, and developing an improved recirculating helmet to replace the MK V MOD 1. By the late 1960s, engineering improvements in demand regulators had reduced breathing resistance on deep dives to acceptable levels. Masks and helmets incorporating the new regulators became commercially available. In 1976, the U.S. Navy approved the MK 1 MOD 0 Lightweight, Mixed-Gas Diving Outfit for dives to 300 fsw on helium-oxygen (Figure 1-16). The MK 1 MOD 0 Diving Outfit incorporated a full face mask (bandmask) featuring a demand open-circuit breathing regulator and a backpack for an emergency gas supply. Surface contact was maintained through an umbilical that included the breathing gas hose, communications cable, lifeline strength member and pneumofathometer hose. The diver was dressed in a dry suit or hot water suit depending on water temperature. The equipment was issued as a lightweight diving outfit in a system with sufficient equipment to support a diving operation employing two working divers and a standby diver. The outfit was used in conjunction with an open diving bell that replaced the traditional diver's stage and added additional safety. In 1990, the MK 1 MOD 0 was replaced by the MK 21 MOD 1 (Superlite 17 B/NS) demand helmet. This is the lightweight rig in use today.

Figure 1-16. MK 1 MOD 0 Diving Outfit.

In 1985, after an extensive development period, the direct replacement for the MK V MOD 1 helmet was approved for Fleet use. The new MK 12 Mixed-Gas Surface-Supplied Diving System (SSDS) was similar to the MK 12 Air SSDS, with the addition of a backpack assembly to allow operation in a semiclosed-circuit mode. The MK 12 system was retired in 1992 after the introduction of the MK 21 MOD 1 demand helmet.

1-4.2 **Diving Bells.** Although open, pressure-balanced diving bells have been used for several centuries, it was not until 1928 that a bell appeared that was capable of maintaining internal pressure when raised to the surface. In that year, Sir Robert H. Davis, the British pioneer in diving equipment, designed the Submersible Decompression Chamber (SDC). The vessel was conceived to reduce the time a diver had to remain in the water during a lengthy decompression.

The Davis SDC was a steel cylinder capable of holding two men, with two inward-opening hatches, one on the top and one on the bottom. A surface-supplied diver was deployed over the side in the normal mode and the bell was lowered to a depth of 60 fsw with the lower hatch open and a tender inside. Surface-supplied air ventilated the bell and prevented flooding. The diver's deep decompression stops were taken in the water and he was assisted into the bell by the tender upon arrival at 60 fsw. The diver's gas supply hose and communications cable were removed from the helmet and passed out of the bell. The lower door was closed and the bell was lifted to the deck where the diver and tender were decompressed within the safety and comfort of the bell.

By 1931, the increased decompression times associated with deep diving and the need for diver comfort resulted in the design of an improved bell system. Davis designed a three-compartment deck decompression chamber (DDC) to which the SDC could be mechanically mated, permitting the transfer of the diver under pressure. The DDC provided additional space, a bunk, food and clothing for the diver's comfort during a lengthy decompression. This procedure also freed the SDC for use by another diving team for continuous diving operations.

The SDC-DDC concept was a major advance in diving safety, but was not applied to American diving technology until the advent of saturation diving. In 1962, E. A. Link employed a cylindrical, aluminum SDC in conducting his first open-sea saturation diving experiment. In his experiments, Link used the SDC to transport the diver to and from the sea floor and a DDC for improved diver comfort. American diving had entered the era of the Deep Diving System (DDS) and advances and applications of the concept grew at a phenomenal rate in both military and commercial diving.

1-4.3 **Saturation Diving.** As divers dove deeper and attempted more ambitious underwater tasks, a safe method to extend actual working time at depth became crucial. Examples of saturation missions include submarine rescue and salvage, sea bed implantments, construction, and scientific testing and observation. These types of operations are characterized by the need for extensive bottom time and, consequently, are more efficiently conducted using saturation techniques.

1-4.3.1 **Advantages of Saturation Diving.** In deep diving operations, decompression is the most time-consuming factor. For example, a diver working for an hour at 200 fsw would be required to spend an additional 3 hours and 20 minutes in the water undergoing the necessary decompression.

However, once a diver becomes saturated with the gases that make decompression necessary, the diver does not need additional decompression. When the blood and tissues have absorbed all the gas they can hold at that depth, the time required for decompression becomes constant. As long as the depth is not increased, additional time on the bottom is free of any additional decompression.

If a diver could remain under pressure for the entire period of the required task, the diver would face a lengthy decompression only when completing the project. For a 40-hour task at 200 fsw, a saturated diver would spend 5 days at bottom pressure and 2 days in decompression, as opposed to spending 40 days making 1-hour dives with long decompression periods using conventional methods.

The U.S. Navy developed and proved saturation diving techniques in its Sealab series. Advanced saturation diving techniques are being developed in ongoing programs of research and development at the Navy Experimental Diving Unit (NEDU), Navy Submarine Medical Research Laboratory (NSMRL), and many institutional and commercial hyperbaric facilities. In addition, saturation diving using Deep Diving Systems (DDS) is now a proven capability.

1-4.3.2 **Bond's Saturation Theory.** True scientific impetus was first given to the saturation concept in 1957 when a Navy diving medical officer, Captain George F. Bond, theorized that the tissues of the body would eventually become saturated with inert gas if exposure time was long enough. Bond, then a commander and the director of the Submarine Medical Center at New London, Connecticut, met with Captain Jacques-Yves Cousteau and determined that the data required to prove the theory of saturation diving could be developed at the Medical Center.

1-4.3.3 **Genesis Project.** With the support of the U.S. Navy, Bond initiated the Genesis Project to test the theory of saturation diving. A series of experiments, first with test animals and then with humans, proved that once a diver was saturated, further extension of bottom time would require no additional decompression time. Project Genesis proved that men could be sustained for long periods under pressure, and what was then needed was a means to put this concept to use on the ocean floor.

1-4.3.4 **Developmental Testing.** Several test dives were conducted in the early 1960s:

- The first practical open-sea demonstrations of saturation diving were undertaken in September 1962 by Edward A. Link and Captain Jacques-Yves Cousteau.
- Link's Man-in-the-Sea program had one man breathing helium-oxygen at 200 fsw for 24 hours in a specially designed diving system.
- Cousteau placed two men in a gas-filled, pressure-balanced underwater habitat at 33 fsw where they stayed for 169 hours, moving freely in and out of their deep-house.
- Cousteau's Conshelf One supported six men breathing nitrogen-oxygen at 35 fsw for 7 days.
- In 1964, Link and Lambertsen conducted a 2-day exposure of two men at 430 fsw.
- Cousteau's Conshelf Two experiment maintained a group of seven men for 30 days at 36 fsw and 90 fsw with excursion dives to 330 fsw.

1-4.3.5 **Sealab Program.** The best known U.S. Navy experimental effort in saturation diving was the Sealab program.

1-4.3.5.1 ***Sealabs I and II.*** After completing the Genesis Project, the Office of Naval Research, the Navy Mine Defense Laboratory and Bond's small staff of volunteers gathered in Panama City, Florida, where construction and testing of the Sealab I habitat began in December 1963.

In 1964, Sealab I placed four men underwater for 10 days at an average depth of 192 fsw. The habitat was eventually raised to 81 fsw, where the divers were transferred to a decompression chamber that was hoisted aboard a four-legged offshore support structure.

In 1965, Sealab II put three teams of ten men each in a habitat at 205 fsw. Each team spent 15 days at depth and one man, Astronaut Scott Carpenter, remained for 30 days (see Figure 1-17).

1-4.3.5.2 ***Sealab III.*** The follow-on seafloor experiment, Sealab III, was planned for 600 fsw. This huge undertaking required not only extensive development and testing of equipment but also assessment of human tolerance to high-pressure environments.

To prepare for Sealab III, 28 helium-oxygen saturation dives were performed at the Navy Experimental Diving Unit to depths of 825 fsw between 1965 and 1968. In 1968, a record-breaking excursion dive to 1,025 fsw from a saturation depth of 825 fsw was performed at the Navy Experimental Diving Unit (NEDU). The culmination of this series of dives was a 1,000 fsw, 3-day saturation dive conducted jointly by the U.S. Navy and Duke University in the hyperbaric chambers at Duke. This was the first time man had been saturated at 1,000 fsw. The Sealab III preparation experiments showed that men could readily perform useful work at pressures up to 31 atmospheres and could be returned to normal pressure without harm.

Figure 1-17. Sealab II.

Figure 1-18. U.S. Navy's First DDS, SDS-450.

Reaching the depth intended for the Sealab III habitat required highly specialized support, including a diving bell to transfer divers under pressure from the habitat to a pressurized deck decompression chamber. The experiment, however, was marred by tragedy. Shortly after being compressed to 600 fsw in February 1969, Aquanaut Berry Cannon convulsed and drowned. This unfortunate accident ended the Navy's involvement with seafloor habitats.

1-4.3.5.3 ***Continuing Research.*** Research and development continues to extend the depth limit for saturation diving and to improve the diver's capability. The deepest dive attained by the U.S. Navy to date was in 1979 when divers from the NEDU completed a 37-day, 1,800 fsw dive in its Ocean Simulation Facility. The world record depth for experimental saturation, attained at Duke University in 1981, is 2,250 fsw, and non-Navy open sea dives have been completed to in excess of 2300 fsw. Experiments with mixtures of hydrogen, helium, and oxygen have begun and the success of this mixture was demonstrated in 1988 in an open-sea dive to 1,650 fsw.

Advanced saturation diving techniques are being developed in ongoing programs of research and development at NEDU, Navy Submarine Medical Research Laboratory (NSMRL), and many institutional and commercial hyperbaric facilities. In addition, saturation diving using Deep Diving Systems (DDS) is now a proven capability.

1-4.4 Deep Diving Systems (DDS).

Experiments in saturation technique required substantial surface support as well as extensive underwater equipment. DDS are a substantial improvement over previous methods of accomplishing deep undersea work. The DDS is readily adaptable to saturation techniques and safely maintains the saturated diver under pressure in a dry environment. Whether employed for saturation or nonsaturation diving, the Deep Diving System totally eliminates long decompression periods in the water where the diver is subjected to extended environmental stress. The diver only remains in the sea for the time spent on a given task. Additional benefits derived from use of the DDS include eliminating the need for underwater habitats and increasing operational flexibility for the surface-support ship.

The Deep Diving System consists of a Deck Decompression Chamber (DDC) mounted on a surface-support ship. A Personnel Transfer Capsule (PTC) is mated to the DDC, and the combination is pressurized to a storage depth. Two or more divers enter the PTC, which is unmated and lowered to the working depth. The interior of the capsule is pressurized to equal the pressure at depth, a hatch is opened, and one or more divers swim out to accomplish their work. The divers can use a self-contained breathing apparatus with a safety tether to the capsule, or employ a mask and an umbilical that provides breathing gas and communications. Upon completing the task, the divers enters the capsule, close the hatch and return to the support ship with the interior of the PTC still at the working pressure. The capsule is hoisted aboard and mated to the pressurized DDC. The divers enter the larger, more comfortable DDC via an entry lock. They remain in the DDC until they must return to the undersea job site. Decompression is carried out comfortably and safely on the support ship.

The Navy developed four deep diving systems: ADS-IV, MK 1 MOD 0, MK 2 MOD 0, and MK 2 MOD 1.

1-4.4.1 **ADS-IV.** Several years prior to the Sealab I experiment, the Navy successfully deployed the Advanced Diving System IV (ADS-IV) (see Figure 1-18). The ADS-IV was a small deep diving system with a depth capability of 450 fsw. The ADS-IV was later called the SDS-450.

1-4.4.2 **MK 1 MOD 0.** The MK 1 MOD 0 DDS was a small system intended to be used on the new ATS-1 class salvage ships, and underwent operational evaluation in 1970. The DDS consisted of a Personnel Transfer Capsule (PTC) (see Figure 1-19), a life-support system, main control console and two deck decompression chambers to handle two teams of two divers each. This system was also used to operationally evaluate the MK 11 UBA, a semiclosed-circuit mixed-gas apparatus, for saturation diving. The MK 1 MOD 0 DDS conducted an open-sea dive to 1,148 fsw in 1975. The MK 1 DDS was not installed on the ATS ships as originally planned, but placed on a barge and assigned to Harbor Clearance Unit Two. The system went out of service in 1977.

Figure 1-19. DDS MK 1 Personnel Transfer Capsule.

Figure 1-20. PTC Handling System, *Elk River*.

1-4.4.3 **MK 2 MOD 0.** The Sealab III experiment required a much larger and more capable deep diving system than the MK 1 MOD 0. The MK 2 MOD 0 was constructed and installed on the support ship *Elk River* (IX-501). With this system, divers could be saturated in the deck chamber under close observation and then transported to the habitat for the stay at depth, or could cycle back and forth between the deck chamber and the seafloor while working on the exterior of the habitat. The bell could also be used in a non-pressurized observation mode. The divers would be transported from the habitat to the deck decompression chamber, where final decompression could take place under close observation.

1-4.4.4 **MK 2 MOD 1.** Experience gained with the MK 2 MOD 0 DDS on board *Elk River* (IX-501) (see Figure 1-20) led to the development of the MK 2 MOD 1, a larger, more sophisticated DDS. The MK 2 MOD 1 DDS supported two four-man teams for long term saturation diving with a normal depth capability of 850 fsw. The diving complex consisted of two complete systems, one at starboard and one at port. Each system had a DDC with a life-support system, a PTC, a main control console, a strength-power-communications cable (SPCC) and ship support. The two systems shared a helium-recovery system. The MK 2 MOD 1 was installed on the ASR 21 Class submarine rescue vessels.

1-5 SUBMARINE SALVAGE AND RESCUE

At the beginning of the 20th century, all major navies turned their attention toward developing a weapon of immense potential—the military submarine. The highly effective use of the submarine by the German Navy in World War I heightened this interest and an emphasis was placed on the submarine that continues today.

The U.S. Navy had operated submarines on a limited basis for several years prior to 1900. As American technology expanded, the U.S. submarine fleet grew rapidly. However, throughout the period of 1912 to 1939, the development of the Navy's F, H, and S class boats was marred by a series of accidents, collisions, and sinkings. Several of these submarine disasters resulted in a correspondingly rapid growth in the Navy diving capability.

Until 1912, U.S. Navy divers rarely went below 60 fsw. In that year, Chief Gunner George D. Stillson set up a program to test Haldane's diving tables and methods of stage decompression. A companion goal of the program was to improve Navy diving equipment. Throughout a 3-year period, first diving in tanks ashore and then in open water in Long Island Sound from the USS *Walkie*, the Navy divers went progressively deeper, eventually reaching 274 fsw.

1-5.1 **USS F-4.** The experience gained in Stillson's program was put to dramatic use in 1915 when the submarine USS F-4 sank near Honolulu, Hawaii. Twenty-one men lost their lives in the accident and the Navy lost its first boat in 15 years of submarine operations. Navy divers salvaged the submarine and recovered the bodies of the crew. The salvage effort incorporated many new techniques, such as using lifting pontoons. What was most remarkable, however, was that the divers completed a major salvage effort working at the extreme depth of 304 fsw, using air as a breathing mixture. The decompression requirements limited bottom time for each dive to about 10 minutes. Even for such a limited time, nitrogen narcosis made it difficult for the divers to concentrate on their work.

The publication of the first U.S. Navy Diving Manual and the establishment of a Navy Diving School at Newport, Rhode Island, were the direct outgrowth of experience gained in the test program and the USS F-4 salvage. When the U.S. entered

World War I, the staff and graduates of the school were sent to Europe, where they conducted various salvage operations along the coast of France.

The physiological problems encountered in the salvage of the USS F-4 clearly demonstrated the limitations of breathing air during deep dives. Continuing concern that submarine rescue and salvage would be required at great depth focused Navy attention on the need for a new diver breathing medium.

1-5.2 **USS S-51.** In September of 1925, the USS S-51 submarine was rammed by a passenger liner and sunk in 132 fsw off Block Island, Rhode Island. Public pressure to raise the submarine and recover the bodies of the crew was intense. Navy diving was put in sharp focus, realizing it had only 20 divers who were qualified to go deeper than 90 fsw. Diver training programs had been cut at the end of World War I and the school had not been reinstituted.

Salvage of the USS S-51 covered a 10-month span of difficult and hazardous diving, and a special diver training course was made part of the operation. The submarine was finally raised and towed to the Brooklyn Navy Yard in New York.

Interest in diving was high once again and the Naval School, Diving and Salvage, was reestablished at the Washington Navy Yard in 1927. At the same time, the Navy brought together its existing diving technology and experimental work by shifting the Experimental Diving Unit (EDU), which had been working with the Bureau of Mines in Pennsylvania, to the Navy Yard as well. In the following years, EDU developed the U.S. Navy Air Decompression Tables, which have become the accepted world standard and continued developmental work in helium-oxygen breathing mixtures for deeper diving.

Losing the USS F-4 and USS S-51 provided the impetus for expanding the Navy's diving ability. However, the Navy's inability to rescue men trapped in a disabled submarine was not confronted until another major submarine disaster occurred.

1-5.3 **USS S-4.** In 1927, the Navy lost the submarine USS S-4 in a collision with the Coast Guard cutter USS *Paulding*. The first divers to reach the submarine in 102 fsw, 22 hours after the sinking, exchanged signals with the men trapped inside. The submarine had a hull fitting designed to take an air hose from the surface, but what had looked feasible in theory proved too difficult in reality. With stormy seas causing repeated delays, the divers could not make the hose connection until it was too late. All of the men aboard the USS S-4 had died. Even had the hose connection been made in time, rescuing the crew would have posed a significant problem.

The USS S-4 was salvaged after a major effort and the fate of the crew spurred several efforts toward preventing a similar disaster. LT C.B. Momsen, a submarine officer, developed the escape lung that bears his name. It was given its first operational test in 1929 when 26 officers and men successfully surfaced from an intentionally bottomed submarine.

1-5.4 **USS *Squalus.*** The Navy pushed for development of a rescue chamber that was essentially a diving bell with special fittings for connection to a submarine deck hatch. The apparatus, called the McCann-Erickson Rescue Chamber, was proven in 1939 when the USS *Squalus*, carrying a crew of 50, sank in 243 fsw. The rescue chamber made four trips and safely brought 33 men to the surface. (The rest of the crew, trapped in the flooded after-section of the submarine, had perished in the sinking.)

The USS *Squalus* was raised by salvage divers (see Figure 1-21). This salvage and rescue operation marked the first operational use of HeO_2 in salvage diving. One of the primary missions of salvage divers was to attach a down-haul cable for the Submarine Rescue Chamber (SRC). Following renovation, the submarine, renamed USS *Sailfish*, compiled a proud record in World War II.

Figure 1-21. Recovery of the *Squalus.*

1-5.5 **USS *Thresher.*** Just as the loss of the USS F-4, USS S-51, USS S-4 and the sinking of the USS *Squalus* caused an increased concern in Navy diving in the 1920s and 1930s, a submarine disaster of major proportions had a profound effect on the development of new diving equipment and techniques in the postwar period. This was the loss of the nuclear attack submarine USS *Thresher* and all her crew in April 1963. The submarine sank in 8,400 fsw, a depth beyond the survival limit of the hull and far beyond the capability of any existing rescue apparatus.

An extensive search was initiated to locate the submarine and determine the cause of the sinking. The first signs of the USS *Thresher* were located and photographed a month after the disaster. Collection of debris and photographic coverage of the wreck continued for about a year.

Two special study groups were formed as a result of the sinking. The first was a Court of Inquiry, which attributed probable cause to a piping system failure. The second, the Deep Submergence Review Group (DSRG), was formed to assess the Navy's undersea capabilities. Four general areas were examined—search, rescue, recovery of small and large objects, and the Man-in-the-Sea concept. The basic recommendations of the DSRG called for a vast effort to improve the Navy's capabilities in these four areas.

1-5.6 **Deep Submergence Systems Project.** Direct action on the recommendations of the DSRG came with the formation of the Deep Submergence Systems Project (DSSP) in 1964 and an expanded interest regarding diving and undersea activity throughout the Navy.

Submarine rescue capabilities have been substantially improved with the development of the Deep Submergence Rescue Vehicle (DSRV) which became operational in 1972. This deep-diving craft is air-transportable, highly instrumented, and capable of diving to 5,000 fsw and rescues to 2,500 fsw.

Three additional significant areas of achievement for the Deep Submergence Systems Project have been that of Saturation Diving, the development of Deep Diving Systems, and progress in advanced diving equipment design.

1-6 SALVAGE DIVING

1-6.1 World War II Era.

1-6.1.1 **Pearl Harbor.** Navy divers were plunged into the war with the Japanese raid on Pearl Harbor. The raid began at 0755 on 7 December 1941; by 0915 that same morning, the first salvage teams were cutting through the hull of the overturned battleship USS *Oklahoma* to rescue trapped sailors. Teams of divers worked to recover ammunition from the magazines of sunken ships, to be ready in the event of a second attack.

The immense salvage effort that followed at Pearl Harbor was highly successful. Most of the 101 ships in the harbor at the time of the attack sustained damage. The battleships, one of the primary targets of the raid, were hardest hit. Six battleships were sunk and one was heavily damaged. Four were salvaged and returned to the fleet for combat duty; the former battleships USS *Arizona* and USS *Utah* could not be salvaged. The USS *Oklahoma* was righted and refloated but sank en route to a shipyard in the U.S.

Battleships were not the only ships salvaged. Throughout 1942 and part of 1943, Navy divers worked on destroyers, supply ships, and other badly needed vessels, often using makeshift shallow water apparatus inside water and gas-filled compartments. In the Pearl Harbor effort, Navy divers spent 16,000 hours underwater during 4,000 dives. Contract civilian divers contributed another 4,000 diving hours.

1-6.1.2 **USS *Lafayette*.** While divers in the Pacific were hard at work at Pearl Harbor, a major challenge was presented to the divers on the East Coast. The interned French passenger liner *Normandie* (rechristened as the USS *Lafayette*) caught fire alongside New York City's Pier 88. Losing stability from the tons of water poured on the fire, the ship capsized at her berth.

The ship had to be salvaged to clear the vitally needed pier. The Navy took advantage of this unique training opportunity by instituting a new diving and salvage school at the site. The Naval Training School (Salvage) was established in September 1942 and was transferred to Bayonne, New Jersey in 1946.

1-6.1.3 **Other Diving Missions.** Salvage operations were not the only missions assigned to Navy divers during the war. Many dives were made to inspect sunken enemy ships and to recover materials such as code books or other intelligence items. One Japanese cruiser yielded not only $500,000 in yen, but also provided valuable information concerning plans for the defense of Japan against the anticipated Allied invasion.

1-6.2 **Vietnam Era.** Harbor Clearance Unit One (HCU 1) was commissioned 1 February 1966 to provide mobile salvage capability in direct support of combat operations in Vietnam. Homeported at Naval Base Subic Bay, Philippines, HCU 1 was dedicated primarily to restoring seaports and rivers to navigable condition following their loss or diminished use through combat action.

Beginning as a small cadre of personnel, HCU 1 quickly grew in size to over 260 personnel, as combat operations in littoral environment intensified. At its peak, the unit consisted of five Harbor Clearance teams of 20 to 22 personnel each and a varied armada of specialized vessels within the Vietnam combat zone.

As their World War II predecessors before them, the salvors of HCU 1 left an impressive legacy of combat salvage accomplishments. HCU 1 salvaged hundreds of small craft, barges, and downed aircraft; refloated many stranded U.S. Military and merchant vessels; cleared obstructed piers, shipping channels, and bridges; and performed numerous underwater repairs to ships operating in the combat zone.

Throughout the colorful history of HCU 1 and her East Coast sister HCU 2, the vital role salvage forces play in littoral combat operations was clearly demonstrated. Mobile Diving and Salvage Unit One and Two, the modern-day descendants of the Vietnam era Harbor Clearance Units, have a proud and distinguished history of combat salvage operations.

1-7 OPEN-SEA DEEP DIVING RECORDS

Diving records have been set and broken with increasing regularity since the early 1900s:

- **1915.** The 300-fsw mark was exceeded. Three U.S. Navy divers, F. Crilley, W.F. Loughman, and F.C. Nielson, reached 304 fsw using the MK V dress.
- **1972.** The MK 2 MOD 0 DDS set the in-water record of 1,010 fsw.
- **1975.** Divers using the MK 1 Deep Dive System descended to 1,148 fsw.

- **1977.** A French dive team broke the open-sea record with 1,643 fsw.
- **1981.** The deepest salvage operation made with divers was 803 fsw when British divers retrieved 431 gold ingots from the wreck of HMS *Edinburgh*, sunk during World War II.
- **Present.** Commercial open water diving operations to over 1,000 fsw.

1-8 SUMMARY

Throughout the evolution of diving, from the earliest breath-holding sponge diver to the modern saturation diver, the basic reasons for diving have not changed. National defense, commerce, and science continue to provide the underlying basis for the development of diving. What has changed and continues to change radically is diving technology.

Each person who prepares for a dive has the opportunity and obligation to take along the knowledge of his or her predecessors that was gained through difficult and dangerous experience. The modern diver must have a broad understanding of the physical properties of the undersea environment and a detailed knowledge of his or her own physiology and how it is affected by the environment. Divers must learn to adapt to environmental conditions to successfully carry out their missions.

Much of the diver's practical education will come from experience. However, before a diver can gain this experience, he or she must build a basic foundation from certain principles of physics, chemistry and physiology and must understand the application of these principles to the profession of diving.

PAGE LEFT BLANK INTENTIONALLY

CHAPTER 2

Underwater Physics

2-1 INTRODUCTION

2-1.1 **Purpose.** This chapter describes the laws of physics as they affect humans in the water.

2-1.2 **Scope.** A thorough understanding of the principles outlined in this chapter is essential to safe and effective diving performance.

2-2 PHYSICS

Humans readily function within the narrow atmospheric envelope present at the earth's surface and are seldom concerned with survival requirements. Outside the boundaries of the envelope, the environment is hostile and our existence depends on our ability to counteract threatening forces. To function safely, divers must understand the characteristics of the subsea environment and the techniques that can be used to modify its effects. To accomplish this, a diver must have a basic knowledge of physics—the science of matter and energy. Of particular importance to a diver are the behavior of gases, the principles of buoyancy, and the properties of heat, light, and sound.

2-3 MATTER

Matter is anything that occupies space and has mass, and is the building block of the physical world. Energy is required to cause matter to change course or speed. The diver, the diver's air supply, everything that supports him or her, and the surrounding environment is composed of matter.

2-3.1 **Elements.** An *element* is the simplest form of matter that exhibits distinct physical and chemical properties. An element cannot be broken down by chemical means into other, more basic forms. Scientists have identified more than 100 elements in the physical universe. Elements combine to form the more than four million substances known to man.

2-3.2 **Atoms.** The *atom* is the smallest particle of matter that carries the specific properties of an element. Atoms are made up of electrically charged particles known as protons, neutrons, and electrons. Protons have a positive charge, neutrons have a neutral charge, and electrons have a negative charge.

2-3.3 **Molecules.** *Molecules* are formed when atoms group together (Figure 2-1). Molecules usually exhibit properties different from any of the contributing atoms. For example, when two hydrogen atoms combine with one oxygen atom, a new substance—water—is formed. Some molecules are active and try to combine with many of the other molecules that surround them. Other molecules are inert and

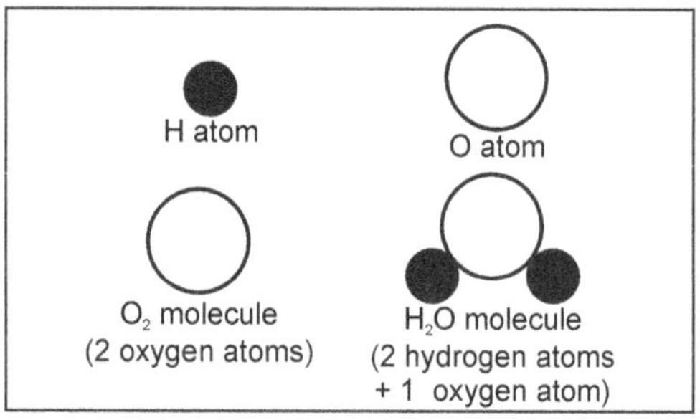

Figure 2-1. Molecules. Two similar atoms combine to form an oxygen molecule while the atoms of two different elements, hydrogen and oxygen, combine to form a water molecule.

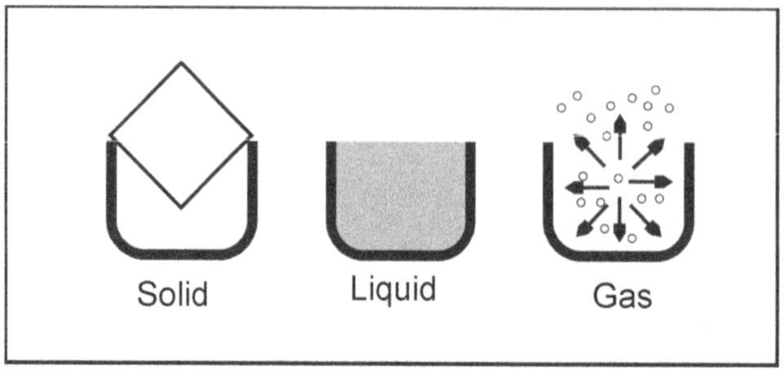

Figure 2-2. The Three States of Matter.

do not naturally combine with other substances. The presence of inert elements in breathing mixtures is important when calculating a diver's decompression obligations.

2-3.4 **The Three States of Matter.** Matter can exist in one of three natural states: solid, liquid, or gas (Figure 2-2). A solid has a definite size and shape. A liquid has a definite volume, but takes the shape of the container. Gas has neither definite shape nor volume, but will expand to fill a container. Gases and liquids are collectively referred to as fluids.

The physical state of a substance depends primarily upon temperature and partially upon pressure. A solid is the coolest of the three states, with its molecules rigidly aligned in fixed patterns. The molecules move, but their motion is like a constant vibration. As heat is added the molecules increase their motion, slip apart from each other and move around; the solid becomes a liquid. A few of the molecules will spontaneously leave the surface of the liquid and become a gas. When the substance reaches its boiling point, the molecules are moving very rapidly in all directions and the liquid is quickly transformed into a gas. Lowering the temperature reverses the sequence. As the gas molecules cool, their motion is reduced and the gas condenses into a liquid. As the temperature continues to fall, the liquid reaches the freezing point and transforms to a solid state.

2-4 MEASUREMENT

Physics relies heavily upon standards of comparison of one state of matter or energy to another. To apply the principles of physics, divers must be able to employ a variety of units of measurement.

2-4.1 **Measurement Systems.** Two systems of measurement are widely used throughout the world. Although the English System is commonly used in the United States, the most common system of measurement in the world is the International System of Units. The International System of Units, or *SI* system, is a modernized metric system designated in 1960 by the General Conference on Weights and Measures. The SI system is decimal based with all its units related, so that it is not necessary to use calculations to change from one unit to another. The SI system changes one of its units of measurement to another by moving the decimal point, rather than by the lengthy calculations necessary in the English System. Because measurements are often reported in units of the English system, it is important to be able to convert them to SI units. Measurements can be converted from one system to another by using the conversion factors in Table 2-10 through 2-18.

2-4.2 **Temperature Measurements.** While the English System of weights and measures uses the Fahrenheit (°F) temperature scale, the Celsius (°C) scale is the one most commonly used in scientific work. Both scales are based upon the freezing and boiling points of water. The freezing point of water is 32°F or 0°C; the boiling point of water is 212°F or 100°C. Temperature conversion formulas and charts are found in Table 2-18.

Absolute temperature values are used when employing the ideal gas laws. The absolute temperature scales are based upon absolute zero. Absolute zero is the lowest temperature that could possibly be reached at which all molecular motion would cease (Figure 2-3).

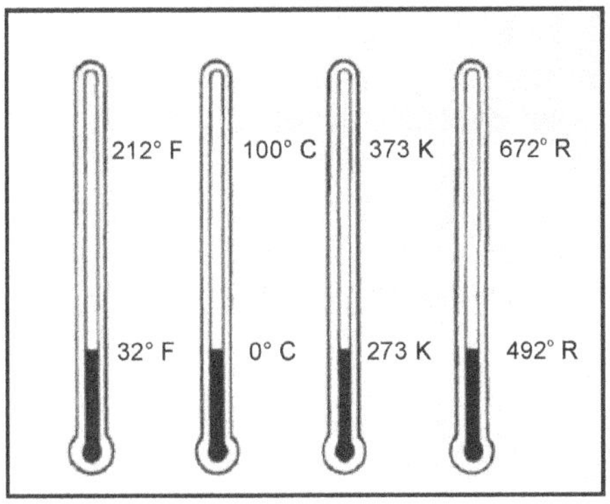

Figure 2-3. Temperature Scales. Fahrenheit, Celsius, Kelvin, and Rankine temperature scales showing the freezing and boiling points of water.

2-4.2.1 **Kelvin Scale.** One example of an absolute temperature scale is the Kelvin scale, which has the same size degrees as the Celsius scale. The freezing point of water is 273°K and boiling point of water is 373°K. Use this formula to convert from Celsius to absolute temperature (Kelvin):

Kelvin (K) = °C + 273.

2-4.2.2 **Rankine Scale.** The Rankine scale is another absolute temperature scale, which has the same size degrees as the Fahrenheit scale. The freezing point of water is 492°R and the boiling point of water is 672°R. Use this formula to convert from Fahrenheit to absolute temperature (degrees Rankine, °R):

°R = °F + 460

2-4.3 **Gas Measurements.** When measuring gas, actual cubic feet (acf) of a gas refers to the quantity of a gas at ambient conditions. The most common unit of measurement for gas in the United States is standard cubic feet (scf). Standard cubic feet relates the quantity measurement of a gas under pressure to a specific condition. The specific condition is a common basis for comparison. For air, the standard cubic foot is measured at 60°F and 14.696 psia.

2-5 ENERGY

Energy is the capacity to do work. The six basic types of energy are mechanical, heat, light, chemical, electromagnetic, and nuclear, and may appear in a variety of forms (Figure 2-4). Energy is a vast and complex aspect of physics beyond the scope of this manual. Consequently, this chapter only covers a few aspects of light, heat, and mechanical energy because of their unusual effects underwater and their impact on diving.

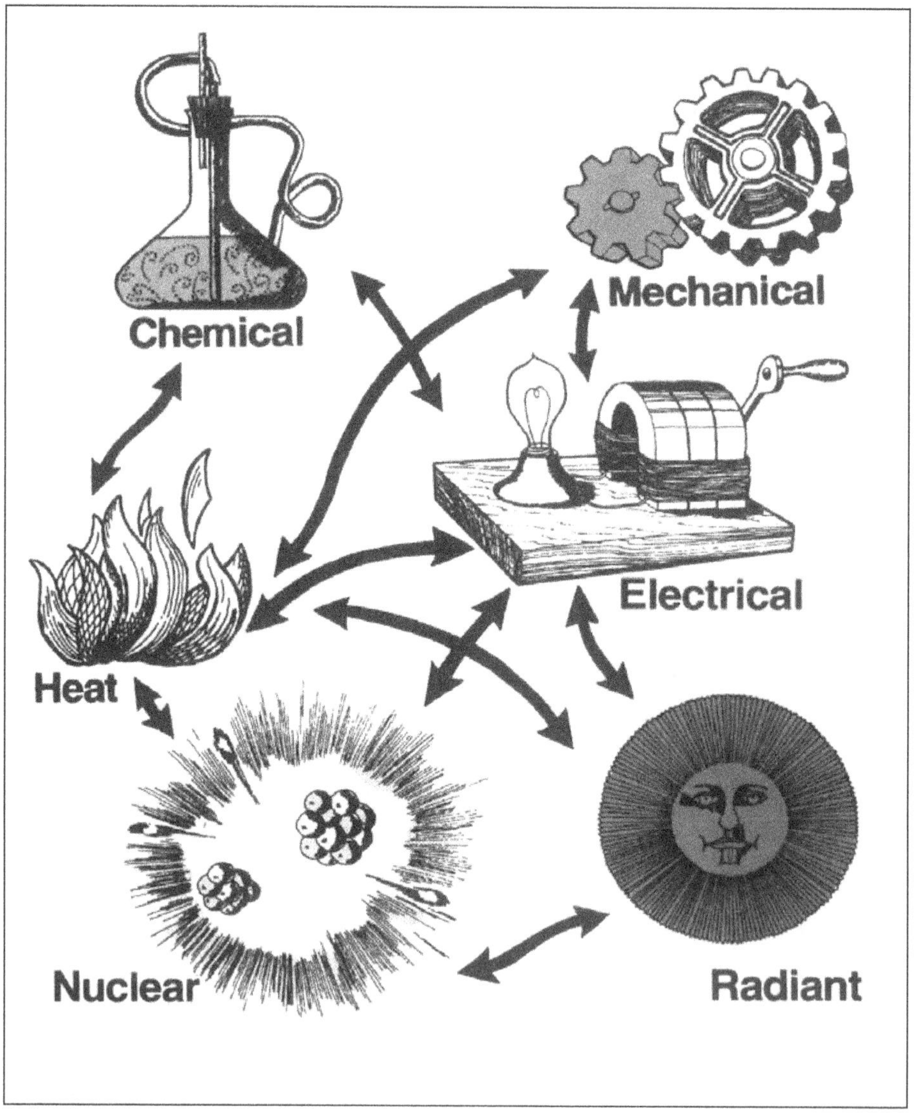

Figure 2-4. The Six Forms of Energy.

2-5.1 **Conservation of Energy.** The Law of the Conservation of Energy, formulated in the 1840s, states that energy in the universe can neither be created nor destroyed. Energy can be changed, however, from one form to another.

2-5.2 **Classifications of Energy.** The two general classifications of energy are potential energy and kinetic energy. Potential energy is due to position. An automobile parked on a hill with its brakes set possesses potential energy. Kinetic energy is energy of motion. An automobile rolling on a flat road possesses kinetic energy while it is moving.

2-6 LIGHT ENERGY IN DIVING

Refraction, turbidity of the water, salinity, and pollution all contribute to the distance, size, shape, and color perception of underwater objects. Divers must understand the factors affecting underwater visual perception, and must realize that distance perception is very likely to be inaccurate.

2-6.1 **Refraction.** Light passing from an object bends as it passes through the diver's faceplate and the air in his mask (Figure 2-5). This phenomenon is called refraction, and occurs because light travels faster in air than in water. Although the refraction that occurs between the water and the air in the diver's face mask produces undesirable perceptual inaccuracies, air is essential for vision. When a diver loses his face mask, his eyes are immersed in water, which has about the same refractive index as the eye. Consequently, the light is not focused normally and the diver's vision is reduced to a level that would be classified as legally blind on the surface.

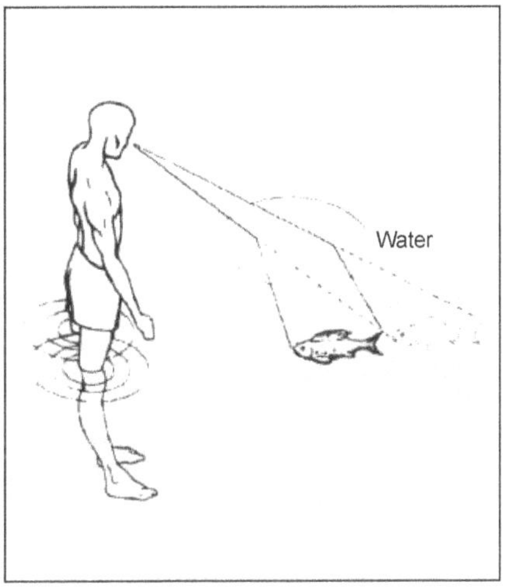

Figure 2-5. Objects Underwater Appear Closer.

Refraction can make objects appear closer than they really are. A distant object will appear to be approximately three-quarters of its actual distance. At greater distances, the effects of refraction may be reversed, making objects appear farther away than they actually are. Reduced brightness and contrast combine with refraction to affect visual distance relationships.

Refraction can also affect perception of size and shape. Generally, underwater objects appear to be about 30 percent larger than they actually are. Refraction effects are greater for objects off to the side in the field of view. This distortion interferes with hand-eye coordination, and explains why grasping objects underwater is sometimes difficult for a diver. Experience and training can help a diver learn to compensate for the misinterpretation of size, distance, and shape caused by refraction.

2-6.2 **Turbidity of Water.** Water turbidity can also profoundly influence underwater vision and distance perception. The more turbid the water, the shorter the distance at which the reversal from underestimation to overestimation occurs. For example, in highly turbid water, the distance of objects at 3 or 4 feet may be overestimated; in moderately turbid water, the change might occur at 20 to 25 feet and in very clear water, objects as far away as 50 to 70 feet might appear closer than they actually are. Generally speaking, the closer the object, the more it will appear to be too close, and the more turbid the water, the greater the tendency to see it as too far away.

2-6.3 **Diffusion.** Light scattering is intensified underwater. Light rays are diffused and scattered by the water molecules and particulate matter. At times diffusion is helpful because it scatters light into areas that otherwise would be in shadow or have no illumination. Normally, however, diffusion interferes with vision and underwater photography because the backscatter reduces the contrast between an object and its background. The loss of contrast is the major reason why vision underwater is so much more restricted than it is in air. Similar degrees of scattering occur in air only in unusual conditions such as heavy fog or smoke.

2-6.4 **Color Visibility.** Object size and distance are not the only characteristics distorted underwater. A variety of factors may combine to alter a diver's color perception. Painting objects different colors is an obvious means of changing their visibility by enhancing their contrast with the surroundings, or by camouflaging them to merge with the background. Determining the most and least visible colors is much more complicated underwater than in air.

Colors are filtered out of light as it enters the water and travels to depth. Red light is filtered out at relatively shallow depths. Orange is filtered out next, followed by yellow, green, and then blue. Water depth is not the only factor affecting the filtering of colors. Salinity, turbidity, size of the particles suspended in the water, and pollution all affect the color-filtering properties of water. Color changes vary from one body of water to another, and become more pronounced as the amount of water between the observer and the object increases.

The components of any underwater scene, such as weeds, rocks, and encrusting animals, generally appear to be the same color as the depth or viewing range increases. Objects become distinguishable only by differences in brightness and not color. Contrast becomes the most important factor in visibility; even very large objects may be undetectable if their brightness is similar to that of the background.

2-7 MECHANICAL ENERGY IN DIVING

Mechanical energy mostly affects divers in the form of sound. Sound is a periodic motion or pressure change transmitted through a gas, a liquid, or a solid. Because liquid is denser than gas, more energy is required to disturb its equilibrium. Once this disturbance takes place, sound travels farther and faster in the denser medium. Several aspects of sound underwater are of interest to the working diver.

2-7.1 Water Temperature and Sound. In any body of water, there may be two or more distinct contiguous layers of water at different temperatures; these layers are known as thermoclines. The colder a layer of water, the greater its density. As the difference in density between layers increases, the sound energy transmitted between them decreases. This means that a sound heard 50 meters from its source within one layer may be inaudible a few meters from its source if the diver is in another layer.

2-7.2 Water Depth and Sound. In shallow water or in enclosed spaces, reflections and reverberations from the air/water and object/water interfaces produce anomalies in the sound field, such as echoes, dead spots, and sound nodes. When swimming in shallow water, among coral heads, or in enclosed spaces, a diver can expect periodic losses in acoustic communication signals and disruption of acoustic navigation beacons. The problem becomes more pronounced as the frequency of the signal increases.

Because sound travels so quickly underwater (4,921 feet per second), human ears cannot detect the difference in time of arrival of a sound at each ear. Consequently, a diver cannot always locate the direction of a sound source. This disadvantage can have serious consequences for a diver or swimmer trying to locate an object or a source of danger, such as a powerboat.

2-7.2.1 **Diver Work and Noise.** Open-circuit SCUBA affects sound reception by producing high noise levels at the diver's head and by creating a screen of bubbles that reduces the effective sound pressure level (SPL). When several divers are working in the same area, the noise and bubbles affect communication signals more for some divers than for others, depending on the position of the divers in relation to the communicator and to each other.

A neoprene wet suit is an effective barrier to sound above 1,000 Hz and it becomes more of a barrier as frequency increases. This problem can be overcome by exposing a small area of the head either by cutting holes at the ears of the suit or by folding a small flap away from the surface.

2-7.2.2 **Pressure Waves.** Sound is transmitted through water as a series of pressure waves. High-intensity sound is transmitted by correspondingly high-intensity pressure waves. A high-pressure wave transmitted from the water surrounding a diver to the open spaces within the body (ears, sinuses, lungs) may increase the pressure within these open spaces, causing injury. Underwater explosions and sonar can create high-intensity sound or pressure waves. Low intensity sonar, such as depth finders and fish finders, do not produce pressure waves intense enough to endanger divers. However, anti-submarine sonar-equipped ships do pulse dangerous, high-intensity pressure waves.

It is prudent to suspend diving operations if a high-powered sonar transponder is being operated in the area. When using a diver-held pinger system, divers are advised to wear the standard ¼-inch neoprene hood for ear protection. Experiments have shown that such a hood offers adequate protection when the ultrasonic pulses are of 4-millisecond duration, repeated once per second for acoustic source levels up to 100 watts, at head-to-source distances as short as 0.5 feet (Pence and Sparks, 1978).

2-7.3 **Underwater Explosions.** An underwater explosion creates a series of waves that are transmitted as hydraulic shock waves in the water, and as seismic waves in the seabed. The hydraulic shock wave of an underwater explosion consists of an initial wave followed by further pressure waves of diminishing intensity. The initial high-intensity shock wave is the result of the violent creation and liberation of a large volume of gas, in the form of a gas pocket, at high pressure and temperature. Subsequent pressure waves are caused by rapid gas expansion in a non-compressible environment, causing a sequence of contractions and expansions as the gas pocket rises to the surface.

The initial high-intensity shock wave is the most dangerous; as it travels outward from the source of the explosion, it loses its intensity. Less severe pressure waves closely follow the initial shock wave. Considerable turbulence and movement of the water in the area of the explosion are evident for an extended time after the detonation.

2-7.3.1 **Type of Explosive and Size of the Charge.** Some explosives have characteristics of high brisance (shattering power in the immediate vicinity of the explosion) with less power at long range, while the brisance of others is reduced to increase their power over a greater area. Those with high brisance generally are used for cutting or shattering purposes, while high-power, low-brisance explosives are used in depth charges and sea mines where the target may not be in immediate contact and the ability to inflict damage over a greater area is an advantage. The high-brisance explosives create a high-level shock and pressure waves of short duration over a limited area. Low brisance explosives create a less intense shock and pressure waves of long duration over a greater area.

2-7.3.2 **Characteristics of the Seabed.** Aside from the fact that rock or other bottom debris may be propelled through the water or into the air with shallow-placed charges, bottom conditions can affect an explosion's pressure waves. A soft bottom tends to dampen reflected shock and pressure waves, while a hard, rock bottom may amplify the effect. Rock strata, ridges and other topographical features of the seabed may affect the direction of the shock and pressure waves, and may also produce secondary reflecting waves.

2-7.3.3 **Location of the Explosive Charge.** Research has indicated that the magnitude of shock and pressure waves generated from charges freely suspended in water is considerably greater than that from charges placed in drill holes in rock or coral.

2-7.3.4 **Water Depth.** At great depth, the shock and pressure waves are drawn out by the greater water volume and are thus reduced in intensity. An explosion near the surface is not weakened to the same degree.

2-7.3.5 **Distance from the Explosion.** In general, the farther away from the explosion, the greater the attenuation of the shock and pressure waves and the less the intensity. This factor must be considered in the context of bottom conditions, depth of water, and reflection of shock and pressure waves from underwater structures and topographical features.

2-7.3.6 **Degree of Submersion of the Diver.** A fully submerged diver receives the total effect of the shock and pressure waves passing over the body. A partially submerged diver whose head and upper body are out of the water, may experience a reduced effect of the shock and pressure waves on the lungs, ears, and sinuses. However, air will transmit some portion of the explosive shock and pressure waves. The head, lungs, and intestines are the parts of the body most vulnerable to the pressure effects of an explosion. A pressure wave of 500 pounds per square inch is sufficient to cause serious injury to the lungs and intestinal tract, and one greater than 2,000 pounds per square inch will cause certain death. Even a pressure wave of 500 pounds per square inch could cause fatal injury under certain circumstances.

2-7.3.7 **Estimating Explosion Pressure on a Diver.** There are various formulas for estimating the pressure wave resulting from an explosion of TNT. The equations vary in format and the results illustrate that the technique for estimation is only an approximation. Moreover, these formulas relate to TNT and are not applicable to other types of explosives.

The formula below (Greenbaum and Hoff, 1966) is one method of estimating the pressure on a diver resulting from an explosion of tetryl or TNT.

$$P = \frac{13{,}000\sqrt[3]{W}}{r}$$

Where:

P = pressure on the diver in pounds per square inch
W = weight of the explosive (TNT) in pounds
r = range of the diver from the explosion in feet

Sample Problem. Determine the pressure exerted by a 45-pound charge at a distance of 80 feet.

1. Substitute the known values.

$$P = \frac{13{,}000\sqrt[3]{45}}{80}$$

2. Solve for the pressure exerted.

$$\begin{aligned} P &= \frac{13{,}000\sqrt[3]{45}}{80} \\ &= \frac{13{,}000 \times 3.56}{80} \\ &= 578.5 \end{aligned}$$

Round up to 579 psi.

A 45-pound charge exerts a pressure of 579 pounds per square inch at a distance of 80 feet.

2-7.3.8 **Minimizing the Effects of an Explosion.** When expecting an underwater blast, the diver shall get out of the water and out of range of the blast whenever possible. If the diver must be in the water, it is prudent to limit the pressure he experiences from the explosion to less than 50 pounds per square inch. To minimize the effects, the diver can position himself with feet pointing toward and head directly away from the explosion. The head and upper section of the body should be out of the water or the diver should float on his back with his head out of the water.

2-8 HEAT ENERGY IN DIVING

Heat is crucial to man's environmental balance. The human body functions within only a very narrow range of internal temperature and contains delicate mechanisms to control that temperature.

Heat is a form of energy associated with and proportional to the molecular motion of a substance. It is closely related to temperature, but must be distinguished from temperature because different substances do not necessarily contain the same heat energy even though their temperatures are the same.

Heat is generated in many ways. Burning fuels, chemical reactions, friction, and electricity all generate heat. Heat is transmitted from one place to another by conduction, convection, and radiation.

2-8.1 **Conduction, Convection, and Radiation.** *Conduction* is the transmission of heat by direct contact. Because water is an excellent heat conductor, an unprotected diver can lose a great deal of body heat to the surrounding water by direct conduction.

Convection is the transfer of heat by the movement of heated fluids. Most home heating systems operate on the principle of convection, setting up a flow of air currents based on the natural tendency of warm air to rise and cool air to fall. A diver seated on the bottom of a tank of water in a cold room can lose heat not only by direct conduction to the water, but also by convection currents in the water. The warmed water next to his body will rise and be replaced by colder water passing along the walls of the tank. Upon reaching the surface, the warmed water will lose heat to the cooler surroundings. Once cooled, the water will sink only to be warmed again as part of a continuing cycle.

Radiation is heat transmission by electromagnetic waves of energy. Every warm object gives off waves of electromagnetic energy, which is absorbed by cool objects. Heat from the sun, electric heaters, and fireplaces is primarily radiant heat.

2-8.2 **Heat Transfer Rate.** To divers, conduction is the most significant means of transmitting heat. The rate at which heat is transferred by conduction depends on two basic factors:

- The difference in temperature between the warmer and cooler material
- The thermal conductivity of the materials

Not all substances conduct heat at the same rate. Iron, helium, and water are excellent heat conductors while air is a very poor conductor. Placing a poor heat conductor between a source of heat and another substance insulates the substance and slows the transfer of heat. Materials such as wool and foam rubber insulate the human body and are effective because they contain thousands of pockets of trapped air. The air pockets are too small to be subject to convective currents, but block conductive transfer of heat.

2-8.3 **Diver Body Temperature.** A diver will start to become chilled when the water temperature falls below a seemingly comfortable 70°F (21°C). Below 70°F, a diver wearing only a swimming suit loses heat to the water faster than his body can replace it. Unless he is provided some protection or insulation, he may quickly experience difficulties. A chilled diver cannot work efficiently or think clearly, and is more susceptible to decompression sickness.

Suit compression, increased gas density, thermal conductivity of breathing gases, and respiratory heat loss are contributory factors in maintaining a diver's body temperature. Cellular neoprene wet suits lose a major portion of their insulating properties as depth increases and the material compresses. As a consequence, it is often necessary to employ a thicker suit, a dry suit, or a hot water suit for extended exposures to cold water.

The heat transmission characteristics of an individual gas are directly proportional to its density. Therefore, the heat lost through gas insulating barriers and respiratory heat lost to the surrounding areas increase with depth. The heat loss is further aggravated when high thermal conductivity gases, such as helium-oxygen, are used for breathing. The respiratory heat loss alone increases from 10 percent of the body's heat generating capacity at one ata (atmosphere absolute), to 28 percent at 7 ata, to 50 percent at 21 ata when breathing helium-oxygen. Under these circumstances, standard insulating materials are insufficient to maintain body temperatures and supplementary heat must be supplied to the body surface and respiratory gas.

2-9 PRESSURE IN DIVING

Pressure is defined as a force acting upon a particular area of matter. It is typically measured in pounds per square inch (psi) in the English system and Newton per square centimeter (N/cm^2) in the System International (SI). Underwater pressure is a result of the weight of the water above the diver and the weight of the atmosphere over the water. There is one concept that must be remembered at all times—any diver, at any depth, must be in pressure balance with the forces at that depth. The body can only function normally when the pressure difference between the forces acting inside of the diver's body and forces acting outside is very small. Pressure, whether of the atmosphere, seawater, or the diver's breathing gases, must always be thought of in terms of maintaining pressure balance.

2-9.1 **Atmospheric Pressure.** Given that one atmosphere is equal to 33 feet of sea water or 14.7 psi, 14.7 psi divided by 33 feet equals 0.445 psi per foot. Thus, for every foot of sea water, the total pressure is increased by 0.445 psi. Atmospheric pressure is constant at sea level; minor fluctuations caused by the weather are usually ignored. Atmospheric pressure acts on all things in all directions.

Most pressure gauges measure differential pressure between the inside and outside of the gauge. Thus, the atmospheric pressure does not register on the pressure gauge of a cylinder of compressed air. The initial air in the cylinder and the gauge are already under a base pressure of one atmosphere (14.7 psi or $10N/cm^2$). The gauge measures the pressure difference between the atmosphere and the increased air pressure in the tank. This reading is called *gauge pressure* and for most purposes it is sufficient.

In diving, however, it is important to include atmospheric pressure in computations. This total pressure is called *absolute pressure* and is normally expressed in units of atmospheres. The distinction is important and pressure must be identified as either gauge (psig) or absolute (psia). When the type of pressure is identified only as psi, it refers to gauge pressure. Table 2-10 contains conversion factors for pressure measurement units.

2-9.2 **Terms Used to Describe Gas Pressure.** Four terms are used to describe gas pressure:

- **Atmospheric.** Standard atmosphere, usually expressed as $10N/cm^2$, 14.7 psi, or one atmosphere absolute (1 ata).
- **Barometric.** Essentially the same as atmospheric but varying with the weather and expressed in terms of the height of a column of mercury. Standard pressure is equal to 29.92 inches of mercury, 760 millimeters of mercury, or 1013 millibars.
- **Gauge.** Indicates the difference between atmospheric pressure and the pressure being measured.
- **Absolute.** The total pressure being exerted, i.e., gauge pressure plus atmospheric pressure.

2-9.3 **Hydrostatic Pressure.** The water on the surface pushes down on the water below and so on down to the bottom where, at the greatest depths of the ocean (approximately 36,000 fsw), the pressure is more than 8 tons per square inch (1,100 ata). The pressure due to the weight of a water column is referred to as hydrostatic pressure.

The pressure of seawater at a depth of 33 feet equals one atmosphere. The absolute pressure, which is a combination of atmospheric and water pressure for that depth, is two atmospheres. For every additional 33 feet of depth, another atmosphere of pressure (14.7 psi) is encountered. Thus, at 99 feet, the absolute pressure is equal to four atmospheres. Table 2-1 and Figure 2-7 shows how pressure increases with depth.

Table 2-1. Pressure Chart.

Depth Gauge Pressure	Atmospheric Pressure	Absolute Pressure
0	One Atmosphere	1 ata (14.7 psia)
33 fsw	+ One Atmosphere	2 ata (29.4 psia)
66 fsw	+ One Atmosphere	3 ata (44.1 psia)
99 fsw	+ One Atmosphere	4 ata (58.8 psia)

The change in pressure with depth is so pronounced that the feet of a 6-foot tall person standing underwater are exposed to pressure that is almost 3 pounds per square inch greater than that exerted at his head.

2-9.4 **Buoyancy.** Buoyancy is the force that makes objects float. It was first defined by the Greek mathematician Archimedes, who established that "Any object wholly or partly immersed in a fluid is buoyed up by a force equal to the weight of the fluid displaced by the object." This is known as Archimedes' Principle and applies to all objects and all fluids.

2-9.4.1 **Archimedes' Principle.** According to Archimedes' Principle, the buoyancy of a submerged body can be established by subtracting the weight of the submerged body from the weight of the displaced liquid. If the total displacement (the weight of the displaced liquid) is greater than the weight of the submerged body, the buoyancy is positive and the body will float or be buoyed upward. If the weight of the body is equal to that of the displaced liquid, the buoyancy is neutral and the body will remain suspended in the liquid. If the weight of the submerged body is greater than that of the displaced liquid, the buoyancy is negative and the body will sink.

The buoyant force on an object is dependent upon the density of the substance it is immersed in (weight per unit volume). Fresh water has a density of 62.4 pounds per cubic foot. Sea water is heavier, having a density of 64.0 pounds per cubic foot. Thus an object is buoyed up by a greater force in seawater than in fresh water, making it easier to float in the ocean than in a fresh water lake.

2-9.4.2 **Diver Buoyancy.** Lung capacity has a significant effect on buoyancy of a diver. A diver with full lungs displaces a greater volume of water and, therefore, is more buoyant than with deflated lungs. Individual differences that may affect the buoyancy of a diver include bone structure, bone weight, and body fat. These differences explain why some individuals float easily while others do not.

A diver can vary his buoyancy in several ways. By adding weight to his gear, he can cause himself to sink. When wearing a variable volume dry suit, he can increase or decrease the amount of air in his suit, thus changing his displacement and thereby his buoyancy. Divers usually seek a condition of neutral to slightly negative buoyancy. Negative buoyancy gives a diver in a helmet and dress a better foothold on the bottom. Neutral buoyancy enhances a SCUBA diver's ability to swim easily, change depth, and hover.

2-10 GASES IN DIVING

Knowledge of the properties and behavior of gases, especially those used for breathing, is vitally important to divers.

2-10.1 **Atmospheric Air.** The most common gas used in diving is atmospheric air, the composition of which is shown in Table 2-2. Any gases found in concentrations different than those in Table 2-2 or that are not listed in Table 2-2 are considered contaminants. Depending on weather and location, many industrial pollutants may be found in air. Carbon monoxide is the most commonly encountered and is often present around air compressor engine exhaust. Care must be taken to exclude the pollutants from the diver's compressed air by appropriate filtering, inlet location, and compressor maintenance. Water vapor in varying quantities is present in compressed air and its concentration is important in certain instances.

Table 2-2. *Components of Dry Atmospheric Air.*

	Concentration	
Component	**Percent by Volume**	**Parts per Million (ppm)**
Nitrogen	78.084	
Oxygen	20.9476	
Carbon Dioxide	0.038	380
Argon	0.0934	
Neon		18.18
Helium		5.24
Krypton		1.14
Xenon		0.08
Hydrogen		0.5
Methane		2.0
Nitrous Oxide		0.5

For most purposes and computations, diving air may be assumed to be composed of 79 percent nitrogen and 21 percent oxygen. Besides air, varying mixtures of oxygen, nitrogen, and helium are commonly used in diving. While these gases are discussed separately, the gases themselves are almost always used in some mixture. Air is a naturally occurring mixture of most of them. In certain types of diving applications, special mixtures may be blended using one or more of the gases with oxygen.

2-10.2 **Oxygen.** Oxygen (O_2) is the most important of all gases and is one of the most abundant elements on earth. Fire cannot burn without oxygen and people cannot survive without oxygen. Atmospheric air contains approximately 21 percent oxygen, which exists freely in a diatomic state (two atoms paired off to make one molecule). This colorless, odorless, tasteless, and active gas readily combines with other elements. From the air we breathe, only oxygen is actually used by the body. The other 79 percent of the air serves to dilute the oxygen. Pure 100 percent oxygen is often used for breathing in hospitals, aircraft, and hyperbaric medical treatment facilities. Sometimes 100 percent oxygen is used in shallow diving operations and certain phases of mixed-gas diving operations. However, breathing pure oxygen under pressure may induce the serious problems of oxygen toxicity.

2-10.3 **Nitrogen.** Like oxygen, nitrogen (N_2) is diatomic, colorless, odorless, and tasteless, and is a component of all living organisms. Unlike oxygen, it will not support life or aid combustion and it does not combine easily with other elements. Nitrogen in the air is inert in the free state. For diving, nitrogen may be used to dilute oxygen. Nitrogen is not the only gas that can be used for this purpose and under some conditions it has severe disadvantages as compared to other gases. Nitrogen narcosis, a disorder resulting from the anesthetic properties of nitrogen breathed under pressure, can result in a loss of orientation and judgment by the diver. For this reason, compressed air, with its high nitrogen content, is not used below a specified depth in diving operations.

2-10.4 **Helium.** Helium (He) is a colorless, odorless, and tasteless gas, but it is monatomic (exists as a single atom in its free state). It is totally inert. Helium is a rare element, found in air only as a trace element of about 5 parts per million (ppm). Helium coexists with natural gas in certain wells in the southwestern United States, Canada, and Russia. These wells provide the world's supply. When used in diving to dilute oxygen in the breathing mixture, helium does not cause the same problems associated with nitrogen narcosis, but it does have unique disadvantages. Among these is the distortion of speech which takes place in a helium atmosphere. The "Donald Duck" effect is caused by the acoustic properties of helium and it impairs voice communications in deep diving. Another negative characteristic of helium is its high thermal conductivity which can cause rapid loss of body and respiratory heat.

2-10.5 **Hydrogen.** Hydrogen (H_2) is diatomic, colorless, odorless, and tasteless, and is so active that it is rarely found in a free state on earth. It is, however, the most abundant element in the visible universe. The sun and stars are almost pure hydrogen. Pure hydrogen is violently explosive when mixed with air in proportions that include a presence of more than 5.3 percent oxygen. Hydrogen has been used in diving (replacing nitrogen for the same reasons as helium) but the hazards have limited this to little more than experimentation.

2-10.6 **Neon.** Neon (Ne) is inert, monatomic, colorless, odorless, and tasteless, and is found in minute quantities in the atmosphere. It is a heavy gas and does not exhibit the narcotic properties of nitrogen when used as a breathing medium. Because it does not cause the speech distortion problem associated with helium and has superior thermal insulating properties, it has been the subject of some experimental diving research.

2-10.7 **Carbon Dioxide.** Carbon dioxide (CO_2) is colorless, odorless, and tasteless when found in small percentages in the air. In greater concentrations it has an acid taste and odor. Carbon dioxide is a natural by-product of animal and human respiration, and is formed by the oxidation of carbon in food to produce energy. For divers, the two major concerns with carbon dioxide are control of the quantity in the breathing supply and removal of the exhaust after breathing. Carbon dioxide can cause unconsciousness when breathed at increased partial pressure. In high concentrations the gas can be extremely toxic. In the case of closed and semiclosed breathing apparatus, the removal of excess carbon dioxide generated by breathing is essential to safety.

2-10.8 **Carbon Monoxide.** Carbon monoxide (CO) is a colorless, odorless, tasteless, and poisonous gas whose presence is difficult to detect. Carbon monoxide is formed as a product of incomplete fuel combustion, and is most commonly found in the exhaust of internal combustion engines. A diver's air supply can be contaminated by carbon monoxide when the compressor intake is placed too close to the compressor's engine exhaust. The exhaust gases are sucked in with the air and sent on to the diver, with potentially disastrous results. Carbon monoxide seriously interferes with the blood's ability to carry the oxygen required for the body to function normally. The affinity of carbon monoxide for hemoglobin is approximately 210 times that of oxygen. Carbon monoxide dissociates from hemoglobin at a much slower rate than oxygen.

2-10.9 **Kinetic Theory of Gases.** On the surface of the earth the constancy of the atmosphere's pressure and composition tend to be accepted without concern. To the diver, however, the nature of the high pressure or hyperbaric, gaseous environment assumes great importance. The basic explanation of the behavior of gases under all variations of temperature and pressure is known as the kinetic theory of gases.

The kinetic theory of gases states: "The kinetic energy of any gas at a given temperature is the same as the kinetic energy of any other gas at the same temperature." Consequently, the measurable pressures of all gases resulting from kinetic activity are affected by the same factors.

The kinetic energy of a gas is related to the speed at which the molecules are moving and the mass of the gas. Speed is a function of temperature and mass is a function of gas type. At a given temperature, molecules of heavier gases move at a slower speed than those of lighter gases, but their combination of mass and speed results in the same kinetic energy level and impact force. The measured impact force, or pressure, is representative of the kinetic energy of the gas. This is illustrated in Figure 2-6.

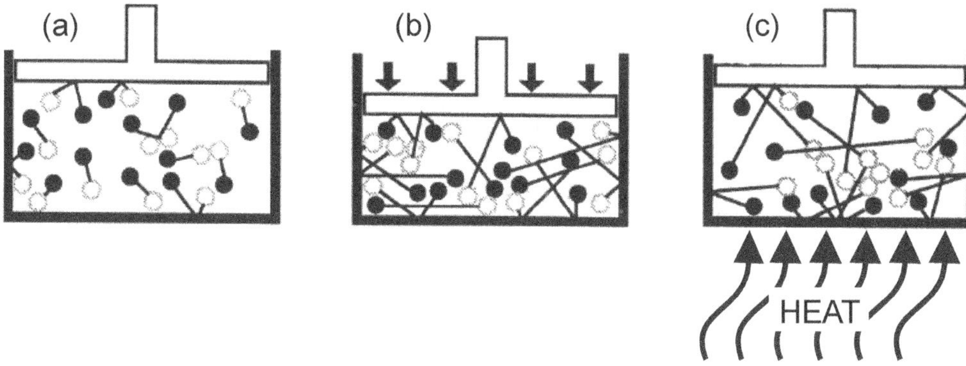

Figure 2-6. Kinetic Energy. The kinetic energy of the molecules inside the container (a) produces a constant pressure on the internal surfaces. As the container volume is decreased (b), the molecules per unit volume (density) increase and so does the pressure. As the energy level of the molecules increases from the addition of thermal energy (heat), so does the pressure (c).

2-11 GAS LAWS

Gases are subject to three closely interrelated factors—temperature, pressure, and volume. As the kinetic theory of gases points out, a change in one of these factors must result in some measurable change in the other factors. Further, the theory indicates that the kinetic behavior of any one gas is the same for all gases or mixtures of gases. Consequently, basic laws have been established to help predict the changes that will be reflected in one factor as the conditions of one or both of the other factors change. A diver needs to know how changing pressure will effect the air in his suit and lungs as he moves up and down in the water. He must be able to determine whether an air compressor can deliver an adequate supply of air to a proposed operating depth. He also needs to be able to interpret the reading on the pressure gauge of his tanks under varying conditions of temperature and pressure. The answers to such questions are calculated using a set of rules called the gas laws. This section explains the gas laws of direct concern to divers.

2-11.1 **Boyle's Law.** Boyle's law states that at constant temperature, the absolute pressure and the volume of gas are inversely proportional. As pressure increases the gas volume is reduced; as the pressure is reduced the gas volume increases. Boyle's law is important to divers because it relates to change in the volume of a gas caused by the change in pressure, due to depth, which defines the relationship of pressure and volume in breathing gas supplies.

The formula for Boyle's law is: $C = P \times V$

Where:

C = a constant
P = absolute pressure
V = volume

Boyle's law can also be expressed as: $P_1V_1 = P_2V_2$

Where:

P_1 = initial pressure
V_1 = initial volume
P_2 = final pressure
V_2 = final volume

When working with Boyle's law, pressure may be measured in atmospheres absolute. To calculate pressure using atmospheres absolute:

$$P_{ata} = \frac{\text{Depth fsw} + 33\text{ fsw}}{33\text{ fsw}} \quad \text{or} \quad P_{ata} = \frac{\text{psig} + 14.7\text{ psi}}{14.7\text{ psi}}$$

Sample Problem 1. An open diving bell with a volume of 24 cubic feet is to be lowered into the sea from a support craft. No air is supplied to or lost from the bell. Calculate the volume of the air in the bell at 99 fsw.

1. Rearrange the formula for Boyle's law to find the final volume (V_2):

$$V_2 = \frac{P_1V_1}{P_2}$$

2. Calculate the final pressure (P_2) at 99 fsw:

$$P_2 = \frac{99\text{ fsw} + 33\text{ fsw}}{33\text{ fsw}}$$
$$= 4\text{ ata}$$

3. Substitute known values to find the final volume:

$$V_2 = \frac{1\text{ata} \times 24\text{ ft}^3}{4\text{ ata}}$$
$$= 6\text{ ft}^3$$

The volume of air in the open bell has been compressed to 6 ft^3 at 99 fsw.

2-11.2 **Charles'/Gay-Lussac's Law.** When working with Boyle's law, the temperature of the gas is a constant value. However, temperature significantly affects the pressure and volume of a gas. Charles'/Gay-Lussac's law describes the physical relationships of temperature upon volume and pressure. Charles'/Gay-Lussac's law states that at a constant pressure, the volume of a gas is directly proportional to the change in the absolute temperature. If the pressure is kept constant and the absolute temperature is doubled, the volume will double. If the temperature decreases, volume decreases. If volume instead of pressure is kept constant (i.e., heating in a rigid container), then the absolute pressure will change in proportion to the absolute temperature.

The formulas for expressing Charles'/Gay-Lussac's law are as follows.

For the relationship between volume and temperature:

$$\frac{V_1}{T_1} = \frac{V_2}{T_2}$$

Where: Pressure is constant

T_1 = initial temperature (absolute)

T_2 = final temperature (absolute)

V_1 = initial volume

V_2 = final volume

And, for the relationship between pressure and temperature:

$$\frac{P_1}{T_1} = \frac{P_2}{T_2}$$

Where: Volume is constant

P_1 = initial pressure (absolute)

P_2 = final pressure (absolute)

T_1 = initial temperature (absolute)

T_2 = final temperature (absolute)

Sample Problem 1. An open diving bell of 24 cubic feet capacity is lowered into the ocean to a depth of 99 fsw. The surface temperature is 80°F, and the temperature at depth is 45°F. From the sample problem illustrating Boyle's law, we know that the volume of the gas was compressed to 6 cubic feet when the bell was lowered to 99 fsw. Apply Charles'/Gay-Lussac's law to determine the volume when it is effected by temperature.

1. Convert Fahrenheit temperatures to absolute temperatures (Rankine):

$$°R = °F + 460$$

$$T_1 = 80°F + 460$$

$$= 540°R$$

$$T_2 = 45°F + 460$$

$$= 505°R$$

2. Transpose the formula for Charles'/Gay-Lussac's law to solve for the final volume (V_2):

$$V_2 = \frac{V_1 T_2}{T_1}$$

3. Substitute known values to solve for the final volume (V_2):

$$V_2 = \frac{6\ \text{ft.}^3 \times 505}{540}$$
$$= 5.61\ \text{ft.}^3$$

The volume of the gas at 99 fsw is 5.61 ft^3.

Sample Problem 2. A 6-cubic-foot flask is charged to 3000 psig and the temperature in the flask room is 72° F. A fire in an adjoining space causes the temperature in the flask room to reach 170° F. What will happen to the pressure in the flask?

1. Convert gauge pressure unit to atmospheric pressure unit:

$$P_1 = 3000\ \text{psig} + 14.7\ \text{psi}$$
$$= 3014.7\ \text{psia}$$

2. Convert Fahrenheit temperatures to absolute temperatures (Rankine):

$$°R = °F + 460$$
$$T_1 = 72°F + 460$$
$$= 532°R$$
$$T_2 = 170°F + 460$$
$$= 630°R$$

3. Transpose the formula for Gay-Lussac's law to solve for the final pressure (P_2):

$$P_2 = \frac{P_1 T_2}{T_1}$$

4. Substitute known values and solve for the final pressure (P_2):

$$P_2 = \frac{3014.7 \times 630}{532}$$
$$= \frac{1,899,261}{532}$$
$$= 3570.03\ \text{psia} - 14.7$$
$$= 3555.33\ \text{psig}$$

The pressure in the flask increased from 3000 psig to 3555.33 psig. Note that the pressure increased even though the flask's volume and the volume of the gas remained the same.

This example also shows what would happen to a SCUBA cylinder that was filled to capacity and left unattended in the trunk of an automobile or lying in direct sunlight on a hot day.

2-11.3 **The General Gas Law.** Boyle, Charles, and Gay-Lussac demonstrated that temperature, volume, and pressure affect a gas in such a way that a change in one factor must be balanced by corresponding change in one or both of the others. Boyle's law describes the relationship between pressure and volume, Charles'/ Gay-Lussac's law describes the relationship between temperature and volume and the relationship between temperature and pressure. The general gas law combines the laws to predict the behavior of a given quantity of gas when any of the factors change.

The formula for expressing the general gas law is: $\frac{P_1V_1}{T_1} = \frac{P_2V_2}{T_2}$

Where:

P_1 = initial pressure (absolute)
V_1 = initial volume
T_1 = initial temperature (absolute)
P_2 = final pressure (absolute)
V_2 = final volume
T_2 = final temperature (absolute)

Two simple rules must be kept in mind when working with the general gas law:

- There can be only one unknown value.
- The equation can be simplified if it is known that a value remains unchanged (such as the volume of an air cylinder) or that the change in one of the variables is of little consequence. In either case, cancel the value out of both sides of the equation to simplify the computations.

Sample Problem 1. Your ship has been assigned to salvage a sunken LCM landing craft located in 130 fsw. An exploratory dive, using SCUBA, is planned to survey the wreckage. The SCUBA cylinders are charged to 2,250 psig, which raises the temperature in the tanks to 140 °F. From experience in these waters, you know that the temperature at the operating depth will be about 40°F. Apply the general gas law to find what the gauge reading will be when you first reach the bottom. (Assume no loss of air due to breathing.)

1. Simplify the equation by eliminating the variables that will not change. The volume of the tank will not change, so V_1 and V_2 can be eliminated from the formula in this problem:

$$\frac{P_1}{T_1} = \frac{P_2}{T_2}$$

2. Calculate the initial pressure by converting the gauge pressure unit to the atmospheric pressure unit:

$$P_1 = 2{,}250 \text{ psig} + 14.7$$
$$= 2{,}264.7 \text{ psia}$$

3. Convert Fahrenheit temperatures to Rankine (absolute) temperatures:

   Conversion formula: °R = °F + 460

$$T_1 = 140° \text{ F} + 460$$
$$= 600° \text{ R}$$
$$T_2 = 40° \text{ F} + 460$$
$$= 500° \text{ R}$$

4. Rearrange the formula to solve for the final pressure (P_2):

$$P_2 = \frac{P_1 T_2}{T_1}$$

5. Fill in known values:

$$P_2 = \frac{2{,}264.7 \text{ psia} \times 500°\text{R}}{600°\text{R}}$$
$$= 1887.25 \text{ psia}$$

6. Convert final pressure (P_2) to gauge pressure:

$$P_2 = 1{,}887.25 \text{ psia} - 14.7$$
$$= 1{,}872.55 \text{ psia}$$

The gauge reading when you reach bottom will be 1,872.55 psig.

Sample Problem 2. During the survey dive for the operation outlined in Sample Problem 1, the divers determined that the damage will require a simple patch. The Diving Supervisor elects to use surface-supplied MK 21 equipment. The compressor discharge capacity is 60 cubic feet per minute, and the air temperature on the deck of the ship is 80°F.

Apply the general gas law to determine whether the compressor can deliver the proper volume of air to both the working diver and the standby diver at the operating depth and temperature.

1. Calculate the absolute pressure at depth (P_2):

$$P_2 = \frac{130 \text{ fsw} + 33 \text{ fsw}}{33 \text{ fsw}}$$
$$= 4.93 \text{ ata}$$

2. Convert Fahrenheit temperatures to Rankine (absolute) temperatures:

   Conversion formula:

$$°R = °F + 460$$
$$T_1 = 80°F + 460$$
$$= 540°R$$
$$T_2 = 40°F + 460$$
$$= 500°R$$

3. Rearrange the general gas law formula to solve for the volume of air at depth (V_2):

$$V_2 = \frac{P_1 V_1 T_2}{P_2 T_1}$$

4. Substitute known values and solve:

$$V_2 = \frac{1 \text{ ata} \times 60 \text{ cfm} \times 500°R}{4.93 \text{ ata} \times 540°R}$$
$$= 11.26 \text{ acfm at bottom conditions}$$

Based upon an actual volume (displacement) flow requirement of 1.4 acfm for a deep-sea diver, the compressor capacity is sufficient to support the working and standby divers at 130 fsw.

Sample Problem 3. Find the actual cubic feet of air contained in a 700-cubic inch internal volume cylinder pressurized to 3,000 psi.

1. Simplify the equation by eliminating the variables that will not change. The temperature of the tank will not change so T_1 and T_2 can be eliminated from the formula in this problem:

$$P_1 V_1 = P_2 V_2$$

2. Rearrange the formula to solve for the initial volume:

$$V_1 = \frac{P_2 V_2}{P_1}$$

Where:

$P_1 =$ 14.7 psi

$P_2 =$ 3,000 psi + 14.7 psi

$V_2 =$ 700 in^3

3. Fill in the known values and solve for V_1:

$$V_1 = \frac{3014.7 \text{ psia} \times 700 \text{ in}^3}{14.7 \text{ psi}}$$
$$= 143{,}557.14 \text{ in}^3$$

4. Convert V_1 to cubic feet:

$$V_1 = \frac{143{,}557.14 \text{ in}^3}{1728 \text{ in}^3} \quad (1728 \text{ in}^3 = 1 \text{ ft}^3)$$
$$= 83.07 \text{ scf}$$

2-12 GAS MIXTURES

If a diver used only one gas for all underwater work, at all depths, then the general gas law would suffice for most of his necessary calculations. However, to accommodate use of a single gas, oxygen would have to be chosen because it is the only one that provides life support. But 100 percent oxygen can be dangerous to a diver as depth and breathing time increase. Divers usually breathe gases in a mixture, either air (21 percent oxygen, 78 percent nitrogen, 1 percent other gases) or oxygen with one of the inert gases serving as a diluent for the oxygen. The human body has a wide range of reactions to various gases under different conditions of pressure and for this reason another gas law is required to help compute the differences between breathing at the surface and breathing under pressure.

2-12.1 **Dalton's Law.** Dalton's law states: "The total pressure exerted by a mixture of gases is equal to the sum of the pressures of each of the different gases making up the mixture, with each gas acting as if it alone was present and occupied the total volume."

In a gas mixture, the portion of the total pressure contributed by a single gas is called the partial pressure (pp) of that gas. An easily understood example is that of a container at atmospheric pressure (14.7 psi). If the container were filled with oxygen alone, the partial pressure of the oxygen would be one atmosphere. If the same container at 1 atm were filled with dry air, the partial pressures of all the constituent gases would contribute to the total partial pressure, as shown in Table 2-3.

If the same container was filled with air to 2,000 psi (137 ata), the partial pressures of the various components would reflect the increased pressure in the same proportion as their percentage of the gas, as illustrated in Table 2-4.

Table 2-3. *Partial Pressure at 1 ata.*

Gas	Percent of Component	Atmospheres Partial Pressure
N_2	78.08	0.7808
O_2	20.95	0.2095
CO_2	.03	0.0003
Other	.94	0.0094
Total	100.00	1.0000

Table 2-4. *Partial Pressure at 137 ata.*

Gas	Percent of Component	Atmospheres Partial Pressure
N_2	78.08	106.97
O_2	20.95	28.70
CO_2	.03	0.04
Other	.94	1.29
Total	100.00	137.00

The formula for expressing Dalton's law is:

$$P_{Total} = pp_A + pp_B + pp_C + \ldots$$

Where: A, B, and C are gases and

$$pp_A = \frac{P_{Total} \times \%Vol_A}{1.00}$$

Another method of arriving at the same conclusion is to use the T formula. When using the T formula, there can be only one unknown value. Then it is merely a case of multiplying across, or dividing up to solve for the unknown value. The T formula is illustrated as:

$$\frac{\text{partial pressure}}{\text{atmosphere(s) absolute} \mid \text{\% volume (in decimal form)}}$$

Sample Problem 1. Use the T formula to calculate oxygen partial pressure given 10 ata and 16 percent oxygen.

1. Fill in the known values:

$$\frac{pp}{10 \mid .16}$$

2. Multiply the pressure by the volume to solve for the oxygen partial pressure (pp):

$$\frac{1.6\ ppO_2}{10 \mid .16}$$

The oxygen partial pressure is 1.6.

Sample Problem 2. What happens to the breathing mixture at the operating depth of 130 fsw (4.93 ata)? The air compressor on the ship is taking in air at the surface, at normal pressure and normal mixture, and sending it to the diver at pressure sufficient to provide the necessary balance. The composition of air is not changed, but the quantity being delivered to the diver is five times what he was breathing on the surface. More molecules of oxygen, nitrogen, and carbon dioxide are all compressed into the same volume at the higher pressure. Use Dalton's law to determine the partial pressures at depth.

1. Calculate the oxygen partial pressure at depth.

$$ppO_2 = .21\ (\text{surface}) \times 4.93\ \text{ata}$$

$$= 1.03\ \text{ata}$$

2. Calculate the nitrogen partial pressure at depth.

$$ppN_2 = .79\ (\text{surface}) \times 4.93\ \text{ata}$$

$$= 3.89\ \text{ata}$$

3. Calculate the carbon dioxide partial pressure at depth.

$$ppCO_2 = .0003\ (\text{surface}) \times 4.93\ \text{ata}$$

$$= .0014\ \text{ata}$$

2-12.1.1 **Expressing Small Quantities of Pressure.** Expressing partial pressures of gases in atmospheres absolute (ata) is the most common method employed in large quantities of pressure. Partial pressures of less than 0.1 atmosphere are usually expressed in millimeters of mercury (mmHg). At the surface, atmospheric pressure is equal to 1 ata or 14.7 psia or 760 mmHg. The formula used to calculate the $ppCO_2$ at 130 fsw in millimeters of mercury is:

$$ppCO_2 = \frac{0.03}{100} \times 4.93 \text{ ata} \times \frac{760\text{mmHg}}{1 \text{ ata}}$$
$$= 1.12\text{mmHg}$$

2-12.1.2 **Calculating Surface Equivalent Value.** From the previous calculations, it is apparent that the diver is breathing more molecules of oxygen breathing air at 130 fsw than he would be if using 100 percent oxygen at the surface. He is also inspiring five times as many carbon dioxide molecules as he would breathing normal air on the surface. If the surface air were contaminated with 2 percent (0.02 ata) carbon dioxide, a level that could be readily accommodated by a normal person at one ata, the partial pressure at depth would be dangerously high—0.0986 ata (0.02 x 4.93 ata). This partial pressure is commonly referred to as a surface equivalent value (sev) of 10 percent carbon dioxide. The formula for calculating the surface equivalent value is:

$$\text{sev} = \frac{\text{pp at depth (in ata)} \times 100\%}{1 \text{ ata}}$$
$$= \frac{0.0986 \text{ ata}}{1 \text{ ata}} \times 100\%$$
$$= 9.86\% \; CO_2$$

2-12.2 **Gas Diffusion.** Another physical effect of partial pressures and kinetic activity is that of gas diffusion. Gas diffusion is the process of intermingling or mixing of gas molecules. If two gases are placed together in a container, they will eventually mix completely even though one gas may be heavier. The mixing occurs as a result of constant molecular motion.

An individual gas will move through a permeable membrane (a solid that permits molecular transmission) depending upon the partial pressure of the gas on each side of the membrane. If the partial pressure is higher on one side, the gas molecules will diffuse through the membrane from the higher to the lower partial pressure side until the partial pressure on sides of the membrane are equal. Molecules are actually passing through the membrane at all times in both directions due to kinetic activity, but more will move from the side of higher concentration to the side of lower concentration.

Body tissues are permeable membranes. The rate of gas diffusion, which is related to the difference in partial pressures, is an important consideration in determining the uptake and elimination of gases in calculating decompression tables.

2-12.3 **Humidity.** Humidity is the amount of water vapor in gaseous atmospheres. Like other gases, water vapor behaves in accordance with the gas laws. However, unlike other gases encountered in diving, water vapor condenses to its liquid state at temperatures normally encountered by man.

Humidity is related to the vapor pressure of water, and the maximum partial pressure of water vapor in the gas is governed entirely by the temperature of the gas.

As the gas temperature increases, more molecules of water can be maintained in the gas until a new equilibrium condition and higher maximum partial pressure are established. As a gas cools, water vapor in the gas condenses until a lower partial pressure condition exists regardless of the total pressure of the gas. The temperature at which a gas is saturated with water vapor is called the *dewpoint*.

In proper concentrations, water vapor in a diver's breathing gas can be beneficial to the diver. Water vapor moistens body tissues, thus keeping the diver comfortable. As a condensing liquid, however, water vapor can freeze and block air passageways in hoses and equipment, fog a diver's faceplate, and corrode his equipment.

2-12.4 **Gases in Liquids.** When a gas comes in contact with a liquid, a portion of the gas molecules enters into solution with the liquid. The gas is said to be *dissolved* in the liquid. Solubility is vitally important because significant amounts of gases are dissolved in body tissues at the pressures encountered in diving.

2-12.5 **Solubility.** Some gases are more soluble (capable of being dissolved) than others, and some liquids and substances are better solvents (capable of dissolving another substance) than others. For example, nitrogen is five times more soluble in fat than it is in water.

Apart from the individual characteristics of the various gases and liquids, temperature and pressure greatly affect the quantity of gas that will be absorbed. Because a diver is always operating under unusual conditions of pressure, understanding this factor is particularly important.

2-12.6 **Henry's Law.** Henry's law states: "The amount of any given gas that will dissolve in a liquid at a given temperature is directly proportional to the partial pressure of that gas." Because a large percentage of the human body is water, the law simply states that as one dives deeper and deeper, more gas will dissolve in the body tissues and that upon ascent, the dissolved gas must be released.

2-12.6.1 **Gas Tension.** When a gas-free liquid is first exposed to a gas, quantities of gas molecules rush to enter the solution, pushed along by the partial pressure of the gas. As the molecules enter the liquid, they add to a state of gas tension. Gas tension is a way of identifying the partial pressure of that gas in the liquid.

The difference between the gas tension and the partial pressure of the gas outside the liquid is called the *pressure gradient*. The pressure gradient indicates the rate at which the gas enters or leaves the solution.

2-12.6.2 **Gas Absorption.** At sea level, the body tissues are equilibrated with dissolved nitrogen at a partial pressure equal to the partial pressure of nitrogen in the lungs. Upon exposure to altitude or increased pressure in diving, the partial pressure of nitrogen in the lungs changes and tissues either lose or gain nitrogen to reach a new equilibrium with the nitrogen pressure in the lungs. Taking up nitrogen in tissues is called *absorption* or *uptake*. Giving up nitrogen from tissues is termed *elimination* or *offgassing*. In air diving, nitrogen absorption occurs when a diver is exposed to an increased nitrogen partial pressure. As pressure decreases, the nitrogen is eliminated. This is true for any inert gas breathed.

Absorption consists of several phases, including transfer of inert gas from the lungs to the blood and then from the blood to the various tissues as it flows through the body. The gradient for gas transfer is the partial pressure difference of the gas between the lungs and blood and between the blood and the tissues.

The volume of blood flowing through tissues is small compared to the mass of the tissue, but over a period of time the gas delivered to the tissue causes it to become equilibrated with the gas carried in the blood. As the number of gas molecules in the liquid increases, the tension increases until it reaches a value equal to the partial pressure. When the tension equals the partial pressure, the liquid is saturated with the gas and the pressure gradient is zero. Unless the temperature or pressure changes, the only molecules of gas to enter or leave the liquid are those which may, in random fashion, change places without altering the balance.

The rate of equilibration with the blood gas depends upon the volume of blood flow and the respective capacities of blood and tissues to absorb dissolved gas. For example, fatty tissues hold significantly more gas than watery tissues and will thus take longer to absorb or eliminate excess inert gas.

2-12.6.3 **Gas Solubility.** The solubility of gases is affected by temperature—the lower the temperature, the higher the solubility. As the temperature of a solution increases, some of the dissolved gas leaves the solution. The bubbles rising in a pan of water being heated (long before it boils) are bubbles of dissolved gas coming out of solution.

The gases in a diver's breathing mixture are dissolved into his body in proportion to the partial pressure of each gas in the mixture. Because of the varied solubility of different gases, the quantity of a particular gas that becomes dissolved is also governed by the length of time the diver is breathing the gas at the increased pressure. If the diver breathes the gas long enough, his body will become saturated.

The dissolved gas in a diver's body, regardless of quantity, depth, or pressure, remains in solution as long as the pressure is maintained. However, as the diver ascends, more and more of the dissolved gas comes out of solution. If his ascent rate is controlled (i.e., through the use of the decompression tables), the dissolved gas is carried to the lungs and exhaled before it accumulates to form significant bubbles in the tissues. If, on the other hand, he ascends suddenly and the pressure is reduced at a rate higher than the body can accommodate, bubbles may form, disrupt body tissues and systems, and produce decompression sickness.

Table 2-5. Symbols and Values.

Symbol	Value
°F	Degrees Fahrenheit
°C	Degrees Celsius
°R	Degrees Rankine
A	Area
C	Circumference
D	Depth of Water
H	Height
L	Length
P	Pressure
r	Radius
T	Temperature
t	Time
V	Volume
W	Width
Dia	Diameter
Dia^2	Diameter Squared
Dia^3	Diameter Cubed
π	3.1416
ata	Atmospheres Absolute
pp	Partial Pressure
psi	Pounds per Square Inch
psig	Pounds per Square Inch Gauge
psia	Pounds per Square Inch Absolute
fsw	Feet of Sea Water
fpm	Feet per Minute
scf	Standard Cubic Feet
BTU	British Thermal Unit
cm^3	Cubic Centimeter
kw hr	Kilowatt Hour
mb	Millibars

Table 2-6. *Buoyancy (In Pounds).*

Fresh Water	(*V cu ft* x 62.4) - *Weight of Unit*
Salt Water	(*V cu ft* x 64) - *Weight of Unit*

Table 2-7. *Formulas for Area.*

Square or Rectangle	A = L x W
Circle	$A = 0.7854 \times Dia^2$ *or* $A = \pi r^2$

Table 2-8. *Formulas for Volumes.*

Compartment	$V = L \times W \times H$
Sphere	$= \pi \times 4/3 \times r^3$ $= 0.5236 \times Dia^3$
Cylinder	$V = \pi \times r^2 \times L$ $= \pi \times 1/4 \times Dia^2 \times L$ $= 0.7854 \times Dia^2 \times L$

Table 2-9. *Formulas for Partial Pressure/Equivalent Air Depth.*

Partial Pressure Measured in psi	$pp = (D + 33\text{ fsw}) \times 0.445\text{ psi} \times \left(\frac{\%V}{100\%}\right)$
Partial Pressure Measured in ata	$pp = \frac{D + 33\text{ fsw}}{33\text{ fsw}} \times \frac{\%V}{100\%}$
Partial Pressure Measured in fsw	$pp = (D + 33\text{ fsw}) \times \frac{\%V}{100\%}$
T formula for Measuring Partial Pressure	$\frac{pp}{ata \mid \%}$
Equivalent Air Depth for N_2O_2 Diving Measured in fsw	$EAD = \left[\frac{(1.0 - O_2\%)(D + 33)}{.79}\right] - 33$
Equivalent Air Depth for N_2O_2 Diving Measured in meters	$EAD = \left[\frac{(1.0 - O_2\%)(M + 10)}{.79}\right] - 10$

Table 2-10. Pressure Equivalents.

Atmos-pheres	Bars	10 Newton Per Square Centimeter	Pounds Per Square Inch	Columns of Mercury at 0°C		Columns of Water* at 15°C			
				Meters	Inches	Meters	Inches	Feet (FW)	Feet (FSW)
1	1.01325	1.03323	14.696	0.76	29.9212	10.337	406.966	33.9139	33.066
0.986923	1	1.01972	14.5038	0.750062	29.5299	10.2018	401.645	33.4704	32.6336
0.967841	0.980665	1	14.2234	0.735559	28.959	10.0045	393.879	32.8232	32.0026
0.068046	0.068947	0.070307	1	0.0517147	2.03601	0.703386	27.6923	2.30769	2.25
1.31579	1.33322	1.35951	19.33369	1	39.37	13.6013	535.482	44.6235	43.5079
0.0334211	0.0338639	0.0345316	0.491157	0.0254	1	0.345473	13.6013	1.13344	1.1051
0.09674	0.09798	0.099955	1.42169	0.073523	2.89458	1	39.37	3.28083	3.19881
0.002456	0.002489	0.002538	0.03609	0.001867	0.073523	0.02540	1	0.08333	0.08125
0.029487	0.029877	0.030466	0.43333	0.02241	0.882271	0.304801	12	1	0.975
0.030242	0.030643	0.031247	0.44444	0.022984	0.904884	0.312616	12.3077	1.02564	1

1. Fresh Water (FW) = 62.4 lbs/ft^3; Salt Water (fsw) = 64.0 lbs/ft^3.
2. The SI unit for pressure is Kilopascal (KPA)—1KG/CM^2 = 98.0665 KPA and by definition 1 BAR = 100.00 KPA @ 4°C.
3. In the metric system, 10 MSW is defined as 1 BAR. Note that pressure conversion from MSW to FSW is different than length conversion; i.e., 10 MSW = 32.6336 FSW and 10 M = 32.8083 feet.

Table 2-11. Volume and Capacity Equivalents.

Cubic Centi-meters	Cubic Inches	Cubic Feet	Cubic Yards	Milliliters	Liters	Pint	Quart	Gallon
1	.061023	3.531×10^{-5}	1.3097×10^{-6}	.999972	9.9997×10^{-4}	2.113×10^{-3}	1.0567×10^{-3}	2.6417×10^{-4}
16.3872	1	5.787×10^{-4}	2.1434×10^{-5}	16.3867	0.0163867	0.034632	0.017316	4.329×10^{-3}
28317	1728	1	0.037037	28316.2	28.3162	59.8442	29.9221	7.48052
764559	46656	27	1	764538	764.538	1615.79	807.896	201.974
1.00003	0.0610251	3.5315×10^{-5}	1.308×10^{-6}	1	0.001	2.1134×10^{-3}	1.0567×10^{-3}	2.6418×10^{-4}
1000.03	61.0251	0.0353154	1.308×10^{-3}	1000	1	2.11342	1.05671	0.264178
473.179	28.875	0.0167101	6.1889×10^{-4}	473.166	0.473166	1	0.5	0.125
946.359	57.75	0.0334201	1.2378×10^{-3}	946.332	0.946332	2	1	0.25
3785.43	231	0.133681	49511×10^{-3}	3785.33	3.78533	8	4	1

Table 2-12. Length Equivalents.

Centi-meters	Inches	Feet	Yards	Meters	Fathom	Kilometers	Miles	Int. Nautical Miles
1	0.3937	0.032808	0.010936	0.01	5.468×10^{-3}	0.00001	6.2137×10^{-5}	5.3659×10^{-6}
2.54001	1	0.08333	0.027778	0.025400	0.013889	2.540×10^{-5}	1.5783×10^{-5}	1.3706×10^{-5}
30.4801	12	1	0.33333	0.304801	0.166665	3.0480×10^{-4}	1.8939×10^{-4}	1.6447×10^{-4}
91.4403	36	3	1	0.914403	0.5	9.144×10^{-4}	5.6818×10^{-4}	4.9341×10^{-4}
100	39.37	3.28083	1.09361	1	0.5468	0.001	6.2137×10^{-4}	5.3959×10^{-4}
182.882	72	6	2	1.82882	1	1.8288×10^{-3}	1.1364×10^{-3}	9.8682×10^{-4}
100000	39370	3280.83	1093.61	1000	546.8	1	0.62137	0.539593
160935	63360	5280	1760	1609.35	80	1.60935	1	0.868393
185325	72962.4	6080.4	2026.73	1853.25	1013.36	1.85325	1.15155	1

Table 2-13. Area Equivalents.

Square Meters	Square Centimeters	Square Inches	Square Feet	Square Yards	Acres	Square Miles
1	10000	1550	10.7639	1.19599	2.471×10^{-4}	3.861×10^{-7}
0.0001	1	0.155	1.0764×10^{-3}	1.196×10^{-4}	2.471×10^{-8}	3.861×10^{-11}
6.4516×10^{-4}	6.45163	1	6.944×10^{-3}	7.716×10^{-4}	1.594×10^{-7}	2.491×10^{-10}
0.092903	929.034	144	1	0.11111	2.2957×10^{-5}	3.578×10^{-8}
0.836131	8361.31	1296	9	1	2.0661×10^{-4}	3.2283×10^{-7}
4046.87	4.0469×10^{7}	6.2726×10^{6}	43560	4840	1	1.5625×10^{-3}
2.59×10^{6}	2.59×10^{10}	4.0145×10^{9}	2.7878×10^{7}	3.0976×10^{6}	640	1

Table 2-14. Velocity Equivalents.

Centimeters Per Second	Meters Per Second	Meters Per Minute	Kilometers Per Hour	Feet Per Second	Feet Per Minute	Miles Per Hour	Knots
1	0.01	0.6	0.036	0.0328083	1.9685	0.0223639	0.0194673
100	1	60	3.6	3.28083	196.85	2.23693	1.9473
1.66667	0.016667	1	0.06	0.0546806	3.28083	0.0372822	0.0324455
27.778	0.27778	16.667	1	0.911343	54.6806	0.62137	0.540758
30.4801	0.304801	18.288	1.09728	1	60	0.681818	0.593365
0.5080	5.080×10^{-3}	0.304801	0.018288	0.016667	1	0.0113636	9.8894×10^{-3}
44.7041	0.447041	26.8225	1.60935	1.4667	88	1	0.870268
51.3682	0.513682	30.8209	1.84926	1.6853	101.118	1.14907	1

Table 2-15. *Mass Equivalents.*

Kilograms	Grams	Grains	Ounces	Pounds	Tons (short)	Tons (long)	Tons (metric)
1	1000	15432.4	35.274	2.20462	1.1023×10^{-3}	9.842×10^{-4}	0.001
0.001	1	15432.4	0.035274	2.2046×10^{-3}	1.1023×10^{-6}	9.842×10^{-7}	0.000001
6.4799×10^{-5}	0.6047989	1	2.2857×10^{-3}	1.4286×10^{-4}	7.1429×10^{-8}	6.3776×10^{-8}	6.4799×10^{-8}
0.0283495	28.3495	437.5	1	0.0625	3.125×10^{-5}	2.790×10^{-5}	2.835×10^{-5}
0.453592	453.592	7000	16	1	0.0005	4.4543×10^{-4}	4.5359×10^{-4}
907.185	907185	1.4×10^{7}	32000	2000	1	0.892857	0.907185
1016.05	1.016×10^{6}	1.568×10^{7}	35840	2240	1.12	1	1.01605
1000	10^{6}	1.5432×10^{7}	35274	2204.62	1.10231	984206	1

Table 2-16. *Energy or Work Equivalents.*

International Joules	Ergs	Foot - Pounds	International Kilowatt Hours	Horse Power Hours	Kilo - Calories	BTUs
1	10^{7}	0.737682	2.778×10^{-7}	$3.7257\ 10^{-7}$	2.3889×10^{-4}	9.4799×10^{-4}
10^{-7}	1	7.3768×10^{-8}	2.778×10^{-14}	3.726×10^{-14}	2.389×10^{-11}	9.4799×10^{-11}
1.3566	1.3556×10^{7}	1	3.766×10^{-7}	5.0505×10^{-7}	3.238×10^{-4}	1.285×10^{-3}
3.6×10^{6}	3.6×10^{13}	2.6557×10^{6}	1	1.34124	860	3412.76
2.684×10^{6}	2.684×10^{13}	1.98×10^{6}	0.745578	1	641.197	2544.48
4186.04	4.186×10^{10}	3087.97	1.163×10^{-3}	1.596×10^{-3}	1	3.96832
1054.87	1.0549×10^{10}	778.155	2.930×10^{-4}	3.93×10^{-4}	0.251996	1

Table 2-17. *Power Equivalents.*

Horse Power	International Kilowatts	International Joules/ Second	Kg-M Second	Foot lbs. Per Second	IT Calories Per Second	BTUs Per Second
1	0.745578	745.578	76.0404	550	178.11	0.7068
1.34124	1	1000	101.989	737.683	238.889	0.947989
1.3412×10^{-3}	0.001	1	0.101988	0.737682	0.238889	9.4799×10^{-4}
0.0131509	9.805×10^{-3}	9.80503	1	7.233	2.34231	9.2951×10^{-3}
1.8182×10^{-3}	1.3556×10^{-3}	1.3556	0.138255	1	0.323837	1.2851×10^{-3}
5.6145×10^{-3}	4.1861×10^{-3}	4.18605	0.426929	3.08797	1	3.9683×10^{-3}
1.41483	1.05486	1054.86	107.584	778.155	251.995	1

Table 2-18. Temperature Equivalents.

Conversion Formulas: $°C = (°F - 32) \times \frac{5}{9}$ $\quad °F = (\frac{9}{5} \times °C) + 32$

°C	°F	°C	°F	°C	°F	°C	°F	°C	°F	°C	°F	°C	°F
-100	-148.0	-60	-76.0	-20	-4.0	20	68.0	60	140.0	100	212.0	140	284.0
-98	-144.4	-58	-72.4	-18	-0.4	22	71.6	62	143.6	102	215.6	142	287.6
-96	-140.8	-56	-68.8	-16	3.2	24	75.2	64	147.2	104	219.2	144	291.2
-94	-137.2	-54	-65.2	-14	6.8	26	78.8	66	150.8	106	222.8	146	294.8
-92	-133.6	-52	-61.6	-12	10.4	28	82.4	68	154.4	108	226.4	148	298.4
-90	-130.0	-50	-58.0	-10	14.0	30	86.0	70	158.0	110	230.0	150	302.0
-88	-126.4	-48	-54.4	-8	17.6	32	89.6	72	161.6	112	233.6	152	305.6
-86	-122.8	-46	-50.8	-6	21.2	34	93.2	74	165.2	114	237.2	154	309.2
-84	-119.2	-44	-47.2	-4	24.8	36	96.8	76	168.8	116	240.8	156	312.8
-82	-115.6	-42	-43.6	-2	28.4	38	100.4	78	172.4	118	244.4	158	316.4
-80	-112.0	-40	-40.0	0	32	40	104.0	80	176.0	120	248.0	160	320.0
-78	-108.4	-38	-36.4	2	35.6	42	107.6	82	179.6	122	251.6	162	323.6
-76	-104.8	-36	-32.8	4	39.2	44	111.2	84	183.2	124	255.2	164	327.2
-74	-101.2	-34	-29.2	6	42.8	46	114.8	86	186.8	126	258.8	166	330.8
-72	-97.6	-32	-25.6	8	46.4	48	118.4	88	190.4	128	262.4	168	334.4
-70	-94.0	-30	-22.0	10	50.0	50	122.0	90	194.0	130	266.0	170	338.0
-68	-90.4	-28	-18.4	12	53.6	52	125.6	92	197.6	132	269.6	172	341.6
-66	-86.8	-26	-14.8	14	57.2	54	129.2	94	201.2	134	273.2	174	345.2
-64	-83.2	-24	-11.2	16	60.8	56	132.8	96	204.8	136	276.8	176	348.8
-62	-79.6	-22	-7.6	18	64.4	58	136.4	98	208.4	138	280.4	178	352.4

Table 2-19. Atmospheric Pressure at Altitude.

	Atmospheric Pressure				
Altitude in Feet	**Atmospheres absolute**	**Millimeters of Mercury**	**Pounds per sq. in. absolute**	**Millibars**	**Kilopascals**
500	0.982	746.4	14.43	995.1	99.51
1000	0.964	732.9	14.17	977.2	97.72
1500	0.947	719.7	13.92	959.5	95.95
2000	0.930	706.7	13.66	942.1	94.21
2500	0.913	693.8	13.42	925.0	92.50
3000	0.896	681.1	13.17	908.1	90.81
3500	0.880	668.7	12.93	891.5	89.15
4000	0.864	656.4	12.69	875.1	87.51
4500	0.848	644.3	12.46	859.0	85.90
5000	0.832	632.4	12.23	843.1	84.31
5500	0.817	620.6	12.00	827.4	82.74
6000	0.801	609.0	11.78	812.0	81.20
6500	0.786	597.7	11.56	796.8	79.68
7000	0.772	586.4	11.34	781.9	78.19
7500	0.757	575.4	11.13	767.1	76.71
8000	0.743	564.5	10.92	752.6	75.26
8500	0.729	553.8	10.71	738.3	73.83
9000	0.715	543.3	10.50	724.3	72.43
9500	0.701	532.9	10.30	710.4	71.04
10000	0.688	522.7	10.11	696.8	69.68

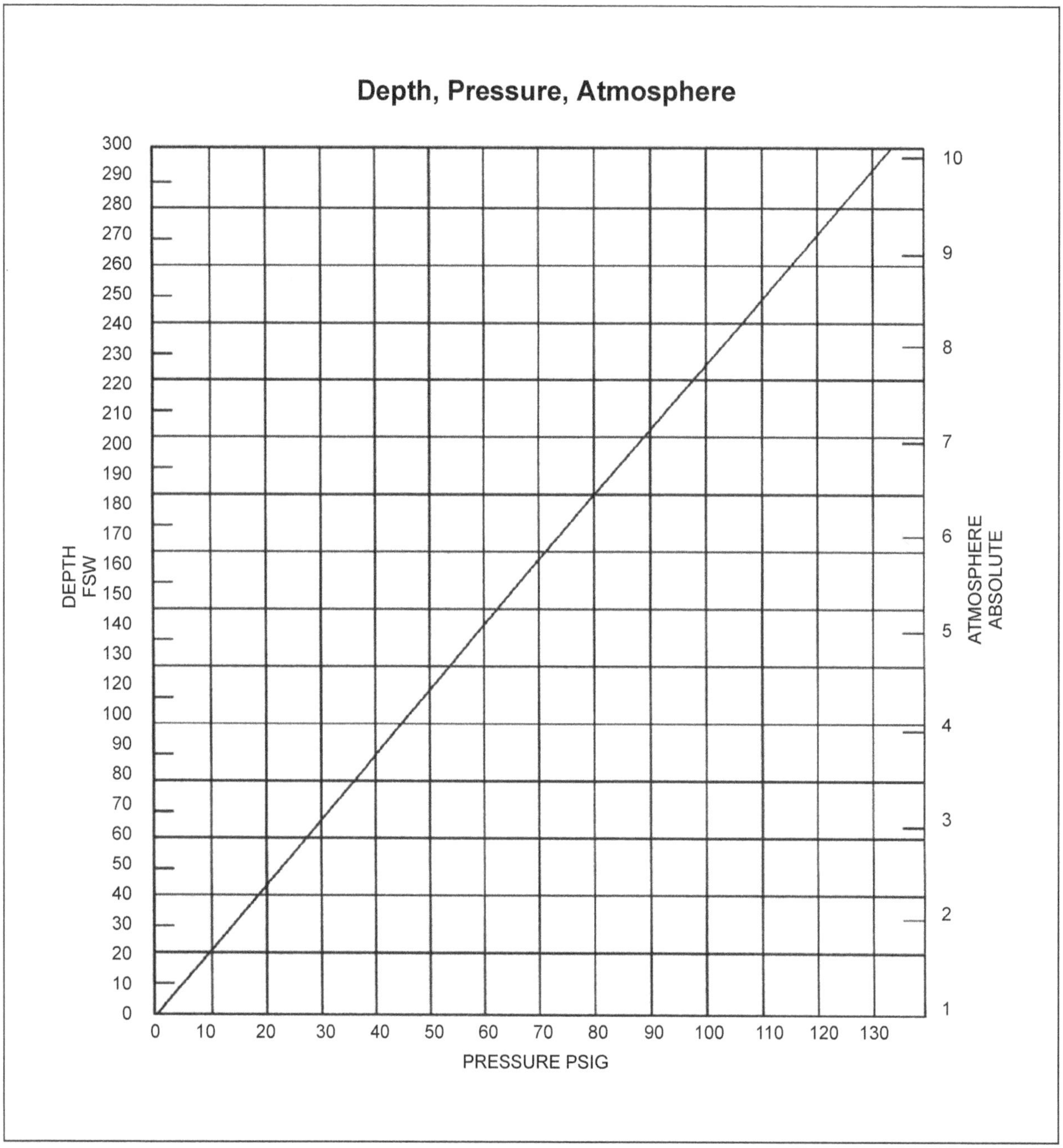

Figure 2-7. Depth, Pressure, Atmosphere Graph.

CHAPTER 3

Underwater Physiology and Diving Disorders

3-1 INTRODUCTION

3-1.1 **Purpose.** This chapter provides basic information on the changes in human anatomy and physiology that occur while working in the underwater environment. It also discusses the diving disorders that result when these anatomical or physiological changes exceed the limits of adaptation.

3-1.2 **Scope.** Anatomy is the study of the structure of the organs of the body. Physiology is the study of the processes and functions of the body. This chapter explains the basic anatomical and physiological changes that occur when diver enters the water and is subject to increased ambient pressure. A diver's knowledge of these changes is as important as his knowledge of diving gear and procedures. When the changes in normal anatomy or physiology exceed the limits of adaptation, one or more pathological states may emerge. These pathological states are called diving disorders and are also discussed in this chapter. Safe diving is only possible when the diver fully understands the fundamental processes at work on the human body in the underwater environment.

3-1.3 **General.** A body at work requires coordinated functioning of all organs and systems. The heart pumps blood to all parts of the body, the tissue fluids exchange dissolved materials with the blood, and the lungs keep the blood supplied with oxygen and cleared of excess carbon dioxide. Most of these processes are controlled directly by the brain, nervous system, and various glands. The individual is generally unaware that these functions are taking place.

As efficient as it is, the human body lacks effective ways of compensating for many of the effects of increased pressure at depth and can do little to keep its internal environment from being upset. Such external effects set definite limits on what a diver can do and, if not understood, can give rise to serious accidents.

3-2 THE NERVOUS SYSTEM

The nervous system coordinates all body functions and activities. The nervous system comprises the brain, spinal cord, and a complex network of nerves that course through the body. The brain and spinal cord are collectively referred to as the *central nervous system* (CNS). Nerves originating in the brain and spinal cord and traveling to peripheral parts of the body form the *peripheral nervous system* (PNS). The peripheral nervous system consists of the cranial nerves, the spinal nerves, and the sympathetic nervous system. The peripheral nervous system is involved in regulating cardiovascular, respiratory, and other automatic body functions. These nerve trunks also transmit nerve impulses associated with sight, hearing, balance, taste, touch, pain, and temperature between peripheral sensors and the spinal cord and brain.

3-3 THE CIRCULATORY SYSTEM

The circulatory system consists of the heart, arteries, veins, and capillaries. The circulatory system carries oxygen, nutrients, and hormones to every cell of the body, and carries away carbon dioxide, waste chemicals, and heat. Blood circulates through a closed system of tubes that includes the lung and tissue capillaries, heart, arteries, and veins.

3-3.1 **Anatomy.** Every part of the body is completely interwoven with intricate networks of extremely small blood vessels called capillaries. The very large surface areas required for ample diffusion of gases in the lungs and tissues are provided by the thin walls of the capillaries. In the lungs, capillaries surround the tiny air sacs (alveoli) so that the blood they carry can exchange gases with air.

3-3.1.1 **The Heart.** The heart (Figure 3-1) is the muscular pump that propels the blood throughout the system. It is about the size of a closed fist, hollow, and made up almost entirely of muscle tissue that forms its walls and provides the pumping action. The heart is located in the front and center of the chest cavity between the lungs, directly behind the breastbone (sternum).

The interior of the heart is divided lengthwise into halves, separated by a wall of tissue called a septum. The two halves have no direct connection to each other. Each half is divided into an upper chamber (the atrium), which receives blood from the veins of its circuit and a lower chamber (the ventricle) which takes blood from the atrium and pumps it away via the main artery. Because the ventricles do most of the pumping, they have the thickest, most muscular walls. The arteries carry blood from the heart to the capillaries; the veins return blood from the capillaries to the heart. Arteries and veins branch and rebranch many times, very much like a tree. Trunks near the heart are approximately the diameter of a human thumb, while the smallest arterial and venous twigs are microscopic. Capillaries provide the connections that let blood flow from the smallest branch arteries (arterioles) into the smallest veins (venules).

3-3.1.2 **The Pulmonary and Systemic Circuits.** The circulatory system consists of two circuits with the same blood flowing through the body. The pulmonary circuit serves the lung capillaries; the systemic circuit serves the tissue capillaries. Each circuit has its own arteries and veins and its own half of the heart as a pump. In complete circulation, blood first passes through one circuit and then the other, going through the heart twice in each complete circuit.

3-3.2 **Circulatory Function.** Blood follows a continuous circuit through the human body. Blood leaving a muscle or organ capillary has lost most of its oxygen and is loaded with carbon dioxide. The blood flows through the body's veins to the main veins in the upper chest (the superior and inferior vena cava). The superior vena cava receives blood from the upper half of the body; the inferior vena cava receives blood from areas of the body below the diaphragm. The blood flows

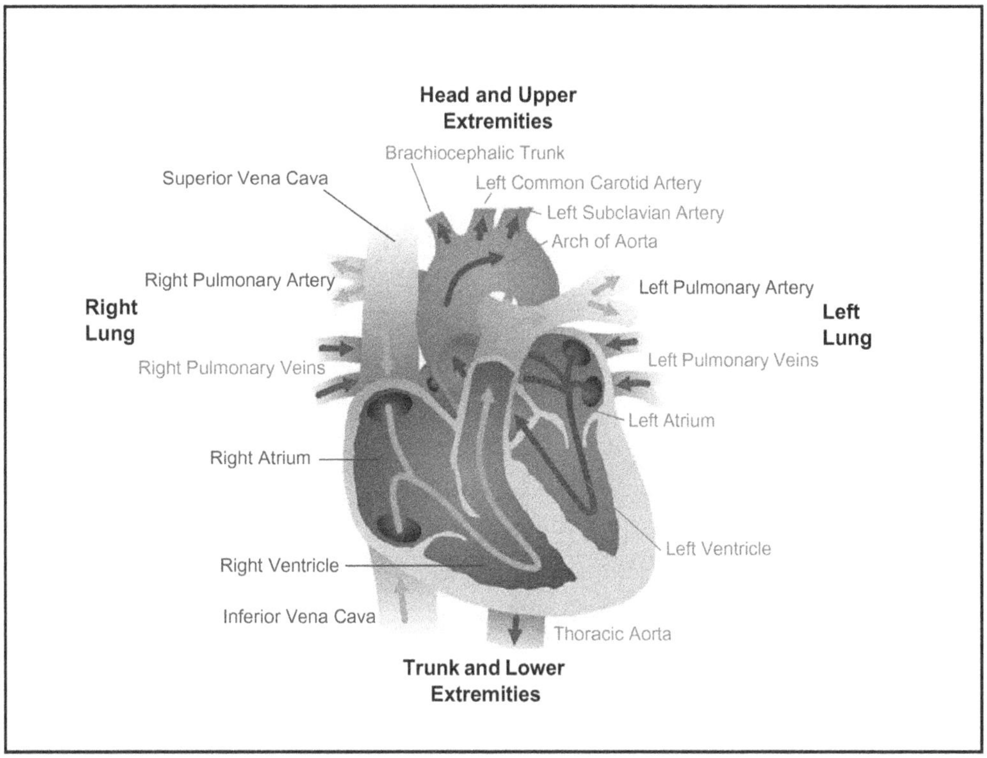

Figure 3-1. The Heart's Components and Blood Flow.

through the main veins into the right atrium and then through the tricuspid valve into the right ventricle.

The next heart contraction forces the blood through the pulmonic valve into the pulmonary artery. The blood then passes through the arterial branchings of the lungs into the pulmonary capillaries, where gas transfer with air takes place. By diffusion, the blood exchanges inert gas as well as carbon dioxide and oxygen with the air in the lungs. The blood then returns to the heart via the pulmonary venous system and enters the left atrium.

The next relaxation finds it going through the mitral valve into the left ventricle to be pumped through the aortic valve into the main artery (aorta) of the systemic circuit. The blood then flows through the arteries branching from the aorta, into successively smaller vessels until reaching the capillaries, where oxygen is exchanged for carbon dioxide. The blood is now ready for another trip to the lungs and back again. Figure 3-2 shows how the pulmonary circulatory system is arranged.

The larger blood vessels are somewhat elastic and have muscular walls. They stretch and contract as blood is pumped from the heart, maintaining a slow but adequate flow (perfusion) through the capillaries.

3-3.3 **Blood Components.** The average human body contains approximately five liters of blood. Oxygen is carried mainly in the red corpuscles (red blood cells). There are approximately 300 million red corpuscles in an average-sized drop of blood.

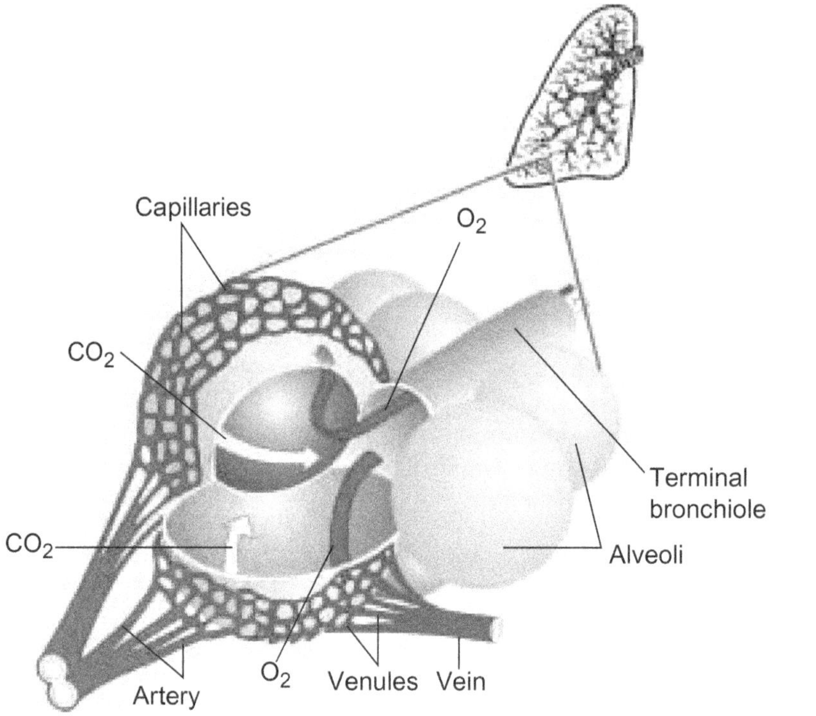

Figure 3-2. Respiration and Blood Circulation. The lung's gas exchange system is essentially three pumps. The thorax, a gas pump, moves air through the trachea and bronchi to the lung's air sacs. These sacs, the alveoli, are shown with and without their covering of pulmonary capillaries. The heart's right ventricle, a fluid pump, moves blood that is low in oxygen and high in carbon dioxide into the pulmonary capillaries. Oxygen from the air diffuses into the blood while carbon dioxide diffuses from the blood into the air in the lungs. The oxygenated blood moves to the left ventricle, another fluid pump, which sends the blood via the arterial system to the systemic capillaries which deliver oxygen to and collect carbon dioxide from the body's cells.

These corpuscles are small, disc-shaped cells that contain hemoglobin to carry oxygen. Hemoglobin is a complex chemical compound containing iron. It can form a loose chemical combination with oxygen, soaking it up almost as a sponge soaks up liquid. Hemoglobin is bright red when it is oxygen-rich; it becomes increasingly dark as it loses oxygen. Hemoglobin gains or loses oxygen depending upon the partial pressure of oxygen to which it is exposed. Hemoglobin takes up about 98 percent of the oxygen it can carry when it is exposed to the normal partial pressure of oxygen in the lungs. Because the tissue cells are using oxygen, the partial pressure (tension) in the tissues is much lower and the hemoglobin gives up much of its oxygen in the tissue capillaries.

Acids form as the carbon dioxide dissolves in the blood. Buffers in the blood neutralize the acids and permit large amounts of carbon dioxide to be carried away to prevent excess acidity. Hemoglobin also plays an important part in transporting carbon dioxide. The uptake or loss of carbon dioxide by blood depends mainly upon the partial pressure (or tension) of the gas in the area where the blood is exposed. For example, in the peripheral tissues, carbon dioxide diffuses into the blood and oxygen diffuses into the tissues.

Blood also contains infection-fighting white blood cells, and platelets, which are cells essential in blood coagulation. Plasma is the colorless, watery portion of the blood. It contains a large amount of dissolved material essential to life. The blood also contains several substances, such as fibrinogen, associated with blood clotting. Without the clotting ability, even the slightest bodily injury could cause death.

3-4 THE RESPIRATORY SYSTEM

Every cell in the body must obtain energy to maintain its life, growth, and function. Cells obtain their energy from oxidation, which is a slow, controlled burning of food materials. Oxidation requires fuel and oxygen. Respiration is the process of exchanging oxygen and carbon dioxide during oxidation and releasing energy and water.

3-4.1 **Gas Exchange.** Few body cells are close enough to the surface to have any chance of obtaining oxygen and expelling carbon dioxide by direct air diffusion. Instead, the gas exchange takes place via the circulating blood. The blood is exposed to air over a large diffusing surface as it passes through the lungs. When the blood reaches the tissues, the small capillary vessels provide another large surface where the blood and tissue fluids are in close contact. Gases diffuse readily at both ends of the circuit and the blood has the remarkable ability to carry both oxygen and carbon dioxide. This system normally works so well that even the deepest cells of the body can obtain oxygen and get rid of excess carbon dioxide almost as readily as if they were completely surrounded by air.

If the membrane surface in the lung, where blood and air come close together, were just an exposed sheet of tissue like the skin, natural air currents would keep fresh air in contact with it. Actually, this lung membrane surface is many times larger than the skin area and is folded and compressed into the small space of the lungs that are protected inside the bony cage of the chest. This makes it necessary to continually move air in and out of the space. The processes of breathing and the exchange of gases in the lungs are referred to as *ventilation* and *pulmonary gas exchange*, respectively.

3-4.2 **Respiration Phases.** The complete process of respiration includes six important phases:

1. Ventilation of the lungs with fresh air
2. Exchange of gases between blood and air in lungs
3. Transport of gases by blood
4. Exchange of gases between blood and tissue fluids
5. Exchange of gases between the tissue fluids and cells
6. Use and production of gases by cells

If any one of the processes stops or is seriously hindered, the affected cells cannot function normally or survive for any length of time. Brain tissue cells, for example, stop working almost immediately and will either die or be permanently injured in a few minutes if their oxygen supply is completely cut off.

The respiratory system is a complex of organs and structures that performs the pulmonary ventilation of the body and the exchange of oxygen and carbon dioxide between the ambient air and the blood circulating through the lungs. It also warms the air passing into the body and assists in speech production by providing air to the larynx and the vocal chords. The respiratory tract is divided into upper and lower tracts.

3-4.3 **Upper and Lower Respiratory Tract.** The upper respiratory tract consists of the nose, nasal cavity, frontal sinuses, maxillary sinuses, larynx, and trachea. The upper respiratory tract carries air to and from the lungs and filters, moistens and warms air during each inhalation.

The lower respiratory tract consists of the left and right bronchi and the lungs, where the exchange of oxygen and carbon dioxide occurs during the respiratory cycle. The bronchi divide into smaller bronchioles in the lungs, the bronchioles divide into alveolar ducts, the ducts into alveolar sacs, and the sacs into alveoli. The alveolar sacs and the alveoli present about 850 square feet of surface area for the exchange of oxygen and carbon dioxide that occurs between the internal alveolar surface and the tiny capillaries surrounding the external alveolar wall.

3-4.4 **The Respiratory Apparatus.** The mechanics of taking fresh air into the lungs (inspiration or inhalation) and expelling used air from the lungs (expiration or exhalation) is diagrammed in Figure 3-3. By elevating the ribs and lowering the diaphragm, the volume of the lung is increased. Thus, according to Boyle's Law, a lower pressure is created within the lungs and fresh air rushes in to equalize this lowered pressure. When the ribs are lowered again and the diaphragm rises to its original position, a higher pressure is created within the lungs, expelling the used air.

3-4.4.1 **The Chest Cavity.** The chest cavity does not have space between the outer lung surfaces and the surrounding chest wall and diaphragm. Both surfaces are covered by membranes; the visceral pleura covers the lung and the parietal pleura lines the chest wall. These pleurae are separated from each other by a small amount of fluid that acts as a lubricant to allow the membranes to slide freely over themselves as the lungs expand and contract during respiration.

3-4.4.2 **The Lungs.** The lungs are a pair of light, spongy organs in the chest and are the main component of the respiratory system (see Figure 3-4). The highly elastic lungs are the main mechanism in the body for inspiring air from which oxygen is extracted for the arterial blood system and for exhaling carbon dioxide dispersed from the venous system. The lungs are composed of lobes that are smooth and shiny on their surface. The lungs contain millions of small expandable air sacs (alveoli) connected to air passages. These passages branch and rebranch like the

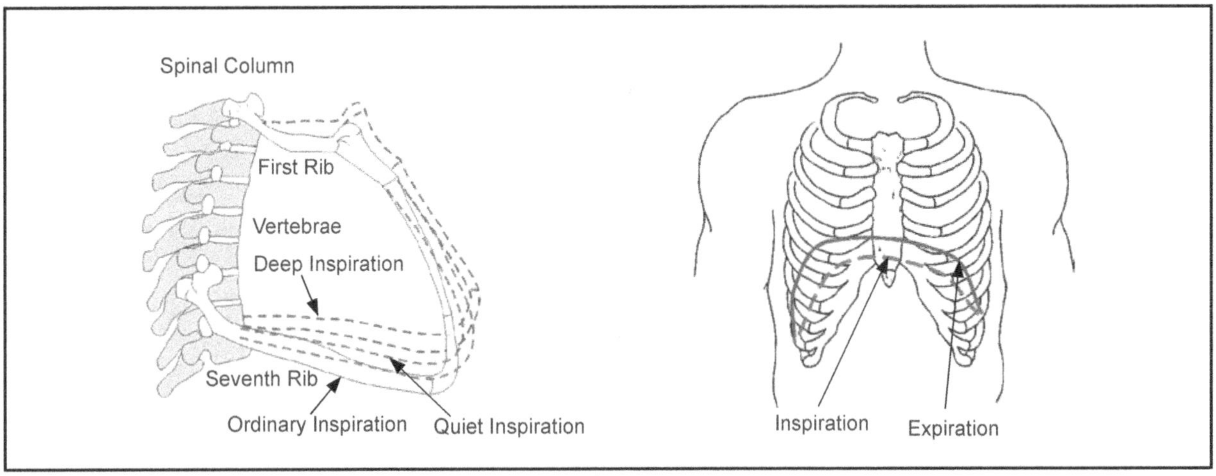

Figure 3-3. Inspiration Process. Inspiration involves both raising the rib cage (left panel) and lowering the diaphragm (right panel). Both movements enlarge the volume of the thoracic cavity and draw air into the lung.

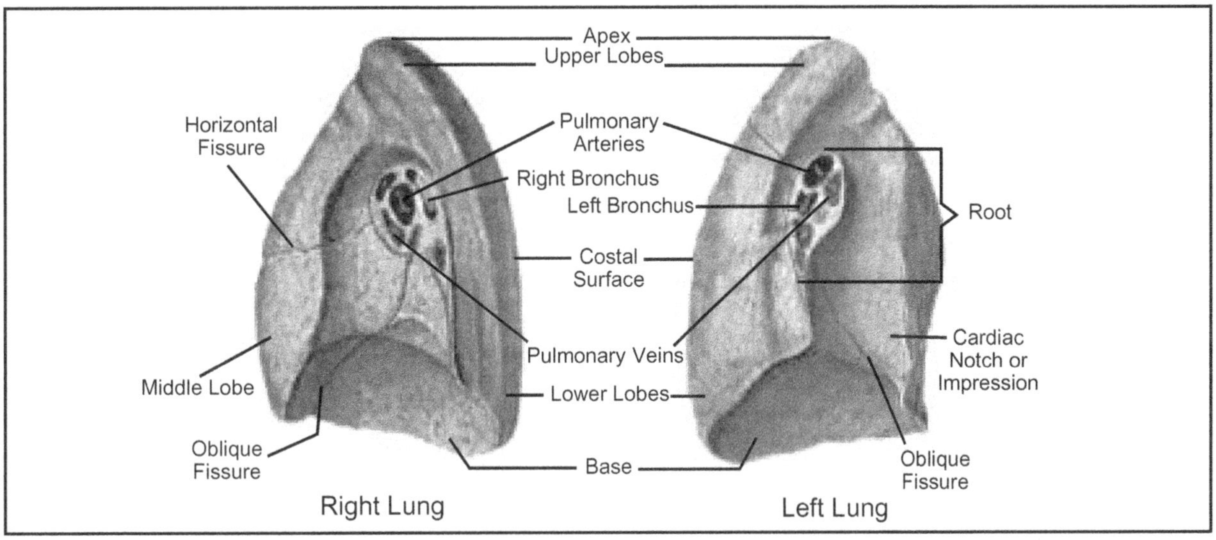

Figure 3-4. Lungs Viewed from Medical Aspect.

twigs of a tree. Air entering the main airways of the lungs gains access to the entire surface of these alveoli. Each alveolus is lined with a thin membrane and is surrounded by a network of very small vessels that make up the capillary bed of the lungs. Most of the lung membrane has air on one side of it and blood on the other; diffusion of gases takes place freely in either direction.

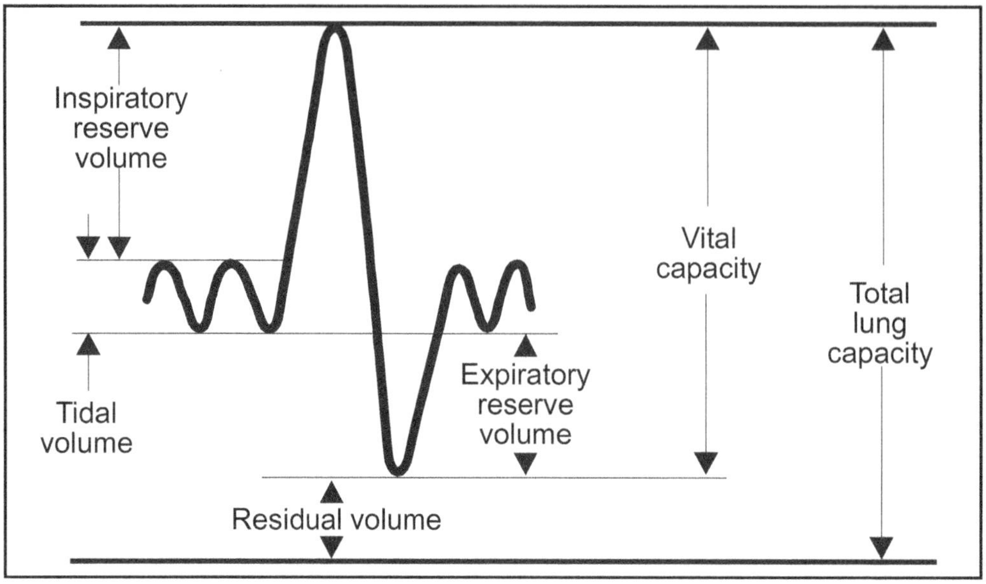

Figure 3-5. Lung Volumes. The heavy line is a tracing, derived from a subject breathing to and from a sealed recording bellows. Following several normal tidal breaths, the subject inhales maximally, then exhales maximally. The volume of air moved during this maximal effort is called the vital capacity. During exercise, the tidal volume increases, using part of the inspiratory and expiratory reserve volumes. The tidal volume, however, can never exceed the vital capacity. The residual volume is the amount of air remaining in the lung after the most forceful expiration. The sum of the vital capacity and the residual volume is the total lung capacity.

3-4.5 **Respiratory Tract Ventilation Definitions.** Ventilation of the respiratory system establishes the proper composition of gases in the alveoli for exchange with the blood. The following definitions help in understanding respiration (Figure 3-5).

Respiratory Cycle. The *respiratory cycle* is one complete breath consisting of an inspiration and exhalation, including any pause between the movements.

Respiratory Rate. The number of complete respiratory cycles that take place in 1 minute is the respiratory rate. An adult at rest normally has a respiratory rate of approximately 12 to 16 breaths per minute.

Total Lung Capacity. The *total lung capacity* (TLC) is the total volume of air that the lungs can hold when filled to capacity. TLC is normally between five and six liters.

Vital Capacity. *Vital capacity* is the volume of air that can be expelled from the lungs after a full inspiration. The average vital capacity is between four and five liters.

Tidal Volume. *Tidal volume* is the volume of air moved in or out of the lungs during a single normal respiratory cycle. The tidal volume generally averages about one-half liter for an adult at rest. Tidal volume increases considerably during physical exertion, and may be as high as 3 liters during severe work.

Respiratory Minute Volume. The *respiratory minute volume* (RMV) is the total amount of air moved in or out of the lungs in a minute. The respiratory minute volume is calculated by multiplying the tidal volume by the respiratory rate. RMV varies greatly with the body's activity. It is about 6 to 10 liters per minute at complete rest and may be over 100 liters per minute during severe work.

Maximal Breathing Capacity and Maximum Ventilatory Volume. The *maximum breathing capacity* (MBC) and *maximum voluntary ventilation* (MVV) are the greatest respiratory minute volumes that a person can produce during a short period of extremely forceful breathing. In a healthy young man, they may average as much as 180 liters per minute (the range is 140 to 240 liters per minute).

Maximum Inspiratory Flow Rate and Maximum Expiratory Flow Rate. The *maximum inspiratory flow rate* (MIFR) and *maximum expiratory flow rate (*MEFR) are the fastest rates at which the body can move gases in and out of the lungs. These rates are important in designing breathing equipment and computing gas use under various workloads. Flow rates are usually expressed in liters per second.

Respiratory Quotient. *Respiratory quotient* (RQ) is the ratio of the amount of carbon dioxide produced to the amount of oxygen consumed during cellular processes per unit time. This value ranges from 0.7 to 1.0 depending on diet and physical exertion and is usually assumed to be 0.9 for calculations. This ratio is significant when calculating the amount of carbon dioxide produced as oxygen is used at various workloads while using a closed-circuit breathing apparatus. The duration of the carbon dioxide absorbent canister can then be compared to the duration of the oxygen supply.

Respiratory Dead Space. *Respiratory dead space* refers to the part of the respiratory system that has no alveoli, and in which little or no exchange of gas between air and blood takes place. It normally amounts to less than 0.2 liter. Air occupying the dead space at the end of expiration is rebreathed in the following inspiration. Parts of a diver's breathing apparatus can add to the volume of the dead space and thus reduce the proportion of the tidal volume that serves the purpose of respiration. To compensate, the diver must increase his tidal volume. The problem can best be visualized by using a breathing tube as an example. If the tube contains one liter of air, a normal exhalation of about one liter will leave the tube filled with used air from the lungs. At inhalation, the used air will be drawn right back into the lungs. The tidal volume must be increased by more than a liter to draw in the needed fresh supply, because any fresh air is diluted by the air in the dead space. Thus, the air that is taken into the lungs (inspired air) is a mixture of fresh and dead space gases.

3-4.6 **Alveolar/Capillary Gas Exchange.** Within the alveolar air spaces, the composition of the air (alveolar air) is changed by the elimination of carbon dioxide from the blood, the absorption of oxygen by the blood, and the addition of water vapor. The air that is exhaled is a mixture of alveolar air and the inspired air that remained in the dead space.

The blood in the capillary bed of the lungs is exposed to the gas pressures of alveolar air through the thin membranes of the air sacs and the capillary walls. With this exposure taking place over a vast surface area, the gas pressure of the blood leaving the lungs is approximately equal to that present in alveolar air.

When arterial blood passes through the capillary network surrounding the cells in the body tissues it is exposed to and equalizes with the gas pressure of the tissues. Some of the blood's oxygen is absorbed by the cells and carbon dioxide is picked up from these cells. When the blood returns to the pulmonary capillaries and is exposed to the alveolar air, the partial pressures of gases between the blood and the alveolar air are again equalized.

Carbon dioxide diffuses from the blood into the alveolar air, lowering its partial pressure, and oxygen is absorbed by the blood from the alveolar air, increasing its partial pressure. With each complete round of circulation, the blood is the medium through which this process of gas exchange occurs. Each cycle normally requires approximately 20 seconds.

3-4.7 **Breathing Control.** The amount of oxygen consumed and carbon dioxide produced increases markedly when a diver is working. The amount of blood pumped through the tissues and the lungs per minute increases in proportion to the rate at which these gases must be transported. As a result, more oxygen is taken up from the alveolar air and more carbon dioxide is delivered to the lungs for disposal. To maintain proper blood levels, the respiratory minute volume must also change in proportion to oxygen consumption and carbon dioxide output.

Changes in the partial pressure (concentration) of oxygen and carbon dioxide (ppO_2 and $ppCO_2$) in the arterial circulation activate central and peripheral chemoreceptors. These chemoreceptors are attached to important arteries. The most important are the carotid bodies in the neck and aortic bodies near the heart. The chemoreceptor in the carotid artery is activated by the $ppCO_2$ in the blood and signals the respiratory center in the brain stem to increase or decrease respiration. The chemoreceptor in the aorta causes the aortic body reflex. This is a normal chemical reflex initiated by decreased oxygen concentration and increased carbon dioxide concentration in the blood. These changes result in nerve impulses that increase respiratory activity. Low oxygen tension alone does not increase breathing markedly until dangerous levels are reached. The part played by chemoreceptors is evident in normal processes such as breathholding.

As a result of the regulatory process and the adjustments they cause, the blood leaving the lungs usually has about the same oxygen and carbon dioxide levels during work that it did at rest. The maximum pumping capacity of the heart (blood circulation) and respiratory system (ventilation) largely determines the amount of work a person can do.

3-4.8 Oxygen Consumption. A diver's oxygen consumption is an important factor when determining how long breathing gas will last, the ventilation rates required to maintain proper helmet oxygen level, and the length of time a canister will absorb carbon dioxide. Oxygen consumption is a measure of energy expenditure and is closely linked to the respiratory processes of ventilation and carbon dioxide production.

Oxygen consumption is measured in liters per minute (l/min) at Standard Temperature (0°C, 32°F) and Pressure (14.7 psia, 1 ata), Dry Gas (STPD). These rates of oxygen consumption are not depth dependent. This means that a fully charged MK 16 oxygen bottle containing 360 standard liters (3.96 scf) of usable gas will last 225 minutes at an oxygen consumption rate of 1.6 liters per minute at any depth, provided no gas leaks from the rig.

Minute ventilation, or respiratory minute volume (RMV), is measured at BTPS (body temperature 37°C/98.6°F, ambient barometric pressure, saturated with water vapor at body temperature) and varies depending on a person's activity level, as shown in Figure 3-6. Surface RMV can be approximated by multiplying the oxygen consumption rate by 25. Although this 25:1 ratio decreases with increasing gas density and high inhaled oxygen concentrations, it is a good rule-of-thumb approximation for computing how long the breathing gas will last.

Unlike oxygen consumption, the amount of gas a diver inhales is depth dependent. At the surface, a diver swimming at 0.5 knot inhales 20 l/min of gas. A SCUBA cylinder containing 71.2 standard cubic feet (scf) of air (approximately 2,000 standard liters) lasts approximately 100 minutes. At 33 fsw, the diver still inhales 20 l/min at BTPS, but the gas is twice as dense; thus, the inhalation would be approximately 40 standard l/min and the cylinder would last only half as long, or 50 minutes. At three atmospheres, the same cylinder would last only one-third as long as at the surface.

Carbon dioxide production depends only on the level of exertion and can be assumed to be independent of depth. Carbon dioxide production and RQ are used to compute ventilation rates for chambers and free-flow diving helmets. These factors may also be used to determine whether the oxygen supply or the duration of the CO_2 absorbent will limit a diver's time in a closed or semi-closed system.

3-5 RESPIRATORY PROBLEMS IN DIVING.

Physiological problems often occur when divers are exposed to the pressures of depth. However, some of the difficulties related to respiratory processes can occur at any time because of an inadequate supply of oxygen or inadequate removal of carbon dioxide from the tissue cells. Depth may modify these problems for the diver, but the basic difficulties remain the same. Fortunately, the diver has normal physiological reserves to adapt to environmental changes and is only marginally aware of small changes. The extra work of breathing reduces the diver's ability to do heavy work at depth, but moderate work can be done with adequate equipment at the maximum depths currently achieved in diving.

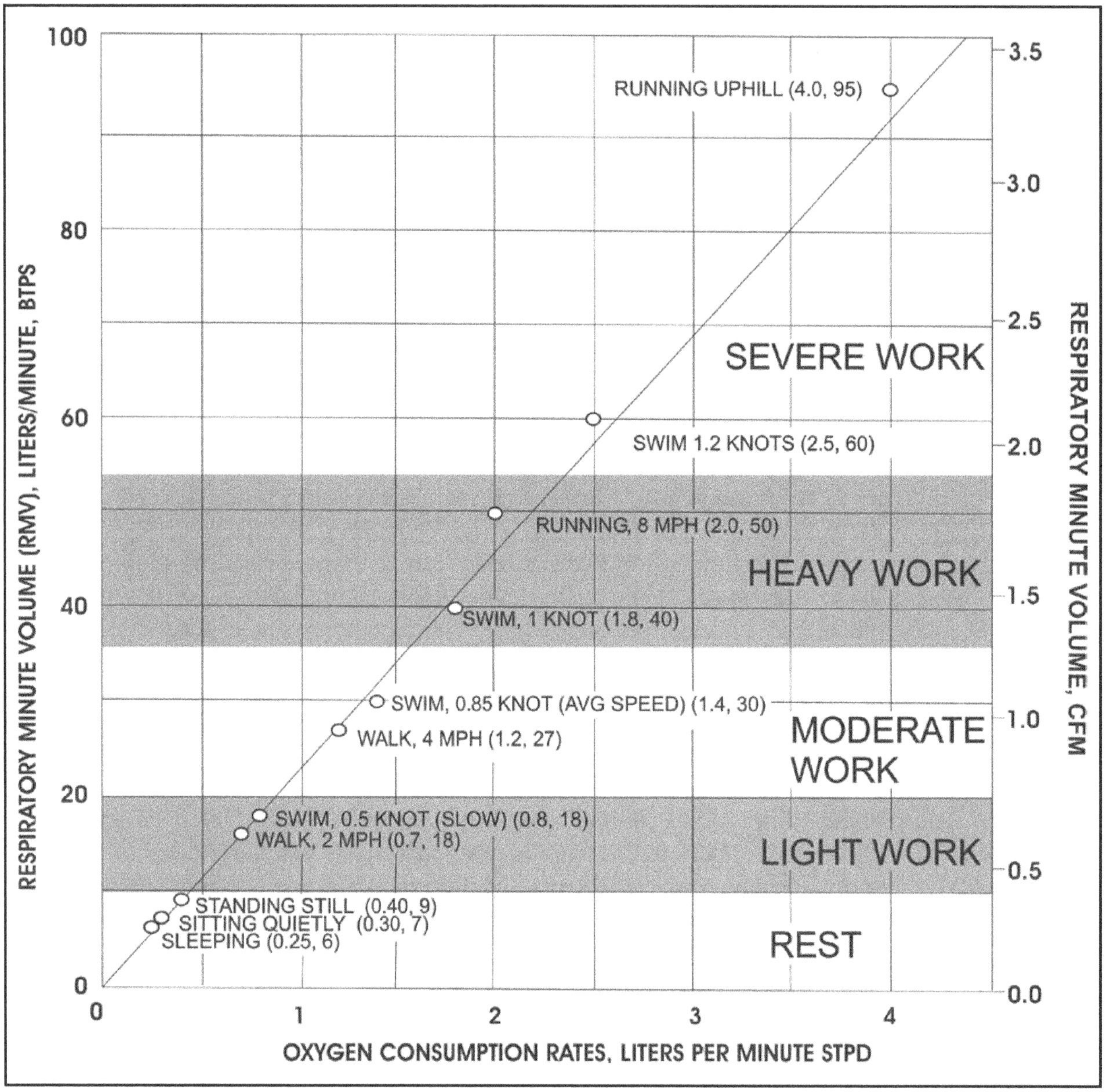

Figure 3-6. Oxygen Consumption and RMV at Different Work Rates.

3-5.1 **Oxygen Deficiency (Hypoxia).** Hypoxia, is an abnormal deficiency of oxygen in the arterial blood. Severe hypoxia will impede the normal function of cells and eventually kill them. The brain is the most vulnerable organ in the body to the effects of hypoxia.

The partial pressure of oxygen (ppO_2) determines whether the amount of oxygen in a breathing medium is adequate. Air contains approximately 21 percent oxygen and provides an ample ppO_2 of about 0.21 ata at the surface. A drop in ppO_2 below 0.16 ata causes the onset of hypoxic symptoms. Most individuals become hypoxic to the point of helplessness at a ppO_2 of 0.11 ata and unconscious at a ppO_2 of 0.10 ata. Below this level, permanent brain damage and eventually death will occur. In diving, a lower percentage of oxygen will suffice as long as the total pressure is sufficient to maintain an adequate ppO_2. For example, 5 percent oxygen gives a ppO_2 of 0.20 ata for a diver at 100 fsw. On ascent, however, the diver would rapidly experience hypoxia if the oxygen percentage were not increased.

3-5.1.1 **Causes of Hypoxia.** The causes of hypoxia vary, but all interfere with the normal oxygen supply to the body. For divers, interference of oxygen delivery can be caused by:

- Improper line up of breathing gases resulting in a low partial pressure of oxygen in the breathing gas supply.
- Partial or complete blockage of the fresh gas injection orifice in a semiclosed-circuit UBA. Failure of the oxygen addition valve in closed circuit rebreathers like the MK 16.
- Inadequate purging of breathing bags in closed-circuit oxygen rebreathers like the MK 25.
- Blockage of all or part of the air passages by vomitus, secretions, water, or foreign objects.
- Collapse of the lung due to pneumothorax.
- Paralysis of the respiratory muscles from spinal cord injury.
- Accumulation of fluid in the lung tissues (pulmonary edema) due to diving in cold water while overhydrated, negative pressure breathing, inhalation of water in a near drowning episode, or excessive accumulation of venous gas bubbles in the lung during decompression. The latter condition is referred to as "chokes". Pulmonary edema causes a mismatch of alveolar ventilation and pulmonary blood flow and decreases the rate of transfer of oxygen across the alveolar capillary membrane.
- Carbon monoxide poisoning. Carbon monoxide interferes with the transport of oxygen by the hemoglobin in red blood cells and blocks oxygen utilization at the cellular level.
- Breathholding. During a breathhold the partial pressure of oxygen in the lung falls progressively as the body continues to consume oxygen. If the breathhold is long enough, hypoxia will occur.

3-5.1.2 **Symptoms of Hypoxia.** The symptoms of hypoxia include:

- Loss of judgment
- Lack of concentration
- Lack of muscle control

- Inability to perform delicate or skill-requiring tasks
- Drowsiness
- Weakness
- Agitation
- Euphoria
- Loss of consciousness

Brain tissue is by far the most susceptible to the effects of hypoxia. Unconsciousness and death can occur from brain hypoxia before the effects on other tissues become very prominent.

There is no reliable warning of the onset of hypoxia. It can occur unexpectedly, making it a particularly serious hazard. A diver who loses his air supply is in danger of hypoxia, but he immediately knows he is in danger and usually has time to do something about it. He is much more fortunate than a diver who gradually uses up the oxygen in a closed-circuit rebreathing rig and has no warning of impending unconsciousness.

When hypoxia develops, pulse rate and blood pressure increase as the body tries to offset the hypoxia by circulating more blood. A small increase in breathing may also occur. A general blueness (cyanosis) of the lips, nail beds, and skin may occur with hypoxia. This may not be noticed by the diver and often is not a reliable indicator of hypoxia, even for the trained observer at the surface. The same signs could be caused by prolonged exposure to cold water.

If hypoxia develops gradually, symptoms of interference with brain function will appear. None of these symptoms, however, are sufficient warning and very few people are able to recognize the mental effects of hypoxia in time to take corrective action.

3-5.1.3 **Treatment of Hypoxia.** A diver suffering from severe hypoxia must be rescued promptly. Treat with basic first aid and 100% oxygen. If a victim of hypoxia is given gas with adequate oxygen content before his breathing stops, he usually regains consciousness shortly and recovers completely. For SCUBA divers, this usually involves bringing the diver to the surface. For surface-supplied mixed-gas divers, it involves shifting the gas supply to alternative banks and ventilating the helmet or chamber with the new gas. Refer to Volume 4 for information on treatment of hypoxia arising in specific operational environments for dives involving semi-closed and closed-circuit rebreathers.

3-5.1.4 **Prevention of Hypoxia.** Because of its insidious nature and potentially fatal outcome, preventing hypoxia is essential. In open-circuit SCUBA and helmets, hypoxia is unlikely unless the supply gas has too low an oxygen content. On mixed-gas operations, strict attention must be paid to gas analysis, cylinder lineups and predive checkout procedures. In closed and semi-closed circuit rebreathers, a malfunction can cause hypoxia even though the proper gases are being used. Electronically controlled, fully closed-circuit Underwater Breathing Apparatus (UBAs), like the MK 16, have oxygen sensors to read out oxygen partial pressure, but divers must be constantly alert to the possibility of hypoxia from a UBA malfunction. **To prevent hypoxia, oxygen sensors should be monitored closely throughout the dive. MK 25 UBA breathing bags should be purged in accordance with Operating Procedures (OPs).** Recently surfaced mixed-gas chambers should not be entered until after they are thoroughly ventilated with air.

3-5.2 **Carbon Dioxide Retention (Hypercapnia).** Hypercapnia is an abnormally high level of carbon dioxide in the blood and body tissues.

3-5.2.1 **Causes of Hypercapnia.** In diving operations, hypercapnia is generally the result of a buildup of carbon dioxide in the breathing supply or an inadequate respiratory minute volume. The principal causes are:

- Excess carbon dioxide levels in compressed air supplies due to improper placement of the compressor inlet.
- Inadequate ventilation of surface-supplied helmets or UBAs.
- Failure of carbon dioxide absorbent canisters to absorb carbon dioxide or incorrect installation of breathing hoses in closed or semi-closed circuit UBAs.
- Inadequate lung ventilation in relation to exercise level. The latter may be caused by skip breathing, increased apparatus dead space, excessive breathing resistance, or increased oxygen partial pressure.

Excessive breathing resistance is an important cause of hypercapnia and arises from two sources: flow resistance and static lung load. Flow resistance results from the flow of dense gas through tubes, hoses, and orifices in the diving equipment and through the diver's own airways. As gas density increases, a larger driving pressure must be applied to keep gas flowing at the same rate. The diver has to exert higher negative pressures to inhale and higher positive pressures to exhale. As ventilation increases with increasing levels of exercise, the necessary driving pressures increase. Because the respiratory muscles can only exert so much effort to inhale and exhale, a point is reached when further increases cannot occur. At this point, metabolically produced carbon dioxide is not adequately eliminated and increases in the blood and tissues, causing symptoms of hypercapnia. Symptoms of hypercapnia usually become apparent when divers attempt heavy work at depths deeper then 120 FSW on air or deeper than 850 FSW on helium-oxygen. At very great depths (1,600-2,000 FSW), shortness of breath and other signs of carbon dioxide toxicity may occur even at rest.

Static lung load is the result of breathing gas being supplied at a different pressure than the hydrostatic pressure surrounding the lungs. For example, when swimming horizontally with a single-hose regulator, the regulator diaphragm is lower than the mouth and the regulator supplies gas at a slight positive pressure once the demand valve has opened. If the diver flips onto his back, the regulator diaphragm is shallower than his mouth and the regulator supplies gas at a slightly negative pressure. Inhalation is harder but exhalation is easier because the exhaust ports are above the mouth and at a slightly lower pressure.

Static lung loading is more apparent in closed and semi-closed circuit underwater breathing apparatus such as the MK 25 and MK 16. When swimming horizontally with the MK 16, the diaphragm on the diver's back is shallower than the lungs and the diver feels a negative pressure at the mouth. Exhalation is easier than inhalation. If the diver flips onto his back, the diaphragm is below the lungs and the diver feels a positive pressure at the mouth. Inhalation becomes easier than exhalation. Static lung load is an important contributor to hypercapnia.

Excessive breathing resistance may cause shortness of breath and a sensation of labored breathing (dyspnea) without any increase in blood carbon dioxide level. In this case, the sensation of shortness of breath is due to activation of pressure and stretch receptors in the airways, lungs, and chest wall rather than activation of the chemoreceptors in the brain stem and carotid and aortic bodies. Usually, both types of activation are present when breathing resistance is excessive.

3-5.2.2 **Symptoms of Hypercapnia.** Hypercapnia affects the brain differently than hypoxia does. However, it can result in similar symptoms. Symptoms of hypercapnia include:

- Increased breathing rate
- Shortness of breath, sensation of difficult breathing or suffocation (dyspnea)
- Confusion or feeling of euphoria
- Inability to concentrate
- Increased sweating
- Drowsiness
- Headache
- Loss of consciousness
- Convulsions

The increasing level of carbon dioxide in the blood stimulates the respiratory center to increase the breathing rate and volume. The pulse rate also often increases. On dry land, the increased breathing rate is easily noticed and uncomfortable enough to warn the victim before the rise in $ppCO_2$ becomes dangerous. This is usually not the case in diving. Factors such as water temperature, work rate, increased breathing resistance, and an elevated ppO_2 in the breathing mixture may produce changes in respiratory drive that mask changes caused by excess carbon dioxide. This is especially true in closed-circuit UBAs, particularly 100-percent oxygen rebreathers. In cases where the ppO_2 is above 0.5 ata, the shortness of breath usually associated with excess carbon dioxide may not be prominent and may go unnoticed by the diver, especially if he is breathing hard because of exertion. In these cases the diver may become confused and even slightly euphoric before losing consciousness. For this reason, a diver must be particularly alert for any marked change in his breathing comfort or cycle (such as shortness of breath or hyperventilation) as a warning of hypercapnia. A similar situation can occur in cold water. Exposure to cold water often results in an increase in respiratory rate. This increase can make it difficult for the diver to detect an increase in respiratory rate related to a buildup of carbon dioxide.

Injury from hypercapnia is usually due to secondary effects such as drowning or injury caused by decreased mental function or unconsciousness. A diver who loses consciousness because of excess carbon dioxide in his breathing medium and does not inhale water generally revives rapidly when given fresh air and usually feels normal within 15 minutes. The after effects rarely include symptoms more serious than headache, nausea, and dizziness. Permanent brain damage and death are much less likely than in the case of hypoxia. If breathing resistance was high, the diver may note some respiratory muscle soreness post-dive.

Excess carbon dioxide also dilates the arteries of the brain. This may partially explain the headaches often associated with carbon dioxide intoxication, though these headaches are more likely to occur following the exposure than during it. The increase in blood flow through the brain, which results from dilation of the arteries, is thought to explain why carbon dioxide excess speeds the onset of CNS oxygen toxicity. Excess carbon dioxide during a dive is also believed to increase the likelihood of decompression sickness, but the reasons are less clear.

The effects of nitrogen narcosis and hypercapnia are additive. A diver under the influence of narcosis will probably not notice the warning signs of carbon dioxide intoxication. Hypercapnia in turn will intensify the symptoms of narcosis.

3-5.2.3 **Treatment of Hypercapnia.** Hypercapnia is treated by:

- Decreasing the level of exertion to reduce CO_2 production
- Increasing helmet and lung ventilation to wash out excess CO_2
- Shifting to an alternate breathing source or aborting the dive if defective equipment is the cause.

Because the first sign of hypercapnia may be unconsciousness and it may not be readily apparent whether the cause is hypoxia or hypercapnia. It is important to rule out hypoxia first because of the significant potential for brain damage in hypoxia. Hypercapnia may cause unconsciousness, but by itself will not injure the brain permanently.

3-5.2.4 **Prevention of Hypercapnia.** In surface-supplied diving, hypercapnia is prevented by ensuring that gas supplies do not contain excess carbon dioxide, by maintaining proper manifold pressure during the dive and by ventilating the helmet frequently with fresh gas. For dives deeper than 150 fsw, helium-oxygen mixtures should be used to reduce breathing resistance. In closed or semiclosed-circuit UBAs, hypercapnia is prevented by carefully filling the CO_2 absorbent canister and limiting dive duration to established canister duration limits. For dives deeper than 150 fsw, helium-oxygen mixtures should be used to reduce breathing resistance.

3-5.3 **Asphyxia.** *Asphyxia* is a condition where breathing stops and both hypoxia and hypercapnia occur simultaneously. Asphyxia will occur when there is no gas to breathe, when the airway is completely obstructed, when the respiratory muscles become paralyzed, or when the respiratory center fails to send out impulses to breathe. Running out of air is a common cause of asphyxia in SCUBA diving. Loss of the gas supply may also be due to equipment failure, for example regulator freeze up. Divers who become unconscious as a result of hypoxia, hypercapnia, or oxygen toxicity may lose the mouthpiece and suffer asphyxia. Obstruction of the airway can be caused by injury to the windpipe, the tongue falling back in the throat during unconsciousness, or the inhalation of water, saliva, vomitus or a foreign body. Paralysis of the respiratory muscles may occur with high cervical spinal cord injury due to trauma or decompression sickness. The respiratory center in the brain stem may become non-functional during a prolonged episode of hypoxia.

3-5.4 **Drowning/Near Drowning.** Drowning is fluid induced asphyxia. *Near drowning* is the term used when a victim is successfully resuscitated following a drowning episode.

3-5.4.1 **Causes of Drowning.** A swimmer or diver can fall victim to drowning because of overexertion, panic, inability to cope with rough water, exhaustion, or the effects of cold water or heat loss. Drowning in a hard-hat diving rig is rare. It can happen if the helmet is not properly secured and comes off, or if the diver is trapped in a head-down position with a water leak in the helmet. Normally, as long as the diver is in an upright position and has a supply of air, water can be kept out of the helmet regardless of the condition of the suit. Divers wearing lightweight or SCUBA gear can drown if they lose or ditch their mask or mouthpiece, run out of air, or inhale even small quantities of water. This could be the direct result of failure of the air supply, or panic in a hazardous situation. The SCUBA diver, because of direct exposure to the environment, can be affected by the same conditions that may cause a swimmer to drown.

3-5.4.2 **Symptoms of Drowning/Near Drowning.**

- Unconsciousness
- Pulmonary edema
- Increased respiratory rate.

3-5.4.3 **Treatment of Near Drowning.**

- Assess airway, breathing, and circulation.
- Rescue breathing should be started as soon as possible, even before the victim is removed from the water.
- Give 100 percent oxygen by mask.
- Call for assistance from qualified medical personnel and transport to nearest medical facility for evaluation.

Victims of near drowning who have no neurological symptoms should be evaluated by a Diving Medical Officer for pulmonary aspiration. Pneumonia is the classic result of near drowning.

3-5.4.4 **Prevention of Near Drowning.** Drowning is best prevented by thoroughly training divers in safe diving practices and carefully selecting diving personnel. A trained diver should not easily fall victim to drowning. However, overconfidence can give a feeling of false security that might lead a diver to take dangerous risks.

3-5.5 **Breathholding and Unconsciousness.** Most people can hold their breath approximately 1 minute, but usually not much longer without training or special preparation. At some time during a breathholding attempt, the desire to breathe becomes uncontrollable. The demand to breathe is signaled by the respiratory center responding to the increasing levels of carbon dioxide in the arterial blood and peripheral chemoreceptors responding to the corresponding fall in arterial oxygen partial pressure. If the breathhold is preceded by a period of voluntary hyperventilation, the breathhold can be much longer. Voluntary hyperventilation lowers body stores of carbon dioxide below normal (a condition known as hypocapnia), without significantly increasing oxygen stores. During the breathhold, it takes an appreciable time for the body stores of carbon dioxide to return to the normal level then to rise to the point where breathing is stimulated. During this time the oxygen partial pressure may fall below the level necessary to maintain consciousness. This is a common cause of breathholding accidents in swimming pools. Extended breathholding after hyperventilation is not a safe procedure.

WARNING **Voluntary hyperventilation is dangerous and can lead to unconsciousness and death during breathhold dives.**

Another hazard of breathhold diving is the possible loss of consciousness from hypoxia during ascent. Air in the lungs is compressed during descent, raising the oxygen partial pressure. The increased ppO_2 readily satisfies the body's oxygen demand during descent and while on the bottom, even though a portion is being consumed by the body. During ascent, the partial pressure of the remaining oxygen is reduced rapidly as the hydrostatic pressure on the body lessens. If the ppO_2 falls below 0.10 ata (10% sev), unconsciousness may result. This danger is further heightened when hyperventilation has eliminated normal body warning signs of carbon dioxide accumulation and allowed the diver to remain on the bottom for a longer period of time. Refer to Chapter 6 for breathhold diving restrictions.

3-5.6 **Involuntary Hyperventilation.** Hyperventilation is the term applied to breathing more than is necessary to keep the body's carbon dioxide tensions at proper level. Hyperventilation may be voluntary (for example, to increase breathholding time) or involuntary. In involuntary hyperventilation, the diver is either unaware that he is breathing excessively, or is unable to control his breathing.

3-5.6.1 **Causes of Involuntary Hyperventilation.** Involuntary hyperventilation can be triggered by fear experienced during stressful situations. It can also be initiated by the slight "smothering sensation" that accompanies an increase in equipment dead space, an increase in static lung loading, or an increase in breathing resistance. Cold water exposure can add to the sensation of needing to breathe faster and deeper. Divers using SCUBA equipment for the first few times are likely to hyperventilate to some extent because of anxiety.

3-5.6.2 **Symptoms of Involuntary Hyperventilation.** Hyperventilation may lead to a biochemical imbalance that gives rise to dizziness, tingling of the extremities, and spasm of the small muscles of the hands and feet. Hyperventilating over a long period, produces additional symptoms such as weakness, headaches, numbness, faintness, and blurring of vision. The diver may experience a sensation of "air hunger" even though his ventilation is more than enough to eliminate carbon dioxide. All these symptoms can be easily confused with symptoms of CNS oxygen toxicity.

3-5.6.3 **Treatment of Involuntary Hyperventilation.** Hyperventilation victims should be encouraged to relax and slow their breathing rates. The body will correct hyperventilation naturally.

3-5.7 **Overbreathing the Rig.** "Overbreathing the Rig" is a special term divers apply to an episode of acute hypercapnia that develops when a diver works at a level greater than his UBA can support. When a diver starts work, or abruptly increases his workload, the increase in respiratory minute ventilation lags the increase in oxygen consumption and carbon dioxide production by several minutes. When the RMV demand for that workload finally catches up, the UBA may not be able to supply the gas necessary despite extreme respiratory efforts on the part of the diver. Acute hypercapnia with marked respiratory distress ensues. Even if the diver stops work to lower the production of carbon dioxide, the sensation of shortness of breath may persist or even increase for a short period of time. When this occurs, the inexperienced diver may panic and begin to hyperventilate. The situation can rapidly develop into a malicious cycle of severe shortness of breath and uncontrollable hyperventilation. In this situation, if even a small amount of water is inhaled, it can cause a spasm of the muscles of the larynx (voice box), called a laryngospasm, followed by asphyxia and possible drowning.

The U.S. Navy makes every effort to ensure that UBA meet adequate breathing standards to minimize flow resistance and static lung loading problems. However, all UBA have their limitations and divers must have sufficient experience to recognize those limitations and pace their work accordingly. Always increase workloads gradually to insure that the UBA can match the demand for increased lung ventilation. If excessive breathing resistance is encountered, slow or stop the pace of work until a respiratory comfort level is achieved. If respiratory distress occurs following an abrupt increase in workload, stop work and take even controlled breaths until the sensation of respiratory distress subsides. If the situation does not improve, abort the dive.

3-5.8 **Carbon Monoxide Poisoning.** The body produces carbon monoxide as a part of the process of normal metabolism. Consequently, there is always a small amount of carbon monoxide present in the blood and tissues. Carbon monoxide poisoning occurs when levels of carbon monoxide in the blood and tissues rise above these normal values due to the presence of carbon monoxide in the diver's gas supply. Carbon monoxide not only blocks hemoglobin's ability to delivery oxygen to the cells, causing cellular hypoxia, but also poisons cellular metabolism directly.

3-5.8.1 **Causes of Carbon Monoxide Poisoning.** Carbon monoxide is not found in any significant quantity in fresh air. Carbon monoxide poisoning is usually caused by a compressor's intake being too close to the exhaust of an internal combustion engine or malfunction of a oil lubricated compressor. Concentrations as low as 0.002 ata (2,000 ppm, or 0.2%) can prove fatal.

3-5.8.2 **Symptoms of Carbon Monoxide Poisoning.** The symptoms of carbon monoxide poisoning are almost identical to those of hypoxia. When toxicity develops gradually the symptoms are:

- Headache
- Dizziness
- Confusion
- Nausea
- Vomiting
- Tightness across the forehead

When carbon monoxide concentrations are high enough to cause rapid onset of poisoning, the victim may not be aware of any symptoms before he becomes unconscious.

Carbon monoxide poisoning is particularly treacherous because conspicuous symptoms may be delayed until the diver begins to ascend. While at depth, the greater partial pressure of oxygen in the breathing supply forces more oxygen into solution in the blood plasma. Some of this additional oxygen reaches the cells and helps to offset the hypoxia. In addition, the increased partial pressure of oxygen forcibly displaces some carbon monoxide from the hemoglobin. During ascent, however, as the partial pressure of oxygen diminishes, the full effect of carbon monoxide poisoning is felt.

3-5.8.3 **Treatment of Carbon Monoxide Poisoning.** The immediate treatment of carbon monoxide poisoning consists of getting the diver to fresh air and seeking medical attention. Oxygen, if available, shall be administered immediately and while transporting the patient to a hyperbaric or medical treatment facility. Hyperbaric oxygen therapy is the definitive treatment of choice and transportation for recompression should not be delayed except to stabilize the serious patient. Divers with severe symptoms (i.e. severe headache, mental status changes, any neurological symptoms, rapid heart rate) should be treated using Treatment Table 6.

3-5.8.4 **Prevention of Carbon Monoxide Poisoning.** Locating compressor intakes away from engine exhausts and maintaining air compressors in the best possible mechanical condition can prevent carbon monoxide poisoning. When carbon monoxide poisoning is suspected, isolate the suspect breathing gas source, and forward gas samples for analysis as soon as possible.

3-6 MECHANICAL EFFECTS OF PRESSURE ON THE HUMAN BODY-BAROTRAUMA DURING DESCENT

Barotrauma, or damage to body tissues from the mechanical effects of pressure, results when pressure differentials between body cavities and the hydrostatic pressure surrounding the body, or between the body and the diving equipment, are not equalized properly. Barotrauma most frequently occurs during descent, but may also occur during ascent. Barotrauma on descent is called squeeze. Barotrauma on ascent is called reverse squeeze.

3-6.1 **Prerequisites for Squeeze.** For squeeze to occur during descent the following five conditions must be met:

- There must be a gas-filled space. Any gas-filled space within the body (such as a sinus cavity) or next to the body (such as a face mask) can damage the body tissues when the gas volume changes because of increased pressure.

- The gas-filled space must have rigid walls. If the walls are collapsible like a balloon, no damage will be done by compression.

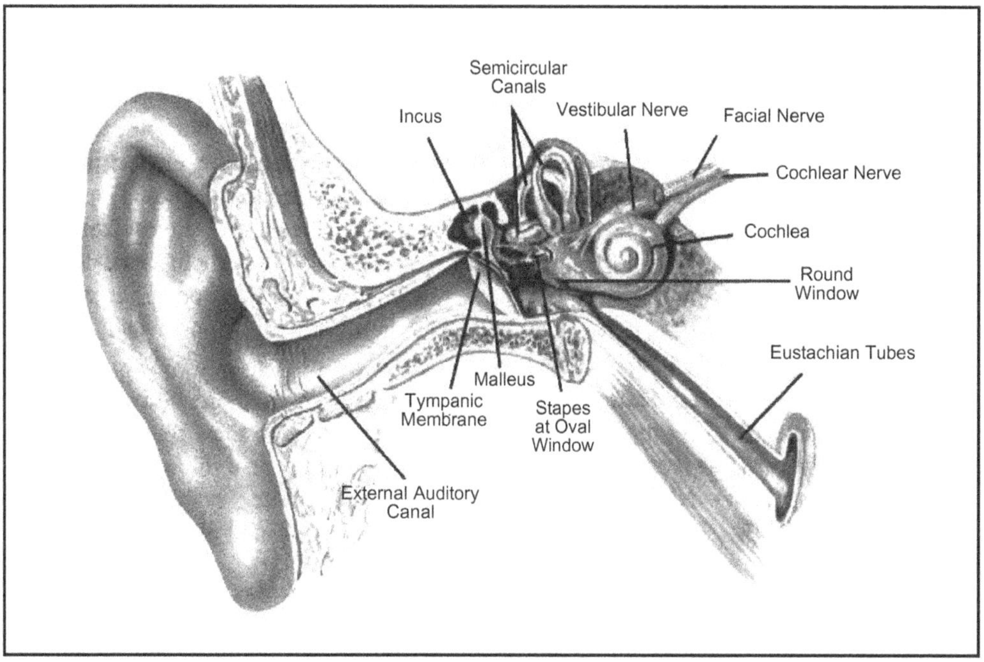

Figure 3-7. Gross Anatomy of the Ear in Frontal Section.

- The gas-filled space must be enclosed. If gas or liquid can freely enter the space as the gas volume changes, no damage will occur.
- The space must have lining membrane with an arterial blood supply and venous drainage that penetrates the space from the outside. This allows blood to be forced into the space to compensate for the change in pressure.
- There must be a change in ambient pressure.

3-6.2 **Middle Ear Squeeze.** Middle ear squeeze is the most common type of barotrauma. The anatomy of the ear is illustrated in Figure 3-7. The eardrum completely seals off the outer ear canal from the middle ear space. As a diver descends, water pressure increases on the external surface of the drum. To counterbalance this pressure, the air pressure must reach the inner surface of the eardrum. This is accomplished by the passage of air through the narrow eustachian tube that leads from the nasal passages to the middle ear space. When the eustachian tube is blocked by mucous, the middle ear meets four of the requirements for barotrauma to occur (gas filled space, rigid walls, enclosed space, penetrating blood vessels).

As the diver continues his descent, the fifth requirement (change in ambient pressure) is attained. As the pressure increases, the eardrum bows inward and initially equalizes the pressure by compressing the middle ear gas. There is a limit to this stretching capability and soon the middle ear pressure becomes lower than the external water pressure, creating a relative vacuum in the middle ear space. This negative pressure causes the blood vessels of the eardrum and lining of the middle ear to first expand, then leak and finally burst. If descent continues, either the eardrum ruptures, allowing air or water to enter the middle ear and equalize the pressure, or blood vessels rupture and cause sufficient bleeding into the middle ear to equalize the pressure. The latter usually happens.

The hallmark of middle ear squeeze is sharp pain caused by stretching of the eardrum. The pain produced before rupture of the eardrum often becomes intense enough to prevent further descent. Simply stopping the descent and ascending a few feet usually brings about immediate relief.

If descent continues in spite of the pain, the eardrum may rupture. When rupture occurs, this pain will diminish rapidly. Unless the diver is in hard hat diving dress, the middle ear cavity may be exposed to water when the ear drum ruptures. This exposes the diver to a possible middle ear infection and, in any case, prevents the diver from diving until the damage is healed. If eardrum rupture occurs, the dive shall be aborted. At the time of the rupture, the diver may experience the sudden onset of a brief but violent episode of vertigo (a sensation of spinning). This can completely disorient the diver and cause nausea and vomiting. This vertigo is caused by violent disturbance of the malleus, incus, and stapes, or by cold water stimulating the balance mechanism of the inner ear. The latter situation is referred to as caloric vertigo and may occur from simply having cold or warm water enter one ear and not the other. The eardrum does not have to rupture for caloric vertigo to occur. It can occur as the result of having water enter one ear canal when swimming or diving in cold water. Fortunately, these symptoms quickly pass when the water reaching the middle ear is warmed by the body. Suspected cases of eardrum rupture shall be referred to medical personnel.

3-6.2.1 **Preventing Middle Ear Squeeze.** Diving with a partially blocked eustachian tube increases the likelihood of middle ear squeeze. Divers who cannot clear their ears on the surface should not dive. Medical personnel shall examine divers who have trouble clearing their ears before diving. The possibility of barotrauma can be virtually eliminated if certain precautions are taken. While descending, stay ahead of the pressure. To avoid collapse of the eustachian tube and to clear the ears, frequent adjustments of middle ear pressure must be made by adding gas through the eustachian tubes from the back of the nose. If too large a pressure difference develops between the middle ear pressure and the external pressure, the eustachian tube collapses as it becomes swollen and blocked. For some divers, the eustachian tube is open all the time so no conscious effort is necessary to clear their ears. For the majority, however, the eustachian tube is normally closed and some action must be taken to clear the ears. Many divers can clear by yawning, swallowing, or moving the jaw around.

Some divers must gently force gas up the eustachian tube by closing their mouth, pinching their nose and exhaling. This is called a Valsalva maneuver. If too large a relative vacuum exists in the middle ear, the eustachian tube collapses and no amount of forceful clearing will open it. If a squeeze is noticed during descent, the diver shall stop, ascend a few feet and gently perform a Valsalva maneuver. If clearing cannot be accomplished as described above, abort the dive.

WARNING **Never do a forceful Valsalva maneuver during descent. A forceful Valsalva maneuver can result in alternobaric vertigo or barotrauma to the inner ear (see below).**

WARNING **If decongestants must be used, check with medical personnel trained in diving medicine to obtain medication that will not cause drowsiness and possibly add to symptoms caused by the narcotic effect of nitrogen.**

3-6.2.2 **Treating Middle Ear Squeeze.** Upon surfacing after a middle ear squeeze, the diver may complain of pain, fullness in the ear, hearing loss, or even mild vertigo. Occasionally, the diver may have a bloody nose, the result of blood being forced out of the middle ear space and into the nasal cavity through the eustachian tube by expanding air in the middle ear. The diver shall report symptoms of middle ear squeeze to the diving supervisor and seek medical attention. Treatment consists of taking decongestants, pain medication if needed, and cessation of diving until the damage is healed. If the eardrum has ruptured antibiotics may be prescribed as well. Never administer medications directly into the external ear canal if a ruptured eardrum is suspected or confirmed unless done in direct consultation with an ear, nose, and throat (ENT) medical specialist.

3-6.3 **Sinus Squeeze.** Sinuses are located within hollow spaces of the skull bones and are lined with a mucous membrane continuous with that of the nasal cavity (Figure 3-8). The sinuses are small air pockets connected to the nasal cavity by narrow passages. If pressure is applied to the body and the passages to any of these sinuses are blocked by mucous or tissue growths, pain will soon be experienced in the affected area. The situation is very much like that described for the middle ear.

3-6.3.1 **Causes of Sinus Squeeze.** When the air pressure in these sinuses is less than the pressure applied to the tissues surrounding these incompressible spaces, the same relative effect is produced as if a vacuum were created within the sinuses: the lining membranes swell and, if severe enough, hemorrhage into the sinus spaces. This process represents nature's effort to balance the relative negative air pressure by filling the space with swollen tissue, fluid, and blood. The sinus is actually squeezed. The pain produced may be intense enough to halt the diver's descent. Unless damage has already occurred, a return to normal pressure will bring about immediate relief. If such difficulty has been encountered during a dive, the diver may often notice a small amount of bloody nasal discharge on reaching the surface.

3-6.3.2 **Preventing Sinus Squeeze.** Divers should not dive if any signs of nasal congestion or a head cold are evident. The effects of squeeze can be limited during a dive by halting the descent and ascending a few feet to restore the pressure balance. If the space cannot be equalized by swallowing or blowing against a pinched-off nose, the dive must be aborted.

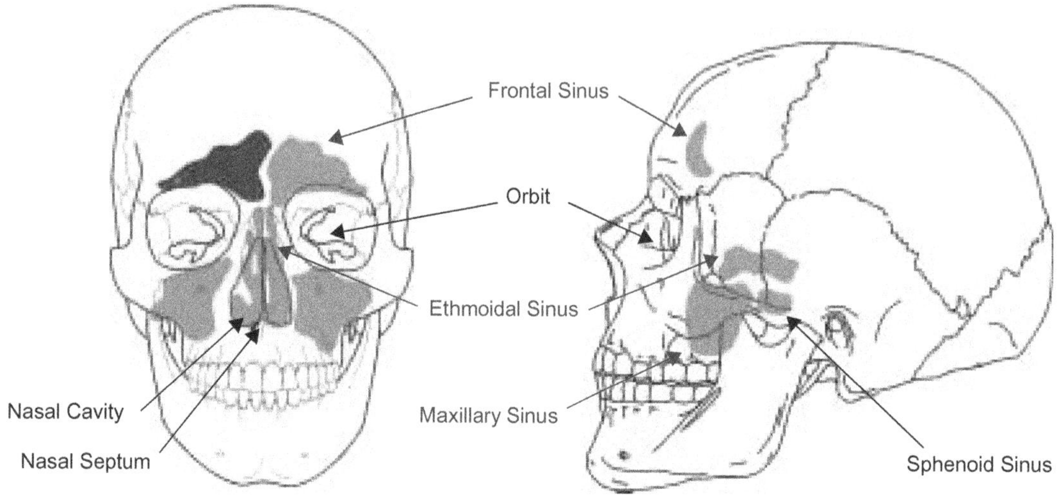

Figure 3-8. Location of the Sinuses in the Human Skull.

3-6.4 **Tooth Squeeze (Barodontalgia).** Tooth squeeze occurs when a small pocket of gas, generated by decay, is lodged under a poorly fitted or cracked filling. If this pocket of gas is completely isolated, the pulp of the tooth or the tissues in the tooth socket can be sucked into the space causing pain. If additional gas enters the tooth during descent and does not vent during ascent, it can cause the tooth to crack or the filling to be dislodged. Prior to any dental work, personnel shall identify themselves as divers to the dentist.

3-6.5 **External Ear Squeeze.** A diver who wears ear plugs, has an infected external ear (external otitis), has a wax-impacted ear canal, or wears a tight-fitting wet suit hood, can develop an external ear squeeze. The squeeze occurs when gas trapped in the external ear canal remains at atmospheric pressure while the external water pressure increases during descent. In this case, the eardrum bows outward (opposite of middle ear squeeze) in an attempt to equalize the pressure difference and may rupture. The skin of the canal swells and hemorrhages, causing considerable pain.

Ear plugs must never be worn while diving. In addition to creating the squeeze, they may be forced deep into the ear canal. When a hooded suit must be worn, air (or water in some types) must be allowed to enter the hood to equalize pressure in the ear canal.

3-6.6 **Thoracic (Lung) Squeeze.** When making a breathhold dive, it is possible to reach a depth at which the air held in the lungs is compressed to a volume somewhat smaller than the normal residual volume of the lungs. At this volume, the chest wall becomes stiff and incompressible. If the diver descends further, the additional pressure is unable to compress the chest walls, force additional blood into the blood vessels in the chest, or elevate the diaphragm further. The pressure in the lung becomes negative with respect to the external water pressure. Injury takes the form of squeeze. Blood and tissue fluids are forced into the lung alveoli and air passages where the air is under less pressure than the blood in the surrounding vessels. This amounts to an attempt to relieve the negative pressure within the lungs by partially filling the air space with swollen tissue, fluid, and blood. Considerable lung damage results and, if severe enough, may prove fatal. If the diver descends still further, death will occur as a result of the collapse of the chest. Breathhold diving shall be limited to controlled, training situations or special operational situations involving well-trained personnel at shallow depths.

A surface-supplied diver who suffers a loss of gas pressure or hose rupture with failure of the nonreturn valve may suffer a lung squeeze, if his depth is great enough, as the surrounding water pressure compresses his chest.

3-6.7 **Face or Body Squeeze.** SCUBA face masks, goggles, and certain types of exposure suits may cause squeeze under some conditions. Exhaling through the nose can usually equalize the pressure in a face mask, but this is not possible with goggles. Goggles shall only be used for surface swimming. The eye and the eye socket tissues are the most seriously affected tissues in an instance of face mask or goggle squeeze. When using exposure suits, air may be trapped in a fold in the garment and may lead to some discomfort and possibly a minor case of hemorrhage into the skin from pinching.

3-6.8 **Inner Ear Barotrauma.** The inner ear contains no gas and therefore cannot be "squeezed" in the same sense that the middle ear and sinuses can. However, the inner ear is located next to the middle ear cavity and is affected by the same conditions that lead to middle ear squeeze. To understand how the inner ear could be damaged as a result of pressure imbalances in the middle ear, it is first necessary to understand the anatomy of the middle and inner ear.

The inner ear contains two important organs, the cochlea and the vestibular apparatus. The cochlea is the hearing sense organ; damage to the cochlea will result in hearing loss and ringing in the ear (tinnitus). The vestibular apparatus is the balance organ; damage to the vestibular apparatus will result in vertigo and unsteadiness.

There are three bones in the middle ear: the malleus, the incus, and the stapes. They are also commonly referred to as the hammer, anvil, and stirrup, respectively (Figure 3-9). The malleus is connected to the eardrum (tympanic membrane) and transmits sound vibrations to the incus, which in turn transmits these vibrations to the stapes, which relays them to the inner ear. The stapes transmits these vibrations to the inner ear fluid through a membrane-covered hole called the oval window. Another membrane-covered hole called the round window connects the inner ear with the middle ear and relieves pressure waves in the inner ear caused by movement of the stapes. When the stapes drives the oval window inward, the round window bulges outward to compensate. The fluid-filled spaces of the inner ear are also connected to the fluid spaces surrounding the brain by a narrow passage called the cochlear aqueduct. The cochlear aqueduct can transmit increases in cerebrospinal fluid pressure to the inner ear. When Valsalva maneuvers are performed to equalize middle ear and sinus pressure, cerebrospinal fluid pressure increases.

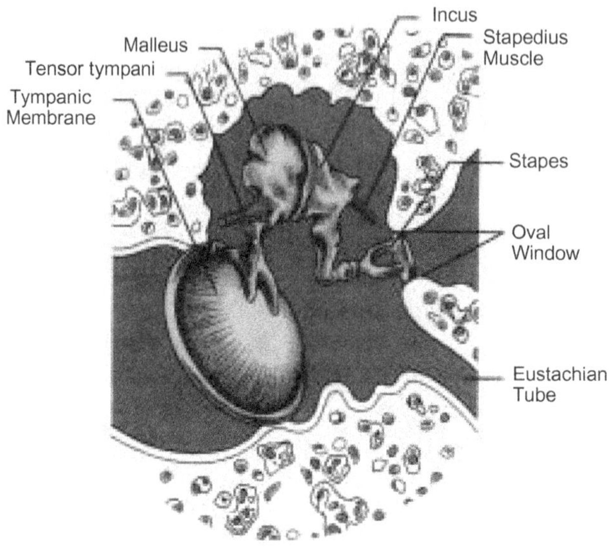

Figure 3-9. Components of the Middle/Inner Ear.

If middle ear pressure is not equalized during descent, the inward bulge of the eardrum is transmitted to the oval window by the middle ear bones. The stapes pushes the oval window inward. Because the inner ear fluids are incompressible, the round window correspondingly bulges outward into the middle ear space. If this condition continues, the round window may rupture spilling inner ear fluids into the middle ear and leading to a condition know as *inner ear barotrauma with perilymph fistula*. Fistula is a medical term for a hole in a membrane; the fluid in the inner ear is called perilymph. Rupture of the oval or round windows may also occur when middle ear pressures are suddenly and forcibly equalized. When equalization is sudden and forceful, the eardrum moves rapidly from a position of bulging inward maximally to bulging outward maximally. The positions of the oval and round windows are suddenly reversed Inner ear pressure is also increased by transmission of the Valsalva-induced increase in cerebrospinal fluid pressure. This puts additional stresses on these two membranes. Either the round or oval window may rupture. Rupture of the round window is by far the most common. The oval window is a tougher membrane and is protected by the footplate of the stapes. Even if rupture of the round or oval window does not occur, the pressure waves induced in the inner ear during these window movements may lead to disruption of the delicate cells involved in hearing and balance. This condition is referred to *inner ear barotrauma without perilymph fistula*.

The primary symptoms of inner ear barotrauma are persistent vertigo and hearing loss. Vertigo is the false sensation of motion. The diver feels that he is moving with respect to his environment or that the environment is moving with respect to him, when in fact no motion is taking place. The vertigo of inner ear barotrauma is generally described as whirling, spinning, rotating, tilting, rocking, or undulating. This sensation is quite distinct from the more vague complaints of dizziness or lightheadedness caused by other conditions. The vertigo of inner ear barotrauma is often accompanied by symptoms that may or may not be noticed depending on the severity of the insult. These include nausea, vomiting, loss of balance, incoordination, and a rapid jerking movement of the eyes, called nystagmus. Vertigo may be accentuated when the head is placed in certain positions. The hearing loss of inner ear barotrauma may fluctuate in intensity and sounds may be distorted. Hearing loss is accompanied by ringing or roaring in the affected ear. The diver may also complain of a sensation of bubbling in the affected ear.

Symptoms of inner ear barotrauma usually appear abruptly during descent, often as the diver arrives on the bottom and performs his last equalization maneuver. However, the damage done by descent may not become apparent until the dive is over. A common scenario is for the diver to rupture a damaged round window while lifting heavy weights or having a bowel movement post dive. Both these activities increase cerebrospinal fluid pressure and this pressure increase is transmitted to the inner ear. The round window membrane, weakened by the trauma suffered during descent, bulges into the middle ear space under the influence of the increased cerebrospinal fluid pressure and ruptures.

All cases of suspected inner ear barotrauma should be referred to an ear, nose and throat (ENT) physician as soon as possible. Treatment of inner ear barotrauma ranges from bed rest with head elevation to exploratory surgery, depending on the severity of the symptoms and whether a perilymph fistula is suspected. Any hearing loss or vertigo occurring within 72 hours of a hyperbaric exposure should be evaluated as a possible case of inner ear barotrauma.

When either hearing loss or vertigo develop after the diver has surfaced, it may be impossible to tell whether the symptoms are caused by inner ear barotrauma, decompression sickness or arterial gas embolism. For the latter two conditions, recompression treatment is mandatory. Although it might be expected that recompression treatment would further damage to the inner ear in a case of barotrauma and should be avoided, experience has shown that recompression is generally not harmful provided a few simple precautions are followed. The diver should be placed in a head up position and compressed slowly to allow adequate time for middle ear equalization. Clearing maneuvers should be gentle. The diver should not be exposed to excessive positive or negative pressure when breathing oxygen on the built-in breathing system (BIBS) mask. Always recompress the diver if there is any doubt about the cause of post-dive hearing loss or vertigo.

CAUTION **When in doubt, always recompress.**

Frequent oscillations in middle ear pressure associated with difficult clearing may lead to a transient vertigo. This condition is called *alternobaric vertigo of descent*. Vertigo usually follows a Valsalva maneuver, often with the final clearing episode just as the diver reaches the bottom. Symptoms typically last less than a minute but can cause significant disorientation during that period. Descent should be halted until the vertigo resolves. Once the vertigo resolves, the dive may be continued.

Alternobaric vertigo is a mild form of inner ear barotrauma in which no lasting damage to the inner ear occurs.

3-7 MECHANICAL EFFECTS OF PRESSURE ON THE HUMAN BODY--BAROTRAUMA DURING ASCENT

During ascent gases expand according to Boyle's Law. If the excess gas is not vented from enclosed spaces, damage to those spaces may result.

3-7.1 **Middle Ear Overpressure (Reverse Middle Ear Squeeze).** Expanding gas in the middle ear space during ascent ordinarily vents out through the eustachian tube. If the tube becomes blocked, pressure in the middle ear relative to the external water pressure increases. To relieve this pressure, the eardrum bows outward causing pain. If the overpressure is significant, the eardrum may rupture. If rupture occurs, the middle ear will equalize pressure with the surrounding water and the pain will disappear. However, there may be a transient episode of intense vertigo as cold water enters the middle ear space.

The increased pressure in the middle ear may also affect the inner ear balance mechanism, leading to a condition called *alternobaric vertigo of ascent*. Alternobaric vertigo occurs when the middle ear space on one side is overpressurized while the other side is equalizing normally. The onset of vertigo is usually sudden and may be preceded by pain in the ear that is not venting excess pressure. Alternobaric vertigo usually lasts for only a few minutes, but may be incapacitating during that time. Relief is usually abrupt and may be accompanied by a hissing sound in the affected ear as it equalizes. Alternobaric vertigo during ascent will disappear immediately if the diver halts his ascent and descends a few feet.

Increased pressure in the middle ear can also produce paralysis of the facial muscles, a condition known as *facial baroparesis*. In some individuals, the facial nerve is exposed to middle ear pressure as it traverses the temporal bone. If the middle ear fails to vent during ascent, the overpressure can shut off the blood supply to the nerve causing it to stop transmitting neural impulses to the facial muscles on the affected side. Generally, a 10 to 30 min period of overpressure is necessary for symptoms to occur. Full function of the facial muscles returns 5-10 min after the overpressure is relieved.

Increased pressure in the middle ear can also cause structural damage to the inner ear, a condition known as *inner ear barotrauma of ascent*. The bulging ear drum pulls the oval window outward into the middle ear space through the action of the middle ear bones. The round window correspondingly bulges inward. This inward deflection can be enhanced if the diver further increases middle ear pressure by performing a Valsalva maneuver. The round window may rupture causing inner ear fluids to spill into the middle ear space. The symptoms of marked hearing loss and sustained vertigo are identical to the symptoms experienced with inner ear barotrauma during descent.

A diver who has a cold or is unable to equalize the ears is more likely to develop reverse middle ear squeeze. There is no uniformly effective way to clear the ears on ascent. Do not perform a Valsalva maneuver on ascent, as this will increase the pressure in the middle ear, which is the direct opposite of what is required. The Valsalva maneuver can also lead to the possibility of an arterial gas embolism. If pain in the ear or vertigo develops on ascent, the diver should halt the ascent, descend a few feet to relieve the symptoms and then continue his ascent at a slower rate. Several such attempts may be necessary as the diver gradually works his way to the surface. If symptoms of sustained hearing loss or vertigo appear during ascent, or shortly after ascent, it may be impossible to tell whether the symptoms are arising from inner ear barotrauma or from decompression sickness or arterial gas embolism. Recompression therapy is always indicated unless there is 100% certainty that the condition is inner ear barotrauma.

3-7.2 **Sinus Overpressure (Reverse Sinus Squeeze).** Overpressure is caused when gas is trapped within the sinus cavity. A fold in the sinus-lining membrane, a cyst, or an outgrowth of the sinus membrane (polyp) may act as a check valve and prevent gas from leaving the sinus during ascent. Sharp pain in the area of the affected sinus results from the increased pressure. The pain is usually sufficient to stop the diver from ascending. Pain is immediately relieved by descending a few feet. From that point, the diver should titrate himself slowly to the surface in a series of ascents and descents just as with a reverse middle ear squeeze.

When overpressure occurs in the maxillary sinus, the blood supply to the infraorbital nerve may be reduced, leading to numbness of the lower eyelid, upper lip, side of the nose, and cheek on the affected side. This numbness will resolve spontaneously when the sinus overpressure is relieved.

3-7.3 **Gastrointestinal Distention.** Divers may occasionally experience abdominal pain during ascent because of gas expansion in the stomach or intestines. This condition is caused by gas being generated in the intestines during a dive, or by swallowing air (aerophagia). These pockets of gas will usually work their way out of the system through the mouth or anus. If not, distention will occur.

If the pain begins to pass the stage of mild discomfort, ascent should be halted and the diver should descend slightly to relieve the pain. The diver should then attempt to gently burp or release the gas anally. Overzealous attempts to belch should be avoided as they may result in swallowing more air. Abdominal pain following fast ascents shall be evaluated by a Diving Medical Officer.

To avoid intestinal gas expansion:

- Do not dive with an upset stomach or bowel.
- Avoid eating foods that are likely to produce intestinal gas.
- Avoid a steep, head-down angle during descent to minimize the amount of air swallowed.

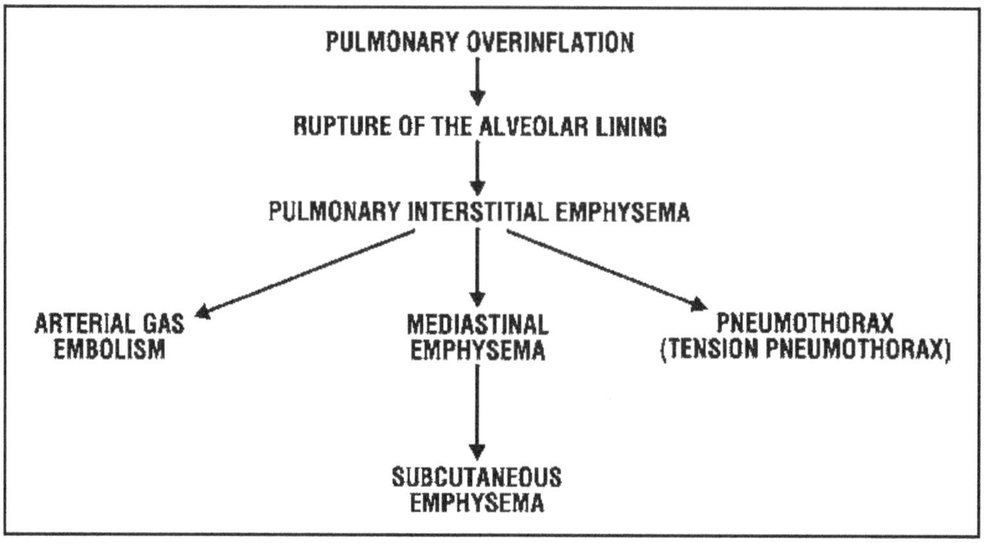

Figure 3-10. Pulmonary Overinflation Syndromes (POIS). Leaking of gas into the pulmonary interstitial tissue causes no symptoms unless further leaking occurs. If gas enters the arterial circulation, potentially fatal arterial gas embolism may occur. Pneumothorax occurs if gas accumulates between the lung and chest wall and if accumulation continues without venting, then tension pneumothorax may result.

3-8 PULMONARY OVERINFLATION SYNDROMES

Pulmonary overinflation syndromes are a group of barotrauma-related diseases caused by the expansion of gas trapped in the lung during ascent (reverse squeeze) or overpressurization of the lung with subsequent overexpansion and rupture of the alveolar air sacs. Excess pressure inside the lung can also occur when a diver presses the purge button on a single-hose regulator while taking a breath. The two main causes of alveolar rupture are:

- Excessive pressure inside the lung caused by positive pressure
- Failure of expanding gas to escape from the lung during ascent

Pulmonary overinflation from expanding gas failing to escape from the lung during ascent can occur when a diver voluntarily or involuntarily holds his breath during ascent. Localized pulmonary obstructions that can cause air trapping, such as asthma or thick secretions from pneumonia or a severe cold, are other causes. The conditions that bring about these incidents are different from those that produce lung squeeze and they most frequently occur during free and buoyant ascent training or emergency ascent from dives made with lightweight diving equipment or SCUBA.

The clinical manifestations of pulmonary overinflation depend on the location where the free air collects. In all cases, the first step is rupture of the alveolus with a collection of air in the lung tissues, a condition known as interstitial emphysema. Interstitial emphysema causes no symptoms unless further distribution of the air occurs. Gas may find its way into the chest cavity or arterial circulation. These conditions are depicted in Figure 3-10.

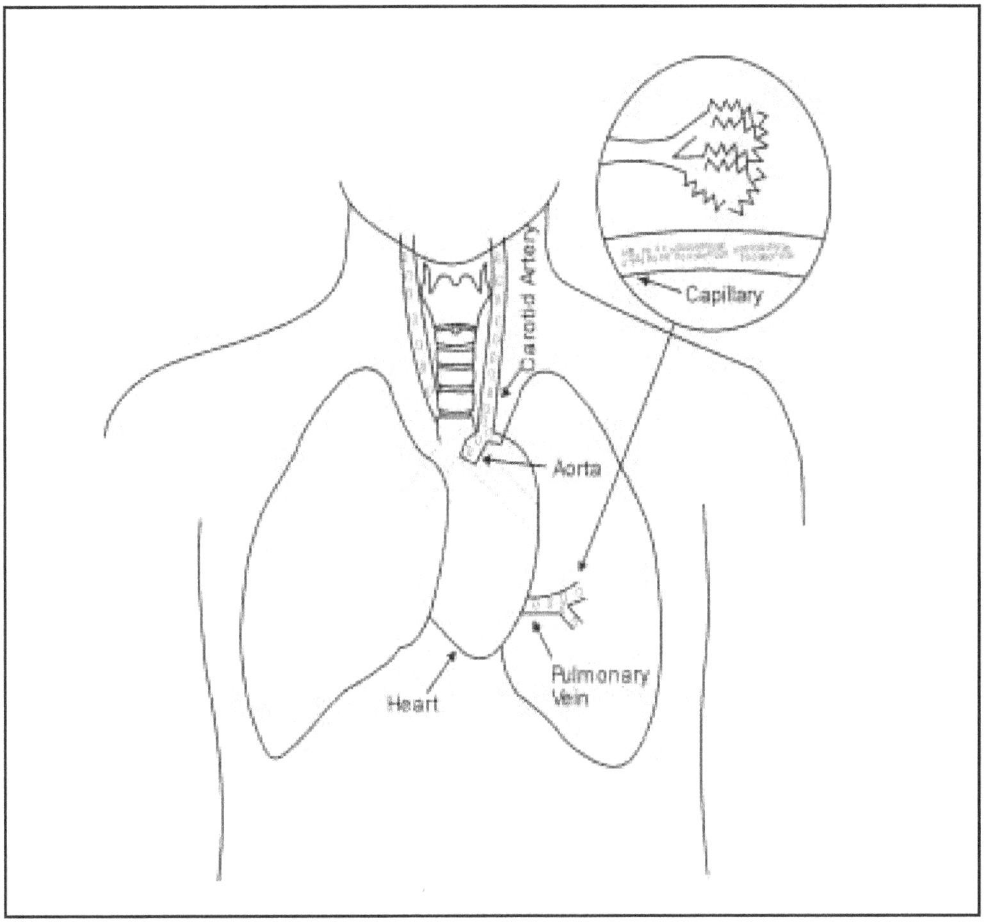

Figure 3-11. Arterial Gas Embolism.

3-8.1 **Arterial Gas Embolism (AGE).** Arterial gas embolism (AGE), sometimes simply called gas embolism, is an obstruction of blood flow caused by gas bubbles (emboli) entering the arterial circulation. Obstruction of the arteries of the brain and heart can lead to death if not promptly relieved (see Figure 3-11).

3-8.1.1 **Causes of AGE.** AGE is caused by the expansion of gas taken into the lungs while breathing under pressure and held in the lungs during ascent. The gas might have been retained in the lungs by choice (voluntary breathholding) or by accident (blocked air passages). The gas could have become trapped in an obstructed portion of the lung that has been damaged from some previous disease or accident; or the diver, reacting with panic to a difficult situation, may breathhold without realizing it. If there is enough gas and if it expands sufficiently, the pressure will force gas through the alveolar walls into surrounding tissues and into the bloodstream. If the gas enters the arterial circulation, it will be dispersed to all organs of the body. The organs that are especially susceptible to arterial gas embolism and that are responsible for the life-threatening symptoms are the central nervous system (CNS) and the heart. In all cases of arterial gas embolism, associated pneumothorax is possible and should not be overlooked. Exhaustion of air supply and the need for an emergency ascent is the most common cause of AGE.

3-8.1.2 **Symptoms of AGE**

- Unconsciousness
- Paralysis
- Numbness
- Weakness
- Extreme fatigue
- Large areas of abnormal sensations (Paresthesias)
- Difficulty in thinking
- Vertigo
- Convulsions
- Vision abnormalities
- Loss of coordination
- Nausea and or vomiting
- Hearing abnormalities
- Sensation similar to that of a blow to the chest during ascent
- Bloody sputum
- Dizziness
- Personality changes
- Loss of control of bodily functions
- Tremors

Symptoms of subcutaneous/medistinal emphysema, pneumothorax and/or pneumopericardium may also be present (see below). In all cases of arterial gas embolism, the possible presence of these associated conditions should not be overlooked.

3-8.1.3 **Treatment of AGE.**

- Basic first aid (ABC)
- 100 percent oxygen

- Immediate recompression
- See Volume 5 for more specific information regarding treatment.

3-8.1.4 **Prevention of AGE.** The risk of arterial gas embolism can be substantially reduced or eliminated by paying careful attention to the following:

- Every diver must receive intensive training in diving physics and physiology, as well as instruction in the correct use of diving equipment. Particular attention must be given to the training of SCUBA divers, because SCUBA operations produce a comparatively high incidence of embolism accidents.
- A diver must never interrupt breathing during ascent from a dive in which compressed gas has been breathed.
- A diver must exhale continuously while making an emergency ascent. The rate of exhalation must match the rate of ascent. For a free ascent, where the diver uses natural buoyancy to be carried toward the surface, the rate of exhalation must be great enough to prevent embolism, but not so great that positive buoyancy is lost. In a uncontrolled or buoyant ascent, where a life preserver, dry suit or buoyancy compensator assists the diver, the rate of ascent may far exceed that of a free ascent. The exhalation must begin before the ascent and must be a strong, steady, and forceful. It is difficult for an untrained diver to execute an emergency ascent properly. It is also often dangerous to train a diver in the proper technique.
- The diver must not hesitate to report any illness, especially respiratory illness such as a cold, to the Diving Supervisor or Diving Medical Personnel prior to diving.

3-8.2 **Mediastinal and Subcutaneous Emphysema.** Mediastinal emphysema, also called pneumomediastinum, occurs when gas is forced through torn lung tissue into the loose mediastinal tissues in the middle of the chest surrounding the heart, the trachea, and the major blood vessels (see Figure 3-12). Subcutaneous emphysema occurs when that gas subsequently migrates into the subcutaneous tissues of the neck (Figure 3-13). Mediastinal emphysema is a pre-requisite for subcutaneous emphysema.

3-8.2.1 **Causes of Mediastinal and Subcutaneous Emphysema.** Mediastinal/subcutaneous emphysema is caused by over inflation of the whole lung or parts of the lung due to:

- Breath holding during ascent
- Positive pressure breathing such as ditch and don exercises
- Drown proofing exercises
- Cough during surface swimming

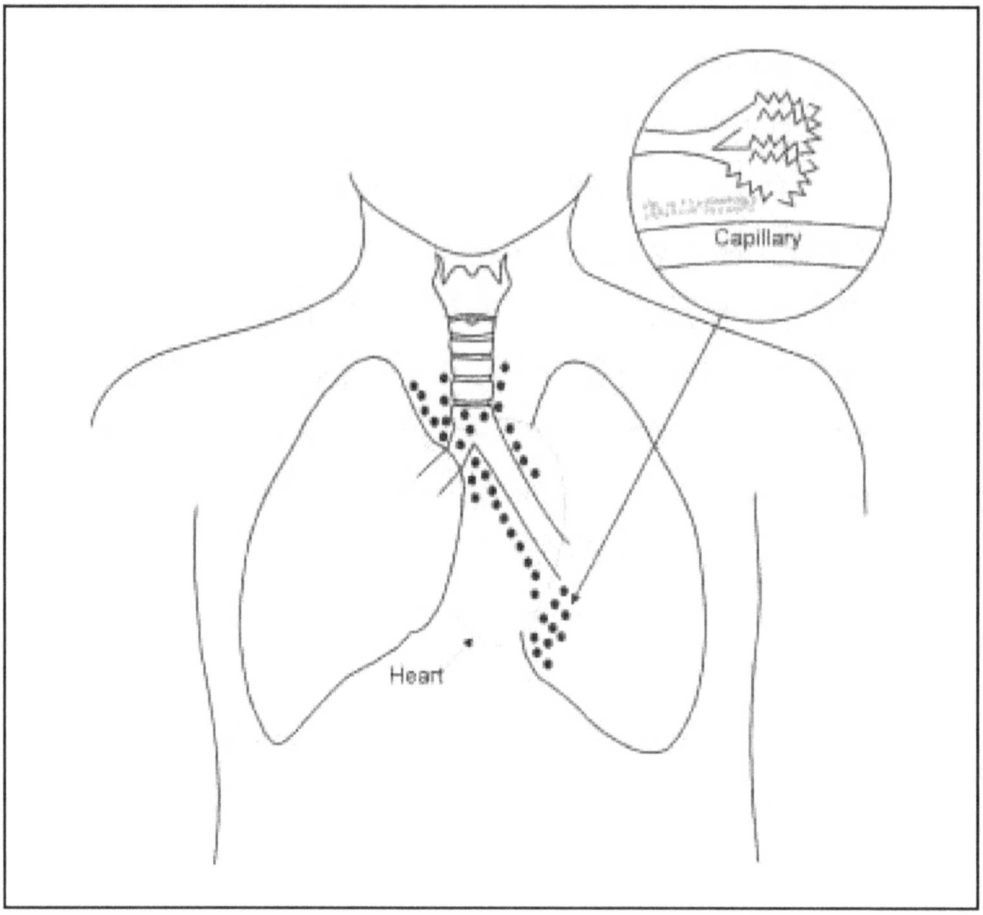

Figure 3-12. Mediastinal Emphysema.

3-8.2.2 **Symptoms of Mediastinal and Subcutaneous Emphysema.** Mild cases are often unnoticed by the diver. In more severe cases, the diver may experience mild to moderate pain under the breastbone, often described as dull ache or feeling of tightness. The pain may radiate to the shoulder or back and may increase upon deep inspiration, coughing, or swallowing. The diver may have a feeling of fullness around the neck and may have difficulty in swallowing. His voice may change in pitch. An observer may note a swelling or apparent inflation of the diver's neck. Movement of the skin near the windpipe or about the collar bone may produce a cracking or crunching sound (crepitation).

3-8.2.3 **Treatment of Mediastinal and Subcutaneous Emphysema.** Suspicion of mediastinal or subcutaneous emphysema warrants prompt referral to medical personnel to rule out the coexistence of arterial gas embolism or pneumothorax. The latter two conditions require more aggressive treatment. Treatment of mediastinal or subcutaneous emphysema with mild symptoms consists of breathing 100 percent oxygen at the surface. If symptoms are severe, shallow recompression may be beneficial. Recompression should only be carried out upon the recommendation of a Diving Medical Officer who has ruled out the occurrence of pneumothorax. Recompression is performed with the diver breathing 100 percent oxygen and using the shallowest depth of relief (usually 5 or 10 feet). An hour of breathing oxygen

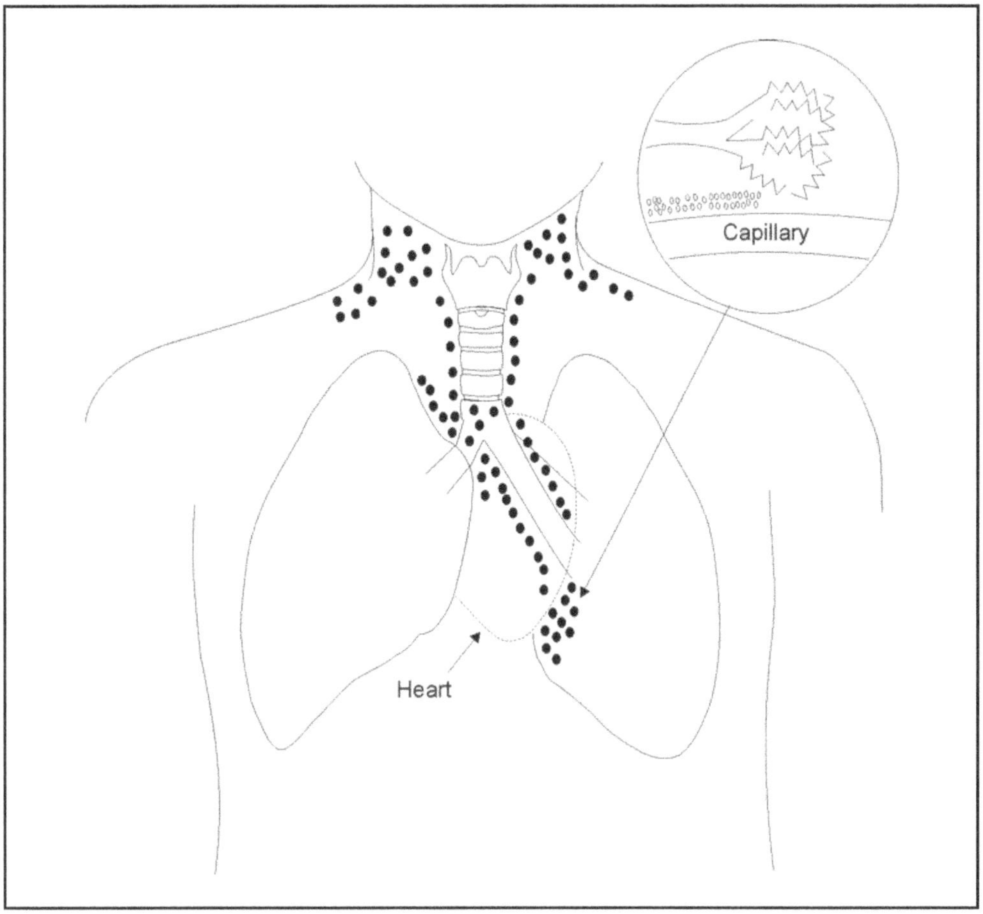

Figure 3-13. Subcutaneous Emphysema.

should be sufficient for resolution, but longer stays may be necessary. Decompression will be dictated by the tender's decompression obligation. The appropriate air table should be used, but the ascent rate should not exceed 1 foot per minute. In this specific case, the delay in ascent should be included in bottom time when choosing the proper decompression table.

3-8.2.4 **Prevention of Mediastinal and Subcutaneous Emphysema.** The strategies for preventing mediastinal/subcutaneous emphysema are identical to the strategies for preventing arterial gas embolism. Breathe normally during ascent. If emergency ascent is required, exhale continuously. Mediastinal/subcutaneous emphysema is particularly common after ditch and don exercises. Avoid positive pressure breathing situations during such exercises. The mediastinal/subcutaneous emphysema that is seen during drown proofing exercises and during surface swimming unfortunately is largely unavoidable.

3-8.3 **Pneumothorax.** A pneumothorax is air trapped in the pleural space between the lung and the chest wall (Figure 3-14).

3-8.3.1 **Causes of Pneumothorax.** A pneumothorax occurs when the lung surface ruptures and air spills into the space between the lung and chest wall. Lung rupture can

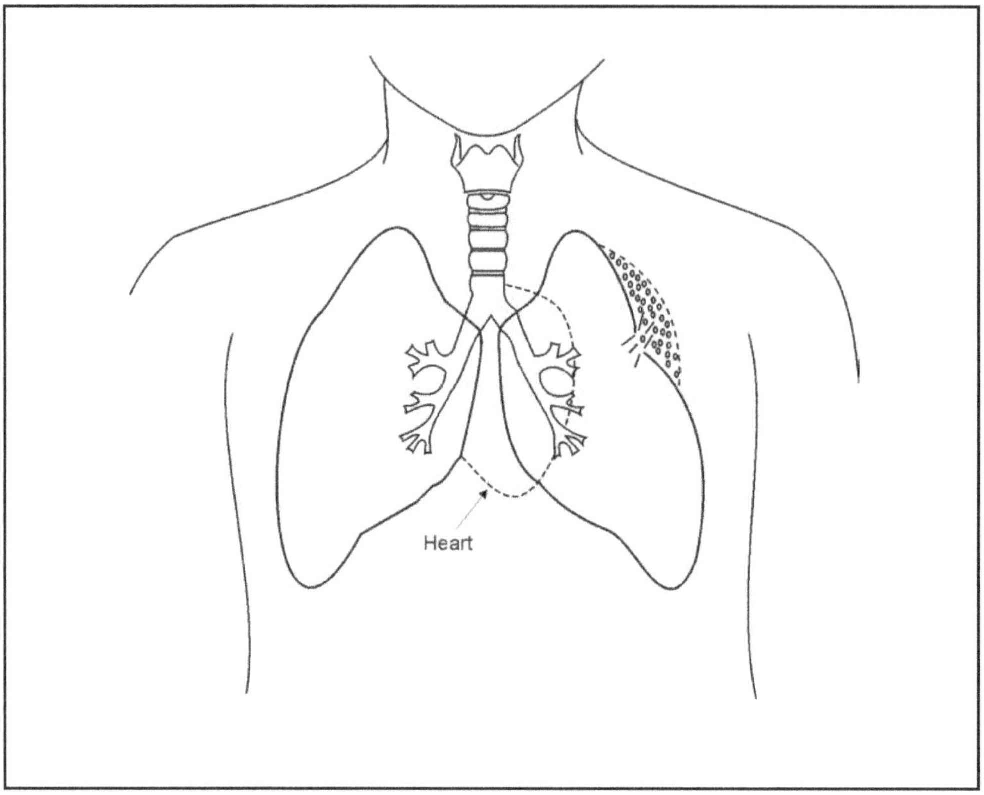

Figure 3-14. Pneumothorax.

result from a severe blow to the chest or from overpressurization of the lung. In its usual manifestation, called a simple pneumothorax, a one-time leakage of air from the lung into the chest partially collapses the lung, causing varying degrees of respiratory distress. This condition normally improves with time as the air is reabsorbed. In severe cases of collapse, the air must be removed with the aid of a tube or catheter.

In certain instances, the damaged lung may allow air to enter but not exit the pleural space. Successive breathing gradually enlarges the air pocket. This is called a tension pneumothorax (Figure 3-15) because of the progressively increasing tension or pressure exerted on the lung and heart by the expanding gas. If uncorrected, this force presses on the involved lung, causing it to completely collapse. The lung, and then the heart, are pushed toward the opposite side of the chest, which impairs both respiration and circulation.

A simple pneumothorax that occurs while the diver is at depth can be converted to a tension pneumothorax by expansion of the gas pocket during ascent. Although a ball valve like mechanism that allows air to enter the pleural cavity but not escape is not present, the result is the same. The mounting tension collapses the lung on the affected side and pushes the heart and lung to the opposite side of the chest.

3-8.3.2 **Symptoms of Pneumothorax.** The onset of a simple pneumothorax is accompanied by a sudden, sharp chest pain, followed by shortness of breath, labored breathing,

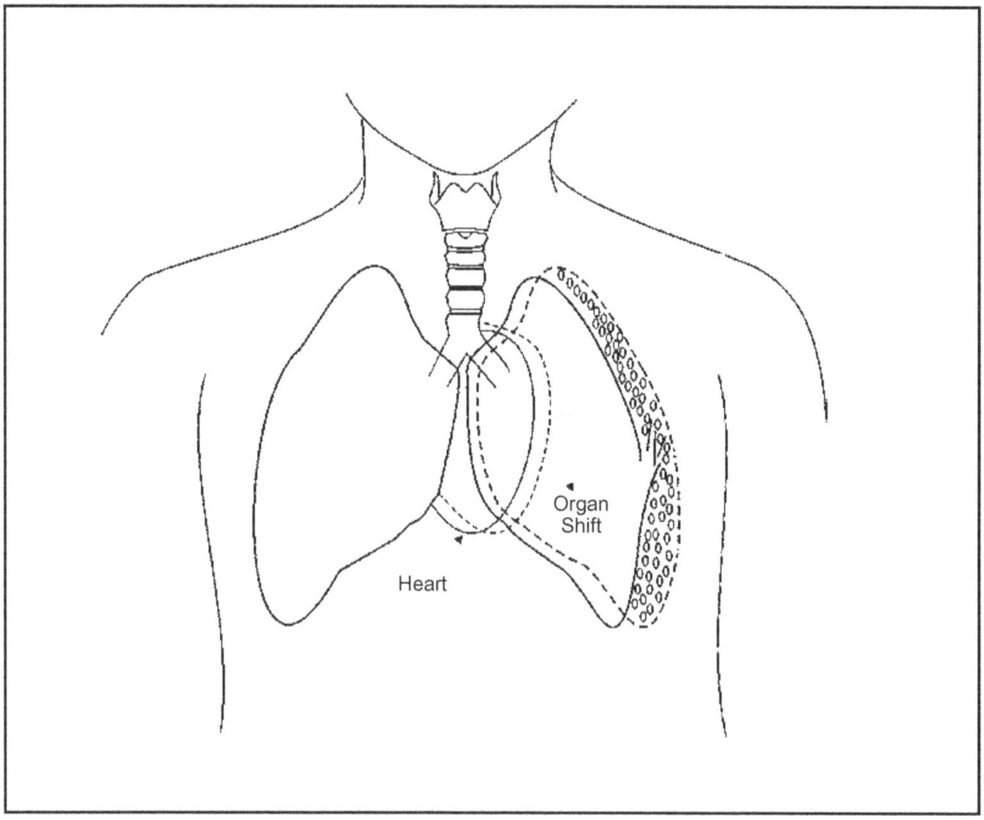

Figure 3-15. Tension Pneumothorax.

rapid heart rate, a weak pulse, and anxiety. The normal chest movements associated with respiration may be reduced on the affected side and breath sounds may be difficult to hear with a stethoscope.

The symptoms of tension pneumothorax are similar to simple pneumothorax, but become progressively more intense over time. As the heart and lungs are displaced to the opposite side of the chest, blood pressure falls along with the arterial oxygen partial pressure. Cyanosis (a bluish discoloration) of the skin appears. If left untreated, shock and death will ensue. Tension pneumothorax is a true medical emergency.

3-8.3.3 **Treatment of Pneumothorax.** A diver believed to be suffering from pneumothorax must be thoroughly examined for the possible co-existence of arterial gas embolism. This is covered more fully in Volume 5.

A small pneumothorax (less than 15%) normally will improve with time as the air in the pleural space is reabsorbed spontaneously. A larger pneumothorax may require active treatment. Mild pneumothorax can be treated by breathing 100 percent oxygen. Cases of pneumothorax that demonstrate cardio-respiratory compromise may require the insertion of a chest tube, largebore intravenous (IV) catheter, or other device designed to remove intrathoracic gas (gas around the lung). Only personnel trained in the use of these and the other accessory devices (one-way valves, underwater suction, etc.) necessary to safety decompress the thoracic cavity should insert them. Divers recompressed for treatment of arterial gas embolism or decompression sickness, who also have a pneumothorax, will experience relief upon recompression. A chest tube or other device with a one-way relief valve may need to be inserted at depth to prevent expansion of the trapped gas during subsequent ascent. A tension pneumothorax should always be suspected if the diver's condition deteriorates rapidly during ascent, especially if the symptoms are respiratory. If a tension pneumothorax is found, recompress to depth of relief until the thoracic cavity can be properly vented. Pneumothorax, if present in combination with arterial gas embolism or decompression sickness, should not prevent immediate recompression therapy. However, a pneumothorax may need to be vented as described before ascent from treatment depth. In cases of tension pneumothorax, this procedure may be lifesaving. Volume 5 fully discusses the treatment of simple and tension pneumothorax.

3-8.3.4 **Prevention of Pneumothorax.** The strategies for avoiding pneumothorax are the same as those for avoiding arterial gas embolism. Breathe normally during ascent. If forced to perform an emergency ascent, exhale continuously

3-9 INDIRECT EFFECTS OF PRESSURE ON THE HUMAN BODY

The conditions previously described occur because of differences in pressure that damage body structures in a direct, mechanical manner. The indirect or secondary effects of pressure are the result of changes in the partial pressure of individual gases in the diver's breathing medium. The mechanisms of these effects include saturation and desaturation of body tissues with dissolved gas and the modification of body functions by abnormal gas partial pressures.

3-9.1 Nitrogen Narcosis.

Nitrogen narcosis is the state of euphoria and exhilaration that occurs when a diver breathes a gas mixture with a nitrogen partial pressure greater than 4 ata.

3-9.1.1 **Causes of Nitrogen Narcosis.** Breathing nitrogen at high partial pressures has a narcotic effect on the central nervous system that causes euphoria and impairs the diver's ability to think clearly. The narcotic effect begins at a nitrogen partial pressure of approximately 4 ata and increases in severity as the partial pressure is increased beyond that point. A nitrogen partial pressure of 8 ata causes very marked impairment; partial pressures in excess of 10 ata may lead to hallucinations and unconsciousness. For a dive on air, narcosis usually appears at a depth of approximately 130 fsw, is very prominent at a depth of 200 fsw, and becomes disabling at deeper depths.

There is a wide range of individual susceptibility to narcosis. There is also some evidence that adaptation occurs on repeated exposures. Some divers, particularly those experienced in deep operations with air, can often work as deep as 200 fsw without serious difficulty. Others cannot.

3-9.1.2 **Symptoms of Nitrogen Narcosis.** The symptoms of nitrogen narcosis include:

- Loss of judgment or skill

- A false feeling of well-being
- Lack of concern for job or safety
- Apparent stupidity
- Inappropriate laughter
- Tingling and vague numbness of the lips, gums, and legs

Disregard for personal safety is the greatest hazard of nitrogen narcosis. Divers may display abnormal behavior such as removing the regulator mouthpiece or swimming to unsafe depths without regard to decompression sickness or air supply.

3-9.1.3 **Treatment of Nitrogen Narcosis.** The treatment for nitrogen narcosis is to bring the diver to a shallower depth where the effects are not felt. The narcotic effects will rapidly dissipate during the ascent. There is no hangover associated with nitrogen narcosis.

3-9.1.4 **Prevention of Nitrogen Narcosis.** Experienced and stable divers may be reasonably productive and safe at depths where others fail. They are familiar with the extent to which nitrogen narcosis impairs performance. They know that a strong conscious effort to continue the dive requires unusual care, time, and effort to make even the simplest observations and decisions. Any relaxation of conscious effort can lead to failure or a fatal blunder. Experience, frequent exposure to deep diving, and training may enable divers to perform air dives as deep as 180-200 fsw, but novices and susceptible individuals should remain at shallower depths or dive with helium-oxygen mixtures.

Helium is widely used in mixed-gas diving as a substitute for nitrogen to prevent narcosis. Helium has not demonstrated narcotic effects at any depth tested by the U.S. Navy. Diving with helium-oxygen mixtures is the only way to prevent nitrogen narcosis. Helium-oxygen mixtures should be considered for any dive in excess of 150 fsw.

3-9.2 Oxygen Toxicity. Exposure to a partial pressure of oxygen above that encountered in normal daily living may be toxic to the body. The extent of the toxicity is dependent upon both the oxygen partial pressure and the exposure time. The higher the partial pressure and the longer the exposure, the more severe the toxicity. The two types of oxygen toxicity experienced by divers are pulmonary oxygen toxicity and central nervous system (CNS) oxygen toxicity.

3-9.2.1 **Pulmonary Oxygen Toxicity.** Pulmonary oxygen toxicity, sometimes called low pressure oxygen poisoning, can occur whenever the oxygen partial pressure exceeds 0.5 ata. A 12 hour exposure to a partial pressure of 1 ata will produce mild symptoms and measurable decreases in lung function. The same effect will occur with a 4 hour exposure at a partial pressure of 2 ata.

Long exposures to higher levels of oxygen, such as administered during Recompression Treatment Tables 4, 7, and 8, may produce pulmonary oxygen toxicity. The symptoms of pulmonary oxygen toxicity may begin with a burning sensation on inspiration and progress to pain on inspiration. During recompression treatments, pulmonary oxygen toxicity may have to be tolerated in patients with severe neurological symptoms to effect adequate treatment. In conscious patients, the pain and coughing experienced with inspiration eventually limit further exposure to oxygen. Unconscious patients who receive oxygen treatments do not feel pain and it is possible to subject them to exposures resulting in permanent lung damage or pneumonia. For this reason, care must be taken when administering 100 percent oxygen to unconscious patients even at surface pressure.

Return to normal pulmonary function gradually occurs after the exposure is terminated. There is no specific treatment for pulmonary oxygen toxicity.

The only way to avoid pulmonary oxygen toxicity completely is to avoid the long exposures to moderately elevated oxygen partial pressures that produce it. However, there is a way of extending tolerance. If the oxygen exposure is periodically interrupted by a short period of time at low oxygen partial pressure, the total exposure time needed to produce a given level of toxicity can be increased significantly. This is the basis for the "air breaks" commonly seen in both decompression and recompression treatment tables.

3-9.2.2 **Central Nervous System (CNS) Oxygen Toxicity.** Central nervous system (CNS) oxygen toxicity, sometimes called high pressure oxygen poisoning, can occur whenever the oxygen partial pressure exceeds 1.3 ata in a wet diver or 2.4 ata in a dry diver. The reason for the marked increase in susceptibility in a wet diver is not completely understood. At partial pressures above the respective 1.3 ata wet and 2.4 ata dry thresholds, the risk of CNS toxicity is dependent on the oxygen partial pressure and the exposure time. The higher the partial pressure and the longer the exposure time, the more likely CNS symptoms will occur. This gives rise to partial pressure of oxygen-exposure time limits for various types of diving.

3-9.2.2.1 ***Factors Affecting the Risk of CNS Oxygen Toxicity.*** A number of factors are known to influence the risk of CNS oxygen toxicity:

Individual Susceptibility. Susceptibility to CNS oxygen toxicity varies markedly from person to person. Individual susceptibility also varies markedly from time to time and for this reason divers may experience CNS oxygen toxicity at exposure times and pressures previously tolerated. Individual variability makes it difficult to set oxygen exposure limits that are both safe and practical.

CO_2 Retention. Hypercapnia greatly increases the risk of CNS toxicity probably through its effect on increasing brain blood flow and consequently brain oxygen levels. Hypercapnia may result from an accumulation of CO_2 in the inspired gas or from inadequate ventilation of the lungs. The latter is usually due to increased breathing resistance or a suppression of respiratory drive by high inspired ppO_2. Hypercapnia is most likely to occur on deep dives and in divers using closed and semi-closed circuit rebreathers.

Exercise. Exercise greatly increases the risk of CNS toxicity, probably by increasing the degree of CO_2 retention. Exposure limits must be much more conservative for exercising divers than for resting divers.

Immersion in Water. Immersion in water greatly increases the risk of CNS toxicity. The precise mechanism for the big increase in risk over comparable dry chamber exposures is unknown, but may involve a greater tendency for diver CO_2 retention during immersion. Exposure limits must be much more conservative for immersed divers than for dry divers.

Depth. Increasing depth is associated with an increased risk of CNS toxicity even though ppO_2 may remain unchanged. This is the situation with UBAs that control the oxygen partial pressure at a constant value, like the MK 16. The precise mechanism for this effect is unknown, but is probably more than just the increase in gas density and concomitant CO_2 retention. There is some evidence that the inert gas component of the gas mixture accelerates the formation of damaging oxygen free radicals. Exposure limits for mixed gas diving must be more conservative than for pure oxygen diving.

Intermittent Exposure. Periodic interruption of high ppO_2 exposure with a 5-15 min exposure to low ppO_2 will reduce the risk of CNS toxicity and extend the total allowable exposure time to high ppO_2. This technique is most often employed in hyperbaric treatments and surface decompression.

Because of these modifying influences, allowable oxygen exposure times vary from situation to situation and from diving system to diving system. In general, closed and semi-closed circuit rebreathing systems require the lowest partial pressure limits, whereas surface-supplied open-circuit systems permit slightly higher limits. Allowable oxygen exposure limits for each system are discussed in later chapters.

3-9.2.2.2 ***Symptoms of CNS Oxygen Toxicity.*** The most serious direct consequence of oxygen toxicity is convulsions. Sometimes recognition of early symptoms may provide sufficient warning to permit reduction in oxygen partial pressure and prevent the onset of more serious symptoms. The warning symptoms most often encountered also may be remembered by the mnemonic VENTIDC:

V: Visual symptoms. Tunnel vision, a decrease in diver's peripheral vision, and other symptoms, such as blurred vision, may occur.

E: Ear symptoms. Tinnitus, any sound perceived by the ears but not resulting from an external stimulus, may resemble bells ringing, roaring, or a machinery-like pulsing sound.

N: Nausea or spasmodic vomiting. These symptoms may be intermittent.

T: Twitching and tingling symptoms. Any of the small facial muscles, lips, or muscles of the extremities may be affected. These are the most frequent and clearest symptoms.

I: Irritability. Any change in the diver's mental status including confusion, agitation, and anxiety.

D: Dizziness. Symptoms include clumsiness, incoordination, and unusual fatigue.

C: Convulsions. The first sign of CNS oxygen toxicity may be convulsions that occur with little or no warning.

Warning symptoms may not always appear and most are not exclusively symptoms of oxygen toxicity. Muscle twitching is perhaps the clearest warning, but it may occur late, if at all. If any of these warning symptoms occur, the diver should take immediate action to lower the oxygen partial pressure.

A convulsion, the most serious direct consequence of CNS oxygen toxicity, may occur suddenly without being preceded by any other symptom. During a convulsion, the individual loses consciousness and his brain sends out uncontrolled nerve impulses to his muscles. At the height of the seizure, all of the muscles are stimulated at once and lock the body into a state of rigidity. This is referred to as the *tonic phase* of the convulsion. The brain soon fatigues and the number of impulses slows. This is the *clonic phase* and the random impulses to various muscles may cause violent thrashing and jerking for a minute or so.

After the convulsive phase, brain activity is depressed and a *postconvulsive (postictal) depression* follows. During this phase, the patient is usually unconscious and quiet for a while, then semiconscious and very restless. He will then usually sleep on and off, waking up occasionally though still not fully rational. The depression phase sometimes lasts as little as 15 minutes, but an hour or more is not uncommon. At the end of this phase, the patient often becomes suddenly alert and complains of no more than fatigue, muscular soreness, and possibly a headache. After an oxygen-toxicity convulsion, the diver usually remembers clearly the events up to the moment when consciousness was lost, but remembers nothing of the convulsion itself and little of the postictal phase.

3-9.2.2.3 ***Treatment of CNS Oxygen Toxicity.*** A diver who experiences the warning symptoms of oxygen toxicity shall inform the Diving Supervisor immediately. The following actions can be taken to lower the oxygen partial pressure:

- Ascend
- Shift to a breathing mixture with a lower oxygen percentage
- In a recompression chamber, remove the mask.

WARNING **Reducing the oxygen partial pressure does not instantaneously reverse the biochemical changes in the central nervous system caused by high oxygen partial pressures. If one of the early symptoms of oxygen toxicity occurs, the diver may still convulse up to a minute or two after being removed from the high oxygen breathing gas. One should not assume**

that an oxygen convulsion will not occur unless the diver has been off oxygen for 2 or 3 minutes.

Despite its rather alarming appearance, the convulsion itself is usually not much more than a strenuous muscular workout for the victim. The possible danger of hypoxia during breathholding in the tonic phase is greatly reduced because of the high partial pressure of oxygen in the tissues and brain. If a diver convulses, the UBA should be ventilated immediately with a gas of lower oxygen content, if possible. If depth control is possible and the gas supply is secure (helmet or full face mask), the diver should be kept at depth until the convulsion subsides and normal breathing resumes. If an ascent must take place, it should be done as slowly as possible to reduce the risk of an arterial gas embolism. A diver surfacing unconscious because of an oxygen convulsion must be treated as if suffering from arterial gas embolism. Arterial gas embolism cannot be ruled out in an unconscious diver.

If the convulsion occurs in a recompression chamber, it is important to keep the individual from thrashing against hard objects and being injured. Complete restraint of the individual's movements is neither necessary nor desirable. The oxygen mask shall be removed immediately. It is not necessary to force the mouth open to insert a bite block while a convulsion is taking place. After the convulsion subsides and the mouth relaxes, keep the jaw up and forward to maintain a clear airway until the diver regains consciousness. Breathing almost invariably resumes spontaneously. Management of CNS oxygen toxicity during recompression therapy is discussed fully in Volume 5.

If a convulsing diver is prevented from drowning or causing other injury to himself, full recovery with no lasting effects can be expected within 24 hours. Susceptibility to oxygen toxicity does not increase as a result of a convulsion, although divers may be more inclined to notice warning symptoms during subsequent exposures to oxygen.

3-9.2.2.4 ***Prevention of CNS Oxygen Toxicity.*** The actual mechanism of CNS oxygen toxicity remains unknown in spite of many theories and much research. Preventing oxygen toxicity is important to divers. When use of high pressures of oxygen is advantageous or necessary, divers should take sensible precautions, such as being sure the breathing apparatus is in good order, observing depth-time limits, avoiding excessive exertion, and heeding abnormal symptoms that may appear. Interruption of oxygen breathing with periodic "air" breaks can extend the exposure time to high oxygen partial pressures significantly. Air breaks are routinely incorporated into recompression treatment tables and some decompression tables.

3-9.3 **Decompression Sickness (DCS).** A diver's blood and tissues absorb additional nitrogen (or helium) from the lungs when at depth. If a diver ascends too fast this excess gas will separate from solution and form bubbles. These bubbles produce mechanical and biochemical effects that lead to a condition known as *decompression sickness*.

3-9.3.1 **Absorption and Elimination of Inert Gases.** The average human body at sea level contains about 1 liter of nitrogen. All of the body tissues are saturated with nitrogen at a partial pressure equal to the partial pressure in the alveoli, about 0.79 ata. If the partial pressure of nitrogen changes because of a change in the pressure or composition of the breathing mixture, the pressure of the nitrogen dissolved in the body gradually attains a matching level. Additional quantities of nitrogen are absorbed or eliminated, depending on the partial pressure gradient, until the partial pressure of the gas in the lungs and in the tissues is equal. If a diver breathes helium, a similar process occurs.

As described by Henry's Law, the amount of gas that dissolves in a liquid is almost directly proportional to the partial pressure of the gas. If one liter of inert gas is absorbed at a pressure of one atmosphere, then two liters are absorbed at two atmospheres and three liters at three atmospheres, etc.

The process of taking up more inert gas is called absorption or saturation. The process of giving up inert gas is called elimination or desaturation. The chain of events is essentially the same in both processes even though the direction of exchange is opposite.

Shading in diagram (Figure 3-16) indicates saturation with nitrogen or helium under increased pressure. Blood becomes saturated on passing through lungs, and tissues are saturated in turn via blood. Those with a large supply (as in A above) are saturated much more rapidly than those with poor blood supply (C) or an unusually large capacity for gas, as fatty tissues have for nitrogen. In very abrupt ascent from depth, bubbles may form in arterial blood or in "fast" tissue (A) even through the body as a whole is far from saturation. If enough time elapses at depth, all tissues will become equally saturated, as shown in lower diagram.

3-9.3.1.1 ***Saturation of Tissues.*** The sequence of events in the process of saturation can be illustrated by considering what happens in the body of a diver taken rapidly from the surface to a depth of 100 fsw (Figure 3-16). To simplify matters, we can say that the partial pressure of nitrogen in his blood and tissues on leaving the surface is roughly 0.8 ata. When the diver reaches 100 fsw, the alveolar nitrogen pressure in his lungs will be about 0.8×4 ata = 3.2 ata, while the blood and tissues remain temporarily at 0.8 ata. The partial pressure difference or gradient between the alveolar air and the blood and tissues is thus 3.2 minus 0.8, or 2.4 ata. This gradient is the driving force that makes the molecules of nitrogen move by diffusion from one place to another. Consider the following 10 events and factors in the diver at 100 fsw:

1. As blood passes through the alveolar capillaries, nitrogen molecules move from the alveolar air into the blood. By the time the blood leaves the lungs, it has reached equilibrium with the new alveolar nitrogen pressure. It now has a nitrogen tension (partial pressure) of 3.2 ata and contains about four times as much nitrogen as before. When this blood reaches the tissues, there is a similar gradient and nitrogen molecules move from the blood into the tissues until equilibrium is reached.

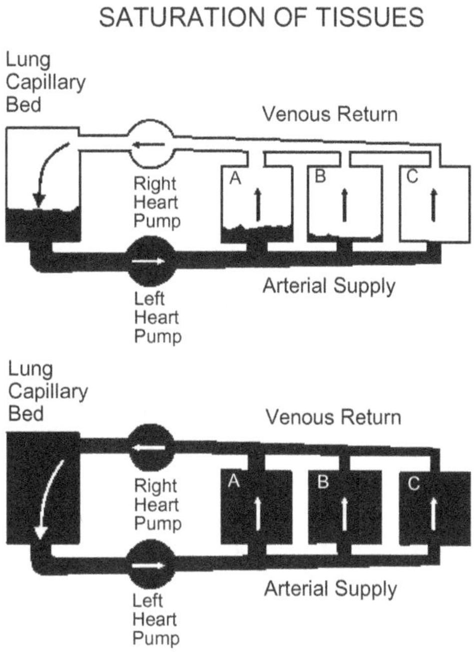

Figure 3-16. Saturation of Tissues. Shading in diagram indicates saturation with nitrogen or helium under increased pressure. Blood becomes saturated on passing through lungs, and tissues are saturated in turn via blood. Those with a large supply (as in A above) are saturated much more rapidly than those with poor blood supply (C) or an unusually large capacity for gas, as fatty tissues have for nitrogen. In very abrupt ascent from depth, bubbles may form in arterial blood or in "fast" tissue (A) even through the body as a whole is far from saturation. If enough time elapses at depth, all tissues will become equally saturated, as shown in lower diagram.

2. The volume of blood in a tissue is relatively small compared to the volume of the tissue and the blood can carry only a limited amount of nitrogen. Because of this, the volume of blood that reaches a tissue over a short period of time loses its excess nitrogen to the tissue without greatly increasing the tissue nitrogen pressure.

3. When the blood leaves the tissue, the venous blood nitrogen pressure is equal to the new tissue nitrogen pressure. When this blood goes through the lungs, it again reaches equilibrium at 3.2 ata.

4. When the blood returns to the tissue, it again loses nitrogen until a new equilibrium is reached.

5. As the tissue nitrogen pressure rises, the blood-tissue gradient decreases, slowing the rate of nitrogen exchange. The rate at which the tissue nitrogen partial pressure increases, therefore, slows as the process proceeds. However, each volume of blood that reaches the tissue gives up some nitrogen which increases the tissue

partial pressure until complete saturation, in this case at 3.2 ata of nitrogen, is reached.

6. Tissues that have a large blood supply in proportion to their own volume have more nitrogen delivered to them in a certain amount of time and therefore approach complete saturation more rapidly than tissues that have a poor blood supply.

7. All body tissues are composed of lean and fatty components. If a tissue has an unusually large capacity for nitrogen, it takes the blood longer to deliver enough nitrogen to saturate it completely. Nitrogen is about five times as soluble (capable of being dissolved) in fat as in water. Therefore, fatty tissues require much more nitrogen and much more time to saturate them completely than lean (watery) tissues do, even if the blood supply is ample. Adipose tissue (fat) has a poor blood supply and therefore saturates very slowly.

8. At 100 fsw, the diver's blood continues to take up more nitrogen in the lungs and to deliver more nitrogen to tissues, until all tissues have reached saturation at a pressure of 3.2 ata of nitrogen. A few watery tissues that have an excellent blood supply will be almost completely saturated in a few minutes. Others, like fat with a poor blood supply, may not be completely saturated unless the diver is kept at 100 fsw for 72 hours or longer.

9. If kept at a depth of 100 fsw until saturation is complete, the diver's body contains about four times as much nitrogen as it did at the surface. Divers of average size and fatness have about one liter of dissolved nitrogen at the surface and about four liters at 100 fsw. Because fat holds about five times as much nitrogen as lean tissues, much of a diver's nitrogen content is in his fatty tissue.

10. An important fact about nitrogen saturation is that the process requires the same length of time regardless of the nitrogen pressure involved. For example, if the diver had been taken to 33 fsw instead of 100, it would have taken just as long to saturate him completely and to bring his nitrogen pressures to equilibrium. In this case, the original gradient between alveolar air and the tissues would have been only 0.8 ata instead of 2.4 ata. Because of this, the amount of nitrogen delivered to tissues by each round of blood circulation would have been smaller from the beginning. Less nitrogen would have to be delivered to saturate him at 33 fsw, but the slower rate of delivery would cause the total time required to be the same.

When any other inert gas, such as helium, is used in the breathing mixture, the body tissues become saturated with that gas in the same process as for nitrogen. However, the time required to reach saturation is different for each gas. This is because the blood and tissue solubilities are different for the different inert gases. Helium, for example, is much less soluble in fat than nitrogen is.

3-9.3.1.2 ***Desaturation of Tissues.*** The process of desaturation is the reverse of saturation (Figure 3-17). If the partial pressure of the inert gas in the lungs is reduced, either through a reduction in the diver's depth or a change in the breathing medium, the new pressure gradient induces the nitrogen to diffuse from the tissues to the blood, from the blood to the gas in the lungs, and then out of the body with the expired breath. Some parts of the body desaturate more slowly than others for the same

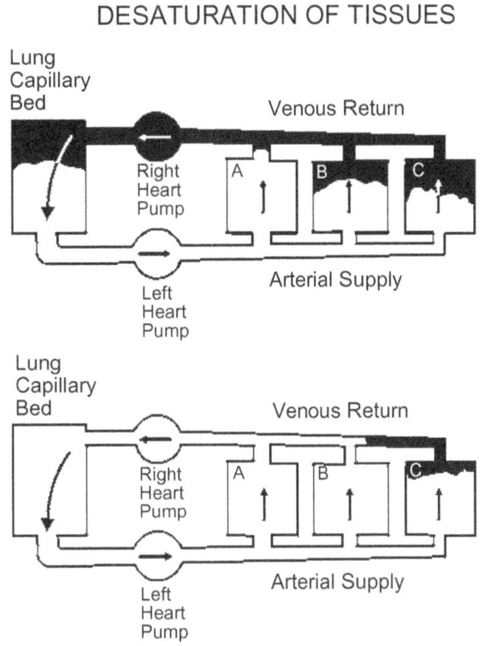

Figure 3-17. Desaturation of Tissues. The desaturation process is essentially the reverse of saturation. When pressure of inert gas is lowered, blood is cleared of excess gas as it goes through the lungs. Blood then removes gas from the tissues at rates depending on amount of blood that flows through them each minute. Tissues with poor blood supply (as in C in upper sketch) or large gas capacity will lag behind and may remain partially saturated after others have cleared (see lower diagram).

reason that they saturate more slowly: poor blood supply or a greater capacity to store inert gas. Washout of excess inert gas from these "slow" tissues will lag behind washout from the faster tissues.

3-9.3.2 **Bubble Formation.** Inert gas may separate from physical solution and form bubbles if the partial pressure of the inert gas in blood and tissues exceeds the ambient pressure by more than a critical amount. During descent and while the diver is on the bottom, blood and tissue inert gas partial pressures increase significantly as tissue saturation takes place, but the inert gas pressure always remains less than the ambient pressure surrounding the diver. Bubbles cannot form in this situation. During ascent the converse is true. Blood and tissue inert gas pressures fall as the tissues desaturate, but blood and tissue inert gas pressures can exceed the ambient pressure if the rate of ascent is faster than the rate at which tissues can equilibrate. Consider an air diver fully saturated with nitrogen at a depth of 100 fsw. All body tissues have a nitrogen partial pressure of 3.2 ata. If the diver were to quickly ascend to the surface, the ambient pressure surrounding his tissues would be reduced to 1 ata. Assuming that ascent was fast enough not to allow for any tissue desaturation, the nitrogen pressure in all the tissues would be 2.2 ata greater than the ambient pressure (3.2 ata - 1 ata). Under this circumstance bubbles can form.

Bubble formation can be avoided if the ascent is controlled in such a way that the tissue inert gas pressure never exceeds the ambient pressure by more than the critical amount. This critical amount, called the *allowable supersaturation*, varies from tissue to tissue and from one inert gas to another. A decompression table shows the time that must be spent at various decompression stops on the way to the surface to allow each tissue to desaturate to the point where its allowable supersaturation is not exceeded.

3-9.3.3 **Direct Bubble Effects.** Bubbles forming in the tissues (autochthonous bubbles) and in the bloodstream (circulating bubbles) may exert their effects directly in several ways:

- Autochthonous bubbles can put pressure on nerve endings, stretch and tear tissue leading to hemorrhage, and increase pressure in the tissue leading to slowing or cessation of incoming blood flow. These are thought to be the primary mechanisms for injury in Spinal Cord, Musculoskeletal, and Inner Ear DCS.

- Venous bubbles can partially or completely block the veins draining various organs leading to reduced organ blood flow (venous obstruction). Venous obstruction in turn leads to tissue hypoxia, cell injury and death. This is one of the secondary mechanisms of injury in Spinal Cord DCS.

- Venous bubbles carried to the lung as emboli (called venous gas emboli or VGE) can partially block the flow of blood through the lung leading to fluid build up (pulmonary edema) and decreased gas exchange. The result is systemic hypoxia and hypercarbia. This is the mechanism of damage in Pulmonary DCS.

- Arterial bubbles can act as emboli blocking the blood supply of almost any tissue leading to hypoxia, cell injury and death. Arterial gas embolism and autochotonous bubble formation are thought be the primary mechanisms of injury in Cerebral (brain) DCS.

The damage done by the direct bubble effect occurs within a relatively short period of time (a few minutes to hours). The primary treatment for these effects is recompression. Recompression will compress the bubble to a smaller diameter, restore blood flow, decrease venous congestion, and improve gas exchange in the lungs and tissues. It also increases the speed at which the bubbles outgas and collapse.

3-9.3.4 **Indirect Bubble Effects.** Bubbles may also exert their effects indirectly because a bubble acts like a foreign body. The body reacts as it would if there were a cinder in the eye or a splinter in the hand. The body's defense mechanisms become alerted and try to eliminate the foreign body. Typical reactions include:

- Blood vessels become "leaky" due to damage to the endothelial lining cells and chemical release. Blood plasma leaks out while blood cells remain inside. The blood becomes thick and more difficult to pump. Organ blood flow is reduced.

- The platelet system becomes active and the platelets gather at the site of the bubble causing a clot to form.
- The injured tissue releases fats that clump together in the bloodstream. These fat clumps act as emboli, causing tissue hypoxia.
- Injured tissues release histamine and histamine-like substances, causing edema, which leads to allergic-type problems of shock and respiratory distress.

Indirect bubble effects take place over a longer period of time than the direct bubble effects. Because the non-compressible clot replaces a compressible bubble, recompression alone is not enough. To restore blood flow and relieve hypoxia, hyperbaric treatment and other therapies are often required.

3-9.3.5 **Symptoms of Decompression Sickness.** Decompression sickness is generally divided into two categories. Type I decompression sickness involves the skin, lymphatic system, muscles and joints and is not life threatening. Type II decompression sickness (also called serious decompression sickness) involves the nervous system, respiratory system, or circulatory system. Type II decompression sickness may become life threatening. Because the treatment of Type I and Type II decompression sickness may be different, it is important to distinguish between these two types. Symptoms of Type I and Type II decompression sickness may be present at the same time.

When the skin is involved, the symptoms are itching or burning usually accompanied by a rash. Involvement of the lymphatic system produces swelling of regional lymph nodes or an extremity. Involvement of the musculoskeletal system produces pain, which in some cases can be excruciating. Bubble formation in the brain can produce blindness, dizziness, paralysis and even unconsciousness and convulsion. When the spinal cord is involved, paralysis and/or loss of feeling occur. Bubbles in the inner ear produce hearing loss and vertigo. Bubbles in the lungs can cause coughing, shortness of breath, and hypoxia, a condition referred to as "the chokes." This condition may prove fatal. A large number of bubbles in the circulation can lead to cardiovascular collapse and death. Unusual fatigue or exhaustion after a dive is probably due to bubbles in unusual locations and the biochemical changes they have induced. While not attributable to a specific organ system, unusual fatigue is a definite symptom of decompression sickness.

3-9.3.5.1 ***Time Course of Symptoms.*** Decompression sickness usually occurs after surfacing. If the dive is particularly arduous or decompression has been omitted, however, the diver may experience decompression sickness before reaching the surface.

After surfacing, there is a latency period before symptoms appear. This may be as short as several minutes to as long as several days. Long, shallow dives are generally associated with longer latencies than deep, short dives. For most dives, the onset of decompression sickness can be expected within several hours of surfacing.

3-9.3.6 **Treating Decompression Sickness.** Treatment of decompression sickness is accomplished by recompression. This involves putting the victim back under pressure to reduce the size of the bubbles to cause them to go back into solution and to supply extra oxygen to the hypoxic tissues. Treatment is done in a recompression chamber, but can sometimes be accomplished in the water if a chamber cannot be reached in a reasonable period of time. Recompression in the water is not recommended, but if undertaken, must be done following specified procedures. Further discussion of the symptoms of decompression sickness and a complete discussion of treatment are presented in Volume 5.

3-9.3.7 **Preventing Decompression Sickness.** Prevention of decompression sickness is generally accomplished by following the decompression tables. However, individual susceptibility or unusual conditions, either in the diver or in connection with the dive, produces a small percentage of cases even when proper dive procedures are followed meticulously. To be absolutely free of decompression sickness under all possible circumstances, the decompression time specified would have to be far in excess of that normally needed. On the other hand, under ideal circumstances, some individuals can ascend safely in less time than the tables specify. This must not be taken to mean that the tables contain an unnecessarily large safety factor. The tables represent the minimum workable decompression time that permits average divers to surface safely from normal working dives without an unacceptable incidence of decompression sickness.

3-10 THERMAL PROBLEMS IN DIVING

The human body functions effectively within a relatively narrow range of internal temperature. The average, or normal, core temperature of 98.6°F (37°C) is maintained by natural mechanisms of the body, aided by artificial measures such as the use of protective clothing or environmental conditioning when external conditions tend toward cold or hot extremes.

Thermal problems, arising from exposure to various temperatures of water, pose a major consideration when planning operational dives and selecting equipment. Bottom time may be limited more by a diver's intolerance to heat or cold than his exposure to increased oxygen partial pressures or the amount of decompression required.

The diver's thermal status will affect the rate of inert gas uptake and elimination. Recent studies suggest divers who are warm on the bottom but cold during decompression may more susceptible to decompression sickness. This may require modification of a diver's decompression schedule. Rewarming before a repetitive dive is as important as accounting for residual nitrogen levels.

3-10.1 **Regulating Body Temperature.** The metabolic processes of the body constantly generate heat. If heat is allowed to build up inside the body, damage to the cells can occur. To maintain internal temperature at the proper level, the body must lose heat equal to the amount it produces.

Heat transfer is accomplished in several ways. The blood, while circulating through the body, picks up excess heat and carries it to the lungs, where some of it is lost with the exhaled breath. Heat is also transferred to the surface of the skin, where much of it is dissipated through a combination of conduction, convection, and radiation. Moisture released by the sweat glands cools the surface of the body as it evaporates and speeds the transfer of heat from the blood to the surrounding air. If the body is working hard and generating greater than normal quantities of heat, the blood vessels nearest the skin dilate to permit more of the heated blood to reach the body surfaces, and the sweat glands increase their activity.

Maintaining proper body temperature is particularly difficult for a diver working underwater. The principal temperature control problem encountered by divers is keeping the body warm. The high thermal conductivity of water, coupled with the normally cool-to-cold waters in which divers operate, can result in rapid and excessive heat loss.

3-10.2 **Excessive Heat Loss (Hypothermia).** Hypothermia is a lowering of the core temperature of the body. Immersion hypothermia is a potential hazard whenever diving operations take place in cool to cold waters. A diver's response to immersion in cold water depends on the degree of thermal protection worn and water temperature. A water temperature of approximately 91°F (33°C) is required to keep an unprotected, resting man at a stable temperature. The unprotected diver will be affected by excessive heat loss and become chilled within a short period of time in water temperatures below 72°F (23°C).

3-10.2.1 **Causes of Hypothermia.** Hypothermia in diving occurs when the difference between the water and body temperature is large enough for the body to lose more heat than it produces. Exercise normally increases heat production and body temperature in dry conditions. Paradoxically, exercise in cold water may cause the body temperature to fall more rapidly. Any movement that stirs the water in contact with the skin creates turbulence that carries off heat (convection). Heat loss is caused not only by convection at the limbs, but also by increased blood flow into the limbs during exercise. Continual movement causes the limbs to resemble the internal body core rather than the insulating superficial layer. These two conflicting effects result in the core temperature being maintained or increased in warm water and decreased in cold water.

3-10.2.2 **Symptoms of Hypothermia.** In mild cases, the victim will experience uncontrolled shivering, slurred speech, imbalance, and/or poor judgment. Severe cases of hypothermia are characterized by loss of shivering, impaired mental status, irregular heartbeat, and/or very shallow pulse or respirations. This is a medical emergency. The signs and symptoms of falling core temperature are given in Table 3-1, though individual responses to falling core temperature will vary. At extremely low temperatures or with prolonged immersion, body heat loss reaches a point at which death occurs.

Table 3-1. *Signs and Symptoms of Dropping Core Temperature.*

Core Temperature °F	°C	Symptoms
98	37	Cold sensations, skin vasoconstriction, increased muscle tension, increased oxygen consumption
97	36	Sporadic shivering suppressed by voluntary movements, gross shivering in bouts, further increase in oxygen consumption, uncontrollable shivering
95	35	Voluntary tolerance limit in laboratory experiments, mental confusion, impairment of rational thought, possible drowning, decreased will to struggle
93	34	Loss of memory, speech impairment, sensory function impairment, motor performance impairment
91	33	Hallucinations, delusions, partial loss of consciousness, shivering impaired
90	32	Heart rhythm irregularities, motor performance grossly impaired
88	31	Shivering stopped, failure to recognize familiar people
86	30	Muscles rigid, no response to pain
84	29	Loss of consciousness
80	27	Ventricular fibrillation (ineffective heartbeat), muscles flaccid
79	26	Death

3-10.2.3 **Treatment of Hypothermia.** To treat mild hypothermia, passive and active rewarming measures may be used and should be continued until the victim is sweating. Rewarming techniques include:

Passive:

- Remove all wet clothing.
- Wrap victim in a blanket (preferably wool).
- Place in an area protected from wind.
- If possible, place in a warm area (i.e. galley).

Active:

- Warm shower or bath.
- Place in a very warm space (i.e., engine room).

To treat severe hypothermia avoid any exercise, keep the victim lying down, initiate only passive rewarming, and immediately transport to the nearest medical treatment facility.

CAUTION **Do not institute active rewarming with severe cases of hypothermia.**

WARNING **CPR should not be initiated on a severely hypothermic diver unless it can be determined that the heart has stopped or is in ventricular fibrillation. CPR should not be initiated in a patient that is breathing.**

3-10.2.4 **Prevention of Hypothermia.** The body's ability to tolerate cold environments is due to natural insulation and a built-in means of heat regulation. Temperature is not uniform throughout the body. It is more accurate to consider the body in terms of an inner core where a constant or uniform temperature prevails and a superficial region through which a temperature gradient exists from the core to the body surface. Over the trunk of the body, the thickness of the superficial layer may be 1 inch (2.5 cm). The extremities become a superficial insulating layer when their blood flow is reduced to protect the core.

Once in the water, heat loss through the superficial layer is lessened by the reduction of blood flow to the skin. The automatic, cold-induced vasoconstriction (narrowing of the blood vessels) lowers the heat conductance of the superficial layer and acts to maintain the heat of the body core. Unfortunately, vasoconstrictive regulation of heat loss has only a narrow range of protection. When the extremities are initially put into very cold water, vasoconstriction occurs and the blood flow is reduced to preserve body heat. After a short time, the blood flow increases and fluctuates up and down for as long as the extremities are in cold water. As circulation and heat loss increase, the body temperature falls and may continue falling, even though heat production is increased by shivering.

Much of the heat loss in the trunk area is transferred over the short distance from the deep organs to the body surface by physical conduction, which is not under any physiological control. Most of the heat lost from the body in moderately cold water is from the trunk and not the limbs.

Hypothermia can be insidious and cause problems without the diver being aware of it. The diver should wear appropriate thermal protection based upon the water temperature and expected bottom time (See Chapter 6). Appropriate dress can greatly reduce the effects of heat loss and a diver with proper dress can work in very cold water for reasonable periods of time. Acclimatization, adequate hydration, experience, and common sense all play a role in preventing hypothermia. Provide the diver and topside personnel adequate shelter from the elements. Adequate predive hydration is essential.

Heat loss through the respiratory tract becomes an increasingly significant factor in deeper diving. Inhaled gases are heated in the upper respiratory tract and more energy is required to heat the denser gases encountered at depth. In fact, a severe respiratory insult can develop if a diver breathes unheated gas while making a deep saturation dive in cold water. Respiratory gas heating is required in such situations.

3-10.3 **Other Physiological Effects of Exposure to Cold Water.** In addition to hypothermia, other responses to exposure to cold water create potential hazards for the diver.

3-10.3.1 **Caloric Vertigo.** The eardrum does not have to rupture for caloric vertigo to occur. Caloric vertigo can occur simply as the result of having water enter the external ear canal on one side but not the other. The usual cause is a tight fitting wet suit hood that allows cold water access to one ear, but not the other. It can also occur when one external canal is obstructed by wax. Caloric vertigo may occur suddenly upon entering cold water or when passing through thermoclines. The effect is usually short lived, but while present may cause significant disorientation and nausea.

3-10.3.2 **Diving Reflex.** Sudden exposure of the face to cold water or immersion of the whole body in cold water may cause an immediate slowing of the heart rate (bradycardia) and intense constriction of the peripheral blood vessels. Sometimes abnormal heart rhythms accompany the bradycardia. This response is known as the *diving reflex*. Removing or losing a facemask in cold water can trigger the diving reflex. It is still not known whether cardiac arrhythmias associated with the diving reflex contribute to diving casualties. Until this issue is resolved, it is prudent for divers to closely monitor each other when changing rigs underwater or buddy breathing.

3-10.3.3 **Uncontrolled Hyperventilation.** If a diver with little or no thermal protection is suddenly plunged into very cold water, the effects are immediate and disabling. The diver gasps and his respiratory rate and tidal volume increase. His breathing becomes so rapid and uncontrolled that he cannot coordinate his breathing and swimming movements. The lack of breathing control makes survival in rough water very unlikely.

3-10.4 **Excessive Heat Gain (Hyperthermia).** Hyperthermia is a raising of the core temperature of the body. Hyperthermia should be considered a potential risk any time air temperature exceeds 90°F or water temperature is above 82°F. An individual is considered to have developed hyperthermia when core temperature rises 1.8°F (1°C) above normal (98.6°F, 37°C). The body core temperature should not exceed 102.2°F (39°C). By the time the diver's core temperature approaches 102°F noticeable mental confusion may be present.

3-10.4.1 **Causes of Hyperthermia.** Divers are susceptible to hyperthermia when they are unable to dissipate their body heat. This may result from high water temperatures, protective garments, rate of work, and the duration of the dive. Predive heat exposure may lead to significant dehydration and put the diver at greater risk of hyperthermia.

3-10.4.2 **Symptoms of Hyperthermia.** Signs and symptoms of hyperthermia can vary among individuals. Since a diver might have been in water that may not be considered hot, support personnel must not rely solely on classical signs and symptoms of heat stress for land exposures. Table 3-2 lists commonly encountered signs and symptoms of heat stress in diving. In severe cases of hyperthermia (severe heat exhaustion or heat stroke), the victim will experience disorientation, tremors, loss of consciousness and/or seizures.

Table 3-2. *Signs of Heat Stress.*

Least Severe	High breathing rate
	Feeling of being hot, uncomfortable
	Low urine output
	Inability to think clearly
	Fatigue
	Light-headedness or headache
	Nausea
	Muscle cramps
	Sudden rapid increase in pulse rate
	Disorientation, confusion
	Exhaustion
	Collapse
Most Severe	Death

3-10.4.3 **Treatment of Hyperthermia.** The treatment of all cases of hyperthermia shall include cooling of the victim to reduce the core temperature. In mild to moderate hyperthermia cooling should be started immediately by removing the victim's clothing, spraying him with a fine mist of lukewarm-to-cool water, and then fanning. This causes a large increase in evaporative cooling. Avoid whole body immersion in cold water or packing the body in ice as this will cause vasoconstriction which will decrease skin blood flow and may slow the loss of heat. Ice packs to the neck, armpit or groin may be used. Oral fluid replacement should begin as soon as the victim can drink and continue until he has urinated pale to clear urine several times. If the symptoms do not improve, the victim shall be transported to a medical treatment facility.

Severe hyperthermia is a medical emergency. Cooling measures shall be started and the victim shall be transported immediately to a medical treatment facility. Intravenous fluids should be administered during transport.

3-10.4.4 **Prevention of Hyperthermia.** Acclimatization, adequate hydration, experience, and common sense all play a role in preventing hyperthermia. Shelter personnel from the sun and keep the amount of clothing worn to a minimum. Adequate predive hydration is essential. Alcohol or caffeine beverages should be avoided since they can produce dehydration. Medications containing antihistamines or aspirin should not be used in warm water diving. Physically fit individuals and those with lower levels of body fat are less likely to develop hyperthermia. Guidelines for diving in warm water are contained in Chapter 6.

Acclimatization is the process where repeated exposures to heat will reduce (but not eliminate) the rise in core temperature. At least 5 consecutive days of acclimatization to warm water diving are needed to see an increased tolerance to heat. Exercise training is essential for acclimation to heat. Where possible, acclimatization should be completed before attempting long duration working dives. Acclimatization should begin with short exposures and light workloads. All support personnel should also be heat acclimatized. Fully acclimatized divers can still develop hyperthermia, however. Benefits of acclimatization begin to disappear in 3 to 5 days after stopping exposure to warm water.

3-11 SPECIAL MEDICAL PROBLEMS ASSOCIATED WITH DEEP DIVING

3-11.1 **High Pressure Nervous Syndrome (HPNS).** High Pressure Nervous Syndrome (HPNS) is a derangement of central nervous system function that occurs during deep helium-oxygen dives, particularly saturation dives. The cause is unknown. The clinical manifestations include nausea, fine tremor, imbalance, incoordination, loss of manual dexterity, and loss of alertness. Abdominal cramps and diarrhea develop occasionally. In severe cases a diver may develop vertigo, extreme indifference to his surroundings and marked confusion such as inability to tell the right hand from the left hand. HPNS is first noted between 400 and 500 fsw and the severity appears to be both depth and compression rate dependent. With slow compression, depth of 1000 fsw may be achieved with relative freedom from HPNS. Beyond 1000 fsw, some HPNS may be present regardless of the compression rate. Attempts to block the appearance of the syndrome have included the addition of nitrogen or hydrogen to the breathing mixture and the use of various drugs. No method appears to be entirely satisfactory.

3-11.2 **Compression Arthralgia.** Most divers will experience pain in the joints during compression on deep dives. This condition is called *compression arthralgia.* The shoulders, knees, writs, and hips are the joints most commonly affected. The fingers, lower back, neck, and ribs may also be involved. The pain may be a constant deep ache similar to Type I decompression sickness, or a sudden, sharp, and intense but short-lived pain brought on my movement of the joint. These pains may be accompanied by "popping" or "cracking" of joints or a dry "gritty" feeling within the joint.

The incidence and intensity of compression arthralgia symptoms are dependent on the depth of the dive, the rate of compression, and individual susceptibility. While primarily a problem of deep saturation diving, mild symptoms may occur with rapid compression on air or helium-oxygen dives as shallow as 100 fsw. In deep helium saturation dives with slower compression rates, symptoms of compression arthralgia usually begins between 200 and 300 fsw, and increase in intensity as deeper depths are attained. Deeper than 600 fsw, compression pain may occur even with extremely slow rates of compression.

Compression joint pain may be severe enough to limit diver activity, travel rate, and depths attainable during downward excursion dives from saturation. Improvement is generally noted during the days spent at the saturation depth but, on occasion, these pains may last well into the decompression phase of the dive until shallower depths are reached. Compression pain can be distinguished from decompression sickness pain because it was present before decompression was started and does not increase in intensity with decreasing depth.

The mechanism of compression pain is unknown, but is thought to result from the sudden increase in inert gas tension surrounding the joints causing fluid shifts that interfere with joint lubrication.

3-12 OTHER DIVING MEDICAL PROBLEMS

3-12.1 **Dehydration.** Dehydration is a concern to divers, particularly in tropical zones. It is defined as an excessive loss of water from the body tissues and is accompanied by a disturbance in the balance of essential electrolytes, particularly sodium, potassium, and chloride.

3-12.1.1 **Causes of Dehydration.** Dehydration usually results from inadequate fluid intake and/or excessive perspiration in hot climates. Unless adequate attention is paid to hydration, there is a significant chance the diver in a hot climate will enter the water in a dehydrated state.

Immersion in water creates a special situation that can lead to dehydration in its own right. The water pressure almost exactly counterbalances the hydrostatic pressure gradient that exists from head to toe in the circulatory system. As a result, blood which is normally pooled in the leg veins is translocated to the chest, causing an increase central blood volume. The body mistakenly interprets the increase in central blood as a fluid excess. A reflex is triggered leading to an increase in urination, a condition called *immersion diuresis*. The increased urine flow leads to steady loss of water from the body and a concomitant reduction in blood volume during the dive. The effects of immersion diuresis are felt when the diver leaves the water. Blood pools once again in the leg veins. Because total blood volume is reduced, central blood volume falls dramatically. The heart may have difficulty getting enough blood to pump. The diver may experience lightheadness or faint while attempting to climb out of the water on a ladder or while standing on the stage. This is the result of a drop in blood pressure as the blood volume shifts to the legs. More commonly the diver will feel fatigued, less alert, and less able to think clearly than normal. His exercise tolerance will be reduced.

3-12.1.2 **Preventing Dehydration.** Dehydration is felt to increase the risk of decompression sickness. Divers should monitor their fluid intake and urine output during diving operations to insure that they keep themselves well hydrated. During the dive itself, there is nothing one can do to block the effects of immersion diuresis. Upon surfacing they should rehydrate themselves as soon as the opportunity presents itself.

3-12.2 **Immersion Pulmonary Edema.** Immersion in water can cause fluid to leak out of the circulation system and accumulate first in the interstitial tissues of the lungs then in the alveoli themselves. This condition is called *immersion pulmonary edema*. The exact mechanism of injury is not know, but the condition is probably related to the increase in central blood volume that occurs during immersion (see description above). Contributing factors include immersion in cold water, negative pressure breathing, and overhydration pre-dive, all of which enhance the increase in central blood volume with immersion. Heavy exercise is also a contributor.

Symptoms may begin on the bottom, during ascent, or shortly after surfacing and consist primarily of cough and shortness of breath. The diver may cough up blood tinged mucus. Chest pain is notably absent. A chest x-ray shows the classic pattern of pulmonary edema seen in heart failure.

A diver with immersion pulmonary edema should be placed on surface oxygen and transported immediately to a medical treatment facility. Signs and symptoms will usually resolve spontaneously over 24 hours with just bed rest and 100% oxygen.

Immersion pulmonary edema is a relatively rare condition, but the incidence appears to be increasing perhaps because of an over-emphasis on the need to hydrate before a dive. Adequate pre-dive hydration is essential, but overhydration is to be avoided. Beyond avoiding overhydration and negative pressure breathing situations, there is nothing the diver can do to prevent immersion pulmonary edema.

3-12.3 **Carotid Sinus Reflex.** External pressure on the carotid artery from a tight fitting neck dam, wet suit, or dry suit can activate receptors in the arterial wall, causing a decrease in heart rate with possible loss of consciousness. Using an extra-tight-fitting dry or wet suit or tight neck dams to decrease water leaks increase the chances of activation of the carotid reflex and the potential for problems.

3-12.4 **Middle Ear Oxygen Absorption Syndrome.** Middle ear oxygen absorption syndrome refers to the negative pressure that may develop in the middle ear following a long oxygen dive. Gas with a very high percentage of oxygen enters the middle ear cavity during an oxygen dive. Following the dive, the tissues of the middle ear slowly absorb the oxygen. If the eustachian tube does not open spontaneously, a negative pressure relative to ambient may result in the middle ear cavity. Symptoms are often noted the morning after a long oxygen dive. Middle ear oxygen absorption syndrome is difficult to avoid but usually does not pose a significant problem because symptoms are generally minor and easily eliminated. There may also be fluid (serous otitis media) present in the middle ear as a result of the differential pressure.

3-12.4.1 **Symptoms of Middle Ear Oxygen Absorption Syndrome.** The diver may notice mild discomfort and hearing loss in one or both ears. There may also be a sense of pressure and a moist, cracking sensation as a result of fluid in the middle ear.

3-12.4.2 **Treating Middle Ear Oxygen Absorption Syndrome.** Equalizing the pressure in the middle ear using a normal Valsalva maneuver or the diver's procedure of choice, such as swallowing or yawning, will usually relieve the symptoms. Discomfort and hearing loss resolve quickly, but the middle ear fluid is absorbed more slowly. If symptoms persist, a Diving Medical Technician or Diving Medical Officer shall be consulted.

3-12.5 Underwater Trauma. Underwater trauma is different from trauma that occurs at the surface because it may be complicated by the loss of the diver's gas supply and by the diver's decompression obligation. If possible, injured divers should be surfaced immediately and treated appropriately. If an injured diver is trapped, the first priority is to ensure sufficient breathing gas is available, then to stabilize the injury. At that point, a decision must be made as to whether surfacing is possible. If the decompression obligation is great, the injury will have to be stabilized until sufficient decompression can be accomplished. If an injured diver must be surfaced with missed decompression, the diver must be treated as soon as possible, realizing that the possible injury from decompression sickness may be as severe or more severe than that from the other injuries.

3-12.6 Blast Injury. Divers frequently work with explosive material or are involved in combat swimming and therefore may be subject to the hazards of underwater explosions. An explosion is the violent expansion of a substance caused by the gases released during rapid combustion. One effect of an explosion is a shock wave that travels outward from the center, somewhat like the spread of ripples produced by dropping a stone into a pool of water. This shock wave moving through the surrounding medium (whether air or water) passes along some of the force of the blast.

A shock wave moves more quickly and is more pronounced in water than in air because of the relative incompressibility of liquids. Because the human body is mostly water and incompressible, an underwater shock wave passes through the body with little or no damage to the solid tissues. However, the air spaces of the body, even though they may be in pressure balance with the ambient pressure, do not readily transmit the overpressure of the shock wave. As a result, the tissues that line the air spaces are subject to a violent fragmenting force at the interface between the tissues and the gas.

The amount of damage to the body is influenced by a number of factors. These include the size of the explosion, the distance from the site, and the type of explosive (because of the difference in the way the expansion progresses in different types of explosives). In general, larger, closer, and slower-developing explosions are more hazardous. The depth of water and the type of bottom (which can reflect and amplify the shock wave) may also have an effect. Under average conditions, a shock wave of 500 psi or greater will cause injury to the lungs and intestinal tract.

The extent of injury is also determined in part by the degree to which the diver's body is submerged. For an underwater blast, any part of the body that is out of the water is not affected. Conversely, for an air blast, greater depth provides more protection. The maximum shock pressure to which a diver should be exposed is 50 psi. The safest and recommended procedure is to have all divers leave the water if an underwater explosion is planned or anticipated. A diver who anticipates a nearby underwater explosion should try to get all or as much of his body as possible out of the water. If in the water, the diver's best course of action is to float face up, presenting the thicker tissues of the back to the explosion.

3-12.7 **Otitis Externa.** Otitis externa (swimmer's ear) is an infection of the ear canal caused by repeated immersion. The water in which the dive is being performed does not have to be contaminated with bacteria for otitis externa to occur. The first symptom of otitis externa is an itching and/or wet feeling in the affected ear. This feeling will progress to local pain as the external ear canal becomes swollen and inflamed. Local lymph nodes (glands) may enlarge, making jaw movement painful. Fever may occur in severe cases. Once otitis externa develops, the diver should discontinue diving and be examined and treated by Diving Medical Personnel.

Unless preventive measures are taken, otitis externa is very likely to occur during diving operations, causing unnecessary discomfort and restriction from diving. *External ear prophylaxis*, a technique to prevent swimmer's ear, should be done each morning, after each wet dive, and each evening during diving operations. External ear prophylaxis is accomplished using a 2 percent acetic acid in aluminum acetate (e.g., Otic Domboro) solution. The head is tilted to one side and the external ear canal gently filled with the solution, which must remain in the canal for 5 minutes. The head is then tilted to the other side, the solution allowed to run out and the procedure repeated for the other ear. The 5-minute duration shall be timed with a watch. If the solution does not remain in the ear a full 5 minutes, the effectiveness of the procedure is greatly reduced.

During prolonged diving operations, the external ear canal may become occluded with wax (cerumen). When this happens, external ear prophylaxis is ineffective and the occurrence of otitis externa will become more likely. The external ear canal can be examined periodically with an otoscope to detect the presence of ear wax. If the eardrum cannot be seen during examination, the ear canal should be flushed gently with water, dilute hydrogen peroxide, or sodium bicarbonate solutions to remove the excess cerumen. Never use swabs or other instruments to remove cerumen; this is to be done only by trained medical personnel. Otitis externa is a particular problem in saturation diving if divers do not adhere to prophylactic measures.

3-12.8 **Hypoglycemia.** Hypoglycemia is an abnormally low blood sugar (glucose) level. Episodes of hypoglycemia are common in diabetics and pre-diabetics, but may also occur in normal individuals. Simply missing a meal tends to reduce blood sugar levels. A few individuals who are otherwise in good health will develop some degree of hypoglycemia if they do not eat frequently. Severe exercise on an empty stomach will occasionally bring on symptoms even in an individual who ordinarily has no abnormality in this respect.

Symptoms of hypoglycemia include unusual hunger, excessive sweating, numbness, chills, headache, trembling, dizziness, confusion, incoordination, anxiety, and in severe cases, loss of consciousness.

If hypoglycemia is present, giving sugar by mouth relieves the symptoms promptly and proves the diagnosis. If the victim is unconscious, glucose should be given intravenously.

The possibility of hypoglycemia increases during long, drawn out diving operations. Personnel have a tendency to skip meals or eat haphazardly during the operation. For this reason, attention to proper nutrition is required. Prior to long, cold, arduous dives, divers should be encouraged to load up on carbohydrates. For more information, see Naval Medical Research Institute (NMRI) Report 89-94.

PAGE LEFT BLANK INTENTIONALLY

CHAPTER 4

Dive Systems

4-1 INTRODUCTION

4-1.1 **Purpose.** The purpose of this chapter is to promulgate general policy for maintaining diving equipment and systems.

4-1.2 **Scope.** This chapter provides general guidance applicable to maintaining all diving equipment and diving systems. Detailed procedures for maintaining diving equipment and systems are found in applicable military and manufacturer's operating and maintenance (O&M) manuals and Planned Maintenance System (PMS) Maintenance Requirement Cards (MRC).

4-2 GENERAL INFORMATION

4-2.1 **Document Precedence.** If a conflict arises between the documents containing the maintenance procedures for diving equipment and systems, the following actions are required:

1. PMS/MRC takes precedence.
2. If PMS/MRC is inadequate or incorrect, the applicable military O&M manual takes precedence. Report inadequate or incorrect PMS via a PMS feedback report in accordance with current PMS instructions.
3. If PMS/MRC and applicable military O&M manual are inadequate or incorrect, the manufacturer's technical manual takes precedence. Report inadequate or incorrect military technical manual information in accordance with procedures in the affected technical manual.

Call NAVSEA or NAVFAC prior to disregarding any required maintenance procedures on certified diving equipment. Failure to do so may compromise certification.

4-2.2 **Equipment Authorized For Navy Use (ANU).** Diving equipment used to conduct diving operations shall be authorized for use by NAVSEA/00C Diving Equipment Authorized For Navy Use (ANU) list or hold a current NAVSEA or NAVFAC system safety certification certificate. Naval Sea Systems Command (Code 00C3B), Supervisor of Diving is the cognizant authority for the NAVSEA/00C ANU list. Surface supplied diving systems, hyperbaric chamber systems, and selected free swimming SCUBA underwater breathing apparatus shall be certified in accordance with *U.S. Navy Diving and Manned Hyperbaric System Safety Certification Manual* (SS521-AA-MAN-010).

4-2.3 **System Certification Authority (SCA).** Naval Sea Systems Command Code 00C4 is SCA for all afloat and portable diving and hyperbaric systems. Naval Facilities

Engineering Command Code OFP-SCA is SCA for all shore-based diving and hyperbaric systems. Naval Sea Systems Command Code 07Q is SCA for submarine-employed Dry Deck Shelters and one atmosphere diving systems.

4-2.4 **Planned Maintenance System.** Diving equipment shall be maintained in accordance with the applicable PMS package. Failure to maintain equipment in accordance with current PMS guidance reduces the equipment reliability and may void the system safety certification for formally certified systems.

4-2.5 **Alteration of Diving Equipment.** Diving equipment shall not be modified or altered from approved configuration unless prior written approval has been granted by the applicable diving equipment technical program manager.

4-2.5.1 **Technical Program Managers for Shore-Based Systems.** Alterations for shore-based systems are managed by Naval Facilities Engineering Command (Code OFP-SCA), who is the cognizant technical authority for the development and approval of alterations to shore-based systems.

4-2.5.2 **Technical Program Managers for Other Diving Apparatus.** The technical program managers for other diving apparatus are:

- MK 16 MOD 0 - NAVSEASYSCOM (PMS NSW)
- MK 16 MOD 1 - NAVSEASYSCOM (PMS-EOD)
- MK 20 - NAVSEASYSCOM (SEA 00C)
- MK 21 - NAVSEASYSCOM (SEA 00C)
- MK 25 - NAVSEASYSCOM (PMS NSW)
- Dry Deck Shelter - NAVSEASYSCOM (PMS 399)

4-2.6 **Operating and Emergency Procedures.** Operating procedures (OPs) are detailed check sheets for operating the diving system and for performing various system-related tasks. All diving and recompression chamber systems shall be operated in accordance with a set of NAVSEA or NAVFAC approved operating procedures (OPs) and Emergency Operating Procedures (EPs) and requires the Commanding Officer's or OIC's signature on the cover page as final review.

4-2.6.1 **Standardized OP/EPs.** Standardized diving equipment such as the Light Weight MK 3 Surface Supplied Diving System, Transportable Recompression Chamber System (TRCS), and class-certified equipment such as the MK 16 and MK 25 Underwater Breathing Apparatus shall be operated per a single set of standardized OP/EPs that are included as part of the system O&M Manual.

Proposed changes/updates to OP/EPs for standardized diving equipment shall be submitted as a formal change proposal to the respective O&M Manual in accordance with directions contained therein.

4-2.6.2 **Non-standardized OP/EPs.** Diving and diving support equipment such as ships, small boats, and unique shore facility surface supplied diving and recompression chamber systems shall be operated in accordance with a single set of standard OP/

EPs that are developed at the command level and approved for use after validation by NAVSEA Code 00C3 or NAVFAC Code OFP-SCA. Proposed changes/updates to OPs/EPs for non-standardized diving equipment shall be submitted to the applicable approval authority. The following addresses are provided to assist in submitting proposed OP/EP changes and updates.

Submit proposed OP/EP changes and updates for afloat, portable diving and recompression chamber systems, and class-certified equipment to:

COMNAVSEASYSCOM (Code 00C3)
1333 Isaac Hull Ave., SE
Washington Navy Yard, DC 20376-1070

Submit proposed OP/EP changes and updates for fixed, shore-based facilities to:

COMNAVFACENGCOM (OFP-SCA)
1322 Patterson Ave., SE
Suite 1000
Washington Navy Yard, DC 20374-5065

4-2.6.3 **OP/EP Approval Process.** Submission of OPs/EPs for approval (if required) must precede the requested on-site survey date by 90 calendar days to allow complete review and resolution of questions. Follow these procedures when submitting OPs/EPs for approval:

- The command shall validate in the forwarding letter that the OPs/EPs are complete and accurate.
- The command must verify that drawings are accurate. Accurate drawings are used as a guide for evaluating OPs/EPs. Fully verified system schematics/drawings with components, gas consoles, manifolds, and valves clearly labeled shall be forwarded with the OPs/EPs.
- Approved OPs/EPs shall have the revision date listed on each page and not have any changes without written NAVSEA/NAVFAC approval.
- The command shall retain system documentation pertaining to DLSS approval, i.e., PSOBs, supporting manufacturing documentation, and OPs/EPs.

4-2.6.4 **Format.** The format for OPs/EPs is as follows:

- System: (Name or description, consistent with drawings)
- Step, Component, Description, Procedure, Location, Initials, Note (read in seven columns)

4-2.6.5 **Example.**

- System: High Pressure Air
- Step/Component/Description/Procedure/Location /Initials /Note

  1. ALP-15/Reducer outlet/Open/Salvage Hold/Initials/Note
  2. ALP-GA-7/Reducer outlet/Record Pressure/Salvage Hold/Initials/Note 1

The operator executing the procedure shall initial the Check column. Hazards and items of particular concern shall be identified in the Note column.

Once NAVSEA or NAVFAC has approved the system OP/EPs, they shall not be changed without specific written approval from NAVSEA or NAVFAC.

4-3 DIVER'S BREATHING GAS PURITY STANDARDS

4-3.1 **Diver's Breathing Air.** Diver's air compressed from ANU or certified diving system sources shall meet the U.S. Military Diver's Breathing Air Standards contained in Table 4-1.

Table 4-1. *U.S. Military Diver's Compressed Air Breathing Purity Requirements for ANU Approved or Certified Sources.*

Constituent	Specification
Oxygen (percent by volume)	20–22%
Carbon dioxide (by volume)	1,000 ppm (max)
Carbon monoxide (by volume)	20 ppm (max)
Total hydrocarbons (as CH_4 by volume)	25 ppm (max)
Odor and taste	Not objectionable
Oil, mist, particulates	5 mg/m^3 (max)

Diver's breathing air may be procured from commercial sources if a source of military diver's air is not readily available. Diver's air procured from commercial sources shall be certified in writing by the vendor as meeting the purity standards of FED SPEC BB-A-1034 Grade A Source I (pressurized container) or Source II (compressor) air. Specifications for this standard are outlined in Table 4-2.

Table 4-2. *Diver's Compressed Air Breathing Requirements if from Commercial Source.*

Constituent	Specification Source I Source II
Oxygen (percent by volume)	20–22%
Carbon dioxide (by volume)	500 ppm (max)
Carbon monoxide (by volume)	10 ppm (max)
Total hydrocarbons [as Methane (CH_4) by volume]	25 ppm (max)
Odor	Not objectionable
Oil, mist, particulates	.005 mg/l (max)
Separated Water	None
Total Water	0.02 mg/l (max)
Halogenated Compounds (by volume):	
Solvents	0.2 ppm (max)
Reference: FED SPEC BB-A-1034 B	

4-3.2 **Diver's Breathing Oxygen.** Oxygen used for breathing at 100-percent concentrations and for mixing of diver's breathing gases shall meet Military Specification MIL-PRF-27210G, Oxygen, Aviators Breathing, Liquid and Gaseous. The purity standards are contained in Table 4-3.

Table 4-3. *Diver's Compressed Oxygen Breathing Purity Requirements.*

Constituent	Specification
General Note: Gaseous and liquid oxygen shall contain not less than 99.5% by volume. The remainder, except for moisture and minor constituents specified below, shall be Argon and Nitrogen.	
Type I Gaseous	
Oxygen (percent by volume)	99.5%
Carbon dioxide (by volume)	10 ppm (max)
Methane (CH_4 by volume)	50 ppm (max)
Acetylene (C_2H_2)	0.1 ppm (max)
Ethylene (C_2H_4)	0.4 ppm (max)
Ethane (C_2H_6 and other hydrocarbons)	6.0 ppm (max)
Nitrous Oxide (N_2O by volume)	4.0 ppm (max)
Halogenated Compounds (by volume):	
Refrigerants	2.0 ppm (max)
Solvents	0.2 ppm (max)
Moisture (water vapor measured by ppm or measured by dew point)	7 ppm (max) $<-82°F$
Odor	Odor free

Table 4-3. Diver's Compressed Oxygen Breathing Purity Requirements (Continued).

Constituent	Specification
Type II Liquid	
Oxygen (percent by volume)	99.5%
Carbon dioxide (by volume)	5 ppm (max)
Methane (CH_4 by volume)	25 ppm (max)
Acetylene (C_2H_2)	0.05 ppm (max)
Ethylene (C_2H_4)	0.2 ppm (max)
Ethane (C_2H_6 and other hydrocarbons)	3.0 ppm (max)
Nitrous Oxide (N_2O by volume)	2.0 ppm (max)
Halogenated Compounds (by volume):	
Refrigerants	1.0 ppm (max)
Solvents	0.10 ppm (max)
Moisture (water vapor measured by ppm or measured by dew point)	7 ppm (max) <–82°F
Odor	Odor free
Reference: Military Specification MIL-PRF-27210G	

4-3.3 **Diver's Breathing Helium.** Helium used for diver's breathing gas shall meet Military Specification, MIL-PRF-27407B Propellant Pressurizing Agent Helium, Type I Gaseous Grade B, Respirable Helium. The purity standards are contained in Table 4-4.

Table 4-4. Diver's Compressed Helium Breathing Purity Requirements.

Constituent	Specification
Helium (percent by volume)	99.997%
Moisture (water vapor)	9 ppm (max)
Dew Point (not greater than)	–78°F
Hydrocarbons (as Methane)	1 ppm (max)
Oxygen	3 ppm (max)
Nitrogen + Argon	5 ppm (max)
Neon	23 ppm (max)
Hydrogen	1 ppm (max)
Reference: Military Specification MIL-PRF-27407B	

4-3.4 **Diver's Breathing Nitrogen.** Nitrogen used for divers breathing gas shall meet Federal Specification A-A-59155 Nitrogen, High Purity, Special Purpose. The purity standards are contained in Table 4-5.

Table 4-5. Diver's Compressed Nitrogen Breathing Purity Requirements.

Class I Oil Free, Type I Gaseous & Type II Liquid		
	Specification/Grade	
Constituent	**A**	**B**
Nitrogen	99.95%	99.50%
Oxygen	0.05%	0.50%
Moisture (water vapor)	.02 mg/l	.02 mg/l
Total Hydrocarbons (as methane by volume)	50 ppm	50 ppm
Odor	None	None
Note: Type I Nitrogen shall not contain any solid particles whose dimensions are greater than 50 microns. A 10 micron or better nominal filter at or close to the cylinder charging manifold will be used.		
Reference: Federal Specification A-A-59155		

4-4 DIVER'S AIR SAMPLING PROGRAM

NAVSEA Code 00C manages the diver's breathing air sampling program in accordance with OPNAVINST 3150.27 (series). The purpose of the air sampling program is to:

- Provide technical support for the operation and maintenance of diver's breathing air compressors and diving air storage systems.
- Provide general guidance concerning use of local commercial air sampling sources, including the evaluation of commercial air sampling capabilities and equipment.
- Perform program management for centrally funded air sampling services as directed by CNO Code N873.
- Collaborate with other government agencies and commercial industry on gas purity standards and sampling procedures related to diver's breathing gases.

4-4.1 **Maintenance Requirements.** Taking periodic air samples is a required maintenance action and shall be performed in accordance with the PMS card(s) applicable to the compressor or system producing diver's breathing air. Each diver breathing-air source in service must be sampled approximately every 6 months (within the interval between 4 and 8 months following the last accomplishment), when contamination is suspected and after system overhaul.

Do not use a compressor that is suspected of producing contaminated air or that has failed an air sample analysis until the cause of the problem has been corrected and a satisfactory air sample analysis has been obtained validating the production of acceptable air.

Diving systems that do not have a high-pressure (HP) air compressor within the scope of certification shall only be charged with air produced by HP air compressors listed on the ANU list and must have all applicable PMS completed up to date, including air sample requirements. Examples of these types of systems include MK 3 LWDS, Roper Cart, and various diving boats. HP banks on these systems need not be sampled unless contamination is suspected.

Air drawn from submarine HP air storage banks for use as diver's breathing air shall be sampled in accordance with the PMS maintenance requirement card applicable to the system, i.e., dry deck shelter system, submarine escape trunk, SCUBA charging station. See paragraph 4-4.2 for additional information on system line-up for sampling compressors where a sampling connection cannot be made immediately downstream from the last air filtration device.

Table 4-1 shows the minimum purity requirements for diving air produced by ANU-approved and certified diving air compressors. Air sampling services may be procured locally from government or commercial air analysis facilities, or may be acquired by utilizing analysis services coordinated via Naval Surface Warfare Center, Panama City, Florida (NSWC-PC).

NOTE **The most recent air sample analysis report shall be maintained on file for each air compressor (by compressor serial number) used to produce diver's breathing air.**

4-4.2 **General Air Sampling Procedures.** The following general information is provided to assist commands in managing air sample analysis programs.

Ensure all applicable PMS has been completed on the compressor and associated filtration system prior to taking an air sample.

- When sampling from HP charging systems, separate samples should be taken from each compressor supplying the system. Samples from the compressors should be taken as close to the compressor as possible but down stream of the last compressor-mounted air treatment device (moisture separator, filter, etc.). Some systems do not have fittings that allow samples to be taken from the system at a location other than the charging connection. In this case, the storage flasks should be isolated from the system, the system purged with air from the compressor to be sampled and the sample taken at the charging connection.

- When sampling from a low-pressure (LP) breathing-air system, separate air samples shall be taken from each LP compressor connected to the system. Samples shall be taken from each LP compressor as close to the compressor as possible, but downstream of the last compressor installed air treatment device (moisture separator, filter, etc.). Some systems do not have fittings that allow samples to be taken at connections other than the diver's manifold. In this case, a HP source should be isolated from the LP system, the system purged with air from the LP compressor to be sampled, and the sample obtained from the diver's manifold.

NOTE **Failure to purge the system line-up of air produced from other compressors or storage flasks will lead to an invalid air sample for the compressor being sampled.**

- Ensure that the compressor being sampled has reached full operating status (proper operating temperature, oil pressure, and air pressure) and is properly lined up to deliver air to the sample kit.
- Ensure that the compressor's intake is clear of any potential sources of contamination (including consideration of ambient smog levels in areas where smog is a problem).
- Follow the procedures on applicable air sample MRC card.
- Follow the instructions for operation of the air sampling kit.

4-4.3 **NSWC-PC Air Sampling Services.** The following applies to centrally funded air sampling services coordinated by NSWC-PC. Due to limited funding, commands are requested to schedule all compressors and associated samples to be taken at the same time. NSWC-PC coordinates air sampling services with a commercial contractor. Commands are not authorized to communicate directly with the commercial contractor. Sampling services are provided at no cost to the command. To request air sampling services, fill out and fax Air Sampling services request to NSWC-PC (Attn: Air Sampling). Telephone numbers are listed in Appendix 1C.

- The user must provide the sample expiration date, the number and type (HP or LP) of samples required, a complete mailing address, user point of contact and phone number. Air sample kits will not be shipped until the required information is received.
- Allow a minimum of 5 working days after submitting a properly filled out request form for delivery of a sampling kit in CONUS. Kits will be sent via commercial air with a prepaid return mailer. Incomplete sample requests cannot be acted on and will result in delay of shipping of sample kit.
- Allow a minimum of 3 weeks after submitting a properly filled out request form for delivery of a sampling kit if overseas. Kits will be sent via certified priority mail for overseas/FPO-APO addressees with prepaid return mailing. Incomplete sample requests cannot be acted on and will result in delay of shipping of sample kit.
- Detailed instructions are included with each sample kit. It is imperative to follow those instructions and the instructions on the applicable compressor air sampling MRC card.
- Air samples shall be taken and returned to NSWC-PC within 5 working days of receipt of the air sample kit to preclude incurring late fees.

- Air sample analysis reports for samples that meet air purity standards will be mailed to the command. Commands will be notified by quickest means possible of any samples that do not meet minimum purity requirements.

- The user will be contacted immediately by phone and/or message by NSWC-PC if the sample fails to meet established purity standards. The user will discontinue use of the air source until cause of contamination is corrected. Corrective action must be taken prior to laboratory retest.

4-4.4 **Local Air Sampling Services.** Commands may use local government (e.g., shipyards, ship repair facilities, government research laboratories) or commercial laboratories to analyze diver's air samples. Commands are required to bear the cost of locally procured air sample services. Local sampling facilities must be able to analyze to U.S. Navy air purity standards.

4-5 DIVING COMPRESSORS

4-5.1 **Equipment Requirements.** Compressors used to supply diving air or transfer oxygen or mixed gases shall be listed in the NAVSEA/00C Authorized for Navy use (ANU) list or be an element of a certified diving system.

4-5.2 **Air Filtration System.** Military diving compressors shall be equipped with an air filtration system that is listed in the NAVSEA/00C Authorized for Navy use (ANU) list or be an element of a certified diving system. The term air filtration system as used here is inclusive, referring collectively to compressed gas system filters, moisture separators, air purification, air cooling, and dehydration equipment.

4-5.3 **Lubrication.** Compressors used to produce military diver's breathing air are normally of oil-lubricated, two-to-five-stage reciprocating type. Oil lubrication:

- Prevents wear between friction surfaces

- Seals close clearances

- Protects against corrosion

- Transfers heat away from heat-producing surfaces

- Transfers minute particles generated from normal system wear to the oil sump or oil filter if so equipped

A malfunctioning oil-lubricated compressor poses a contamination risk to the diver's air supply. Contamination may occur due to excess oil mist being passed out of the compressor due to excess clearances, broken parts, or overfilling the oil sump.

Gaseous hydrocarbons and carbon monoxide may also be produced should a compressor overheat to the point of causing combustion of the lubricating oil and/or gaskets and other soft goods found in the compressor. Compressor overheating may be caused by a number of events including, but not limited to: loss of cooling water or air flow, low lube oil level, malfunction of stage unloader or relief valves, friction from broken or excessively worn parts, and/or compressor operation at an RPM above its rated capacity.

Diver's air filtration systems are designed to work with compressors operating under normal conditions, and cannot be relied on to filter or purify air from a malfunctioning compressor.

WARNING **Do not use a malfunctioning compressor to pump diver's breathing air or charge diver's air storage flasks as this may result in contamination of the diver's air supply.**

Lubricants used in diver's air compressors shall conform to MIL-PRF-17331 (2190 TEP) for normal operations, or MIL-PRF-17672 (2135TH) for cold weather operations. Where the compressor manufacturer specifically recommends the use of a synthetic base oil in their compressor for production of breathing air, that manufacturer recommended synthetic base oil may be used in lieu of MIL-PRF-17331 or MIL-PRF-17672 oil. Oil shall be changed out on compressors in strict accordance with the PMS requirements applicable to that compressor.

4-6 DIVING GAUGES

4-6.1 **Selecting Diving System Gauges.** Select a gauge whose full scale reading approximates 130 percent to 160 percent of the maximum operating pressure of the system. Following this guideline, a gauge with a full scale reading of 4,000 or 5,000 psi would be satisfactory for installation in a system with a maximum operating pressure of 3,000 psi.

Selecting gauge accuracy and precision should be based on the type of system and how the gauge will be used. For example, a high level of precision is not required on air bank pressure gauges where only relative values are necessary to determine how much air is left in the bank or when to shut down the charging compressor. However, considerable accuracy (¼ of 1 percent of full scale for saturation diving operations and 1 percent of full scale for surface supplied operations) is required for gauges that read diver depth (pneumofathometers and chamber depth gauges). Depth gauge accuracy is critical to selecting the proper decompression or treatment table.

Many gauges are provided with a case blowout plug on the rear surface. The blowout plug protects the operator in the event of Bourdon tube failure, when case overpressurization could otherwise result in explosion of the gauge lens. The plug must not be obstructed by brackets or other hardware.

All diving system gauges should be provided with gauge isolation valves and calibration fittings. If a gauge fails during an operation, the isolation valve closes to prevent loss of system pressure.

4-6.2 **Calibrating and Maintaining Gauges.** All installed gauges and portable gauges (tank pressure gauges, submersible tank pressure gauges, and gauges in small portable test sets) in use must be calibrated or compared in accordance with the Planned Maintenance System schedule unless a malfunction requires repair and calibration sooner. Programs such as the Shipboard Gauge Calibration Program as outlined in the NAVSEA Instruction 4734.1 (series) provide authority for a command to calibrate its own gauges. Calibrated gauges not in use should be kept in a clean, dry, vibration-free environment.

Calibration and comparison data must include the date of the last satisfactory check, the date the next calibration is due, and the activity accomplishing the calibration.

Gauges are delicate instruments and can be damaged by vibration, shock, or impact. They should be mounted in locations that minimize these factors and should always be mounted to gauge boards, panels, or brackets. The piping connection should not be the sole support for the gauge. A gauge can be severely damaged by rapid pulsations of the system when the fluid pressure is being measured. When this condition exists, a gauge snubber should be installed between the isolation valve and the gauge to protect the instrument. Most gauges are not waterproof and are not designed for use in a marine environment. Enclosures of transparent acrylic plastic, such as lucite, can be used to protect the gauges from water and salt spray. However, the enclosure must have vent passages to allow the atmospheric pressure to act on the gauge sensing element.

4-6.3 **Helical Bourdon Tube Gauges.** Manufacturers make two basic types of helical Bourdon tube gauges for use on recompression chambers and for surface-supplied diving systems. One is a caisson gauge with two ports on the back. The reference port, which is capped, is sealed with ambient air pressure or is piped to the exterior of the pressure chamber. The sensing port is left open to interior pressure. The other gauge is the standard exterior gauge.

Both are direct-drive instruments employing a helical Bourdon tube as the sensing element. The gauges are accurate to ¼ of 1 percent of full scale pressure at all dial points. With no gears or linkages, the movement is unaffected by wear, and accuracy and initial calibration remains permanent.

A comparative check in lieu of recalibration should be made in accordance with the Planned Maintenance System. A dial adjustment screw on the front face of the gauge provides for zero-point adjustment and special set pressure. Dial readout units of measure can be in pounds per square inch (psi) and/or feet of seawater (fsw).

4-7 COMPRESSED GAS HANDLING AND STORAGE

Handling and storing compressed gas are inherent parts of virtually all diving activities, whether conducted with SCUBA or surface supplied diving equipment. It is imperative that divers be familiar with the safety aspects of handling compressed gas. Diver's compressed gas shall be stored in military standard (MIL-STD) or DOT approved cylinders or ASME flasks applicable to the type and pressure levels of the compressed gas being stored.

Compressed gas shall be transported in cylinders meeting Department of Transportation (DOT) regulations applicable to the compressed gas being handled. DOT approved cylinders bear a serial number, DOT inspection stamp, a pressure rating, the date of last hydrostatic test, are equipped with applicable cylinder valve, and are appropriately color coded.

Refer to the following references for more detailed information on compressed gas handling and storage:

- *Industrial Gases, Generating, Handling and Storage*, NAVSEA Technical Manual S9086-SX-STM-000/CH-550.
- *American and Canadian Standard Compressed-Gas Cylinder Valve Outlet and Inlet Connections* (ANSI-B57.1 and CSA-B96).
- *American National Standard Method of Marking Portable Compressed-Gas Containers to Identify the Material Contained* (Z48.1).
- *Guide to the Preparation of Precautionary Labeling and Marking of Compressed Gas Cylinders* (CGA Pamphlet C-7).

PAGE LEFT BLANK INTENTIONALLY

CHAPTER 5

Dive Program Administration

5-1 INTRODUCTION

5-1.1 **Purpose.** The purpose of this chapter is to promulgate general policy for maintaining and retaining command smooth diving logs, personal diving logs, personal diving records, diving mishap reports, and failure analysis reports.

5-1.2 **Scope.** The record keeping and reporting instructions outlined in this chapter pertain to command smooth diving logs, individual diving logs, personal diving records, diving mishap reports, and failure analysis reports.

5-2 OBJECTIVES OF THE RECORD KEEPING AND REPORTING SYSTEM

There are five objectives in the diving record keeping and reporting system.

1. Establish a comprehensive operational record for each diving command. The Command Smooth Diving Log is a standardized operational record prepared in accordance with established military practice. This record establishes the diving history for each diving command and constitutes the basic operational record requirement under normal, uneventful circumstances.

2. Gather data for safety and trend analysis. Information about current diving operations conducted in the Navy, the incidence of Hyperbaric Treatments, and diving mishaps is provided to the Naval Safety Center through the Diving Reporting System and by message as required in OPNAVINST 5102.1 (series) via the Web Enabled Safety System (WESS). This information enables the Safety Center to identify safety-related problems associated with operating procedures and training.

3. Provide data for a personal record. OPNAVINST 3150.27 (series) requires each diver to maintain a personal diving log/history.

4. Report information about diving mishaps and casualties in accordance with the requirements of OPNAVINST 5102.1 (series) via WESS. Complete and accurate information enables the command to take appropriate action and prevent reoccurrence.

5. Report information about equipment deficiencies to the responsible technical agencies through the Failure Analysis Report (FAR) system.

5-3 RECORD KEEPING AND REPORTING DOCUMENTS

The documents established to meet the objectives of the record keeping and reporting system are:

- Command Smooth Diving Log (Figure 5-1)

- Dive/Jump Reporting System (DJRS)
- Diver's Personal Dive Record (diskette or hard copy)
- Diving Mishap/Hyperbaric Treatment/Death Report, Symbol OPNAV 5102/5 (via WESS)
- Diving Mishaps reported in accordance with OPNAVINST 5102.1 (series) via WESS
- Equipment Accident/Incident Information Sheet (Figure 5-2)
- Diving Life Support Equipment Failure Analysis Report (FAR) for surface-supplied diving systems, and open-circuit SCUBA (NAVSEA Form 10560/4) (Figure 5-3). FARS may be reported via the on-line reporting system at www.supsalv.org.
- Failure Analysis Report (NAVSEA Form 10560/1) (Figure 5-4) or Failure Analysis or Inadequacy Report. FARS maybe reported via the on-line reporting system at www.supsalv.org.

5-4 COMMAND SMOOTH DIVING LOG

The Command Smooth Diving Log is a chronological record of all dives conducted at that facility or command. It contains information on dives by personnel attached to the reporting command and dives by personnel temporarily attached to the command, such as personnel on TAD/TDY.

Dives conducted while temporarily assigned to another diving command shall be recorded in the host command's Smooth Diving Log. Additionally, record the dive in the Dive/Jump Reporting System (DJRS) of the host command.

The OPNAVINST 3150.27 (series) requires commands to retain the official diving log for 3 years. The minimum data items in the Command Smooth Diving Log include:

- Date of dive
- Purpose of the dive
- Identification of divers and standby divers
- Times left and reached surface, bottom time
- Depth
- Decompression time
- Air and water temperature
- Signatures of Diving Supervisor or Diving Officer/Master Diver

U.S. NAVY COMMAND SMOOTH DIVING LOG

Start Date ____

End Date ____

This log must be maintained in accordance with the *U.S. Navy Diving Manual, Volume 1,* (NAVSEA).

Figure 5-1. U.S. Navy Diving Log (sheet 1 of 2).

<table>
<tr><th colspan="9">COMMAND SMOOTH DIVING LOG</th></tr>
<tr><td colspan="2">Date</td><td colspan="4">Geographic Location</td><td colspan="3">Air Temp (°F)</td></tr>
<tr><td colspan="3">Equipment Used</td><td colspan="3">Dress</td><td colspan="3">Wave Height (ft)</td></tr>
<tr><td colspan="3">Breathing Medium</td><td colspan="3">Platform</td><td colspan="3">Water Temp (°F)</td></tr>
<tr><td colspan="6">Breathing Medium Source</td><td colspan="3">Current (kts.)</td></tr>
<tr><td colspan="3">Depth of Dive (fsw)</td><td colspan="3">Bottom Type</td><td colspan="3">Bottom Vis (ft)</td></tr>
<tr><td>Diver</td><td>LS</td><td>RB</td><td>LB</td><td>RS</td><td>TBT</td><td>TDT</td><td>TTD</td><td>Sched Used</td></tr>
<tr><td></td><td></td><td></td><td></td><td></td><td></td><td></td><td></td><td></td></tr>
<tr><td></td><td></td><td></td><td></td><td></td><td></td><td></td><td></td><td></td></tr>
<tr><td></td><td></td><td></td><td></td><td></td><td></td><td></td><td></td><td></td></tr>
<tr><td colspan="6">Purpose of Dive, Tools Used, etc.</td><td colspan="3">Repet Group</td></tr>
<tr><td colspan="6"></td><td colspan="3">Surface Interval</td></tr>
<tr><td colspan="6"></td><td colspan="3">New Repet Group</td></tr>
<tr><td colspan="6"></td><td colspan="3">RNT</td></tr>
<tr><td colspan="9">Dive Comments</td></tr>
<tr><td colspan="9"></td></tr>
<tr><td colspan="9"></td></tr>
<tr><td colspan="9"></td></tr>
<tr><td colspan="9">Signature (Diving Supervisor)</td></tr>
<tr><td colspan="9">Signature (Diving Officer/Master Diver)</td></tr>
</table>

Figure 5-1. U.S. Navy Diving Log (sheet 2 of 2).

EQUIPMENT ACCIDENT/INCIDENT INFORMATION SHEET

GENERAL

Unit point of contact________________ Position________________

Command UIC____________ Date__________ Time of occurrence____________

____

EQUIPMENT (indicate type of all equipment worn/used) Contributing factor______________

UBA: SCUBA__________ MK21__________ MK20__________

MK 16__________ LAR V__________ KM37__________

Other (specify)______________________

Suit type: Dry__________ Wet__________ Hot water__________

Other dress: Gloves__________ Booties__________ Fins__________

Mask__________ Snorkel__________ Knife__________

Weight belt (indicate weight)______________________

Depth gauge______________ Last calibration date______________

Buoyancy compensator/life preserver:______________________

Inflated at scene:__________ Partially__________ Operational __________

Inflation mode: Oral__________ CO_2__________ Independent supply__________

Cylinders: Number worn________ Size (cu ft)________ Valve type__________

Gas mix__________ Aluminum________ Steel__________

Surface pressure: Before______________ After______________

Regulator:____________ Last PMS date__________ Functional at scene?__________

Submersible pressure gauge:______________ Functional at scene?__________

CONDITIONS Location______________________

____

Depth________fsw Visibility________ft. Current________Knots sea state________(0-9)

Air temp__________°F Water temp: at surface__________°F at depth__________°F

Bottom type (mud, sand, coral, etc.)______________________

DIVE TIME

Bottom____________ Decompression____________ Total dive time____________

Was equipment operating and maintenance procedure a contributing factor?

(Explain):______________________

Is there contributory error in O&M Manual or 3M System?

(Explain):______________________

OTHER CONTRIBUTING FACTORS______________________

Figure 5-2. Equipment Accident/Incident Information Sheet. (sheet 1 of 2).

EQUIPMENT ACCIDENT/INCIDENT INFORMATION SHEET

Pertaining to UBA involved, fill in blanks with data required by items 1 through 9.

	KM 37 ↓	MK 21 ↓	MK 20 MOD 0 ↓	SCUBA ↓	MK 16 ↓	MK 25 ↓	OTHER ↓
1. Number of turns to secure topside gas umbilical supply:				N/A	N/A	N/A	
2. Number of turns to secure valve on emergency gas supply (EGS):				Reserve Up/Down	N/A	N/A	
3. Number of turns to secure gas supply at mask/helmet:				N/A	Mouthpiece Valve: Surface ______ Dive ______	Mouthpiece Valve: Surface ______ Dive ______	
4. Number of turns to secure gas bottle:	N/A	N/A	N/A	Air Bottle ______	O_2 ______ Diluent ______	O_2 Bottle ______	
5. Bottle Pressure:	EGS ______ psig	EGS ______ psig	EGS ______ psig	______ psig	O_2 ______ psig Diluent ______ psig	______ psig	
6. Gas Mixture:	Primary % ______ EGS % ______	Primary % ______ EGS % ______		N/A	Diluent N_2O_2 ______ HeO_2 ______	N/A	
7. Data/color of electronic display:	N/A	N/A	N/A	N/A	Primary ______ Secondary ______ ______ ______	N/A	
8. Battery voltage level:	N/A	N/A	N/A	N/A	Primary ______ Secondary ______	N/A	
9. Condition of canister:	N/A	N/A	N/A	N/A			

Note: If UBA involved is not listed above, provide information on separate sheet.

Figure 5-2. Equipment Accident/Incident Information Sheet. (sheet 2 of 2).

5-5 RECOMPRESSION CHAMBER LOG

The Recompression Chamber Log is the official chronological record of procedures and events for an entire dive. It is mandatory that all U.S. Navy diving activities maintain a Recompression Chamber Log. The log shall be legibly maintained in a narrative style. The Diving Officer, Master Diver, and Diving Supervisor shall review and sign the log daily or at the end of their watches. The Recompression Chamber Log must be retained for 3 years after the date of the dive. The minimum data items in the Recompression Chamber Log include:

- Date of dive
- Purpose of the dive
- Identification of diver(s)/patients(s)
- Identification of tender(s)
- Time left surface
- Time reached treatment depth
- Time reached stop
- Time left stop
- Depth/time of relief
- Change in symptoms
- Recompression chamber air temperature (if available)
- Oxygen and Carbon Dioxide % (if available)
- Medicine given
- Fluid administered
- Fluid void
- Signatures of Diving Officer, Master Diver, or Diving Supervisor

NAVAL SEA SYSTEMS COMMAND

DIVING LIFE SUPPORT EQUIPMENT FAILURE ANALYSIS REPORT (FAR)

1. Reporting Activity Name	2. Unit Identification Code ________ FAR Serial No. ________	3. Date Problem Detected (Mo/Day/Yr) ________

4. Reporting Activity Name Name: ________ Telephone: ________	5. Classification (NCSC Use Only) 1 ☐ 2 ☐ 3 ☐ 4 ☐ 5 ☐
6. Equipment Name	Equipment Serial Number
7. Item Name	Part Number or Federal Stock Number

8. Failure

A. How Discovered: ☐ Predive ☐ Operating ☐ Postdive ☐ PMS ☐ Other (Block 11)

B. Type: ☐ Malfunction ☐ Broken or Damaged ☐ Tech Documentation ☐ Other (Block 11)

C. Cause: ☐ Normal Wear ☐ Human Error ☐ Design Flaw ☐ PMS ☐ Wrong Parts ☐ Other (Block 11)

9. Spare Parts   Part Contract Number (   )

☐ Not Available from Stock ☐ Improperly Packaged ☐ Defective on Receipt

☐ Defective on Trial ☐ Quality Deficiency Report (QDR) Submitted ☐ Other (Block 11)

10. Correction ☐ Repaired Parts ☐ Replaced Parts ☐ Other (Block 11)

Date Completed: Man Hours: Est. Cost of Parts:.

11. Comments (Reference Block Numbers)

NAVSEA 10560/4 (1086) (Front)   **S/N 0116-LF-105-6020**

Figure 5-3. Failure Analysis Report (NAVSEA Form 10560/4).

FAILURE ANALYSIS REPORT

(See SS600-AH-MMA-010 for Information Concerning Use of This Form)

Disposition: Maintain the Original of This Form in Auditable Fashion With the UBA for the Entire Period Between NAVSEA Certification Surveys. Forward Copies 1-3 (Self Mailers) to the Addresses as Shown on the Bottom Right-Hand Corner and Back of the Forms.

1. Name of Reporting Activity	Unit Identification Code	2. Report Category (*Check Applicable Block*) ☐ Safety ☐ Routine	3. Report Serial Number 4. Date Discovered	
5. Deficiency Category (*Check One*) ☐ Equipment ☐ Publication	6. UBA Serial Number	7. Point of Contact for Activity	Autovon No.	Commercial No. ( )

8. Reason for Report (*Check Applicable Block*)

☐ Failure/Failure Suspected or Malfunction ☐ Damage Due to Improper Maintenance/Operation/Test ☐ Damage or Defective on Receipt ☐ Other (*Explain in Item 15*)

9. When Discovered (*Check Applicable Block*)

☐ Predive ☐ Postdive ☐ PMS ☐ During Operations ☐ Other (*Explain Here or in Item 15*)

10. System, Subsystem, or Component(s) Affected	11. Reentry Control Form No. (*Attach Copy*)

12. Description of Failure/Trouble/Discrepancy

13. Cause of Failure/Trouble/Discrepancy, If Known

14. Corrective Action Taken

15. Comments or Recommendations for Prevention or Elimination of Problems

16. Signature of Preparer	Rank/Rate	Date Signed	17. Signature, Approving Official	Rank/Rate	Date Approved

NAVSEA 10560/1 (12-84)

Figure 5-4. Failure Analysis Report. (NAVSEA Form 10560/1).

5-6 DIVER'S PERSONAL DIVE LOG

Although specific Navy Divers Personal Logbooks are no longer required, each Navy trained diver is still required to maintain a record of his dives in accordance with the OPNAVINST 3150.27 (series). The best way for each diver to accomplish this is to keep a copy of each Diving Log Form in a binder or folder. The Diving Log Form is generated by the Diver Reporting System (DRS) software. These forms, when signed by the Diving Supervisor and Diving Officer, are an acceptable record of dives that may be required to justify special payments made to you as a diver and may help substantiate claims made for diving-related illness or injury. If an individual desires a hard copy of the dives, the diver's command can generate a report using the DRS or by submitting a written request to the Naval Safety Center.

5-7 DIVING MISHAP/CASUALTY REPORTING

Specific instructions for diving mishap, casualty, and hyperbaric treatment are provided in OPNAVINST 5102.1 (series). The Judge Advocate General (JAG) Manual provides instructions for investigation and reporting procedures required in instances when the mishap may have occurred as a result of procedural or personnel error. Diving equipment status reporting instructions related to diving accidents/incidents are specified in this chapter.

5-8 EQUIPMENT FAILURE OR DEFICIENCY REPORTING

The Failure Analysis Report (FAR) system provides the means for reporting, tracking and resolving material failures or deficiencies in diving life-support equipment (DLSE). The FAR was developed to provide a rapid response to DLSE failures or deficiencies. It is sent directly to the configuration manager, engineers, and technicians who are qualified to resolve the deficiency. FAR Form 10560/4 (stock number 0116-LF-105-6020) covers all DLSE not already addressed by other FARs or reporting systems. For example, the MK 21 MOD 1, MK 20 MOD 0 mask, and all open-circuit SCUBA are reportable on this FAR form; the UBAs MK 16 and MK 25 are reportable on a FAR or a Failure Analysis or Inadequacy Report (FAR) in accordance with their respective technical manuals. When an equipment failure or deficiency is discovered, the Diving Supervisor or other responsible person shall ensure that the FAR is properly prepared and distributed. Refer to paragraph 5-10 for additional reporting requirements for an equipment failure suspected as the cause of a diving accident.

An electronic version of the FAR form is also available on-line at http://www.supsalv.org. Click on Diving or 00C3 Diving. When the next screen appears, click on Failure Analysis Reporting. Follow the instructions and submit the form.

5-9 U.S. NAVY DIVE REPORTING SYSTEM (DRS)

The Dive Reporting System (DRS) is a computer-based method of recording and reporting dives required by the OPNAVINST 3150.27 (series), and replaces reporting on DD Form 2544. The computer software provides all diving commands with a computerized record of dives.

The DRS makes it easy for commands to submit diving data to the Naval Safety Center. The computer software allows users to enter dive data, transfer data to the Naval Safety Center, and to generate individual diver and command reports. The DRS was designed for all branches of the U.S. Armed Services and can be obtained through:

Commander, Naval Safety Center
Attention: Code 37
375 A Street
Norfolk, VA 23511-4399

5-10 ACCIDENT/INCIDENT EQUIPMENT INVESTIGATION REQUIREMENTS

An *accident* is an unexpected event that culminates in loss of or serious damage to equipment or loss of consciousness, injury, or death to personnel. An *incident* is an unexpected event that degrades safety and increases the probability of an accident.

The number of diving accidents/incidents involving U.S. Navy divers is small when compared to the total number of dives conducted each year. The mishaps that do occur, however, must receive a thorough review to identify the cause and determine corrective measures to prevent further diving mishaps.

This section expands on the OPNAVINST 5102.1 (series) that requires expeditious reporting and investigation of diving related mishaps. The accident/incident equipment status reporting procedures in this chapter apply, in general, to all diving mishaps when malfunction or inadequate equipment performance, or unsound equipment operating and maintenance procedures are a factor.

In many instances a Diving Life Support Equipment Failure Analysis Report (FAR) may also be required. The primary purpose of this requirement is to identify any material deficiency that may have contributed to the mishap. Any suspected malfunction or deficiency of life support equipment will be thoroughly investigated by controlled testing at the Navy Experimental Diving Unit (NEDU). NEDU has the capability to perform engineering investigations and full unmanned testing of all Navy diving equipment under all types of pressure and environmental conditions. Depth, water turbidity, and temperature can be duplicated for all conceivable U.S. Navy dive scenarios.

Contact NAVSEA/00C3 to assist diving units with investigations and data collection following a diving mishap. 00C3 will assign a representative to inspect the initial condition of equipment and to pick up or ship all pertinent records and equipment to NEDU for full unmanned testing. Upon receiving the defective equipment, NEDU will conduct unmanned tests as rapidly as possible and will then return the equipment to the appropriate activity.

NOTE **Do not tamper with equipment without first contacting NAVSEA/00C3 for guidance.**

5-11 REPORTING CRITERIA

The diving and diving related accident/incident equipment status requirements set forth in this chapter are mandatory for all U.S. Navy diving units in each of the following circumstances:

- In all cases when an accident/incident results in a fatality or serious injury.
- When an accident/incident occurs and a malfunction or inadequate performance of the equipment may have contributed to the accident/incident.

5-12 ACTIONS REQUIRED

U.S. Navy diving units shall perform the following procedure when a diving accident/incident or related mishap meets the criteria stated in paragraph 5-11.

1. Immediately secure and safeguard from tampering all diver-worn and ancillary/support equipment that may have contributed to the mishap. This equipment should also include, but is not limited to, the compressor, regulator, depth gauge, submersible pressure gauge, diver dress, buoyancy compensator/life preserver, weight belt, and gas supply (SCUBA, emergency gas supply, etc.).

2. Expeditiously report circumstances of the accident/incident via WESS. Commands without WESS access should report by message (see OPNAVINST 5102.1 (series) for format requirements) to:

   - NAVSAFECEN NORFOLK VA//JJJ// with information copies to CNO WASHINGTON DC//N773// COMNAVSEASYSCOM WASHINGTON DC//00C// and NAVXDIVINGU PANAMA CITY FL//JJJ//.
   - If the accident/incident is MK 16 MOD 1 related, also send information copies to PEO LMW WASHINGTON DC//PMS-EOD// and NAVEODTECHDIV INDIAN HEAD MD//70//.
   - If the accident/incident is MK 16 MOD 0 related, also send information copies to PEO LMW WASHINGTON DC//PMS-NSW//.
   - If the accident/incident occurs at a shore-based facility, contact NAVFAC SCA, also send information copies to NFESC EAST COAST DET WASHINGTON DC//55//.

3. Expeditiously prepare a **separate, written report** of the accident/incident. The report shall include:

   - A completed Equipment Accident/Incident Information Sheet (Figure 5-2)
   - A sequential narrative of the mishap including relevant details that might not be apparent in the data sheets

4. The data sheets and the written narrative shall be mailed by traceable registered mail to:

   Commanding Officer
   Navy Experimental Diving Unit
   321 Bullfinch Road
   Panama City, Florida 32407-7015
   Attn: Code 03, Test & Evaluation

5. Package a certified copy of all pertinent 3M records and deliver to NAVSEA/00C3 on-scene representative.

NOTE **Call NAVSEA/NEDU/NAVFAC with details of the mishap or incident whenever possible. Personal contact may prevent loss of evidence vital to the evaluation of the equipment.**

5-12.1 **Technical Manual Deficiency/Evaluation Report.** If the accident/incident is believed to be solely attributable to unsound operating and maintenance procedures, including publications, submit a NAVSEA (user) Technical Manual Deficiency/Evaluation Report (TMDER) and request guidance from NEDU to ascertain if shipment of all or part of the equipment is necessary.

5-12.2 **Shipment of Equipment.** To expedite delivery, SCUBA, MK 16 and EGS bottles shall be shipped separately in accordance with current DOT directives and command procedures for shipment of compressed gas cylinders. Cylinders shall be forwarded in their exact condition of recovery (e.g., empty, partially filled, fully charged). If the equipment that is believed to be contributory to the accident/incident is too large to ship economically, contact NEDU to determine alternate procedures.

PAGE LEFT BLANK INTENTIONALLY

APPENDIX 1A

Safe Diving Distances from Transmitting Sonar

1A-1 INTRODUCTION

The purpose of this appendix is to provide guidance regarding safe diving distances and exposure times for divers operating in the vicinity of ships transmitting with sonar. Table 1A-1 provides guidance for selecting Permissible Exposure Limits Tables; Table 1A-2 provides additional guidance for helmeted divers. Tables 1A-3 through 1A-5 provide specific procedures for diving operations involving AN/SQS-23, -26, -53, -56; AN/BSY-1, -2; and AN/BQQ-5 sonars. Table 1A-6 provides procedures for diving operations involving AN/SQQ-14, -30, and -32. Section 1A-5 provides guidance and precautions concerning diver exposure to low-frequency sonar (160-320Hz). Contact NAVSEA Supervisor of Diving (00C3B) for guidance on other sonars. This appendix has been substantially revised from Safe Diving Distances from Transmitting Sonar (NAVSEAINST 3150.2 Series) and should be read in its entirety.

1A-2 BACKGROUND

Chapter 18 of OPNAVINST 5100.23 Series is the basic instruction governing hearing conservation and noise abatement, but it does not address exposure to waterborne sound. Tables 1A-3 through 1A-6 are derived from experimental and theoretical research conducted at the Naval Submarine Medical Research Laboratory (NSMRL) and Naval Experimental Diving Unit (NEDU). This instruction provides field guidance for determining safe diving distances from transmitting sonar. This instruction supplements OPNAVINST 5100.23 Series, and should be implemented in conjunction with OPNAVINST 5100.23 Series by commands that employ divers.

The Sound Pressure Level (SPL), not distance, is the determining factor for establishing a Permissible Exposure Limit (PEL). The exposure SPLs in Tables 1A-3 through 1A-6 are based upon the sonar equation and assume omni-directional sonar and inverse square law spreading. Any established means may be used to estimate the SPL at a dive site, and that SPL may be used to determine a PEL. When the exposure level is overestimated, little damage, except to working schedules, will result. Any complaints of excessive loudness or ear pain for divers require that corrective action be taken. Section 1A-5 provides guidance for diver exposure to low-frequency active sonar (LFA), which should be consulted if exposure to LFA is either suspected or anticipated.

This appendix does not preclude the operation of any sonar in conjunction with diving operations, especially under operationally compelling conditions. It is based upon occupational safety and health considerations that should be implemented for routine diving operations. It should be applied judiciously under special operational circumstances. The guidance in Tables 1A-3 through 1A-6 is intended to facilitate the successful integration of operations.

1A-3 ACTION

Commanding Officers or Senior Officers Present Afloat are to ensure that diving and sonar operations are integrated using the guidance given by this appendix. Appropriate procedures are to be established within each command to effect coordination among units, implement safety considerations, and provide efficient operations using the guidance in Tables 1A-3 though 1A-6.

1A-4 SONAR DIVING DISTANCES WORKSHEETS WITH DIRECTIONS FOR USE

1A-4.1 **General Information/Introduction.** Permissible Exposure Limits (PEL) in minutes for exposure of divers to sonar transmissions are given in Tables 1A-3 through 1A-6.

1A-4.1.1 **Effects of Exposure.** Tables 1A-3 through 1A-5 are divided by horizontal double lines. Exposure conditions above the double lines should be avoided for routine operations. As Sound Pressure Level (SPL) increases above 215 dB for hooded divers, slight visual-field shifts (probably due to direct stimulation of the semicircular canals), fogging of the face plate, spraying of any water within the mask, and other effects may occur. In the presence of long sonar pulses (one second or longer), depth gauges may become erratic and regulators may tend to free-flow. Divers at Naval Submarine Medical Research Laboratory experiencing these phenomena during controlled research report that while these effects are unpleasant, they are tolerable. Similar data are not available for un-hooded divers but visual-field shifts may occur for these divers at lower levels. If divers need to be exposed to such conditions, they must be carefully briefed and, if feasible, given short training exposures under carefully controlled conditions. Because the probability of physiological damage increases markedly as sound pressures increase beyond 200 dB at any frequency, exposure of divers above 200 dB is prohibited unless full wet suits and hoods are worn. Fully protected divers (full wet suits and hoods) must not be exposed to SPLs in excess of 215 dB at any frequency for any reason.

1A-4.1.2 **Suit and Hood Characteristics.** There is some variation in nomenclature and characteristics of suits and hoods used by divers. The subjects who participated in the Naval Submarine Medical Research Laboratory experiments used 3/8-inch nylon-lined neoprene wet suits and hoods. Subsequent research has shown that 3/16-inch wet suit hoods provide about the same attenuation as 3/8-inch hoods. Hoods should be well fitted and cover the skull completely including cheek and chin areas. The use of wet-suit hoods as underwater ear protection is strongly recommended.

1A-4.1.3 **In-Water Hearing vs. In-Gas Hearing.** A distinction is made between in-water hearing and in-gas hearing. In-water hearing occurs when the skull is directly in contact with the water, as when the head is bare or covered with a wet-suit hood. In-gas hearing occurs when the skull is surrounded by gas as in the MK 21 diving helmet. In-water hearing occurs by bone conduction—sound incident anywhere on the skull is transmitted to the inner ear, bypassing the external and middle ear. In-gas hearing occurs in the normal way—sound enters the external ear canal and stimulates the inner ear through the middle ear.

1A-4.2 **Directions for Completing the Sonar Diving Distances Worksheet.** Follow the steps listed below to determine Permissible Exposure Limits (PELs) for the case when the actual dB Sound Pressure Level (SPL) at the dive site is unknown. Figure 1A-1 is a worksheet for computing the safe diving distance/exposure time. Figures 1A-2 through 1A-5 are completed worksheets using example problems. Work through these example problems before applying the worksheet to your particular situation.

Step 1. **Diver Dress.** Identify the type of diving equipment—wet-suit un-hooded; wet-suit hooded; helmeted. Check the appropriate entry on step 1 of the worksheet.

Step 2. **Sonar Type(s).** Identify from the ship's Commanding Officer or representative the type(s) of sonar that will be transmitting during the period of time the diver is planned to be in the water. Enter the sonar type(s) in step 2 of the worksheet.

Step 3. **PEL Table Selection.** Use the Table 1A-1 to determine which PEL table you will use for your calculations. For swimsuit diving use wet suit un-hooded tables. Check the table used in step 3 of the worksheet.

Table 1A-1. *PEL Selection Table.*

	SONAR		
DIVER DRESS:	**All except AN/SQQ -14, - 30, -32**	**AN/SQQ -14, -30, -32**	**Unknown Sonar**
Wet suit - Un-hooded	Table 1A-3	Table 1A-6	Start at 1000 yards and move in to diver comfort
Wet suit - Hooded	Table 1A-4	Table 1A-6	Start at 600 yards and move in to diver comfort
Helmeted	Table 1A-5	No restriction	Start at 3000 yards and move in to diver comfort

For guidance for sonars not addressed by this instruction, contact NAVSEA (00C32).

NOTE **If the type of sonar is unknown, start diving at 600–3,000 yards, depending on diving equipment (use greater distance if helmeted), and move in to limits of diver comfort.**

Step 4. **Distance to Sonar.** Determine the distance (yards) to the transmitting sonar from place of diver's work. Enter the range in yards in step 4 of the worksheet.

SONAR SAFE DIVING DISTANCE/EXPOSURE TIME WORKSHEET

1. Diver dress: Wet Suit - Un-hooded ______
   Wet Suit - Hooded ______
   Helmeted ______

2. Type(s) of sonar: ____________________

3. PEL Table 1A-3 ___; 1A-4 _____; 1A-5 ___; 1A-6 ___

4. Range(s) to sonar (yards): ____________________

5. Estimated SPL at range(s) in step 3 (from table/column in step 3): ________________

   Reminder: If range is between two values in the table, use the shorter range. If the SPL is measured at the dive site, use the measured value.

6. Depth Reduction ______ dB

   Reminder: 0 if <u>not</u> helmeted, see table in instructions if helmeted.

7. Corrected SPL (Step 5 minus Step 6) ____________________________

8. Estimated PEL at SPL (from table/column in step 3 of the appendix): ____________________

9. Duty Cycle Known: Yes _______ (do step 9); No _____ (stop)

   Adjusted PEL for actual duty cycle

   Actual DC % = 100 × ____ sec. (pulse length / ___ sec. (pulse repetition period)
   Actual DC % = ____
   Adjusted PEL = PEL (from step 8) ____ min. × 20 / actual duty cycle (%) ___ = ____ min.

   PEL1 = _____ minutes; PEL2 = _____ minutes

   Reminder: Do not adjust the PEL if duty cycle is unknown.

10. Multiple Sonars: Yes ____ (do step 10); No ____ (stop)

    Sonar 1: DT1 = ______ (Desired dive duration)

    PEL1 = _______ (from Step 8 or 9, as applicable)

    DT1/PEL1 = ____________ .

    Sonar 2: DT1 = ______ (Desired dive duration)

    PEL1 = _______ (from Step 8 or 9, as applicable)

    DT1/PEL1 = ____________ .

    ND = ______ + _______ = _______ (This is less than 1.0, so dive is acceptable and may proceed.)

    Reminder: The Noise Dose must not exceed a value of 1.0.

Figure 1A-1. Sonar Safe Diving Distance/Exposure Time Worksheet.

NOTE **If range is between two values in the table, use the shorter range. This will insure that the SPL is not underestimated and that the PEL is conservative.**

Step 5. **Estimated SPL.** In the PEL selection table (Table 1A-1) determined in step 3 of the worksheet (Figure 1A-1), locate the diving distance (range) in the appropriate sonar equipment column. Read across to the leftmost column to find the SPL in dB. For ranges intermediate to those shown use the shorter range. Enter this SPL value in step 5 of the worksheet. If the SPL value in dB can be determined at the dive site, enter the measured SPL value in step 5.

Step 6. **Helmeted Dive Depth Reduction.**

If the diver dress is not helmeted, enter 0 in step 6 of the worksheet and go to step 7 of these instructions.

Helmeted divers experience reduced sensitivity to sound pressure as depth increases. The reductions listed in Table 1A-2 may be subtracted from the SPLs for helmeted divers in Table 1A-5. Enter the reduction in step 6 of the worksheet. If the depth is between two values in the table, use the lesser reduction since that value will produce a conservative PEL.

Table 1A-2. *Depth Reduction Table.*

Depth (FSW)	Reduction (dB)	Depth (FSW)	Reduction (dB)
9	1	98	6
19	2	132	7
33	3	175	8
50	4	229	9
71	5	297	10

Step 7. **Corrected SPL.** The corrected SPL equals the Estimated SPL from step 5 minus the reduction in dB from step 6. Enter the corrected SPL in step 7 of the worksheet.

Step 8. **PEL Determination.** Go to the SPL in the appropriate table and read one column right to find the PEL for the SPL shown in step 7 of the worksheet. Enter in step 8 of the worksheet.

Step 9. **Duty Cycle/Adjusted PEL Calculation.** Tables 1A-3 through 1A-6 assume a transmit duty cycle of 20 percent. Duty cycle (DC) is the percentage of time in a given period that the water is being insonified (sonar transmitting). Sonar operators may use various means of computing DC that are valid for the purpose of this instruction. If the actual duty cycle is different from 20 percent, PELs may be extended or shortened proportionally. Use step 9 of the worksheet to calculate and enter the corrected PEL.

The formula for duty cycle is:

DC = 100 × Pulse length (sec.) / Pulse Repetition Period (sec.)

The formula for the adjusted PEL is:

Adjusted PEL = PEL × 20 / actual duty cycle; **Equation 1**

Example Problem. An un-hooded wet suited diver is 16 yards from an AN/SQQ-14 sonar transmitting a 500 msec pulse (.5 seconds) every 10 seconds.

Solution. The actual duty cycle (DC) % is:

Actual DC % = 100 × .5 / 10 = 5 percent.

Locate the PEL from the table (which is for a 20% duty cycle). Compute the adjusted PEL as:

Using worksheet step 9, Adjusted PEL = PEL (from step 8) 170 × 20/5=680 minutes.

If variable duty cycles are to be used, select the greatest percent value.

Step 10. **Multiple Sonar/Noise Dose Calculation.** When two or more sonars are operating simultaneously, or two or more periods of noise exposure of different values occur, the combined effects must be considered. In the following formula, **ND is the daily noise dose and must not exceed a value of 1.0**, DT is the dive (exposure) time (left surface to reach surface), and PEL is the PEL for each noise exposure condition computed as described above:

ND = DT1/PEL1 + DT2/PEL2 + ... DTn/PELn; **Equation 2**

Note: DT1/PEL1 is for the first sonar, DT2/PEL2 is for the second sonar, up to the total number of sonars in use.

To use the worksheet, go through the steps 1-9 for each sonar, entering the appropriate values in each step of the worksheet. Enter the PELs into the worksheet step 10. There is room for two sonars in the worksheet. If more than two are being used, follow the same format and continue the calculations in the white space at the end of the worksheet.

Example Problem. A hooded wet suited diver is 100 yards from a transmitting AN/SQS-53A sonar and a transmitting AN/SQS-23 sonar for fifteen minutes.

Solution.
DT1 = 15 minutes
PEL1 (for SQS-53A) = 50 minutes
DT1/PEL1 = 15/50 = .3

DT2 = 15 minutes
PEL2 (for SQS-23) = 285 minutes
DT2/PEL2 = 15/285 = .05

ND = .3 + .05 = .35
This is less than 1.0 and therefore is acceptable.

Example 1: *You are planning a routine dive for 160 minutes using wet-suited divers without hoods at a dive site 17 yards from an AN/SQQ-14 sonar. The duty cycle for the AN/SQQ-14 sonar is unknown. Is this dive permitted? Provide justification for your decision.*

SONAR SAFE DIVING DISTANCE/EXPOSURE TIME WORKSHEET

1. Diver dress: Wet Suit - Un-hooded __X__
   Wet Suit - Hooded _____
   Helmeted ______
2. Type(s) of sonar: AN/SQQ-14
3. PEL Table 1A-3 __; 1A-4 __; 1A-5 __; 1A-6 _X_
4. Range(s) to sonar (yards): 17
5. Estimated SPL at range(s) in step 3 (from table/column in step 3): SPL = 198 dB

   Reminder: If range is between two values in the table, use the shorter range. If the SPL is measured at the dive site, use the measured value.
6. Depth Reduction __0__ dB

   Reminder: 0 if not helmeted, see table in instructions if helmeted.
7. Corrected SPL (Step 5 minus Step 6) SPL1 198 – 0 = 198 dB
8. Estimated PEL at SPL (from table/column in step 3 of the appendix): PEL1 = 170 minutes
9. Duty Cycle Known: Yes ______ (do step 9); No __X__ (stop)
   Adjusted PEL for actual duty cycle
   Actual DC % = 100 × ______ sec. (pulse length / ______ sec. (pulse repetition period)
   Actual DC % = ______
   Adjusted PEL = PEL (from step 8) ___ min. × 20 / actual duty cycle (%) ___ = ___ min.

   Reminder: Do not adjust the PEL if duty cycle is unknown.
10. Multiple Sonars: Yes _____ (do step 10); No _X__ (stop)

    Sonar 1: DT1 = ____ (Desired dive duration)
    PEL1 = ____ (from Step 8 or 9, as applicable)
    DT1/PEL1 = _______ .

    Sonar 2: DT1 = ____ (Desired dive duration)
    PEL1 = ____ (from Step 8 or 9, as applicable)
    DT1/PEL1 = _______ .

    ND = ____ + _____ = ____ (This is less than 1.0, so dive is acceptable and may proceed.)

    Reminder: The Noise Dose must not exceed a value of 1.0.

The dive time of 160 minutes is permitted because the PEL is 171 minutes.

Figure 1A-2. Sonar Safe Diving Distance/Exposure Time Worksheet (Completed Example).

Example 2: *You are planning a routine dive for 75 minutes using wet-suited divers without hoods at a dive site which is 1000 yards from an AN/SQQ-23 sonar. The SPL was measures at 185 dB. The duty cycle for the AN/SQS-23 sonar is unknown. Is this dive permitted? Provide justification for your decision.*

SONAR SAFE DIVING DISTANCE/EXPOSURE TIME WORKSHEET

1. Diver dress: Wet Suit - Un-hooded __X__
   Wet Suit - Hooded _____
   Helmeted ______
2. Type(s) of sonar: AN/SQS-23
3. PEL Table 1A-3 _X_; 1A-4 __; 1A-5 __; 1A-6 __
4. Range(s) to sonar (yards): 1000
5. Estimated SPL at range(s) in step 3 (from table/column in step 3): SPL = 185 dB

   Reminder: If range is between two values in the table, use the shorter range.
   If the SPL is measured at the dive site, use the measured value.
6. Depth Reduction __0__ dB

   Reminder: 0 if not helmeted, see table in instructions if helmeted.
7. Corrected SPL (Step 5 minus Step 6) SPL1 185 – 0 = 185 dB
8. Estimated PEL at SPL (from table/column in step 3 of the appendix): PEL1 = 170 minutes
9. Duty Cycle Known: Yes ______ (do step 9); No __X__ (stop)
   Adjusted PEL for actual duty cycle
   Actual DC % = 100 × ______ sec. (pulse length / ______ sec. (pulse repetition period)
   Actual DC % = ______
   Adjusted PEL = PEL (from step 8) ___ min. × 20 / actual duty cycle (%) ___ = ___ min.

   Reminder: Do not adjust the PEL if duty cycle is unknown.
10. Multiple Sonars: Yes _____ (do step 10); No __X__ (stop)

    Sonar 1: DT1 = ____ (Desired dive duration)
    PEL1 = ____ (from Step 8 or 9, as applicable)
    DT1/PEL1 = _______ .

    Sonar 2: DT1 = ____ (Desired dive duration)
    PEL1 = ____ (from Step 8 or 9, as applicable)
    DT1/PEL1 = _______ .

    ND = ____ + _____ = ____ (This is less than 1.0, so dive is acceptable and may proceed.)
    Reminder: The Noise Dose must not exceed a value of 1.0..

The dive time of 75 minutes is permitted because the PEL is 170 minutes.

Figure 1A-3. Sonar Safe Diving Distance/Exposure Time Worksheet (Completed Example).

Example 3: *You are planning a 98 fsw dive for 35 minutes using the MK 21 at a dive site which is 3000 yards from an AN/SQS-53C sonar. The duty cycle for the AN/SQS-53C sonar is unknown. Is this dive permitted? Provide justification for your decision.*

SONAR SAFE DIVING DISTANCE/EXPOSURE TIME WORKSHEET

1. Diver dress: Wet Suit - Un-hooded ______
   Wet Suit - Hooded _____
   Helmeted __X___
2. Type(s) of sonar: AN/SQS-53C
3. PEL Table 1A-3 ___; 1A-4 __; 1A-5 _X_; 1A-6 __
4. Range(s) to sonar (yards): 3000
5. Estimated SPL at range(s) in step 3 (from table/column in step 3): SPL1 = 181 dB

   Reminder: If range is between two values in the table, use the shorter range. If the SPL is measured at the dive site, use the measured value.
6. Depth Reduction __6__ dB

   Reminder: 0 if not helmeted, see table in instructions if helmeted.
7. Corrected SPL (Step 5 minus Step 6) SPL1 181 – 6 = 175 dB
8. Estimated PEL at SPL (from table/column in step 3 of the appendix): PEL1 = 50 minutes
9. Duty Cycle Known: Yes ______ (do step 9); No __X__ (stop)
   Adjusted PEL for actual duty cycle
   Actual DC % = 100 × ______ sec. (pulse length / ______ sec. (pulse repetition period)
   Actual DC % = ______
   Adjusted PEL = PEL (from step 8) ___ min. × 20 / actual duty cycle (%) ___ = ___ min.

   Reminder: Do not adjust the PEL if duty cycle is unknown.
10. Multiple Sonars: Yes _____ (do step 10); No __X__ (stop)

    Sonar 1: DT1 = ____ (Desired dive duration)
    PEL1 = ____ (from Step 8 or 9, as applicable)
    DT1/PEL1 = _______ .

    Sonar 2: DT1 = ____ (Desired dive duration)
    PEL1 = ____ (from Step 8 or 9, as applicable)
    DT1/PEL1 = _______ .

    ND = ____ + _____ = ____ (This is less than 1.0, so dive is acceptable and may proceed.)
    Reminder: The Noise Dose must not exceed a value of 1.0.

The dive time of 35 minutes is permitted because the PEL is 50 minutes.

Figure 1A-4. Sonar Safe Diving Distance/Exposure Time Worksheet (Completed Example).

Example 4: *You are planning a routine dive for 120 minutes using wet-suited divers with hoods at a dive site which is 200 yards from an AN/SQS-53A sonar and 120 yards from an AN/SQS-23 sonar. The AN/SQS-53A sonar is transmitting an 800 msec pulse (0.8 sec) every 20 seconds. The duty cycle for the AN/SQS-23 sonar is unknown. Is this dive permitted? Provide justification for your decision.*

SONAR SAFE DIVING DISTANCE/EXPOSURE TIME WORKSHEET

1. Diver dress: Wet Suit - Un-hooded ______
   Wet Suit - Hooded __X__
   Helmeted ______
2. Type(s) of sonar: AN/SQS-53A and AN/SQS-23
3. PEL Table 1A-3 ___; 1A-4 _X_; 1A-5 ___; 1A-6 ___
4. Range(s) to sonar (yards): 200 (from SQS-53A); 120 (from SQS-23)
5. Estimated SPL at range(s) in step 3 (from table/column in step 3): SPL1 = 201; SPL2 = 196 (per reminder, use SPL for 112 yard range)

   Reminder: If range is between two values in the table, use the shorter range. If the SPL is measured at the dive site, use the measured value.
6. Depth Reduction __0__ dB

   Reminder: 0 if not helmeted, see table in instructions if helmeted.
7. Corrected SPL (Step 5 minus Step 6) SPL1 201 – 0 = 201 dB; SPL2 196 – 0 = 196 dB;
8. Estimated PEL at SPL (from table/column in step 3 of the appendix): PEL1 = 143 min; PEL 2 = 339 min
9. Duty Cycle Known: Yes __X__ (do step 9); No _____ (stop)
   Adjusted PEL for actual duty cycle
   Actual DC % = 100 × __0.8__ sec. (pulse length / __20__ sec. (pulse repetition period)
   Actual DC % = __4__
   Adjusted PEL = PEL (from step 8) 143 min. × 20 / actual duty cycle (%) 4 = 715 min.
   PEL1 = 715 minutes; PEL2 = 339 minutes

   Reminder: Do not adjust the PEL if duty cycle is unknown.
10. Multiple Sonars: Yes __X__ (do step 10); No ____ (stop)

    Sonar 1: DT1 = 120 (Desired dive duration)
    PEL1 = 715 (from Step 8 or 9, as applicable)
    DT1/PEL1 = 120/715 = 0.17 .

    Sonar 2: DT1 = 120 (Desired dive duration)
    PEL1 = 339 (from Step 8 or 9, as applicable)
    DT1/PEL1 = 120/339 = .35 .

    ND = 0.17 + 0.35 = 0.52 (This is less than 1.0, so dive is acceptable and may proceed.)
    Reminder: The Noise Dose must not exceed a value of 1.0.

The dive time of 120 minutes is permitted because the ND is less than 1.0.

Figure 1A-5. Sonar Safe Diving Distance/Exposure Time Worksheet (Completed Example).

Table 1A-3. Wet Suit Un-Hooded.

Permissible Exposure Limit (PEL) within a 24-hour period for exposure to AN/SQS-23, -26, -53, -56, AN/BSY-1, -2 and AN/BQQ-5 sonars, including versions and upgrades. Exposure conditions shown above the double line should be avoided except in cases of compelling operational necessity.

Estimated Ranges in yards for given SPL and PEL for sonar.

SPL (dB)	PEL (MIN)	BSY-1 SQS-53C	BQQ-5 BSY-2 SQS-26CX(U) SQS-53A, SQS-53B SQS-56(U)	SQS-23 SQS-26AX SQS-26BX, SQS-26CX SQS-56	
200	13	316	224	71	A
199	15	355	251	79	V E
198	18	398	282	89	O X
197	21	447	316	100	I P
196	25	501	355	112	D O
195	30	562	398	126	S
194	36	631	447	141	T U
193	42	708	501	158	H R
192	50	794	562	178	I E
191	60	891	631	200	S
190	71	1,000	708	224	
189	85	1,122	794	251	
188	101	1,259	891	282	
187	120	1,413	1,000	316	
186	143	1,585	1,122	355	
185	170	1,778	1,259	398	
184	202	1,995	1,413	447	
183	240	2,239	1,585	501	
182	285	2,512	1,778	562	
181	339	2,818	1,995	631	
180	404	3,162	2,239	708	
179	480	3,548	2,512	794	
178	571	3,981	2,818	891	
177	679	4,467	3,162	1,000	
176	807	5,012	3,548	1,122	
175	960	5,623	3,981	1,259	

All ranges and SPLs are nominal.

*SPL is measured in dB/1 μPA at the dive site. To convert SPL for sound levels referenced to mbar, subtract 100 dB from tabled levels.

(U) = upgrade

Table 1A-4. *Wet Suit Hooded.*

Permissible Exposure Limit (PEL) within a 24-hour period for exposure to AN/SQS-23, -26, -53, -56, AN/BSY-1, -2, and AN/BQQ-5 sonar, including versions and upgrades. Exposure conditions shown above the double line should be avoided except in cases of compelling operational necessity.

Estimated Ranges in yards for given SPL and PEL for sonar.

SPL (dB)	PEL (MIN)	BSY-1 SQS-53C	BQQ-5 BSY-2 SQS-26CX(U) SQS-53A, SQS-53B SQS-56(U)	SQS-23 SQS-26AX SQS-26BX, SQS-26CX SQS-56		
215	13	56	40	13	A	
214	15	63	45	14	V	E
213	18	71	50	16	O	X
212	21	79	56	18	I	P
211	25	89	63	20	D	O
210	30	100	71	22		S
209	36	112	79	25	T	U
208	42	126	89	28	H	R
207	50	141	100	32	I	E
206	60	158	112	35	S	
205	71	178	126	40		
204	85	200	141	45		
203	101	224	158	50		
202	120	251	178	56		
201	143	282	200	63		
200	170	316	224	71		
199	202	355	251	79		
198	240	398	282	89		
197	285	447	316	100		
196	339	501	355	112		
195	404	562	398	126		
194	480	631	447	141		
193	571	708	501	158		
192	679	794	562	178		
191	807	891	631	200		
190	960	1,000	708	224		

All ranges and SPLs are nominal.

*SPL is measured in dB/1 μPA at the dive site. To convert SPL for sound levels referenced to mbar, subtract 100 dB from tabled levels.

(U) = upgrade

Table 1A-5. *Helmeted.*

Permissible Exposure Limit (PEL) within a 24-hour period for exposure to AN/SQS-23, -26, -53, -56, AN/BSY-1, -2, and AN/BQQ-5 sonar, including versions and upgrades. Exposure conditions shown above the double line should be avoided except in cases of compelling operational necessity.

Estimated Ranges in yards for given SPL and PEL for sonar.

SPL (dB)	PEL (MIN)	BSY-1 SQS-53C	BQQ-5 BSY-2 SQS-26CX(U) SQS-53A, SQS-53B SQS-56(U)	SQS-23 SQS-26AX SQS-26BX, SQS-26CX SQS-56	
183	13	2,239	1,585	501	AVOID THIS EXPOSURE
182	15	2,512	1,778	562	
181	18	2,818	1,995	631	
180	21	3,162	2,239	708	
179	25	3,548	2,512	794	
178	30	3,981	2,818	891	
177	36	4,467	3,162	1,000	
176	42	5,012	3,548	1,122	
175	50	5,623	3,981	1,259	
174	60	6,310	4,467	1,413	
173	71	7,079	5,012	1,585	
172	85	7,943	5,623	1,778	
171	101	8,913	6,310	1,995	
170	120	10,000	7,079	2,239	
169	143	11,220	7,943	2,512	
168	170	12,589	8,913	2,818	
167	202	14,125	10,000	3,162	
166	240	15,849	11,220	3,548	
165	285	17,783	12,589	3,981	
164	339	19,953	14,125	4,467	
163	404	22,387	15,849	5,012	
162	480	25,119	17,783	5,623	
161	571	28,184	19,953	6,310	
160	679	31,623	22,387	7,079	
159	807	35,481	25,119	7,943	
158	960	39,811	28,184	8,913	

All ranges and SPLs are nominal.

*SPL is measured in dB/1 μPA at the dive site. To convert SPL for sound levels referenced to mbar, subtract 100 dB from tabled levels.

(U) = upgrade

Table 1A-6. *Permissible Exposure Limit (PEL) Within a 24-hour Period for Exposure to AN/SQQ-14, -30, -32 Sonars.*

Estimated Ranges in yards for given SPL and PEL for sonar.

WET SUIT UN-HOODED		
SPL (dB)	PEL (MIN)	Range (yards)
200	120	13
199	143	14
198	170	16
197	202	18
196	240	20
195	285	22
194	339	25
193	404	28
192	480	32
191	571	35
190	679	40
189	807	45
188	960	50
WET SUIT HOODED		
SPL (dB)	**PEL (MIN)**	**Range (yards)**
215	120	2
214	143	3
213	170	3
212	202	3
211	240	4
210	285	4
209	339	4
208	404	5
207	480	6
206	571	6
205	679	7
204	807	8
203	960	9

Dry suit helmeted divers: no restriction for these sonars. All ranges and SPLs are nominal.

*SPL is measured in dB/1 μPA at the dive site. To convert SPL for sound levels referenced to mbar, subtract 100 dB from tabled levels.

1A-5 GUIDANCE FOR DIVER EXPOSURE TO LOW-FREQUENCY SONAR (160–320 Hz)

If possible, you should avoid diving in the vicinity of low-frequency sonar (LFS). LFS generates a dense, high-energy pulse of sound that can be harmful at higher power levels. Because a variety of sensations may result from exposure to LFS, it is necessary to inform divers when exposure is likely and to brief them regarding possible effects; specifically, that they can expect to hear and feel it. Sensations may include mild dizziness or vertigo, skin tingling, vibratory sensations in the throat and abdominal fullness. Divers should also be briefed that voice communications are likely to be affected by the underwater sound to the extent that line pulls or other forms of communication may become necessary. Annoyance and effects on communication are less likely when divers are wearing a hard helmet (MK 21) diving rig. For safe distance guidance, contact NAVSEA (00C3). Telephone numbers are listed in Volume 1, Appendix C.

1A-6 GUIDANCE FOR DIVER EXPOSURE TO ULTRASONIC SONAR (250 KHz AND GREATER)

The frequencies used in ultrasonic sonars are above the human hearing threshold. The primary effect of ultrasonic sonar is heating. Because the power of ultrasonic sonar rapidly falls off with distance, a safe operating distance is 10 yards or greater. Dive operations may be conducted around this type of sonar provided that the diver does not stay within the sonar's focus beam. The diver may finger touch the transducer's head momentarily to verify its operation as long as the sonar is approached from the side.

APPENDIX 1B
References

References	Subject
BUMEDINST 6200.15	Suspension of Diving During Pregnancy
BUMEDINST 6320.38	Clinical Use of Recompression Chambers for Non-Diving Illnesses: Policy for
Manual of the Medical Department, Article 15-66	Medical Examinations
MILPERSMAN Article 1220	Military Personnel Manual
NAVEDTRA 10669-C	Hospital Corpsman 3 & 2
NAVFAC P-990	UCT Conventional Inspection and Repair Techniques
NAVFAC P-991	Expedient Underwater Repair Techniques
NAVFAC P-992	UCT Arctic Operations Manual
NAVMED P-5010	Manual of Naval Preventive Medicine
NAVSEA 10560 ltr, Ser 00C34/3160 of 27 Sept 01	UBA Canister Duration
NAVSEA/00C ANU, www.navsea.navy.mil/sea00c/doc/anu_disc.html	Authorized for Navy Use
NAVSEA (SS521-AA-MAN-010)	U.S. Navy Diving and Manned Hyperbaric System Safety Certification Manual
NAVSEA Technical Manual (S0600-AA-PRO-010)	Underwater Ship Husbandry Manual
NAVSEA Technical Manual (SS500-HK-MMO-010)	MK 3 MOD 0 Light Weight Diving System Operating and Maintenance
NAVSEA Technical Manual (SS500-AW-MMM-010)	MK 6 MOD 0 Transportable Recompression Chamber System Operating and Maintenance
NAVSEA Technical Manual (SS600-AA-MMA-010)	MK 16 MOD 0 Operating and Maintenance
NAVSEA Technical Manual (SS600-AQ-MMO-010)	MK 16 MOD 1 Operating and Maintenance
NAVSEA Technical Manual (SS-600-A3-MMO-010)	MK 25 MOD 2 UBA Operating and Maintenance
NAVSEA Technical Manual (S9592-B1-MMO-010)	Fly Away Dive System (FADS) III Air System Operating and Maintenance
NAVSEA Technical Manual (SS9592-B2-MMO-010)	Fly Away Dive System (FADS) III Mixed Gas System (FMGS) Operating and Maintenance
NAVSEA Technical Manual (S9592-AN-MMO-010)	Emergency Breathing System Type I Operating and Maintenance
NAVSEA Technical Manual (0938-LP-011-4010)	Nuclear Powered Submarine Atmosphere Control Manual
NAVSEA Technical Manual (S9592-AY-MMO-020)	MK 5 MOD 0 Flyaway Recompression Chamber (FARCC)
NAVSEA Technical Manual (SS500-B1-MMO-010)	Standard Navy Double-Lock Recompression Chamber System
NAVSEA Technical Manual (SH700-A2-MMC-010)	Emergency Hyperbaric Stretcher Operations and Maintenance
NAVSEA Technical Manual (SS521-AJ-PRO-010)	Guidance for Diving in Contaminated Waters

Naval Ships Technical Manual, Chapter 74, Vol. 1 (S9086-CH-STM-010)	Welding and Allied Processes
Naval Ships Technical Manual, Chapter 74, Vol. 3 (S9086-CH-STM-030)	Gas Free Engineering
Naval Ships Technical Manual, Chapter 262 (S9086-H7-STM-010)	Lubricating Oils, Greases, Specialty Lubricants, and Lubrication Systems
Naval Ships Technical Manual, Chapter 550 (S9086-SX-STM-010)	Industrial Gases, Generating, Handling, and Storage
NAVSEA Operation & Maintenance Instruction (0910-LP-001-6300)	Fly Away Diving System Filter/Console
NAVSEA Operation & Maintenance Instruction (0910-LP-001-1500)	Fly Away Diving System Diesel Driven Compressor Unit EX 32 MOD 0, PN 5020559
Naval Safety Center Technical Manual	Guide to Extreme Cold Weather
NAVSEA Technical Manual (S0300-A5-MAN-010)	Polar Operations Manual
Office of Naval Research Technical Manual	Guide to Polar Diving
ASTM G-88-90	Standard Guide for Designing Systems for Oxygen Service
ASTM G-63-92	Standard Guide for Evaluating Nonmetallic Materials for Oxygen Service
ASTM G-94-92	Standard Guide for Evaluating Metals for Oxygen Service
FED SPEC BB-A-1034 B	Diver's Compressed Air Breathing Standard
FED SPEC A-A-59503	Compressed Nitrogen Standard
MIL-D -16791	Detergents, General Purpose (Liquid, Nonionic)
MIL-PRF-27210G	Oxygen, Aviators Breathing, Liquid and Gaseous
MIL-PRF-27407B	Propellant Pressurizing Agent Helium, Type I Gaseous Grade B
MIL-STD-438	Schedule of Piping, Valves and Fittings, and Associated Piping Components for Submarine Service
MIL-STD-777	Schedule of Piping, Valves and Fittings, and Associated Piping Components for Naval Surface Ships
MIL-STD-1330	Cleaning and Testing of Shipboard Oxygen, and Nitrogen Systems Helium, Helium - Oxygen
OPNAVINST 3120.32C CH-1	Equipment Tag-Out Bill
OPNAVINST 3150.27 Series	Navy Diving Program
OPNAVINST 5100.19C, Appendix A-6	Navy Occupational Safety and Health (NAVOSH) Program Manual for Forces Afloat
OPNAVINST 5100.23	Navy Occupational Safety and Health (NAVOSH) Afloat Program Manual
OPNAVINST 5102.1C CH-1	Mishap Investigation and Reporting
OPNAVINST 8023.2C CH-1	U.S. Navy Explosives Safety Policies, Requirements, and Procedures (Department of the Navy Explosives Safety Policy Manual)
OSHA 29 CFR Part 1910 Subpart T, PG 6-36	Commercial Diving Operations
MIL-PRF-17331	Lubricant (2190 TEP)

MIL-PRF-17672	Lubricant (2135 TH)
ANSI-B57.1 and CSA-B96	American and Canadian Standard Compressed-Gas Cylinder Valve Outlet and Inlet Connections
Z48.1	American National Standard Method of Marking Portable Compressed-Gas Containers to Identify the Material Contained
CGA Pamphlet C-7	Guide to the Preparation of Precautionary Labeling and Marking of Compressed Gas Cylinders

PAGE LEFT BLANK INTENTIONALLY

APPENDIX 1C

Telephone Numbers

Command	Department	Telephone	Fax
Naval Surface Warfare Center - Panama City, Florida (NSWC-PC)	Diver Life Support (Fleet Support & Air Sampling	(850) 234-4482 DSN: 436-4482	(850) 234-4775
BUMED M3B42		(202) 762-3444	
National Oceanic and Atmospheric Administration (NOAA)	HAZMAT	(206) 526-6317	(206) 526-6329
Naval Sea Systems Command (COMNAVSEASYSCOM)		(202) 781-XXXX DSN: 326-XXXX	(202) 781-4588
00C	Director	(202) 781-0731	
00C1	Finance	(202) 781-0648	
00C2	Salvage	(202) 781-2736	
00C3	Diving	(202) 781-0934	
00C4	Certification	(202) 781-0927	
00C5	Husbandry	(202) 781-3453	
Naval Sea Systems Command Code 07Q	Deep Submergence Systems Certification	(202) 781-1467 (202) 781-1336	
NAVFAC Ocean Facilities Program	(Code OFP)	(202) 433-5596 DSN 288-5596.	(202) 433-2280

PAGE LEFT BLANK INTENTIONALLY

APPENDIX 1D

List of Acronyms

ABS	Acrylonitrile Butadiene Styrene
ACF	Actual Cubic Feet
ACFM	Actual Cubic Feet per Minute
ACGIH	American Conference of Governmental Industrial Hygienists
ACLS	Advanced Cardiac Life Support
ADS	Advance Diving System
AGE	Arterial Gas Embolism
ALSS	Auxiliary Life-Support System
AM	Amplitude Modulated
ANU	Authorized for Navy Use List
AQD	Additional Qualification Designator
ARD	Audible Recall Device
AS	Submarine Tender
ASDS	Advanced SEAL Delivery System
ASRA	Air Supply Rack Assembly
ASME	American Society of Mechanical Engineers
ATA	Atmosphere Absolute
ATP	Ambient Temperature and Pressure
ATS	Active Thermal System
BC	Buoyancy Compensator
BCLS	Basic Cardiac Life Support
BIBS	Built-In Breathing System
BPM	Breaths per Minute

BTPS	Body Temperature, Ambient Pressure
BTU	British Thermal Unit
CDO	Command Duty Officer
CCTV	Closed-Circuit Television
CGA	Compressed Gas Association
CNO	Chief of Naval Operations
CNS	Central Nervous System
CONUS	Continental United States
COSAL	Coordinated Shipboard Allowance List
CPR	Cardiopulmonary Resuscitation
CRS	Chamber Reducing Station
CSMD	Combat Swimmer Multilevel Dive
CUMA	Canadian Underwater Minecountermeasures Apparatus
CWDS	Contaminated Water Diving System
DATPS	Divers Active Thermal Protection System
DC	Duty Cycle
DCS	Decompression Sickness
DDC	Deck Decompression Chamber
DDS	Deep Diving System
DDS	Dry Deck Shelter
DHMLS	Divers Helmet Mounted Lighting System
DLSE	Diving Life-Support Equipment
DLSS	Divers Life Support System
DMO	Diving Medical Officer
DMS	Dive Monitoring System
DMT	Diving Medical Technician

DOT	Department of Transportation
DRS	Dive Reporting System
DSI	Diving Systems International
DSM	Diving System Module
DSRG	Deep Submergence Review Group
DSRV	Deep Submergence Rescue Vehicle
DSSP	Deep Submergence System Project
DT	Dive Time or Descent Time
DT/DG	Dive Timer/Depth Gauge
DUCTS	Divers Underwater Color Television System
DV	Diver
DPV	Diver Propulsion Vehicle
EAD	Equivalent Air Depth
EBA	Emergency Breathing Apparatus
EBS I	Emergency Breathing System I
EDWS	Enhanced Diver Warning System
EEHS	Emergency Evacuation Hyperbaric Stretcher
EGS	Emergency Gas Supply
ENT	Ear, Nose, and Throat
EOD	Explosive Ordnance Disposal
EPs	Emergency Procedures
ESDS	Enclosed Space Diving System
ESDT	Equivalent Single Dive Time
ESSM	Emergency Ship Salvage Material
FADS III	Flyaway Air Dive System III
FAR	Failure Analysis Report

FARCC	Flyaway Recompression Chamber
FED SPEC	Federal Specifications
FFM	Full Face Mask
FFW	Feet of Fresh Water
FMGS	Flyaway Mixed-Gas System
FPM	Feet per Minute
FSW	Feet of Sea Water
FV	Floodable Volume
GFI	Ground Fault Interrupter
GPM	Gallons per Minute
HBO2	Hyperbaric Oxygen
HOSRA	Helium-Oxygen Supply Rack Assembly
HP	High Pressure
HPNS	High Pressure Nervous Syndrome
HSU	Helium Speech Unscrambler
ICCP	Impressed-Current Cathodic Protection
IDV	Integrated Divers Vest
IL	Inner Lock
ILS	Integrated Logistics Support
ISIC	Immediate Senior in Command
JAG	Judge Advocate General
J/L	Joules per Liter, Unit of Measure for Work of Breathing
KwHr	Kilowatt Hour
LB	Left Bottom
LCM	Landing Craft, Medium
LFA	Low Frequency Acoustic

LFS	Low Frequency Sonar
LP	Low Pressure
LPM	Liters per Minute
LS	Left Surface
LSS	Life Support System or Life Support Skid
LWDS	Light Weight Diving System
MBC	Maximal Breathing Capacity
MCC	Main Control Console
MD	Maximum Depth
MDSU	Mobile Diving and Salvage Unit
MDV	Master Diver
MEFR	Maximum Expiratory Flow Rate
MEV	Manual Exhaust Valve
MFP	Minimum Flask Pressure
MGCCA	Mixed-Gas Control Console Assembly
MIFR	Maximum Inspiratory Flow Rate
MIL-STD	Military Standard
MMP	Minimum Manifold Pressure
MP	Medium Pressure
MRC	Maintenance Requirement Card
MSW	Meters of Sea Water
MVV	Maximum Ventilatory Volume
NAVEDTRA	Naval Education Training
NAVFAC	Naval Facilities Engineering Command
NAVMED	Naval Medical Command
NAVSEA	Naval Sea Systems Command

ND	Noise Dose
NDSTC	Naval Diving and Salvage Training Center
NEC	Navy Enlisted Classification
NEDU	Navy Experimental Diving Unit
NEURO	Neurological Examination
NID	Non-Ionic Detergent
NITROX	Nitrogen-Oxygen
NMRI	Navy Medical Research Institute
NOAA	National Oceanic and Atmospheric Administration
NO-D	No Decompression
NPC	Naval Personnel Command
NRV	Non Return Valve
NSMRL	Navy Submarine Medical Research Laboratory
NSN	National Stock Number
NSTM	Naval Ships Technical Manual or NAVSEA Technical Manual
NSWC-PC	Naval Surface Warfare Center - Panama City
O&M	Operating and Maintenance
OBP	Over Bottom Pressure
OCEI	Ocean Construction Equipment Inventory
OIC	Officer in Charge
OJT	On the Job Training
OL	Outer Lock
OOD	Officer of the Deck
OPs	Operating Procedures
OSF	Ocean Simulation Facility
OSHA	Occupational Safety and Health Administration

PEL	Permissible Exposure Limit
PMS	Planned Maintenance System
PNS	Peripheral Nervous System
PP	Partial Pressure
PPCO2	Partial Pressure Carbon Dioxide
PPM	Parts per Million
PPO2	Partial Pressure Oxygen
PSI	Pounds per Square Inch
PSIA	Pounds per Square Inch Absolute
PSIG	Pounds per Square Inch Gauge
PSOB	Pre-Survey Outline Booklet
PTC	Personnel Transfer Capsule
PTS	Passive Thermal System
QA	Quality Assurance
RB	Reached Bottom
RCC	Recompression Chamber
REC	Re-Entry Control
RMV	Respiratory Minute Ventilation
RNT	Residual Nitrogen Time
ROV	Remotely Operated Vehicle
RQ	Respiratory Quotient
RS	Reached Surface
RSP	Render Safe Procedure
SAD	Safe Ascent Depth
SCA	System Certification Authority
SCF	Standard Cubic Feet

SCFM	Standard Cubic Feet per Minute
SCFR	Standard Cubic Feet Required
SCSCs	System Certification Survey Cards
SCUBA	Self Contained Underwater Breathing Apparatus
SDRW	Sonar Dome Rubber Window
SDS	Saturation Diving System
SDV	SEAL Delivery Vehicle
SEAL	Sea, Air, and Land
SET	Surface Equivalent Table
SEV	Surface Equivalent (percent or pressure)
SI	Surface Interval or System International
SLED	Sea Level Equivalent Depth
SLM	Standard Liters per Minute (short version used in formulas)
SLPM	Standard Liters per Minute
SNDB	Standard Navy Dive Boat
SOC	Scope of Certifications
SPL	Sound Pressure Level
SRDRS	Submarine Rescue and Diver Recompression System
SSB	Single Side Band
SSDS	Surface Supplied Diving System
STEL	Safe Thermal Exposure Limits
STP	Standard Temperature and Pressure
STPD	Standard Temperature and Pressure, Dry Gas
SUR D	Surface Decompression
SUR D AIR	Surface Decompression Using Air
SUR D O2	Surface Decompression Using Oxygen

T-ARS	Auxiliary Rescue/Salvage Ship
T-ATF	Fleet Ocean Tug
TBT	Total Bottom Time
TDCS	Tethered Diver Communication System
TDT	Total Decompression Time
TL	Transfer Lock
TLC	Total Lung Capacity
TLD	Thermal Luminescence Dosimeter
TLV	Threshold Limit Values
TM	Technical Manual
TMDER	Technical Manual Deficiency Evaluation Report
TRC	Transportable Recompression Chamber
TRCS	Transportable Recompression Chamber System
TTD	Total Time of Dive
UBA	Underwater Breathing Apparatus
UCT	Underwater Construction Team
UDM	Underwater Decompression Monitor
UQC	Underwater Sound Communications
UWSH	Underwater Ship Husbandry
VENTIDC	Vision Ear Nausea Twitching Irritability Dizziness Convulsions
VTA	Volume Tank Assembly
VVDS	Variable Volume Dry Suit
WOB	Work of Breathing
YDT	Diving Tender

PAGE LEFT BLANK INTENTIONALLY

VOLUME 2

Air Diving Operations

U.S. NAVY DIVING MANUAL

PAGE LEFT BLANK INTENTIONALLY

Volume 2 - Table of Contents

Chap/Para | Page

Chap/Para | Page

Chap/Para | Page

Chap/Para		Page

Chap/Para Page

Chap/Para | Page

Volume 2 - List of Illustrations

PAGE LEFT BLANK INTENTIONALLY

Volume 2 - List of Tables

PAGE LEFT BLANK INTENTIONALLY

CHAPTER 6

Operational Planning and Risk Management

6-1 INTRODUCTION

6-1.1 **Purpose.** Diving operations are inherently risky. This chapter provides a general guide for planning diving operations. All Naval activities shall apply the Operational Risk Management (ORM) process in planning operations and training to optimize operational capability and readiness in accordance with OPNAV INSTRUCTION 3500.39 (series). Correct application of these techniques will reduce mishaps and associated costs resulting in more efficient use of resources. ORM is a decision making tool used by personnel at all levels to increase operational effectiveness by identifying, assessing, and managing risks. Proper application of ORM minimizes risks to acceptable levels, commensurate with mission accomplishment. The amount of risk we will accept in war is much greater than that we should accept in peace, but the ORM process remains the same.

6-1.2 **Scope.** This chapter outlines a comprehensive planning process to effectively plan and execute diving operations in support of military operations. The planning worksheets and checklists contained in this chapter are examples of U.S. Navy material. They may be used as provided or modified locally to suit specific needs.

6-2 MISSION OBJECTIVE AND OPERATIONAL TASKS

A clear and concise statement of the mission objective shall be established. If the officer planning the operation is unclear about the urgency of the mission objective, he or she shall obtain clarification from the tasking authority to determine acceptable risks.

Example: Locate, recover, and deliver lost anchor to USS SMITH at Pier A.

This section outlines the primary diving functions that may be identified in an operational task. These functions may be incorporated singly or in conjunction with others. Each task shall be identified and placed in the context of an overall schedule or job profile. Work items that must be coordinated with other support teams shall also be identified. The availability of outside assistance, including assistance for possible emergencies, from a diving unit or other sources must be coordinated in advance.

6-2.1 **Underwater Ship Husbandry (UWSH).** UWSH is the inspection, maintenance, and repair of Navy hulls and hull appendages while the hulls are waterborne. UWSH includes tasks such as patching, plugging, attaching cofferdams, waterborne hull cleaning, underwater weld repair to ship's hulls and appendages, propeller replacement, underwater hull inspection, and nondestructive testing (Figure 6-1).

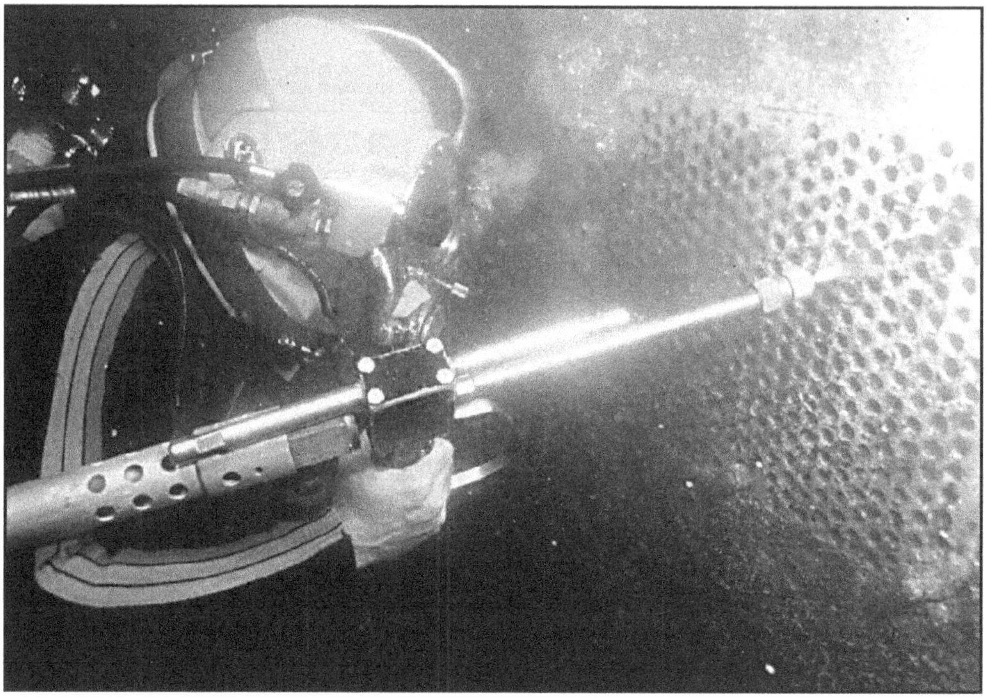

Figure 6-1. Underwater Ship Husbandry Diving.

6-2.1.1 **Objective of UWSH Operations.** The objective of all UWSH operations is to provide a permanent repair without dry-docking the ship. When a permanent repair is not possible, temporary repairs are performed to allow the ship to operate until its next scheduled drydocking where permanent repairs can be accomplished.

6-2.1.2 **Repair Requirements.** All UWSH repairs shall follow strict Quality Assurance (QA) procedures to ensure underwater systems are properly repaired. Divers shall work closely with all other repair activities to ensure procedures comply with prescribed ship design and maintenance specifications. All relevant technical manuals shall be made available for dive planning, and individual diver background and expertise shall be considered when assembling dive teams. The *NAVSEA Underwater Ship Husbandry Manual* (S0600-AA-PRO-010) provides general guidance and specific procedures to accomplish many underwater repairs.

6-2.1.3 **Diver Training and Qualification Requirements.** Many UWSH training requirements and qualifications are task specific. General training may be accomplished by:

- Formalized instruction as in First or Second Class Dive School
- NAVSEA-sponsored training, e.g., Sonar Dome Rubber Window (SDRW) Repair
- On the Job Training (OJT)
- Personnel Qualification Standards (PQS)

6-2.1.4 **Training Program Requirements.** A proper training program should result in permanent repairs meeting the same tolerances and QA requirements as if performed in dry-dock. If there are any questions as to the qualifications required for a permanent repair, divers should consult with their command repair department or contact NAVSEA 00C5.

6-2.1.5 **Ascent Training and Operations.** Ascent operations are conducted by qualified divers or combat swimmers. These operations require the supervision of an Ascent Supervisor but operational conditions preclude the use of instructors. Ascent training is distinctly different from ascent operations as performed by Navy Special Warfare groups.

No ascent training may be conducted unless fully qualified instructors are present, recompression chamber is available within 10 minutes, Diving Medical Technician is on station, and a Diving Medical Officer is able to provide immediate response to an accident.

6-2.2 **Salvage/Object Recovery.** In a salvage or object-recovery operation, divers work to recover sunken or wrecked naval craft, submersibles, downed aircraft, human remains, or critical items of equipment to help determine the cause of a mishap. Salvaged items may include classified or sensitive materials (Figure 6-2).

6-2.3 **Search Missions.** Underwater searches are conducted to locate underwater objects or subsurface geological formations. Searches can be performed by various methods depending on the undersea terrain and purpose of the mission. Because using divers for an unaided visual search over a large area is time consuming and labor intensive, this type of search operation should incorporate the use of sidescan sonar and other search equipment whenever possible. Remotely Operated Vehicles (ROVs) may be used to extend searches into deep waters and areas that are particularly dangerous for a diver. A reconnaissance dive may be conducted prior to other scheduled dives to gather information that can save in-water time and identify any special hazards of the dive mission.

6-2.4 **Explosive Ordnance Disposal.** Divers perform Explosive Ordnance Disposal tasks including recovering, identifying, disarming, and disposing of explosive devices that must be cleared from harbors, ships, and sea-lanes (Figure 6-3). Diving in the vicinity of ordnance combines the risks of diving and the explosive hazards of the ordnance. EOD divers shall accomplish diving to investigate, render safe, or dispose of explosive ordnance found underwater, regardless of type or fusing. Refer to Chapter 18 for more information on EOD operations.

6-2.5 **Security Swims.** Security swims are employed to search for underwater explosives or other devices that may have been attached to ships or piers. All qualified divers may conduct ship security swims. Once a task is identified as involving ordnance disposal, the area shall be marked. If EOD qualified personnel are not on site they shall be requested. Only EOD personnel may attempt to handle or dispose of underwater explosives.

Figure 6-2. Salvage Diving. Surface-supplied divers on an aircraft recovery mission.

Figure 6-3. Explosive Ordnance Disposal Diving. An EOD diver using handheld sonar to locate objects underwater.

6-2.6 **Underwater Construction.** Underwater construction is the construction, inspection, repair, and removal of in-water facilities in support of military operations. An in-water facility can be defined as a fixed harbor, waterfront, or ocean structure located in or near the ocean. Pipelines, cables, sensor systems, and fixed/advanced-base structures are examples of in-water facilities (Figure 6-4).

6-2.6.1 **Diver Training and Qualification Requirements.** Seabee divers are specifically trained in the special techniques used to accomplish underwater construction tasks.

6-2.6.2 **Equipment Requirements.** Tools and equipment used include common underwater tools in addition to specialized ocean construction equipment. Specific tools and components for large ocean engineering projects are maintained in the Ocean Construction Equipment Inventory (OCEI) located at St. Julian Creek, Norfolk, Virginia.

6-2.6.3 **Underwater Construction Planning Resources.** References for underwater construction planning can be found in:

- *UCT Conventional Inspection and Repair Techniques Manual* NAVFAC P-990
- *Expedient Underwater Repair Techniques* NAVFAC P-991
- *UCT Arctic Operations Manual* NAVFAC P-992
- *Design and Installation of Nearshore Ocean Cable Protection Systems* FPO-1-78(3)

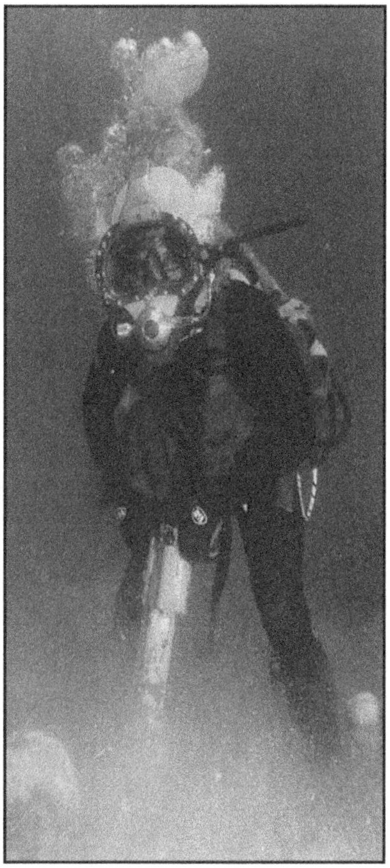

Figure 6-4. Underwater Construction Diving.

For more information on ocean construction, commands should consult NAVFAC Ocean Facilities Program.

6-2.7 **Demolition Missions.** Diving operations may include demolition duties to remove man-made structures such as barriers, sunken naval craft, and damaged piers. Blasting, freeing, flattening, or cutting with explosives define demolition operations. Divers may also be assigned to destroy natural formations, such as reefs, bars, and rock structures that interfere with transportation routes. All personnel involved in handling explosives shall be qualified in accordance with the OPNAVINST 8023.2 series.

6-2.8 **Combat Swimmer Missions.** Combat swimmers conduct reconnaissance and neutralization of enemy ships, shore-based installations, and personnel. Some missions may require an underwater approach to reach coastal installations undetected. Reconnaissance missions and raids may expose the combat swimmers to additional risk but may be necessary to advance broader warfare objectives.

6-2.9 **Enclosed Space Diving.** Divers are often required to work in enclosed or confined spaces. Using surface-supplied Underwater Breathing Apparatus (UBA) (MK 20 MOD 0, MK 21 MOD 1, KM-37, or EXO BR MS), divers may enter submarine ballast tanks, mud tanks, or cofferdams, which may be in either a flooded or dry condition. Access to these spaces is normally restrictive, making it difficult for the diver to enter and exit. Enclosed space diving shall be supported by a surface-supplied air system. Refer to Section 8-11.4 for more information on the hazards of enclosed space diving.

6-3 GENERAL PLANNING AND ORM PROCESS

A successful diving mission is the direct outcome of careful, thorough planning. The nature of each operation determines the scope of the planning effort, but certain general considerations apply to every operation.

- Bottom Time. Bottom time is always at a premium. Developing measures to conserve bottom time or increase diver effectiveness is critical for success.
- Preplanning. An operation that is delayed due to unanticipated problems may fail. Preplanning the use of the time available to accomplish specific objectives is a prerequisite to success.
- Equipment. Selecting the correct equipment for the job is critical to success.
- Environmental Conditions. Diving operational planners must plan for safely mitigating extreme environmental conditions. Personnel and support facility safety shall be given the highest priority.
- Diver Protection. It is critical to protect divers from all anticipated hazards. Application of the ORM process will identify hazards prior to the operation.
- Emergency Assistance. It is critical to coordinate emergency assistance from outside sources before the operation begins.
- Weather. Because diving operations are weather dependent, dive planning shall allow for worst-case scenarios.

6-3.1 Concept of ORM:

- ORM is a decision making tool used by people at all levels to increase operational effectiveness by anticipating hazards and reducing the potential for loss, thereby increasing the probability of successful mission.
- Increases our ability to make informed decisions by providing the best baseline of knowledge and experience available.
- Minimizes risks to acceptable levels, commensurate with mission accomplishment. The amount of risk we will take in war is much greater than that we should be willing to take in peace, but the process is the same. Applying the ORM process will reduce mishaps, lower costs, and provide for more efficient use of resources.

6-3.2 Risk Management Terms:

- Hazard – A condition with potential to cause personal injury or death, property damage, or mission degradation.
- Risk – An expression of possible loss in terms of severity and probability.

- Risk Assessment – The process of detecting hazards and assessing associated risks.
- ORM – The process of dealing with risk associated within military operations, which includes risk assessment, risk decision-making and implementation of effective risk controls.

6-3.3 **ORM Process.** The five step process is:

1. Identify Hazards – Begin with an outline or chart of the major steps in the operation (operational analysis). Next, conduct a Preliminary Hazard Analysis by listing all of the hazards associated with each step in the operational analysis along with possible causes for those hazards.

2. Assess Hazards – For each hazard identified, determine the associated degree of risk in terms of probability and severity. Although not required; the use of a matrix may be helpful in assessing hazards.

3. Make Risk Decisions – First, develop risk control options. Start with the most serious risk first and select controls that will reduce the risk to a minimum consistent with mission accomplishment. With selected controls in place, decide if the benefit of the operation outweighs the risk. If risk outweighs benefit or if assistance is required to implement controls, communicate with higher authority in the chain of command.

4. Implement Controls – The following measures can be used to eliminate hazards or reduce the degree of risk. These are listed by order of preference:
   - Administrative Controls – Controls that reduce risks through specific administrative actions, such as:
   - Providing suitable warnings, markings, placards, signs, and notices.
   - Establishing written policies, programs, instructions and standard operating procedures (SOP).
   - Training personnel to recognize hazards and take appropriate precautionary measures.
   - Limiting the exposure to hazard (either by reducing the number or personnel/assets or the length of time they are exposed).
   - Engineering Controls – Controls that use engineering methods to reduce risks by design, material selection or substitution when technically or economically feasible.
   - Personal Protective Equipment – Serves as a barrier between personnel and hazard. It should be used when other controls do not reduce the hazard to an acceptable level.

5. Supervise – conduct follow-up evaluations of the controls to ensure they remain in place and have the desired effect. Monitor for changes, which may require further ORM. Take corrective action when necessary.

6-4 COLLECT AND ANALYZE DATA

Information pertinent to the mission objective shall be collected, organized, and analyzed to determine what may affect successful accomplishment of the objective. This process aids in:

- Planning for contingencies
- Developing the dive plan
- Selecting diving technique, equipment, and diver personnel
- Identifying potential hazards and the need for any special emergency procedures

6-4.1 **Information Gathering.** The size of the operation, the diving site location, and the prevailing environmental conditions influence the extent and type of information that must be gathered when planning an operation. Some operations are of a recurring nature; so much of the required information is readily available. An example of a recurring operation is removing a propeller from a particular class of ship. However, even for a standard operation, the ship may have been modified or special environmental conditions may exist, requiring a change in procedure or special tools. Potential changes in task requirements affecting work procedures should not be overlooked during planning.

6-4.2 **Planning Data.** Many operations require that detailed information be collected in advance. For example, when planning to salvage a sunken or stranded vessel, the diving team needs to know the construction of the ship, the type and location of cargo, the type and location of fuel, the cause of the sinking or stranding, and the nature and degree of damage sustained. Such information can be obtained from ship's plans, cargo manifests and loading plans, interviews with witnesses and survivors, photographs, and official reports of similar accidents.

6-4.3 **Object Recovery.** Operations involving the recovery of an object from the bottom require knowledge of the dimensions and weight of the object. Other useful information includes floodable volume, established lifting points, construction material, length of time on the bottom, probable degree of embedment in mud or silt, and the nature and extent of damage. This data helps determine the type of lift to be used (e.g., boom, floating crane, lifting bags, pontoons), indicates whether high-pressure hoses are needed to jet away mud or silt, and helps determine the disposition of the object after it is brought to the surface. Preliminary planning may find the object too heavy to be placed on the deck of the support ship, indicating the need for a barge and heavy lifting equipment.

6-4.3.1 **Searching for Objects or Underwater Sites.** When the operation involves searching for an object or underwater site, data gathered in advance helps to limit the search area. There are numerous planning data sources available to help supervisors collect data for the operation (see Figure 6-5).

PLANNING DATA SOURCES		
■ Aircraft Drawings ■ Cargo Manifest ■ Coastal Pilot Publications ■ Cognizant Command ■ Communications Logs ■ Construction Drawings ■ Current Tables ■ Diving Advisory Messages ■ DRT Tracks ■ DSV/DSRV Observations ■ Electronic Analysis ■ Equipment Operating Procedures (OPs) ■ Equipment Operation and Maintenance Manuals ■ Eyewitnesses ■ Flight or Ship Records ■ Flight Plan ■ Hydrographic Publications	■ Light Lists ■ Local Yachtsmen/Fishermen ■ LORAN Readings ■ Magnetometer Plots ■ Navigation Text (Dutton's/Bowditch) ■ Navigational Charts ■ NAVOCEANO Data ■ Notices to Mariners ■ OPORDERS ■ Photographs ■ Radar Range and Bearings ■ RDF Bearings ■ ROV Video and Pictures ■ Sailing Directions ■ Salvage Computer Data ■ Ship's Curves of Forms ■ Ship's Equipment ■ Ship's Logs and Records	■ Ship's Personnel ■ Ships Drawings (including docking plan) ■ Side-Scan Sonar Plots ■ SINS Records ■ SITREP ■ Sonar Readings and/or Charts ■ TACAN Readings ■ Technical Reference Books ■ Test Records ■ Tide Tables ■ Underwater Work Techniques ■ USN Diving Manual Reference List ■ USN Instructions ■ USN Ship Salvage Manual ■ Visual Bearings ■ Weather Reports

Figure 6-5. Planning Data Sources.

For example, information useful in narrowing the search area for a lost aircraft includes the aircraft's last known heading, altitude, and speed; radar tracks plotted by ships and shore stations; tape recordings and radio transmissions; and eyewitness accounts. Once a general area is outlined, a side scan sonar system can be used to locate the debris field, and an ROV can identify target items located by the side scan sonar. Once the object of the search has been found, the site should be marked, preferably with an acoustic transponder (pinger) and/or a buoy. If time and conditions permit, preliminary dives by senior, experienced members of the team can be of great value in verifying, refining, and analyzing the data to improve the dive plan. This method saves diver effort for recovering items of interest.

6-4.4 **Data Required for All Diving Operations.** Data involving the following general categories shall be collected and analyzed for all diving operations:

- Surface conditions
- Underwater conditions
- Equipment and personnel resources
- Assistance in emergencies

6-4.4.1 **Surface Conditions.** Surface conditions in the operating area affect both the divers and the topside team members. Surface conditions are influenced by location, time of year, wind, waves, tides, current, cloud cover, temperature, visibility, and the presence of other ships. Completing the Environmental Assessment Worksheet (Figure 6-6) helps ensure that environmental factors are not overlooked during planning. For an extensive dive mission, a meteorological detachment may be requested from the local or regional meteorological support activity.

6-4.4.1.1 ***Natural Factors.*** Normal conditions for the area of operations can be determined from published tide and current tables, sailing directions, notices to mariners, and special charts that show seasonal variations in temperature, wind, and ocean currents. Weather reports and long-range weather forecasts shall be studied to determine if conditions will be acceptable for diving. Weather reports shall be continually monitored while an operation is in progress.

NOTE **Diving shall be discontinued if sudden squalls, electrical storms, heavy seas, unusual tide or any other condition exists that, in the opinion of the Diving Supervisor, jeopardizes the safety of the divers or topside personnel.**

6-4.4.1.2 ***Sea State.*** A significant factor is the sea state (Figure 6-7). Wave action can affect everything from the stability of the moor to the vulnerability of the crew to seasickness or injury. Unless properly moored, a ship or boat drifts or swings around an anchor, fouling lines and dragging divers. Because of this, any vessel being used to support surface-supplied or tended diving operations shall be secured by at least a two-point moor. Exceptions to diving from a two-point moor may occur when moored alongside a pier or another vessel that is properly anchored, or when a ship is performing diving during open ocean transits and cannot moor due to depth. A three- or four-point moor, while more difficult to set, may be preferred depending on dive site conditions.

Divers are not particularly affected by the action of surface waves unless operating in surf or shallow waters, or if the waves are exceptionally large. Surface waves may become a serious problem when the diver enters or leaves the water and during decompression stops near the surface.

6-4.4.1.3 ***Tender Safety.*** Effective dive planning shall provide for extreme temperatures that may be encountered on the surface. Normally, such conditions are a greater problem for tending personnel than for a diver. Any reduction in the effectiveness of the topside personnel may endanger the safety of a diver. Tending personnel shall guard against:

- Sunburn and windburn
- Hypothermia and frostbite
- Heat exhaustion

6-4.4.1.4 ***Windchill Factor.*** In cold, windy weather, the windchill factor shall be considered. Exposure to cold winds greatly increases dangers of hypothermia and all types of cold injury. For example, if the actual temperature is 35°F and the wind velocity is

ENVIRONMENTAL CHECKLIST

Date:____________

Surface

Atmosphere
Visibility ____________
Sunrise (set) ____________
Moonrise (set) ____________
Temperature (air) ____________
Humidity ____________
Barometer ____________
Precipitation ____________
Cloud Description ____________
Percent Cover ____________
Wind Direction ____________
Wind Force (knots) ____________
Other:
____
____
____

Sea Surface
Sea State ____________
Wave Action:
Height ____________
Length ____________
Direction ____________
Current:
Direction ____________
Velocity ____________
Type ____________
Surf. Visibility ____________
Surf. Water Temp. ____________
Local Characteristics
____
____
____

Subsurface

Underwater & Bottom
Depth ____________
Water Temperature:
________ depth ________
________ depth ________
________ depth ________
________ bottom ________
Thermoclines ____________
____

Curent:
Direction ____________
Source ____________
Velocity ____________
Pattern ____________
Tides:
High Water ______/______ Time
Low Water ______/______ Time
Ebb Dir. ______ Vel. ______
Flood Dir. ______ Vel. ______

Visibility
Underwater
ft ________ at ________ depth
ft ________ at ________ depth
ft ________ at ________ depth
Bottom
ft ________ at ________ depth
Bottom Type:
____

Obstructions:
____
____
____

Marine Life:
____
____
____

Other Data:
____
____

NOTE: A meteorological detachment may be requested from the local meteorological support activity.

Figure 6-6. Environmental Assessment Worksheet. The Environmental Assessment Worksheet indicates categories of data that might be gathered for an operation. Planners may develop an assessment methodology to suit the particular situation. The data collected is vital for effective operations planning, and is also of value when filing Post Salvage Reports.

Sea State	Description	Wind Force (Beaufort)	Wind Description	Wind Range (knots)	Wind Velocity (knots)	Average Wave Height (ft)
0	Sea like a mirror.	0	Calm	<1	0	0
	Ripples with the appearance of scales are formed, but without foam crests.	1	Light Air	5-3	2	0.05
1	Small wavelets still short but more pronounced; crests have a glassy appearance but do not break.	2	Light Breeze	4-6	5	0.18
2	Large wavelets, crests begin to break. Foam of glassy appearance, perhaps scattered whitecaps.	3	Gentle Breeze	7-10	8.5 10	0.6 0.88
3	Small waves, becoming longer; fairly frequent whitecaps.	4	Moderate Breeze	15-16	12 13.5 14 16	1.4 1.8 2.0 2.9
4	Moderate waves, taking a more pronounced long form; many whitecaps are formed. Chance of some spray.	5	Fresh Breeze	17-21	18 19 20	3.8 4.3 5.0
5	Large waves begin to form; white foam crests are more extensive everywhere. Some spray.	6	Strong Breeze	22-27	22 24 24.5 26	6.4 7.9 8.2 9.6
6	Sea heaps up and white foam from breaking waves begins to be blown in streaks along the direction of the wind. Spindrift begins.	7	Moderate Gale	28-33	28 30 30.5 32	11 14 14 16
7	Moderately high waves of greater length; edges of crests break into spindrift. The foam is blown in well marked streaks along the direction of the wind. Spray affects visibility.	8	Fresh Gale	34-40	34 36 37 38 40	19 21 23 25 28
8	High waves. Dense streaks of foam along the direction of the wind. Sea begins to roll. Visibility affected.	9	Strong Gale	45-47	42 44 46	31 36 40
9	Very high waves with long overhanging crests. Foam is in great patches and is blown in dense white streaks along the direction of the wind. The surface of the sea takes on a white appearance. The rolling of the sea becomes heavy and shock-like. Visibility is affected.	10	Whole Gale	48-55	48 50 51.5 52 54	44 49 52 54 59
	Exceptionally high waves. The sea is completely covered with long white patches of foam along the direction of the wind. Everywhere the edges of the wave crests are blown into froth. Visibility seriously affected.	11	Storm	56-63	56 59.5	64 73
	Air filled with foam and spray. Sea completely white with driving spray. Visibility seriously affected.	12	Hurricane	64-71	>64	>80

Figure 6-7. Sea State Chart.

35 mph, the windchill factor is equivalent to 5°F (Figure 6-8). For information on ice and cold water diving operations, refer to Chapter 11.

6-4.4.1.5 ***Surface Visibility.*** Variations in surface visibility are important. Reduced visibility may seriously hinder or force postponement of diving operations. For operations to be conducted in a known fog belt, the diving schedule should allow for delays because of low visibility. Diver and support crew safety is the prime consideration when determining whether surface visibility is adequate. For example, a surfacing diver might not be able to find his support craft, or the diver and the craft itself might be in danger of being hit by surface traffic. A proper radar reflector for small craft should be considered.

6-4.4.2 **Depth.** Depth is a major factor in selecting both diving personnel and apparatus and influences the decompression profile for any dive. Operations in deep waters may also call for special support equipment such as underwater lights, cameras, ROV, etc.

Depth must be carefully measured and plotted over the general area of the operation to get an accurate depth profile of the dive site. Soundings by a ship-mounted fathometer are reasonably accurate but shall be verified by either a lead-line sounding, a pneumofathometer (Figure 6-9), or a high resolution sonar (bottom finder or fish finder). Depth readings taken from a chart should only be used as an indication of probable depth.

6-4.4.3 **Type of Bottom.** The type of bottom may have a significant effect upon a diver's ability to move and work efficiently and safely. Advance knowledge of bottom conditions is important in scheduling work, selecting dive technique and equipment, and anticipating possible hazards. The type of bottom is often noted on the chart for the area, but conditions can change within just a few feet.

Independent verification of the type of bottom should be obtained by sample or observation. Figure 6-10 outlines the basic types of bottoms and the characteristics of each.

6-4.4.4 **Tides and Currents.** The basic types of currents that affect diving operations are:

- River or Major Ocean Currents. The direction and velocity of normal river, ocean, and tidal currents will vary with time of the year, phase of the tide, configuration of the bottom, water depth, and weather. Tide and current tables show the conditions at the surface only and should be used with caution when planning diving operations. The direction and velocity of the current beneath the surface may be quite different than that observed on the surface.

- Ebb Tides. Current produced by the ebb and flow of the tides may add to or subtract from any existing current.

- Undertow or Rip Current. Undertow or rip currents are caused by the rush of water returning to the sea from waves breaking along a shoreline. Rip currents will vary with the weather, the state of the tide, and the slope of the bottom.

Actual Air Temp °F (°C)	Wind MPH 5	10	15	20	25	30	35	40
	Equivalent Chill Temperature °F (°C)							
40 (4)	35 (2)	30 (-1)	25 (-4)	20 (-7)	15 (-9)	10 (-12)	10 (-12)	10 (-12)
35 (2)	30 (-1)	20 (-7)	15 (-9)	10 (-12)	10 (-12)	5 (-15)	5 (-15)	0 (-17)
30 (-1)	25 (-4)	15 (-9)	10 (-12)	5 (-15)	0 (-17)	0 (-17)	0 (-17)	-5 (-21)
25 (-4)	20 (-7)	10 (-12)	0 (-17)	0 (-17)	-5 (-21)	-10 (-23)	-10 (-23)	-15 (-26)
20 (-7)	15 (-9)	5 (-15)	-5 (-21)	-10 (-23)	-15 (-26)	-20 (-29)	-20 (-29)	-20 (-29)
15 (-9)	10 (-12)	0 (-17)	-10 (-23)	-15 (-26)	-20 (-29)	-25 (-32)	-25 (-32)	-30 (-34)
10 (-12)	5 (-15)	-10 (-23)	-20 (-29)	-25 (-32)	-30 (-34)	-30 (-34)	-30 (-34)	-35 (-37)
5 (-15)	0 (-17)	-15 (-26)	-25 (-32)	-30 (-34)	-35 (-37)	-40 (-40)	-40 (-40)	-45 (-43)
0 (-17)	-5 (-15)	-20 (-24)	-30 (-34)	-35 (-37)	-45 (-43)	-55 (-46)	-50 (-46)	-55 (-48)
-5 (-21)	-10 (-23)	-25 (-32)	-40 (-40)	-45 (-43)	-50 (-46)	-65 (-54)	-60 (-51)	-60 (-51)
-10 (-23)	-15 (-26)	-35 (-37)	-45 (-43)	-50 (-46)	-60 (-54)	-70 (-57)	-65 (-54)	-70 (-57)
-15 (-26)	-20 (-29)	-40 (-40)	-50 (-46)	-60 (-51)	-65 (-54)	-70 (-57)	-75 (-60)	-75 (-60)
-20 (-29)	-25 (-32)	-45 (-43)	-60 (-51)	-65 (-54)	-75 (-60)	-80 (-62)	-85 (-65)	-90 (-68)
-25 (-32)	-30 (-34)	-50 (-46)	-65 (-45)	-75 (-60)	-80 (-62)	-85 (-65)	-90 (-68)	-95 (-71)
-30 (-34)	-35 (-37)	-60 (-51)	-70 (-57)	-80 (-62)	-90 (-68)	-95 (-71)	-100 (-73)	-100 (-73)
-35 (-37)	-40 (-40)	-65 (-54)	-80 (-62)	-85 (-65)	-95 (-71)	-100 (-73)	-105 (-76)	-110 (-79)
-40 (-40)	-45 (-43)	-70 (-57)	-85 (-65)	-95 (-71)	-105 (-76)	-110 (-79)	-115 (-82)	-115 (-82)
-45 (-43)	-50 (-46)	-75 (-60)	-90 (-68)	-100 (-73)	-110 (-79)	-115 (-82)	-120 (-85)	-125 (-87)
-50 (-46)	-55 (-48)	-80 (-62)	-100 (-73)	-110 (-79)	-120 (-85)	-125 (-87)	-130 (-90)	-130 (-90)
-55 (-48)	-60 (-51)	-90 (-68)	-105 (-76)	-115 (-82)	-125 (-87)	-130 (-90)	-135 (-93)	-140 (-96)
-60 (-51)	-70 (-57)	-95 (-71)	-110 (-79)	-120 (-85)	-135 (-93)	-140 (-96)	-145 (-98)	-150(-101)

LITTLE DANGER

INCREASING DANGER (flesh may freeze within one minute)

GREAT DANGER (flesh may freeze within 20 seconds)

Figure 6-8. Equivalent Wind Chill Temperature Chart.

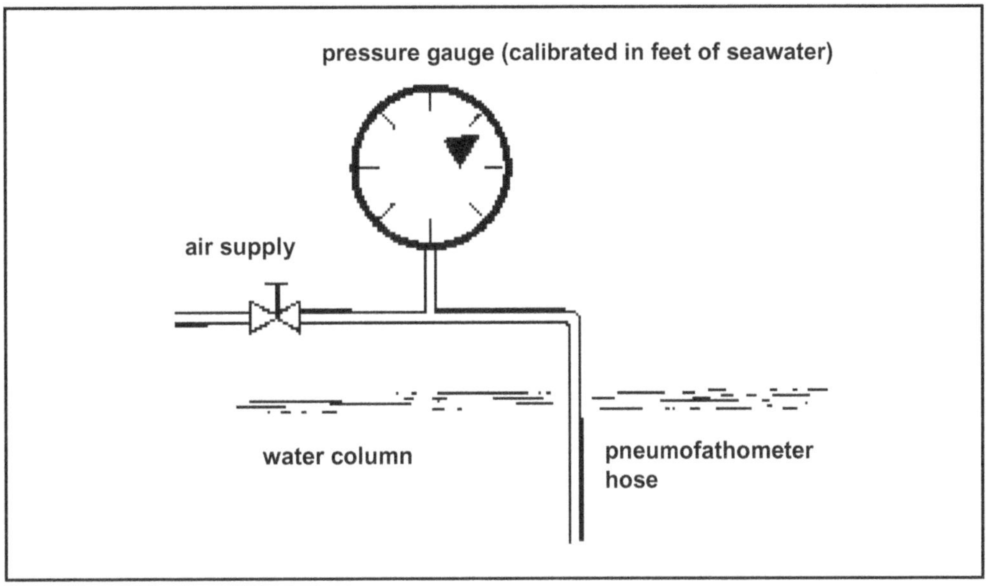

Figure 6-9. Pneumofathometer. The pneumofathometer hose is attached to a diver or weighted object and lowered to the depth to be measured. Water is forced out of the hose by pressurized air until a generally constant reading is noted on the pressure gauge. The air supply is secured, and the actual depth (equal to the height of the water column displaced by the air) is read on the gauge.

These currents may run as fast as two knots and may extend as far as one-half mile from shore. Rip currents, not usually identified in published tables, can vary significantly from day to day in force and location.

- Surface Current Generated by Wind. Wind-generated surface currents are temporary and depend on the force, duration, and fetch of the wind. If the wind has been blowing steadily for some time, this current should be taken into consideration especially when planning surface swims and SCUBA dives.

6-4.4.4.1 ***Equipment Requirements for Working in Currents.*** A diver wearing a surface-supplied outfit, such as the MK 21 SSDS with heavy weights, can usually work in currents up to 1.5 knots without undue difficulty. A diver supplied with an additional weighted belt may be able to accomplish useful work in currents as strong as 2.5 knots. A SCUBA diver is severely handicapped by currents greater than 1.0 knot. If planning an operation in an area of strong current, it may be necessary to schedule work during periods of slack water to minimize the tidal effect.

6-5 IDENTIFY OPERATIONAL HAZARDS

Underwater environmental conditions have a major influence on the selection of divers, diving technique, and the equipment to be used. In addition to environmental hazards, a diver may be exposed to operational hazards that are not unique to the diving environment. This section outlines the environmental and operational hazards that may impact an operation.

TYPE	CHARACTERISTICS	VISIBILITY	DIVER MOBILITY ON BOTTOM
Rock	Smooth or jagged, minimum sediment	Generally unrestricted by dive movement	Good, exercise care to prevent line snagging and falls from ledges
Coral	Solid, sharp and jagged, found in tropical waters only	Generally unrestricted by diver movement	Good, exercise care to prevent line snagging and falls from ledges
Gravel	Relatively smooth, granular base	Generally unrestricted by diver movement	Good, occasional sloping bottoms of loose gravel impair walking and cause instability
Shell	Composed principally of broken shells mixed with sand or mud	Shell-sand mix does not impair visibility when moving over bottom. Shell-mud mix does impair visibility. With higher mud concentrations, visibility is increasingly impaired.	Shell-sand mix provides good stability. High mud content can cause sinking and impaired movement
Sand	Common type of bottom, packs hard	Generally unrestricted by diver movement	Good
Mud and Silt	Common type of bottom, composed of varying amounts of silt and clay, commonly encountered in river and harbor areas	Poor to zero. Work into the current to carry silt away from job site, minimize bottom disturbance. Increased hazard presented by unseen wreckage, pilings, and other obstacles.	Poor, can readily cause diver entrapment. Crawling may be required to prevent excessive penetration, fatiguing to diver.

Figure 6-10. Bottom Conditions and Effects Chart.

6-5.1 **Underwater Visibility.** Underwater visibility varies with depth and turbidity. Horizontal visibility is usually quite good in tropical waters; a diver may be able to see more than 100 feet at a depth of 180 fsw. Horizontal visibility is almost always less than vertical visibility. Visibility is poorest in harbor areas because of river silt, sewage, and industrial wastes flowing into the harbor. Agitation of the bottom caused by strong currents and the passage of large ships can also affect visibility. The degree of underwater visibility influences selection of dive technique and can greatly increase the time required for a diver to complete a given task. For example, a diving team preparing for harbor operations should plan for extremely limited visibility, possibly resulting in an increase in bottom time, a longer period on station for the diving unit, and a need for additional divers on the team.

6-5.2 **Temperature.** Figure 6-11 illustrates how water temperature can affect a diver's performance, and is intended as a planning guide. A diver's physical condition, amount of body fat, and thermal protection equipment determine how long exposure to extreme temperatures can be endured safely. In cold water, ability to concentrate and work efficiently will decrease rapidly. Even in water of moderate temperature (60–70°F, 15.5–21.5°C), the loss of body heat to the water can quickly bring on diver exhaustion.

6-5.3 **Warm Water Diving.** Warm water diving is defined as those diving operations that occur in water temperatures exceeding 88° F. During recent studies at the Navy Experimental Diving Unit, physiological limits have been developed for diving operations in water temperatures up to 99°F. Diving in water temperatures above 99°F should not be attempted without first contacting NAVSEA 00C.

6-5.3.1 **Operational Guidelines and Safety Precautions.** These guidelines are based on data collected from heat acclimated divers dressed in UDT swim trunks and t-shirts who were well rested, calorically replete, well hydrated, and had no immediate heat exposure prior to starting exercise. Exercise rate for the divers replicated a moderate swimming effort. Conditions that contribute to thermal loading such as heavy work rates, significant pre/post dive activities, and various diver dress (dive skins/wetsuits/dry suits) can reduce exposure limits appreciably. Guidelines for exposure limits are based on diver dress and water temperatures. The following precautions apply to all warm water diving operations above 88°F:

- Weight losses up to 15 lbs (or 6-8% of body weight) due to fluid loss may occur and mental and physical performance can be affected. Divers should hydrate fully (approximately 500 ml or 17 oz) two hours before diving. Fluid loading in excess of the recommended 500 ml may cause life-threatening pulmonary edema and should not be attempted.

- Hydrating with water or a glucose/electrolyte beverage should occur as soon as possible after diving. Approximately 500 ml should be replaced for each hour of diving.

- Exposure limits represent maximum cumulative exposure over a 12 hour period. Divers should be hydrated and calorically replete to baseline weight, rested, and kept in a cool environment for at least 12 hours before a repeat exposure to warm water is deemed safe.

NOTE **The following are the general guidelines for warm water diving. Specific UBAs may have restrictions greater than the ones listed below; refer to the appropriate UBA Operations and Maintenance manual. The maximum warm water dive time exposure limit shall be the lesser of the approved UBA operational limits, canister duration limits, oxygen bottle duration or the diver physiological exposure limit.**

- A diver working at a moderate rate e.g. swimming at 0.8 kts or less:

  88°–94°F - limited to canister/O2 bottle duration or diver aerobic endurance

  94°–97°F - limited to three hours based on physiological limits.

  97°–99°F - limited to one hour based on physiological limits.

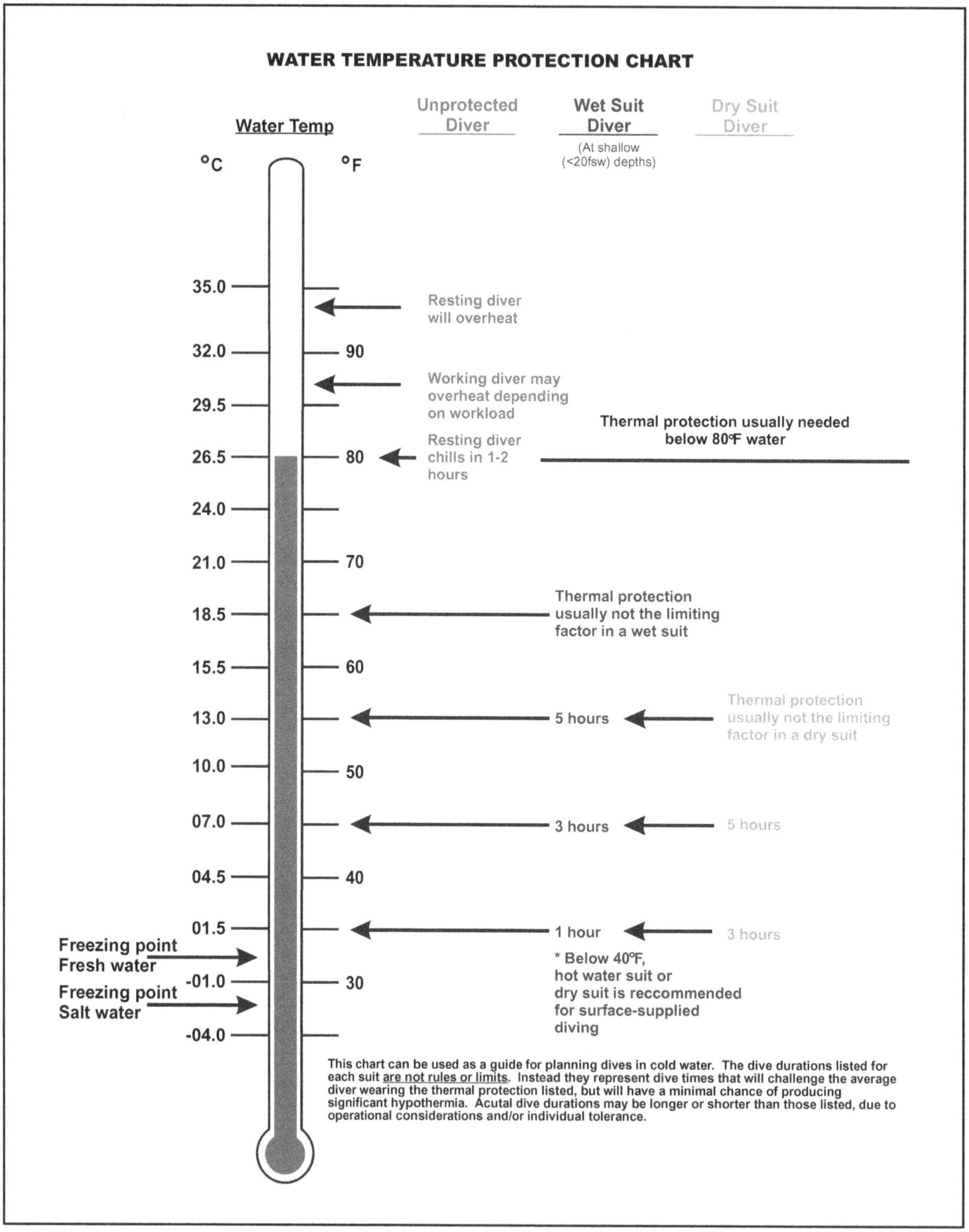

Figure 6-11. Water Temperature Protection Chart.

NOTE **In cases of SDV and DDS operations, thermal loading may change during the course of the mission. Exposure times should be reduced and fluids replaced during the dive when possible.**

- A resting diver e.g. during decompression:

  88°–94°F - limited to canister duration.

  94°–97°F - limited to canister duration.

  97°–99°F - limited to two hours based on physiological limits.

6-5.3.2 **Mission Planning Factors.** The following mission planning factors may mitigate thermal loading and allow greatest utilization of the exposure limits:

1. Conduct diving operations at night, dusk, or dawn to reduce heat stress incurred from sun exposure and high air temperatures.
2. Avoid wearing a hood with a dive skin to allow evaporative cooling.
3. When possible avoid wearing dive skin or anti-chafing dress. Although the effect of various diver dress is not known, it is expected that safe exposure durations at temperatures above 96°F will be less.
4. Follow the guidelines in paragraph 3-10.4 regarding acclimatization. Reduce the intensity of the diving for five days immediately prior to the diving operation.
5. Ensure divers maintain physical conditioning during periods of warm water diving.
6. Methods of cooling the diver should be employed whenever possible. These include using hot water suits to supply cold water to the diver and the use of ice vests.

Mission planning should also include recognition and management of heat stress injuries as part of pre-dive training and briefing. The diver and topside personnel shall be particularly alert for the symptoms of heat stress. Further guidance is contained in paragraph 3-10.4.4 (Excessive Heat - Hyperthermia), paragraph 3-12.1 (Dehydration), and Figure 3-6 (Oxygen Consumption and RMV at Different Work Rates).

6-5.4 **Contaminated Water.** When planning for contaminated water diving, medical personnel should be consulted to ensure proper pre-dive precautions are taken and post-dive monitoring of divers is conducted. In planning for operations in polluted waters, protective clothing and appropriate preventative medical procedures shall be taken. Diving equipment shall be selected that gives the diver maximum protection consistent with the threat. Resources outside the scope of this manual may be required to deal with nuclear, biological, or chemical contaminants. Resources and technical advice for dealing with contaminated water diving conditions are available in the *Guidance for Diving in Contaminated Waters,* SS521-AJ-PRO-010, or contact NAVSEA 00C3.

6-5.5 **Chemical Contamination.** Oil leaking from underwater wellheads or damaged tanks can foul equipment and seriously impede a diver's movements. Toxic materials or volatile fuels leaking from barges or tanks can irritate the skin and corrode equipment. Diving units should not conduct the dive until the contaminant has been identified, the safety factors evaluated, and a process for decontamination set up. Divers operating in waters where a chemical or chemical warfare threat is known or suspected shall evaluate the threat and protect themselves as appropriate. The MK 21 UBA with a double exhaust and a dry suit dress assembly affords limited protection for diving in polluted and contaminated water. Refer to the *MK 21 UBA NAVSEA Technical Manual*, S6560-AG-OMP-010, for more information on using the MK 21 UBA with a dry suit assembly.

6-5.6 **Biological Contamination.** A diver working near sewer outlets may be exposed to biological hazards. SCUBA divers are especially vulnerable to ear and skin infections when diving in waters that contain biological contamination. Divers may also inadvertently take polluting materials into the mouth, posing both physiological and psychological problems. External ear prophylaxis should be provided to diving personnel to prevent ear infections.

6-5.7 **Altitude Diving.** Divers may be required to dive in bodies of water at higher altitudes. Planning shall address the effects of the atmospheric pressures that may be much lower than those at sea level. Air Decompression Tables and Surface-Supplied Helium-Oxygen Tables are authorized for use at altitudes up to 300 feet above sea level without corrections (see paragraphs 9-13 and 14-6). Transporting divers out of the diving area, which may include movement into even higher elevations either overland or by plane, requires special consideration and planning. The Diving Supervisor shall be alert for symptoms of hypoxia and decompression sickness after the dive due to the lower oxygen partial pressure and atmospheric pressure.

6-5.8 **Underwater Obstacles.** Various underwater obstacles, such as wrecks or discarded munitions, offer serious hazards to diving. Wrecks and dumping grounds are often noted on charts, but the actual presence of obstacles might not be discovered until an operation begins. This is a good reason for scheduling a preliminary inspection dive before a final work schedule and detailed dive plan is prepared.

6-5.9 **Electrical Shock Hazards.** Electrical shock may occur when using electric welding or power equipment. All electrical equipment shall be in good repair and be inspected before diving. Although equipped with test buttons, electrical Grounds Fault Interrupters (GFI) often do not provide any indication when the unit has experienced an internal component failure in the fault circuitry. Therefore, GFI component failure during operation (subsequent to testing the unit) may go unnoticed. Although this failure alone will not put the diver at risk, the GFI will not protect the diver if he is placed in contact with a sufficiently high fault current. The following is some general information concerning GFIs:

- GFIs are required when line voltage is above 7.5 VAC or 30 VDC.

- GFIs shall be capable of tripping within 20 milliseconds (ms) after detecting a maximum leakage current of 30 milliamps (ma).

CAUTION **GFIs require an established reference ground in order to function properly. Cascading GFIs could result in loss of reference ground; therefore, GFIs or equipment containing built-in GFIs should not be plugged into an existing GFI circuit.**

In general, three independent actions must occur simultaneously to electrically shock a diver:

- The GFI must fail.
- The electrical equipment which the diver is operating must experience a ground fault.
- The diver must place himself in the path between the fault and earth ground.

6-5.9.1 **Reducing Electrical Shock Hazards.** The only effective means of reducing electrical shock hazards are to ensure:

- Electrical equipment is properly maintained.
- All electrical devices and umbilicals are inspected carefully before all operations.
- Electrical umbilicals are adequately protected to reduce the risk of being abraded or cut when pulled over rough or sharp objects.
- Personnel are offered additional protection through the use of rubber suits (wet, dry, or hot-water) and rubber gloves.
- GFI circuits are tested at regular intervals throughout the operation using built-in test circuits.

Divers operating with remotely operated vehicles (ROVs) should take similar precautions to ensure the ROV electrical system offers the required protection. Many new ROVs use extremely high voltages which make these protective actions even more critical to diver safety.

6-5.9.2 **Securing Electrical Equipment.** The Ship Repair Safety Checklist for Diving requires underwater electrical equipment to be secured while divers are working over the side. While divers are in the water:

- Ship impressed current cathodic protection (ICCP) systems must be secured, tagged out, and confirmed secured before divers may work on an ICCP device such as an anode, dielectric shield, or reference cell.
- When divers are required to work close to an active ICCP anode and there is a risk of contact with the anode, the system must also be secured.
- In situations other than those described above, the ICCP should remain active.

- Divers working within 15 feet of active systems must wear a full dry suit, unisuit, or wet suit with hood and gloves.
- All other underwater electrical equipment shall be secured while divers are working over the side.

6-5.10 **Explosions.** Explosions may be set off in demolition tasks intentionally, accidentally, or as the result of enemy action. When working with or near explosives, the procedures outlined in SWO 60-AA-MMA-010 shall be followed. Divers should stay clear of old or damaged munitions. Divers should get out of the water when an explosion is imminent.

WARNING **Welding or cutting torches may cause an explosion on penetration of gas-filled compartments, resulting in serious injury or death.**

6-5.11 **Sonar.** Appendix 1A provides guidance regarding safe diving distances and exposure times for divers operating in the vicinity of ships transmitting with sonar.

6-5.12 **Nuclear Radiation.** Radiation may be encountered as the result of an accident, proximity to weapons or propulsion systems, weapons testing, or occasionally natural conditions. Radiation exposure can cause serious injury and illness. Safe tolerance levels have been set and shall not be exceeded. These levels may be found in the *Radiological Control Manual*, NAVSEA 0389-LP-660-6542. Local instructions may be more stringent and in such case shall be followed. Prior to diving, all dive team members shall be thoroughly knowledgeable of the local/command radiological control requirements. When required divers shall have a Thermal Luminescence Dosimeter (TLD) or similar device and be apprised of the locations of items such as the reactor compartment, discharges, etc.

6-5.13 **Marine Life.** Certain marine life, because of its aggressive or venomous nature, may be dangerous to man. Some species of marine life are extremely dangerous, while some are merely an uncomfortable annoyance. Most dangers from marine life are largely overrated because most underwater animals leave man alone. All divers should be able to identify the dangerous species that are likely to be found in the area of operation and should know how to deal with each. Refer to Appendix 5C for specific information about dangerous marine life, including identification factors, dangerous characteristics, injury prevention, and treatment methods.

6-5.14 **Vessels and Small Boat Traffic.** The presence of other ships is often a serious problem. It may be necessary to close off an area or limit the movement of other ships. A local Notice to Mariners should be issued. At any time that diving operations are to be conducted in the vicinity of other ships, they shall be properly notified by International Code signal flags (Figure 6-12). An operation may have to be conducted in an area with many small boats operated by people with varied levels of seamanship and knowledge of Nautical Rules of the Road. The diving team should assume that these operators are not acquainted with diving signals and take the precautions required to ensure that these vessels remain clear of the

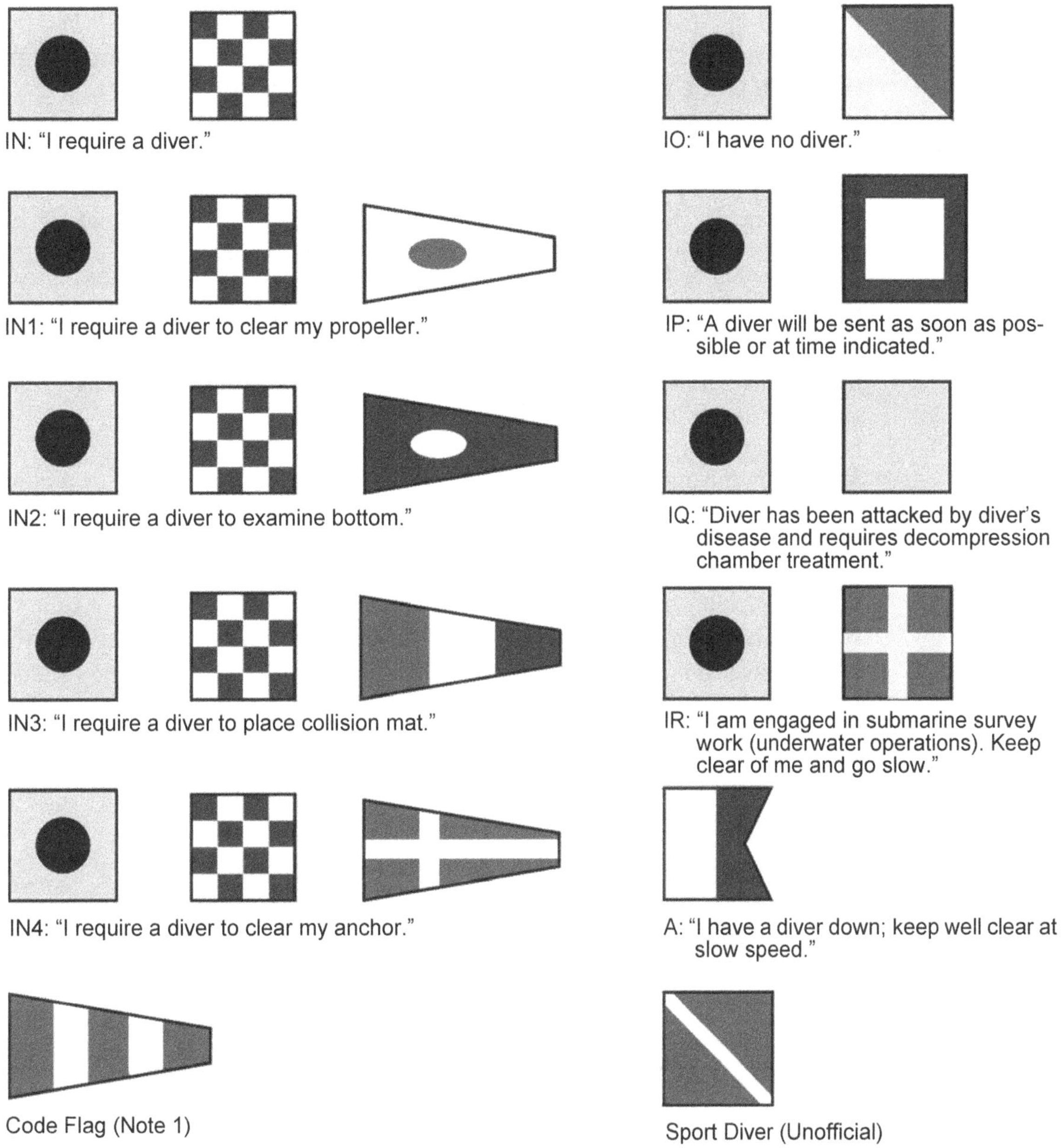

General Note: Rule 27 of Navigation Rules-International-Inland of March 1999 states the lights and shapes that must be displayed when engaged in diving operations.

Note 1: International Signal Code – All signals must be preceded by the code flag to signify that they are international signals. (Do not use code flag in inland waters.)

Figure 6-12. International Code Signal Flags.

diving area. Hazards associated with vessel traffic are intensified under conditions of reduced visibility.

NOTE: When small civilian boats are in the area, use the civilian Sport Diver flag (red with white diagonal stripe) as well as "Code Alpha."

6-5.15 **Territorial Waters.** Diving operations conducted in the territorial waters of other nations shall be properly coordinated prior to diving. Diving units must be alert to the presence of foreign intelligence-collection ships and the potential for hostile action when diving in disputed territorial waters or combat zones.

6-5.16 **Emergency Equipment.** The Diving Safety and Planning Checklist (see Figure 6-19) lists operational steps and equipment required to safely conduct diving operations. The following minimum emergency equipment will be available on-station for every diving operation:

- Communications equipment capable of reaching help in the event of an emergency
- A completely stocked first aid kit
- Portable oxygen supply with sufficient capacity to reach either the recompression chamber or the planned evacuation location listed in the Emergency Assistance Checklist (Figure 6-22)
- Resuscitator or Bag-mask (to provide rescue breathing)
- A means of extracting and transporting an unconscious diver (e.g., litter, stretcher, mesh stretcher, backboard)

If unable to comply due to operational restrictions (limited space, DDS operations, saturation diving), this equipment will be as close as practical to the diving operations and ready for immediate use.

6-6 SELECT DIVING TECHNIQUE

The four main types of air diving equipment used in U.S. Navy diving operations are (Figure 6-13):

1. Open-circuit SCUBA
2. MK 20 MOD 0 Full Face Mask surface-supplied or open-circuit SCUBA
3. MK 21 MOD 1, KM-37 surface-supplied gear
4. EXO BR MS Full Face Mask surface-supplied or open-circuit SCUBA

6-6.1 **Factors to Consider when Selecting the Diving Technique.** When selecting the technique to be used for a dive, the following factors must be considered:

- Duration and depth of the dive
- Type of work to be performed

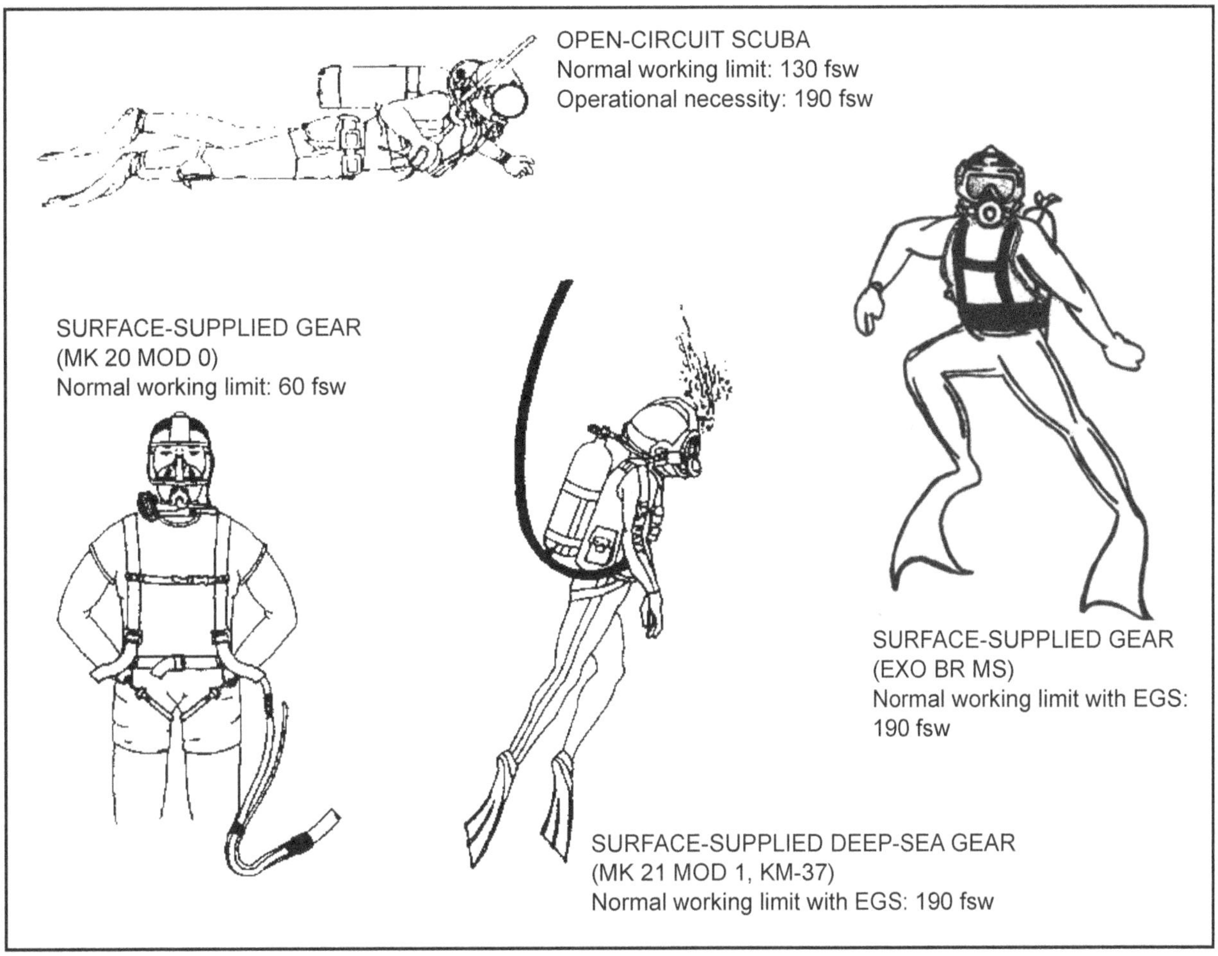

Figure 6-13. Air Diving Techniques. A choice of four air diving techniques are available: open circuit SCUBA, surface-supplied gear (MK 20 MOD 0), surface-supplied deep-sea gear (MK 21 MOD 1 and KM-37), and surface-supplied deep sea gear (EXO BR MS).

- Environmental conditions
- Time constraints

A dive of extended length, even in shallow water, may require an air supply exceeding that which could be provided by SCUBA. Specific depth limits have been established for each type of diving gear and shall not be exceeded without specific approval of the Chief of Naval Operations in accordance with the OPNAVINST 3150.27 series (see Figure 6-14).

The increase of air consumption with depth limits open-circuit SCUBA to 130 fsw for reasonable working dives. The hazards of nitrogen narcosis and decompression further limit open-circuit SCUBA to 190 fsw even for short duration dives. Surface-supplied equipment is generally preferred between 130 and 190 fsw, although open-circuit SCUBA may be used under some circumstances. Decompression SCUBA dives and SCUBA dives deeper than 130 fsw may be conducted when dictated by operational necessity and with the specific approval of the Commanding Officer

NORMAL AND MAXIMUM LIMITS FOR AIR DIVING		
Depth fsw (meters)	**Limit for Equipment**	**Notes**
60 (18)	MK 21 MOD 1, KM-37 diving equipment, maximum working limit without Emergency Gas Supply (EGS)	a
60 (18)	MK 20 MOD 0 equipment surface-supplied	a
60 (18)	Maximum depth for standby SCUBA diver using a single cylinder with less than 100 SCF capacity	
100 (30)	Open-circuit SCUBA with less than 100 SCF cylinder capacity	b
130 (40)	Open-circuit SCUBA, normal working limit	b
190 (58)	Open-circuit SCUBA, maximum working limit with Commanding Officer's or Officer-in-Charge's permission	b, d
190 (58)	MK 21 MOD 1, KM-37 and EXO BR MS (air) diving equipment with EGS, normal working limit	c, d, e
285 (87)	MK 21 MOD 1, KM-37 and EXO BR MS (air) diving equipment with EGS, maximum working limit, exceptional exposure with authorization from the Chief of Naval Operations (N873)	c, d, e

General Operating Notes (Apply to all):

1. These limits are based on a practical consideration of working time versus decompression time and oxygen-tolerance limits. These limits shall not be exceeded except by specific authorization from the Chief of Naval Operations (N873).
2. Do not exceed the limits for exceptional exposures for the Air Decompression Table.
3. In an emergency, any operable recompression chamber may be used for treatment if deemed safe to use by a DSWS qualified Chamber Supervisor.

Specific Notes:

a. When diving in an enclosed space, EGS must be used by each diver.

b. Under normal circumstances, do not exceed the limits of the No-Decompression Table. Dives requiring decompression may be made if considered necessary with approval by the Commanding Officer or Officer-in-Charge of the diving command. The total time of a SCUBA dive (including decompression) shall not exceed the duration of the apparatus in use, disregarding any reserves.

c. A Diving Medical Officer is required on the dive station for all air dives deeper than 190 fsw and for exceptional exposure dives.

d. All planned decompression dives deeper than 130 fsw require a certified recompression chamber on site. **An on-site chamber is defined as a certified and ready chamber accessible within 30 minutes of the dive site by available transportation.**

e. Exceptional exposure dives have a significantly higher probability of DCS and CNS oxygen toxicity.

Figure 6-14. Normal and Maximum Limits for Air Diving.

or the Officer-in-Charge. All open-circuit SCUBA dives deeper than 100 fsw shall employ cylinders having a capacity of at least 100 cubic feet.

In some operations there may be no clear-cut choice of which diving technique to use. Selecting a diving technique may depend upon availability of equipment or trained personnel. The following comparison of SCUBA and surface-supplied techniques highlights the significant differences between the methods and outlines the effect these differences will have on planning.

6-6.2 **Breathhold Diving Restrictions.** Breathhold diving shall be confined to tactical and work situations that cannot be effectively accomplished by the use of underwater breathing apparatus and applicable diver training situations such as SCUBA pool phase and shallow water obstacle/ordnance clearance. Breathhold diving includes the practice of taking two or three deep breaths prior to the dive. The diver shall terminate the dive and surface at the first sign of the urge to breathe. Hyperventilation (excessive rate and depth of breathing prior to a dive, as differentiated from two or three deep breaths prior to a dive) shall not be practiced because of the high possibility of causing unconsciousness under water.

6-6.3 **Operational Characteristics of SCUBA.** The term SCUBA refers to open-circuit air SCUBA unless otherwise noted. The main advantages of SCUBA are mobility, depth flexibility and control, portability, and reduced requirement for surface support. The main disadvantages are limited depth, limited duration, lack of voice communications (unless equipped with a through-water communications system), limited environmental protection, remoteness from surface assistance, and the negative psychological and physiological problems associated with isolation and direct exposure to the underwater environment.

6-6.3.1 **Mobility.** The SCUBA diver is not hindered by bulky or heavy equipment and can cover a considerable distance, with an even greater range through the use of diver propulsion vehicles (DPVs), moving freely in any direction. However, the SCUBA diver shall be able to ascend directly to the surface in case of emergency.

WARNING SCUBA equipment is not authorized for use in enclosed space diving.

6-6.3.2 **Buoyancy.** SCUBA equipment is designed to have nearly neutral buoyancy when in use, permitting the diver to change or maintain depth with ease. This allows the SCUBA diver to work at any level in the water column.

6-6.3.3 **Portability.** The portability and ease with which SCUBA can be employed are distinct advantages. SCUBA equipment can be transported easily and put into operation with minimum delay. SCUBA offers a flexible and economical method for accomplishing a range of tasks.

6-6.3.4 **Operational Limitations.** Divers shall adhere to the operational limitations contained in Figure 6-14. Bottom time is limited by the SCUBA's fixed air supply, which is depleted more rapidly when diving deep or working hard.

6-6.3.5 **Environmental Protection.** The SCUBA diver is not as well protected from cold or from contact with marine plants and animals as a diver in surface-supplied gear, and is more easily swept along by current.

6-6.4 **Operational Characteristics of SSDS.** Surface-supplied diving systems can be divided into two major categories: lightweight full face mask (MK 20 and EXO 26-BR), and deep-sea (MK 21 and KM-37) gear.

6-6.4.1 **Mobility.** Surface-supplied gear allows the diver almost as much mobility as SCUBA. The primary use for deep-sea gear is bottom work in depths up to 190 fsw.

6-6.4.2 **Buoyancy.** The buoyancy associated with SSDS varies with the diving dress selected. Variable Volume Dry Suit (VVDS) provides the greatest buoyancy control (see paragraph 7-3.1.2), making it a desirable technique for working on muddy bottoms, conducting jetting or tunneling, or working where the reaction forces of tools are high.

6-6.4.3 **Operational Limitations.** Divers using surface-supplied gear are restricted to the operational limitations described in Figure 6-14. Additional limitations of using surface-supplied gear include additional topside support personnel and lengthy predive and postdive procedures.

6-6.4.4 **Environmental Protection.** Surface-supplied diving systems can offer the diver increased thermal protection when used with a Hot Water or VVDS. The MK 21 helmet can increase protection of the diver's head. Deep sea gear (MK 21 MOD 1, KM-37) should be used for jobs involving underwater rigging, heavy work, use of certain underwater tools, and any situation where more physical protection is desired. Because the diver's negative buoyancy is easily controlled, an SSDS allows diving in areas with strong currents.

6-7 SELECT EQUIPMENT AND SUPPLIES

6-7.1 **Equipment Authorized for Navy Use.** Equipment procured for use in the U.S. Navy has been tested under laboratory and field conditions to ensure that it will perform according to design specifications. A vast array of equipment and tools is available for use in diving operations. The NAVSEA/00C Diving Equipment Authorized for U.S. Navy Use (ANU) list identifies much of this equipment and categorizes diving equipment authorized for U.S. Navy use.

6-7.2 **Air Supply.** The quality of diver's breathing air is vitally important. Air supplies provided to the diver in tanks or through a compressor shall meet five basic criteria.

1. Air shall conform to standards for diving air purity found in paragraph 4-3 and paragraph 4-4.
2. Flow to the diver must be sufficient. Refer to the appropriate equipment operations and maintenance manual for flow requirements.
3. Adequate overbottom pressure shall be maintained at the dive station.

4. Adequate air supply shall be available to support the duration and depth of the dive (see paragraph 7-4.1 for SCUBA; paragraph 8-2.2 for MK 21).
5. A secondary air supply shall be available for surface-supplied diving.

6-7.3 **Diving Craft and Platforms.** Regardless of the technique being supported, craft used for diving operations shall:

- Be seaworthy
- Include required lifesaving and other safety gear
- Have a reliable engine (unless it is a moored platform or barge)
- Provide ample room for the divers to dress
- Provide adequate shelter and working area for the support crew
- Be able to carry safely all equipment required for the operation
- Have a well-trained crew

Other support equipment—including barges, tugs, floating cranes, or vessels and aircraft for area search—may be needed, depending on the type of operation. The need for additional equipment should be anticipated as far in advance as possible.

6-7.4 **Deep-Sea Salvage/Rescue Diving Platforms.**

- **Auxiliary Rescue/Salvage Ship (T-ARS) (Safeguard Class).** The mission of the T-ARS ship is to assist disabled ships, debeach stranded vessels, fight fires alongside other ships, lift heavy objects, recover submerged objects, tow other vessels, and perform manned diving operations. The T-ARS class ships carry a complement of divers to perform underwater ship husbandry tasks and salvage operations as well as underwater search and recovery. This class of vessel is equipped for all air diving techniques. Onboard equipment allows diving with air to a depth of 190 fsw.

- **Submarine Tender (AS).** U.S. submarine tenders are designed specifically for servicing nuclear-powered submarines. Submarine tenders are fitted with a recompression chamber used for hyperbaric treatments. Submarine tenders support underwater ship husbandry and maintenance and security swims.

- **Fleet Ocean Tug (T-ATF).** T-ATFs are operated by the Military Sealift Command. Civilian crews are augmented with military communications and diving detachments. In addition to towing, these large ocean-going tugs serve as salvage and diving platforms.

- **Diving Tender (YDT).** These vessels are used to support shallow-water diving operations. Additionally, a wide variety of Standard Navy Dive Boats (SNDB), LCM-8, LCM-6, 50-foot work boats, and other yard craft have been fitted with surface-supplied dive systems.

6-7.5 **Small Craft.** SCUBA operations are normally conducted from small craft. These can range in size and style from an inflatable rubber raft with an outboard engine to a small landing craft. If divers are operating from a large ship or diving float, a small boat must be ready as a rescue craft in the event a surfacing diver is in trouble some distance from the support site. A small boat used by SCUBA divers must be able to slip its moorings quickly and move to a diver needing assistance.

6-8 SELECT AND ASSEMBLE THE DIVING TEAM

When planning diving assignments and matching the qualifications and experience of diving personnel to specific requirements of the operation, a thorough knowledge of the duties, responsibilities, and relationships of the various members of the diving team is essential. The diving team may include the Diving Officer, Master Diver, Diving Supervisor, Diving Medical Officer, divers qualified in various techniques and equipment, support personnel (tenders—qualified divers if possible), recorder, and medical personnel, as indicated by the type of operation (Figure 6-15). Other members of the ship's company, when properly instructed, provide support in varying degrees in such roles as boat crew, winch operators, and line handlers.

6-8.1 **Manning Levels.** The size of the diving team may vary with the operation, depending upon the type of equipment being used, the number of divers needed to complete the mission, and the depth. Other factors, such as weather, planned length of the mission, the nature of the objective, and the availability of various resources will also influence the size of the team. The minimum number of personnel required on station for each particular type of diving equipment is provided in Figure 6-16. Minimum levels as determined by ORM shall be maintained; levels must be increased as necessary to meet anticipated operational conditions and situations.

Figure 6-15. MK 21 Dive Requiring Two Divers. The team consists of one Diving Supervisor, two divers, a standby diver, one tender per diver, comms and logs, console operator, and extra personnel (as required).

MINIMUM MANNING LEVELS FOR AIR DIVING

	Open circuit SCUBA Operations		Surface-Supplied Operations
	Single Diver	Buddy Pair	
Diving Supervisor	1	1	1
Comms and Logs	(a)	(a)	(a)
Console Operator			(a)
Diver	1	2	1
Standby Diver	1	1	1
Diver Tender (b, c)	1(b)		1(b)
Standby Diver Tender	(c)	(c)	1
Total	4(d)	4	5(e)

WARNING

These are the minimum personnel levels required. ORM may require these personnel levels be increased so the diving operations can be conducted safely. See Paragraph 6-1.1 and 6-9.1

NOTES:

(a) Diving Supervisor may perform/assign Comms/Logs or Console Operator positions as necessary or required by the system/operations/mission.

(b) See paragraph 6-8.8.5.2 for Tender Qualifications.

(c) If the standby diver is deployed, the Diving Supervisor shall tend the standby diver.

(d) The diver will be tended or have a witness float attached, see paragraph 7-3.1.7. A tender is required when the diver does not have free access to the surface, see paragraph 7-8.2 for further guidance. During mission essential open circuit SCUBA operations, minimum-manning level may be reduced to three qualified divers at the Diving Supervisor's discretion.

(e) Although five is the minimum number of personnel for the MK III and Extreme Lightweight Dive System (XLDS) operations, six or more is highly recommended based on mission requirements and ORM.

Figure 6-16. Minimum Personnel Levels for Air Diving Stations.

6-8.2 **Commanding Officer.** The ultimate responsibility for the safe and successful conduct of all diving operations rests with the Commanding Officer. The Commanding Officer's responsibilities for diving operations are defined and the provisions of U.S. Navy Regulations and other fleet, force, or command regulations confirm specific authority. To ensure diving operations are efficiently conducted, the Commanding Officer delegates appropriate authority to selected members of the command who, with subordinate personnel, make up the diving team.

6-8.3 **Command Diving Officer.** The Command Diving Officer's primary responsibility is the safe conduct of all diving operations within the command. The Command Diving Officer will become thoroughly familiar with all command diving techniques and have a detailed knowledge of all applicable regulations and is responsible for all operational and administrative duties associated with the command diving program. The Command Diving Officer is designated in writing by the Commanding Officer and must be a qualified diver. In the absence of a commissioned officer or a Master Diver, a senior enlisted diving supervisor may be assigned as the Command Diving Officer. On submarines the senior qualified diver may be assigned Command Diving Officer.

6-8.4 **Watchstation Diving Officer.** The Watchstation Diving Officer must be a qualified diver and is responsible to the Commanding Officer for the safe and successful conduct of the diving operation. The Watchstation Diving Officer provides overall supervision of diving operations, ensuring strict adherence to procedures and precautions. A qualified Diving Officer or Master Diver may be assigned this watchstation. The Watchstation Diving Officer must be designated in writing by the Commanding Officer.

Figure 6-17. Master Diver Supervising Recompression Treatment.

6-8.5 Master Diver

6-8.5.1 **Master Diver Responsibilities.** The Master Diver is the most qualified person to supervise air and mixed-gas dives (using SCUBA and surface-supplied diving equipment) and recompression treatments (Figure 6-17). He is directly responsible to the Commanding Officer, via the Diving Officer, for the safe conduct of all phases of diving operations. The Master Diver manages preventive and corrective maintenance on diving equipment, support systems, salvage machinery, handling systems, and submarine rescue equipment. The Master Diver, who also ensures that divers are trained in emergency procedures, conducts training and requalification of divers attached to the command. The Master Diver recommends to the Commanding Officer, via the Diving Officer, which enlisted divers are qualified to serve as Diving Supervisors. The Master Diver oversees the efforts of the Diving Supervisor and provides advice and technical expertise. If circumstances warrant, the Master Diver shall relieve the Diving Supervisor and assume control of the dive station. In the absence of a Diving Officer, the Master Diver can assume the duties and responsibilities of the Diving Officer.

6-8.5.2 **Master Diver Qualifications.** The Master Diver has completed Master Diver evaluation course (CIN A-433-0019) successfully and is proficient in the operation of Navy-approved underwater breathing equipment, support systems, and recompression chambers. He is also trained in diagnosing and treating diving injuries and illnesses. The Master Diver is thoroughly familiar with operating and emergency procedures for diving systems, and possesses a working knowledge of gas mixing and analysis, computations, salvage theory and methods, submarine rescue procedures, towing, and underwater ship husbandry. The Master Diver shall possess a comprehensive knowledge of the scope and application of all Naval instructions and publications pertaining to diving, and shall ensure that logs and reports are maintained and submitted as required.

6-8.6 **Diving Supervisor.** While the Master Diver is in charge of the overall diving operation, the Diving Supervisor is in charge of the actual diving operation for a particular dive or series of dives. Diving operations shall not be conducted without the presence of the Diving Supervisor. The Diving Supervisor has the authority and responsibility to discontinue diving operations in the event of unsafe diving conditions.

6-8.6.1 **Pre-dive Responsibilities.** The Diving Supervisor shall be included in preparing the operational plans. The Diving Supervisor shall consider contingencies, determine equipment requirements, recommend diving assignments, and establish back-up requirements for the operation. The Diving Supervisor shall be familiar with all divers on the team and shall evaluate the qualifications and physical fitness of the divers selected for each particular job. The Diving Supervisor inspects all equipment and conducts pre-dive briefings of personnel.

6-8.6.2 **Responsibilities While Operation is Underway.** While the operation is underway, the Diving Supervisor monitors progress; debriefs divers; updates instructions to subsequent divers; and ensures that the Master Diver, Diving Officer, Commanding Officer, and other personnel as necessary are advised of progress and of any changes to the original plan. The Diving Supervisor should not hesitate to call upon the technical advice and expertise of the Master Diver during the conduct of the dive operation.

6-8.6.3 **Post-dive Responsibilities.** When the mission has been completed, the Diving Supervisor gathers appropriate data, analyzes the results of the mission, prepares reports to be submitted to higher authority, and ensures that required records are completed. These records may range from equipment logs to individual diving records.

6-8.6.4 **Diving Supervisor Qualifications.** The Diving Supervisor may be commissioned or enlisted depending on the size of the operation and the availability of qualified personnel. When qualifying a Diving Supervisor, selection is based on knowledge of diving technique, experience, level of training, and the competence of the available personnel. Regardless of rank, the Diving Supervisor shall be a qualified diver of demonstrated ability and experience. The Diving Supervisor shall be designated in writing by the Commanding Officer. Diving Supervisors under instruction shall stand their watches under the supervision of a qualified Diving Supervisor.

6-8.7 **Diving Medical Officer.** The Diving Medical Officer recommends the proper course of medical action during medical emergencies. The Diving Medical Officer provides on-site medical care for divers as conditions arise and ensures that diving personnel receive proper attention before, during, and after dives. The Diving Medical Officer may modify recompression treatment tables, with the specific concurrence of the Commanding Officer. A Diving Medical Officer is required on site for all air dives deeper than 190 fsw, or for planned exceptional exposure dives. A DMO must be consulted at some point during an actual recompression chamber treatment prior to the release of the patient.

6-8.8 Diving Personnel

6-8.8.1 **Diving Personnel Responsibilities.** While working, the diver shall keep topside personnel informed of conditions on the bottom, progress of the task, and of any developing problems that may indicate the need for changes to the plan or a call for assistance from other divers. To ensure safe conduct of the dive, the diver shall always obey a signal from the surface and repeat all commands when using voice communications. The diver is responsible for the diving gear worn and shall ensure that it is complete and in good repair.

6-8.8.2 **Diving Personnel Qualifications.** Military divers shall be qualified and designated in accordance with instructions issued by the Naval Personnel Command (NPC) or as appropriate by USMC, U.S. Army, or U.S. Air Force orders. Civilian divers under military cognizance must be qualified in accordance with OPNAV 3150.27 (Series). The diving team selected for an operation shall be qualified for the positions manned, diving technique used, the equipment involved, and for diving to the depth required. The DSWS NAVEDTRA 43245 Series Personnel Qualification Standard (PQS) is required for Navy Diver, and equivalent Navy civilian divers. All other Military Divers qualifying to operate or supervise diving systems and equipment contained in the NAVEDTRA 43245 Series PQS, should use the current NAVEDTRA 43245 Series PQS Watch Stations as a guide for qualification, in an effort to standardize DOD qualifications and ensure safe conduct of diving operations. Diving personnel assigned to Navy Experimental Diving Unit (NEDU) and Naval Submarine Medical Research Laboratory (NSMRL) are exempt from such requirements as they are assigned as experimental test subjects and may be employed in experimental dive profiles as required within approved test protocols.

Formal training is required for all designated U.S. Military and DOD civilian employee divers. The Center for EOD and Diving (CENEODDIVE) is authorized to designate fleet units to train personnel in specific critical diving skill sets (HEO_2, Saturation, MK-16 Mod 0 and Mod 1, and MK 25). Commands performing these local qualifications must be designated in writing. Qualifications will be conducted using curricula and materials provided and controlled by CENEODDIVE. Commands conducting local qualifications must have a Master Diver qualified in the equipment being trained and who holds NEC 9502 or an equivalent instructor qualification as determined by CENEODDIVE.

6-8.8.3 **Standby Diver.** A standby diver with a tender is required for all diving operations. The standby diver need not be equipped with the same equipment as the primary diver (except as otherwise specified), but shall have equivalent depth and operational capabilities. SCUBA shall not be used for the standby diver for surface-supplied diving operations.

Figure 6-18. Standby Diver.

6-8.8.3.1 ***Standby Diver Qualifications.*** The standby diver is a fully qualified diver, assigned for back-up or to provide emergency assistance, and is ready to enter the water immediately. For surface-supplied operations, the standby diver shall be dressed to the following points, MK 20 or MK 21 MOD 1, KM-37, with strain relief connected to the harness. Under certain conditions, the Diving Supervisor may require that the helmet be worn. A standby SCUBA diver shall don all equipment and be checked by the Diving Supervisor. The standby diver may then remove the mask and fins and have them ready to don immediately for quick deployment. For safety reasons at the discretion of the Diving Supervisor, the standby diver may remove the tank. The standby diver receives the same briefings and instructions as the working diver, monitors the progress of the dive, and is fully prepared to respond if called upon for assistance. The SCUBA standby diver shall be equipped with an octopus rig.

6-8.8.3.2 ***Deploying the Standby Diver as a Worker Diver.*** The standby diver may be deployed as a working diver provided all of the following conditions are met:

1. Surface-supplied no-decompression dive of 60 fsw or less.

2. Same job/location, e.g., working on port and starboard propellers on the vessel:

   - Prior to deploying the standby diver, the work area shall be determined to be free of hazards (i.e., suctions, discharges) by the first diver on the job site.

- When working in ballast tanks or confined spaces, the standby diver may deploy as a working diver, but both divers shall be tended by a third diver who is outside the confined space.

NOTE **The standby diver shall remain on deck ready for deployment when salvage operations diving is being done.**

6-8.8.4 **Buddy Diver.** A buddy diver is the diver's partner for a SCUBA operation. The buddy divers are jointly responsible for the assigned mission. Each diver keeps track of depth and time during the dive. Each diver shall watch out for the safety and well-being of his buddy and shall be alert for symptoms of nitrogen narcosis, decompression sickness, and carbon dioxide build-up. A diver shall keep his buddy within sight and not leave his buddy alone except to obtain additional assistance in an emergency. If visibility is limited, a buddy line shall be used to maintain contact and communication. If SCUBA divers get separated and cannot locate each other, both divers shall surface immediately.

6-8.8.5 **Diver Tender**

6-8.8.5.1 ***Diver Tender Responsibilities.*** The tender is the surface member of the diving team who works closely with the diver on the bottom. At the start of a dive, the tender checks the diver's equipment and topside air supply for proper operation and dresses the diver. Once the diver is in the water, the tender constantly tends the lines to eliminate excess slack or tension (certain UWSH tasking may preclude this requirement, e.g., working in submarine ballast tanks, shaft lamination, dry habitat welding, etc.). The tender exchanges line-pull signals with the diver, keeps the Diving Supervisor informed of the line-pull signals and amount of diving hose/tending line over the side, and remains alert for any signs of an emergency.

6-8.8.5.2 ***Diver Tender Qualifications.*** The tender should be a qualified diver. When circumstances require the use of a non-diver as a tender, the Diving Supervisor shall ensure that the tender has been thoroughly instructed in the required duties. If a substitute tender shall be employed during an operation, the Diving Supervisor must make certain that the substitute is adequately briefed before assuming duties.

6-8.8.6 **Recorder.** The recorder shall be a qualified diver. The recorder maintains worksheets, fills out the diving log for the operation, and records the diver's descent time, depth of dive, and bottom time. The recorder reports to the Diving Supervisor the ascent time, first stop, and time required at the decompression stop. In SCUBA operations, the Diving Supervisor may assume the duties of the recorder. The recorder is required to have on hand a copy of the U.S. Navy Decompression Table being used. When decompression begins, the schedule selected by the Diving Supervisor is recorded on the chart and log. The recorder keeps all members of the team advised of the decompression requirements of the divers.

6-8.8.7 **Medical Personnel.** Diving Medical Officers and Diving Medical Technicians are given special training in hyperbaric medicine and in diving. They provide medical advice and treatment to diving personnel. They also instruct members of the diving team in first aid procedures and participate in diving operations when the presence of diving medical personnel is indicated, as when particularly hazardous operations are being conducted.

Diving medical personnel evaluate the fitness of divers before operations begin and are prepared to handle any emergencies which might arise. They also observe the condition of other support personnel and are alert for signs of fatigue, overexposure, and heat exhaustion.

There are no hard and fast rules for deciding when a medication would preclude a diver from diving. In general, topical medications, antibiotics, birth control medication, and decongestants that do not cause drowsiness would not restrict diving. Diving Medical Personnel should be consulted to determine if any other drugs would preclude diving.

6-8.8.8 **Other Support Personnel.** Other support personnel may include almost any member of the command when assigned to duties that support diving operations. Some personnel need specific indoctrination. Small-Boat operators shall understand general diving procedures, know the meanings of signals, and be aware of the mission objectives. Other personnel, such as winch operators or deck crew, might interact with the operation directly, but only when under the control of the Diving Supervisor. Engineering personnel may be directed to secure overboard discharges and lock the shafts; a sonar operator might be required to secure equipment and put a Do Not Energize tag on the power switch (see Figure 6-20 for a detailed Ship Repair Safety Checklist).

The Officer of the Deck (OOD) or Command Duty Officer (CDO) is responsible to the Commanding Officer for the operation and safety of the ship and crew during the watch. He shall be concerned with the activities of the diving team. The OOD/CDO shall stay informed of the progress of the operation, of any changes to the original plan, and shall be notified as far in advance as possible of any special requirements. The Officer of the Deck or Command Duty Officer shall be alert for any shifting of the moor or changing weather/sea conditions. He shall inform the Diving Officer and/or Diving Supervisor of any changes in these conditions.

6-8.8.9 **Cross-Training and Substitution.** Each member of the diving team should be qualified to act in any position on the team. Because it is probable that substitutions will be made at some point during a lengthy mission, dive plans and diving schedules should organize personnel and work objectives so that experienced personnel will always be available on site. All personnel who participate in the operation should be included in initial briefings.

6-8.8.10 **Physical Condition.** Diving candidates shall meet the specific physical requirements for divers set forth by the Commander Naval Medical Command and pass a physical screening test as outlined in MILPERSMAN Article 1220.100. Once qualified, the diver is responsible for maintaining good health and top physical condition.

Reference NAVMEDCOMINST 6200.15 (series) to provide guidance on suspension of diving duty of pregnant servicewomen.

Medical personnel assigned to a diving unit shall evaluate the day-to-day condition of each diver and the Diving Supervisor shall verify the fitness of each diver immediately before a dive. Any symptom such as cough, nasal congestion, apparent fatigue, emotional stress, skin or ear infection is reason for placing the diver on the binnacle list until the problem is corrected.

Physical condition is often best judged by the diver who is obligated to report to the Diving Supervisor when not feeling fit to dive. A diver who, for any reason, does not want to make a dive should not be forced. A diver who regularly declines diving assignments shall be disqualified as a diver.

6-8.8.11 **Underwater Salvage or Construction Demolition Personnel.** Underwater salvage demolition training is provided at the Naval Diving and Salvage Training Center in both the Second class and First class Diver curriculum. Demolition diving personnel shall be qualified in accordance with the requirements of OPNAVINST 8023.2 (series).

6-8.8.12 **Blasting Plan.** The Master Diver or senior qualified diver is responsible for providing the Commanding Officer with a comprehensive and written blasting plan. At a minimum, the blasting plan contains:

- Demolition team organization
- Work description with alternatives
- Range standard operating procedures
- Prefiring procedures
- Postfiring procedures
- Area security plan
- Misfire procedures
- Personnel and equipment casualty procedures
- Blasting sequence of events

All demolition operations shall be conducted using approved operating and safety procedures. Qualified demolition personnel shall ensure the operation does not proceed until receiving specific approval from the diving supervisor and shall take charge of all misfires, ensuring they are handled in accordance with the approved plan.

6-8.8.13 **Explosive Handlers.** All divers who handle explosives shall be trained and certified in accordance with the OPNAVINST 8023.2 (series).

6-8.9 **OSHA Requirements for U.S. Navy Civilian Diving.** U.S. Navy Civilian Divers are governed by the provisions of the U.S. Navy Diving Program, yet they must also comply with U.S. Government Occupational Safety and Health Administration

(OSHA) diving standards, delineated in 29 CFR Part 1910 Subpart T; Subj: Commercial Diving Operations. U.S. Navy Civilian Divers are identified as all permanent Navy employees who have formally trained at an approved U.S. Navy diving school as either a SCUBA diver, Second Class diver, or First Class diver. Commercial divers contracted by the Navy who are not permanent government employees are not subject to these provisions.

Most directives of the U.S. Navy Diving Program provide parallel requirements, or are similar enough not to be considered of substantive difference. Several requirements of OSHA do, however, exceed those delineated for U.S. Navy divers and must be identified to ensure compliance by USN civilian divers to both standards. Therefore, the following restrictions, in addition to all other requirements addressed in this manual, apply to USN civilian divers:

6-8.9.1 **SCUBA Diving (Air) Restriction.**

1. SCUBA diving shall not be conducted.
   - To depths deeper than 130 fsw
   - To depths deeper than 100 fsw unless a recompression chamber is on station
2. All SCUBA cylinder manifolds shall be equipped with a manual reserve (J valve), or an independent reserve cylinder gas supply with a separate regulator.
3. A SCUBA cylinder submersible pressure gauge shall be worn by each diver.

6-8.9.2 **Surface Supplied Air Diving Restrictions.**

1. Surface Supplied air diving shall not be conducted to depths greater than 190 fsw.
2. Dives shall be limited to in-water decompression times of less than 120 minutes.
3. An emergency gas supply (come-home bottle) is required for any dive greater than 60 fsw planned decompression dives or for which direct access to the surface is not available.

6-8.9.3 **Mixed Gas Diving Restrictions.** All mixed gas diving shall be limited to:

- A maximum depth of 220 fsw
- Less than 120 minutes total in-water decompression time
- Having a recompression chamber on station

6-8.9.4 **Recompression Chamber Requirements.**

1. An on-station recompression chamber is defined as a certified and ready chamber on the dive station.

2. A recompression chamber shall be on-station for all planned decompression dives or dives deeper than 100 fsw.

3. Civilian divers shall remain at the location of a manned recompression chamber for 1 hour after surfacing from a dive that requires a recompression chamber on station.

6-9 ORGANIZE AND SCHEDULE OPERATIONS

6-9.1 **Task Planning and Scheduling.** All phases of an operation are important. A common failure when planning an operation is to place excessive emphasis on the actual dive phases, while not fully considering pre-dive and post-dive activities. Another failure is to treat operations of a recurring nature with an indifference to safety that comes with over-familiarity. In developing a detailed task-by-task schedule for an operation, the following points shall be considered.

- The schedule shall allocate sufficient time for preparation, transit to the site, rendezvous with other vessels or units, and establishing a secure mooring.
- Bottom time is always at a premium, and all factors that shall affect bottom time shall be carefully considered. These include depth, decompression, number of divers available, support craft size, and surface and underwater environmental conditions.
- The number and profile of repetitive dives in a given time period are limited. This subject is discussed in Chapter 9.
- Plans may include the option to work night and day; however, there is an increased risk of a diving mishap from fatigue.
- The level of personnel support depends on the diving techniques selected (see Minimum Manning Levels, Figure 6-16).
- In planning tasks, non-diving topside support personnel shall be selected carefully, especially those who are not members of the diving team.
- Any schedule must be flexible to accommodate unexpected complications, delays, and changing conditions.
- The Diving Supervisor shall anticipate difficulties and be prepared to either overcome them or find alternative methods to circumvent them.
- If divers have been inactive and operating conditions permit, work-up dives should be conducted in-water or in the recompression chamber.

6-9.2 **Post-dive Tasks.** A diving operation is completed when the objective has been met, the diving team demobilized, and records and reports are filed. Time shall be allocated for:

- Recovering, cleaning, inspecting, maintaining, repairing, and stowing all equipment
- Disposing materials brought up during the operation
- Debriefing divers and other team members
- Analyzing the operation, as planned and as actually carried out
- Restocking expended materials
- Ensuring the readiness of the team to respond to the next assignment

6-10 BRIEF THE DIVING TEAM

6-10.1 **Establish Mission Objective.** The Master Diver or the Diving Supervisor shall brief the team on the overall mission and the aspects of the operation necessary to safely achieve the objective. Major points of discussion include:

1. Clear, brief statement of the mission objective
2. Dominant factors that may determine mission outcome (i.e., environment, enemy/friendly actions, and hazards)
3. All tasks required to accomplish the mission
4. Time factors that may prevail
5. Any changes or augmentations of the dive plan

Prior to starting a dive mission or dive day, coordination with other commands and/or shipboard departments shall be accomplished.

6-10.2 **Identify Tasks and Procedures.** A briefing may be elaborate or simple. For complex operations, briefing with charts, slides, and diagrams may be required. For most operations, the briefing need not be complex and may be an informal meeting. The briefing shall present a breakdown of the dive objective, primary tasks, diving procedures, and related work procedures for the mission or dive day. Prompt debriefing of divers returning to the surface provides the Diving Supervisor with information that may influence or alter the next phase of the operation. Divers should be questioned about the progress of the work, bottom conditions and anticipated problems. They should also be asked for suggestions for immediate changes.

6-10.3 **Review Diving Procedures.** Diving and work procedures to be used for the task at hand shall be reviewed during the briefing. The Diving Safety and Planning Checklist (Figure 6-19), Ship Repair Safety Checklist for Diving (Figure 6-20) and the Surface-Supplied Diving Operations Pre-dive Checklist (Figure 6-21) support control of diving operations. These checklists may be tailored to specific missions and environmental circumstances.

6-10.4 **Assignment of Personnel.** All personnel assignments shall be reviewed and verified to ensure properly trained personnel are assigned to operations.

6-10.5 **Assistance and Emergencies.** In any diving operation, three types of assistance may be required:

1. Additional equipment, personnel, supplies, or services
2. Clarification, authorization, or decisions from higher command
3. Emergency assistance in the event of an accident or serious illness

Unexpected developments or emergency situations may be accompanied by confusion. The source and availability of any needed assistance and the method for obtaining it as quickly as possible, shall be determined in advance. The location of the nearest recompression chamber shall be identified and the chamber operators notified before the operation begins. The sources of emergency transportation, military or civilian, shall be established and alerted and the nearest Diving Medical Officer should be located and notified. Arrangements must be made to ensure a 24-hour availability for emergency assistance.

If emergency transportation is required by civilian Emergency Medical Services (EMS) sources, a Memorandum of Agreement or Diving Protocol should be established in advance and those casualty response agreements incorporated into the Command Diving Bill.

When a recompression chamber is required by Figure 6-14, the chamber shall be currently certified and within 30 minutes' travel time from the dive site. If a recompression chamber is required in an emergency, a non-certified chamber may be used if the DSWS qualified Chamber Supervisor is of the opinion that it is safe to operate.

Figure 6-22 is a suggested format for the Emergency Assistance Checklist that shall be completed and posted at the diving station to provide necessary information so that any member of the team could take prompt action.

6-10.6 **Notification of Ship's Personnel.** In the event of a diving casualty or mishap on dive station, calm must be maintained. Maintain silence on the side and take orders from the Diving Officer, Master Diver, and/or Diving Supervisor.

6-10.7 **Fouling and Entrapment.** Fouling and entrapment are more common with surface-supplied gear than SCUBA because of the ease with which the umbilicals can become entangled. Divers shall be particularly careful and watch their own umbilicals and those of their partners as well.

The surface-supplied diver may become fouled more easily, but will usually have an ample air supply while working to get free. The SCUBA diver may have no other recourse but to remove the gear and make a free ascent. If trapped, the SCUBA diver must face the possibility of running out of air before being able to work free.

The first and most important action that a trapped diver can take is to stop and think. The diver shall remain calm, analyze the situation, and carefully try to work free. Panic and overexertion are the greatest dangers to the trapped diver. If the situation cannot be resolved readily, help should be obtained. A new umbilical can be provided to the surface-supplied diver; the SCUBA diver can be given a new apparatus or may be furnished air by the dive partner.

Once the diver has been freed and returns to the surface, the diver shall be examined and treated, bearing in mind the following considerations:

- The diver will probably be overtired and emotionally exhausted.
- The diver may be suffering from or approaching hypothermia.
- The diver may have a physical injury.
- A SCUBA diver may be suffering from asphyxia. If a free ascent has been made, gas embolism may have developed.
- Significant decompression time may have been missed.

6-10.8 **Equipment Failure.** With well-maintained equipment that is thoroughly inspected and tested before each dive, operational failure is rarely a problem. When a failure does occur, the correct procedures will depend upon the type of equipment and dive. As with most emergencies, the training and experience of the diver and the diving team will be the most important factor in resolving the situation safely.

6-10.8.1 **Loss of Gas Supply.** Usually, when a diver loses breathing gas it should be obvious almost immediately. Some diving apparatus configurations may have an emergency gas supply (EGS). When breathing gas is interrupted, the dive shall be aborted and the diver surfaced as soon as possible. Surfacing divers may be suffering from hypoxia, hypercapnia, missed decompression, or a combination of the three, and should be treated accordingly.

6-10.8.2 **Loss of Communications.** If audio communications are lost with surface-supplied gear, the system may have failed or the diver could be in trouble. If communications are lost:

1. Use line-pull signals at once. Depth, current, bottom or work site conditions may interfere.
2. Check the rising bubbles of air. A cessation or marked decrease of bubbles could be a sign of trouble.
3. Listen for sounds from the diving helmet. If no sound is heard, the circuit is probably out of order. If the flow of bubbles seems normal, the diver may be all right.

DIVING SAFETY AND PLANNING CHECKLIST

(Sheet 1 of 4)

STEPS IN PLANNING OF DIVING OPERATIONS

Detailed, advanced planning is the foundation of diving safety.

A. ANALYZE THE MISSION FOR SAFETY.

- __ Ensure mission objective is defined.
- __ Determine that non-diving means of mission accomplishment have been considered and eliminated as inappropriate.
- __ Coordinate emergency assistance.
- __ Review relevant Naval Warfare Publications (NWP) and OPNAV instructions.

B. IDENTIFY AND ANALYZE POTENTIAL HAZARDS.

__ **Natural Hazards:**

1. Atmospheric:
   - __ Exposure of personnel to extreme conditions
   - __ Adverse exposure of equipment and supplies to elements
   - __ Delays or disruption caused by weather
2. Surface:
   - __ Sea sickness
   - __ Water entry and exit
   - __ Handling of heavy equipment in rough seas
   - __ Maintaining location in tides and currents
   - __ Ice, flotsam, kelp, and petroleum in the water
   - __ Delays or disruption caused by sea state
3. Underwater and Bottom:
   - __ Depth which exceeds diving limits or limits of available equipment
   - __ Exposure to cold temperatures
   - __ Dangerous marine life
   - __ Tides and currents
   - __ Limited visibility
   - __ Bottom obstructions
   - __ Ice (underwater pressure ridges, loss of entry hole, loss of orientation, etc.)
   - __ Dangerous bottom conditions (mud, drop-offs, etc.)

__ **On-Site Hazards:**

- __ Local marine traffic or other conflicting naval operations
- __ Other conflicting commercial operations
- __ High-powered, active sonar
- __ Radiation contamination and other pollution (chemical, sewer outfalls, etc.)

__ **Mission Hazards:**

- __ Decompression sickness
- __ Communications problems
- __ Drowning
- __ Other trauma (injuries)
- __ Hostile action

__ **Object Hazards:**

- __ Entrapment and entanglement
- __ Shifting or working of object
- __ Explosives or other ordnance

Figure 6-19. Diving Safety and Planning Checklist (sheet 1 of 4).

DIVING SAFETY AND PLANNING CHECKLIST

(Sheet 2 of 4)

C. SELECT EQUIPMENT, PERSONNEL and EMERGENCY PROCEDURES.

__ **Diving Personnel:**

- __ 1. Assign a complete and properly qualified Diving Team.
- __ 2. Assign the right man to the right task.
- __ 3. Verify that each member of the Diving Team is properly trained and qualified for the equipment and depths involved.
- __ 4. Determine that each man is physically fit to dive, paying attention to:
  - __ general condition and any evidence of fatigue
  - __ record of last medical exam
  - __ ears and sinuses
  - __ severe cold or flu
  - __ use of stimulants or intoxicants
- __ 5. Observe divers for emotional readiness to dive:
  - __ motivation and professional attitude
  - __ stability (no noticeably unusual or erratic behavior)

__ **Diving Equipment:**

- __ 1. Verify that diving gear chosen and diving techniques are adequate and authorized for mission and particular task.
- __ 2. Verify that equipment and diving technique are proper for depth involved.
- __ 3. Verify that life support equipment has been tested & approved for U.S. Navy use.
- __ 4. Determine that all necessary support equipment and tools are readily available and are best for accomplishing job efficiently and safely.
- __ 5. Determine that all related support equipment such as winches, boats, cranes, floats, etc. are operable, safe and under control of trained personnel.
- __ 6. Check that all diving equipment has been properly maintained (with appropriate records) and is in full operating condition.

__ **Provide for Emergency Equipment:**

- __ 1. Obtain suitable communications equipment with sufficient capability to reach outside help; check all communications for proper operation.
- __ 2. Verify that a recompression chamber is ready for use, or notify the nearest command with one that its use may be required within a given timeframe.
- __ 3. Verify that a completely stocked first aid kit is at hand.
- __ 4. If oxygen will be used as standby first aid, verify that the tank is full and properly pressurized, and that masks, valves, and other accessories are fully operable.
- __ 5. If a resuscitator will be used, check apparatus for function.
- __ 6. Check that fire-fighting equipment is readily available and in full operating condition.
- __ 7. Verify that emergency transportation is either standing by or on immediate call.

__ **Establish Emergency Procedures:**

- __ 1. Know how to obtain medical assistance immediately.
- __ 2. For each potential emergency situation, assign specific tasks to the diving team and support personnel.
- __ 3. Complete and post Emergency Assistance Checklist; ensure that all personnel are familiar with it.
- __ 4. Verify that an up-to-date copy of U.S. Navy Decompression Tables is available.
- __ 5. Ensure that all divers, boat crews and other support personnel understand all diver hand signals.
- __ 6. Predetermine distress signals and call-signs.

Figure 6-19. Diving Safety and Planning Checklist (sheet 2 of 4).

DIVING SAFETY AND PLANNING CHECKLIST

(Sheet 3 of 4)

__ 7. Ensure that all divers have removed anything from their mouths on which they might choke during a dive (gum, dentures, tobacco).

__ 8. Thoroughly drill all personnel in Emergency Procedures, with particular attention to cross-training; drills should include:

Emergency recompression	Rapid undressing	Fire
First aid	Rapid dressing	Embolism
Restoration of breathing	Near-drowning	Electric shock
Blowup	Entrapment	Lost diver

D. ESTABLISH SAFE DIVING OPERATIONAL PROCEDURES

__ **Complete Planning, Organization, and Coordination Activities:**

__ 1. Ensure that other means of accomplishing mission have been considered before deciding to use divers.

__ 2. Ensure that contingency planning has been conducted.

__ 3. Carefully state goals and tasks of each mission and develop a flexible plan of operations (Dive Plan).

__ 4. Completely brief the diving team and support personnel (paragraph 6-7).

__ 5. Designate a Master Diver or properly qualified Diving Supervisor to be in charge of the mission.

__ 6. Designate a recorder/timekeeper and verify that he understands his duties and responsibilities.

__ 7. Determine the exact depth at the job-site through the use of a lead line, pneumofathometer, or commercial depth sounder.

__ 8. Verify existence of an adequate supply of compressed air available for all planned diving operations **plus an adequate reserve for emergencies.**

__ 9. Ensure that no operations or actions on part of diving team, support personnel, technicians, boat crew, winch operators, etc., take place without the knowledge of and by the direct command of the Diving Supervisor.

__ 10. All efforts must be made through planning, briefing, training, organization, and other preparations to minimize bottom time. Water depth and the condition of the diver (especially fatigue), rather than the amount of work to be done, shall govern diver's bottom time.

__ 11. Current decompression tables shall be on hand and shall be used in all planning and scheduling of diving operations.

__ 12. Instruct all divers and support personnel not to cut any lines until approved by the Diving Supervisor.

__ 13. Ensure that ship, boat, or diving craft is securely moored and in position to permit safest and most efficient operations (exceptions are emergency and critical ship repairs).

__ 14. Verify that, when using surface-supplied techniques, the ship, boat, or diving craft has at least a two-point moor.

__ 15. Ensure that, when conducting SCUBA operations in hazardous conditions, a boat can be quickly cast off and moved to a diver in distress.

__ **Perform Diving Safety Procedures, Establish Safety Measures:**

__ 1. Ensure that each diver checks his own equipment in addition to checks made by tenders, technicians or other support personnel.

__ 2. Designate a standby diver for all diving operations; standby diver shall be dressed to the necessary level and ready to enter the water if needed.

__ 3. Assign buddy divers, when required, for all SCUBA operations.

Figure 6-19. Diving Safety and Planning Checklist (sheet 3 of 4).

DIVING SAFETY AND PLANNING CHECKLIST

(Sheet 4 of 4)

__ 4. Take precautions to prevent divers from being fouled on bottom. If work is conducted inside a wreck or other structure, assign a team of divers to accomplish task. One diver enters wreck, the other tends his lines from point of entry.

__ 5. When using explosives, take measures to ensure that no charge shall be fired while divers are in water.

__ 6. Use safety procedures as outlined in relevant Naval publications for all U/W cutting and welding operations.

__ 7. Brief all divers and deck personnel on the planned decompression schedules for each particular dive. Check provisions for decompressing the diver.

__ 8. Verify that ship, boat, or diving craft is displaying proper signals, flags, day shapes, or lights to indicate diving operations are in progress. (Consult publications governing International or Inland Rules, International/Inland local signals, and Navy communications instructions.)

__ 9. Ensure that protection against harmful marine life has been provided. (See Appendix 5C.)

__ 10. Check that the quality of diver's air supply is periodically and thoroughly tested to ensure purity.

__ 11. Thoroughly brief boat crew.

__ 12. Verify that proper safety and operational equipment is aboard small diving boats or craft.

__ **Notify Proper Parties that Dive Operations Are Ready to Commence:**

__ 1. Diving Officer

__ 2. Commanding Officer

__ 3. Area Commander

__ 4. Officer of the Deck/Day

__ 5. Command Duty Officer or Commanding Officer of ships alongside

__ 6. Bridge, to ensure that ship's personnel shall not:

- __ turn the propeller or thrusters
- __ get underway
- __ activate active sonar or other electronics
- __ drop heavy items overboard
- __ shift the moor

__ 7. Ship Duty Officer, to ensure that ship's personnel shall not:

- __ activate sea discharges or suctions
- __ operate bow or stern-planes or rudder
- __ operate vents or torpedo shutters
- __ turn propellers

__ 8. Other Interested Parties and Commands:

- __ Harbor Master/Port Services Officer
- __ Command Duty Officers
- __ Officers in tactical command
- __ Cognizant Navy organizations
- __ U.S. Coast Guard (if broadcast warning to civilians is required)

__ 9. Notify facilities having recompression chambers and sources of emergency transportation that diving operations are underway and their assistance may be needed.

Figure 6-19. Diving Safety and Planning Checklist (sheet 4 of 4).

SHIP REPAIR SAFETY CHECKLIST FOR DIVING

(Sheet 1 of 2)

When diving operations will involve underwater ship repairs, the following procedures and safety measures are required in addition to the Diving Safety Checklist.

SAFETY OVERVIEW

A. The Diving Supervisor shall advise key personnel of the ship undergoing repair:

1. OOD	4. OODs of ships alongside
2. Engineering Officer	5. Squadron Operations (when required)
3. CDO	6. Combat Systems Officer (when required)

B. The Diving Supervisor shall request that OOD/Duty Officer of ship being repaired ensure that appropriate equipment is secured and tagged out.

C. The Diving Supervisor shall request that OOD/Duty Officer advise him when action has been completed and when diving operations may commence.

D. When ready, the diving Supervisor shall request that the ship display appropriate diving signals and pass a diving activity advisory over the 1MC every 30 minutes. For example, "There are divers working over the side. Do not operate any equipment, rotate screws, cycle rudder, planes or torpedo shutters, take suction from or discharge to sea, blow or vent any tanks, activate sonar or underwater electrical equipment, open or close any valves, or cycle trash disposal unit before checking with the Diving Supervisor."

E. The Diving Supervisor shall advise the OOD/Duty Officer when diving operations commence and when they are concluded. At conclusion, the ship will be requested to pass the word on the 1MC, "Diving operations are complete. Carry out normal work routine."

F. Diving within 50 feet of an active sea suction (located on the same side of the keel) that is maintaining a suction of 50 gpm or more, is not authorized unless considered as an emergency repair and is authorized by the Commanding Officers of both the repair activity and tended vessel. When it is determined that the sea suction is maintaining a suction of less than 50 gpm and is less than 50 feet, or maintaining a suction of more than 50 gpm and is less than 50 feet but on the opposite side of the keel, the Diving Supervisor shall determine if the sea suction is a safety hazard to the divers prior to conducting any diving operation. In all cases the Diving Supervisor shall be aware of the tend of the diver's umbilical to ensure that it will not cross over or become entrapped by an active sea suction. Diving on 688 and 774 class submarines do not present a hazard to divers when ASW pumps are operating in slow speed and MSW pumps are operating in super slow speed. Diver tag-out procedures must be completed in accordance with the TUMS and SORM to ensure ASW pumps are not operated in fast speed and MSW pumps are not operated in either fast or slow speeds. Divers must be properly briefed on location of suctions and current status of equipment.

NOTIFY KEY PERSONNEL.

1.	OOD	______________________	(signature)
2.	Engineering Officer	______________________	(signature)
3.	CDO	USS ______________________	(signature)
4.	OOD	USS ______________________	
	OOD	USS ______________________	
	OOD	USS ______________________	
	OOD	USS ______________________	
5.	Squadron Operations	______________________	
6.	Port Services Officer	______________________	

(Diving Supervisor (Signature))

Figure 6-20. Ship Repair Safety Checklist for Diving (sheet 1 of 2).

SHIP REPAIR SAFETY CHECKLIST FOR DIVING

(Sheet 2 of 2)

TAG OUT EQUIPMENT

TAG OUT	SIGNATURE AND RATE
Rudder	______________
Anchors	______________
Planes	______________
Torpedo tube shutters	______________
Trash disposal unit	______________
Tank blows	______________
Tank vents	______________
Shaft(s) locked	______________
Sea suctions	______________
Sea discharges	______________
U/W electrical equipment	______________
Sonars	______________
Other U/W equipment	______________

USS ______________

(name of ship)

CDO______________

(signature of CDO)

Figure 6-20. Ship Repair Safety Checklist for Diving (sheet 2 of 2).

SURFACE-SUPPLIED DIVING OPERATIONS PREDIVE CHECKLIST

(Sheet 1 of 3)

CAUTION

This checklist is an overview intended for use with the detailed Operating Procedures (OPs) from the appropriate equipment O&M technical manual.

A. Basic Preparation:

___1. Verify that a recompression chamber is onsite for all decompression dives deeper than 130 fsw.

___2. Verify that proper signals indicating underwater operations being conducted are displayed correctly.

___3. Ensure that all personnel concerned, or in the vicinity, are informed of diving operations.

___4. Determine that all valves, switches, controls, and equipment components affecting diving operation are tagged-out to prevent accidental shut-down or activation.

___5. Verify that diving system and recompression chamber are currently certified or granted a Chief of Naval Operations (CNO) waiver to operate.

B. Equipment Protection:

___1. Assemble all members of the diving team and support personnel (winch operators, boat crew, watchstanders, etc.) for a predive briefing.

___2. Assemble and lay out all dive equipment, both primary equipment and standby spares for diver (or standby diver), including all accessory equipment and tools.

___3. Check all equipment for superficial wear, tears, dents, distortion, or other discrepancies.

___4. Check all masks, helmets, view ports, faceplates, seals, and visors for damage.

___5. Check all harnesses, laces, strain reliefs, and lanyards for wear; renew as needed.

C. MK 21 MOD1/KM-37:

___ Ensure that all Operating Procedures (OPs) have been completed in accordance with *UBA MK 21 MOD 1,* NAVSEA S6560-AG-OMP-010, or *KM-37 Technical Manual.*

D. MK 20 MOD 0:

___ Ensure that all Operating Procedures (OPs) have been completed in accordance with *UBA MK 20 MOD 0 Technical Manual*, NAVSEA SS600-AK-MMO-010.

E. General Equipment:

___1. Check that all accessory equipment – tools, lights, special systems, spares, etc., – are on site and in working order. In testing lights, tests should be conducted with lights submerged in water and extinguished before removal, to prevent overheating and failure.

___2. Erect diving stage or attach diving ladder. In the case of the stage, ensure that the screw pin shackle connecting the stage line is securely fastened with the shackle pin seized with wire or a safety shackle is used to help prevent opening.

F. Preparing the Diving System:

___1. Check that a primary and suitable back-up air supply is available with a capacity in terms of purity, volume, and supply pressure to completely service all divers including decompression, recompressions and accessory equipment throughout all phases of the planned operation.

___2. Verify that all diving system operating procedures have been conducted to properly align the dive system.

___3. Ensure that qualified personnel are available to operate and stand watch on the dive system.

Figure 6-21. Surface-Supplied Diving Operations Predive Checklist (sheet 1 of 3).

SURFACE-SUPPLIED DIVING OPERATIONS PREDIVE CHECKLIST

(Sheet 2 of 3)

___4. Compressors:

- ___ a. Determine that sufficient fuel, coolant, lubricants, and antifreeze are available to service all components throughout the operation. All compressors should be fully fueled, lubricated, and serviced (with all spillage cleaned up completely).
- ___ b. Verify that all diving system operating procedures have been conducted properly to align the dive system.
- ___ c. Check maintenance and repair logs to ensure the suitability of the compressor (both primary and back-up) to support the operation.
- ___ d. Verify that all compressor controls are properly marked and any remote valving is tagged with "Divers Air Supply - Do Not Touch" signs.
- ___ e. Ensure that compressor is secure in diving craft and shall not be subject to operating angles, caused by roll or pitch, that will exceed 15 degrees from the horizontal.
- ___ f. Verify that oil in the compressor is an approved type. Check that the compressor oil does not overflow Fill mark; contamination of air supply could result from fumes or oil mist.
- ___ g. Check that compressor exhaust is vented away from work areas and, specifically, does not foul the compressor intake.
- ___ h. Check that compressor intake is obtaining a free and pure suction without contamination. Use pipe to lead intake to a clear suction if necessary.
- ___ i. Check all filters, cleaners and oil separators for cleanliness IAW PMS.
- ___ j. Bleed off all condensed moisture from filters and from the bottom of volume tanks. Check all manifold drain plugs, and that all petcocks are closed.
- ___ k. Check that all belt-guards are properly in place on drive units.
- ___ l. Check all pressure-release valves, check valves and automatic unloaders.
- ___ m. Verify that all supply hoses running to and from compressor have proper leads, do not pass near high-heat areas such as steam lines, are free of kinks and bends, and are not exposed on deck in such a way that they could be rolled over, damaged, or severed by machinery or other means.
- ___ n. Verify that all pressure supply hoses have safety lines and strain reliefs properly attached.

H. Activate the Air Supply in accordance with approved OPs.

___1. Compressors:

- ___ a. Ensure that all warm-up procedures are completely followed.
- ___ b. Check all petcocks, filler valves, filler caps, overflow points, bleed valves, and drain plugs for leakage or malfunction of any kind.
- ___ c. Verify that there is a properly functioning pressure gauge on the air receiver and that the compressor is meeting its delivery requirements.

___2. Cylinders:

- ___ a. Gauge all cylinders for proper pressure.
- ___ b. Verify availability and suitability of reserve cylinders.
- ___ c. Check all manifolds and valves for operation.
- ___ d. Activate and check delivery.

___3. For all supply systems, double check "Do Not Touch" tags (tags outs).

Figure 6-21. Surface-Supplied Diving Operations Predive Checklist (sheet 2 of 3).

SURFACE-SUPPLIED DIVING OPERATIONS PREDIVE CHECKLIST

(Sheet 3 of 3)

I. Diving Hoses:

___1. Ensure all hoses have a clear lead and are protected from excessive heating and damage.
___2. Check hose in accordance with PMS.
___3. Ensure that the hose (or any length) has not been used in a burst test program. No hose length involved in such a program shall be part of an operational diving hose.
___4. Check that hoses are free of moisture, packing material, or chalk.
___5. Soap test hose connections after connection to air supply and pressurization.
___6. Ensure umbilical boots are in good condition.

J. Test Equipment with Activated Air Supply in accordance with approved OPs.

___1. Hook up all air hoses to helmets, masks and chamber; make connections between back-up supply and primary supply manifold.
___2. Verify flow to helmets and masks.
___3. Check all exhaust and non-return valves.
___4. Hook up and test all communications.
___5. Check air flow from both primary and back-up supplies to chamber.

K. Recompression Chamber Checkout (Predive only):

___1. Check that chamber is completely free and clear of all combustible materials.
___2. Check primary and back-up air supply to chamber and all pressure gauges.
___3. Check that chamber is free of all odors or other "contaminants."
___4. Hook up and test all communications.
___5. Check air flow from both primary and back-up supplies to chamber.

Final Preparations:

___1. Verify that all necessary records, logs, and timesheets are on the diving station.
___2. Check that appropriate decompression tables are readily at hand.
___3. Place the dressing bench in position, reasonably close to the diving ladder or stage, to minimize diver travel.

Figure 6-21. Surface-Supplied Diving Operations Predive Checklist (sheet 3 of 3).

EMERGENCY ASSISTANCE CHECKLIST

____
Location

____
Name/Phone Number

____
Response Time

AIR TRANSPORTATION

____
Location

____
Name/Phone Number

____
Response Time

SEA TRANSPORTATION

____
Location

____
Name/Phone Number

____
Response Time

HOSPITAL

____
Location

____
Name/Phone Number

____
Response Time

DIVING MEDICAL OFFICER

____
Location

____
Name/Phone Number

____
Response Time

____
Location

____
Name/Phone Number

____
Response Time

COMMUNICATIONS

____
Location

____
Name/Phone Number

____
Response Time

DIVING UNITS

____
Location

____
Name/Phone Number

____
Response Time

COMMAND

____
Location

____
Name/Phone Number

____
Response Time

EMERGENCY CONSULTATION
Duty Phone Numbers 24 Hours a Day
Navy Experimental Dive Unit (NEDU)

Commercial	(850) 234-4351
	(850) 230-3100
DSN	436-4351

Navy Diving Salvage and Training Center (NDSTC)

Commercial	(850) 234-4651
DSN	436-4651

Figure 6-22. Emergency Assistance Checklist.

4. If sounds are heard and the diver does not respond to signals, assume the diver is in trouble.
5. Have divers already on the bottom investigate, or send down the standby diver to do so.

6-10.9 **Lost Diver.** In planning for an operation using SCUBA, lost diver procedures shall be included in the dive plan and dive brief. Losing contact with a SCUBA diver can be the first sign of a serious problem. If contact between divers is lost, each diver shall surface. If the diver is not located quickly, or not found at the surface, the Diving Supervisor shall initiate search procedures immediately. At the same time, medical personnel should be notified and the recompression chamber team alerted.

A lost diver is often disoriented and confused and may have left the operating area. Nitrogen narcosis or other complications involving the breathing mixture, which can result in confusion, dizziness, anxiety, or panic, are common in recovered lost divers. The diver may harm the rescuers unknowingly. When the diver is located, the rescuer should approach with caution to prevent being harmed and briefly analyze the stricken diver's condition.

If the diver is found unconscious, attempts should be made to resupply breathing gas and restore consciousness. If this cannot be accomplished, the diver shall be brought to the surface immediately. Gas Embolism may occur during ascent and significant decompression may be missed and immediate recompression may be required. If it is possible to provide the diver with an air supply such as a single-hose demand SCUBA, the rescuer should do so during the ascent.

6-10.10 **Debriefing the Diving Team.** After the day's diving has been completed (or after a shift has finished work if the operation is being carried on around the clock), all members of the diving team should be brought together for a short debriefing of the day's activities. This offers all personnel a chance to provide feedback to the Diving Supervisor and other members of the team. This group interaction can help clarify any confusion that may have arisen because of faulty communications, lack of dive site information, or misunderstandings from the initial briefing.

6-11 AIR DIVING EQUIPMENT REFERENCE DATA

There are several diving methods which are characterized by the diving equipment used. The following descriptions outline capabilities and logistical requirements for various air diving systems.

SCUBA General Characteristics

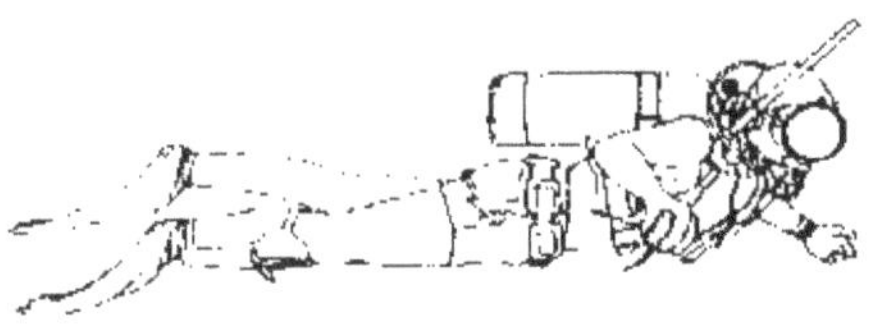

Principle of Operation:

Self contained, open-circuit demand system

Minimum Equipment:

1. Open-circuit SCUBA with J-valve or submersible pressure gauge
2. Life preserver/buoyancy compensator
3. Weight belt (if required)
4. Dive knife
5. Face mask
6. Swim fins
7. Submersible wrist watch
8. Depth gauge

Principal Applications:

1. Shallow water search
2. Inspection
3. Light repair and recovery

Advantages:

1. Rapid deployment
2. Portability
3. Minimum support requirements
4. Excellent horizontal and vertical mobility
5. Minimum bottom disturbances

Disadvantages:

1. Limited endurance (depth and duration)
2. Limited physical protection
3. Influenced by current
4. Lack of voice communication (unless equipped with a through-water communications system or full face mask)

Restrictions:

Work limits:

1. Normal 130 fsw
2. Maximum 190 fsw with Commanding Officer or Officer-in-Charge's permission
3. 100 fsw using SCUBA cylinder(s) with less than 100 SCF
4. Standby diver with at least 100 SCF cylinder capacity for dives deeper than 60 fsw
5. Within no-decompression limits
6. Current - 1 knot maximum. Current greater than 1 knot, requires ORM analysis. As a minimum the divers(s) must be tended or have a witness float.

Operational Considerations:

1. Standby diver required
2. Small craft is mandatory for diver recovery during open-ocean diving, when diving off of a large platform or when the diver is untended and may be displaced from dive site, e.g., during a bottom search in a strong current or a long duration swim.
3. Moderate to good visibility preferred
4. Ability to free ascend to surface required (see paragraph 7-8.2)

Figure 6-23. SCUBA General Characteristics.

MK 20 MOD 0 General Characteristics

Principle of Operation:

Surface-supplied, open-circuit lightweight system

Minimum Equipment:

1. MK 20 MOD 0 mask
2. Harness
3. Weight belt (as required)
4. Dive knife
5. Swim fins or boots
6. Surface umbilical

Principal Applications:

Diving in mud tanks and enclosed spaces

Advantages:

1. Unlimited by air supply
2. Good horizontal mobility
3. Voice and/or line-pull signal capabilities

Disadvantages:

1. Limited physical protection

Restrictions:

1. Depth limits: 60 fsw
2. Current - Above 1.5 knots requires extra weights
3. Enclosed space diving requires an Emergency Gas Supply (EGS) with 50 to 150 foot whip and second-stage regulator.

Operational Considerations:

1. Adequate air supply system required
2. Standby diver required

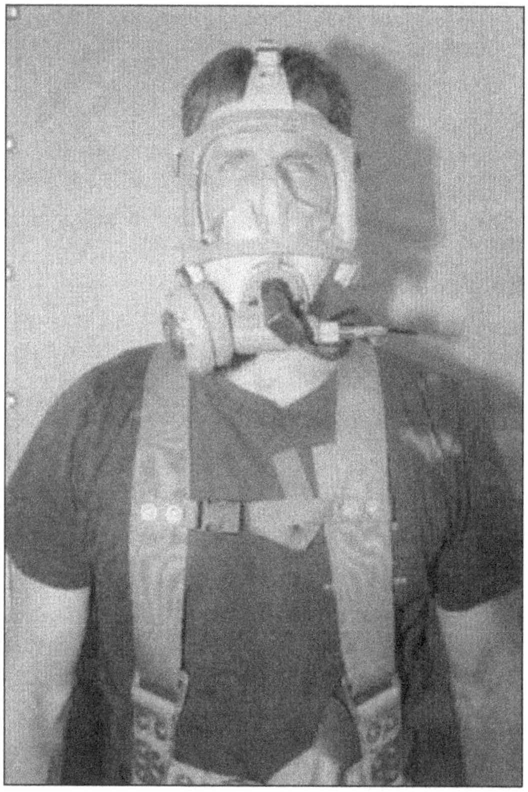

MK 20 MOD 0 Helmet

Figure 6-24. MK 20 MOD 0 General Characteristics.

MK 21 MOD 1, KM-37 General Characteristics

Principle of Operation:

Surface-supplied, open-circuit system

Minimum Equipment:

1. MK 21 MOD 1, KM-37 Helmet
2. Harness
3. Weight belt (if required)
4. Dive knife
5. Swim fins or boots
6. Surface umbilical
7. EGS bottle deeper than 60 fsw

Principal Applications:

1. Search
2. Salvage
3. Inspection
4. Underwater Ships Husbandry and enclosed space diving

Advantages:

1. Unlimited by air supply
2. Head protection
3. Good horizontal mobility
4. Voice and/or line pull signal capabilities
5. Fast deployment

Disadvantages:

1. Limited mobility

Restrictions:

1. Depth limits: 190 fsw
2. Emergency air supply (EGS) required deeper than 60 fsw or diving inside a wreck or enclosed space
3. Current - Above 1.5 knots requires extra weights
4. Enclosed space diving requires an Emergency Gas Supply (EGS).

Operational Considerations:

1. Adequate air supply system required
2. Standby diver required

MK 21 MOD 1, KM-37 Helmet.

Figure 6-25. MK 21 MOD 1, KM-37 General Characteristics.

EXO BR MS Characteristics

Principle of Operation:

Surface-supplied, open-circuit system
Self contained, open-circuit demand system

Minimum Equipment:

1. EXO BR MS Full Face Mask
2. Manifold Block (except for SCUBA and ship husbandry enclosed spaces)
3. Harness
4. Weight belt (if required)
5. Dive knife
6. Swim fins or boots
7. Surface umbilical
8. EGS bottle deeper than 60 fsw

Principal Applications:

1. Search
2. Salvage
3. Inspection
4. Underwater Ships Husbandry and enclosed space diving

Advantages:

1. Unlimited by air supply
2. Good horizontal mobility
3. Voice and/or line pull signal capabilities
4. Fast deployment

Disadvantages:

1. Limited physical protection

Restrictions:

1. Depth limits: 190 fsw
2. Emergency air supply (EGS) required deeper than 60 fsw or diving inside a wreck or enclosed space
3. Current - Above 1.5 knots requires extra weights
4. Enclosed space diving requires an Emergency Gas Supply (EGS) with 50 to 150 foot whip and second stage regulator.

Operational Considerations:

1. Adequate air supply system required
2. Standby diver required

EXO BR MS Full Face Mask.

Figure 6-26. EXO BR MS Characteristics.

CHAPTER 7

SCUBA Air Diving Operations

7-1 INTRODUCTION

7-1.1 **Purpose.** The purpose of this chapter is to familiarize divers with standard and emergency procedures when diving with SCUBA equipment.

7-1.2 **Scope.** This chapter covers the use of open-circuit SCUBA, which is normally deployed in operations not requiring decompression. Decompression diving using open-circuit air SCUBA may be undertaken only if no other option exists and only with the concurrence of the Commanding Officer or Officer-in-Charge (OIC). Closed-circuit underwater breathing apparatus is the preferred method of performing SCUBA decompression dives. Operation of open-circuit, closed-circuit, and semiclosed-circuit systems designed for use with mixed-gas or oxygen is covered in Volume 4.

7-2 REQUIRED EQUIPMENT FOR SCUBA OPERATIONS

At a minimum, each diver must be equipped with the following items to safely conduct an open-circuit SCUBA dive:

- Open-circuit SCUBA.
- Face mask.
- Life preserver/buoyancy compensator.*
- Weight belt and weights as required.**
- Knife.**
- Swim fins.
- Submersible pressure gauge or Reserve J-valve.
- Submersible wrist watch. Only one is required when diving in pairs with a buddy line.**
- Depth gauge. **
- Octopus. ***

* During the problem-solving pool phase of SCUBA training, CO_2 cartridges may be removed and replaced with plugs or expended cartridges that are painted International Orange.

** These items are not required for the pool phase of SCUBA training.

*** At Commanding Officers discretion based on ORM

7-2.1 **Equipment Authorized for Navy Use.** Only diving equipment that has been certified or authorized for use by the NAVSEA/00C ANU list shall be used in a Navy dive. However, many items, such as hand tools, which are not specifically listed in the ANU list or do not fit under the scope of certification and are deemed valuable to the success of the dive, can be used. A current copy must be maintained by all diving activities. Look for the ANU on the SUPSALV website.

7-2.2 **Open-Circuit SCUBA.** All open-circuit SCUBA authorized for Navy use employ a demand system that supplies air each time the diver inhales. The basic open-circuit SCUBA components are:

- Demand regulator assembly
- One or more air cylinders
- Cylinder valve and manifold assembly
- Backpack or harness

7-2.2.1 **Demand Regulator Assembly.** The demand regulator assembly is the central component of the open-circuit system. The regulator delivers air to the diver after reducing the high-pressure air in the cylinder to a pressure that can be used by the diver. There are two stages in a typical system (Figure 7-1).

7-2.2.1.1 ***First Stage.*** In the regulator's first stage, high-pressure air from the cylinder passes through a regulator that reduces the pressure of the air to a predetermined level over ambient pressure. Refer to the regulator technical manual for the specific setting.

7-2.2.1.2 ***Second Stage.*** In the second stage of a regulator, a movable diaphragm is linked by a lever to the low-pressure valve, which leads to a low-pressure chamber. When the air pressure in the low-pressure chamber equals the ambient water pressure, the diaphragm is in the center position and the low-pressure valve is closed. When the diver inhales, the pressure in the low-pressure chamber is reduced, causing the diaphragm to be pushed inward by the higher ambient water pressure. The diaphragm actuates the low-pressure valve which opens, permitting air to flow to the diver. The greater the demand, the wider the low-pressure valve is opened, thus allowing more air flow to the diver. When the diver stops inhaling, the pressure on either side of the diaphragm is again balanced and the low-pressure valve closes. As the diver exhales, the exhausted air passes through at least one check valve and vents to the water.

7-2.2.1.3 ***Single Hose Regulators.*** In the single-hose, two-stage demand regulator the first stage is mounted on the cylinder valve assembly. The second-stage assembly includes the mouthpiece and a valve to exhaust exhaled air directly into the water. The two stages are connected by a length of low-pressure hose, which passes over the diver's right shoulder. The second stage has a purge button, which when activated allows low-pressure air to flow through the regulator and the mouthpiece, forcing out any water which may have entered the system. The principal disadvantages of the single-hose unit are an increased tendency to freeze up in very cold water and the exhaust of air in front of the diver's mask. While

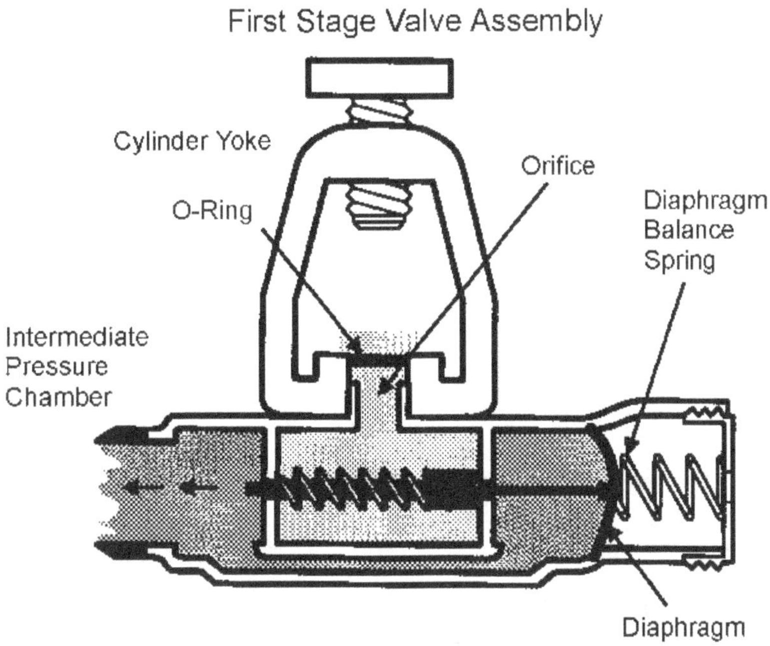

First Stage. High pressure air flows through the orifice of the first stage into the intermediate chamber. When the pressure in the intermediate chamber reaches ambient plus diaphragm balance spring set pressure, the first stage assembly closes.

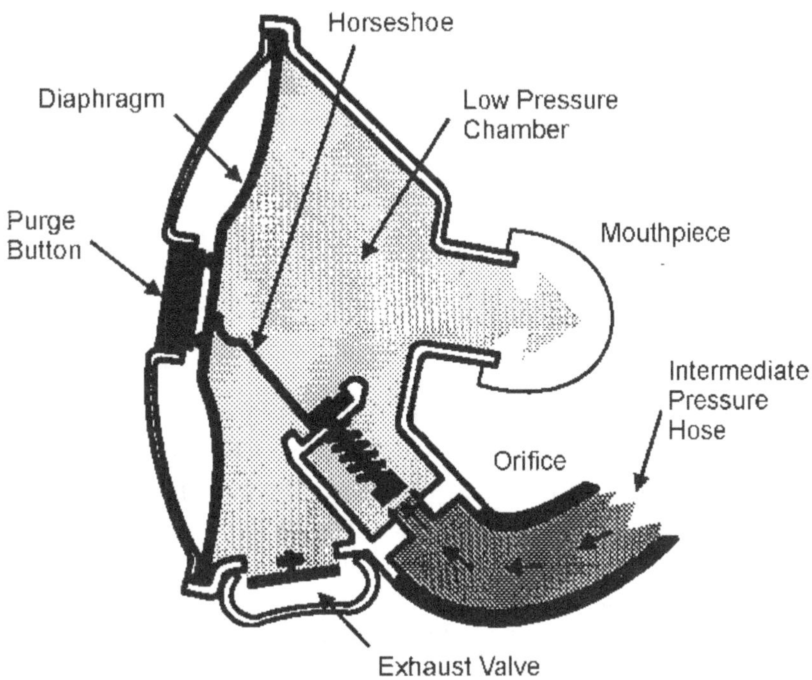

Second Stage. Upon inhalation the second stage diaphragm moves inward and the horseshoe lever opens the second stage valve assembly. Intermediate pressure air from the hoses is throttled across the orifice and fills the low pressure chamber to ambient pressure and flow is provided to the diver. Upon exhalation the diaphragm is pushed outward and the second stage is closed. Expired air is dumped from the low pressure chamber to the surrounding water through the exhaust valve.

Figure 7-1. Schematic of Demand Regulator.

the Navy PMS system provides guidance for repairing and maintaining SCUBA regulators, the manufacturer's service manual should be followed for specific procedures.

7-2.2.1.4 ***Full Face Mask.*** The AGA/Divator-IIG/MK20 full face mask may be used with an approved single-hose first-stage regulator with an octopus, to the maximum approved depth of the regulator, as indicated in the NAVSEA/00C ANU list (Figure 7-2).

Figure 7-2. Full Face Mask.

7-2.2.1.5 ***Mouthpiece.*** The size and design of SCUBA mouthpieces differ between manufacturers, but each mouthpiece provides relatively watertight passageways for delivering breathing air into the diver's mouth. The mouthpiece should fit comfortably with slight pressure from the lips.

7-2.2.1.6 ***Octopus.*** An octopus is an additional single hose second stage regulator connected to the diver's first stage regulator and may be used in case the diver's primary second stage regulator fails or for buddy breathing. The octopus must be an ANU approved second stage regulator. Hose length and designation markings are at the discretion of the diving supervisor. An octopus is mandatory for the standby divers. Use of an octopus is the preferred method to accomplish buddy breathing (see paragraph 7-7.7). During predive inspection, the diver shall breathe the octopus to ensure it is working properly.

7-2.2.2 **Cylinders.** SCUBA cylinders (tanks or bottles) are designed to hold high pressure compressed air. Because of the extreme stresses imposed on a cylinder at these pressures, all cylinders used in SCUBA diving must be inspected and tested periodically. Seamless steel or aluminum cylinders which meet Department of Transportation (DOT) specifications (DOT 3AA, DOT 3AL, DOT SP6498, and

DOT E6498) are approved for Navy use. Each cylinder used in Navy operations must have identification symbols stamped into the shoulder (Figure 7-3).

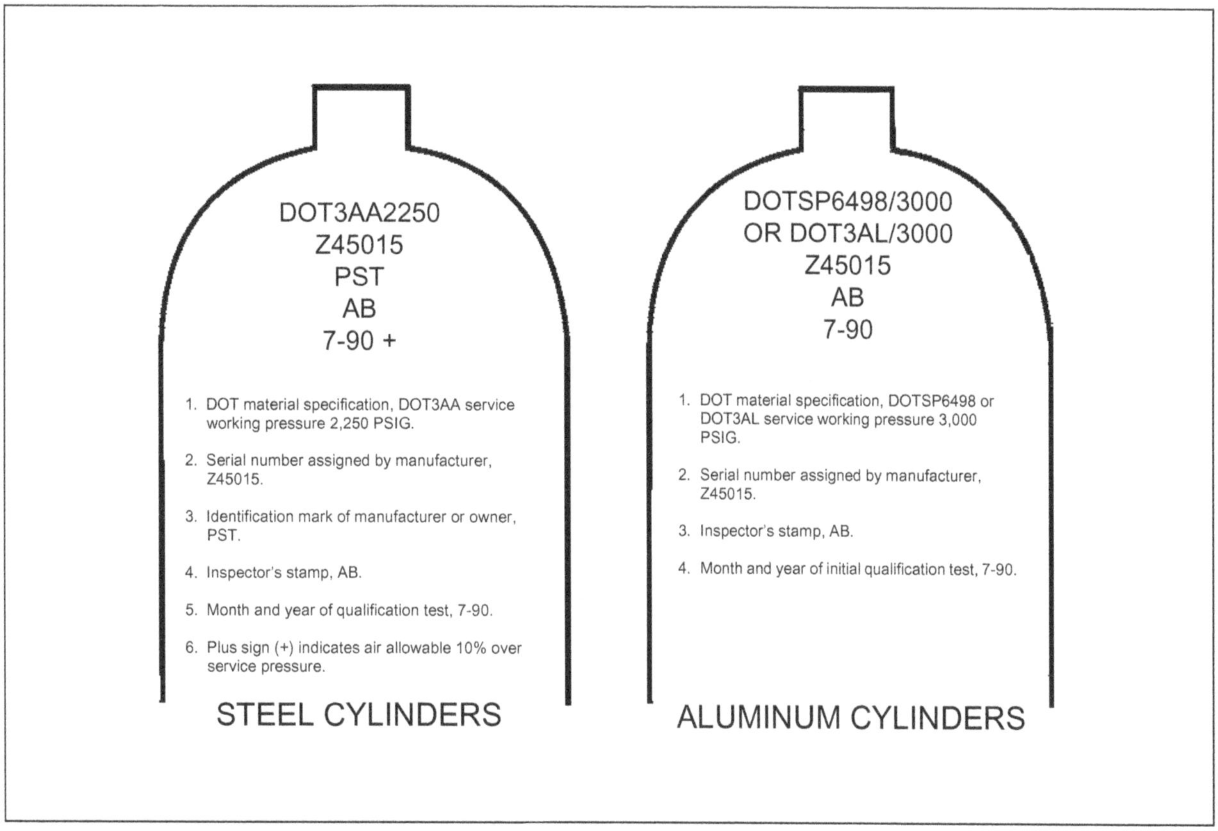

Figure 7-3. Typical Gas Cylinder Identification Markings.

7-2.2.2.1 ***Sizes of Approved SCUBA Cylinders.*** Approved SCUBA cylinders are available in several sizes and one or two cylinders may be worn to provide the required quantity of air for the dive. The volume of a cylinder, expressed in actual cubic feet or cubic inches, is a measurement of the internal volume of the cylinder. The capacity of a cylinder, expressed in standard cubic feet or liters, is the amount of gas (measured at surface conditions) that the cylinder holds when charged to its rated pressure. Table 7-1 lists the sizes of some standard SCUBA cylinders. Refer to the NAVSEA/00C ANU list for a list of approved SCUBA cylinders.

7-2.2.2.2 ***Inspection Requirements.*** Open-circuit SCUBA cylinders must be visually inspected at least once every 12 months and every time water or particulate matter is suspected in the cylinder. Cylinders containing visible accumulations of corrosion must be cleaned before being placed into service. Commercially available steel and aluminum SCUBA cylinders, as specified in the NAVSEA/00C ANU list, which meet DOT specifications, as well as SCUBA cylinders designed to Navy specifications, must be visually inspected at least annually and must be hydrostatically tested at least every five years in accordance with DOT regulations and Compressed Gas Association (CGA) pamphlets C-1 and C-6.

Table 7-1. *Sample SCUBA Cylinder Data.*

Open-Circuit Cylinder Description (Note 1)	Rated Working Pressure (PSIG)	Floodable Volume (Cu.Ft.)	Absolute Air Capacity at Rated Pressure (Cu.Ft.)	Reserve Pressure
Steel 72	2,250	0.420	64.7	500
Steel 100	3,500	0.445	106.4	500
Steel 120	3,500	0.526	125.7	500
Aluminum 50	3,000	0.281	48.5	500
Aluminum 63	3,000	0.319	65.5	500
Aluminum 80	3,000	0.399	81.85	500
Aluminum 100	3,300	0.470	105.9	500

Note 1: Fifty cubic feet is the minimum size SCUBA cylinder authorized. SEAL teams are authorized smaller cylinders for special operations.

Note 2: For EGS Cylinder requirements refer to paragraph 8-2.2.1.

7-2.2.2.3 ***Guidelines for Handling Cylinders.*** General safety regulations governing the handling and use of compressed gas cylinders aboard Navy ships are contained in NAVSEA 0901-LP-230-0002, NSTM Chapter 550, "Compressed Gas Handling." Persons responsible for handling, storing, and charging SCUBA cylinders must be familiar with these regulations. Safety rules applying to SCUBA cylinders are contained in paragraph 7-4.5. Because SCUBA cylinders are subject to continuous handling and because of the hazards posed by a damaged unit, close adherence to the rules is mandatory.

7-2.2.3 **Cylinder Valves and Manifold Assemblies.** Cylinder valves and manifolds make up the system that passes the high-pressure air from the cylinders to the first-stage regulator. The cylinder valve serves as an on/off valve and is sealed to the tank by a straight-threaded male connection containing a neoprene O-ring on the valve's body.

7-2.2.3.1 ***Blowout Plugs and Safety Discs.*** The cylinder valve contains a high-pressure blowout plug or safety disc plug in the event of excessive pressure buildup. When a dual manifold is used, two blowout plugs or safety disc plugs are installed as specified by the manufacturers' technical manual.

For standard diving equipment, a safety disc plug similar to new issue equipment is recommended. The safety disc plug and safety disc are not always identified by a National Stock Number (NSN), but are available commercially.

7-2.2.3.2 ***Manifold Connectors.*** If two or more cylinders are to be used together, a manifold unit is needed to provide the necessary interconnection. Most manifolds incorporate an O-ring as a seal, but some earlier models may have a tapered (pipe) thread design. One type will not connect with the other type.

7-2.2.3.3 ***Pressure Gauge Requirements.*** A cylinder valve with an air reserve (J valve) is preferred. When a cylinder valve without an air reserve (K valve) is used, the SCUBA regulator must be equipped with a submersible pressure gauge to indicate pressure contents of the cylinder. The dive must be terminated when the cylinder pressure reaches 500 psi for a single cylinder or 250 psi for twin manifold cylinders. The air reserve mechanism alerts the diver that the available air supply is almost exhausted and provides the diver with sufficient reserve air to reach the surface. The air reserve mechanism contains a spring-loaded check valve. When it becomes increasingly difficult to obtain a full breath, the diver must reach over the left shoulder and push down the reserve lever, opening the reserve valve to make the remaining air available.

Dive planning should not extend bottom time by including the use of reserve air. The diver should never assume that the reserve air supply will be provided. When the resistance to breathing becomes obvious, the diver should notify the dive partner that the air supply is low and both should start for the surface immediately. **The dive must be terminated when either diver shifts to reserve air.**

7-2.2.4 **Backpack or Harness.** A variety of backpacks or harnesses, used for holding the SCUBA on the diver's back, have been approved for Navy use. The backpack may include a lightweight frame with the cylinder(s) held in place with clamps or straps. The usual system for securing the cylinder to the diver uses shoulder and waist straps. All straps must have a quick-release feature, easily operated by either hand, so that the diver can remove the cylinder and leave it behind in an emergency.

7-2.3 Minimum Equipment.

7-2.3.1 **Face Mask.** The face mask protects the diver's eyes and nose from the water. Additionally, it provides maximum visibility by putting a layer of air between the diver's eyes and the water.

Face masks are available in a variety of shapes and sizes for diver comfort. To check for proper fit, hold the mask in place with one hand and inhale gently through the nose. The suction produced should hold the mask in place. Don the mask with the head strap properly adjusted, and inhale gently through the nose. If the mask seals, it should provide a good seal in the water.

Some masks are equipped with a one-way purge valve to aid in clearing the mask of water. Some masks have indentations at the nose or a neoprene nose pad to allow the diver to block the nostrils to equalize the pressure in the ears and sinuses. Several models are available for divers who wear eyeglasses. One type provides a prescription-ground faceplate, while another type has special holders for separate lenses. All faceplates must be constructed of tempered or shatterproof safety glass because faceplates made of ordinary glass can be hazardous. Plastic faceplates are generally unsuitable as they fog too easily and are easily scratched.

The size or shape of the faceplate is a matter of personal choice, but the diver should use a mask that provides a wide, clear range of vision.

7-2.3.2 **Life Preserver.** The principal functions of the life preserver are to assist a diver in rising to the surface in an emergency and to keep the diver on the surface in face-up position (Figure 7-4). The low-pressure inflation device on the preserver may be actuated by the diver, or by a dive partner should the diver be unconscious or otherwise incapacitated.

All models used by the Navy must be authorized by NAVSEA/00C Authorized for Navy Use List and have a manual inflation device in addition to the low pressure inflation device. With the exception of the UDT (9C-4220-00-276-8929), an overinflation valve or relief valve is required to ensure against possible rupture of the life preserver on ascent. Some ANU models are available commercially while others may be procured through the Navy supply system. In selecting a life preserver for a specific task, the individual technical manuals should be consulted. The use of certain closed and semi-closed UBAs will require the wearing of a life preserver.

Figure 7-4. Life Preserver.

The life preserver must be sturdy enough to resist normal wear and tear, and of sufficient volume to raise an unconscious diver safely from maximum dive depth to the surface.

Most life preservers currently in use employ carbon dioxide (CO_2) cartridges to provide inflation in an emergency. The cartridges must be the proper size for the life preserver. Cartridges must be weighed prior to use, in accordance with the planned maintenance system (PMS) for the life preserver, to ensure the actual weight is in compliance with the weight tolerance for the cartridge cylinder. Carbon dioxide cartridges used with commercially available life preservers with low-pressure inflators do not have the weight stamped on the cartridge cylinder. The actual weight of these cartridges must be inscribed on the cartridge, and be within the tolerance for weight.

7-2.3.3 **Buoyancy Compensator.** When a life preserver is not required by a specific UBA, a buoyancy compensator may be used at the Diving Supervisor's discretion. When selecting a buoyancy compensator, a number of factors must be considered. These factors include: type of wet suit, diving depth, breathing equipment characteristics, nature of diving activity, accessory equipment, and weight belt. A list of approved buoyancy compensators is contained in the NAVSEA/00C Authorized for Navy Use List (ANU).

As a buoyancy compensating device, the compensator can be inflated by a low-pressure inflator connected to the first-stage regulator, or an oral inflation tube. Any buoyancy compensator selected for Navy use must have an over-pressure relief valve. The compensator is used in conjunction with the diver weights to control buoyancy in the water column by allowing the diver to increase displacement through inflation of the device, or to decrease displacement by venting. Training and practice under controlled conditions are required to master the buoyancy compensation technique. Rapid, excessive inflation can cause excessive buoyancy and uncontrolled ascent. The diver must systematically vent air from the compensator during ascent to maintain proper control. Weights installed in a vest type buoyancy compesator must be jettisonable.

Refer to the appropriate technical manual for complete operations and maintenance instructions for the equipment. At the dive supervisor's discretion, when using a variable volume dry suit (VVDS), a Buoyancy Compensator is not required.

CAUTION **Prior to use of VVDS as a buoyancy compensator, divers must be thoroughly familiar with its use.**

7-2.3.4 **Weight Belt.** SCUBA is designed to have nearly neutral buoyancy. With full tanks, a unit tends to have negative buoyancy, becoming slightly positive as the air supply is consumed. Most divers are positively buoyant and need to add extra weight to achieve a neutral or slightly negative status. This extra weight is furnished by a weighted belt worn outside of all other equipment and strapped so that it can easily released in the event of an emergency.

Each diver may select the style and size of belt and weights that best suit the diver. A number of different models are available. A weight belt shall meet certain basic standards: the buckle must have a quick-release feature, easily operated by either hand; the weights (normally made of lead) should have smooth edges so as not to chafe the diver's skin or damage any protective clothing, and the belt should be made of rot- and mildew-resistant fabric, such as nylon webbing.

7-2.3.5 **Knife.** Several types of knives are available. For EOD and other special missions, a nonmagnetic knife designed for use when diving near magnetic-influence mines is used.

Knives may have single- or double-edged blades with chisel or pointed tips. The most useful knife has one sharp edge and one saw-toothed edge. All knives must be kept sharp.

The knife must be carried in a suitable scabbard and worn on the diver's life preserver, hip, thigh, or calf. The knife must be readily accessible, must not interfere with body movement, and must be positioned so that it will not become fouled while swimming or working. The scabbard should hold the knife with a positive but easily released lock.

The knife and scabbard must not be secured to the weight belt. If the weights are released in an emergency, the knife may be also dropped unintentionally.

7-2.3.6 **Swim Fins.** Swim fins increase the efficiency of the diver, permitting faster swimming over longer ranges with less expenditure of energy. Swim fins are made of a variety of materials and styles.

Each feature—flexibility, blade size, and configuration—contributes to the relative power of the fin. A large blade will transmit more power from the legs to the water, provided the legs are strong enough to use a larger blade. Small or soft blades should be avoided. Ultimately, selection of blade type is a matter of personal preference based on the diver's strength and experience.

7-2.3.7 **Wrist Watch.** Analog diver's watches must be waterproof, pressure proof, and equipped with a rotating bezel outside the dial that can be set to indicate the elapsed time of a dive. A luminous dial with large numerals is also necessary. Additional features such as automatic winding, nonmagnetic components, and stop watch action are available. Digital watches, with a stop watch feature to indicate the elapsed time of a dive, are also approved for Navy use.

7-2.3.8 **Depth Gauge.** The depth gauge measures the pressure created by the water column above the diver and is calibrated to provide a direct reading of depth in feet of sea water. It must be designed to be read under conditions of limited visibility. The gauge mechanism is delicate and should be handled with care. Accurate depth determination is important to a diver's safety. The accuracy of a gauge must be checked in accordance with the planned maintenance system or whenever a malfunction is suspected. This can be done by taking the gauge to a known depth and checking the reading, or by placing it in a recompression chamber or test pressure chamber for depth comparison.

7-3 OPTIONAL EQUIPMENT FOR SCUBA OPERATIONS

The requirements of a specific diving operation determine which items of optional diving equipment may be necessary. This section lists some of the equipment that may be used.

- Protective clothing
  - Wet suit
  - Variable volume dry suit
  - Gloves
  - Hoods
  - Boots or hard-soled shoes

- Whistle
- Slate and pencil
- Tools and light
- Signal flare
- Tool bag
- Acoustic beacons
- Lines and floats
- Wrist compass
- Witness float
- Snorkel
- Submersible cylinder pressure gauge (see note)
- Chem light and strobe light

NOTE **Submersible cylinder pressure gauge is required when using K valve**

7-3.1 **Protective Clothing.** A diver needs some form of protection from cold water, from heat loss during long exposure in water of moderate temperature, from chemical or bacterial pollution in the water, and from the hazards posed by marine life and underwater obstacles. Wet suit, or a dry suit with or without thermal underwear in Figure 7-5 can provide protection.

7-3.1.1 **Wet Suits.** The wet suit is a form-fitting suit, usually made of closed-cell neoprene. The suit traps a thin layer of water next to the diver's skin, where it is warmed by the diver's body. Wet suits are available in thicknesses of 1/8-, 3/16-, 3/8-, and 1/2-inch, with the thickest providing better insulation. The selection of the type of wet suit used is left to each diver. Standard size suits are available at most commercial diving shops. Proper fit is critical in the selection of a wet suit. The suit must not restrict the diver's movements. A custom-fitted suit is recommended. The performance of a suit depends upon suit thickness, water temperature, and water depth.

7-3.1.2 **Dry Suits.** The Variable Volume Dry Suit (VVDS) has proven to be effective in keeping divers warm in near-freezing water. It is typically constructed of 1/4-inch closed-cell neoprene with nylon backing on both sides. Boots are provided as an integral part of the suit, but the hood and three finger gloves are usually separate. The suit is entered by means of a water- and pressure-proof zipper. Inflation is controlled using inlet and outlet valves, which are fitted into the suit. Air is supplied from a pressure reducer on an auxiliary cylinder or from the emergency gas supply or the SCUBA bottle. About 0.2 actual cubic foot of air is required for normal inflation. Because of this inflation, slightly more weight than would be used with a wet suit must be carried. Normally, thermal underwear can be worn under the suit for insulation.

Wet or dry suits can be worn with hoods, gloves, boots, or hard-soled shoes depending upon conditions. If the diver will be working under conditions where the suit may be easily torn or punctured, the diver should be provided with additional protection such as coveralls or heavy canvas chafing gear.

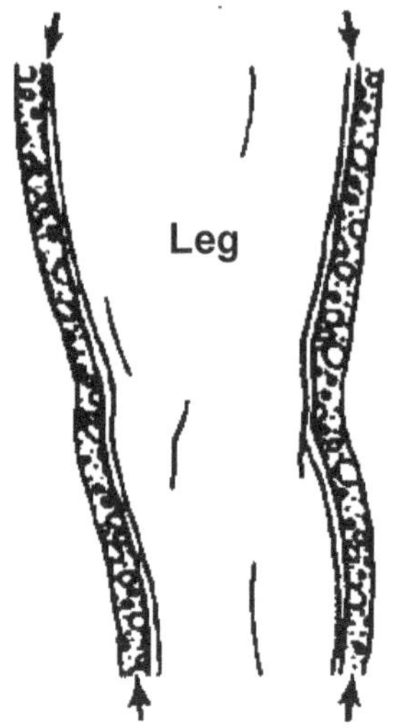

Figure 7-5. Protective Clothing.

7-3.1.3 **Gloves.** Gloves are an essential item of protective clothing. They can be made of leather, cloth, or rubber, depending upon the degree and type of protection required. Gloves shield the hands from cuts and chafing, and provide protection from cold water. Some styles are designed to have insulating properties but may limit the diver's dexterity.

7-3.1.4 **Writing Slate.** A rough-surfaced sheet of acrylic makes an excellent writing slate for recording data, carrying or passing instructions, and communicating between divers. A grease pencil or graphite pencil should be attached to the slate with a lanyard.

7-3.1.5 **Signal Flare.** A signal flare is used to attract attention if the diver has surfaced away from the support crew. Any waterproof flare that can be carried and safely ignited by a diver can be used, but the preferred type is the MK 99 MOD 3 (NSN 1370-01-177-4072; pouch is NSN 1370-01-194-0844). These are day-or-night flares that give off a heavy orange smoke for day time and a brilliant red light at night. Each signal lasts for approximately 45 seconds and will withstand submersion up to depths of 200 fsw without adverse effects. A hexagon shaped end cap marked SMOKE is threaded into the smoke assembly and a round shaped end cap with eight grooves marked FLARE is threaded onto the flare assembly. Also available are the MK 131 MOD 0 (NSN 1370-01-252-0318) and MK 132 MOD 0 (NSN 1370-01-252-0317). The MK 131 is for day time distress signaling while the MK 132 is for night. The only difference between the MK 99 and the MK 131/132, other than the fact that the MK 99 is a combined day/night signal flare which gives off yellow smoke and light, is that the MK 99 satisfies magnetic effect limits of MIL-M-19595 for explosive ordinance disposal (EOD) usage. Flares should be handled with care. For safety, each diver should carry a maximum of two flares. All divers/combat swimmers engaged in submarine Dry Deck Shelter operations should stow flares in hangar prior to reentering the host submarine.

7-3.1.6 **Acoustic Beacons.** Acoustic beacons or pingers are battery-operated devices that emit high-frequency signals when activated. The devices may be worn by divers to aid in keeping track of their position or attached to objects to serve as fixed points of reference. The signals can be picked up by hand-held sonar receivers, which are used in the passive or listening mode, at ranges of up to 1,000 yards. The hand-held sonar enables the search diver to determine the direction of the signal source and swim toward the pinger using the heading noted on a compass.

7-3.1.7 **Lines and Floats.** A lifeline should be used when it is necessary to exchange signals, keep track of the diver's location, or operate in limited visibility. There are three basic types of lifelines: the tending line, the float line, and the buddy line.

A single diver will be tended with either a tending line or a float line. When direct access to the surface is not available a tending line is mandatory. A float line may not be used.

The float line reaches from the diver to a suitable float on the surface. This float can be a brightly painted piece of wood, an empty sealed plastic bottle, a life ring, or any similar buoyant, visible object. An inner tube with a diving flag attached makes an excellent float and provides a hand-hold for a surfaced diver. If a pair of divers are involved in a search, the use of a common float gives them a rendezvous point. Additional lines for tools or other equipment can be tied to the float. A buddy line, 6 to 10 feet long, is used to connect the diver partners at night or when visibility is poor.

Any line used in SCUBA operations should be strong and have neutral or slightly positive buoyancy. Nylon, Dacron, and manila are all suitable materials. Always attach a lifeline to the diver, never to a piece of equipment that may be ripped away or may be removed in an emergency.

7-3.1.8 **Snorkel.** A snorkel is a simple breathing tube that allows a diver to swim on the surface for long or short distances face-down in the water. This permits the diver to search shallow depths from the surface, conserving the SCUBA air supply. When snorkels are used for skin diving, they are often attached to the face mask with a lanyard or rubber connector to the opposite side of the regulator.

7-3.1.9 **Compass.** Small magnetic compasses are commonly used in underwater navigation. Such compasses are not highly accurate, but can be valuable when visibility is poor. Submersible wrist compasses, watches, and depth gauges covered by NAVSUPINST 5101.6 (series) are items controlled by the Nuclear Regulatory Commission and require leak testing and reporting every 6 months.

7-3.1.10 **Submersible Cylinder Pressure Gauge.** The submersible cylinder pressure gauge provides the diver with a continual read-out of the air remaining in the cylinder(s). Various submersible pressure gauges suitable for Navy use are commercially available. Most are equipped with a 2- to 3-foot length of high-pressure rubber hose with standard fittings, and are secured directly into the first stage of the regulator. When turning on the cylinder air, the diver should turn the face of the gauge away in the event of a blowout. When worn, the gauge and hose should be tucked under a shoulder strap or otherwise secured to avoid its entanglement with bottom debris or other equipment. The gauge must be calibrated in accordance with the equipment planned maintenance system.

7-4 AIR SUPPLY

An important early step in any SCUBA dive is computing the air supply requirement. The air supply requirement is a function of the expected duration of the dive at a specific working depth. The duration of the air supply in the SCUBA cylinders depends on the depth at which the air is delivered. Air consumption rate increases with depth.

7-4.1 **Duration of Air Supply.** The duration of the air supply of any given cylinder or combination of cylinders depends upon:

- The diver's consumption rate, which varies with the diver's work rate,
- The depth of the dive, and
- The capacity and minimum pressure of the cylinder(s).

Temperature is usually not significant in computing the duration of the air supply, unless the temperature conditions are extreme. When diving in extreme temperature conditions, Charles'/Gay-Lussac's law must be applied.

There are three steps in calculating how long a diver's air supply will last:

1. Calculate the diver's consumption rate by using this formula:

$$C = \frac{D + 33}{33} \times RMV$$

Where:

C	=	Diver's consumption rate, standard cubic feet per minute (scfm)
D	=	Depth, fsw
RMV	=	Diver's Respiratory Minute Volume, actual cubic feet per minute (acfm) (from Figure 3-6)

2. Calculate the available air capacity provided by the cylinders. The air capacity must be expressed as the capacity that will actually be available to the diver, rather than as a total capacity of the cylinder. The formula for calculating the available air capacity is:

$$V_a = \frac{P_c - P_m}{14.7} \times FV \times N$$

Where:

P_c = Measured cylinder pressure, psig

P_m = Minimum pressure of cylinder, psig

FV = Floodable Volume (scf)

N = Number of cylinders

V_a = Capacity available (scf)

3. Calculate the duration of the available capacity (in minutes) by using this formula:

$$\text{Duration} = \frac{V_a}{C}$$

Where:

V_a = Capacity available, scf

C = Consumption rate, scfm

Sample Problem. Determine the duration of the air supply of a diver doing moderate work at 70 fsw using twin 72-cubic-foot steel cylinders charged to 2,250 psig.

1. Calculate the diver's consumption rate in scfm. According to Figure 3-6, the diver's consumption rate at depth is 1.4 acfm.

$$C = \frac{D + 33}{33} \times RMV$$

$$= \frac{70 + 33}{33} \times 1.4$$

$$= 4.37 \text{ scfm}$$

2. Calculate the available air capacity provided by the cylinders. Table 7-1 contains the cylinder data used in this calculation:

- Floodable Volume = 0.420 scf
- Rated working pressure = 2250 psig
- Reserve pressure for twin 72-cubic-foot cylinders = 250 psig

$$V_a = \frac{P_c - P_m}{14.7} \times FV \times N$$
$$= \frac{2250-250}{14.7} \times 0.420 \times 2$$
$$= 114 \text{ scf}$$

3. Calculate the duration of the available capacity.

$$\text{Duration} = \frac{V_a}{C}$$
$$= \frac{114 \text{ scf}}{4.37 \text{ scfm}}$$
$$= 26 \text{ minutes}$$

The total time for the dive, from initial descent to surfacing at the end of the dive, is limited to 26 minutes.

7-4.2 **Compressed Air from Commercial Sources.** Compressed air meeting the established standards can usually be obtained from Navy sources. In the absence of appropriate Navy sources, air may be procured from commercial sources. Usually, any civilian agency or firm which handles compressed oxygen can provide pure compressed air. Air procured from commercial sources must meet the requirements of Grade A Source I or Source II air as specified by FED SPEC BB-A-1034B. Refer to Table 4-2 in Chapter 4 for the air purity requirements.

7-4.3 Methods for Charging SCUBA Cylinders.

NOTE Paragraph 7-4.5 addresses safety precautions for charging and handling cylinders.

SCUBA cylinders shall be charged only with air that meets diving air purity standards. A diving unit can charge its own cylinders by one of two accepted methods: (1) by cascading or transferring air from banks of large cylinders into the SCUBA tanks; or (2) by using a high-pressure air compressor. Cascading is the fastest and most efficient method for charging SCUBA tanks. The NAVSEA/00C ANU list lists approved high-pressure compressors and equipment authorized for SCUBA air sources.

The normal cascade system consists of supply flasks connected together by a manifold and feeding into a SCUBA high-pressure whip. This whip consists of a SCUBA yoke fitting, a pressure gauge, and a bleed valve for relieving the pressure in the lines after charging a cylinder. A cascade system, with attached whip, is shown in Figure 7-6.

SCUBA charging lines shall be fabricated using SAE 100R7 hose for 3,000 psi service and SAE 100R8 hose for 5,000 psi service. The service pressure of the

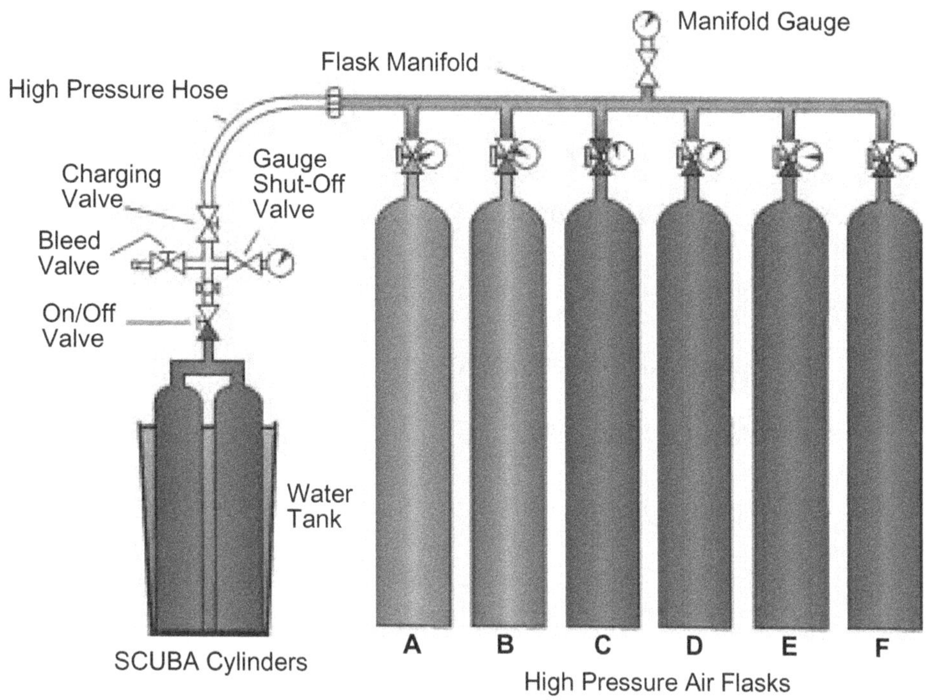

Figure 7-6. Cascading System for Charging SCUBA Cylinders.

SCUBA charging lines shall be no greater than the working pressure of the hose used.

The working pressure of a hose is determined as one-fourth of its burst pressure. While this criteria for working pressure was developed based on the characteristics of rubber hose, it has also been determined to be appropriate for use with the plastic hoses cited above.

Fleet units using charging lines shall not exceed the rated working pressure of the hose. If the charging line working pressure rating does not meet service requirements, restrict the service pressure of the hose to its working pressure and initiate replacement action immediately.

The use of strain reliefs made from cable, chain, 21-thread, or 3/8-inch nylon, married at a minimum of every 18 inches and at the end of the hose, is a required safety procedure to prevent whipping in the event of hose failure under pressure. Marrying cord shall be 1/8-inch nylon or material of equivalent strength. Tie wraps, tape, and marlin are not authorized for this purpose.

7-4.4 **Operating Procedures for Charging SCUBA Tanks.** Normally, SCUBA tanks are charged using the following operating procedures (OPs), which may be tailored to each unit:

1. Determine that the cylinder is within the hydrostatic test date.

2. Check the existing pressure in the SCUBA cylinder with an accurate pressure gauge.

3. Attach the cylinder to the yoke fitting on the charging whip, and attach the safety strain relief.

4. For safety and to dissipate heat generated in the charging process, when facilities are available, immerse the SCUBA cylinder in a tank of water while it is being filled. A 55-gallon drum is a suitable container for this purpose.

5. Tighten all fittings in the system.

6. Close the bleed valve.

7. Place reserve mechanism lever in the open (lever down) position.

8. Open the cylinder (on/off) valve. This valve is fully opened with about two turns on the handle, counter-clockwise. However, the valve must not be used in a fully open position as it may stick or be stripped if force is used to open a valve that is incorrectly believed to be closed. The proper procedure is to open the valve fully and then close or back off one-quarter to one-half turn. This will not impede the flow of air.

9. Open the supply flask valve.

10. Slowly open the charging valve. The sound of the air flowing into the SCUBA cylinder is noticeable. The operator will control the flow so that the pressure in the cylinder increases at a rate not to exceed 400 psig per minute. If unable to submerge SCUBA cylinders during charging, the charging rate must not exceed 200 psig per minute. The rate of filling must be controlled to prevent overheating; the cylinder must not be allowed to become too hot to touch.

11. Monitor the pressure gauge carefully. When the reading reaches the rated pressure for the SCUBA cylinder, close the valve on the first cylinder and take a reading.

12. Close the charging valve.

13. Close the on/off valve on the SCUBA cylinder.

14. Ensure that all valves in the system are firmly closed.

15. Let the SCUBA cylinder cool to room temperature. Once the cylinder is cool, the pressure will have dropped and you may need to top off the SCUBA cylinder.

7-4.4.1 **Topping off the SCUBA Cylinder.** Follow this procedure to top off a SCUBA cylinder:

1. Open the on/off valve on the SCUBA cylinder.
2. Select a supply flask with higher pressure than the SCUBA rated limit.
3. Open the supply valve on the flask.
4. Throttle the charging valve to bring the SCUBA cylinder up to the rated limit.
5. Close all valves.
6. Open the bleed valve and depressurize the lines.
7. When air has stopped flowing through the bleed valve, disconnect the SCUBA cylinder from the yoke fitting.
8. Reset the reserve mechanism (lever in up position).

In the absence of high-pressure air systems, large-volume air compressors can be used to charge SCUBA cylinders directly. However, few compressors can deliver air in sufficient quantity at the needed pressure for efficient operation. Small compressors should be used only if no other suitable source is available.

If a suitable compressor is available, the basic charging procedure will be the same as that outlined for cascading except that the compressor will replace the bank of cylinders. Special considerations that apply when using air compressors are:

- The compressor must be listed in the NAVSEA/00C ANU list if it is not part of a certified system.
- The compressor must deliver air that meets the established purity standards.
- The compressor shall be equipped with ANU particulate filters. Chemically active filters are not authorized.
- An engine-driven compressor must always be mounted so there is no danger of taking in exhaust fumes from the engine, stack gas, or other contaminated air from local sources.
- Only approved diving compressor lubricants are to be used in accordance with PMS procedures or manufacturer's recommendations.

Additional information on using air compressors is found in paragraph 8-7.2.2.

7-4.5 **Safety Precautions for Charging and Handling Cylinders.** The following safety rules apply to charging and handling SCUBA cylinders:

- Carry cylinders by holding the valve and body of the cylinder. Avoid carrying a cylinder by the backpack or harness straps as the quick-release buckle can be accidentally tripped or the straps may fail.

- Do not attempt to fill any cylinder if the hydrostatic test date has expired or if the cylinder appears to be substandard. Dents, severe rusting, bent valves, frozen reserve mechanisms, or evidence of internal contamination (e.g., water scales or rust) are all signs of unsuitability. See CGA Pamphlet C-6, Standards for Visual Inspection of Compressed Gas Cylinders.

- Always use gauges to measure cylinder pressure. Never point the dial of a gauge to which pressure is being applied toward the operators face.

- Never work on a cylinder valve while the cylinder is charged.

- Make sure that the air reserve mechanism is open (lever down) before charging.

- Use only compressed air for filling conventional SCUBA cylinders. Never fill SCUBA cylinders with oxygen. Air is color-coded black, while oxygen is color-coded green.

- Tighten all fittings before pressurizing lines.

- When fully charged, close the air reserve (lever up). Mark the filled tank to indicate the pressure to which it was charged.

- Handle charged cylinders with care. If a charged cylinder is damaged or if the valve is accidentally knocked loose, the cylinder tank can become an explosive projectile. A cylinder charged to 2,000 psi has enough potential energy to propel itself for some distance, tearing through any obstructions in its way.

- Store filled cylinders in a cool, shaded area. Never leave filled cylinders in direct sunlight.

- Cylinders should always be properly secured aboard ship or in a diving boat.

7-5 PREDIVE PROCEDURES

Predive procedures for SCUBA operations include equipment preparation, diver preparation, and conducting a predive inspection before the divers enter the water.

7-5.1 **Equipment Preparation.** Prior to any dive, all divers must carefully inspect their own equipment for signs of deterioration, damage, or corrosion. The equipment must be tested for proper operation. Predive preparation procedures must be standardized, not altered for convenience, and must be the personal concern of each diver.

7-5.1.1 **Air Cylinders.**

- Inspect air cylinder exteriors and valves for rust, cracks, dents, and any evidence of weakness.
- Inspect O-ring.
- Verify that the reserve mechanism is closed (lever in up position) signifying a filled cylinder ready for use.
- Gauge the cylinders according to the following procedure:

  1. Attach pressure gauge to O-ring seal face of the on/off valve.
  2. Close gauge bleed valve and open air reserve mechanism (lever in down position). Slowly open the cylinder on/off valve, keeping a cloth over the face of the gauge.
  3. Read pressure gauge. The cylinder must not be used if the pressure is not sufficient to complete the planned dive.
  4. Close the cylinder on/off valve and open the gauge bleed valve.
  5. When the gauge reads zero, remove the gauge from the cylinder.
  6. Close the air reserve mechanism (lever in up position).
  7. If the pressure in cylinders is 50 psi or greater over rating, open the cylinder on/off valve to bleed off excess and regauge the cylinder.

7-5.1.2 **Harness Straps and Backpack.**

- Check for signs of rot and excessive wear.
- Adjust straps for individual use and test quick-release mechanisms.
- Check backpack for cracks and other unsafe conditions.

7-5.1.3 **Breathing Hoses.**

- Check the hoses for cracks and punctures.
- Test the connections of each hose at the regulator and mouthpiece assembly by tugging on the hose.
- Check the clamps for corrosion and damage; replace as necessary and in accordance with PMS procedures.

7-5.1.4 **Regulator.**

1. Ensure over-bottom pressure of first stage regulator has been set to a minimum of 135 psig or in accordance with manufacturer's recommendations within the past year.

2. Attach regulator to the cylinder manifold, ensuring that the O-ring is properly seated.

3. Crack the cylinder valve open and wait until the hoses and gauges have equalized.

4. Next open the cylinder valve completely and then close (back off) one-quarter turn.

5. Check for any leaks in the regulator by listening for the sound of escaping air. If a leak is suspected, determine the exact location by submerging the valve assembly and the regulator in a tank of water and watch for escaping bubbles. Frequently the problem can be traced to an improperly seated regulator and is corrected by closing the valve, bleeding the regulator, detaching and reseating. If the leak is at the O-ring and reseating does not solve the problem, replace the O-ring and check again for leaks.

7-5.1.5 **Life Preserver/Buoyancy Compensator (BC).**

- Orally inflate preserver to check for leaks and then squeeze out all air. The remaining gas should be removed after entry into the water by rolling onto the back and depressing the oral inflation tube just above the surface. Never suck the air out, as it may contain excessive carbon dioxide.
- Inspect the carbon dioxide cartridges to ensure they have not been used (seals intact) and are the proper size for the vest being used and for the depth of dive.
- The cartridges shall be weighed in accordance with the Planned Maintenance System.
- The firing pin should not show wear and should move freely.
- The firing lanyards and life preserver straps must be free of any signs of deterioration.
- When the life preserver inspection is completed, place it where it will not be damaged. Life preservers should never be used as a buffer, cradle, or cushion for other gear.

7-5.1.6 **Face Mask.**

- Check the seal of the mask and the condition of the head strap.
- Check for cracks in the skirt and faceplate.

7-5.1.7 **Swim Fins.**

- Check straps for signs of cracking.
- Inspect blades for signs of cracking.

7-5.1.8 **Dive Knife.**

- Test the edge of the knife for sharpness.
- Ensure the knife is fastened securely in the scabbard.
- Verify that the knife can be removed from the scabbard without difficulty, but will not fall out.

7-5.1.9 **Snorkel.**

- Inspect the snorkel for obstructions.
- Check the condition of the mouthpiece.

7-5.1.10 **Weight Belt.**

- Check the condition of the weight belt.
- Make sure that the proper number of weights are secure and in place.
- Verify that the quick-release buckle is functioning properly.

7-5.1.11 **Submersible Wrist Watch.**

- Ensure wrist watch is wound and set to the correct time.
- Inspect the pins and strap of the watch for wear.

7-5.1.12 **Depth Gauge and Compass.**

- Inspect pins and straps.
- If possible, check compass with another compass.
- Make comparative checks on depth gauges to ensure depth gauges read zero fsw on the surface.

7-5.1.13 **Miscellaneous Equipment.**

- Inspect any other equipment that will be used on the dive as well as any spare equipment that may be needed during the dive including spare regulators, cylinders, and gauges.
- Check all protective clothing, lines, tools, flares, and other optional gear.

7-5.2 **Diver Preparation and Brief.** When the divers have completed inspecting and testing their equipment, they shall report to the Diving Supervisor. The divers shall be given a predive briefing of the dive plan. This briefing is critical to the success and safety of any diving operation and shall be concerned with only the dive about to begin. All personnel directly involved in the dive should be included in the briefing. Minimum items to be covered are:

- Dive objectives
- Time and depth limits for the dive
- Task assignments
- Buddy assignments
- Work techniques and tools
- Phases of the dive
- Route to the work site
- Special signals
- Anticipated conditions
- Anticipated hazards
- Emergency procedures (e.g., unconscious diver, trapped diver, loss of air, aborted dive, injured diver, lost diver, etc.)

When the Diving Supervisor determines all requirements for the dive have been met, the divers may dress for the dive.

7-5.3 **Donning Gear.** Although SCUBA divers should be able to put on all gear themselves, the assistance of a tender is encouraged. Dressing sequence is important as the weight belt must be outside of all backpack harness straps and other equipment in order to facilitate its quick release in the event of an emergency. The following is the recommended dressing sequence to be observed:

1. Protective clothing. Ensure adequate protection is provided with a wet suit.
2. Booties and hood.
3. Dive knife.
4. Life preserver, with inflation tubes in front and the actuating lanyards exposed and accessible.
5. SCUBA. Most easily donned with the tender holding the cylinders in position while the diver fastens and adjusts the harness. The SCUBA should be worn centered on the diver's back as high up as possible but not high enough to interfere with head movement. All quick-release buckles must be positioned so that they can be reached by either hand. All straps must be pulled snug so the cylinders are held firmly against the body. The ends of the straps must hang free so the quick-release feature of the buckles will function. If the straps are too long, they should be cut and the ends whipped with small line or a plastic sealer. At this time, the cylinder on/off valve should be opened fully and then

backed off one-quarter to one-half turn. Ensure buoyancy compensator whip is connected to the buoyancy compensator.

6. Accessory equipment (diving wrist watch, depth gauge, snorkel).

7. Weight belt.

8. Gloves.

9. Swim fins.

10. Face mask or full face mask.

7-5.4 **Predive Inspection.** The divers must report to the Diving Supervisor for a final inspection. During this final predive inspection the Diving Supervisor must:

1. Ensure that the divers are physically and mentally ready to enter the water.

2. Verify that all divers have all minimum required equipment (SCUBA, face mask, life preserver or buoyancy compensator, weight belt, dive knife, scabbard, swim fins, watch and depth gauge). When diving SCUBA and a buddy line is used, only one depth gauge and one watch per dive team is required.

3. Verify that the cylinders have been gauged and that the available volume of air is sufficient for the planned duration of the dive.

4. Ensure that all quick-release buckles and fastenings can be reached by either hand and are properly rigged for quick release.

5. Verify that the weight belt is outside of all other belts, straps, and equipment and will not become pinched under the bottom edge of the cylinders.

6. Verify that the life preserver or buoyancy compensator is not constrained and is free to expand, and that all air has been evacuated.

7. Check position of the knife to ensure that it will remain with the diver no matter what equipment is left behind.

8. Ensure that the cylinder valve is open fully and backed off one-quarter to one-half turn.

9. Ensure that the hose supplying air passes over the diver's right shoulder and the exhaust hose on the double-hose unit passes over the left shoulder. Double-hose regulators are attached so that the exhaust ports face up when the tank is standing upright.

10. With mouthpiece or full face mask in place, breathe in and out for several breaths, ensuring that the demand regulator and check valves are working correctly.

11. With a single-hose regulator, depress and release the purge button at the mouthpiece and listen for any sound of leaking air. Breathe in and out several times ensuring valves are working correctly.

12. Give the breathing hoses and mouthpiece a final check; ensure that none of the connections have been pulled open during the process of dressing.

13. Check that the air reserve mechanism lever is up (closed position).

14. Conduct a brief final review of the dive plan.

15. Verify that dive signals are displayed and personnel and equipment are ready to signal other vessels in the event of an emergency.

7-6 WATER ENTRY AND DESCENT

The divers are now ready to enter the water, where their SCUBA shall be given another brief inspection by their dive partners or tenders prior to descent.

7-6.1 **Water Entry.** There are several ways to enter the water, with the choice usually determined by the nature of the diving platform (Figure 7-7). Whenever possible, entry should be made by ladder, especially in unfamiliar waters. Several basic rules apply to all methods of entry:

- Look before jumping or pushing off from the platform or ladder.
- Tuck chin into chest and hold the cylinders with one hand to prevent the manifold from hitting the back of the head.
- Hold the mask in place with the fingers and the mouthpiece in place with the heel of the hand.

7-6.1.1 **Step-In Method.** The step-in method is the most frequently used, and is best used from a stable platform or vessel. The divers should simply take a large step out from the platform, keeping legs in an open stride. They should try to enter the water with a slightly forward tilt of the upper body so that the force of entry will not cause the cylinder to hit the back of the head.

7-6.1.2 **Rear Roll Method.** The rear roll is the preferred method for entering the water from a small boat. A fully outfitted diver standing on the edge of a boat would upset the stability of the craft and would be in danger of falling either into the boat or into the water. To execute a rear roll, the diver sits on the gunwale of the boat, facing inboard. With chin tucked in and one hand holding the mask and mouthpiece in place, the diver rolls backward, basically moving through a full backward somersault.

Front jump or step-in. On edge of platform, one hand holding face mask and regulator, the other holding the cylinders, the diver takes a long step forward, keeping his legs astride.

Rear roll. The diver, facing inboard, sits on the gunwale. With chin tucked in, holding his mask, mouthpiece, and cylinders, the diver rolls backwards, basically completing a full backward somersault.

Side roll. Tender assists diver in taking a seated position. Tender stands clear as diver holds mask and cylinders and rolls into the water.

Front roll. Diver sits on edge of platform with a slight forward lean to offset the weight of the cylinders. Holding his mask and cylinders, the diver leans forward.

Figure 7-7. SCUBA Entry Techniques.

7-6.1.3 **Entering the Water from the Beach.** Divers working from the beach choose their method of entry according to the condition of the surf and the slope of the bottom. If the water is calm and the slope gradual, the divers can walk out, carrying their swim fins until they reach water deep enough for swimming. In a moderate to high surf, the divers, wearing swim fins, should walk backwards into the waves until they have enough depth for swimming. They should gradually settle into the waves as the waves break around them.

Rear step-in. The diver steps backward pushing himself away with his feet.

Figure 7-7. SCUBA Entry Techniques (continued).

7-6.2 **Pre-descent Surface Check.** Once in the water, and before descending to operating depth, the divers make a final check of their equipment. They must:

- Make a breathing check of the SCUBA. Breathing should be easy, with no resistance and no evidence of water leaks.
- Visually check dive partner's equipment for leaks, especially at all connection points (i.e., cylinder valve, hoses at regulator and mouthpiece).
- Check partner for loose or entangled straps.
- Check face mask seal. A small amount of water may enter the mask upon the diver's entry into the water. The mask may be cleared through normal methods (see paragraph 7-7.2).
- Check buoyancy. SCUBA divers should strive for neutral buoyancy. When carrying extra equipment or heavy tools, the divers might easily be negatively buoyant unless the weights are adjusted accordingly.
- If wearing a dry suit, check for leaks. Adjust suit inflation for proper buoyancy.
- Orient position with the compass or other fixed reference points.

When satisfied that all equipment checks out properly, the divers report their readiness to the Diving Supervisor. The Diving Supervisor directs the divers to zero their watches and bottom time begins. The Diving Supervisor gives a signal to descend and the divers descend below the surface.

7-6.3 **Surface Swimming.** The diving boat should be moored as near to the dive site as possible. While swimming, dive partners must keep visual contact with each other and other divers in the group. They should be oriented to their surroundings to avoid swimming off course. The most important factor in surface swimming with SCUBA is to maintain a relaxed pace to conserve energy. The divers should keep their masks on and breathe through the snorkel. When surface swimming with a SCUBA regulator, hold the mouthpiece so that air does not free-flow from the system.

Divers should use only their legs for propulsion and employ an easy kick from the hips without lifting the swim fins from the water. Divers can rest on their backs and still make headway by kicking. Swimming assistance can be gained by partially inflating the life preserver or buoyancy compensator. However, the preserver must be deflated again before the dive begins.

7-6.4 **Descent.** The divers may swim down or they may use a descending line to pull themselves down. The rate of descent will generally be governed by the ease with which the divers will be able to equalize the pressure in their ears and sinuses, but it should never exceed 75 feet per minute. If either diver experiences difficulty in clearing, both divers must stop and ascend until the situation is resolved. If the problem persists after several attempts to equalize, the dive shall be aborted and both divers shall return to the surface. When visibility is poor, the divers should extend an arm to ward off any obstructions.

Upon reaching the operating depth, the divers must orient themselves to their surroundings, verify the site, and check the underwater conditions. If conditions appear to be radically different from those anticipated and seem to pose a hazard, the dive should be aborted and the conditions reported to the Diving Supervisor. The dive should be aborted if the observed conditions call for any major change in the dive plan. The divers should surface, discuss the situation with the Diving Supervisor, and modify the dive plan.

7-7 UNDERWATER PROCEDURES

In a SCUBA dive, bottom time is at a premium because of a limited supply of air. Divers must pace their work, conserve their energy, and take up each task or problem individually. At the same time they must be flexible. They must be ready to abort the dive at any time they feel that they can no longer progress toward the completion of their mission or when conditions are judged unsafe. The divers must be alert for trouble at all times and must monitor the condition of the dive partner constantly.

7-7.1 **Breathing Technique.** When using SCUBA for the first time, a novice diver is likely to experience anxiety and breathe more rapidly and deeply than normal. The diver must learn to breathe in an easy, slow rhythm at a steady pace. The rate of work should be paced to the breathing cycle, rather than changing the breathing to support the work rate. If a diver is breathing too hard, he should pause in the work until breathing returns to normal. If normal breathing is not restored soon, the diver must signal the dive partner and break off the operation, and together they should ascend to the surface.

Some divers, knowing that they have a limited air supply, will attempt to conserve air by holding their breath. One common technique is to skip-breathe: to insert an unnatural, long pause between each breath.

WARNING **Skip-breathing may lead to hypercapnia and is prohibited.**

Increased breathing resistance results from the design of the equipment and increased air density. For normal diving, a marked increase of breathing resistance should not occur until the primary air supply has been almost depleted. This increase in breathing resistance is a signal to the diver to activate the reserve air supply and to begin an ascent with the partner immediately. When equipped with a submersible bottle gauge, the diver shall monitor his air supply pressure and must terminate the dive whenever bottle pressure is reduced to 500 psi for a single bottle or 250 psi for a set of double bottles.

7-7.2 **Mask Clearing.** Some water seepage into the face mask is a normal condition and is often useful in defogging the lens. From time to time the quantity may build to a point that it must be removed. On occasion, a mask may become dislodged and flooded. To clear a flooded mask not equipped with a purge valve, the diver should roll to the side or look upward, so that the water will collect at the side or bottom of the mask. Using either hand, the diver applies a firm direct pressure on the opposite side or top of the mask and exhales firmly and steadily through the nose. The water will be forced out under the skirt of the mask. When the mask has a purge valve, the diver tilts his head so that the accumulated water covers the valve, presses the mask against the face and then exhales firmly and steadily through the nose. The increased pressure in the mask will force the water through the valve. Occasionally, more than one exhalation will be required (see Figure 7-8).

7-7.3 **Hose and Mouthpiece Clearing.** The mouthpiece and the breathing hoses can become flooded if the mouthpiece is accidentally pulled from the mouth. With a single-hose SCUBA this is not a serious problem since the hose (carrying air at medium pressure) will not flood and the mouthpiece can be cleared quickly by depressing the purge button as the mouthpiece is being replaced.

To clear a double-hose SCUBA regulator that has flooded, the diver, swimming in a horizontal position, should grasp the mouthpiece. The diver should then blow into the mouthpiece, forcing any water trapped in it out through the regulator's exhaust ports. The diver should carefully take a shallow breath. If water is still trapped in the mouthpiece, the diver should blow through it once more and resume normal breathing. If the diver is out of breath, he should roll over onto his back and the regulator will free flow.

7-7.4 **Swimming Technique.** In underwater swimming, all propulsion comes from the action of the legs. The hands are used for maneuvering. The leg kick should be through a large, easy arc with main thrust coming from the hips. The knees and

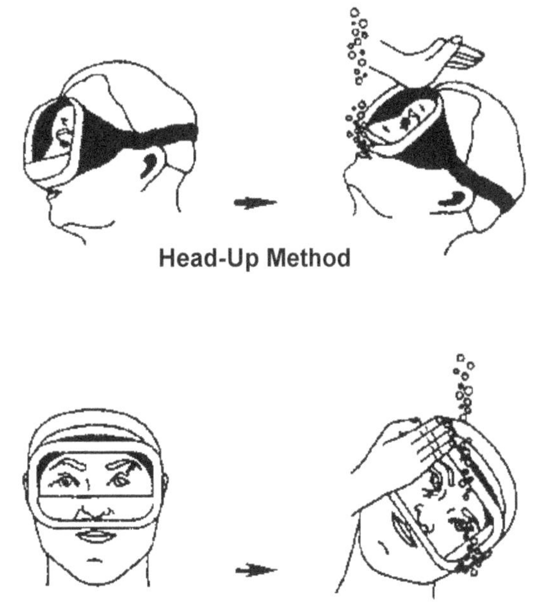

Figure 7-8. Clearing a Face Mask. To clear a flooded face mask, push gently on the upper or side portion of the mask and exhale through the nose into the mask. As water is forced out, tilt the head backward or sideway until the mask is clear.

ankles should be relaxed. The rhythm of the kick should be maintained at a level that will not tire the legs unduly or bring on muscle cramps.

7-7.5 **Diver Communications.** Some common methods of diver communications are: through-water communication systems, hand signals, slate boards, and line-pull signals. Communication between the surface and a diver can be best accomplished with through-water voice communications. However, when through-water communications are not available, hand signals or line-pull signals can be used.

7-7.5.1 **Through-Water Communication Systems.** Presently, several types of through-water communication systems are available for SCUBA diving operations. Acoustic systems provide one-way, topside-to-diver communications. The multidirectional audio signal is emitted through the water by a submerged transducer. Divers can hear the audio signal without signal receiving equipment. Amplitude Modulated (AM) and Single Sideband (SSB) systems provide round-robin, diver-to-diver, diver-to-topside, and topside-to-diver communications. Both the AM and SSB systems require transmitting and receiving equipment worn by the divers. AM systems provide a stronger signal and better intelligibility, but are restricted to line-of-sight use. SSB systems provide superior performance in and around obstacles. Before any through-water communication system is used, consult the NAVSEA/00C Authorized for Navy Use (ANU) list.

7-7.5.2 **Hand and Line-Pull Signals.** Navy divers shall only use hand signals that have been approved for Navy diving use. Figure 7-9 presents the U.S. Navy approved hand signals. Under certain conditions, special signals applicable to a specific mission may be devised and approved by the Diving Supervisor. If visibility is poor, the dive partners may be forced to communicate with line-pull signals on a buddy line. Line-pull signals are discussed in Table 8-3. Hand signals and line-pull signals should be delivered in a forceful, exaggerated manner so that there is no ambiguity and no doubt that a signal is being given. Every signal must be acknowledged.

7-7.6 **Buddy Diver Responsibilities.** The greatest single safety practice in Navy SCUBA operations is the use of the buddy system. Dive partners operating in pairs are responsible for both the assigned task and each other's safety. The basic rules for buddy diving are:

- Always maintain contact with the dive partner. In good visibility, keep the partner in sight. In poor visibility, use a buddy line.
- Know the meaning of all hand and line-pull signals.
- If a signal is given, it must be acknowledged immediately. Failure of a dive partner to respond to a signal must be considered an emergency.
- Monitor the actions and apparent condition of the dive partner. Know the symptoms of diving ailments. If at any time the dive partner appears to be in distress or is acting in an abnormal manner, determine the cause immediately and take appropriate action.
- Never leave a partner unless the partner has become trapped or entangled and cannot be freed without additional assistance. If surface assistance must be sought, mark the location of the distressed diver with a line and float or other locating device. Do not leave a partner if voice communications or line-pull signals are being used; contact the surface and await assistance or instructions.
- Establish a lost-diver plan for any dive. If partner contact is broken, follow the plan.
- If one member of a dive team aborts a dive, for whatever reason, the other member also aborts and both must surface.
- Know the proper method of buddy breathing.

7-7.7 **Buddy Breathing Procedure.** If a diver runs out of air or the SCUBA malfunctions, air may be shared with the dive partner. The preferred method of buddy breathing is the use of an octopus. As an alternative, the two divers may face each other and alternately breathe from the same mouthpiece while ascending. Buddy breathing may be used in an emergency and must be practiced so that each diver will be thoroughly familiar with the procedure.

1. The distressed diver should remain calm and signal the partner by pointing to SCUBA mouthpiece.

	Meaning/Signal	Comment
	**STOP** Clenched fist.	
	**SOMETHING IS WRONG** Hand flat, fingers together, palm out, thumb down then hand rocking back and forth on axis of forearm.	This is the opposite of Okay. The signal does not indicate an emergency.
 	**I AM OKAY or ARE YOU OKAY?** Thumb and forefinger making a circle with three remaining fingers extended (if possible).	Divers wearing mittens may not be able to extend three remaining fingers distinctly. Short range use.
 	**OKAY ON THE SURFACE (CLOSE)** Right hand raised overhead giving Okay signal with fingers. **OKAY ON THE SURFACE (DISTANT)** Both hands touching overhead with both arms bent at 45° angle.	Given when diver is close to pickup boat. Given when diver is at a distance from the pickup boat.
	**DISTRESS or HELP or PICK ME UP** Hand waving overhead (diver may also thrash hand in water).	Indicates immediate aid is required.
	**WHAT TIME? or WHAT DEPTH?** Diver points to either watch or depth gauge.	When indicating time, this signal is commonly used for bottom time remaining.
	**GO DOWN or GOING DOWN** Two fingers up, two fingers and thumb against palm.	
	**GO UP or GOING UP** Four fingers pointing up, thumb against palm.	
 	**I'M OUT OF AIR** Hand slashing or chopping at throat. **I NEED TO BUDDY BREATHE** Fingers pointing to mouth or regulator.	Indicates signaler is out of air. Signaler's regulator may be in or out of mouth.

Figure 7-9. SCUBA Hand Signals (page 1 of 3).

	Meaning/Signal	Comment
	**COME HERE** Hand to chest, repeated.	
	**ME or WATCH ME** Finger to chest, repeated.	
	**OVER, UNDER, or AROUND** Fingers together and arm moving in and over, under, or around movement.	Diver signals intention to move over, under, or around an object.
	**LEVEL OFF or HOW DEEP?** Fingers and thumb spread out and hand moving back and forth in a level position.	
	**GO THAT WAY** Fist clenched with thumb pointing up, down, right, or left.	Indicates which direction to swim.
	**WHICH DIRECTION?** Fingers clenched, thumb and hand rotating right and left.	
	**EAR TROUBLE** Diver pointing to either ear.	Divers should ascend a few feet. If problem continues, both divers must surface.
	**I'M COLD** Both arms crossed over chest.	
	**TAKE IT EASY OR SLOW DOWN** Hand extended, palm down, in short up-and-down motion.	
	**YOU LEAD, I'LL FOLLOW** Index fingers extended, one hand forward of the other.	

Figure 7-9. SCUBA Hand Signals (page 2 of 3).

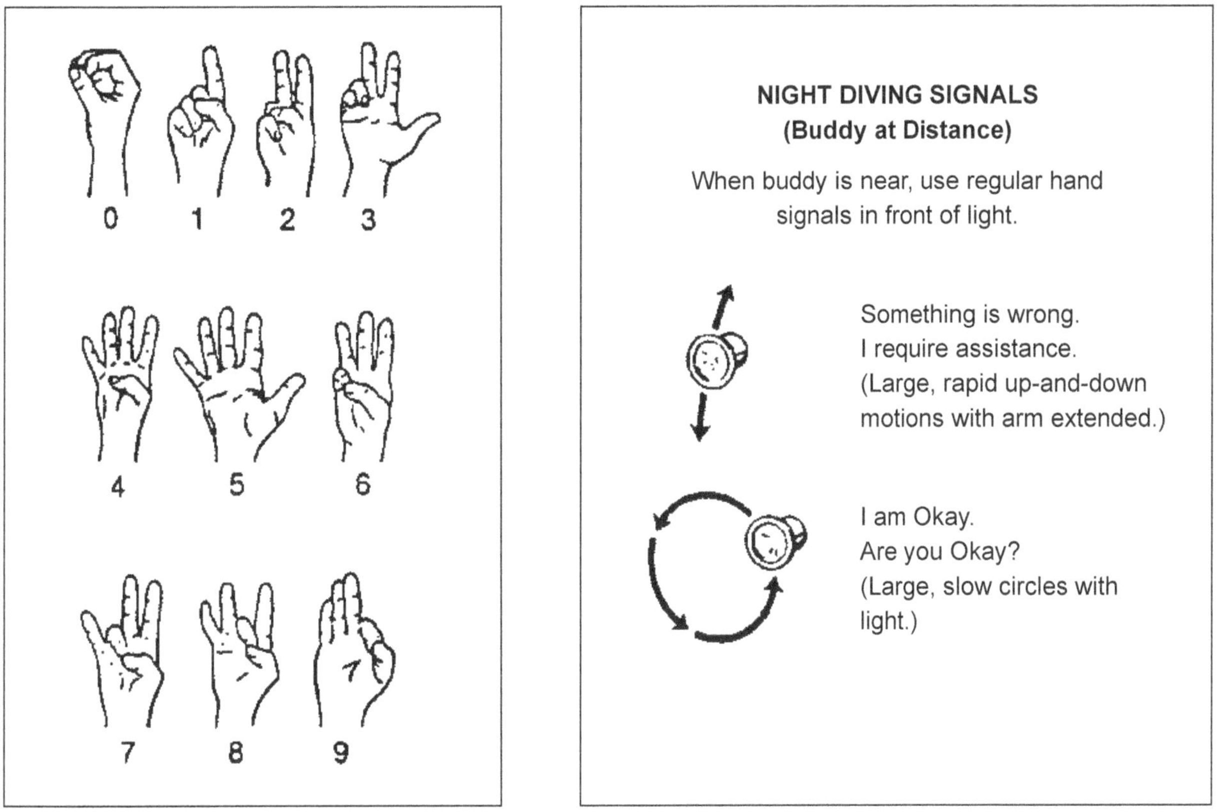

Figure 7-9. SCUBA Hand Signals (page 3 of 3).

2. The partner and the distressed diver should hold on to each other by grasping a strap or the free arm. The divers must be careful not to drift away from each other. The partner gives his octopus to the distressed diver. If an octopus is not available, proceed to step 3.

3. The partner must make the first move by taking a breath and passing the mouthpiece to the distressed diver. The distressed diver must not grab for the dive partner's mouthpiece. The dive partner guides it to the distressed diver's mouth. Both divers maintain direct hand contact on the mouthpiece.

4. The mouthpiece may have flooded during the transfer. In this case, clear the mouthpiece by using the purge button (if single-hose) or by exhaling into the mouthpiece before a breath can be taken. If using a double-hose regulator, the mouthpiece should be kept slightly higher than the regulator so that free-flowing air will help keep the mouthpiece clear.

5. The distressed diver should take two full breaths (exercising caution in the event that all of the water has not been purged) and guide the mouthpiece back to the partner. The partner should then purge the mouthpiece as necessary and take two breaths.

6. The divers should repeat the breathing cycle and establish a smooth rhythm. No attempt should be made to surface until the cycle is stabilized and the proper signals have been exchanged.

WARNING **During ascent, the diver without the mouthpiece must exhale to offset the effect of decreasing pressure on the lungs which could cause an air embolism.**

7-7.8 Tending.

7-7.8.1 **Tending with a Surface or Buddy Line.** When a diver is being tended by a line from the surface or a buddy line, several basic considerations apply.

- Lines should be kept free of slack.
- Line signals must be given in accordance with the procedures given in Table 8-3.
- Any signals via the line must be acknowledged immediately by returning the same signal.
- The tender should signal the diver with a single pull every 2 or 3 minutes to determine that the diver is all right. A return signal of one pull indicates that the diver is all right.
- If the diver fails to respond to line-pull signals after several attempts, the standby diver must investigate immediately.
- The diver must be particularly aware of the possibilities for the line becoming snagged or entangled.

7-7.8.2 **Tending with No Surface Line.** If a surface line is not being used, the tender must keep track of the general location of the divers by observing the bubble tracks or the float or locating device (such as a pinger or strobe light). When tending a single diver, the tender shall continually monitor the diver float for diver location and line pull signals.

7-7.9 **Working with Tools.** The near-neutral buoyancy of a SCUBA diver poses certain problems when working with tools. A diver is at a disadvantage when applying leverage with tools. When applying force to a wrench, for example, the diver is pushed away and can apply very little torque. If both sides of the work are accessible, two wrenches—one on the nut and one on the bolt—should be used. By pulling on one wrench and pushing on the other, the counter-force permits most of the effort to be transmitted to the work. When using any tool that requires leverage or force (including pneumatic power tools), the diver should be braced with feet, a free hand, or a shoulder.

NOTE **When using externally powered tools with SCUBA, the diver must have voice communications with the Diving Supervisor.**

Any tools to be used should be organized in advance. The diver should carry as few items as possible. If many tools are required, a canvas tool bag should be used to lower them to the diver as needed. Further guidelines for working underwater are provided in the *U.S. Navy Underwater Ship Husbandry Manual* (NAVSEA S0600-AA-PRO-010). Authorized power tools are listed in the NAVSEA/00C ANU list.

7-7.10 **Adapting to Underwater Conditions.** Through careful and thorough planning, the divers can be properly prepared for the underwater conditions at the diving site and be provided with appropriate auxiliary equipment, protective clothing, and tools. However, the diver may have to employ the following techniques to offset the effects of certain underwater conditions:

- Stay 2 or 3 feet above a muddy bottom; use a restricted kick and avoid stirring up the mud. A diver should be positioned so that the current will carry away any clouds of mud.
- Avoid coral or rocky bottoms, which may cause cuts and abrasions.
- Avoid abrupt changes of depth.
- Do not make excursions away from the dive site unless the excursions have been included in the dive plan.
- Be aware of the peculiar properties of light underwater. Depth perception is altered so that an object appearing to be 3 feet away is actually 4 feet away, and objects appear larger than they actually are.
- Be aware of unusually strong currents, particularly rip currents near a shoreline. If caught in a rip current, relax and ride along with it until it diminishes enough to swim clear.
- If practical, swim against a current to approach a job site. The return swim with the current will be easier and will offset some of the fatigue caused by the job.
- Stay clear of lines or wires that are under stress.

7-8 ASCENT PROCEDURES

When it is time to return to the surface, either diver may signal the end of the dive. When the signal has been acknowledged, the divers shall ascend to the surface together at a rate not to exceed 30 feet per minute. For a normal ascent, the divers will breathe steadily and naturally. Divers must never hold their breath during ascent, because of the danger of an air embolism. While ascending, divers must keep an arm extended overhead to watch for obstructions and should spiral slowly while rising to obtain a full 360 degree scan of the water column.

7-8.1 **Emergency Free-Ascent Procedures.** If a diver is suddenly without air or if the SCUBA is entangled and the dive partner cannot be reached quickly, a free ascent must be made. Guidelines for a free ascent are:

1. Drop any tools or objects being carried by hand.
2. Abandon the weight belt.
3. If the SCUBA has become entangled and must be abandoned, actuate the quick-release buckles on the waist, chest, shoulder, and crotch straps. Slip an arm out of one shoulder strap and roll the SCUBA off the other arm. An alternate method is to flip the SCUBA over the head and pull out from underneath. Ensure that the hoses do not wrap around or otherwise constrict the neck. The neck straps packed with some single-hose units can complicate the overhead procedure and should be disconnected from the unit and not used.
4. If the reason for the emergency ascent is a loss of air, drop all tools and the weight belt and actuate the life preserver to surface immediately. Do not drop the SCUBA unless it is absolutely necessary.
5. If a diver is incapacitated or unconscious and the dive partner anticipates difficulty in trying to swim the injured diver to the surface, the partner should activate the life preserver or inflate the buoyancy compensator. The weight belt may have to be released also. However, the partner should not lose direct contact with the diver.
6. Exhale continuously during ascent to let the expanding air in the lungs escape freely.

7-8.2 **Ascent From Under a Vessel.** When underwater ship husbandry tasks are required, surface-supplied lightweight equipment is preferred. SCUBA diving is permitted under floating hulls; however, a tending line to the SCUBA diver must be provided. In the event of casualty and the lack of immediate assistance by the dive partner, the SCUBA diver will be able to return to the surface using the tending line. Ships are often moored against closed-face piers or heavy camels and care must be exercised to ensure that the tending line permits a clear path for emergency surfacing of the diver.

Due to the unique nature of EOD operations involving limpet search and neutralization, the use of tending lines is not practical and is not required. During EOD limpet mine training, the use of tending lines is required.

SCUBA dive plans on deep-draft ships should restrict diving operations to one quadrant of the hull at a time. This theoretical quartering of the ship's hull will minimize potential diver disorientation caused by multiple keel crossings or fore and aft confusion.

When notified of a lost diver, a search shall be conducted by a tended diver in the area where the lost diver was last seen.

Predive briefs must include careful instruction on life preserver use when working under a hull to prevent panic blowup against the hull. Life preservers should not be fully inflated until after the diver passes the turn of the bilge.

7-8.3 **Decompression.** Open-circuit SCUBA dives are normally planned as no-decompression dives. Open-circuit SCUBA dives requiring decompression may be made only when considered absolutely necessary and authorized by the Commanding Officer or Officer in Charge (OIC). Under this unique situation, the following provides guidance for SCUBA decompression diving.

The Diving Supervisor shall determine the required bottom time for each dive. Based upon the time and depth of the dive, the required decompression profile from the tables presented in Chapter 9 shall be computed. The breathing supply required to support the total time in the water must then be calculated. If the air supply is not sufficient, a backup SCUBA will have to be made available to the divers. The backup unit can be strapped to a stage or tied off on a descent line which also has been marked to indicate the various decompression stops to be used.

When the divers have completed the assigned task, or have reached the maximum allowable bottom time prescribed in the dive plan, they must ascend to the stage or the marked line and signal the surface to begin decompression. With the stage being handled from the surface, the divers will be taken through the appropriate stops while the timekeeper controls the progress. Before each move of the stage, the tender will signal the divers to prepare for the lift and the divers will signal back when prepared. When using a marked line, the tender will signal when each stop has been completed, at which point the divers will swim up, signaling their arrival at the next stop. Stop times will always be regulated by the Dive Supervisor.

In determining the levels for the decompression stops, the sea state on the surface must be taken into consideration. If large swells are running, the stage or marker line will be constantly rising and falling with the movements of the surface-support craft. The depth of each decompression stop should be calculated so that the divers' chests will never be brought above the depths prescribed for the stops in the decompression tables.

In the event of an accidental surfacing or an emergency, the Diving Supervisor will have to determine if decompression should be resumed in the water or if the services of a recompression chamber are required. The possibility of having to make such a choice should be anticipated during the planning stages of the operation (Chapter 1 and Chapter 5).

7-8.4 **Surfacing and Leaving the Water.** When approaching the surface, divers must not come up under the support craft or any other obstruction. They should listen for the sound of propellers and delay surfacing until satisfied that there is no obstruction. On the surface, the diver should scan immediately in all directions and check the location of the support craft, other divers, and any approaching surface traffic. If they are not seen by the support craft, they should attempt to signal the support craft with hand signals, whistle, or flare.

As the divers break the surface, the tender and other personnel in the support craft must keep them in sight constantly and be alert for any signs of trouble. While one diver is being taken aboard the support craft, attention must not be diverted from the divers remaining in the water. The dive is completed when all divers are safely aboard.

Usually, getting into the boat will be easier if the divers remove the weight belts and SCUBA and then hand them to the tenders. If the boat has a ladder, swim fins should also be removed. Without a ladder, the swim fins will help to give the diver an extra push to get aboard. A small boat may be boarded over the side or over the stern depending on the type of craft and the surface conditions. As each diver comes aboard a small boat or a raft, other personnel in the boat should remain seated.

7-9 POSTDIVE PROCEDURES

The Diving Supervisor should debrief each returning diver while the experience of the dive is still fresh. The Diving Supervisor should determine if the assigned tasks were completed, if any problems were encountered, if any changes to the overall dive plan are indicated and if the divers have any suggestions for the next team.

When satisfied with their physical condition, the divers' first responsibility after the dive is to check their equipment for damage and get it properly cleaned and stowed. Each diver is responsible for the immediate postdive maintenance and proper disposition of the equipment used during the dive. The Planned Maintenance System provides direction for postdive maintenance.

CHAPTER 8

Surface Supplied Air Diving Operations

8-1 INTRODUCTION

8-1.1 **Purpose.** Surface supplied air diving includes those forms of diving where air is supplied from the surface to the diver by a flexible hose. The Navy Surface Supplied Diving Systems (SSDS) are used primarily for operations to 190 feet of seawater (fsw).

8-1.2 **Scope.** This chapter identifies the required equipment and procedures for using surface supplied Underwater Breathing Apparatus (UBA) diving equipment.

8-2 MK 21 MOD 1, KM-37

The MK 21 MOD 1 and KM-37 are open circuit, demand, diving helmets (Figure 8-1). The maximum working depth for air diving operations using the MK 21 MOD 1 and KM-37 UBAs is 190 fsw. The MK 21 MOD 1 and KM-37 UBAs may be used up to 60 fsw without an Emergency Gas Supply (EGS). An EGS is mandatory at depths deeper than 60 fsw and when diving inside a wreck or enclosed space. An EGS may be required for dives shallower than 60 fsw. The decision on EGS use will be based on Operational Risk Management (ORM). The Diving Supervisor may elect to use an EGS that can be man-carried or located outside the wreck or enclosed space and connected to the diver with a 50 to 150 foot whip. Planned air dives below 190 fsw require CNO approval.

Figure 8-1. MK 21 MOD 1 SSDS.

8-2.1 **Operation and Maintenance.** To ensure safe and reliable service, all surface supplied UBAs must be maintained and repaired in accordance with PMS procedures and the operation and maintenance manual.

The following technical manuals are for use with surface supplied UBAs MK-21 MOD 1 and KM-37:

- MK 21 MOD 1, NAVSEA S6560-AG-OMP-010, Technical Manual, Operation and Maintenance Instructions,

- KM-37 Surface Supported Diving System.

8-2.2 **Air Supply.** Air for the MK 21 MOD 1, KM-37 system is supplied from the surface by either an air compressor or a bank of high pressure air flasks as described in paragraph 8-7.2.3

8-2.2.1 **Emergency Gas Supply Requirements.** The EGS system consists of an adequately charged ANU approved SCUBA cylinder with either a K- or J- valve (with reserve turned down) and a first stage regulator set at manufacturer's recommended pressure, but not lower than 135 psig. A relief valve set at 180 ± 5 psig over bottom pressure must be installed on the first stage regulator to prevent rupture of the low pressure hose should the first stage regulator fail. The flexible low pressure hose from the first stage regulator attaches to the emergency supply valve on the helmet sideblock. The emergency breathing supply valve provides an air supply path parallel to the nonreturn valve and permits attachment of the EGS hose. A submersible pressure gauge is also required on the first stage regulator.

An adequately charged SCUBA cylinder is defined as the pressure that provides sufficient air to bring the diver to his first decompression stop or the surface for no-decompression dives. It is assumed that this will give topside personnel enough time to perform required emergency procedures to restore umbilical air to the diver.

For enclosed space diving an extended EGS whip 50 to 150 feet in length may be used. If the diving scenario requires the EGS topside, adjust the first stage regulator to 135 psi over bottom pressure.

NOTE For open water dives 60 fsw and shallower, an EGS may be required based upon the application of ORM.

Sample Problem 1. Determine the minimum EGS cylinder pressure required for a MK-21 MOD 1, KM-37 dive to 190 fsw for five minutes.

1. To calculate the EGS cylinder pressure, you must first determine the amount of gas required to get the diver back to the stage and leave bottom plus the gas required for ascent to the first decompression stop. The formula for calculating gas required is:

$$V_r = \frac{D+33}{33} \times C \times T$$

Where:

V_r = Capacity required (scf)
D = Depth (fsw)
C = Consumption rate in acfm per diver from Table 8-2
T = Time (minutes)

Air required while on the bottom: For this example, if the time to get the diver to the stage and leave bottom is 3 minutes, then:

$$\text{Bottom } V_r = \frac{190+33}{33} \times 1.4 \times 3$$
$$= 28.38 \text{ scf}$$

Air required for ascent to reach the first stop: For this example, you need to determine ascent time and average depth. Ascent time is 7 minutes (rounded up from 6 minutes 20 seconds) from 190 fsw to the surface at 30 feet per minute. Average depth is calculated as follows:

$$\text{average depth} = \frac{190}{2} = 95 \text{ fsw}$$

$$\text{Ascent } V_r = \frac{95+33}{33} \times 0.75 \times 7$$
$$= 20.36 \text{ scf}$$
$$\text{Total } V_r = 28.38 + 20.36$$
$$= 48.74 \text{ scf}$$

2. The next step is to convert the required scf to an equivalent cylinder pressure in psig. In this example, we are using an 80 ft^3 aluminum cylinder to support this dive. Refer to Table 7-1 for cylinder data used in this calculation:

$$\text{psig required} = \frac{V_r}{FV} \times 14.7 + P_m$$

Where:

FV = Floodable Volume (scf) = 0.399 scf

14.7 = Atmospheric Pressure (psi)

P_m = Minimum cylinder pressure

Minimum Cylinder Pressure = First stage regulator setting + bottom pressure at final stop: [135 psig + (0 fsw x 0.445 psi)] = 135 psig

$$= \frac{48.74}{0.399} \times 14.7 + 135$$
$$= 1930.68 \text{ (round to 2000 psig)}$$

8-2.2.2 **Flow Requirements.** When the MK 21 MOD 1, KM-37 system is used, the air supply system must be able to provide an average sustained flow of 1.4 acfm to the diver. The air consumption of divers using the MK 21 MOD 1, KM-37 varies between 0.75 and 1.5 acfm when used in a demand mode, with occasional faceplate and mask clearing. When used in a free-flow mode, greater than eight acfm is consumed.

NOTE **When planning a dive, calculations are based on 1.4 acfm for descent and bottom phase and 0.75 acfm for ascent and decompression phase.**

To satisfactorily support the MK 21 MOD 1, KM-37 system, the air supply must:

- Replenish the air consumed from the system (average rate of flow)
- Replenish the air at a rate sufficient to maintain the required pressure
- Provide the maximum rate of flow required by the diver

8-2.2.3 **Pressure Requirements.** Because the MK 21 MOD 1 and KM-37 helmets are demand type UBAs, the regulators have an optimum overbottom pressure that ensures the lowest possible breathing resistance and reduces the possibility of overbreathing the regulator (demanding more air than is available). For those systems not capable of sustaining 165 psi overbottom due to design limitations, 135 psi overbottom is acceptable. Table 8-1 shows the MK 21 MOD 1 and KM-37 overbottom pressure requirements.

Table 8-1. *MK 21 MOD 1 and KM-37 Overbottom Pressure Requirements.*

Dive Depth	Pressure in psig		
	Minimum	Desired	Maximum
0-60 fsw	90*	135	165
61-130 fsw	135	135	165
131-190 fsw	165**	165	165

* Not approved for use with a double exhaust kit installed. Instead use a minimum of 135 psig.

** For diver life support systems not capable of sustaining 165 psig over bottom due to system design limitations, 135 psig is authorized.

This ensures that the air supply will deliver air at a pressure sufficient to overcome bottom seawater pressure and the pressure drop that occurs as the air flows through the hoses and valves of the mask.

Sample Problem 1. Determine the air supply manifold pressure required to dive the MK 21 MOD 1, KM-37 system to 175 fsw.

1. Determine the bottom pressure at 175 fsw:

   Bottom pressure at 175 fsw = 175 × .445 psi
   = 77.87 psig (round to 78)

2. Determine the overbottom pressure for the MK 21 MOD 1, KM-37 system (see Table 8-1). Because the operating depth is 175 fsw, the overbottom pressure is 165 psig.

3. Calculate the minimum manifold pressure (MMP) by adding the bottom pressure to the overbottom pressure:

   $$\text{MMP} = 78\,\text{psig} + 165\ \text{psig}$$
   $$= 243\ \text{psig}$$

   The minimum manifold pressure for a 175 fsw dive must be 243 psig.

Sample Problem 2. Determine if air from a bank of high pressure flasks is capable of supporting two MK 21 MOD 1, KM-37 divers and one standby diver at a depth of 130 fsw for 30 minutes. There are 5 flasks in the bank; only 4 are on line. Each flask has a floodable volume of 8 cubic feet and is charged to 3,000 psig.

NOTE **When planning a dive, calculations are based on 1.4 acfm for descent and bottom phase and 0.75 acfm for ascent and decompression phase.**

1. Calculate minimum manifold pressure (MMP).

$$\begin{aligned} \text{MMP (psig)} &= (0.445\text{D}) + 135 \text{ psig} \\ &= (0.445 \times 130) + 135 \text{ psig} \\ &= 192.85 \text{ psig} \end{aligned}$$

   Round up to 193 psig

2. Calculate standard cubic feet (scf) of air available. The formula for calculating the scf of air available is:

$$\text{scf available} = \frac{\text{P}_\text{f} - (\text{P}_\text{mf} + \text{MMP})}{14.7} \times \text{FV} \times \text{N}$$

   Where:
   P_f = Flask pressure = 3,000 psig
   P_{mf} = Minimum flask pressure = 200 psig
   MMP = 193 psig
   FV = Floodable Volume of flask = 8 scf
   N = Number of flasks = 4

$$\begin{aligned} \text{scf available} &= \frac{3000 - (200 + 193)}{14.7} \times 8 \times 4 \\ &= 5675.10 \text{ scf (round down to 5675)} \end{aligned}$$

3. Calculate scf of air required to make the dive. You will need to calculate the air required for the bottom time, the air required for each decompression stop, and the air required for the ascent. The formula for calculating the air required is:

$$\text{scf required} = \frac{\text{D} + 33}{33} \times \text{C} \times \text{N} \times \text{T}$$

   Where:
   D = Depth (fsw)
   C = Consumption rate in acfm needed per diver from Table 8-2
   N = Number of divers
   T = Time at depth (minutes)

   Bottom time: 30 minutes

$$\text{scf required} = \frac{130+33}{33} \times 1.4 \times 3 \times 30$$
$$= 622.36 \text{ scf}$$

Decompression stops: A dive to 130 fsw for 30 minutes requires the following decompression stops:

- 34 minutes at 20 fsw

$$\text{scf required} = \frac{20+33}{33} \times 0.75 \times 3 \times 34$$
$$= 122.86 \text{ scf}$$

Ascent time: 5 minutes (rounded up from 4 minutes 20 seconds) from 130 fsw to the surface at 30 feet per minute.

$$\text{average depth} = \frac{130}{2} = 65 \text{ fsw}$$
$$\text{scf required} = \frac{65+33}{33} \times 0.75 \times 3 \times 5$$
$$= 33.41 \text{ scf}$$
$$\text{total air required} = 622.36 + 122.86 + 33.41$$
$$= 778.63 \text{ scf (round to 779 scf)}$$

4. Calculate the air remaining at the completion of the dive to see if there is sufficient air in the air supply flasks to make the dive. Example shows scf for air only.

$$\text{scf remaining} = \text{scf available} - \text{scf required}$$
$$= 5675 \text{ scf} - 779 \text{ scf}$$
$$= 4896 \text{ scf}$$

More than sufficient air is available in the air supply flasks to make this dive.

NOTE **Planned air usage estimates will vary from actual air usage. The air requirements for a standby diver must also be taken into account for all diving operations. The Diving Supervisor must note initial volume/ pressure and continually monitor consumption throughout dive. If actual consumption exceeds planned consumption, the Diving Supervisor may be required to curtail the dive in order to ensure there is adequate air remaining in the primary air supply to complete decompression.**

8-3 MK 20 MOD 0

The MK 20 MOD 0 is a surface-supplied UBA consisting of a full face mask, diver communications components, equipment harness, and an umbilical assembly (Figure 8-2). One of its primary uses is in enclosed spaces, such as submarine ballast tanks. The MK 20 MOD 0 is authorized for use to a depth of 60 fsw with surface-supplied air and must have an Emergency Gas Supply when used for enclosed space diving.

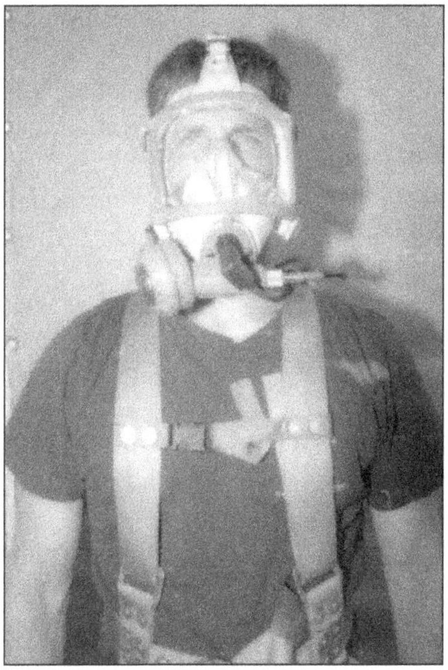

Figure 8-2. MK 20 MOD 0 UBA.

8-3.1 **Operation and Maintenance.** Safety considerations and working procedures are covered in Chapter 6. NAVSEA SS600-AK-MMO-010 *Technical Manual, Operations and Maintenance Instruction Manual* is the technical manual for the MK 20 MOD 0. To ensure safe and reliable service, the MK 20 MOD 0 system must be maintained and repaired in accordance with PMS procedures and the MK 20 MOD 0 operation and maintenance manual.

8-3.2 **Air Supply.** Air for the MK 20 MOD 0 system is supplied from the surface by either an air compressor or a bank of high-pressure flasks as described in paragraph 8-7.2.3.

8-3.2.1 **EGS Requirements for MK 20 MOD 0 Enclosed-Space Diving.** In order to ensure a positive emergency air supply to the diver when working in a ballast tank, mud tank, or confined space, an Emergency Gas Supply (EGS) assembly must be used. As a minimum, the EGS assembly consists of:

- An adequately charged ANU approved SCUBA cylinder with either a K- or J-valve.
- An approved SCUBA regulator set at manufacturer's recommended pressure, but not lower than 135 psi, with an extended EGS whip 50 to 150 feet in length. If the diving scenario dictates leaving the EGS topside, adjust the first stage regulator to 150 psig.
- An approved submersible pressure gauge.

The SCUBA cylinder may be left on the surface and the EGS whip may be married to the diver's umbilical, or it may be secured at the opening of the enclosed space being entered. The diver may then enter the work space with the extended EGS whip trailing. The second stage regulator of the EGS must be securely attached to the diver's harness before entering the work space so that the diver has immediate access to the EGS regulator in an emergency.

An adequately charged SCUBA cylinder is defined as the pressure that provides sufficient air to bring the diver to his first decompression stop or the surface for no-decompression dives. It is assumed that this will give topside personnel enough time to perform required emergency procedures. See paragraph 8-2.2.1 for calculating minimum cylinder pressure.

8-3.2.2 **EGS Requirements for MK 20 MOD 0 Open Water Diving.** When conducting open water dives, the diving supervisor may use a MK 20 designated ANU approved cylinder with the DSI sideblock assembly as an emergency air source.

NOTE **For open water dives 60 fsw and shallower, an EGS may be required based upon the application of ORM.**

8-3.2.3 **Flow Requirements.** The MK 20 MOD 0 requires a breathing gas flow of 1.4 acfm and an overbottom pressure of 90 psig. Flow and pressure requirement calculations are identical to those for the MK 21 MOD 1, KM-37 (see paragraph 8-2.2.3).

8-4 EXO BR MS

8-4.1 **EXO BR MS.** The EXO BR MS is a commercial-off-the-shelf, full face mask, manufactured by Kirby Morgan Dive Systems, which is used for surface-supplied diving. It is authorized for use to 190 fsw on air. An Emergency Gas Supply (EGS) is mandatory at depths deeper than 60 fsw and when diving inside an enclosed space. The Diving Supervisor may elect to use an EGS that can be man-carried or located outside the enclosed space and connected to the diver with a 50-150 foot whip. Conducting air dives below 190 fsw requires CNO approval.

8-4.2 **Operations and Maintenance.** The technical manual for the EXO BR MS is the Kirby Morgan Operations & Maintenance Manual, EXO BR MS Balanced Regulator Full Face Mask Military Standard (DSI Part #100-036). To ensure safe and reliable service, the EXO BR MS must be maintained and repaired in accordance with PMS procedures and the technical manual.

8-4.3 **Air Supply.** For surface-supplied diving, air for the EXO BR MS is supplied from the surface by either an air compressor or a bank of high-pressure flasks as described in paragraph 8-7.2.3.

8-4.4 **EGS Requirements for EXO BR MS.** The EGS system consists of adequately charged ANU approved cylinder with either a K- or J- valve and an approved first stage regulator set at manufacturer's recommended pressure but no lower than 135 psi over bottom pressure. The intermediate hose of the first stage is coupled to the emergency gas supply valve on the manifold block assembly. A relief valve set at 180 +/-5 psi over bottom pressure must be installed on the first stage regulator to prevent rupture of the low pressure hose should the first stage regulator fail. The flexible low pressure hose from the first stage regulator attaches to the emergency supply valve on the manifold block. A submersible pressure gauge is also required on the first stage regulator.

When diving enclosed spaces during ship husbandry operations, the use of an approved second stage regulator with extended EGS whip 50 to 150 feet in length is permissible. The manifold block is not used and the diver's umbilical is connected directly to the low pressure high flow hose from the mask. The SCUBA cylinder may be left on the surface or secured at the opening of the enclosed space. The second stage regulator of the EGS must be securely attached to the diver so the diver has immediate access to the EGS regulator in an emergency. If the diving scenario dictates leaving the EGS topside, adjust the first stage regulator to 150 psig. When diving in submarine ballast tanks, the mask and umbilical may be left up inside the ballast tank adjacent to the opening with the extended EGS whip trailing the diver.

An adequately charged SCUBA cylinder is defined as the pressure that provides sufficient air to bring the diver to his first decompression stop or the surface for no-decompression dives. It is assumed that this will give topside personnel enough time to perform required emergency procedures. See paragraph 8-2.2.1 for calculating minimum cylinder pressure.

NOTE **For open water dives 60 fsw and shallower, an EGS may be required based upon the application of ORM.**

8-4.5 **Flow and Pressure Requirements.** The EXO BR MS requires a breathing gas flow of 1.4 acfm. For dives shallower than 130 fsw, the overbottom pressure shall be 135-165 psi. For those systems which cannot maintain 135 psi overbottom pressure when diving shallower than 60 fsw, 90 psi is permissible. For dives 130-190 fsw, the overbottom pressure shall be 165-225 psi. Flow and pressure calculations are identical to those for the MK21 MOD 1, KM-37 (see paragraph 8-2.2.3).

8-5 PORTABLE SURFACE-SUPPLIED DIVING SYSTEMS

8-5.1 **MK 3 MOD 0 Lightweight Dive System (LWDS).** The MK 3 MOD 0 LWDS is a portable, self-contained, surface-supplied diver life-support system (DLSS). The MK 3 MOD 0 LWDS can be arranged in three different configurations and may be deployed pierside or from a variety of support platforms. Each LWDS includes a control console assembly, volume tank assembly, medium-pressure air compressor (optional), and stackable compressed-air rack assemblies, each consisting of three high-pressure composite flasks (0.935 cu ft floodable volume each). Each flask holds 191 scf of compressed air at 3,000 psi. Set-up and operating procedures for the LWDS are found in the Operating and Maintenance Instructions for Lightweight Dive System (LWDS) MK 3 MOD 0, SS500-HK-MMO-010.

8-5.1.1 **MK 3 MOD 0 Configuration 1.** Air is supplied by a medium-pressure diesel driven compressor unit supplying primary air to the divers at 18 standard cubic feet per minute (scfm) with secondary air being supplied by one air-rack assembly. Total available secondary air is 594 scf. See Figure 8-3.

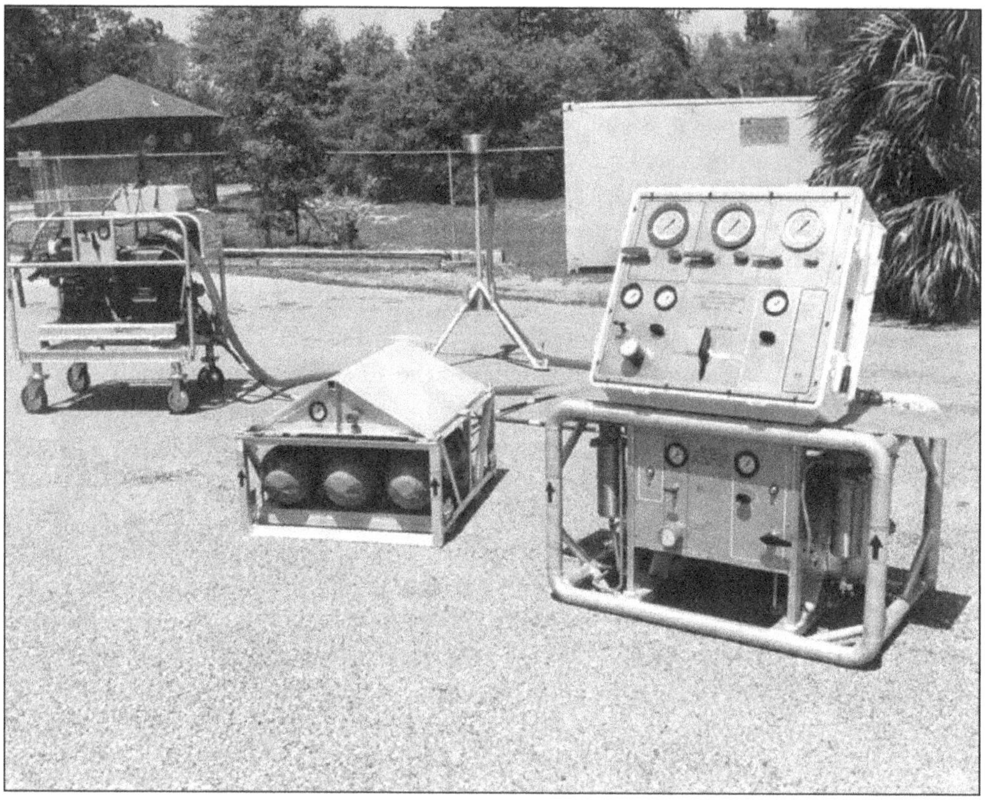

Figure 8-3. MK 3 MOD 0 Configuration 1.

8-5.1.2 **MK 3 MOD 0 Configuration 2.** Primary air is supplied to the divers using three flask rack assemblies. Secondary air is supplied by one flask rack assembly. Total available primary air is 1,782 scf at 3,000 psi. Total available secondary air is 594 scf. See Figure 8-4.

8-5.1.3 **MK 3 MOD 0 Configuration 3.** Primary air is supplied to the divers using three flask rack assemblies. Secondary air is supplied by two flask rack assemblies. Total available primary air is 1,782 scf. Total available secondary air is 1,188 scf. See Figure 8-5.

8-5.2 **MK 3 MOD 1 Lightweight Dive System.** This system is identical to the MK 3 MOD 0 LWDS except that the control console and volume tank have been modified to support 5,000 psi operations for use with the Air Supply Rack Assembly (ASRA). With appropriate adapters the system can still be used to support normal LWDS operations. See Figure 8-6.

8-5.3 **ROPER Diving Cart.** The ROPER diving cart is a trailer-mounted diving system, designed to support one working and one standby diver in underwater operational tasks performed by Ship Repair Activities to 60 fsw (Figure 8-7). The system is self-contained, transportable, and certifiable in accordance with *U.S. Navy Diving and Hyperbaric System Safety Certification Manual*, NAVSEA SS521-AA-MAN-010. The major components/subsystems mounted within the cart body are:

Figure 8-4. MK 3 MOD 0 Configuration 2.

Figure 8-5. MK 3 MOD 0 Configuration 3.

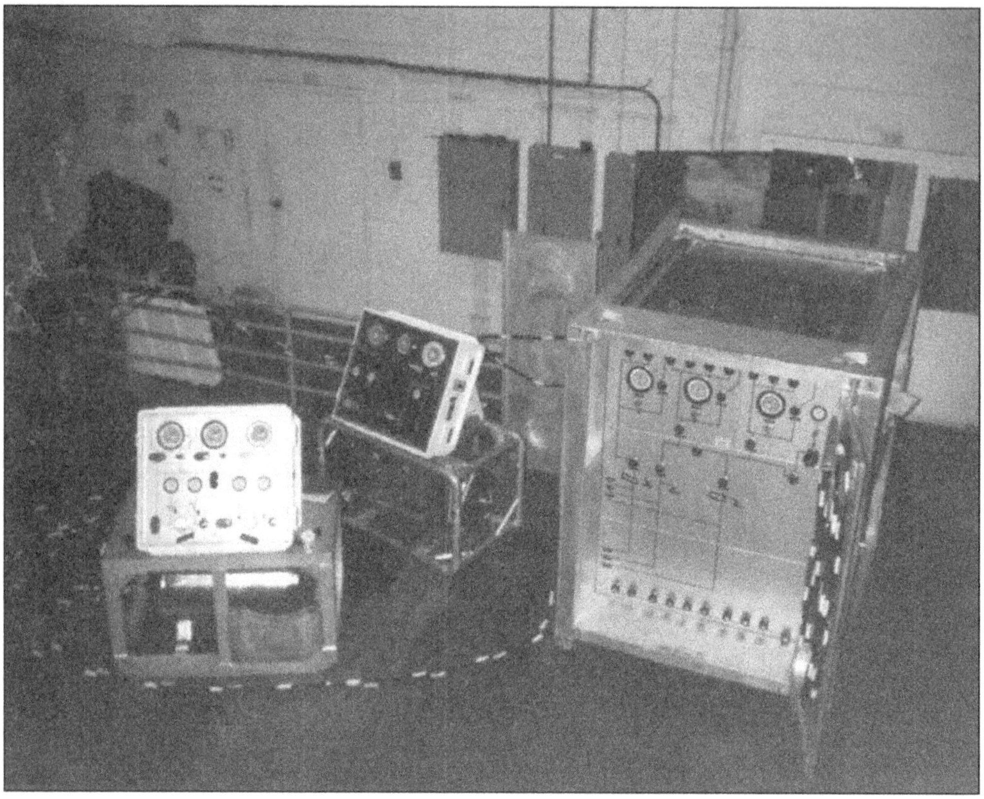

Figure 8-6. Flyaway Dive System (FADS) III.

Figure 8-7. ROPER Cart.

- **Diving control station.** A single operator controls and monitors the air supply and operates the communication system.

- **Power distribution system.** External power for communications and control station lighting.

- **Intercommunication system (AC/DC).** Provides communications between divers and the diving control station.

- **Air supply system.** Primary air source of two 6 cu ft, 3,000 psi air flasks; secondary air source of a single 1.52 cu ft, 3,000 psi air flask; and a SCUBA charging station.

Detailed information and operating instructions are covered in *Operations and Maintenance Instructions for Ready Operational Pierside Emergency Repair (ROPER) Diving Cart*, SS500-AS-MMA-010.

8-5.4 Flyaway Dive System (FADS) III.

The FADS III is a portable, self-contained, surface-supplied diver life-support system designed to support dive missions to 190 fsw (Figure 8-6). Compressed air at 5,000 psi is contained in nine 3.15 cu ft floodable volume composite flasks vertically mounted in an Air Supply Rack Assembly (ASRA). The ASRA will hold 9600 scf of compressed air at 5,000 psi. Compressed air is provided by a 5,000 psi air compressor assembly which includes an air purification system. The FADS III also includes a control console assembly and a volume tank assembly. Three banks of two, three, and four flasks allow the ASRA to provide primary and secondary air to the divers as well as air to support chamber operations. Set-up and operating procedures for the FADS III are found in the *Operating and Maintenance Technical Manual for Fly Away Dive System (FADS) III Air System*, S9592-B1-MMO-010.

8-5.5 Oxygen Regulator Console Assembly (ORCA).

The purpose of the Oxygen Regulator Console Assembly (ORCA) is to provide 100% oxygen to the diver's umbilical's for use during in-water oxygen decompression. It is designed to be used with any currently certified Divers Life Support System (DLSS). The ORCA requires separate oxygen supplies. The ORCA consists of a valve control system and pressure regulator. The valve control system contains isolation, bleed, control valves, gauges, and a high-pressure oxygen pressure regulator to simultaneously provide low-pressure oxygen to up to three divers. When not using the oxygen reducer, system piping is installed to allow a straight pass-through of diver's breathable gas from any compatible diver air supply system. (See Figures 8-8 and 8-9).

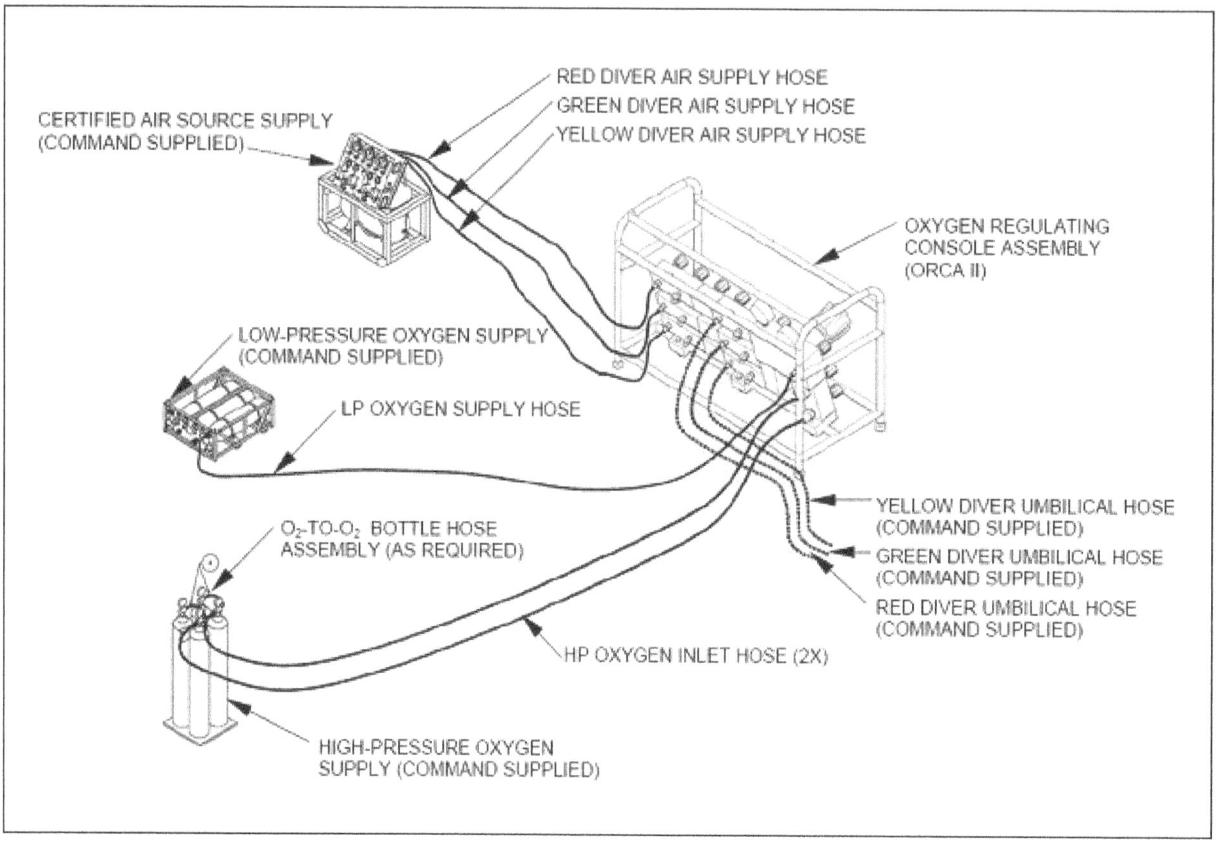

Figure 8-8. Oxygen Regulator Control Assembly (ORCA) II Schematic.

Figure 8-9. Oxygen Regulator Control Assembly (ORCA) II.

8-6 ACCESSORY EQUIPMENT FOR SURFACE-SUPPLIED DIVING

Accessory equipment that is often useful in surface-supplied diving operations includes the following items:

- **Lead Line.** The lead line is used to measure depth.
- **Descent Line.** The descent line guides the diver to the bottom and is used to pass tools and equipment. A 3-inch double-braid line is recommended, to prevent twisting and to facilitate easy identification by the diver on the bottom. In use, the end of the line may be fastened to a fixed underwater object, or it may be anchored with a weight heavy enough to withstand the current.
- **Circling Line.** The circling line is attached to the bottom end of the descent line. It is used by the diver as a guide in searching and for relocating the descent line.
- **Stage.** Constructed to carry one or more divers, the stage is used to put divers into the water and to bring them to the surface, especially when decompression stops must be made. The stage platform is made in an open grillwork pattern to reduce resistance from the water and may include seats. Guides for the descent line, several eyebolts for attaching tools, and steadying lines or weights are provided. The frames of the stages may be collapsible for easy storage. A safety shackle or screw-pin shackle seized with wire or with a cotter pin must be used to connect the stage to the lifting line when raising or lowering. Stages must be weight tested in accordance with PMS.
- **Stage Line.** Used to raise and lower the stage, the stage line is to be 3-inch double braid, or 3/8-inch wire rope minimum, taken to a capstan or run off a winch and davit.
- **Diving Ladder.** The diving ladder is used to enter the water from a vessel.
- **Weights.** Cast iron or lead weights are used to weight the descent line.
- **Tool Bag.** The tool bag is used to carry tools.
- **Stopwatches.** Stopwatches are used to time the total dive time, decompression stop time, travel time, etc.

8-7 SURFACE AIR SUPPLY SYSTEMS

The diver's air supply may originate from an air compressor, a bank of high-pressure air flasks, or a combination of both.

8-7.1 **Requirements for Air Supply.** Regardless of the source, the air must meet certain established standards of purity, must be supplied in an adequate volume for breathing, and must have a rate of flow that properly ventilates the helmet or mask. The air must also be provided at sufficient pressure to overcome the bottom water pressure and the pressure losses due to flow through the diving hose, fittings, and valves. The air supply requirements depend upon specific factors of each dive such as depth, duration, level of work, number of divers being supported, and type of diving system being used.

8-7.1.1 **Air Purity Standards.** Air taken directly from the atmosphere and pumped to the diver may not meet established purity standards. It may be contaminated by engine exhaust. Initially pure air may become contaminated while passing through a faulty air compressor system. For this reason, all divers' air must be periodically sampled and analyzed to ensure the air meets purity standards. Refer to Table 4-1 for compressed air purity requirements.

To meet these standards, specially designed compressors must be used with the air supplied passed through a highly efficient filtration system. Air taken from any machinery space, or downwind from the exhaust of an engine or boiler, must be considered to be contaminated. For this reason, care must be exercised in the placement and operation of diving air compressors to avoid such conditions. Intake piping or ducting must be provided to bring uncontaminated air to the compressor. The outboard end of this piping must be positioned to eliminate sources of contamination. To ensure that the source of diver's breathing air satisfactorily meets the standards established above, it must be checked at intervals not to exceed 8 months, in accordance with the PMS.

8-7.1.2 **Air Supply Flow Requirements.** The required flow for demand breathing equipment such as the MK 21 MOD 1, KM-37 or the MK 20 MOD 0 must meet the diver's flow requirements. The flow requirements for respiration in a demand system are based upon the average rate of air flow demanded by the divers under normal working conditions. The maximum instantaneous (peak) rate of flow under severe work conditions is not a continuous requirement, but rather the highest rate of airflow attained during the inhalation part of the breathing cycle. The diver's requirement varies with the respiratory demands of the diver's work level.

8-7.1.3 **Supply Pressure Requirements.** In order to supply the diver with an adequate flow of air, the air source must deliver air at sufficient pressure to overcome the bottom seawater pressure and the pressure drop that is introduced as the air flows through the hoses and valves of the system. Table 8-2 shows the values for air consumption and minimum over-bottom pressures required for surface-supplied UBAs.

Table 8-2. *Primary Air System Requirements.*

System	Minimum Manifold Pressure (MMP)	AIR CONSUMPTION: Average Over Descent and Bottom Phase (acfm)	AIR CONSUMPTION: Average Over Ascent and Decompression Phase (acfm)
MK 21 MOD 1, KM-37 EXO BR MS	(Depth in fsw × 0.445) + 90 to 165 psi, depending on the depth of the dive	1.4 (Note 1)	0.75
MK 20 MOD 0	(Depth in fsw × 0.445) + 90 psi	1.4	0.75

Note 1: The manifold supply pressure requirement is 90 psig over-bottom pressure for depths to 60 fsw, and 135 psig over-bottom pressure for depths from 61-130 fsw. For dives from 131-190 fsw, 165 psig over-bottom pressure shall be used.

8-7.1.4 **Water Vapor Control.** A properly operated air supply system should never permit the air supplied to the diver to reach its dewpoint. Controlling the amount of water vapor (humidity) in the supplied air is normally accomplished by one or both of the following methods:

- **Compression/Expansion.** As high-pressure air expands across a pressure reducing valve, the partial pressure of the water vapor in the air is decreased. Since the expansion takes place at essentially a constant temperature (isothermal), the partial pressure of water vapor required to saturate the air remains unchanged. Therefore, the relative humidity of the air is reduced.

- **Cooling.** Cooling the air prior to expanding it raises its relative humidity, permitting some of the water to condense. The condensed liquid may then be drained from the system.

8-7.1.5 **Standby Diver Air Requirements.** Air supply requirements cannot be based solely on the calculated continuing needs of the divers who are initially engaged in the operation. There must be an adequate reserve to support a standby diver should one be needed.

8-7.2 **Primary and Secondary Air Supply.** All surface-supplied diving systems must include a primary and a secondary air supply in accordance with the *General Specification for the Design, Construction, and Repair of Diving and Hyperbaric Equipment*, NAVSEA TS500-AU-SPN-010. The primary supply must be able to support the air flow and pressure requirements for the diving equipment designated (Table 8-2). The capacity of the primary supply must meet the consumption rate of the designated number of divers for the full duration of the dive (bottom time plus decompression time). The maximum depth of the dive, the number of divers, and the equipment to be used must be taken into account when sizing the supply. The secondary supply must be sized to be able to support recovery of all divers using the equipment and dive profile of the primary supply if the primary supply sustains a casualty at the worst-case time (for example, immediately prior to completion of planned bottom time of maximum dive depth, when decompression obligation is greatest). Primary and secondary supplies may be either high-pressure (HP) bank-supplied or compressor-supplied.

8-7.2.1 **Requirements for Operating Procedures and Emergency Procedures.** Operating procedures (OPs) and emergency procedures (EPs) must be available to support operation of the system and recovery from emergency situations. OPs and EPs are required to be NAVSEA or NAVFAC approved in accordance with paragraph 4-2.6.3. Should the surface-supplied diving system be integrated with a recompression chamber, an air supply allowance for chamber requirements (Volume 5) must be made.

All valves and electrical switches that directly influence the air supply shall be labeled:

"DIVER'S AIR SUPPLY - DO NOT TOUCH"

Banks of flasks and groups of valves require only one central label at the main stop valve.

A volume tank is required when operating directly from a low pressure air compressor. The volume tank maintains the air supply should the primary supply source fail, providing time to actuate a secondary air supply. It also absorbs pressure pulsations resulting from the compressor operation. A volume tank may also be required when the volume tank is an integral part of the system design such as a Lightweight Dive System. When operating from a high-pressure air source, a volume tank is not required if the pressure reducer has been proven to withstand significant pressure cycling caused by use of UBA demand regulators.

8-7.2.2 **Air Compressors.** Many air supply systems used in Navy diving operations include at least one air compressor as a source of air. To properly select such a compressor, it is essential that the diver have a basic understanding of the principles of gas compression. The NAVSEA/00C ANU list contains guidance for Navy-approved compressors for divers' air systems. See Figure 8-10.

8-7.2.2.1 ***Reciprocating Air Compressors.*** Reciprocating air compressors are the only compressors authorized for use in Navy air diving operations. Low pressure (LP) models can provide rates of flow sufficient to support surface-supplied air diving or recompression chamber operations. High-pressure models can charge high-pressure air banks and SCUBA cylinders.

8-7.2.2.2 ***Compressor Capacity Requirements.*** Air compressors must meet the flow and pressure requirements outlined in paragraph 8-7.1.2 and paragraph 8-7.1.3. Normally, reciprocating compressors have their rating (capacity in cubic feet per minute and delivery pressure in psig) stamped on the manufacturer's identification plate. This rating is usually based on inlet conditions of 70°F (21.1°C), 14.7 psia barometric pressure, and 36 percent relative humidity (an air density of 0.075 pound per cubic foot). If inlet conditions vary, the actual capacity either increases or decreases from rated values. If not provided directly, capacity will be provided by conducting a compressor output test (see Topside Tech Notes, Volume II

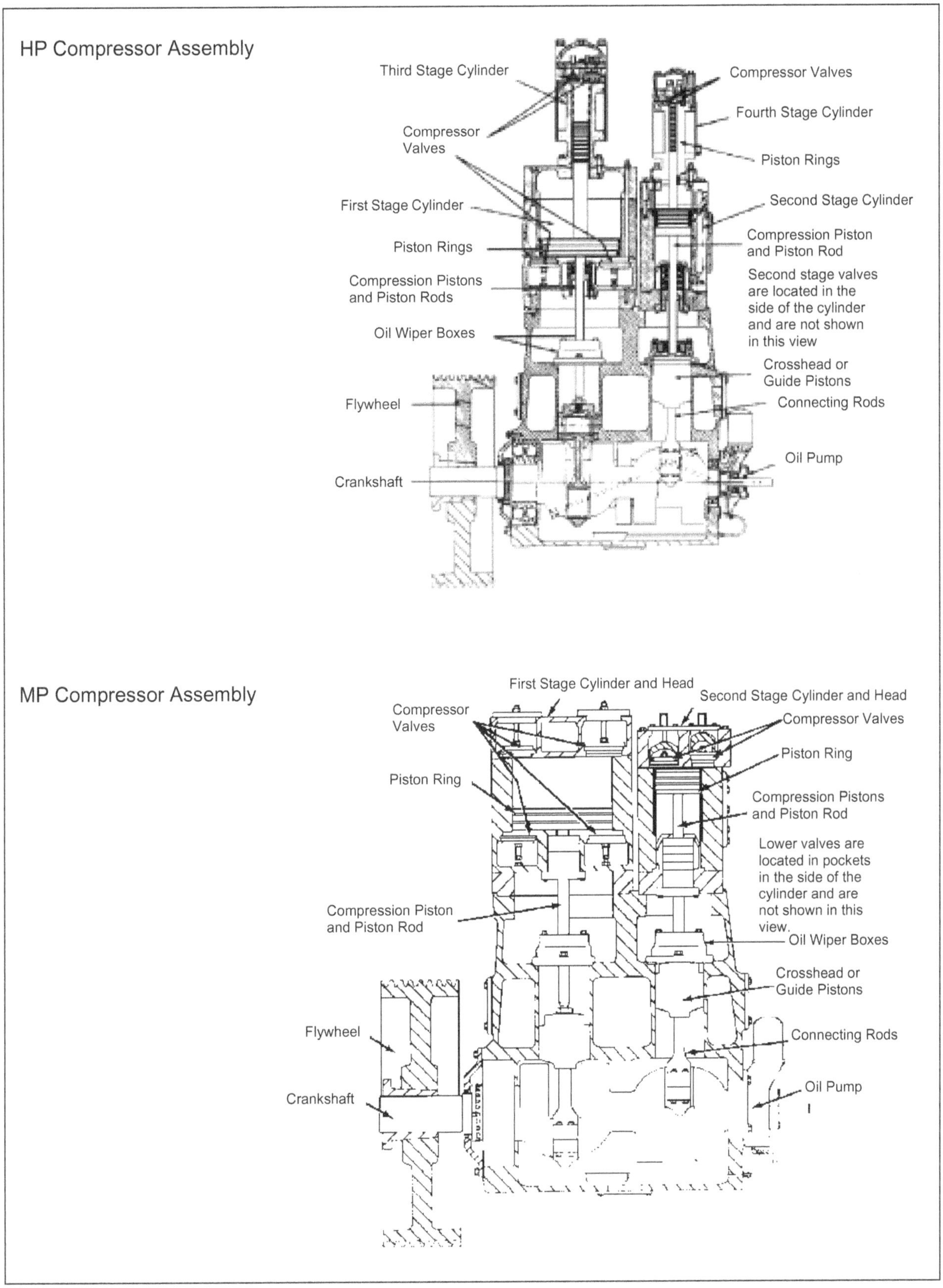

Figure 8-10. HP Compressor Assembly (top); MP Compressor Assembly (bottom).

Compressors/Process Instruction NAVSEA-00C4-PI-004, Compressor Capacity Testing). Since the capacity is the volume of air at defined atmospheric conditions, compressed per unit of time, it is affected only by the first stage, as all other stages only increase the pressure and reduce temperature. All industrial compressors are stamped with a code, consisting of at least two, but usually four to five, numbers that specify the bore and stroke.

The actual capacity of the compressor will always be less than the displacement because of the clearance volume of the cylinders. This is the volume above the piston that does not get displaced by the piston during compression. Compressors having a first stage piston diameter of four inches or larger normally have an actual capacity of about 85 percent of their displacement. The smaller the first stage piston, the lower the percentage capacity, because the clearance volume represents a greater percentage of the cylinder volume.

8-7.2.2.3 ***Lubrication.*** Reciprocating piston compressors are either oil lubricated or water lubricated. The majority of the Navy's diving compressors are lubricated by petroleum or synthetic oil. In these compressors, the lubricant:

- Prevents wear between friction surfaces
- Seals close clearances
- Protects against corrosion
- Transfers heat away from heat-producing surfaces
- Transfers minute particles generated from normal system wear to the oil sump or oil filter if so equipped

8-7.2.2.4 ***Lubricant Specifications.*** Unfortunately, the lubricant vaporizes into the air supply and, if not condensed or filtered out, will reach the diver. Lubricants used in air diving compressors must conform to military specifications MIL-PRF-17331 (2190 TEP) for normal operations, or MIL-PRF-17672 (2135 TH) for cold weather operations. Where the compressor manufacturer specifically recommends using a synthetic base oil, the recommended oil may be used in lieu of MIL-PRF-17331 or MIL-PRF-17672 oil.

8-7.2.2.5 ***Maintaining an Oil-Lubricated Compressor.*** Using an oil-lubricated compressor for diving is contingent upon proper maintenance to limit the amount of oil introduced into the diver's air (see *Topside Tech Notes*, March 1997). When using any lubricated compressor for diving, the air must be checked for oil contamination. Diving operations shall be aborted at the first indication that oil is in the air being delivered to the diver. An immediate air analysis must be conducted to determine whether the amount of oil present exceeds the maximum permissible level in accordance with table Table 4-1.

It should be noted that air in the higher stages of a compressor has a greater amount of lubricant injected into it than in the lower stages. It is recommended that the compressor selected for a diving operation provide as close to the required pressure for that operation as possible. A system that provides excessive pressure contributes to the buildup of lubricant in the air supply.

8-7.2.2.6 ***Intercoolers.*** Intercoolers are heat exchangers that are placed between the stages of a compressor to control the air temperature. Water, flowing through the heat exchanger counter to the air flow, serves both to remove heat from the air and to cool the cylinder walls. Intercoolers are frequently air cooled. During the cooling process, water vapor is condensed out of the air into condensate collectors. The condensate must be drained periodically during operation of the compressor, either manually or automatically.

8-7.2.2.7 ***Filters.*** As the air is discharged from the compressor, it passes through a moisture separator and an approved filter to remove lubricant, aerosols, and particulate contamination before it enters the system. Approved filters are listed in the NAVSEA/00C ANU list.

8-7.2.2.8 ***Pressure Regulators.*** A back-pressure regulator will be installed downstream of the compressor discharge. A compressor only compresses air to meet the supply pressure demand. If no demand exists, air is simply pumped through the compressor at atmospheric pressure. Systems within the compressor, such as the intercoolers, are designed to perform with maximum efficiency at the rated pressure of the compressor. Operating at any pressure below this rating reduces the efficiency of the unit. Additionally, compression reduces water vapor from the air. Reducing the amount of compression increases the amount of water vapor in the air supplied to the diver.

The air supplied from the compressor expands across the pressure regulator and enters the air banks or volume tank. As the pressure builds up in the air banks or volume tank, it eventually reaches the relief pressure of the compressor, at which time the excess air is simply discharged to the atmosphere. Some electrically-driven compressors are controlled by pressure switches installed in the volume tank or HP flask. When the pressure reaches the upper limit, the electric motor is shut off. When sufficient air has been drawn from the volume tank or HP flask to lower its pressure to some lower limit, the electric motor is restarted.

Any diving air compressor, if not permanently installed, must be firmly secured in place. Most portable compressors are provided with lashing rings for this purpose.

8-7.2.3 **High-Pressure Air Cylinders and Flasks.** HP air cylinders and flasks are vessels designed to hold air at pressures over 600 psi. Any HP vessel to be used as a diving air supply unit must bear appropriate Department of Transportation (DOT), American Society of Mechanical Engineers (ASME), or military symbols certifying that the cylinders or flasks meet high-pressure requirements.

A complete air supply system includes the necessary piping and manifolds, HP filter, pressure reducing valve, and a volume tank. An HP gauge must be located ahead of the reducing valve and an LP gauge must be connected to the pressure reducing valve and a volume tank (when required).

In using this type of system, one section must be kept in reserve. The divers take air from the volume tank in which the pressure is regulated to conform to the air supply requirements of the dive. The duration of the dive is limited to the length of time the banks can provide air before being depleted to 200 psi over minimum manifold pressure. This minimum pressure of 200 psi must remain in each flask or cylinder.

As in SCUBA operations, the quantity of air that can be supplied by a system using cylinders or flasks is determined by the initial capacity of the cylinders or flasks and the depth of the dive. The duration of the air supply must be calculated in advance and must include a provision for decompression.

Sample calculations for dive duration, based on bank air supply, are presented in Sample Problem 1 in paragraph 8-2.2.3 for the MK 21 MOD 1, KM-37. The sample problems in this chapter do not take the secondary air system requirements into account. The secondary air system must be able to provide air in the event of failure of the primary system per *General Specification for the Design, Construction, and Repair of Diving and Hyperbaric Equipment*, NAVSEA TS500-AU-SPN-010. In the MK 21 sample problem, this would mean decompressing three divers with a 30-minute bottom time using 0.75 acfm per diver. An additional requirement must be considered if the same air system is to support a recompression chamber. Refer to Chapter 21 for information on the additional capacity required to support a recompression chamber.

8-8 DIVER COMMUNICATIONS

The surface-supplied diver has two means of communicating with the surface, depending on the type of equipment used. If the diver is using a surface-supplied UBA, both voice communications and line-pull signals are available. Voice communications are used as the primary means of communication. Line-pull signals are used only as a backup. Diver-to-diver communications are available through topside intercom, diver-to-diver hand signals, or slate boards.

8-8.1 **Diver Intercommunication Systems.** The major components of the intercommunication system include the diver's earphones and microphone, the communication cable to each diver, the surface control unit, and the tender's speaker and microphone. The system is equipped with an external power cord and can accept 115 VAC or 12 VDC. The internal battery is used for backup power requirements. It should not be used as the primary power source unless an external power source is not available.

The intercom system is operated by a designated phone talker at the diving station. The phone talker monitors voice communications and keeps an accurate log of significant messages. All persons using the intercom system should lower the pitch of their voices and speak slowly and distinctly. The conversation should be kept brief and simple, using standard diving terminology. Divers must repeat verbatim all directions and orders received from topside.

Approved Navy diver communication systems are compatible with all surface-supplied UBAs. This is a surface/underwater system that allows conference communications between the tender and up to three divers.

The divers' voices are continuously monitored on the surface. All communications controls are located at the surface. The topside supervisor speaks with any or all of the divers by exercising the controls on the front panel. It is necessary for a phone talker to monitor and control the underwater communications system at all times.

8-8.2 **Line-Pull Signals.** A line-pull signal consists of one pull or a series of sharp, distinct pulls on the umbilical that are strong enough to be felt by the diver (Figure 8-11). All slack must be taken out of the umbilical before the signal is given.

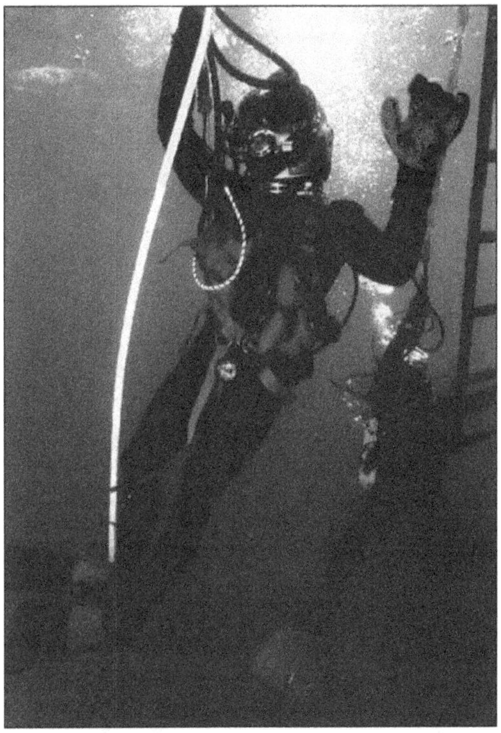

Figure 8-11. Communicating with Line-Pull Signals.

The line-pull signal code (Table 8-3) has been established through many years of experience. Standard signals are applicable to all diving operations; special signals may be arranged between the divers and Diving Supervisor to meet particular mission requirements. Most signals are acknowledged as soon as they are received. This acknowledgment consists of replying with the same signal. If a signal is not properly returned by the diver, the surface signal is sent again. A continued absence of confirmation is assumed to mean one of three things: the line has become fouled, there is too much slack in the line, or the diver is in trouble.

If communications are lost, the Diving Supervisor must be notified immediately and steps taken to identify the problem. The situation is treated as an emergency (see paragraph 6-10.8.2).

There are three line-pull signals that are not answered immediately. Two of these, from diver to tender, are "Haul me up" and "Haul me up immediately." Acknowledgment consists of initiation of the action. The other signal, from the tender to diver, is "Come up." This signal is not acknowledged until the diver is ready to leave the bottom. If for some reason the diver cannot respond to the order, the diver must communicate the reason via the voice intercom system or through the line-pull signal meaning "I understand," followed (if necessary) by an appropriate emergency signal. A special group of searching signals is used by the tender to direct a diver in moving along the bottom. These signals are duplicates of standard

Table 8-3. Line-Pull Signals.

From Tender to Diver		Searching Signals (Without Circling Line)	
1 Pull	"Are you all right?" When diver is descending, one pull means "Stop."	7 Pulls	"Go on (or off) searching signals."
2 Pulls	"Going Down." During ascent, two pulls mean "You have come up too far; go back down until we stop you."	1 Pull	"Stop and search where you are."
3 Pulls	"Stand by to come up."	2 Pulls	"Move directly away from the tender if given slack; move toward the tender if strain is taken on the life line."
4 Pulls	"Come up."	3 Pulls	"Face your umbilical, take a strain, move right."
2-1 Pulls	"I understand" or "Talk to me."	4 Pulls	"Face your umbilical, take a strain, move left."
3-2 Pulls	"Ventilate."		
4-3 Pulls	"Circulate."		
From Diver to Tender		**Searching Signals (With Circling Line)**	
1 Pull	"I am all right." When descending, one pull means "Stop" or "I am on the bottom."	7 Pulls	"Go on (or off) searching signals."
2 Pulls	"Lower" or "Give me slack."	1 Pull	"Stop and search where you are."
3 Pulls	"Take up my slack."	2 Pulls	"Move away from the weight."
4 Pulls	"Haul me up."	3 Pulls	"Face the weight and go right."
2-1 Pulls	"I understand" or "Talk to me."	4 Pulls	"Face the weight and go left."
3-2 Pulls	"More air."		
4-3 Pulls	"Less air."		
Special Signals From the Diver		**Emergency Signals From the Diver**	
1-2-3 Pulls	"Send me a square mark."	2-2-2 Pulls	"I am fouled and need the assistance of another diver."
5 Pulls	"Send me a line."	3-3-3 Pulls	"I am fouled but can clear myself."
2-1-2 Pulls	"Send me a slate."	4-4-4 Pulls	"Haul me up immediately."
ALL EMERGENCY SIGNALS SHALL BE ANSWERED AS GIVEN EXCEPT 4-4-4			

line-pull signals, but their use is indicated by an initial seven-pull signal to the diver that instructs the diver to interpret succeeding signals as searching signals. When the tender wants to revert to standard signals, another seven-pull signal is sent to the diver which means searching signals are no longer in use. Only the tender uses searching signals; all signals initiated by the diver are standard signals. To be properly oriented for using searching signals, the diver must face the line (either the lifeline or the descent line, if a circling line is being employed).

8-9 PREDIVE PROCEDURES

The predive activities for a surface-supplied diving operation involve many people and include inspecting and assembling the equipment, activating the air supply systems, and dressing the divers.

8-9.1 **Predive Checklist.** A comprehensive predive checklist is developed to suit the requirements of the diving unit and of the particular operation. This is in addition to the general Diver Safety and Planning Checklist (Figure 6-19) and suggested Predive Checklist (Figure 6-21).

8-9.2 **Diving Station Preparation.** The diving station is neatly organized with all diving and support equipment placed in an assigned location. Deck space must not be cluttered with gear; items that could be damaged are placed out of the way (preferably off the deck). A standard layout pattern should be established and followed.

8-9.3 **Air Supply Preparation.** The primary and secondary air supply systems are checked to ensure that adequate air is available. Air compressors of the divers' air system are started and checked for proper operation. The pressure in the accumulator tanks is checked. If HP air cylinders are being used, the manifold pressure is checked. If a compressor is being used as a secondary air supply, it is started and kept running throughout the dive. The air supply must meet purity standards (see paragraph 8-7.1.1).

8-9.4 **Line Preparation.** Depth soundings are taken and descent line, stage, stage lines, and connections are checked, with decompression stops properly marked.

8-9.5 **Recompression Chamber Inspection and Preparation.** If available, the recompression chamber is inspected and all necessary equipment and a copy of appropriate recompression treatment tables are placed on hand at the chamber. Two stopwatches and the decompression tables are also required. Adequate air supply for immediate pressurization of the chamber is verified and the oxygen supply system is charged and made ready for operation in accordance with Chapter 21.

8-9.6 **Predive Inspection.** When the Diving Supervisor is satisfied that all equipment is on station and in good operating condition, the next step is to dress the divers.

8-9.7 **Donning Gear.** Dressing the divers is the responsibility of the tender.

8-9.8 **Diving Supervisor Predive Checklist.** The Diving Supervisor must always use a predive checklist prior to putting divers in the water. This checklist must be tailored by the unit to the specific equipment and systems being used. Chapter 6 contains typical predive checklists for surface-supplied equipment. Refer to the appropriate operations and maintenance manual for detailed checklists for specific equipment.

8-10 WATER ENTRY AND DESCENT

Once the predive procedures have been completed, the divers are ready to enter the water. There are several ways to enter the water, with the choice usually determined by the nature of the diving platform. Regardless of the method of entry, the divers should look before entering the water. Three methods for entering the water are the:

- Ladder method
- Stage method
- Step-in method

8-10.1 **Predescent Surface Check.** In the water and prior to descending to operating depth, the diver makes a final equipment check.

- The diver immediately checks for leaks in the suit or air connections.
- If two divers are being employed, both divers perform as many checks as possible on their own rigs and then check their dive partner's rig. The tender or another diver can be of assistance by looking for any telltale bubbles.
- A communications check is made and malfunctions or deficiencies not previously noted are reported at this time.

When satisfied that the divers are ready in all respects to begin the dive, they notify the Diving Supervisor and the tenders move the divers to the descent line. When in position for descent, the diver adjusts for negative buoyancy and signals readiness to the Diving Supervisor.

8-10.2 **Descent.** Descent may be accomplished with the aid of a descent line or stage. Topside personnel must ensure that air is being supplied to the diver in sufficient quantity and at a pressure sufficient to offset the effect of the steadily increasing water pressure.

While descending, the diver adjusts the air supply so that breathing is easy and comfortable. The diver continues to equalize the pressure in the ears as necessary during descent and must be on guard for any pain in the ears or sinuses, or any other warning signals of possible danger. If any such indications are noted, the descent is halted. The difficulty may be resolved by ascending a few feet to regain a pressure balance; if this is not effective, the diver is returned to the surface.

Some specific guidelines for descent are as follows:

- With a descent line, the diver locks the legs around the line and holds on to the line with one hand.
- In a current or tideway, the diver descends with back to the flow in order to be held against the line and not be pulled away. If the current measures more than 1.5 knots, the diver wears additional weights or descends on a weighted stage, so that descent is as nearly vertical as possible.
- When the stage is used for descent, it is lowered with the aid of a winch and guided to the site by a shackle around the descent line. The diver stands in the center of the stage, maintaining balance by holding on to the side bails. Upon reaching the bottom, the diver exits the stage as directed by the Diving Supervisor.
- The maximum allowable rate of descent, by any method, normally should not exceed 75 feet per minute (fpm), although such factors as the diver's ability to clear the ears, currents and visibility, and the need to approach an unknown bottom with caution may render the actual rate of descent considerably less.

- The diver signals arrival on the bottom and quickly checks bottom conditions. Conditions that are radically different than expected are reported to the Diving Supervisor. If there is any doubt about the safety of the diver or the diver's readiness to operate under the changed conditions, the dive is aborted.

- A diver should thoroughly ventilate at subsequent intervals as the diver feels necessary and as directed from the surface. On dives deeper than 100 fsw, the diver may not notice the CO_2 warning symptoms because of nitrogen narcosis. It is imperative that the Diving Supervisor monitors the divers ventilation.

8-11 UNDERWATER PROCEDURES

8-11.1 **Adapting to Underwater Conditions.** Through careful and thorough planning, the divers can be properly prepared for the underwater conditions at the diving site. The diver will employ the following techniques to adapt to underwater conditions:

- Upon reaching the bottom and before leaving the area of the stage or descent line, the diver checks equipment and makes certain that the air supply is adequate.

- The diver becomes oriented to the bottom and the work site using such clues as the lead of the umbilical, natural features on the bottom, and the direction of current. However, bottom current may differ from the surface current. The direction of current flow may change significantly during the period of the dive. If the diver has any trouble in orientation, the tender can guide the diver by using the line-pull searching signals.

The diver is now ready to move to the work site and begin the assignment.

8-11.2 **Movement on the Bottom.** Divers should follow these guidelines for movement on the bottom areas:

- Before leaving the descent line or stage, ensure that the umbilical is not fouled.

- Loop one turn of the lifeline and air hose over an arm; this acts as a buffer against a sudden surge or pull on the lines.

- Proceed slowly and cautiously to increase safety and to conserve energy.

- If obstructions are encountered, pass over the obstruction, not under or around. If you pass around an obstruction, you must return by the same side to avoid fouling lines.

- When using a Variable Volume Dry Suit, buoyancy adjustments to aid in movement, avoid bouncing along the bottom; all diver movements are controlled.

- If the current is strong, stoop or crawl to reduce body area exposed to the current.

- When moving on a rocky or coral bottom, make sure lines do not become fouled on outcroppings, guarding against tripping and getting feet caught in crevices. Watch for sharp projections that can cut hoses, diving dress, or unprotected hands. The tender is particularly careful to take up any slack in the diver's umbilical to avoid fouling.

- Avoid unnecessary movements that stir up the bottom and impair visibility.

CAUTION **When diving with a Variable Volume Dry Suit, avoid overinflation and be aware of the possibility of blowup when breaking loose from mud. It is better to call for aid from the standby diver than to risk blowup.**

- Mud and silt may not be solid enough to support your weight. Many hours may be spent working under mud without unreasonable risk. Demand regulators may not function well when covered by mud or heavy silt. If it is anticipated that the diver may become covered by mud, as in a jetting or tunneling operations, the diver should keep the helmet steady-flow valve slightly open. The primary hazard with mud bottoms comes from the concealment of obstacles and dangerous debris.

8-11.3 **Searching on the Bottom.** If appropriate electronic searching equipment is not available, it may be necessary to use unaided divers to conduct the search. Procedures for searching on the bottom with unaided divers are:

1. A diver search of the bottom can be accomplished with a circling line, using the descent line as the base point of the search. The first sweep is made with the circling line held taut at a point determined by the range of visibility. If possible, the descent line should be in sight or, if visibility is limited, within reach. The starting point is established by a marker, a line orientation with the current or the light, signals from topside, or a wrist compass. After a full 360-degree sweep has been made, the diver moves out along the circling line another increment (roughly double the first) and makes a second sweep in the opposite direction to avoid twisting or fouling the lifeline and air hose.

2. If the object is not found when the end of the circling line has been reached, the base point (the descent line) is shifted. Each base point in succession should be marked by a buoy to avoid unnecessary duplication in the search. If the search becomes widespread, many of the marker buoys can be removed, leaving only those marking the outer limits of the area.

3. If the diver is unable to make a full circle around the descent line because of excessive current or obstructions, the search patterns are adjusted accordingly.

4. A linear search pattern (Jack-Stay) can be established by laying two large buoys and setting a line between them. A diving launch, with a diver on the bottom, can follow along the line from buoy to buoy, coordinating progress with the diver who is searching to each side of the established base line. These buoys may be readjusted to enlarge search areas.

5. Once the object of a search is located, it is marked. The diver can secure the circling line to the object as an interim measure, while waiting for a float line to be sent down.

8-11.4 **Enclosed Space Diving.** Divers are often required to work in an enclosed or confined space. Enclosed space diving shall be supported by a surface-supplied air system and use a surface-supplied UBA.

8-11.4.1 **Enclosed Space Hazards.** The interior of sunken ships, barges, submarine ballast tanks, mud tanks, sonar domes, and cofferdams is hazardous due to limited access, poor visibility, and slippery surfaces. Enclosed spaces may be dry or flooded, and dry spaces may contain a contaminated atmosphere.

NOTE **When a diver is working in an enclosed or confined space with the exception of submarine ballast tanks, the Diving Supervisor shall have the diver tended by another diver at the access opening. Ultimately, the number of tending divers deployed depends on the situation and the good judgement of the Diving Officer, Master Diver, or Diving Supervisor on the site.**

8-11.4.2 **Enclosed Space Safety Precautions.** Because of the hazards involved in enclosed space operations, divers must rigorously adhere to the following warnings.

WARNING **During enclosed space diving, all divers shall be outfitted with a MK 21 MOD 1, KM-37, MK 20 MOD 0, or EXO BR MS that includes a diver-to-diver and diver-to-topside communications system and an EGS for the diver inside the space.**

WARNING **For submarine ballast tanks, the divers shall not remove their diving equipment until the atmosphere has been flushed twice with air from a compressed air source meeting the requirements of Chapter 4, or the submarine L.P. blower, and tests confirm that the atmosphere is safe for breathing. Tests of the air in the enclosed space shall be conducted hourly. Testing shall be done in accordance with NSTM 074, Volume 3, Gas Free Engineering (S9086-CH-STM-030/CH-074) for forces afloat, and NAVSEA S-6470-AA-SAF-010 for shore-based facilities. If the divers smell any unusual odors they shall immediately don their EGS.**

WARNING **If the diving equipment should fail, the diver shall immediately switch to the EGS and abort the dive.**

8-11.5 **Working Around Corners.** When working around corners where the umbilical is likely to become fouled or line-pull signals may be dissipated, a second diver (tending diver) may be sent down to tend the lines of the first diver at the obstruction and to pass along any line-pull signals. Line-pull signals are used when audio communications are lost, and are passed on the first diver's lines; the tending diver uses his own lines only for signals directly pertaining to his own situation.

8-11.6 Working Inside a Wreck. When working inside a wreck, the same procedure of deploying tending divers is followed. This technique applies to the tending divers as well: every diver who penetrates a deck level has another tending diver at that level, or levels, above. Ultimately, the number of tending divers deployed depends on the situation and the good judgment of the Diving Officer, Master Diver, or Diving Supervisor on the site. Obviously, an operation requiring penetration through multiple deck levels requires detailed advanced planning in order to provide for the proper support of the number of divers required. MK 21 MOD 1, KM-37 and MK 20 MOD 0 are the only equipment approved for working inside a wreck. The diver enters a wreck feet first and never uses force to gain entry through an opening.

8-11.7 Working With or Near Lines or Moorings. When working with or near lines or moorings, observe the following rules:

- Stay away from lines under strain.
- Avoid passing under lines or moorings if at all possible; avoid brushing against lines or moorings that have become encrusted with barnacles.
- If a line or mooring is to be shifted, the diver is brought to the surface and, if not removed from the water, moved to a position well clear of any hazard.
- If a diver must work with several lines (messengers, float lines, lifting lines, etc.) each should be distinct in character (size or material) or marking (color codes, tags, wrapping).
- Never cut a line unless the line is positively identified.
- When preparing to lift heavy weights from the bottom, the lines selected must be strong enough and the surface platform must be positioned directly over the object to be raised. Prior to the lift, make sure the diver is clear of the lift area or leaves the water.

8-11.8 Bottom Checks. Bottom checks are conducted after returning to the stage or descent line and prior to ascent. The checks are basically the same for each rig.

1. Ensure all tools are ready for ascent.
2. Check that all umbilicals and lines are clear for ascent.
3. Assess and report your condition (level of fatigue, remaining strength, physical aches or pains, etc.) and mental acuity.

8-11.9 Job Site Procedures. The range of diving jobs is wide and varied. Many jobs follow detailed work procedures and require specific predive training to ensure familiarity with the work. The *Underwater Ship Husbandry Manual*, S0600-AA-PRO-010, presents guidance for most commonly encountered jobs, such as replacement and repair of propellers, propeller blades, auxiliary propulsion motors, and sonar domes.

8-11.9.1 **Underwater Ship Husbandry Procedures.** Due to the complexity of ships' underwater systems and the sophistication of newly developed repair techniques, specific procedures were developed to provide guidance in the underwater repair and maintenance of U.S. Navy ships. These procedures are located in individually bound chapters of the *Underwater Ship Husbandry Manual* (S0600-AA-PRO-010). Chapter 1 of the manual is the Index and User Guide, which provides information on the subsequent chapters of the manual.

8-11.9.2 **Working with Tools.** Underwater work requires appropriate tools and materials, such as cement, foam plastic, and patching compounds. Many of these are standard hand tools (preferably corrosion-resistant) and materials; others are specially designed for underwater work. A qualified diver will become familiar with the particular considerations involved in working with these various tools and materials in an underwater environment. Hands-on training experience is the only way to get the necessary skills. Consult the appropriate operations and maintenance manuals for the use techniques of specific underwater tools. In working with tools the following basic rules always apply:

- Never use a tool that is not in good repair. If a cutting tool becomes dulled, return it to the surface for sharpening.
- Do not overburden the worksite with unnecessary tools, but have all tools that may be needed readily available.
- Tools are secured to the diving stage by lanyard, carried in a tool bag looped over the diver's arm, or lowered on the descent line using a riding shackle and a light line for lowering. Prior to ascent or descent, secure power to all tools. Attach lanyards to all tools, connectors, shackles, and shackle pins.
- Using the diving stage as a worksite permits organization of tools while providing for security against loss. The stage also gives the diver leverage and stability when applying force (as to a wrench), or when working with a power tool that transmits a force back through the diver.
- Tying a hogging line to the work also gives the diver leverage while keeping him close to his task without continually having to fight a current.

8-11.10 **Safety Procedures.** The best safety factors are a positive, confident attitude about diving and careful advance planning for emergencies. A diver in trouble underwater should relax, avoid panic, communicate the problem to the surface, and carefully think through the possible solutions to the situation. Topside support personnel should implement emergency job-site procedures as indicated in Chapter 6. In all situations, the Diving Supervisor should ensure that common sense and good seamanship prevail to safely resolve each emergency.

Emergency procedures are covered specifically for each equipment in its appropriate operations and maintenance manual and in general in Chapter 6. However, there are a number of situations a diver is likely to encounter in the normal range of activity which, if not promptly solved, can lead to full-scale emergencies. These situations and the appropriate action to be taken follow.

8-11.10.1 **Fouled Umbilical Lines.** As soon as a diver discovers that the umbilical has become fouled, the diver must stop and examine the situation. Pulling or tugging without a plan may only serve to complicate the problem and could lead to a severed hose. The Diving Supervisor is notified if possible (the fouling may prevent transmission of line-pull signals). If the lines are fouled on an obstruction, retracing steps should free them. If the lines cannot be cleared quickly and easily, the standby diver is sent down to assist. The standby diver is sent down as normal procedure, should communications be interrupted and the tender be unable to haul the diver up. The standby diver, using the first diver's umbilical (as a descent line), should be able to trace and release the lines. If it is impossible to free the first diver, the standby diver should signal for a replacement umbilical.

8-11.10.2 **Fouled Descent Lines.** If the diver becomes fouled with the descent line and cannot be easily cleared, it is necessary to haul the diver and the line to the surface, or to cut the weight free of the line and attempt to pull it free from topside. If the descent line is secured to an object or if the weight is too heavy, the diver may have to cut the line before being hauled up. For this reason, a diver should not descend on a line that cannot be cut.

WARNING **If job conditions call for using a steel cable or a chain as a descent line, the Diving Officer must approve such use.**

8-11.10.3 **Falling.** When working at mid-depth in the water column, the diver should keep a hand on the stage or rigging to avoid falling. The diver avoids putting an arm overhead in a dry suit; air leakage around the edges of the cuffs may change the suit buoyancy and increase the possibility of a fall in the water column.

8-11.10.4 **Damage to Helmet and Diving Dress.** If a leak occurs in the helmet, the diver's head is lowered and the air pressure slightly increased to prevent water leakage. A leak in the diving suit only requires remaining in an upright position; water in the suit does not endanger breathing.

8-11.11 **Tending the Diver.** Procedures for tending the diver follow.

1. Before the dive, the tender carefully checks the diving dress with particular attention to the nonreturn valve, air control valve, helmet locking device, intercom system, helmet seal, and harness.

2. When the diver is ready, the tenders dress and assist the diver to the stage or ladder or water's edge, always keeping a hand on the umbilical.

3. The primary tender and a backup tender as required are always on station to assist the diver. As the diver enters the water, the tenders handle the umbilical, using care to avoid sharp edges. The umbilical must never be allowed to run

free or be belayed around a cleat or set of bitts. Pay out of the umbilical is at a steady rate to permit the diver to descend smoothly. If a stage is being used, the descent rate is coordinated with the winch operator or line handlers.

4. Throughout the dive the tender keeps slack out of the line while not holding it too tautly. Two or three feet of slack permits the diver freedom of movement and prevents the diver from being pulled off the bottom by surging of the support craft or the force of current acting on the line. The tender occasionally checks the umbilical to ensure that movement by the diver has not resulted in excessive slack. Excessive slack makes signaling difficult and increases the possibility of fouling the umbilical.

5. The tender monitors the umbilical by feel and the descent line by sight for any line-pull signals from the diver. If an intercom is not being used, or if the diver is silent, the tender periodically verifies the diver's condition by line-pull signal. If the diver does not answer, the signal is repeated; if still not answered, the Diving Supervisor is notified. If communications are lost, the situation is treated as an emergency (see paragraph 6-10.8.2 for loss-of-communication procedures).

8-11.12 **Monitoring the Diver's Movements.** The Diving Supervisor and designated members of the dive team constantly monitor the diver's progress and keep track of his relative position.

- **Supervisor Actions.**

  1. Follow the bubble trail, while considering current(s). If the diver is searching the bottom, bubbles move in a regular pattern. If the diver is working in place, bubbles do not shift position. If the diver has fallen, the bubbles may move rapidly off in a straight line.

  2. Monitor the pneumofathometer pressure gauge to keep track of operating depth. If the diver remains at a constant depth or rises, the gauge provides a direct reading, without the need to add air. If the diver descends, the hose must be cleared and a new reading made.

- **Tender Actions.** Feel the pull of the umbilical.

- **Additional Personnel Actions.** Monitor the gauges on the supply systems for any powered equipment. For example, the ammeter on an electric welding unit indicates a power drain when the arc is in use; the gas pressure gauges for a gas torch registers the flow of fuel. A change in pressure and flow of the hydraulic power unit indicates tool use.

8-12 ASCENT PROCEDURES

Follow these ascent procedures when it is time for the divers to return to the surface:

1. To prepare for a normal ascent, the diver clears the job site of tools and equipment. These can be returned to the surface by special messenger lines sent down the descent line. If the diver cannot find the descent line and needs a special line, this can be bent onto his umbilical and pulled down by the diver. The diver must be careful not to foul the line as it is laid down. The tender then pulls up the slack. This technique is useful in shallow water, but not practical in deep dives.

2. If possible, the diving stage is positioned on the bottom. If some malfunction such as fouling of the descent line prevents lowering the stage to the bottom, the stage should be positioned below the first decompression stop if possible. Readings from the pneumofathometer are the primary depth measurements.

3. If ascent is being made using the descent line or the stage has been positioned below the first decompression stop, the tender signals the diver "Standby to come up" when all tools and extra lines have been cleared away. The diver acknowledges the signal. The diver, however, does not pull up. The tender lifts the diver off the bottom when the diver signals "Ready to come up," and the tender signals "Coming up. Report when you leave the bottom." The diver so reports.

4. If, during the ascent, while using a descent line, the diver becomes too buoyant and rises too quickly, the diver checks the ascent by clamping his legs on the descent line.

5. The rate of ascent is a critical factor in decompressing the diver. Ascent must be carefully controlled at 30 feet per minute by the tender. The ascent is monitored with the pneumofathometer. As the diver reaches the stage and climbs aboard, topside is notified of arrival. The stage is then brought up to the first decompression stop. Refer to Chapter 9 for decompression procedures, including an explanation of the tables.

6. While ascending and during the decompression stops, the diver must be satisfied that no symptoms of physical problems have developed. If the diver feels any pain, dizziness, or numbness, the diver immediately notifies topside. During this often lengthy period of ascent, the diver also checks to ensure that his umbilical is not fouled.

7. Upon arrival at the surface, topside personnel, timing the movement as dictated by any surface wave action, coordinate bringing the stage and umbilical up and over the side.

8. If the diver exits the water via the ladder, the tenders provide assistance. The diver will be tired, and a fall back into the water could result in serious injury. Under no conditions is any of the diver's gear to be removed before the diver is firmly on deck.

8-13 SURFACE DECOMPRESSION

8-13.1 **Disadvantages of In-Water Decompression.** Decompression in the water column is time consuming, uncomfortable, and inhibits the ability of the support vessel to get underway. Delay could also present other problems for the support vessel: weather, threatened enemy action, or operating schedule constraints. In-water decompression delays medical treatment, when needed, and increases the possibility of severe chilling and accident. For these reasons, decompression is often accomplished in a recompression chamber on the support ship (Figure 8-12). Refer to Chapter 9 for surface decompression procedures.

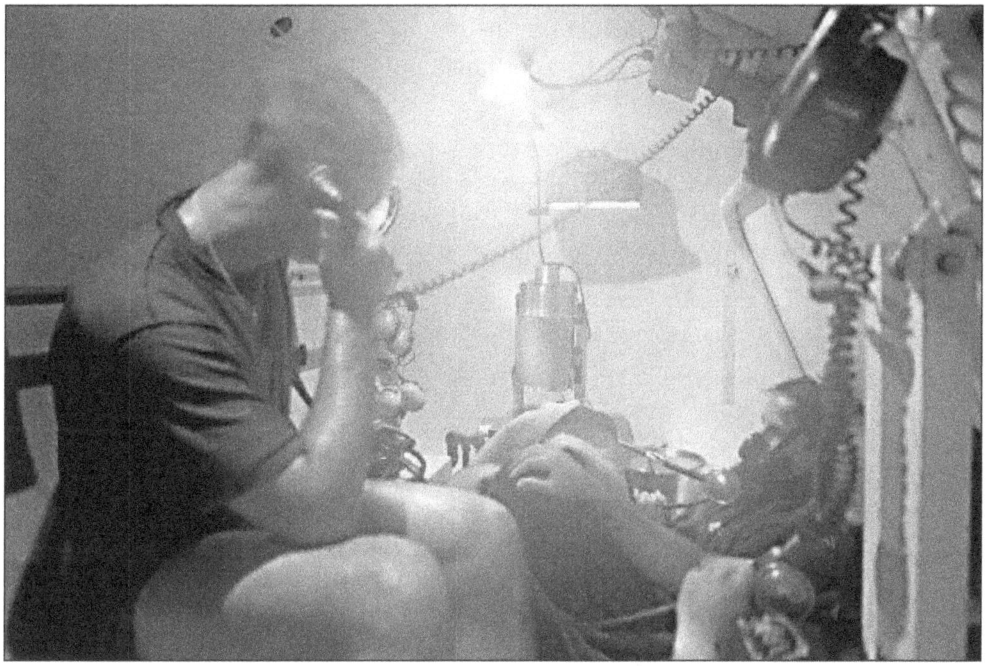

Figure 8-12. Surface Decompression.

8-13.2 **Transferring a Diver to the Chamber.** When transferring a diver from the water to the chamber, the tenders are allowed no more than 3½ minutes to undress the diver. A tender or diving medical personnel, as required by the nature of the dive or the condition of the diver, must be in the chamber with any necessary supplies prior to arrival of the diver. The time factor is critical and delays cannot be tolerated. Undressing a diver for surface decompression should be practiced until a smooth, coordinated procedure is developed.

8-14 POSTDIVE PROCEDURES

Postdive procedures are planned in advance to ensure personnel are carefully examined for any possible injury or adverse effects and equipment is inspected, maintained, and stowed in good order.

8-14.1 **Personnel and Reporting.** Immediate postdive activities include any required medical treatment for the diver and the recording of mandatory reports.

- Medical treatment is administered for cuts or abrasions. The general condition of the diver is monitored until problems are unlikely to develop. The Diving Supervisor resets the stopwatch after the diver reaches the surface and remains alert for irregularities in the diver's actions or mental state. The diver must remain within 30 minutes' travel time of the diving unit for at least 2 hours after surfacing.

- Mandatory records and reports are covered in Chapter 5. Certain information is logged as soon as the diving operations are completed, while other record keeping is scheduled when convenient. The Diving Supervisor is responsible for the diving log, which is kept as a running account of the dive. The diver is responsible for making appropriate entries in the personal diving record. Other personnel, as assigned, are responsible for maintaining equipment usage logs.

8-14.2 **Equipment.** A postdive checklist, tailored to the equipment used, is followed to ensure equipment receives proper maintenance prior to storage. Postdive maintenance procedures are contained in the equipment operation and maintenance manual and the planned maintenance system package.

CHAPTER 9
Air Decompression

9-1 INTRODUCTION

9-1.1 **Purpose.** This chapter discusses the decompression requirements for air diving operations.

9-1.2 **Scope.** The decompression procedures contained in this Chapter are new. They replace the air decompression procedures that have been in use by the Navy for more than fifty years. These new procedures are safer, more flexible, and provide more operational capability than the older procedures. The primary improvement in safety results from the use of oxygen during decompression to accelerate elimination of excess nitrogen from the body. All but the shortest decompressions are performed either with oxygen breathing in the water or with surface decompression on oxygen.

9-2 THEORY OF DECOMPRESSION

As a diver descends, the partial pressure of nitrogen in his lungs rises above the partial pressure of nitrogen dissolved in his tissues. This pressure difference causes nitrogen to be transported from the lungs to the tissues via the bloodstream. Transport to a given tissue will continue as long as the partial pressure of nitrogen in the lungs is higher than the partial pressure of nitrogen in that tissue. The process will stop when the tissue has absorbed enough nitrogen to raise its partial pressure to a value equal to that in the lungs. Different tissues absorb nitrogen at different rates. A tissue with a high blood flow, like the brain, will come into equilibrium with the partial pressure of nitrogen in the lungs faster than a tissue with low blood flow, like muscle or tendon. The total amount of nitrogen absorbed by a tissue will be greater the deeper the dive and the longer the bottom time, until the tissue becomes saturated.

As a diver ascends, the process is reversed. The partial pressure of nitrogen in the tissues comes to exceed that in the lungs. During ascent, nitrogen is transported back from the tissues to the lungs by the circulation. The ascent rate must be carefully controlled to allow time for this process to occur and not allow the tissue nitrogen partial pressure to exceed the ambient pressure by too great an amount. The more the tissue nitrogen partial pressure exceeds the ambient pressure during ascent, the more likely nitrogen bubbles will form in tissues and blood, causing decompression sickness.

To reduce the possibility of decompression sickness, special decompression schedules have been developed for air diving. These schedules take into consideration the amount of nitrogen absorbed by the body at various depths and times. Other considerations are the extent to which the tissue nitrogen partial pressure can exceed the ambient pressure without excessive bubble formation and the different nitrogen elimination rates associated with the various body tissues. Because of its operational simplicity, staged decompression is used for air decompression. Staged decompression requires decompression stops in the water at various depths for specific periods of time.

Years of scientific study, calculations, animal and human experimentation, and extensive field experience have all contributed to the air decompression tables. While the tables contain the best information available, the tables tend to be less accurate as dive depth and bottom time increase. To ensure maximum diver safety, the tables must be strictly followed. Deviations from established decompression procedures are not permitted except in an emergency and with the guidance and recommendation of a Diving Medical Officer (DMO) with the Commanding Officer or Officer-in-Charge's approval.

9-3 AIR DECOMPRESSION DEFINITIONS

The following terms must be understood before using the air decompression tables.

9-3.1 **Descent Time.** *Descent time* is the total elapsed time from the time the diver leaves the surface to the time he reaches the bottom. Descent time is rounded up to the next whole minute for charting purposes.

9-3.2 **Bottom Time.** *Bottom time* is the total elapsed time from the time the diver leaves the surface to the time he leaves the bottom. Bottom time is measured in minutes and is rounded up to the next whole minute.

9-3.3 **Total Decompression Time.** The *total decompression time* is the total elapsed time from the time the diver leaves the bottom to the time he arrives on the surface. This time is also frequently called the *total ascent time*. The two terms are synonymous and can be used interchangeably.

9-3.4 **Total Time of Dive.** The *total time of dive* is the total elapsed time from the time the diver leaves the surface to the time he arrives back on the surface.

9-3.5 **Deepest Depth.** The *deepest depth* is the deepest depth recorded on the depth gauge during a dive.

9-3.6 **Maximum Depth.** *Maximum depth* is the deepest depth obtained by the diver after correction of the depth gauge reading for error. When conducting SCUBA operations, the diver's depth gauge is considered error free. The diver's maximum depth is the deepest depth gauge reading. When conducting surface-supplied diving operations using a pneumofathometer to measure depth, maximum depth is the deepest reading on the pneumofathometer gauge plus the pneumofathometer correction factor (Table 9-1). Maximum depth is the depth used to enter the decompression tables.

9-3.7 **Stage Depth.** *Stage depth* is the pneumofathometer reading taken when the divers are on the stage just prior to leaving the bottom. Stage depth is used to compute the distance and travel time to the first stop, or to the surface if no stops are required.

9-3.8 **Decompression Table.** A *decompression table* is a structured set of decompression schedules, or limits, usually organized in order of increasing bottom times and depths.

9-3.9 **Decompression Schedule.** A *decompression schedule* is a specific decompression procedure for a given combination of depth and bottom time as listed in a decompression table. It is normally indicated as feet/minutes.

9-3.10 **Decompression Stop.** A *decompression stop* is a specified depth where a diver must remain for a specified length of time (stop time) during ascent.

9-3.11 **No-Decompression (No "D") Limit.** The maximum time a diver can spend at a given depth and still ascend directly to the surface at the prescribed travel rate without taking decompression stops.

9-3.12 **No-Decompression Dive.** A dive that does not require a diver to take decompression stops during ascent to the surface.

9-3.13 **Decompression Dive.** A dive that does require a diver to take decompression stops during ascent to the surface.

9-3.14 **Surface Interval.** In the context of repetitive diving, the *surface interval* is the time a diver spends on the surface between dives. It begins as soon as the diver surfaces and ends as soon as he starts his next descent. In the context of surface decompression, the *surface interval* is the total elapsed time from when the diver leaves the 40 fsw water stop to the time he arrives at 50 fsw in the recompression chamber.

9-3.15 **Residual Nitrogen.** *Residual nitrogen* is the excess nitrogen gas still dissolved in a diver's tissues after surfacing. This excess nitrogen is gradually eliminated during the surface interval. If a second dive is performed before all the residual nitrogen has been eliminated, the residual nitrogen must be considered in computing the decompression requirements of the second dive.

9-3.16 **Single Dive.** A *single dive* is any dive conducted after all the residual nitrogen from prior dives has been eliminated from the tissues.

9-3.17 **Repetitive Dive.** A *repetitive dive* is any dive conducted while the diver still has some residual nitrogen in his tissues from a prior dive.

9-3.18 **Repetitive Group Designator.** The *repetitive group designator* is a letter used to indicate the amount of residual nitrogen remaining in the diver's body following a previous dive.

9-3.19 **Residual Nitrogen Time.** *Residual nitrogen time* is the time that must be added to the bottom time of a repetitive dive to compensate for the nitrogen still in solution in a diver's tissues from a previous dive. Residual nitrogen time is expressed in minutes.

9-3.20 **Equivalent Single Dive.** A repetitive dive is converted to its single dive equivalent before entering the decompression tables to determine the decompression requirement. The depth of the equivalent single dive is equal to the depth of the repetitive dive. The bottom time of the equivalent single dive is equal to the sum of the residual nitrogen time and the actual bottom time of the repetitive dive.

9-3.21 **Equivalent Single Dive Time.** The *equivalent single dive time* is the sum of the residual nitrogen time and the bottom time of a repetitive dive. Equivalent single dive time is used to select the decompression schedule for a repetitive dive. This time is expressed in minutes.

9-3.22 **Surface Decompression.** Surface decompression is a technique where some of the decompression stops in the water are skipped. These stops are made up by compressing the diver back to depth in a recompression chamber on the surface.

9-3.23 **Exceptional Exposure Dive.** An *exceptional exposure dive* is one in which the risk of decompression sickness, oxygen toxicity, and/or exposure to the elements is substantially greater than on a normal working dive. Planned exceptional exposure dives require CNO approval.

9-4 DIVE CHARTING AND RECORDING

Chapter 5 provides information for maintaining a Command Diving Log and a personal dive log and for reporting individual dives to the Naval Safety Center. In addition to these records, every Navy dive may be recorded on a diving chart similar to Figure 9-1. The diving chart is a convenient means of collecting the dive data, which in turn will be transcribed into the dive log. Abbreviations that may be used in the diving chart and Command Diving Log are:

- LS - Left Surface
- RB - Reached Bottom
- LB - Left Bottom
- R - Reached a decompression stop
- L - Left a decompression stop
- RS - Reached Surface
- TBT - Total Bottom Time (computed from leaving the surface to leaving the bottom)
- TDT - Total Decompression Time (computed from leaving the bottom to reaching the surface)

Date:	Type of Dive: AIR HeO_2		
Diver 1:	Diver 2:	Standby:	
Rig: PSIG: O_2%:	Rig: PSIG: O_2%:	Rig: PSIG: O_2%:	
Diving Supervisor:	Chartman:	Bottom Mix:	

EVENT	STOP TIME	CLOCK TIME
LS or 20 fsw		
RB		
LB		
R 1st Stop		
190 fsw		
180 fsw		
170 fsw		
160 fsw		
150 fsw		
140 fsw		
130 fsw		
120 fsw		
110 fsw		
100 fsw		
90 fsw		
80 fsw		
70 fsw		
60 fsw		
50 fsw		
40 fsw		
30 fsw		
20 fsw		
RS		
RB CHAMBER		
50 fsw chamber		
40 fsw chamber		
30 fsw chamber		
RS CHAMBER		

TDT	TTD

EVENT	TIME/DEPTH
Descent Time (Water)	
Stage Depth (fsw)	
Maximum Depth (fsw)	
Total Bottom Time	
Table/Schedule	
Time to 1st Stop (Actual)	
Time to 1st Stop (Planned)	
Delay to 1st Stop	
Travel/Shift/Vent Time	
Ascent Time-Water/SurD (Actual)	
Undress Time-SurD (Actual)	
Descent Chamber-SurD (Actual)	
Total SurD Surface Interval	
Ascent Time–Chamber (Actual)	

HOLDS ON DESCENT	
DEPTH	PROBLEM

DELAYS ON ASCENT	
DEPTH	PROBLEM

DECOMPRESSION PROCEDURES USED
AIR ☐ In-water Air decompression ☐ In-water Air/O_2 decompression ☐ $SurDO_2$ HeO_2 ☐ In-water HeO_2/O_2 decompression ☐ $SurDO_2$
REPETITIVE GROUP:

Remarks:

Figure 9-1. Diving Chart.

■ TTD - Total Time of Dive (computed from leaving the surface to reaching the surface)

Figure 9-2 illustrates these abbreviations in conjunction with a dive profile.

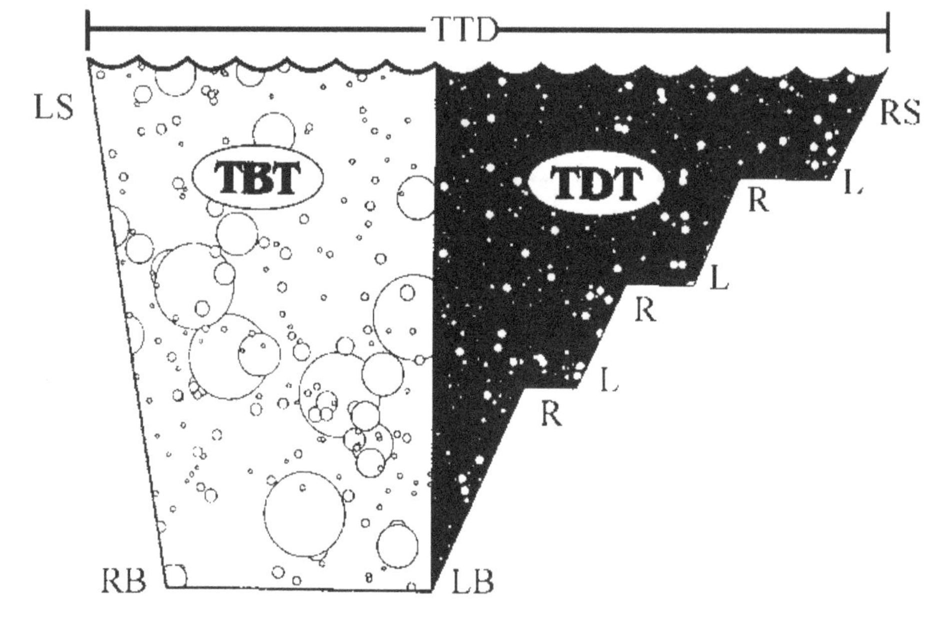

Figure 9-2. Graphic View of a Dive with Abbreviations.

9-5 THE AIR DECOMPRESSION TABLES

Six Tables are required to perform the full spectrum of air dives

- **No-Decompression Limits and Repetitive Group Designation Table for No-Decompression Air Dives.** This Table gives the no-decompression limits and the repetitive group designators for dives that do not require decompression stops.

- **Air Decompression Table.** This Table gives the decompression schedules and repetitive group designators for dives that require decompression stops.

- **Residual Nitrogen Timetable for Repetitive Air Dives.** This Table allows the diver to determine his Residual Nitrogen Time when performing a repetitive dive.

- **Sea Level Equivalent Depth Table.** This Table allows the diver to correct the sea level decompression tables for use at altitude.

- **Repetitive Groups Associated with Initial Ascent to Altitude Table.** This Table allows the diver to adjust his decompression if he is not fully equilibrated at altitude.

- **Required Surface Interval Before Ascent to Altitude After Diving.** This Table tells the diver when it is safe to fly or ascend to higher altitude after a dive.

9-6 GENERAL RULES FOR THE USE OF AIR DECOMPRESSION TABLES

9-6.1 **Selecting the Decompression Schedule.** To select the proper decompression schedule, record the bottom time and the maximum depth attained by the diver. Enter the table at the exact or next greater depth and at the exact or next longer bottom time.

When using a pneumofathometer to measure depth, first correct the observed depth reading by adding the pneumofathometer correction factor shown in Table 9-1. Ensure the pneumofathometer is located at mid-chest level.

Table 9-1. *Pneumofathometer Correction Factors.*

Pneumofathometer Depth	Correction Factor
0-100 fsw	+1 fsw
101-200 fsw	+2 fsw
201-300 fsw	+4 fsw
301-400 fsw	+7 fsw

Example: The diver's pneumofathometer reads 145 fsw. In the depth range of 101–200 fsw, the pneumofathometer underestimates the diver's actual depth by 2 fsw. To determine the diver's actual depth, add 2 fsw to the pneumofathometer reading. The diver's actual depth is 147 fsw.

9-6.2 **Descent Rate.** The descent rate on an air dive is not critical, but in general it should not exceed 75 fsw/min.

9-6.3 **Ascent Rate.** The ascent rate from the bottom to the first decompression stop, between decompression stops, and from the last decompression stop to the surface is 30 fsw/min (20 seconds per 10 fsw). Minor variations in the rate of ascent between 20 and 40 fsw/min are acceptable. For surface decompression, the ascent rate from the 40 fsw water stop to the surface is 40 fsw/min.

9-6.4 **Decompression Stop Time.** For in-water decompression on air, the time at the first decompression stop begins when the diver arrives at the stop and ends when he leaves the stop. For all subsequent stops, the stop time begins when the diver leaves the previous stop and ends when he leaves the stop. In other words, ascent time between stops is included in the subsequent stop time. The same rules apply to in-water decompression on air/oxygen with the exception of the first stop on oxygen. The time at the first oxygen stop begins when all divers are confirmed on oxygen and ends when the divers leave the stop.

9-6.5 **Last Water Stop.** The last water stop for all in-water decompressions is 20 fsw.

9-6.6 **Eligibility for Surface Decompression.** A diver is eligible for surface decompression upon completion of the 40 fsw water stop. If a 40 fsw stop is not required by the decompression schedule, the diver may ascend directly to the surface without decompression stops and begin surface decompression.

9-7 NO-DECOMPRESSION LIMITS AND REPETITIVE GROUP DESIGNATION TABLE FOR NO-DECOMPRESSION AIR DIVES

The No-Decompression Table (Table 9-7) gives the maximum time that can be spent at a given depth without the need for decompression stops during the subsequent ascent to the surface. This table is sometimes called the "no-stop" table. At depths of 20 fsw and shallower, there is no limit on the amount of time that can be spent at depth. Deeper than 20 fsw, the time that can be spent is limited. For example, at 60 fsw, any dive longer than 60 minutes will require decompression stops.

The No-Decompression Table also provides the repetitive group designators for dives that fall within the no-decompression limits. Even though no decompression stops are required during ascent, the diver still surfaces with some residual nitrogen in his tissues. This residual nitrogen needs to be accounted for if a repetitive dive is planned.

If a diver exceeds the limits given in the No-Decompression Table, then the decompression stop requirement must be calculated using Table 9-9.

For each depth listed in the No-Decompression Table, the corresponding no-decompression limit is indicated in the second column. This limit is the maximum bottom time that a diver may spend at that depth and still return to the surface without taking decompression stops. To find the no-decompression limit, enter the table at the depth equal to or next greater than the maximum depth of the dive. Follow that row to the second column to obtain the no-decompression limit.

The columns to the right of the no-decompression limit column contain the repetitive group designators for dives with bottom times equal to or shorter than the no-decompression limit. A repetitive group designator must be assigned to a diver subsequent to every dive, even a no-decompression dive.

To find the repetitive group designator following a no-decompression dive:

1. Enter the table at the depth equal to or next greater than the maximum depth of the dive.

2. Follow that row to the right to the bottom time equal to or next greater than the actual bottom time of the dive.

3. Follow the column up to obtain the repetitive group designator.

Example: Divers conduct a brief inspection of a worksite located at a depth of 74 fsw. Bottom time is 10 min. What is the no-decompression limit for a dive to 74 fsw? What is the repetitive group designator following this 10-minute dive?

Enter the No-Decompression Table at the next greater depth, 80 fsw. Follow the row horizontally to the second column. The no-decompression limit at 80 fsw is 39 min. The divers could spend up to 39 min at this depth and still ascend to the surface without decompression stops. Continue reading horizontally to the right to the bottom time that is next greater than the actual bottom time. This is 12 min. Read vertically up the column to obtain the repetitive group designator for this 10-min dive. The repetitive group designator is C. If the divers had spent the full 39 min allowed at 74 fsw, the repetitive group designator would have been J. This dive is illustrated in Figure 9-3.

9-7.1 **Optional Shallow Water No-Decompression Table.** Appendix 2A contains an expanded version of Table 9-7 and Table 9-8 covering the depth range of 30–50 fsw in one-foot increments. In this depth range, a small change in the diver's maximum depth can make a substantial difference in the allowable no-decompression time. For example, at 35 fsw the no-decompression limit is 232 minutes; at 40 fsw it is only 163 minutes, more than an hour less. When the diver's maximum depth is accurately known at the beginning of the dive, for example in ballast tank dives, or when continuous depth recording is available, for example with a decompression computer, the expanded table can be used to maximize no-decompression time. These optional tables are most suited to ship husbandry diving, but can be used in other shallow air diving applications as well.

9-8 THE AIR DECOMPRESSION TABLE

The Air Decompression Table, Table 9-9, combines three modes of decompression into one table. These modes are: (1) in-water decompression on air, (2) in-water decompression on air and oxygen, and (3) surface decompression on oxygen.

9-8.1 **In-Water Decompression on Air.** This mode of decompression is used when the entire decompression will be conducted on air. The top row labeled "Air" under each depth/bottom time entry gives the decompression schedule for in-water air decompression. Enter the table at the depth that is exactly equal to or next deeper than the diver's maximum depth. Select the schedule for the bottom time that is exactly equal to or next longer than the diver's actual bottom time. Read across the row to obtain the required decompression stop times. The last decompression stop is taken at 20 fsw. The total ascent time is given in the next column. The repetitive group designator upon surfacing is given in the last column.

Example: A diver makes a surface-supplied air dive to 78 fsw for 47 minutes. What is the required decompression?

Enter the Air Decompression Table at the next deeper depth, 80 fsw, and the next longer bottom time, 50 min. Read across the row labeled "Air". A 17 min decompression stop at 20 fsw is required. The diver ascends from 78 to 20 fsw at 30 fsw/min, spends 17 min at 20 fsw, then ascends to the surface at 30 fsw/min.

1313		
Date: 4 Sept 07	Type of Dive: (AIR, circled) HeO_2	
Diver 1: ND1Hooper	Diver 2: ND1 Patterson	Standby: ND2 Webb
Rig: MK 21 PSIG: 3000 O_2%:	Rig: MK 21 PSIG: 3000 O_2%:	Rig: MK 21 PSIG: 3000 O_2%:
Diving Supervisor: NDC Degitz	Chartman: NDC Palmer	Bottom Mix:

EVENT	STOP TIME	CLOCK TIME
LS or 20 fsw		1300
RB		1301
LB		1310
R 1st Stop		
190 fsw		
180 fsw		
170 fsw		
160 fsw		
150 fsw		
140 fsw		
130 fsw		
120 fsw		
110 fsw		
100 fsw		
90 fsw		
80 fsw		
70 fsw		
60 fsw		
50 fsw		
40 fsw		
30 fsw		
20 fsw		
RS		1313
RB CHAMBER		
50 fsw chamber		
40 fsw chamber		
30 fsw chamber		
RS CHAMBER		

TDT	TTD
:02::30	:13

EVENT	TIME/DEPTH
Descent Time (Water)	:01
Stage Depth (fsw)	73
Maximum Depth (fsw)	73+1=74
Total Bottom Time	:10
Table/Schedule	80/12 No D
Time to 1st Stop (Actual)	:02::30
Time to 1st Stop (Planned)	:02::26
Delay to 1st Stop	::04
Travel/Shift/Vent Time	
Ascent Time-Water/SurD (Actual)	
Undress Time-SurD (Actual)	
Descent Chamber-SurD (Actual)	
Total SurD Surface Interval	
Ascent Time–Chamber (Actual)	

HOLDS ON DESCENT	
DEPTH	PROBLEM

DELAYS ON ASCENT	
DEPTH	PROBLEM

DECOMPRESSION PROCEDURES USED

AIR

- ☐ In-water Air decompression
- ☐ In-water Air/O_2 decompression
- ☐ $SurDO_2$

HeO_2

- ☐ In-water HeO_2/O_2 decompression
- ☐ $SurDO_2$

REPETITIVE GROUP: C

Remarks:

Figure 9-3. Completed Air Diving Chart: No-Decompression Dive.

The repetitive group designator for this dive is "M". This dive is illustrated in Figure 9-4.

If the bottom time of a dive is less than the first bottom time listed for its depth in the Air Decompression Table, decompression stops are not required. The divers may ascend directly to the surface at 30 fsw/min. Refer to the No-Decompression Table, Table 9-7, to obtain the repetitive group designator for a no-decompression dive.

If the Air Decompression Table does not list a repetitive group designator for a dive, no repetitive dives deeper than 20 fsw are permitted following this dive. The diver must have an 18-hour surface interval before making another dive deeper than 20 fsw.

9-8.2 **In-Water Decompression on Air and Oxygen.** This mode of decompression is used when the decompression will be conducted partly on air and partly on 100% oxygen. The bottom row labeled "Air/O_2" under each depth/bottom time entry in Table 9-9 gives the decompression schedule for in-water air/oxygen decompression. Enter the table at the depth that is exactly equal to or next deeper than the diver's maximum depth. Select the schedule for the bottom time that is exactly equal to or next longer than the diver's actual bottom time. Read across the Air/O_2 row to obtain the required decompression stop times. The diver follows the air schedule to 30 fsw (or 20 fsw if there is no 30 fsw stop), then shifts from air to 100% oxygen. The oxygen stop times are shown in bold print. Oxygen stop time begins when all divers are confirmed on oxygen. If more than 30 minutes must be spent on oxygen, a 5 min air break is required every 30 minutes. Upon completion of the 20 fsw oxygen stop time, the diver surfaces at 30 fsw/min while continuing to breathe 100% oxygen. The total ascent time, including air breaks, is given in the next column. The repetitive group designator upon surfacing is given in the last column and is the same as the repetitive group designator for an air decompression dive.

All decompression stops deeper than 30 fsw are done on air. Decompression stops on oxygen commence at 20 or 30 fsw in accordance with Table 9-9. Stops on oxygen are in bold type in Table 9-9.

Current USN surface-supplied air diving systems (FADS III, LWDS, DSM, etc.) require the use of an Oxygen Regulator Console Assembly (ORCA) to deliver oxygen to the diver in the water. The Fly-Away Mixed Gas Diving System (FMGS), which can be used to conduct air dives as well as mixed gas dives, is capable of providing oxygen to the diver without the addition of an ORCA.

9-8.2.1 **Procedures for Shifting to 100% Oxygen at 30 or 20 fsw.** Upon arrival at the first oxygen stop, ventilate each diver with oxygen following these steps:

1. Align the ORCA or FMGS to supply 100% oxygen to the diver.

2. Ventilate each diver for 20 seconds. Divers may be vented simultaneously or sequentially.

1007		
Date: 4 Sept 07	Type of Dive: (AIR circled) AIR HeO_2	
Diver 1: ND1 Hedrick	Diver 2: HM2 Tyau	Standby: ND2 Parsons
Rig: MK 21 PSIG: 3000 O_2%:	Rig: MK 21 PSIG: 3000 O_2%:	Rig: MK 21 PSIG: 3000 O_2%:
Diving Supervisor: NDCM Wiggins	Chartman: NDC Kriese	Bottom Mix:

EVENT	STOP TIME	CLOCK TIME
LS or 20 fsw		0900
RB		0902
LB		0947
R 1st Stop		0949
190 fsw		
180 fsw		
170 fsw		
160 fsw		
150 fsw		
140 fsw		
130 fsw		
120 fsw		
110 fsw		
100 fsw		
90 fsw		
80 fsw		
70 fsw		
60 fsw		
50 fsw		
40 fsw		
30 fsw		
20 fsw	:17	1006
RS		1007
RB CHAMBER		
50 fsw chamber		
40 fsw chamber		
30 fsw chamber		
RS CHAMBER		

TDT	TTD
:20	1:07

EVENT	TIME/DEPTH
Descent Time (Water)	:02
Stage Depth (fsw)	77
Maximum Depth (fsw)	77+1=78
Total Bottom Time	:47
Table/Schedule	80/50
Time to 1st Stop (Actual)	:01::58
Time to 1st Stop (Planned)	:01::54
Delay to 1st Stop	::04
Travel/Shift/Vent Time	
Ascent Time-Water/SurD (Actual)	::45
Undress Time-SurD (Actual)	
Descent Chamber-SurD (Actual)	
Total SurD Surface Interval	
Ascent Time–Chamber (Actual)	

HOLDS ON DESCENT

DEPTH	PROBLEM

DELAYS ON ASCENT

DEPTH	PROBLEM

DECOMPRESSION PROCEDURES USED

AIR
- ☑ In-water Air decompression
- ☐ In-water Air/O_2 decompression
- ☐ $SurDO_2$

HeO_2
- ☐ In-water HeO_2/O_2 decompression
- ☐ $SurDO_2$

REPETITIVE GROUP: M

Remarks:

Figure 9-4. Completed Air Diving Chart: In-water Decompression on Air.

3. Verify that the oxygen monitoring device on the ORCA or FMGS, if one is present, shows 100% oxygen being delivered to the diver.

The Air Diving Chart has a space to enter the "Travel/Shift/Vent" time. For dives in which the first stop is at 40 fsw or deeper, the travel/shift/vent time includes the 20 second ascent from 40 to 30 fsw as well as the time required to shift the console to oxygen, vent the divers, and confirm that the divers are on oxygen. For dives in which the first stop is an oxygen stop at 30 or 20 fsw, the travel/shift/vent time only includes the time required to shift the console, vent the divers, and confirm that they are on oxygen. The travel time to the stop is not included. The travel/shift/vent time is recorded as minutes and seconds. The travel/shift/vent time should be under 3 minutes.

9-8.2.2 **Air Breaks at 30 and 20 fsw.** At the 30 fsw and 20 fsw water stops, the diver breathes oxygen for 30 min periods separated by 5 min air breaks. The air breaks do not count toward required decompression time. When an air break is required, shift the ORCA or FMGS to air for 5 minutes then back to 100% oxygen. Ventilation of the divers is not required. For purposes of timing air breaks, begin clocking oxygen time when all divers are confirmed on oxygen. If the total oxygen stop time is 35 minutes or less, an air break is not required at 30 minutes. If the final oxygen period is 35 minutes or less, a final air break at the 30-min mark is not required. In either case, surface the diver on 100% oxygen upon completion of the oxygen time.

Example: A diver makes a surface-supplied air dive to 145 fsw for 39 min. What is the required decompression on air and oxygen?

1. Enter the Air Decompression Table at the next deeper depth, 150 fsw, and the next longer bottom time, 40 min.

2. Read across the row labeled "Air/O_2." A 4-min decompression stop on air at 40 fsw is required.

3. The diver ascends from 145 to 40 fsw at 30 fsw/min, spends 4 min on air at 40 fsw, and then ascends to 30 fsw at 30 fsw/min.

4. Upon arrival at 30 fsw, the diver shifts to 100% oxygen. The diver spends a total of 9 min at 30 fsw after the shift to oxygen.

5. The diver ascends on oxygen to 20 fsw and spends a total of 34 minutes on oxygen at 20 fsw. The 20 second ascent time from 30 to 20 fsw is included in the 34-min stop time. A five minute air break is required 21 minutes into the 20 fsw stop. The diver takes the air break then completes the remaining 13 minutes of oxygen required at 20 fsw.

6. Upon completion of the 20 fsw stop time, the diver ascends to the surface on 100% oxygen at 30 fsw/min. The total ascent time, including the air break is 56 minutes 40 seconds, not counting the time required to shift the divers to oxygen at 30 fsw. The repetitive group designator for this dive is "Z".

This dive is illustrated in Figure 9-5.

1138		
Date: 5 Sept 07	Type of Dive: (AIR circled) HeO_2	
Diver 1: ND1 Poulan	Diver 2: HM2 Montgomery	Standby: NDC Miller
Rig: KM-37 PSIG: 2900 O_2%:	Rig: KM-37 PSIG: 2900 O_2%:	Rig: KM-37 PSIG: 2900 O_2%:
Diving Supervisor: NDCM Westling	Chartman: ND1 Slappy	Bottom Mix:

EVENT	STOP TIME	CLOCK TIME
LS or 20 fsw		1000
RB		1002
LB		1039
R 1st Stop		1043
190 fsw		
180 fsw		
170 fsw		
160 fsw		
150 fsw		
140 fsw		
130 fsw		
120 fsw		
110 fsw		
100 fsw		
90 fsw		
80 fsw		
70 fsw		
60 fsw		
50 fsw		
40 fsw	:04 (Air)	1047
30 fsw	:02+:09 (O_2)	1058
20 fsw	:21+:05+:13(O_2)	1137
RS		1138
RB CHAMBER		
50 fsw chamber		
40 fsw chamber		
30 fsw chamber		
RS CHAMBER		

TDT	TTD
:59	1:38

EVENT	TIME/DEPTH
Descent Time (Water)	:02
Stage Depth (fsw)	143
Maximum Depth (fsw)	143+2=145
Total Bottom Time	:39
Table/Schedule	150/40
Time to 1st Stop (Actual)	:03::30
Time to 1st Stop (Planned)	:03::26
Delay to 1st Stop	::04
Travel/Shift/Vent Time	:02
Ascent Time-Water/SurD (Actual)	::40
Undress Time-SurD (Actual)	
Descent Chamber-SurD (Actual)	
Total SurD Surface Interval	
Ascent Time–Chamber (Actual)	

HOLDS ON DESCENT	
DEPTH	PROBLEM

DELAYS ON ASCENT	
DEPTH	PROBLEM

DECOMPRESSION PROCEDURES USED

AIR

- ☐ In-water Air decompression
- ☑ In-water Air/O_2 decompression
- ☐ $SurDO_2$

HeO_2

- ☐ In-water HeO_2/O_2 decompression
- ☐ $SurDO_2$

REPETITIVE GROUP: Z

Remarks:

Figure 9-5. Completed Air Diving Chart: In-water Decompression on Air and Oxygen.

9-8.3 **Surface Decompression on Oxygen ($SurDO_2$).** Surface decompression is a technique for fulfilling all or a portion of a diver's decompression obligation in a recompression chamber instead of in the water. Surface decompression reduces the time a diver must spend in the water. Surface decompression offers many advantages that enhance the diver's safety. Shorter exposure time in the water keeps divers from chilling to a dangerous level when diving in cold water. Inside the recompression chamber, the divers can be maintained at a constant pressure, unaffected by the surface conditions of the sea. Once divers have been recovered into the recompression chamber, a second dive team can begin descent, provided the recompression chamber and the surface-supplied diving system have separate air supplies. This greatly speeds up operations.

To decompress the diver using the Surface Decompression on Oxygen mode, follow the in-water air decompression schedule (top row) through the end of the 40 fsw water stop, then initiate surface decompression following the rules given below. If there is no 40 fsw water stop in the air schedule, surface the diver without taking any stops. In either case, start timing the surface interval when the diver leaves 40 fsw. The required time on oxygen in the recompression chamber is shown in the next to last column of the Table. Oxygen time is divided into periods. Each period is 30 minutes long; each half-period is 15 minutes long. The first 15 minutes is always spent at 50 fsw in the chamber; the remainder of the oxygen time is taken at 40 fsw. If the schedule requires only one half of an oxygen period, the diver spends 15 minutes breathing oxygen at 50 fsw in the chamber, then surfaces at 30 fsw/min. The repetitive group designator for a surface decompression dive is shown in the last column of the Table and is the same as the repetitive group designator for an air decompression dive.

9-8.3.1 **Surface Decompression on Oxygen Procedure**

1. Complete any required decompression stops on air 40 fsw and deeper.

2. Upon completion of the 40 fsw stop, bring the diver to the surface at 40 fsw/min. If a 40 fsw water stop is not required, bring the diver from the bottom to 40 fsw at 30 fsw/min and then from 40 fsw to the surface at 40 fsw/min. Once the diver is on the surface, tenders have approximately 3 and a half minutes to remove the breathing apparatus and diving dress and assist the diver into the recompression chamber.

3. Place the diver and a tender in the recompression chamber. The job of the tender is to monitor the diver closely for signs of decompression sickness and CNS oxygen toxicity during the subsequent recompression. When two divers undergo surface decompression simultaneously, the dive supervisor may elect not to use an inside tender. In this case, both divers will carefully monitor each other in addition to being closely observed by topside personnel.

4. Compress the diver on air to 50 fsw at a maximum compression rate of 100 fsw/min. The surface interval is the elapsed time from the time the diver leaves the 40 fsw water stop to the time the diver arrives at 50 fsw in the chamber. A normal surface interval should not exceed 5 minutes.

WARNING The interval from leaving 40 fsw in the water to arriving at 50 fsw in the chamber cannot exceed 5 minutes without incurring a penalty. (See **paragraph 9-12.6.**)

5. Upon arrival at 50 fsw, place the diver on 100 percent oxygen by mask. Instruct the diver to strap the mask on tightly to ensure a good oxygen seal.

6. In the chamber, have the diver breathe oxygen for the number of 30-minute periods and 15-min half periods indicated in the next to last column of the Air Decompression Table. The first period consists of 15 minutes on oxygen at 50 fsw followed by 15 minutes on oxygen at 40 fsw. Periods 2–4 are spent at 40 fsw. If more than 4 periods are required, the remaining periods are spent at 30 fsw. Ascent from 50 fsw to 40 fsw and from 40 fsw to 30 fsw is at 30 fsw/min. Ascent time from 50 to 40 fsw is included in the first oxygen period. Ascent from 40 to 30 fsw, if required, should take place during an air break.

7. Interrupt oxygen breathing with a 5-min air break after every 30 minutes on oxygen. This air time is considered dead time. Oxygen time begins when the diver is confirmed to be on oxygen at 50 fsw.

8. When the last oxygen breathing period has been completed, return the diver to breathing chamber air.

9. Ascend to the surface at 30 fsw/min.

Example: A surface-supplied diver makes an air dive to a maximum depth of 118 fsw for 65 minutes. The intent is to decompress the diver using the surface decompression on oxygen mode. What is the proper decompression?

1. Enter the Air Decompression Table at the next deeper depth, 120 fsw, and the next longer bottom time, 70 min.

2. Read across the row labeled "Air." A 12-min decompression stop on air at 40 fsw is required. Continue reading across the row to the column labeled "Chamber O_2 Periods." Two and one half chamber oxygen periods are required.

3. The diver ascends from 118 to 40 fsw at 30 fsw/min, spends 12 minutes on air at 40 fsw, and then ascends to the surface at 40 fsw/min. Surfacing takes 1 minute.

4. Upon surfacing the diver is undressed as quickly as possible, placed in the recompression chamber, and recompressed on air to 50 fsw. The total time from leaving 40 fsw in the water to arriving at 50 fsw in the chamber normally should not exceed 5 minutes.

5. Upon arrival at 50 fsw, the diver goes on 100% oxygen by mask and breathes oxygen for 15 minutes. Time on oxygen begins when the diver goes on the oxygen mask.

6. After 15 minutes on oxygen at 50 fsw, the diver ascends to 40 fsw at 30 fsw/min while continuing to breathe oxygen from the mask. Ascent to 40 fsw takes 20 seconds. The diver continues to breathe oxygen at 40 fsw for an additional 14 min and 40 seconds. This ends the first 30-min oxygen period and the diver takes a 5-min air break.

7. Upon completion of the air break, the diver resumes oxygen breathing by mask for another 30-minute period. This ends the second 30-min oxygen period and the diver takes a second 5-min air break.

8. Upon completion of the second air break, the diver resumes oxygen breathing for 15 minutes, the remaining one-half period of oxygen required.

9. Upon completion of this last half period of oxygen, the diver goes off the oxygen mask and breathes chamber air. The diver is brought to the surface at 30 fsw/min while breathing air.

10. No repetitive group designator is shown for this dive. The diver must wait 18 hours before making another dive.

This dive is illustrated in Figure 9-6.

9-8.3.2 Surface Decompression from 30 and 20 fsw

The diving supervisor can initiate surface decompression at any point during in-water decompression at 30 or 20 fsw, if desired. Surface decompression may become desirable if sea conditions are deteriorating, the diver feels ill, or some other contingency arises. Surface decompression may be initiated regardless of whether the divers are decompressing on air or oxygen. The diving supervisor may elect to prescribe the full number of chamber oxygen periods listed in the surface decompression schedule or elect to reduce that number of periods to take credit for the time already spent on air or oxygen in the water.

1. If surface decompression is elected before the divers have been shifted to oxygen, take the full number of chamber oxygen periods prescribed by the table.

2. If surface decompression is elected after divers have switched to oxygen, compute the number of chamber oxygen periods required by multiplying the remaining oxygen time at the stops by 1.1, dividing the total by 30 minutes, then rounding the result up to the next highest half period. One half period (15 minutes at 50 fsw) is the minimum requirement.

   Example: The supervisor elects to surface decompress when the diver has a remaining oxygen time of 5 minutes at 30 fsw and 33 minutes at 20 fsw. The total remaining oxygen time is 38 minutes. The number of 30-min $SurDO_2$ periods required is $(1.1 \times 38) / 30 = 1.39$. This number is rounded up to 1.5.

3. If surface decompression is elected while the divers are decompressing on air, first convert the remaining air time at the stops to the equivalent remaining oxygen time at the stops, then convert this remaining oxygen time to the number of chamber oxygen periods required as shown above.

1252		
Date: 5 Sept 07	Type of Dive: (AIR) HeO_2	
Diver 1: ND1 Chaisson	Diver 2: ND2 Hutcheson	Standby: ND1 Collins
Rig: KM-37 PSIG: 2900 O_2%:	Rig: KM-37 PSIG: 2900 O_2%:	Rig: KM-37 PSIG: 2900 O_2%:
Diving Supervisor: NDCM Orns	Chartman: ND1 Saurez	Bottom Mix:

EVENT	STOP TIME	CLOCK TIME
LS or 20 fsw		1000
RB		1002
LB		1105
R 1st Stop		1108
190 fsw		
180 fsw		
170 fsw		
160 fsw		
150 fsw		
140 fsw		
130 fsw		
120 fsw		
110 fsw		
100 fsw		
90 fsw		
80 fsw		
70 fsw		
60 fsw		
50 fsw		
40 fsw	:12 (Air)	1120
30 fsw		
20 fsw		
RS		1121
RB CHAMBER		1125
50 fsw chamber	:15	1140
40 fsw chamber	:15+:5+:30+:5+:15	1250
30 fsw chamber		
RS CHAMBER		1252

TDT	TTD
1:47	2:52

EVENT	TIME/DEPTH
Descent Time (Water)	:02
Stage Depth (fsw)	116
Maximum Depth (fsw)	116+2=118
Total Bottom Time	:65
Table/Schedule	120/70
Time to 1st Stop (Actual)	:02::32
Time to 1st Stop (Planned)	:02::32
Delay to 1st Stop	
Travel/Shift/Vent Time	
AscentTime-Water/SurD (Actual)	:01
Undress Time-SurD (Actual)	:03::20
Descent Chamber-SurD (Actual)	::40
Total SurD Surface Interval	:05
Ascent Time–Chamber (Actual)	:01::20

HOLDS ON DESCENT	
DEPTH	PROBLEM

DELAYS ON ASCENT	
DEPTH	PROBLEM

DECOMPRESSION PROCEDURES USED

AIR

- ☐ In-water Air decompression
- ☐ In-water Air/O_2 decompression
- ☑ SurDO_2

HeO_2

- ☐ In-water HeO_2/O_2 decompression
- ☐ SurDO_2

REPETITIVE GROUP: No Repet, must wait 18 hours.

Remarks:

Figure 9-6. Completed Air Diving Chart: Surface Decompression on Oxygen.

- For a diver at 30 fsw: First compute the *air/oxygen trading ratio* at 30 fsw by dividing the 30 fsw air stop time listed in the table by the 30-fsw oxygen time. Next divide the remaining air time at 30 fsw by the air/oxygen trading ratio to determine the equivalent remaining oxygen time at 30 fsw. Add the oxygen time shown in the table at 20 fsw to the equivalent remaining oxygen time at 30 fsw to obtain the total remaining oxygen time. Compute the number of chamber oxygen periods required by multiplying the remaining oxygen time at the stops by 1.1, dividing the total by 30 minutes, then rounding the result up to the next highest half period. One half period (15 minutes at 50 fsw) is the minimum requirement.

- For a diver at 20 fsw: Compute the air/oxygen trading ratio at 20 fsw by dividing the 20 fsw air stop time listed in the table by the 20-fsw oxygen time. Divide the remaining air time at 20 fsw by the air/oxygen trading ratio to obtain the equivalent remaining oxygen time. Compute the number of chamber oxygen periods required by multiplying the remaining oxygen time at the stops by 1.1, dividing the total by 30 minutes, then rounding the result up to the next highest half period. One half period (15 minutes at 50 fsw) is the minimum requirement.

Example: A diver is decompressing on a schedule that calls for a single 50 min stop on air at 20 fsw. The corresponding 20-fsw oxygen stop time is 27 min. After 20 minutes on air at 20 fsw, the diving supervisor elects to surface decompress the diver. The air/oxygen trading ratio at 20 fsw is 50/27 = 1.85, i.e., every 1.85 minutes spent air at 20 fsw is the equivalent of 1 minute spent on oxygen at 20 fsw. The remaining time on air at 20 fsw is 50 – 20 = 30 minutes. The equivalent remaining oxygen time at 20 fsw is 30/1.85 = 16.2 minutes. This remaining oxygen time is rounded up to the next whole minute, 17 min. The number of 30-min SurDO_2 periods required is (1.1 × 17) / 30 = 0.62. This number is rounded up to 1.0.

9-8.4 Selection of the Mode of Decompression

Figure 9-7 provides guidance for selecting the best mode of decompression for a given dive.

In-water decompression on air is the most suitable mode for dives that do not require more than 15 min of total decompression stop time. Most dives will fall in this category. In-water decompression on air avoids the additional logistic burden of bringing an ORCA and/or a recompression chamber to the dive station.

In-water decompression on air and oxygen is strongly recommended whenever the total decompression stop time on air exceeds 15 min and surface decompression on oxygen is not a viable alternative. Surface decompression may not be possible either because a recompression chamber is not available on the dive station or the short surface interval associated with surface decompression does not allow enough time for

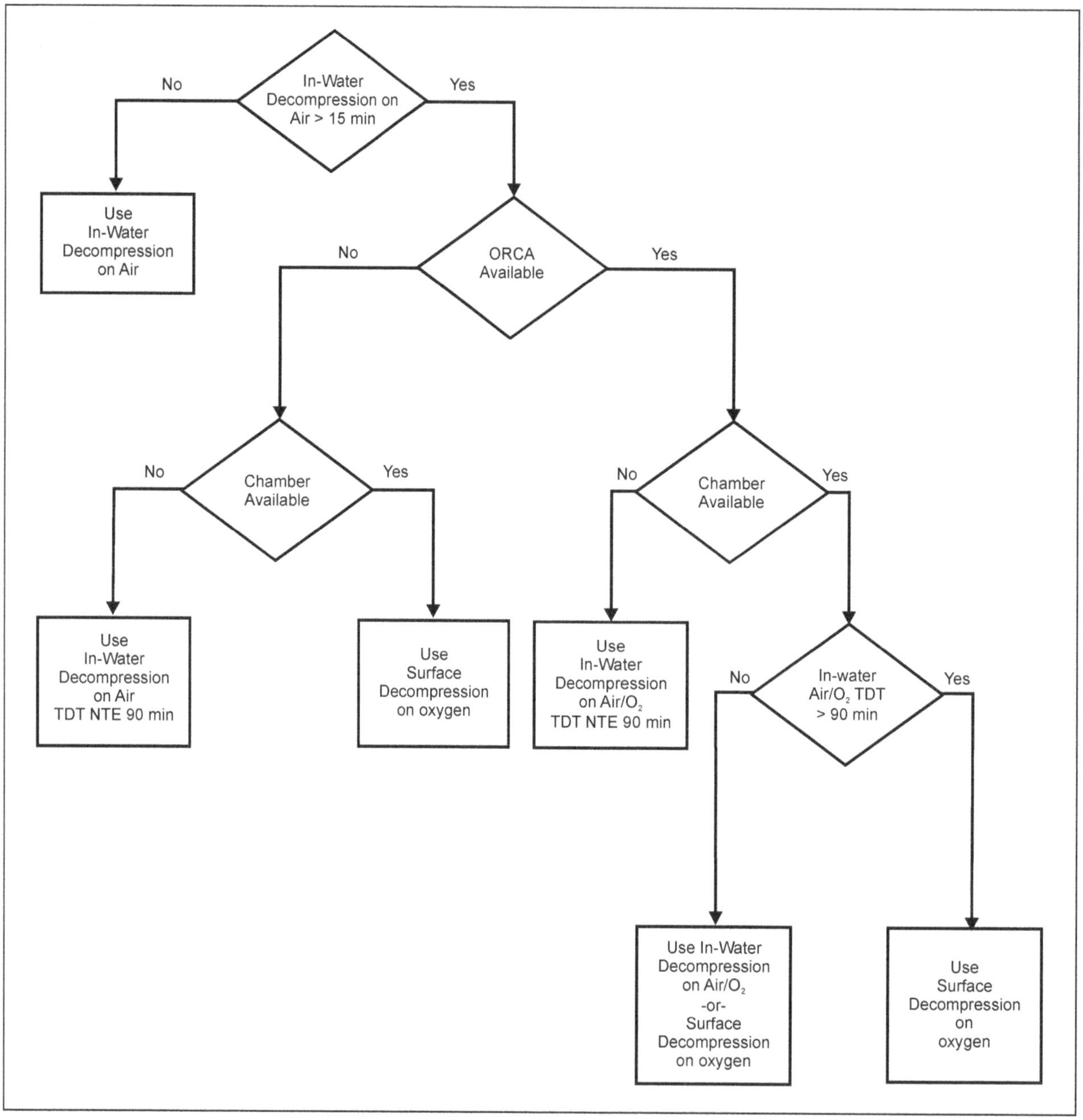

Figure 9-7. Decompression Mode Selection Flowchart.

diver decontamination following a contaminated water dive. In-water decompression on air and oxygen is most suitable for dives that do not require more than 90 min of total air and oxygen time in the water. Longer times increase the risk of CNS oxygen toxicity and exposure to the elements. If the total air/oxygen decompression time in the water is greater than 90 min, surface decompression on oxygen is required unless CNO permission to conduct exceptional exposure dives is obtained.

9-9 REPETITIVE DIVES

During the surface interval after an air dive, the quantity of residual nitrogen in the diver's body will gradually be reduced to its normal value. If the diver makes a second dive before the residual nitrogen has been dissipated (a repetitive dive), he must consider his residual nitrogen level when planning for the second dive.

The procedures for conducting a repetitive dive are summarized in Figure 9-8. Upon completing the first dive, the diver is assigned a repetitive group designator from either the Air Decompression Table or the No-Decompression Table. This designator tells the diver how much residual nitrogen he has upon surfacing from the first dive. A diver in Group A has the lowest amount of residual nitrogen; a diver in Group Z has the highest. As nitrogen passes out of the diver's body during the surface interval, the repetitive group designation changes to a lower letter group to reflect the lower quantity of residual nitrogen. The top half of Table 9-8 allows the repetitive group designator to be determined at any time during the surface interval. The lower half of Table 9-8 gives the Residual Nitrogen Time (RNT) corresponding to the repetitive group designator at the end of the surface interval and the depth of the repetitive dive. The residual nitrogen time is the time a diver would have had to spend at the depth of the repetitive dive to absorb the amount of nitrogen he has left over from the previous dive. The residual nitrogen time is added to the bottom time of the repetitive dive to obtain the Equivalent Single Dive Time (ESDT). The decompression schedule for the repetitive dive is obtained by entering either the Air Decompression Table or the No-Decompression Table at the depth of the repetitive dive and the equivalent single dive time.

9-9.1 **Repetitive Dive Procedure.** To use the repetitive dive procedure described below, the interval on the surface between dives must be at least 10 minutes. If the surface interval between dives is less than 10 minutes, add the bottom time of the two dives and enter the decompression table at the deeper of the two depths.

To determine the decompression schedule for a repetitive dive when the surface interval is greater than 10 minutes:

1. Obtain the repetitive group designator from the Air Decompression Table or the No-Decompression Table upon surfacing from the first dive.

2. Using the repetitive group designator, enter the top half of Table 9-8 on the diagonal. Table 9-8 is the Residual Nitrogen Timetable for Repetitive Air Dives.

3. Read horizontally across the row to locate the time interval that includes the diver's surface interval. The times are expressed in hours and minutes (e.g., 2:21 = 2 hours 21 minutes). Each time interval has a minimum time (top limit) and a maximum time (bottom limit). The time spent on the surface must be between or equal to the limits of the selected interval. If the surface interval exceeds the longest time shown in the row, the dive is not a repetitive dive. No correction for residual nitrogen is required.

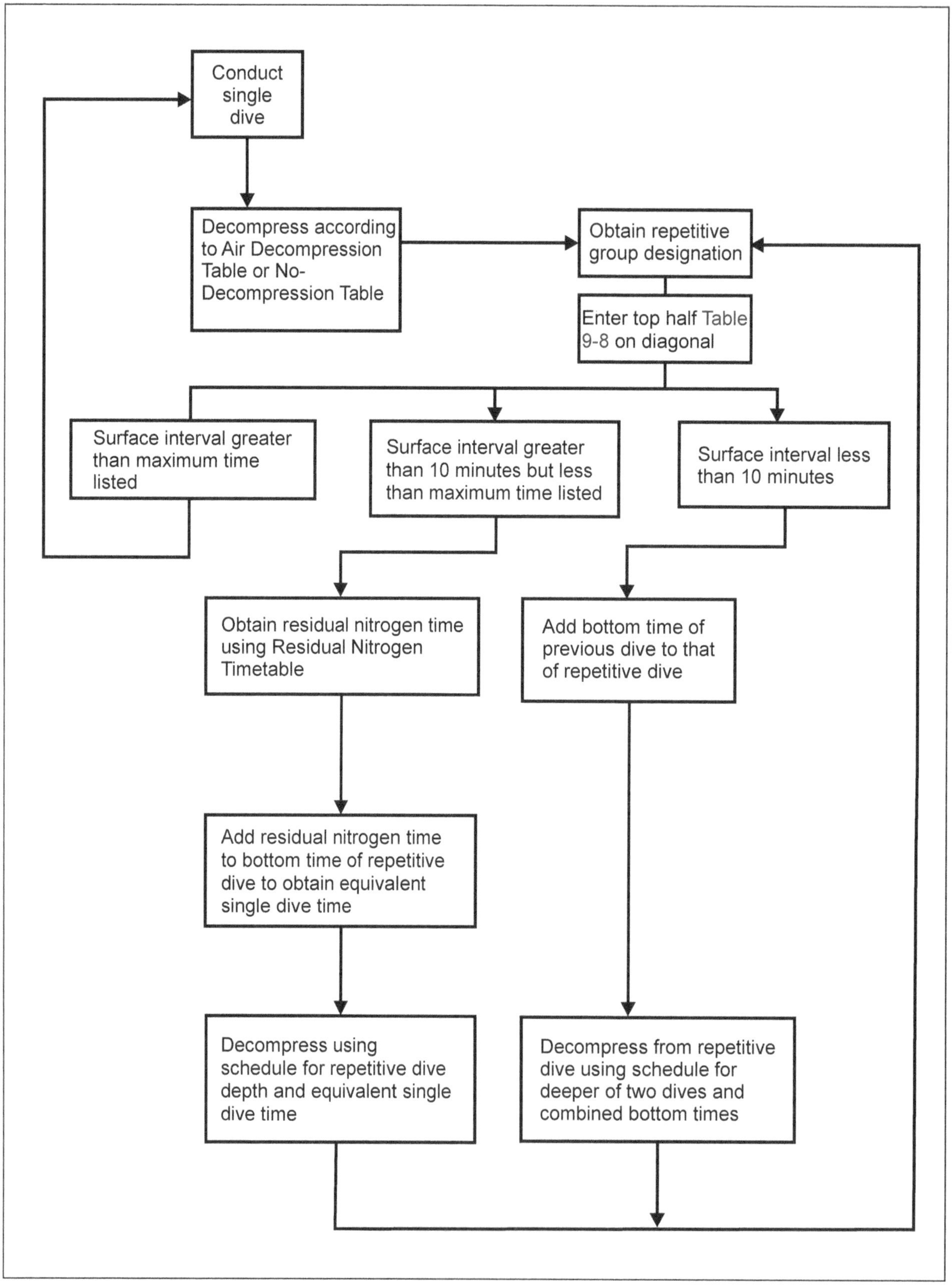

Figure 9-8. Repetitive Dive Flow Chart.

4. Read vertically down the column to obtain the repetitive group designator at the end of the surface interval.

5. Continue down the same column to the depth row that is exactly equal or next deeper than the depth of the repetitive dive. The time given at the intersection of the column and row is the residual nitrogen time in minutes.

6. Add the residual nitrogen time to the actual bottom time of the repetitive dive to get the Equivalent Single Dive Time (ESDT).

7. Enter the Air Decompression Table or No-Decompression Table at the depth that is exactly equal to or next deeper than the actual depth of the repetitive dive. Select the schedule that is exactly equal to or next longer than the Equivalent Single Dive Time. Follow the prescribed decompression to the surface.

8. At depths of 10, 15, and 20 fsw, some of the higher repetitive groups do not have a defined residual nitrogen time. These groups are marked with a double asterisk in the lower half of Table 9-8. The RNT is undefined because the tissue nitrogen loading associated with those repetitive groups is higher than the nitrogen loading that could be achieved even if the diver were to remain at those depths for an infinite period of time. A diver entering the dive in one of those higher groups marked by a double asterisk can still perform a repetitive dive at 10, 15 or 20 fsw because the no-decompression time at those depths is unlimited. An RNT time is not required to make the dive. If a subsequent repetitive dive to a deeper depth is planned, however, the diver will need a repetitive group at the end of the shallow dive in order to continue using the RNT table. If a double asterisk is encountered in Table 9-8, assume that the repetitive group remains unchanged during the course of the dive at 10, 15, or 20 fsw.

   Example: A diver surfaces from a dive in repetitive Group N. Thirty minutes later, he makes a dive to 20 fsw. The diver begins the 20 fsw dive in Group N. The RNT time for Group N at 20 fsw is undefined. This is not a problem because the no-decompression time at 20 fsw is unlimited. Regardless of his starting repetitive group, the diver can spend any amount of time at 20 fsw without incurring a decompression obligation. If a subsequent dive deeper than 20 fsw is planned, the diver should assume that he surfaced from the 20 fsw dive in Group N regardless of the duration of the 20 fsw dive.

9. If a repetitive group is not shown in the decompression schedule, repetitive dives deeper than 20 fsw are not allowed following a dive on that schedule. The diver must remain on the surface for at least 18 hours before making another dive deeper than 20 fsw.

10. Do not perform repetitive dives that require the use of Exceptional Exposure decompression schedules.

Always use the Repetitive Dive Worksheet, shown in Figure 9-9, when determining the decompression schedule for a repetitive dive.

REPETITIVE DIVE WORKSHEET

Date:

1st DIVE							
Max Depth							
Bottom Time							
Table & Schedule						REPET Group	
Surface Interval						New Group	
2nd DIVE							
Max Depth		MD + ESDT = Table & Schedule					
Bottom Time	+	RNT	=	ESDT	=	Table & Schedule	REPET Group
	+		=		=		
Ensure the RNT Exception Rule does not apply							
Surface Interval						New Group	
3rd DIVE							
Max Depth		MD + ESDT = Table & Schedule					
Bottom Time	+	RNT	=	ESDT	=	Table & Schedule	REPET Group
	+		=		=		
Ensure the RNT Exception Rule does not apply							
Surface Interval						New Group	
4th DIVE							
Max Depth		MD + ESDT = Table & Schedule					
Bottom Time	+	RNT	=	ESDT	=	Table & Schedule	REPET Group
	+		=		=		
Ensure the RNT Exception Rule does not apply							
Surface Interval						New Group	

Figure 9-9. Repetitive Dive Worksheet.

Example: A repetitive dive is planned to 98 fsw for an estimated bottom time of 15 minutes. The previous dive was to a depth of 101 fsw (100 fsw + 1 fsw pneumofathometer correction factor) and had a bottom time of 48 minutes. Decompression was conducted using the in-water air/oxygen option. The diver's surface interval is 6 hours 26 minutes (6:26). What is the proper decompression schedule for the repetitive dive?

1. Enter the Air Decompression Table at a depth of 110 fsw and a bottom time of 50 minutes. Read across the row to obtain the repetitive group designator upon surfacing from the first dive. The repetitive group designator is Z.

2. Move to the Residual Nitrogen Timetable for Repetitive Air Dives, Table 9-8.

3. Enter the top half of the table on the diagonal line at Z.

4. Read horizontally across the line until reaching the time interval that includes the diver's surface interval of 6 hours 26 minutes. The diver's surface interval falls within the limits of the 6:07/6:58 column.

5. Read vertically down the 6:07/6:58 column. The repetitive group designator at the end of the surface interval is I.

6. Continue to read down the column until reaching the depth that is exactly equal or next deeper than the depth of the repetitive dive. This is 100 fsw. The residual nitrogen time is 30 minutes.

7. Add the 30 minutes of residual nitrogen time to the estimated bottom time of 15 minutes to obtain the single equivalent dive time of 45 minutes.

8. The diver will be decompressed on the 100 fsw/45 min schedule in the Air Decompression Table.

Figure 9-10 depicts the dive profile for the first dive, Figure 9-11 shows the completed Repetitive Dive Worksheet, and Figure 9-12 shows the dive profile for the repetitive dive.

9-9.2 **RNT Exception Rule.** In some cases, the residual nitrogen time given in Table 9-8 may be longer than needed to provide adequate decompression on the repetitive dive. This situation is most likely to occur when the surface interval between the dives is short. After determining the decompression requirement for the repetitive dive using the procedure in paragraph 9-9.1, the diver should recalculate the requirement by summing the bottom times of the two dives and taking the deepest depth. If the resultant table and schedule produces a longer no-decompression time or a shorter decompression time than the procedure in paragraph 9-9.1, the table and schedule with the lesser decompression obligation may be used. This alternative method of determining the table and schedule is referred to as the RNT Exception Rule.

1025			
Date: 5 Sept 07	Type of Dive: (AIR) HeO_2		
Diver 1: NDCM Boyd	Diver 2: NDC Parson	Standby: ND3 Jones	
Rig: KM-37 PSIG: 2900 O_2%:	Rig: KM-37 PSIG: 2900 O_2%:	Rig: KM-37 PSIG: 2900 O_2%:	
Diving Supervisor: NDCM Mariano	Chartman: ND1 Peters	Bottom Mix:	

EVENT	STOP TIME	CLOCK TIME
LS or 20 fsw		0900
RB		0902
LB		0948
R 1st Stop		0951
190 fsw		
180 fsw		
170 fsw		
160 fsw		
150 fsw		
140 fsw		
130 fsw		
120 fsw		
110 fsw		
100 fsw		
90 fsw		
80 fsw		
70 fsw		
60 fsw		
50 fsw		
40 fsw		
30 fsw		
20 fsw	:02+:31	1024
RS		1025
RB CHAMBER		
50 fsw chamber		
40 fsw chamber		
30 fsw chamber		
RS CHAMBER		

TDT	TTD
:37	1:25

EVENT	TIME/DEPTH
Descent Time (Water)	:02
Stage Depth (fsw)	100
Maximum Depth (fsw)	100+1=101
Total Bottom Time	:48
Table/Schedule	110/50
Time to 1st Stop (Actual)	:02::46
Time to 1st Stop (Planned)	:02::40
Delay to 1st Stop	::06
Travel/Shift/Vent Time	:02
Ascent Time-Water/SurD (Actual)	::40
Undress Time-SurD (Actual)	
Descent Chamber-SurD (Actual)	
Total SurD Surface Interval	
Ascent Time–Chamber (Actual)	

HOLDS ON DESCENT

DEPTH	PROBLEM

DELAYS ON ASCENT

DEPTH	PROBLEM

DECOMPRESSION PROCEDURES USED

AIR
- ☐ In-water Air decompression
- ☑ In-water Air/O_2 decompression
- ☐ $SurDO_2$

HeO_2
- ☐ In-water HeO_2/O_2 decompression
- ☐ $SurDO_2$

REPETITIVE GROUP: Z

Remarks:

Figure 9-10. Completed Air Diving Chart: First Dive of Repetitive Dive Profile.

REPETITIVE DIVE WORKSHEET

Date: 5 Sept 07

1st DIVE							
Max Depth	100+1=101						
Bottom Time	:48						
Table & Schedule	110/50					REPET Group	Z
Surface Interval	6:26					New Group	I
2nd DIVE							
Max Depth	97+1=98	MD + ESDT = Table & Schedule					
Bottom Time	+	RNT	=	ESDT	=	Table & Schedule	REPET Group
:15	+	:30	=	:45	=	100/45	N
Ensure the RNT Exception Rule does not apply							
Surface Interval						New Group	
3rd DIVE							
Max Depth		MD + ESDT = Table & Schedule					
Bottom Time	+	RNT	=	ESDT	=	Table & Schedule	REPET Group
	+		=		=		
Ensure the RNT Exception Rule does not apply							
Surface Interval						New Group	
4th DIVE							
Max Depth		MD + ESDT = Table & Schedule					
Bottom Time	+	RNT	=	ESDT	=	Table & Schedule	REPET Group
	+		=		=		
Ensure the RNT Exception Rule does not apply							
Surface Interval						New Group	

Figure 9-11. Completed Repetitive Dive Worksheet.

1731		
Date: 5 Sept 07	Type of Dive: (AIR circled) HeO_2	
Diver 1: NDCM Boyd	Diver 2: NDC Parson	Standby: CWO5 Armstrong
Rig: KM-37 PSIG: 2900 O_2%:	Rig: KM-37 PSIG: 2900 O_2%:	Rig: KM-37 PSIG: 2900 O_2%:
Diving Supervisor: NDCM Mariano	Chartman: CWO4 Perna	Bottom Mix:

EVENT	STOP TIME	CLOCK TIME
LS or 20 fsw		1651
RB		1653
LB		1706
R 1st Stop		1709
190 fsw		
180 fsw		
170 fsw		
160 fsw		
150 fsw		
140 fsw		
130 fsw		
120 fsw		
110 fsw		
100 fsw		
90 fsw		
80 fsw		
70 fsw		
60 fsw		
50 fsw		
40 fsw		
30 fsw		
20 fsw	:02+:19	1730
RS		1731
RB CHAMBER		
50 fsw chamber		
40 fsw chamber		
30 fsw chamber		
RS CHAMBER		

TDT	TTD
:25	:40

EVENT	TIME/DEPTH
Descent Time (Water)	:02
Stage Depth (fsw)	97
Maximum Depth (fsw)	97+1=98
Total Bottom Time	:15+:30=:45
Table/Schedule	100/45
Time to 1st Stop (Actual)	:02::35
Time to 1st Stop (Planned)	:02::34
Delay to 1st Stop	::01
Travel/Shift/Vent Time	:02
Ascent Time-Water/SurD (Actual)	::45
Undress Time-SurD (Actual)	
Descent Chamber-SurD (Actual)	
Total SurD Surface Interval	
Ascent Time–Chamber (Actual)	

HOLDS ON DESCENT

DEPTH	PROBLEM

DELAYS ON ASCENT

DEPTH	PROBLEM

DECOMPRESSION PROCEDURES USED

AIR

- ☐ In-water Air decompression
- ☑ In-water Air/O_2 decompression
- ☐ $SurDO_2$

HeO_2

- ☐ In-water HeO_2/O_2 decompression
- ☐ $SurDO_2$

REPETITIVE GROUP: N

Remarks:

Figure 9-12. Completed Air Diving Chart: Second Dive of Repetitive Dive Profile.

Example: A diver makes an air dive to 60 fsw for 40 minutes and plans to make a repetitive air dive to 56 fsw for 20 minutes after a 30-minute surface interval. Determine the table and schedule for the repetitive dive.

The diver surfaces from the first dive in repetitive group H. After 30 minutes on the surface he remains in repetitive group H. The depth of the repetitive dive is rounded up to the next deeper depth in Table 9-8, 60 fsw. The residual nitrogen time for a group H diver at 60 fsw is 46 minutes. The equivalent single dive time of the repetitive dive is 20 + 46 = 66 minutes. The 60 fsw/70 min schedule calls for a 7 min stop on air at 20 fsw. The alternative table and schedule for the repetitive dive is 60 fsw (deepest of the two depths) and 60 minutes (sum of the 40 and 20-minute bottom times). The 60 fsw / 60 min schedule does not require decompression stops. The diver uses the 60 fsw / 60 min schedule for the repetitive dive under the RNT exception rule.

Example: A diver makes a dive to 100 fsw for 25 minutes and plans to make a repetitive dive to 60 fsw for 20 minutes after a 30-minute surface interval. Determine the table and schedule for the repetitive dive.

The diver surfaces from the first dive in repetitive group H. After 30 minutes on the surface, he remains in repetitive group H. The residual nitrogen time for group H at 60 fsw is 46 minutes. The equivalent single dive time of the repetitive dive is 20 + 46 = 66 minutes. The 60 fsw / 70 min schedule calls for a 7 min stop on air at 20 fsw. The alternative table and schedule for the repetitive dive is 100 fsw (deepest of the two depths) and 45 minutes (sum of 25 and 20-minute bottom times). The 100 fsw / 45 min schedule calls for 36 minutes on air at 20 fsw. The diver uses the shorter 60 fsw / 70 min schedule under the provisions of paragraph 9-9.1.

The RNT exception rule can be applied to a series of repetitive dives. The table and schedule for the next dive in the series is determined first using the procedure in paragraph 9-9.1, then by adding the bottom times of all the repetitive dives in the series and taking the deepest depth. Whichever table and schedule produces the shorter decompression time or the longer no-decompression time is the table and schedule to be used for the repetitive dive.

9-9.3 **Repetitive Air-MK 16 Dives.** The repetitive group designators for air diving and MK 16 MOD 0 and MOD 1 diving are defined identically. This means that it is possible to perform a repetitive dive on air following either a MK 16 MOD 0 or a MK 16 MOD 1 nitrogen-oxygen dive using the existing tables. To perform a repetitive dive on air following a nitrogen-oxygen dive on either the MK 16 MOD 0 or MOD 1, take the following steps:

WARNING **These procedures cannot be used to make repetitive dives on air following MK 16 helium-oxygen dives.**

1. Obtain the repetitive group designator on surfacing from the MK 16 dive from Table 17-6 or 17-9 (MOD 0) or from Table 18-9 or 18-11 (MOD 1).

2. Using the MK 16 repetitive group designator, enter the top half of Table 9-8 on the diagonal. From this point the procedure is identical to making a repetitive dive on air following an air dive.

3. Read across the row to the appropriate surface interval, then down to the depth of the repetitive dive on air to obtain the residual nitrogen time.

4. Add the residual nitrogen time to the bottom time of the repetitive air dive to obtain the Equivalent Single Dive Time.

5. Enter the Air Decompression Table or No-Decompression Table at the depth that is exactly equal to or next deeper than the actual depth of the repetitive dive. Select the schedule that is exactly equal to or next longer than the Equivalent Single Dive Time. Follow the prescribed decompression to the surface.

6. The RNT exception rule can be applied to repetitive MK 16/Air dives. First compute the Equivalent Air Depth of the MK 16 dive. Then add the bottom times of the two dives and take the deeper of the equivalent air depth of the MK 16 dive or the actual depth of the air dive. If the resultant table and schedule results in less decompression time than the procedure above, the RNT exception rule may be invoked and the shorter schedule used. Equivalent Air Depth for MOD 0 dive = (Depth of MOD 0 dive – 18 fsw)/0.79. Equivalent Air Depth for MOD 1 dive = (Depth of MOD 1 dive – 36 fsw)/0.79.

9-9.4 **Order of Repetitive Dives.** From the decompression standpoint, the most efficient way to perform repetitive dives is to perform the deepest dive first and the shallowest dive last. This pattern yields the most bottom time for the least decompression time. There is no prohibition on performing repetitive dives in the reverse order, i.e., shallowest dive first and deepest dive last, or in any random order if the operational situation requires it. It is just that patterns other than deep to shallow are not the most efficient in terms of decompression.

Example: A diver plans to perform two dives separated by a 30-min surface interval. One dive is to 100 fsw for 20 min. The second dive is to 60 fsw for 20 min. Which dive should be performed first?

Following the normal pattern of deep to shallow, the diver does the 100 fsw dive first. He surfaces in repetitive group G and remains in Group G during the surface interval. The RNT for Group G at 60 fsw is 40 min. The Equivalent Single Dive Time of the 60 fsw dive therefore is 60 min (40 + 20). A 60 fsw/60 min dive is right at the no-decompression limit. No decompression is required for either dive.

Following the reverse pattern of shallow to deep, the diver does the 60 fsw dive first. He surfaces in repetitive Group D and remains in Group D during the surface interval. The RNT for Group D at 100 fsw is 14 min. The Equivalent Single Dive Time of the 100 fsw dive therefore is 34 min (14 + 20). The diver decompresses on the 100 fsw/35 min schedule. A 15 min decompression stop at 20 fsw is required.

With the normal pattern, the diver achieved 40 minutes of bottom time without having to decompress. With the reverse pattern the diver required 15 min of decompression stop time for the same 40 minutes of bottom time.

9-10 EXCEPTIONAL EXPOSURE DIVES

Exceptional exposure dives are those dives in which the risk of decompression sickness, oxygen toxicity, and/or exposure to the elements is substantially greater than on normal working dives. These exceptional exposure schedules are intended to be used only in emergencies such as diver entrapment. Exceptional exposures should not be planned in advance except under the most unusual operational circumstances. The Commanding Officer must carefully assess the need for planned exceptional exposure diving and prior CNO approval for such diving is required.

Exceptional exposure dives are defined by the required decompression time for the decompression mode selected. The following air dives are considered exceptional exposure.

- Any dive deeper than 190 fsw.
- Any in-water decompression dive with a total decompression time on air or air/oxygen greater than 90 minutes.
- Any SurDO_2 dive with a chamber oxygen time greater than 120 minutes (4 oxygen periods).

NOTE **The Commanding Officer must have CNO approval to conduct planned exceptional exposure dives.**

9-11 VARIATIONS IN RATE OF ASCENT

The following rules for correcting for variations in rate of ascent apply to all the tables given in this chapter. The normal rate of ascent to the first stop and between subsequent stops is 30 fsw/min. Minor variations in the rate of travel between 20 and 40 fsw/min are acceptable and do not require correction.

9-11.1 **Travel Rate Exceeded.** If the rate of ascent is greater than 40 fsw/min, stop the ascent, allow the watches to catch up, and then continue ascent.

9-11.2 **Early Arrival at the First Decompression Stop.** If the divers arrive early at the first decompression stop:

1. Begin timing the first stop when the required travel time has been completed.
2. If the first stop is an oxygen stop, shift the divers to oxygen upon arrival at the stop. Begin stop time when the divers are confirmed on oxygen and the required travel time has been completed.

9-11.3 Delays in Arriving at the First Decompression Stop

- **Delay up to 1 minute.** A delay of up to one minute in reaching the first decompression stop can be ignored.

- **Delay greater than 1 minute, deeper than 50 fsw.** Round up the delay time to the next whole minute and add it to the bottom time. Recompute the decompression schedule. If no change in schedule is required, continue on the planned decompression. If a change in schedule is required and the new schedule calls for a decompression stop deeper than the diver's current depth, perform any missed deeper stops at the diver's current depth. Do not go deeper.

  Example: Divers make a dive to 115 fsw. Stage depth is 113 fsw. Bottom time is 55 minutes. According to the 120 fsw / 55 min decompression schedule, the first decompression stop is 30 fsw. During ascent, the divers were delayed at 100 fsw for 3 minutes 27 seconds and it actually took 6 min 13 seconds to reach the 30-foot decompression stop. Determine the new decompression schedule.

  The total delay is 3 minutes 27 seconds. Round this delay time up to the next whole minute, 4 minutes, and add the rounded up delay to the bottom time. The new bottom time is 59 minutes. Re-compute the decompression schedule using a 60-min bottom time and continue decompression according to the new decompression schedule, 120 fsw / 60 min. This dive is illustrated in Figure 9-13.

- **Delay greater than 1 minute, shallower than 50 fsw.** If a delay in ascent greater than 1 minute occurs shallower than 50 fsw, round the delay time up to the next whole minute and add the delay time to the diver's first decompression stop.

  Example: Divers made a dive to 113 fsw. Bottom time was 60 minutes. According to the Air Decompression Table, the first decompression stop is at 30 fsw. During ascent, the divers were delayed at 40 fsw and it actually took 6 minutes 20 seconds to reach the 30-fsw stop. Determine the new decompression schedule.

  If the divers had maintained an ascent rate of 30 fsw/min, the correct ascent time would have been 2 minutes 46 seconds. Because it took 6 minutes 20 seconds to reach the 30-fsw stop, there was a delay of 3 minutes 34 seconds (6 minutes 20 seconds minus 2 minutes 46 seconds). Therefore, increase the length of the 30-fsw decompression stop by 3 minutes 34 seconds, rounded up to 4 minutes. Instead of 14 minutes on oxygen at 30 fsw, the divers must spend 18 minutes on oxygen. This dive is illustrated in Figure 9-14.

9.11.4 Delays in Leaving a Stop or Between Decompression Stops.

- **Delay less than 1 minute leaving an air stop.** When the delay in leaving an air stop is less than 1 minute, disregard the delay. Resume the normal decompression when the delay is over.

- **Delay less than 1 minute between air stops.** If the delay between stops is less than 1 minute, disregard the delay.

1503		
Date: 22 Oct 07	Type of Dive: (AIR) HeO_2	
Diver 1: ND1 Schlabach	Diver 2: ND2 Hedrick	Standby: HM2 Montgomery
Rig: MK 21 PSIG: O_2%:	Rig: MK 21 PSIG: O_2%:	Rig: MK 21 PSIG: O_2%:
Diving Supervisor: NDC Blanton	Chartman: LT Slappy	Bottom Mix:

EVENT	STOP TIME	CLOCK TIME
LS or 20 fsw		1300
RB		1302
LB		1355
R 1st Stop		1402
190 fsw		
180 fsw		
170 fsw		
160 fsw		
150 fsw		
140 fsw		
130 fsw		
120 fsw		
110 fsw		
100 fsw	:3::27	1359
90 fsw		
80 fsw		
70 fsw		
60 fsw		
50 fsw		
40 fsw		
30 fsw	:02+:14	1418
20 fsw	:16+:05+:23	1502
RS		1503
RB CHAMBER		
50 fsw chamber		
40 fsw chamber		
30 fsw chamber		
RS CHAMBER		

TDT	TTD
1:08	2:03

EVENT	TIME/DEPTH
Descent Time (Water)	:02
Stage Depth (fsw)	113
Maximum Depth (fsw)	113+2=115
Total Bottom Time	:55+:04=:59
Table/Schedule	120/60
Time to 1st Stop (Actual)	:06::13
Time to 1st Stop (Planned)	:02::46
Delay to 1st Stop	:03::27
Travel/Shift/Vent Time	:02
Ascent Time-Water/SurD (Actual)	:45
Undress Time-SurD (Actual)	
Descent Chamber-SurD (Actual)	
Total SurD Surface Interval	
Ascent Time–Chamber (Actual)	

HOLDS ON DESCENT

DEPTH	PROBLEM

DELAYS ON ASCENT

DEPTH	PROBLEM
100	fouled

DECOMPRESSION PROCEDURES USED

AIR

- ☐ In-water Air decompression
- ☑ In-water Air/O_2 decompression
- ☐ $SurDO_2$

HeO_2

- ☐ In-water HeO_2/O_2 decompression
- ☐ $SurDO_2$

REPETITIVE GROUP: Z

Remarks: Diver fouled at 100 fsw for :3::27. Rounded up to :4 add to BT, Re-compute T/S

Figure 9-13. Completed Air Diving Chart: Delay in Ascent deeper than 50 fsw.

1711		
Date: 22 Oct 07	Type of Dive: (AIR circled) HeO_2	
Diver 1: ND1 Bauer	Diver 2: ND2 Brown	Standby: HM2 Seymour
Rig: MK 21 PSIG: O_2%:	Rig: MK 21 PSIG: O_2%:	Rig: MK 21 PSIG: O_2%:
Diving Supervisor: NDC Poulan	Chartman: CDR Daubon	Bottom Mix:

EVENT	STOP TIME	CLOCK TIME	EVENT	TIME/DEPTH
LS or 20 fsw		1500	Descent Time (Water)	:02
RB		1502	Stage Depth (fsw)	113
LB		1600	Maximum Depth (fsw)	113+2=115
R 1st Stop		1607	Total Bottom Time	:60
190 fsw			Table/Schedule	120/60
180 fsw			Time to 1st Stop (Actual)	:06::20
170 fsw			Time to 1st Stop (Planned)	:02::46
160 fsw			Delay to 1st Stop	:03::34
150 fsw			Travel/Shift/Vent Time	:02
140 fsw			Ascent Time-Water/SurD (Actual)	:45
130 fsw			Undress Time-SurD (Actual)	
120 fsw			Descent Chamber-SurD (Actual)	
110 fsw			Total SurD Surface Interval	
100 fsw			Ascent Time–Chamber (Actual)	
90 fsw			**HOLDS ON DESCENT**	
80 fsw			**DEPTH**	**PROBLEM**
70 fsw				
60 fsw				
50 fsw				
40 fsw	:03::34	1606	**DELAYS ON ASCENT**	
30 fsw	:02+:04+:14	1626	**DEPTH**	**PROBLEM**
20 fsw	:12+:05+:27	1710	40	fouled
RS		1711		
RB CHAMBER				
50 fsw chamber			**DECOMPRESSION PROCEDURES USED**	
40 fsw chamber				
30 fsw chamber				
RS CHAMBER				

TDT	TTD
1:11	2:11

DECOMPRESSION PROCEDURES USED

AIR

- ☐ In-water Air decompression
- ☑ In-water Air/O_2 decompression
- ☐ $SurDO_2$

HeO_2

- ☐ In-water HeO_2/O_2 decompression
- ☐ $SurDO_2$

REPETITIVE GROUP: Z

Remarks: Diver fouled at 40 fsw for :3::34. Rounded up to :04 add to 1st stop.

Figure 9-14. Completed Air Diving Chart: Delay in Ascent Shallower than 50 fsw.

- **Delay greater than 1 minute leaving an air stop or between air stops deeper than 50 fsw.** Add the delay to the bottom time and recalculate the required decompression. If a new schedule is required, pick up the new schedule at the present stop or subsequent stop if delay occurs between stops. Ignore any missed stops or time deeper than the depth at which the delay occurred.

- **Delay greater than 1 minute leaving an air stop or between air stops shallower than 50 fsw.** Ignore the delay. Resume the normal schedule upon completion of the delay.

- **Delay leaving an oxygen stop at 30 fsw or delay between oxygen stops at 30 and 20 fsw.** Subtract any delay in leaving the 30 fsw oxygen stop or any delay during travel from 30 to 20 fsw on oxygen from the subsequent 20-fsw oxygen stop time. If the delay causes the total time on oxygen deeper than 20 fsw to exceed 30 minutes, shift the diver to air at the 30-minute mark. When the problem has been resolved, shift the diver back to oxygen and resume decompression. Ignore any time spent on air.

  Example: The diver's decompression schedule calls for a 20 min stop on oxygen at 30 fsw and a 40 min stop on oxygen at 20 fsw. The diver has a 15 min delay leaving the 30-fsw stop due to a stage malfunction.

  The first 10 minutes of the delay can be spent on oxygen at 30 fsw, giving a total oxygen time of 30 minutes at 30 fsw. The diver should then be shifted to air for the remaining 5 minutes of the delay. When the problem is resolved, switch the diver back to oxygen at 30 fsw and ascend to 20 fsw to begin the 20-fsw stop time. The 20-fsw stop time is reduced from 40 to 30 minutes because of the extra 10 minutes spent on oxygen at 30 fsw. The 5-min air break is ignored.

- **Delay in leaving the 20-fsw oxygen stop.** Delays leaving the 20-fsw oxygen stop can be ignored. However, do not leave divers on oxygen longer than 30 minutes as described in paragraph 9-8.2.2. Shift the divers to air and remain on air until travel to the surface is possible.

- **Delay in Travel from 40 fsw to the Surface for Surface Decompression.** Disregard any delays in travel from 40 fsw to the surface during surface decompression unless the diver exceeds the allowed 5-minute surface interval. If the diver exceeds the 5-minute surface interval, follow the guidance in paragraph 9-12.6.

9-12 EMERGENCY PROCEDURES

In air diving, specific procedures are used in emergency situations. The following paragraphs detail these emergency procedures.

9-12.1 Bottom Time in Excess of the Table

In the rare instance of diver entrapment or umbilical fouling, bottom time may exceed the longest bottom time listed in the table for the diver's depth. When it is foreseen the bottom time will exceed the longest listed value, immediately contact the Navy Experimental Diving Unit for advice on how to decompress. If the Navy Experimental Diving Unit cannot be contacted in time, take the following action:

1. If available, use the U.S. Navy Thalmann Algorithm Dive Planner to compute the decompression requirement.

2. Read down to deeper depths in the Air Decompression Table until a depth is found that has a schedule that is equal to or longer than the bottom time. The Air Decompression Table contains longer schedules at various depths especially for this purpose.

Example: A diver is trapped on the bottom at a depth of 155 fsw. By the time he is freed, the bottom time is 100 min. The longest schedule in the 160 fsw table is 80 min. Read down to the 170 fsw table. The 120 min schedule is longer than the diver's bottom time. Decompress the diver on the 170 fsw / 120 minute schedule.

9-12.2 Loss of Oxygen Supply in the Water

If the diver cannot be shifted to oxygen at 30 or 20 fsw:

1. Have the diver continue to breathe air while the problem is investigated.

2. If the problem can be corrected quickly, ventilate the diver with oxygen as soon as the gas supply is restored. Consider any time spent on air as dead time. Remain on oxygen at the stop for the full stop time listed in the table.

3. If the problem cannot be corrected, initiate surface decompression or continue decompression in the water on air. In this situation, the surface interval for surface decompression is the time from leaving the in-water stop to reaching the 50-fsw stop in the recompression chamber.

If the oxygen supply is lost during the 30 or 20-fsw water stops after the diver has shifted to oxygen:

1. Shift the diver back to air.

2. If the problem can be corrected quickly, re-ventilate the diver with oxygen and resume the schedule at the point of interruption. Consider any time spent on air as dead time.

3. If the problem cannot be corrected and a recompression chamber is available on the dive station, initiate surface decompression. Compute the number of chamber oxygen periods required by multiplying the remaining oxygen time at the stops by 1.1, dividing the total by 30 minutes, then rounding the result up to the next highest half period. One half period (15 minutes at 50 fsw) is the minimum requirement.

   Example: The oxygen supply is lost permanently when the diver has a remaining oxygen time of 5 minutes at 30 fsw and 33 minutes at 20 fsw. The total remaining oxygen time is 38 minutes. The number of 30-min $SurDO_2$ periods required is $(1.1 \times 38) / 30 = 1.39$. This number is rounded up to 1.5.

4. If the problem cannot be corrected and a recompression chamber is not available on the dive station, continue decompression on air in the water. Compute the remaining stop time on air at the depth of the loss by multiplying the remaining stop time on oxygen at that depth by the ratio of the air stop time to the oxygen time at that depth.

   Example: The oxygen supply is lost permanently when the diver has a remaining oxygen time of 10 minutes at 20 fsw. His decompression schedule calls for either 140 minutes on air at 20 fsw or 34 minutes on oxygen at 20 fsw. The ratio of air stop time to oxygen time at the 20-fsw stop is 140/34 = 4.12. His remaining time on air at 20 fsw is 10 × 4.12 = 41.2 minutes. Round this time up to 42 minutes.

   If the shift to air occurs at 30 fsw, compute the remaining stop time on air at 30 fsw as shown above, then take the full 20-fsw air stop as prescribed in the Air Decompression Table.

9-12.3 Contamination of Oxygen Supply with Air

It will be difficult to detect mixing of air with the oxygen supply during oxygen decompression in the water as no voice change will occur as it does in helium-oxygen diving. On shifting to oxygen, the ORCA operator should verify that the ORCA is properly lined up and that the oxygen monitor, if one is present, indicates 100% oxygen going to the diver's umbilical. The diver should monitor his EGS pressure gauge periodically to ensure that there is no drop in pressure.

If the operator discovers that the ORCA is improperly lined up, take the following action:

1. Align the ORCA properly.
2. Re-ventilate each diver with oxygen for 20 seconds.
3. Restart oxygen time. Consider any time spent on contaminated oxygen as dead time.

9-12.4 CNS Oxygen Toxicity Symptoms (Non-convulsive) at 30 or 20 fsw Water Stop

Most divers will easily tolerate the oxygen exposures prescribed by these Tables. CNS oxygen toxicity symptoms, if they do develop, are most likely to occur near the end of the 20-fsw oxygen stop. Nausea is the most likely symptom.

If the diver develops symptoms of CNS toxicity at the 30- or 20-fsw water stops, take the following action:

1. If a recompression chamber is available on the dive station, initiate surface decompression. Shift the console to air during travel to the surface. Compute the number of chamber oxygen periods required by multiplying the remaining

oxygen time at the stops by 1.1, dividing the total by 30 minutes, then rounding the result up to the next highest half period. One half period (15 minutes at 50 fsw) is the minimum requirement.

2. If a recompression chamber is not available on the dive station and the event occurs at 30 fsw, bring the divers up 10 fsw and shift to air to reduce the partial pressure of oxygen. Shift the console as the divers are traveling to 20 fsw. Ventilate both divers with air upon arrival at 20 fsw. Ventilate the affected diver first. Complete the decompression on air at 20 fsw. Compute the 20-fsw stop time as follows: Multiply the missed stop time on oxygen at 30 fsw by the ratio of the air to oxygen stop time at 30 fsw to obtain the equivalent missed air time at 30 fsw. Add this time to the 20-fsw air stop time shown in the Air Decompression Table.

3. If a recompression chamber is not available on the dive station and the event occurs at 20 fsw, shift the console to air, ventilate both divers, affected diver first, and complete the decompression in the water at 20 fsw on air. Compute the remaining stop time on air at 20 fsw by multiplying the remaining stop time on oxygen at 20 fsw by the ratio of the air stop time to the oxygen time at 20 fsw.

   Example: After 10 minutes on oxygen at 30 fsw, a diver has a non-convulsive CNS oxygen toxicity symptom. A recompression chamber is not available on the dive station. The diver is immediately brought up to 20 fsw and ventilated with air. His decompression schedule calls for 28 minutes on air at 30 fsw and 175 minutes on air at 20 fsw. The oxygen stop time at 30 fsw is 14 minutes. The missed oxygen time at 30 fsw is 4 minutes (14 – 10). The ratio of air to oxygen time at 30 fsw is 28/14 = 2.0. The missed air time at 30 fsw therefore is 4 × 2.0 = 8 minutes. The required air decompression time at 20 fsw is 183 minutes (8 + 175).

   Example: After 24 minutes on oxygen at 20 fsw, a diver has a non-convulsive CNS oxygen toxicity symptom. A recompression chamber is not available on the dive station. The diver is shifted to air with 10 min of oxygen time remaining at 20 fsw. His decompression schedule calls for either 140 minutes on air at 20 fsw or 31 minutes on oxygen at 20 fsw. The ratio of air stop time to oxygen time at the 20-fsw stop is 140/31 = 4.52. His remaining time on air at 20 fsw is 10 × 4.52 = 45.2 minutes. Round this time up to 46 minutes.

9-12.5 Oxygen Convulsion at the 30- or 20-fsw Water Stop

If symptoms progress to an oxygen convulsion despite the above measures, or if a convulsion occurs suddenly without warning, take the following action.

1. Shift both divers to air if this action has not already been taken.

2. Have the unaffected diver ventilate himself and then ventilate the stricken diver.

3. If only one diver is in the water, launch the standby diver immediately and have him ventilate the stricken diver.

4. Hold the divers at depth until the tonic-clonic phase of the convulsion has subsided. The tonic-clonic phase of a convulsion generally lasts 1–2 minutes.

5. At the end of the tonic-clonic phase, have the dive partner or standby diver ascertain whether the diver is breathing. The presence or absence of breath sounds will usually be audible over the diver communication system.

6. If the diver appears not to be breathing, have the dive partner or standby diver attempt to reposition the head to open the airway. Airway obstruction will be the most common reason why an unconscious diver fails to breathe.

7. If the diver is breathing, hold him at depth until he is stable, then surface decompress. Compute the number of chamber oxygen periods required by multiplying the remaining oxygen time at the stops by 1.1, dividing the total by 30 min, then rounding the result up to the next highest half period. One half period (15 minutes at 50 fsw) is the minimum requirement.

8. If surface decompression is not feasible, continue decompression on air in the water. Compute the remaining stop time on air at the depth of the incident by multiplying the remaining stop time on oxygen at that depth by the ratio of the air stop time to the oxygen time at that depth. If the shift to air occurs at 30 fsw, compute the remaining stop time on air at 30 fsw, then take the full 20-fsw air stop as prescribed in the Air Decompression Table.

9. If it is not possible to verify that the affected diver is breathing, leave the unaffected diver at the stop to complete decompression, and surface the affected diver and the standby diver at 30 fsw/min. The standby diver should attempt to maintain an open airway on the stricken diver during ascent. On the surface, the affected diver should receive any necessary airway support and be immediately recompressed and treated for arterial gas embolism in accordance with Figure 20-1.

9-12.6 **Surface Interval Greater than 5 Minutes.** If the time from leaving 40 fsw in the water to the time of arrival at 50 fsw in the chamber during surface decompression exceeds 5 minutes, take the following action:

1. If the surface interval is more than 5 minutes but less than or equal to 7 minutes, increase the time on oxygen at 50 fsw from 15 to 30 minutes, i.e., add one-half oxygen period to the 50 fsw chamber stop. Ascend to 40 fsw during the subsequent air break. The 15-min penalty is considered a part of the normal surface decompression procedure, not an emergency procedure.

   Example: Divers are decompressing on a SurDO_2 schedule that requires 1.5 oxygen breathing periods. It took 6 minutes and 20 seconds to travel from 40 fsw to the surface, undress the diver, and recompress to 50 fsw in the chamber. The divers are placed on oxygen at 50 fsw in the chamber. They will breathe oxygen at 50 fsw for the 15 minutes (one-half period) required by the original schedule plus an additional 15 minutes to compensate for exceeding the normal 5-min surface interval. Upon completion of 30 minutes on oxygen at 50 fsw, they will remove the BIBS to initiate a 5-minute air break and ascend from 50 fsw to 40 fsw at 30 fsw/min while breathing air. After 5 minutes on air, the divers will breathe oxygen for 30 minutes to complete the oxygen time required at 40 fsw on the original schedule. After 30 minutes on oxygen at 40 fsw, the divers will remove the BIBS and ascend to the surface at 30 fsw/min breathing air. Because the divers exceeded the normal 5-minute surface interval, the total number of oxygen periods is increased from 1.5 to 2.0.

2. If the surface interval is greater than 7 minutes, continue compression to a depth of 60 fsw. Treat the divers on Treatment Table 5 if the original schedule required 2 or fewer oxygen periods in the chamber. Treat the divers on Treatment Table 6 if the original schedule required 2.5 or more oxygen periods in the chamber.

3. On rare occasions a diver may not be able to reach 50 fsw in the chamber because of difficulty equalizing middle ear pressure. In this situation, an alternative procedure for surface decompression on oxygen may be used. Compress the diver to the deepest depth he can attain initially. This will usually be less than 20 fsw. Begin oxygen breathing at that depth. Continue attempts to gradually compress the diver deeper. If the in-water air or air/oxygen decompression schedule required only a 20-fsw water stop, attempt to compress the diver to 20 fsw. If the in-water air or air/oxygen decompression schedule required a 30-fsw water stop, attempt to compress the diver to 30 fsw. In either case, double the number of chamber oxygen periods indicated in the table and have the diver take these periods at whatever depth he is able to attain. Oxygen time starts when the diver initially goes on oxygen. Interrupt oxygen breathing every 60 minutes with a 15-min air break. The air break does not count toward the total oxygen time. Upon completion of the oxygen breathing periods, surface the diver at 30 fsw/min. Carefully observe the diver post-dive for the onset of decompression sickness. This "safe way out" procedure is not intended to be used in place of normal surface decompression procedures. Repetitive diving is not allowed following a dive in which the "safe way out" procedure is used.

9-12.7 **Decompression Sickness During the Surface Interval.** If symptoms of Type I decompression sickness occur during travel from 40 fsw to the surface during surface decompression or during the surface undress phase, compress the diver to 50 fsw following normal surface decompression procedures. Delay neurological exam until the diver reaches the 50-fsw stop and is on oxygen. If Type I symptoms resolve during the 15 minute 50-fsw stop, the surface interval was 5 minutes or less, and no neurological signs are found, increase the 50 fsw oxygen time from 15 to 30 minutes as outlined above, then continue normal decompression for the schedule of the dive. Ascend from 50 to 40 fsw during the subsequent air break.

If Type I symptoms do not resolve during the 15 minute 50-fsw stop or symptoms resolve but the surface interval was greater than 5 minutes, compress the diver to 60 fsw on oxygen. Treat the diver on Treatment Table 5 if the original schedule required 2 or fewer oxygen periods in the chamber. Treat the diver on Treatment Table 6 if the original schedule required 2.5 or more oxygen periods in the chamber. Treatment table time starts upon arrival at 60 fsw. Follow the guidelines for treatment of decompression sickness given in Chapter 20, Volume 5.

If symptoms of Type II decompression sickness occur during travel from 40 fsw to the surface, during the surface undress phase, or the neurological examination at 50 fsw is abnormal, compress the diver to 60 fsw on oxygen. Treat the diver on Treatment Table 6. Treatment table time starts upon arrival at 60 fsw. Follow the guidelines for treatment of decompression sickness given in Chapter 20, Volume 5.

Table 9-2 summarizes the guidance for managing an extended surface interval and for managing Type I decompression sickness during the surface interval.

Table 9-2. *Management of Extended Surface Interval and Type I Decompression Sickness during the Surface Interval.*

<table>
<tr><th>Surface Interval
(Note 1)</th><th>Asymptomatic Diver</th><th>Symptomatic Diver
(Type I DCS)</th></tr>
<tr><td>5 min or less</td><td>Follow original schedule</td><td>Increase O_2 time at 50 fsw from 15 to 30 min (Note 2)</td></tr>
<tr><td>Greater than 5 min but less than or equal to 7 min</td><td>Increase O_2 time at 50 fsw from 15 to 30 min</td><td rowspan="2">Treatment Table 5 if 2 or fewer SurDO_2 periods

Treatment Table 6 if more than 2 SurDO_2 periods</td></tr>
<tr><td>Greater than 7 min</td><td>Treatment Table 5 if 2 or fewer SurDO_2 periods

Treatment Table 6 if more than 2 SurDO_2 periods</td></tr>
</table>

Notes:

1. Surface interval is the time from leaving the 40-fsw water stop to arriving at the 50-fsw chamber stop.
2. Type I symptoms must completely resolve during the first 15 minutes at 50 fsw and a full neurological examination at 50 fsw must be normal. If symptoms do not resolve within 15 min, treat the diver on Treatment Tables 5 or 6 as indicated for surface intervals longer than 5 min.
3. If Type II symptoms are present at any time during the surface interval or the neurological examination at 50 fsw is abnormal, treat the diver on Treatment Table 6.

9-12.8 Loss of Oxygen Supply in the Chamber

For loss of oxygen supply in the chamber, have the diver breathe chamber air. If the loss is temporary, return the diver to oxygen breathing. Consider any time spent on air as dead time.

If the loss of the oxygen supply is permanent, complete decompression in the chamber on 50% nitrogen 50% oxygen (preferred) or on air. If 50% nitrogen 50% oxygen is available, multiply the remaining oxygen time by two to obtain the equivalent chamber decompression time on 50/50. Air breaks are not required when breathing 50/50. Diver may remove mask briefly (e.g., for drinking fluids). Consider any time spent on air as dead time. If chamber air is the only gas available, multiply the remaining chamber time on oxygen by the ratio of the water stop times on air at 30 and 20 fsw to the oxygen time at those depths to obtain the equivalent chamber decompression time on air. Allocate 10% of the equivalent air or 50/50 nitrogen-oxygen time to the 40-fsw stop, 20% to the 30-fsw stop, and 70% to the 20-fsw stop. If the diver is at 50 fsw when the loss occurs, ascend to 40 fsw and begin the stop time. If the loss occurred at 30 fsw, allocate 30% of the equivalent air or nitrogen-oxygen time to the 30-fsw stop and 70% to the 20-fsw stop. Round the stop times to the nearest whole minute. Surface the divers upon completion of the 20-fsw stop.

Example: A SurDO_2 schedule calls for two 30-min oxygen periods in the chamber. The chamber oxygen supply is lost permanently after 28 minutes on oxygen at 50 and 40 fsw. Chamber air is the only gas available. The remaining oxygen time is $(2 \times 30) - 28 = 32$ minutes. The original decompression schedule calls for 52 and 140 minute in-water air decompression stops at 30 and 20 fsw for a total air stop time of 192 minutes. The corresponding oxygen stop times are 13 and 34 minutes, for a total of oxygen stop time of 47 min. The ratio of air stop time to oxygen stop time is $192/47 = 4.08$. The remaining chamber air time is $32 \times 4.08 = 131$ minutes. This time is allocated as follows: 13 min at 40 fsw (131×0.1), 26 min at 30 fsw (131×0.2), and 92 min at 20 fsw (131×0.7).

9-12.9 CNS Oxygen Toxicity in the Chamber

At the first sign of CNS oxygen toxicity, the diver should be removed from oxygen and allowed to breathe chamber air. Fifteen minutes after all symptoms have completely subsided, resume oxygen breathing at the point of interruption. If symptoms develop again, or if the first symptom is a convulsion, take the following action:

1. Remove the mask.

2. After all symptoms have completely subsided, decompress 10 feet at a rate of 1 fsw/min. For a convulsion, begin travel when the patient is fully relaxed and breathing normally.

3. Resume oxygen breathing at the shallower depth at the point of schedule interruption.

4. If another oxygen symptom occurs after ascending 10 fsw, complete decompression on chamber air. Compute the remaining chamber time on air as shown in paragraph 9-12.8 above. If the diver is at 40 fsw, allocate 10% of the remaining air time to the 40-fsw stop, 20% to the 30-fsw stop, and 70% to the 20-fsw stop. If the diver is at 30 fsw, allocate 30% of the remaining time to the 30-fsw stop and 70% to the 20-fsw stop. Round the stop times to the nearest whole minute. Surface the divers upon completion of the 20-fsw stop.

9-12.10 Asymptomatic Omitted Decompression

Certain emergencies, such as uncontrolled ascents, an exhausted air supply, or bodily injury may interrupt or prevent required decompression. If the diver shows symptoms of decompression sickness or arterial gas embolism, immediate treatment using the appropriate recompression treatment table is essential. Even if the diver shows no symptoms, omitted decompression must be addressed in some manner to avert later difficulty.

Omitted decompression may or may not be planned. Planned omitted decompression results when a condition develops at depth that will require the diver to surface before completing all of the decompression stops and when there is time to consider all available options, ready the recompression chamber, and alert all personnel as to the planned evolution. Equipment malfunctions, diver injury, or sudden severe storms are examples of these situations. In unplanned omitted decompression, the diver suddenly appears on the surface without warning or misses decompression for some unforeseen reason.

Table 9-3 summarizes management of asymptomatic omitted decompression.

Table 9-3. *Management of Asymptomatic Omitted Decompression.*

Deepest Decompression Stop Omitted	Surface Interval (Note 1)	Action	
		Chamber Available (Note 2)	**No Chamber Available**
None	Any	Observe on surface for 1 hour	
20 or 30 fsw	Less than 1 min	Return to depth of stop. Increase stop time by 1 min. Resume decompression according to original schedule.	
	1 to 7 min	Use Surface Decompression Procedure (Note 3)	Return to depth of stop. Multiply 30 and/or 20 fsw air or O_2 stop times by 1.5.
	Greater than 7 min	Treatment Table 5 if 2 or fewer $SurDO_2$ periods Treatment Table 6 If more than 2 $SurDO_2$ periods	
Deeper than 30 fsw	Any	Treatment Table 6 (Note 4)	Descend to depth of first stop. Follow the schedule to 30 fsw. Switch to O_2 at 30 fsw if available. Multiply 30 and 20 fsw air or O_2 stops by 1.5.

Notes:

1. For surface decompression, surface interval is the time from leaving the stop to arriving at depth in the chamber.
2. Using a recompression chamber is strongly preferred over in-water recompression for returning a diver to pressure. Compress to depth as fast as possible not to exceed 100 fsw/min.
3. For surface intervals greater than 5 minutes but less than or equal to 7 minutes, increase the oxygen time at 50 fsw from 15 to 30 minutes.
4. If a diver missed a stop deeper than 50 fsw, compress to 165 fsw and start Treatment Table 6A.

9-12.10.1 No-Decompression Stops Required

If a diver makes an uncontrolled ascent to the surface at a rate greater than 30 fsw/min, but the dive itself is within no-decompression limits, the diver should be observed on the surface for one hour to ensure that symptoms of decompression sickness or arterial gas embolism do not develop. Recompression is not necessary unless symptoms develop.

9-12.10.2 **Omitted Decompression Stops at 30 and 20 fsw**

If the diver omits some or all of the decompression time at 30 and/or 20 fsw, take the following action:

1. If the diver is on the surface for less than one minute, return the diver to depth of the stop from which he came. Increase that stop time by one minute. Resume decompression according to the original schedule.

2. If the diver is on the surface for 1 to 5 minutes and a recompression chamber is available on dive station, place the diver in the recompression chamber and complete the decompression using surface decompression. If the diver was on oxygen at the time of the omission, compute the number of chamber oxygen periods required by multiplying the remaining oxygen time at the stops by 1.1, dividing the total by 30 min, then rounding the result up to the next highest half period. If the diver was on air at the time of the omission, first compute the equivalent remaining oxygen time at the stop as shown in paragraph 9-8.3.2. If the omission occurred at 20 fsw, use this remaining oxygen time to compute the number of oxygen periods as shown above. If the omission occurred at 30 fsw, compute the remaining oxygen time at 30 fsw, then add the oxygen time shown in the decompression table at 20 fsw to get the total remaining oxygen time. Use the total remaining oxygen time to compute the number of oxygen periods. In all instances, one half period (15 minutes at 50 fsw) is the minimum requirement.

3. If the diver is on the surface for more than 5 minutes but less than or equal to 7 minutes and a recompression chamber is available on the dive station, place the diver in the recompression chamber and complete the decompression using surface decompression as outlined in paragraph 2 above. Increase the time on oxygen at 50 fsw from 15 to 30 minutes.

4. If the diver is on the surface for more than 7 minutes and a recompression chamber is available on site, treat the diver with Treatment Table 5 if the surface decompression schedule for that dive required two or fewer oxygen periods in the chamber. Treat on Treatment Table 6 if the surface decompression schedule for that dive required 2.5 or more oxygen periods in the chamber.

5. If the diver is on the surface for more than 1 minute and a recompression chamber is not available, return the diver to the depth of the omitted stop. Complete decompression in the water by multiplying the 30- and/or 20-fsw air or oxygen stops by 1.5.

9-12.10.3 **Omitted Decompression Stops Deeper than 30 fsw**

If the diver omits part or all of a decompression stop at 40 fsw or deeper and a recompression chamber is available on site, treat the diver with Treatment Table 6. If a recompression chamber is not available on site, return the diver to the depth of the first decompression stop. Follow the original decompression schedule to 30 fsw. At 30 fsw, shift the diver to oxygen if it is available. Complete decompression from 30 fsw by multiplying the 30- and 20-fsw air or oxygen stops by 1.5.

9-12.11 Decompression Sickness in the Water.

In rare instances, decompression sickness may develop in the water during prolonged decompression on air or air/oxygen. The predominant symptom will usually be joint pain but more serious manifestations such as numbness, weakness, hearing loss, and vertigo may also occur. Decompression sickness is most likely to appear at the shallow stops just prior to surfacing. Some cases, however, have occurred during ascent to the first stop or shortly thereafter.

Managing decompression sickness in the water will be difficult in the best of circumstances. Only general guidance can be presented here. Management decisions must be made on site, taking in account all known factors. The advice of a Diving Medical Officer should be sought whenever possible.

9-12.11.1 **Diver Remaining in the Water.** If the diver indicates that he has decompression sickness but feels he can remain in the water:

1. Dispatch the standby diver to assist. Continue to decompress the other divers according to the original schedule.
2. If the diver is decompressing on air at 30 or 20 fsw, switch the diver to 100% oxygen if available.
3. Have the diver descend 10 fsw. If significant relief of symptoms is not obtained, have the diver descend an additional 10 fsw, but no deeper than 40 fsw if the diver is on oxygen.
4. Remain at treatment depth for at least 30 minutes.
5. If the diver is on air, resume decompression from treatment depth by multiplying subsequent air or oxygen stop times in the Air Decompression Table by 1.5. If recompression went deeper than the depth of the first stop on the original air decompression schedule, insert intervening stops in 10 fsw increments between the treatment depth and the original first stop depth equal to 1.5 times the original first stop time.
6. If the diver is undergoing treatment on oxygen at 40 fsw, return to the surface by multiplying the 30 and 20-fsw oxygen stop times by 1.5. If the original schedule did not call for a 30-fsw oxygen stop, insert a 30-fsw oxygen stop with a stop time equal to the 20-fsw stop time.
7. If the diver is undergoing treatment on oxygen at 30 fsw, return to the surface by multiplying the 20-fsw oxygen stop time by 1.5.
8. If the diver is symptom-free upon surfacing, place the diver on oxygen, transport to the nearest recompression chamber, and treat on Treatment Table 5. This requirement may be waived for dives conducted in remote locations that do not have recompression chambers within a reasonable travel distance. If the diver is not symptom-free upon surfacing, transport the diver to the nearest chamber and treat on Treatment Table 6.

9. If a recompression chamber is available on the dive station, the diving supervisor may elect to forego treatment with in-water recompression and surface the diver for treatment in the recompression chamber or treat the diver in the water for 30 minutes to relieve symptoms, then surface the diver for further treatment in the recompression chamber. In either case, the surface interval should be 5 minutes or less, and the diver should be considered to have Type II decompression sickness, even if the symptoms are Type I. After completing recompression treatment, observe the diver for at least 6 hours. If any symptoms recur, treat as a recurrence of Type II symptoms.

9-12.11.2 **Diver Leaving the Water.** If the diver indicates that he has decompression sickness and feels he cannot safely remain in the water:

1. Surface the diver at a moderate rate (not to exceed 30 fsw/min).

2. If a recompression chamber is on site, recompress the diver immediately. Guidance for treatment table selection and use is given in Chapter 20.

3. If a recompression chamber is not on site, follow the management guidance given in Volume 5.

9-13 DIVING AT ALTITUDE

Because of the reduced atmospheric pressure, dives conducted at altitude require more decompression than identical dives conducted at sea level. The air decompression tables, therefore, cannot be used as written. Some organizations calculate specific decompression tables for use at each altitude. An alternative approach is to correct the altitude dive to obtain the equivalent sea level dive, then determine the decompression requirement using standard tables. This procedure is commonly known as the "Cross Correction" technique and always yields a sea level dive that is deeper than the actual dive at altitude. A deeper sea level equivalent dive provides the extra decompression needed to offset effects of diving at altitude.

9-13.1 **Altitude Correction Procedure.** To apply the "Cross Correction" technique, two corrections must be made for altitude diving. First, the actual dive depth must be corrected to determine the sea level equivalent depth. Second, the decompression stops in the sea level equivalent depth table must be corrected for use at altitude. Strictly speaking, ascent rate should also be corrected, but this third correction can safely be ignored.

9-13.1.1 **Correction of Dive Depth.** The depth of the sea level equivalent dive is determined by multiplying the depth of the dive at altitude by the ratio of the atmospheric pressure at sea level to the atmospheric pressure at altitude.

$$\text{Equivalent Depth (fsw)} = \text{Altitude Depth (fsw)} \times \frac{\text{Pressure at Sea Level}}{\text{Pressure at Altitude}}$$

Example: A diver makes a dive to 60 fsw at an altitude of 5000 feet. The atmospheric pressure measured at 5000 feet is 843 millibars (0.832 ATA). Atmospheric pressure at sea level is assumed to be 1013 millibars (1.000 ATA). Sea level equivalent depth is then:

$$\text{Equivalent Depth (fsw)} = 60\ \text{fsw} \times \frac{1.000\ \text{ATA}}{0.832\ \text{ATA}} = 72.1\ \text{fsw}$$

9-13.1.2 **Correction of Decompression Stop Depth.** The depth of the corrected stop at altitude is calculated by multiplying the depth of a sea level equivalent stop by the ratio of the atmospheric pressure at altitude to the atmospheric pressure at sea level. [Note: this ratio is the inverse of the ratio in the formula above.]

$$\text{Altitude Stop Depth (fsw)} = \text{Sea Level Stop Depth (fsw)} \times \frac{\text{Pressure at Altitude}}{\text{Pressure at Sea Level}}$$

Example: A diver makes a dive at an altitude of 5000 feet. An equivalent sea level dive requires a decompression stop at 20 fsw. Stop depth used at altitude is then:

$$\text{Altitude Stop Depth (fsw)} = 20\ \text{fsw} \times \frac{0.832\ \text{ATA}}{1.000\ \text{ATA}} = 16.6\ \text{fsw}$$

To simplify calculations, Table 9-4 gives corrected sea level equivalent depths and equivalent stop depths for dives from 10–190 fsw and for altitudes from 1,000 to 10,000 feet in 1,000 foot increments. For exact calculations, refer to Chapter 2, Table 2-19 for atmospheric pressure at altitude.

WARNING **Table 9-4 cannot be used when diving with equipment that maintains a constant partial pressure of oxygen such as the MK 16 MOD 0 and the MK 16 MOD 1. Consult NAVSEA 00C for specific guidance when diving the MK 16 at altitudes greater than 1000 feet.**

9-13.2 **Need for Correction.** No correction is required for dives conducted at altitudes between sea level and 300 feet. The additional risk associated with these dives is minimal. At altitudes between 300 and 1000 feet, correction is required for dives deeper than 145 fsw (actual depth). At altitudes above 1000 feet, correction is required for all dives.

9-13.3 **Depth Measurement at Altitude.** The preferred method for measuring depth at altitude is a mechanical or electronic gauge that can be re-zeroed at the dive site. Once re-zeroed, no further correction of the reading is required.

When using a recompression chamber for decompression, zero the chamber depth gauges before conducting surface decompression.

Table 9-4. *Sea Level Equivalent Depth (fsw).*

Actual Depth (fsw)	Altitude (feet)									
	1000	2000	3000	4000	5000	6000	7000	8000	9000	10000
10	10	15	15	15	15	15	15	15	15	15
15	15	20	20	20	20	20	20	25	25	25
20	20	25	25	25	25	25	30	30	30	30
25	25	30	30	30	35	35	35	35	35	40
30	30	35	35	35	40	40	40	45	45	45
35	35	40	40	45	45	45	50	50	50	60
40	40	45	45	50	50	50	55	55	60	60
45	45	50	55	55	55	60	60	70	70	70
50	50	55	60	60	70	70	70	70	70	80
55	55	60	70	70	70	70	80	80	80	80
60	60	70	70	70	80	80	80	90	90	90
65	65	70	80	80	80	90	90	90	100	100
70	70	80	80	90	90	90	100	100	100	110
75	75	90	90	90	100	100	100	110	110	110
80	80	90	90	100	100	100	110	110	120	120
85	85	100	100	100	110	110	120	120	120	130
90	90	100	110	110	110	120	120	130	130	140
95	95	110	110	110	120	120	130	130	140	140
100	100	110	120	120	130	130	130	140	140	150
105	105	120	120	130	130	140	140	150	150	160
110	110	120	130	130	140	140	150	150	160	160
115	115	130	130	140	140	150	150	160	170	170
120	120	130	140	140	150	150	160	170	170	180
125	125	140	140	150	160	160	170	170	180	190
130	130	140	150	160	160	170	170	180	190	190
135	135	150	160	160	170	170	180	190	190	200
140	140	160	160	170	170	180	190	190	200	210
145	145	160	170	170	180	190	190	200	210	
150	160	170	170	180	190	190	200	210		
155	170	170	180	180	190	200	210			
160	170	180	180	190	200	200				
165	180	180	190	200	200					
170	180	190	190	200						
175	190	190	200							
180	190	200	210							
185	200	200								
190	200									
Table Water Stops	Equivalent Stop Depths (fsw)									
10	10	9	9	9	8	8	8	7	7	7
20	19	19	18	17	17	16	15	15	14	14
30	29	28	27	26	25	24	23	22	21	21
40	39	37	36	35	33	32	31	30	29	28
50	48	47	45	43	42	40	39	37	36	34
60	58	56	54	52	50	48	46	45	43	41

Note: ▬▬ = Exceptional Exposure Limit

Most mechanical depth gauges carried by divers have a sealed one-atmosphere reference and cannot be adjusted for altitude; thus they will read low throughout a dive at altitude. A correction factor of 1 fsw for every 1000 feet of altitude should be added to the reading of a sealed reference gauge before entering Table 9-4.

Pneumofathometers can be used at altitude. Add the pneumofathometer correction factor (Table 9-1) to the depth reading before entering Table 9-4. The pneumofathometer correction factors are unchanged at altitude.

A sounding line or fathometer may be used to measure the depth if a suitable depth gauge is not available. These devices measure the linear distance below the surface of the water, not the water pressure. Though fresh water is less dense than sea water, all dives will be assumed to be conducted in sea water, thus no corrections will be made based on water salinity. Enter Table 9-4 directly with the depth indicated on the line or fathometer.

9-13.4 **Equilibration at Altitude.** Upon ascent to altitude, two things happen. The body off-gases excess nitrogen to come into equilibrium with the lower partial pressure of nitrogen in the atmosphere. It also begins a series of complicated adjustments to the lower partial pressure of oxygen. The first process is called equilibration; the second is called acclimatization. Approximately twelve hours at altitude is required for equilibration. A longer period is required for full acclimatization.

If a diver begins a dive at altitude within 12 hours of arrival, the residual nitrogen left over from sea level must be taken into account. In effect, the initial dive at altitude can be considered a repetitive dive, with the first dive being the ascent from sea level to altitude. Table 9-5 gives the repetitive group associated with an initial ascent to altitude. Using this group and time at altitude before diving, enter the Residual Nitrogen Timetable for Repetitive Air Dives (Table 9-8) to determine the new repetitive group designator associated with that period of equilibration. Determine the sea level equivalent depth for your planned dive using Table 9-4. From your new repetitive group and sea level equivalent depth, determine the residual nitrogen time associated with the dive. Add this time to the actual bottom time of the dive. If the diver has spent enough time at altitude to desaturate beyond repetitive group A in Table 9-8, no addition of residual nitrogen time to bottom time is needed. The diver is "clean."

Example: A diver ascends rapidly to 6000 feet in a helicopter and begins a dive to 100 fsw 90 minutes later. How much residual nitrogen time should be added to the dive?

From Table 9-5, the repetitive group upon arrival at 6000 feet is Group E. During 90 minutes at altitude, the diver will desaturate to Group D. From Table 9-4, the sea level equivalent depth for a 100 fsw dive is 130 fsw. From Table 9-8, the residual nitrogen time for a 130 fsw dive in Group D is 11 minutes. The diver should add 11 minutes to the bottom time.

Table 9-5. Repetitive Groups Associated with Initial Ascent to Altitude.

Altitude (feet)	Repetitive Group
1000	A
2000	A
3000	B
4000	C
5000	D
6000	E
7000	F
8000	G
9000	H
10000	I

Table 9-5 can also be used when a diver who is fully equilibrated at one altitude ascends to and dives at a higher altitude. Enter Table 9-5 with the difference between the two altitudes to determine the initial repetitive group.

Example: Divers equilibrated at a base camp altitude of 6000 feet fly by helicopter to the dive site at 10,000 feet. The difference between the altitudes is 4000 feet. From Table 9-5, the initial repetitive group to be used at 10,000 feet is Group C.

WARNING **Altitudes above 10,000 feet can impose serious stress on the body resulting in significant medical problems while the acclimatization process takes place. Ascents to these altitudes must be slow to allow acclimatization to occur and prophylactic drugs may be required to prevent the occurrence of altitude sickness. These exposures should always be planned in consultation with a Diving Medical Officer. Commands conducting diving operations above 10,000 feet may obtain the appropriate decompression procedures from NAVSEA 00C.**

9-13.5 **Diving at Altitude Worksheet.** Figure 9-15 is a diving at altitude worksheet. To determine Sea Level Equivalent Depth (SLED) and corrected decompression stops for an altitude dive, follow these steps:

9-13.5.1 Corrections for Depth of Dive at Altitude and In-Water Stops.

Line 1. Determine the dive site altitude by referring to a map or measuring the barometric pressure. From Table 9-4, enter the altitude in feet that is equal to or next greater than the altitude at the dive site.

Line 2. Enter the actual depth of the dive in feet of sea water.

NOTE **Refer to paragraph 9-13.3 to correct divers' depth gauge readings to actual depths at altitude.**

Date: ____________

DIVING AT ALTITUDE WORKSHEET

Actual Dive Site Altitude____________ feet

1. Altitude from Table 9-4 ________ feet
2. Actual Depth of Dive (Corrected per Section 9-13.3) ________ fsw
3. Sea Level Equivalent Depth from Table 9-4 ________ SLED
4. Repetitive Group from Table 9-5 ________
5. Time at Altitude ________ hrs ________ min
6. New Repetitive Group Designator from Table 9-8 ________
7. Residual Nitrogen Time ________ min
8. Planned Bottom Time + ________ min
9. Equivalent Single Dive Time = ________ min
10. Decompression Mode

    □ No-Decompression □ In-water Air/Oxygen Decompression

    □ In-water Air Decompression □ Surface Decompression Using Oxygen

11. Table/Schedule ________ / ________
12. Decompression Schedule

Sea Level Stop Depth	Altitude Stop Depth	Water Stop Time	Chamber Stop Time
60 fsw	________ fsw	________ min	
50 fsw	________ fsw	________ min	________ min *
40 fsw	________ fsw	________ min	________ min *
30 fsw	________ fsw	________ min	________ min *
20 fsw	________ fsw	________ min	

13. Repetitive Group Designator ______

* Chamber stops on $SurDO_2$ will be at 50, 40, and 30 fsw

Figure 9-15. Diving at Altitude Worksheet.

Line 3. Read Table 9-4 vertically down the Actual Depth Column. Select a depth that is equal to or next greater than the actual depth. Reading horizontally, select the Sea Level Equivalent Depth corresponding to an altitude equal to or next greater than that of your dive site.

9-13.5.2 **Corrections for Equilibration.**

Line 4. Enter the Repetitive Group upon arrival at altitude from Table 9-5 for the altitude listed on Line 1.

Line 5. Record the time in hours and minutes spent equilibrating at altitude prior to the dive. If the equilibration time is longer than the time needed to desaturate beyond Repetitive Group A in Table 9-8, proceed to Step 7 and enter zero.

Line 6. Using Table 9-8, determine the Repetitive Group at the end of the pre-dive equilibration interval.

Line 7. Using Table 9-8, determine the Residual Nitrogen Time for the new repetitive group designator from Line 6 and the Sea Level Equivalent Depth from Line 3.

Line 8. Enter the planned bottom time.

Line 9. Add the bottom time and the residual nitrogen time to obtain the Equivalent Single Dive Time.

Line 10. Select the mode of decompression to be used, e.g., in-water air/oxygen.

Line 11. Enter the Schedule from the Air Decompression Table using the Sea Level Equivalent Depth from Line 3 and the Equivalent Single Dive Time from Line 9.

Line 12. Using the lower section of Table 9-4, read down the Table Water Stops column on the left to the decompression stop(s) given in the Sea Level Equivalent Depth Table/Schedule. Read horizontally to the altitude column. Record the corresponding altitude stop depths on the worksheet.

NOTE **For surface decompression dives on oxygen, the chamber stops are not adjusted for altitude. Enter the same depths as at sea level. Keeping chamber stop depths the same as sea level provides an extra decompression benefit for the diver on oxygen.**

Line 13. Record the Repetitive Group Designator at the end of the dive.

NOTE **Follow all decompression table procedures for ascent and descent.**

Example: Five hours after arriving at an altitude of 7750 feet, divers make a 60-minute air dive to a gauge depth of 75 fsw. Depth is measured with a pneumofathometer having a non-adjustable gauge with a fixed reference pressure of one atmosphere. Surface decompression with oxygen will be used for decompression. What is the proper decompression schedule?

The altitude is first rounded up to 8000 feet. A depth correction of +8 fsw must be added to the maximum depth recorded on the fixed reference gauge. A pneumofathometer correction factor of + 1 fsw must also be added. The diver's actual depth is 84 fsw. Table 9-4 is entered at an actual depth of 85 fsw. The Sea Level Equivalent Depth for 8000 feet of altitude is 120 fsw. The repetitive group upon arrival at altitude from Table 9-5 is Group G. This decays to Group B during the five hours at altitude pre-dive. The residual nitrogen time for Group B at 120 fsw is 7 minutes. The Equivalent Single Dive Time therefore is 67 minutes. The appropriate schedule from the Air Decompression Table is 120 fsw for 70 minutes. By the schedule, a 12-minute water stop on air at 40 fsw is required followed by two and one half oxygen periods in the chamber. The water stop is taken at a depth of 30 fsw. The chamber stops are taken at depths of 50 and 40 fsw.

Figure 9-16 shows the filled-out Diving at Altitude Worksheet for this dive. Figure 9-17 shows the filled-out Diving Chart.

9-13.6 **Repetitive Dives.** Repetitive dives may be conducted at altitude. The procedure is identical to that at sea level, with the exception that the sea level equivalent dive depth is always used to replace the actual dive depth. Figure 9-18 is a Repetitive Dive at Altitude Worksheet.

Example: Fourteen hours after ascending to an altitude of 7750 feet, divers make an 82-fsw 50-minute MK 21 dive using in-water air/oxygen decompression. Depth is measured with a pneumofathometer having a depth gauge adjustable for altitude. After two hours and ten minutes on the surface, they make a second dive to 79 fsw for 18 minutes and decompress using surface decompression on oxygen. What is the proper decompression schedule for the second dive?

The altitude is first rounded up to 8000 feet. For the first dive, a depth correction of +1 fsw must be added to the 82 fsw pneumofathometer reading. The divers' actual depth on the first dive is 83 fsw. Table 9-4 is entered at an actual depth of 85 fsw. The Sea Level Equivalent Depth for the first dive is 120 fsw. The repetitive group designation upon completion of the 50 minute dive is Group Z. This decays to Group N during the 2 hour 10 minute surface interval.

The actual depth of the second dive is 80 fsw (79 fsw plus a 1 fsw pneumofathometer correction factor). Table 9-4 is entered at an actual depth of 80 fsw. The Sea Level Equivalent Depth for the second dive is 110 fsw. The residual nitrogen time for Group N at 110 fsw is 42 minutes. The equivalent single dive time therefore is 60 minutes. The appropriate surface decompression schedule is 110 fsw for 60 minutes. This schedule does not require any water stops. The divers spend 60 minutes on oxygen (2 oxygen periods) at 50 and 40 fsw in the recompression chamber.

Date: 23 Oct 07

DIVING AT ALTITUDE WORKSHEET

Actual Dive Site Altitude 7,750 feet

1. Altitude from Table 9-4 — 8,000 feet
2. Actual Depth of Dive (Corrected per Section 9-13.3) — 75+8+1=84 fsw
3. Sea Level Equivalent Depth from Table 9-4 — 120 SLED
4. Repetitive Group from Table 9-5 — G
5. Time at Altitude — 5 hrs ______ min
6. New Repetitive Group Designator from Table 9-8 — B
7. Residual Nitrogen Time — 7 min
8. Planned Bottom Time — + 60 min
9. Equivalent Single Dive Time — = 67 min
10. Decompression Mode

    ☐ No-Decompression ☐ In-water Air/Oxygen Decompression

    ☐ In-water Air Decompression ☑ Surface Decompression Using Oxygen
11. Table/Schedule 120/70
12. Decompression Schedule

Sea Level Stop Depth	Altitude Stop Depth	Water Stop Time	Chamber Stop Time
60 fsw	______ fsw	______ min	
50 fsw	______ fsw	______ min	15 min*
40 fsw	30 fsw	12 min	15+5+30+5+15 min*
30 fsw	______ fsw	______ min	______ min*
20 fsw	______ fsw	______ min	

13. Repetitive Group Designator ______

* Chamber stops on SurDO_2 will be at 50, 40, and 30 fsw

Figure 9-16. Completed Diving at Altitude Worksheet.

1246 *ALTITUDE 8000*		
Date: 5 Sept 07	Type of Dive: (AIR) HeO_2	
Diver 1: ND1 Chaisson	Diver 2: ND2 Hutcheson	Standby: ND1 Collins
Rig: MK-37 PSIG: 2900 O_2%:	Rig: MK-37 PSIG: 2900 O_2%:	Rig: MK-37 PSIG: 2900 O_2%:
Diving Supervisor: NDCM Orns	Chartman: ND1 Saurez	Bottom Mix:

EVENT	STOP TIME	CLOCK TIME
LS or 20 fsw		1000
RB		1002
LB		1100
R 1st Stop		1102
190 fsw		
180 fsw		
170 fsw		
160 fsw		
150 fsw		
140 fsw		
130 fsw		
120 fsw		
110 fsw		
100 fsw		
90 fsw		
80 fsw		
70 fsw		
60 fsw		
50 fsw		
40 fsw		
30 fsw	:12(AIR)	1114
20 fsw		
RS		1115
RB CHAMBER		1119
50 fsw chamber	:15	1134
40 fsw chamber	:15+:5+:30+:5+:15	1244
30 fsw chamber		
RS CHAMBER		1246

TDT	TTD
1:46	2:46

EVENT	TIME/DEPTH
Descent Time (Water)	:02
Stage Depth (fsw)	84
Maximum Depth (fsw)	75+8+1=84
Total Bottom Time	:60+:07=:67
Table/Schedule	120/70 SLED
Time to 1st Stop (Actual)	:01::32
Time to 1st Stop (Planned)	:01::30
Delay to 1st Stop	::02
Travel/Shift/Vent Time	
Ascent Time-Water/SurD (Actual)	:01
Undress Time-SurD (Actual)	:03::20
Descent Chamber-SurD (Actual)	::40
Total SurD Surface Interval	:05
Ascent Time–Chamber (Actual)	:01::20

HOLDS ON DESCENT

DEPTH	PROBLEM

DELAYS ON ASCENT

DEPTH	PROBLEM

DECOMPRESSION PROCEDURES USED

AIR

- ☐ In-water Air decompression
- ☐ In-water Air/O_2 decompression
- ☑ $SurDO_2$

HeO_2

- ☐ In-water HeO_2/O_2 decompression
- ☐ $SurDO_2$

REPETITIVE GROUP: No repet

Remarks:

Figure 9-17. Completed Air Diving Chart: Dive at Altitude.

REPETITIVE DIVE AT ALTITUDE WORKSHEET

Date: ______________

1. PREVIOUS DIVE

Decompression Mode

_____ minutes ☐ No-Decompression ☐ In-water Air/Oxygen Decompression

_____ SLED ☐ In-water Air Decompression ☐ Surface Decompression Using Oxygen

_____ Repetitive Group Letter Designator

2. SURFACE INTERVAL

_____ hours ______ minutes on surface

_____ repetitive group from item 1 above

_____ new repetitive group letter designator from Residual Nitrogen Timetable

3. RESIDUAL NITROGEN TIME FOR REPETITIVE DIVE

Altitude from Table 9-4 _________ feet

Actual Depth of Dive (corrected per section 9-13.3) _________ fsw

Sea Level Equivalent Depth of repetitive dive from Table 9-4 _________ SLED

_____ new repetitive group letter designator from item 2 above

______ minutes, residual nitrogen time from Residual Nitrogen Timetable

4. EQUIVALENT SINGLE DIVE TIME

_____ minutes, residual nitrogen time from item 3 above

\+ _____ minutes, actual bottom time of repetitive dive

= _____ minutes, equivalent single dive time

5. DECOMPRESSION FOR REPETITIVE DIVE

_____ SLED of repetitive dive

_____ minutes, equivalent single dive time from item 4 above

Decompression Mode (check one)

☐ No-Decompression ☐ In-water Air/Oxygen Decompression

☐ In-water Air Decompression ☐ Surface Decompression Using Oxygen

___________ schedule used (depth/time)

Sea Level Stop Depth	Altitude Stop Depth	Water Stop Time	Chamber Stop Time
60 fsw	_________ fsw	_________ min	
50 fsw	_________ fsw	_________ min	_________ min*
40 fsw	_________ fsw	_________ min	_________ min*
30 fsw	_________ fsw	_________ min	_________ min*
20 fsw	_________ fsw	_________ min	

13. Repetitive Group Letter Designator ______

* Chamber stops on $SurDO_2$ will be at 50, 40, and 30 fsw

Figure 9-18. Repetitive Dive at Altitude Worksheet.

Figure 9-19 shows the filled-out Repetitive Dive at Altitude Worksheet for these two dives. Figure 9-20 and Figure 9-21 show the filled-out Diving Charts for the first and second dives.

9-14 ASCENT TO ALTITUDE AFTER DIVING / FLYING AFTER DIVING

Leaving the dive site may require temporary ascent to a higher altitude. For example, divers may drive over a mountain pass at higher altitude or leave the dive site by air. Ascent to altitude after diving increases the risk of decompression sickness because of the additional reduction in atmospheric pressure. The higher the altitude, the greater the risk. (Pressurized commercial airline flights are addressed in Note 3 of Table 9-6).

Table 9-6 gives the surface interval (hours:minutes) required before making a further ascent to altitude. The surface interval depends on the planned increase in altitude and the highest repetitive group designator obtained in the previous 24-hour period. Enter the table with the highest repetitive group designator obtained in the previous 24-hour period. Read the required surface interval from the column for the planned change in altitude.

Example: A diver surfaces from a 60 fsw for 60 minutes no-decompression dive at sea level in Repetitive Group K. After a surface interval of 6 hours 10 minutes, the diver makes a second dive to 30 fsw for 20 minutes placing him in Repetitive Group F. He plans to fly home in a commercial aircraft in which the cabin pressure is controlled at 8000 feet. What is the required interval before flying?

The planned increase in altitude is 8000 feet. Because the diver has made two dives in the previous 24-hour period, you must use the highest repetitive group designator of the two dives. Enter Table 9-6 at 8000 feet and read down to Repetitive Group K. The diver must wait 15 hours 35 minutes after completion of the second dive before flying.

Example: Upon completion of a dive at an altitude of 4000 feet, the diver plans to ascend to 7500 feet in order to cross a mountain pass. The diver's repetitive group upon surfacing is Group G. What is the required surface interval before crossing the pass?

The planned increase in altitude is 3500 feet. Enter Table 9-6 at 4000 feet and read down to Repetitive Group I. The diver must delay 2 hours and 45 minutes before crossing the pass.

Example: Upon completion of a dive at 2000 feet, the diver plans to fly home in an un-pressurized aircraft at 5000 feet. The diver's repetitive group designator upon surfacing is Group K. What is the required surface interval before flying?

The planned increase in altitude is 3000 feet. Enter Table 9-6 at 3000 feet and read down to Repetitive Group K. The diver must delay 3 hours 47 minutes before taking the flight.

REPETITIVE DIVE AT ALTITUDE WORKSHEET Date: 23 Oct 07

1. **PREVIOUS DIVE**

Decompression Mode

50 minutes ☐ No-Decompression ☑ In-water Air/Oxygen Decompression

120 SLED ☐ In-water Air Decompression ☐ Surface Decompression Using Oxygen

Z Repetitive Group Letter Designation

2. **SURFACE INTERVAL**

2 hours 10 minutes on surface

Z repetitive group from item 1 above

N new repetitive group letter designator from Residual Nitrogen Timetable

3. **RESIDUAL NITROGEN TIME FOR REPETITIVE DIVE**

Altitude from Table 9-4 8,000 feet

Actual Depth of Dive (corrected per section 9-13.3) 79+1=80 fsw

Sea Level Equivalent Depth of repetitive dive from Table 9-4 110 SLED

N new repetitive group letter designator from item 2 above

42 minutes, residual nitrogen time from Residual Nitrogen Timetable

4. **EQUIVALENT SINGLE DIVE TIME**

42 minutes, residual nitrogen time from item 3 above

\+ 18 minutes, actual bottom time of repetitive dive

= 60 minutes, equivalent single dive time

5. **DECOMPRESSION FOR REPETITIVE DIVE**

110 SLED of repetitive dive

60 minutes, equivalent single dive time from item 4 above

Decompression Mode (check one)

☐ No-Decompression ☐ In-water Air/Oxygen Decompression

☐ In-water Air Decompression ☑ Surface Decompression Using Oxygen

110/60 schedule used (depth/time)

Sea Level Stop Depth	Altitude Stop Depth	Water Stop Time	Chamber Stop Time
60 fsw	____ fsw	____ min	
50 fsw	____ fsw	____ min	15 min*
40 fsw	____ fsw	____ min	15+3+30 min*
30 fsw	____ fsw	____ min	____ min*
20 fsw	____ fsw	____ min	

13. Repetitive Group Letter Designator ____

* Chamber stops on $SurDO_2$ will be at 50, 40, and 30 fsw

Figure 9-19. Completed Repetitive Dive at Altitude Worksheet.

1038 *ALTITUDE 8000*		
Date: 23 Oct 07	Type of Dive: (AIR) circled / HeO_2	
Diver 1: ND1 Sullivan	Diver 2: ND2 Schleef	Standby: ND2 Bartley
Rig: MK-21 PSIG: 2900 O_2%:	Rig: MK-21 PSIG: 2900 O_2%:	Rig: MK-21 PSIG: 2900 O_2%:
Diving Supervisor: NDCM Van Horn	Chartman: ND2 Bradley	Bottom Mix:

EVENT	STOP TIME	CLOCK TIME
LS or 20 fsw		0900
RB		0902
LB		0950
R 1st Stop		0952
190 fsw		
180 fsw		
170 fsw		
160 fsw		
150 fsw		
140 fsw		
130 fsw		
120 fsw		
110 fsw		
100 fsw		
90 fsw		
80 fsw		
70 fsw		
60 fsw		
50 fsw		
40 fsw		
30 fsw	:02+:05	0959
20 fsw	:15+:05+:18	1037
RS		1038
RB CHAMBER		
50 fsw chamber		
40 fsw chamber		
30 fsw chamber		
RS CHAMBER		

TDT	TTD
:48	1:38

EVENT	TIME/DEPTH
Descent Time (Water)	:02
Stage Depth (fsw)	82
Maximum Depth (fsw)	82+1=83
Total Bottom Time	:50
Table/Schedule	120/50 SLED
Time to 1st Stop (Actual)	:01::44
Time to 1st Stop (Planned)	:01::44
Delay to 1st Stop	
Travel/Shift/Vent Time	:02
AscentTime-Water/SurD (Actual)	::45
Undress Time-SurD (Actual)	
Descent Chamber-SurD (Actual)	
Total SurD Surface Interval	
Ascent Time–Chamber (Actual)	

HOLDS ON DESCENT

DEPTH	PROBLEM

DELAYS ON ASCENT

DEPTH	PROBLEM

DECOMPRESSION PROCEDURES USED

AIR
- ☐ In-water Air decompression
- ☑ In-water Air/O_2 decompression
- ☐ $SurDO_2$

HeO_2
- ☐ In-water HeO_2/O_2 decompression
- ☐ $SurDO_2$

REPETITIVE GROUP: Z

Remarks:

Figure 9-20. Completed Air Diving Chart: First Dive of Repetitive Dive Profile at Altitude.

1420 *ALTITUDE 8000*		
Date: 23 Oct 07	Type of Dive: (AIR) HeO_2	
Diver 1: ND1 Sullivan	Diver 2: ND2 Schleef	Standby: ND2 Bartley
Rig: MK-21 PSIG: 2900 O_2%:	Rig: MK-21 PSIG: 2900 O_2%:	Rig: MK-21 PSIG: 2900 O_2%:
Diving Supervisor: NDCM Van Horn	Chartman: ND2 Bradley	Bottom Mix:

EVENT	STOP TIME	CLOCK TIME
LS or 20 fsw		1248
RB		1250
LB		1306
R 1st Stop		1309
190 fsw		
180 fsw		
170 fsw		
160 fsw		
150 fsw		
140 fsw		
130 fsw		
120 fsw		
110 fsw		
100 fsw		
90 fsw		
80 fsw		
70 fsw		
60 fsw		
50 fsw		
40 fsw		
30 fsw		
20 fsw		
RS		1309
RB CHAMBER		1313
50 fsw chamber	:15	1328
40 fsw chamber	:15+:5+:30	1418
30 fsw chamber		
RS CHAMBER		1420

TDT	TTD
1:14	1:32

EVENT	TIME/DEPTH
Descent Time (Water)	:02
Stage Depth (fsw)	79
Maximum Depth (fsw)	79+1=80
Total Bottom Time	:18 + :42 = 60
Table/Schedule	110/60 SLED
Time to 1st Stop (Actual)	:02::40
Time to 1st Stop (Planned)	:02::38
Delay to 1st Stop	::02
Travel/Shift/Vent Time	
AscentTime-Water/SurD (Actual)	:01
Undress Time-SurD (Actual)	:02::30
Descent Chamber-SurD (Actual)	::50
Total SurD Surface Interval	:04::20
Ascent Time–Chamber (Actual)	:01::20

HOLDS ON DESCENT

DEPTH	PROBLEM

DELAYS ON ASCENT

DEPTH	PROBLEM

DECOMPRESSION PROCEDURES USED

AIR

- ☐ In-water Air decompression
- ☐ In-water Air/O_2 decompression
- ☑ $SurDO_2$

HeO_2

- ☐ In-water HeO_2/O_2 decompression
- ☐ $SurDO_2$

REPETITIVE GROUP: Z

Remarks:

Figure 9-21. Completed Air Diving Chart: Second Dive of Repetitive Dive Profile at Altitude.

Table 9-6. *Required Surface Interval Before Ascent to Altitude After Diving.*

Repetitive Group Designator	Increase in Altitude (feet)									
	1000	2000	3000	4000	5000	6000	7000	8000	9000	10000
A	0:00	0:00	0:00	0:00	0:00	0:00	0:00	**0:00**	0:00	0:00
B	0:00	0:00	0:00	0:00	0:00	0:00	0:00	**0:00**	0:00	1:42
C	0:00	0:00	0:00	0:00	0:00	0:00	0:00	**0:00**	1:48	6:23
D	0:00	0:00	0:00	0:00	0:00	0:00	0:00	**1:45**	5:24	9:59
E	0:00	0:00	0:00	0:00	0:00	0:00	1:37	**4:39**	8:18	12:54
F	0:00	0:00	0:00	0:00	0:00	1:32	4:04	**7:06**	10:45	15:20
G	0:00	0:00	0:00	0:00	1:19	3:38	6:10	**9:13**	12:52	17:27
H	0:00	0:00	0:00	1:06	3:10	5:29	8:02	**11:04**	14:43	19:18
I	0:00	0:00	0:56	2:45	4:50	7:09	9:41	**12:44**	16:22	20:58
J	0:00	0:41	2:25	4:15	6:19	8:39	11:11	**14:13**	17:52	22:27
K	0:30	2:03	3:47	5:37	7:41	10:00	12:33	**15:35**	19:14	23:49
L	1:45	3:18	5:02	6:52	8:56	11:15	13:48	**16:50**	20:29	25:04
M	2:54	4:28	6:12	8:01	10:06	12:25	14:57	**18:00**	21:38	26:14
N	3:59	5:32	7:16	9:06	11:10	13:29	16:02	**19:04**	22:43	27:18
O	4:59	6:33	8:17	10:06	12:11	14:30	17:02	**20:05**	23:43	28:19
Z	5:56	7:29	9:13	11:03	13:07	15:26	17:59	**21:01**	24:40	29:15
Exceptional Exposure	Wait 48 hours before ascent									

NOTE 1 **When using Table 9-6, use the highest repetitive group designator obtained in the previous 24-hour period.**

NOTE 2 **Table 9-6 may only be used when the maximum altitude achieved is 10,000 feet or less. For ascents above 10,000 feet, consult NAVSEA 00C for guidance.**

NOTE 3 **The cabin pressure in commercial aircraft is maintained at a constant value regardless of the actual altitude of the flight. Though cabin pressure varies somewhat with aircraft type, the nominal value is 8,000 feet. For commercial flights, use a final altitude of 8,000 feet to compute the required surface interval before flying.**

NOTE 4 **No surface interval is required before taking a commercial flight if the dive site is at 8,000 feet or higher. In this case, flying results in an increase in atmospheric pressure rather than a decrease.**

NOTE 5 **For ascent to altitude following a non-saturation helium-oxygen dive, wait 12 hours if the dive was a no-decompression dive. Wait 24 hours if the dive was a decompression dive.**

Table 9-7. *No-Decompression Limits and Repetitive Group Designators for No-Decompression Air Dives.*

Depth (fsw)	No-Stop Limit	Repetitive Group Designation															
		A	B	C	D	E	F	G	H	I	J	K	L	M	N	O	Z
10	Unlimited	57	101	158	245	426	*										
15	Unlimited	36	60	88	121	163	217	297	449	*							
20	Unlimited	26	43	61	82	106	133	165	205	256	330	461	*				
25	595	20	33	47	62	78	97	117	140	166	198	236	285	354	469	595	
30	371	17	27	38	50	62	76	91	107	125	145	167	193	223	260	307	371
35	232	14	23	32	42	52	63	74	87	100	115	131	148	168	190	215	232
40	163	12	20	27	36	44	53	63	73	84	95	108	121	135	151	163	
45	125	11	17	24	31	39	46	55	63	72	82	92	102	114	125		
50	92	9	15	21	28	34	41	48	56	63	71	80	89	92			
55	74	8	14	19	25	31	37	43	50	56	63	71	74				
60	60	7	12	17	22	28	33	39	45	51	57	60					
70	48	6	10	14	19	23	28	32	37	42	47	48					
80	39	5	9	12	16	20	24	28	32	36	39						
90	30	4	7	11	14	17	21	24	28	30							
100	25	4	6	9	12	15	18	21	25								
110	20	3	6	8	11	14	16	19	20								
120	15	3	5	7	10	12	15										
130	10	2	4	6	9	10											
140	10	2	4	6	8	10											
150	5	2	3	5													
160	5		3	5													
170	5			4	5												
180	5			4	5												
190	5			3	5												

* Highest repetitive group that can be achieved at this depth regardless of bottom time.

Table 9-8. *Residual Nitrogen Time Table for Repetitive Air Dives.*

Locate the diver's repetitive group designation from his previous dive along the diagonal line above the table. Read horizontally to the interval in which the diver's surface interval lies.

Next, read vertically downward to the new repetitive group designation. Continue downward in this same column to the row that represents the depth of the repetitive dive. The time given at the intersection is residual nitrogen time, in minutes, to be applied to the repetitive dive.

* Dives following surface intervals longer than this are not repetitive dives. Use actual bottom times in the Air Decompression Tables to compute decompression for such dives.

Repetitive Group at Beginning of Surface Interval

Group at beginning	Z	O	N	M	L	K	J	I	H	G	F	E	D	C	B	A
A																:10 – 2:20 *
B															:10 – 1:16	1:17 – 3:36 *
C														:10 – :55	:56 – 2:11	2:12 – 4:31 *
D													:10 – :52	:53 – 1:47	1:48 – 3:03	3:04 – 5:23 *
E												:10 – :52	:53 – 1:44	1:45 – 2:39	2:40 – 3:55	3:56 – 6:15 *
F											:10 – :52	:53 – 1:44	1:45 – 2:37	2:38 – 3:31	3:32 – 4:48	4:49 – 7:08 *
G										:10 – :52	:53 – 1:44	1:45 – 2:37	2:38 – 3:29	3:30 – 4:23	4:24 – 5:40	5:41 – 8:00 *
H									:10 – :52	:53 – 1:44	1:45 – 2:37	2:38 – 3:29	3:30 – 4:21	4:22 – 5:16	5:17 – 6:32	6:33 – 8:52 *
I								:10 – :52	:53 – 1:44	1:45 – 2:37	2:38 – 3:29	3:30 – 4:21	4:22 – 5:13	5:14 – 6:08	6:09 – 7:24	7:25 – 9:44 *
J							:10 – :52	:53 – 1:44	1:45 – 2:37	2:38 – 3:29	3:30 – 4:21	4:22 – 5:13	5:14 – 6:06	6:07 – 7:00	7:01 – 8:16	8:17 – 10:36 *
K						:10 – :52	:53 – 1:44	1:45 – 2:37	2:38 – 3:29	3:30 – 4:21	4:22 – 5:13	5:14 – 6:06	6:07 – 6:58	6:59 – 7:52	7:53 – 9:09	9:10 – 11:29 *
L					:10 – :52	:53 – 1:44	1:45 – 2:37	2:38 – 3:29	3:30 – 4:21	4:22 – 5:13	5:14 – 6:06	6:07 – 6:58	6:59 – 7:50	7:51 – 8:44	8:45 – 10:01	10:02 – 12:21 *
M				:10 – :52	:53 – 1:44	1:45 – 2:37	2:38 – 3:29	3:30 – 4:21	4:22 – 5:13	5:14 – 6:06	6:07 – 6:58	6:59 – 7:50	7:51 – 8:42	8:43 – 9:37	9:38 – 10:53	10:54 – 13:13 *
N			:10 – :52	:53 – 1:44	1:45 – 2:37	2:38 – 3:29	3:30 – 4:21	4:22 – 5:13	5:14 – 6:06	6:07 – 6:58	6:59 – 7:50	7:51 – 8:42	8:43 – 9:34	9:35 – 10:29	10:30 – 11:45	11:46 – 14:05 *
O		:10 – :52	:53 – 1:44	1:45 – 2:37	2:38 – 3:29	3:30 – 4:21	4:22 – 5:13	5:14 – 6:06	6:07 – 6:58	6:59 – 7:50	7:51 – 8:42	8:43 – 9:34	9:35 – 10:27	10:28 – 11:21	11:22 – 12:37	12:38 – 14:58 *
Z	:10 – :52	:53 – 1:44	1:45 – 2:37	2:38 – 3:29	3:30 – 4:21	4:22 – 5:13	5:14 – 6:06	6:07 – 6:58	6:59 – 7:50	7:51 – 8:42	8:43 – 9:34	9:35 – 10:27	10:28 – 11:19	11:20 – 12:13	12:14 – 13:30	13:31 – 15:50 *

Repetitive Group at the End of the Surface Interval

Dive Depth	Z	O	N	M	L	K	J	I	H	G	F	E	D	C	B	A
10	**	**	**	**	**	**	**	**	**	**	**	427	246	159	101	58
15	**	**	**	**	**	**	**	**	450	298	218	164	122	89	61	37
20	**	**	**	**	**	462	331	257	206	166	134	106	83	62	44	27
25	†	†	470	354	286	237	198	167	141	118	98	79	63	48	34	21
30	372	308	261	224	194	168	146	126	108	92	77	63	51	39	28	18
35	245	216	191	169	149	132	116	101	88	75	64	53	43	33	24	15
40	188	169	152	136	122	109	97	85	74	64	55	45	37	29	21	13
45	154	140	127	115	104	93	83	73	64	56	48	40	32	25	18	12
50	131	120	109	99	90	81	73	65	57	49	42	35	29	23	17	11
55	114	105	96	88	80	72	65	58	51	44	38	32	26	20	15	10
60	101	93	86	79	72	65	58	52	46	40	35	29	24	19	14	9
70	83	77	71	65	59	54	49	44	39	34	29	25	20	16	12	8
80	70	65	60	55	51	46	42	38	33	29	25	22	18	14	10	7
90	61	57	52	48	44	41	37	33	29	26	22	19	16	12	9	6
100	54	50	47	43	40	36	33	30	26	23	20	17	14	11	8	5
110	48	45	42	39	36	33	30	27	24	21	18	16	13	10	8	5
120	44	41	38	35	32	30	27	24	22	19	17	14	12	9	7	5
130	40	37	35	32	30	27	25	22	20	18	15	13	11	9	6	4
140	37	34	32	30	27	25	23	21	19	16	14	12	10	8	6	4
150	34	32	30	28	26	23	21	19	17	15	13	11	9	8	6	4
160	32	30	28	26	24	22	20	18	16	14	13	11	9	7	5	4
170	30	28	26	24	22	21	19	17	15	14	12	10	8	7	5	3
180	28	26	25	23	21	19	18	16	14	13	11	10	8	6	5	3
190	26	25	23	22	20	18	17	15	14	12	11	9	8	6	5	3

Residual Nitrogen Times (Minutes)

** Residual Nitrogen Time cannot be determined using this table (see paragraph 9-9.1 subparagraph 8 for instructions).

† Read vertically downward to the 30 fsw repetitive dive depth. Use the corresponding residual nitrogen times to compute the equivalent single dive time. Decompress using the 30 fsw air decompression table.

Table 9-9. *Air Decompression Table.*
(DESCENT RATE 75 FPM—ASCENT RATE 30 FPM)

Bottom Time (min)	Time to First Stop (M:S)	Gas Mix	DECOMPRESSION STOPS (FSW) Stop times (min) include travel time, except first air and first O_2 stop									Total Ascent Time (M:S)	Chamber O_2 Periods	Repet Group
			100	90	80	70	60	50	40	30	20			
30 FSW														
371	1:00	AIR									0	1:00	0	Z
		AIR/O_2									**0**	1:00		
380	0:20	AIR									5	6:00	0.5	Z
		AIR/O_2									**1**	2:00		
In-Water Air/O_2 Decompression or SurDO_2 Recommended ----------														
420	0:20	AIR									22	23:00	0.5	Z
		AIR/O_2									**5**	6:00		
480	0:20	AIR									42	43:00	0.5	
		AIR/O_2									**9**	10:00		
540	0:20	AIR									71	72:00	1	
		AIR/O_2									**14**	15:00		
Exceptional Exposure: In-Water Air Decompression ------------ In-Water Air/O_2 Decompression or SurDO_2 Required ----------														
600	0:20	AIR									92	93:00	1	
		AIR/O_2									**19**	20:00		
660	0:20	AIR									120	121:00	1	
		AIR/O_2									**22**	23:00		
720	0:20	AIR									158	159:00	1	
		AIR/O_2									**27**	28:00		
35 FSW														
232	1:10	AIR									0	1:10	0	Z
		AIR/O_2									**0**	1:10		
240	0:30	AIR									4	5:10	0.5	Z
		AIR/O_2									**2**	3:10		
In-Water Air/O_2 Decompression or SurDO_2 Recommended ----------														
270	0:30	AIR									28	29:10	0.5	Z
		AIR/O_2									**7**	8:10		
300	0:30	AIR									53	54:10	0.5	Z
		AIR/O_2									**13**	14:10		
330	0:30	AIR									71	72:10	1	Z
		AIR/O_2									**18**	19:10		
360	0:30	AIR									88	89:10	1	
		AIR/O_2									**22**	23:10		
Exceptional Exposure: In-Water Air Decompression ------------ In-Water Air/O_2 Decompression or SurDO_2 Required ----------														
420	0:30	AIR									134	135:10	1.5	
		AIR/O_2									**29**	30:10		
480	0:30	AIR									173	174:10	1.5	
		AIR/O_2									**38**	44:10		
540	0:30	AIR									228	229:10	2	
		AIR/O_2									**45**	51:10		
600	0:30	AIR									277	278:10	2	
		AIR/O_2									**53**	59:10		
660	0:30	AIR									314	315:10	2.5	
		AIR/O_2									**63**	69:10		
720	0:30	AIR									342	343:10	3	
		AIR/O_2									**71**	82:10		

Table 9-9. *Air Decompression Table (Continued).*
(DESCENT RATE 75 FPM—ASCENT RATE 30 FPM)

Bottom Time (min)	Time to First Stop (M:S)	Gas Mix	DECOMPRESSION STOPS (FSW) Stop times (min) include travel time, except first air and first O_2 stop 100	90	80	70	60	50	40	30	20	Total Ascent Time (M:S)	Chamber O_2 Periods	Repet Group
40 FSW														
163	1:20	AIR									0	1:20	0	O
		AIR/O_2									**0**	1:20		
170	0:40	AIR									6	7:20	0.5	O
		AIR/O_2									**2**	3:20		
180	0:40	AIR									14	15:20	0.5	Z
		AIR/O_2									**5**	6:20		
In-Water Air/O_2 Decompression or SurDO_2 Recommended														
190	0:40	AIR									21	22:20	0.5	Z
		AIR/O_2									**7**	8:20		
200	0:40	AIR									27	28:20	0.5	Z
		AIR/O_2									**9**	10:20		
210	0:40	AIR									39	40:20	0.5	Z
		AIR/O_2									**11**	12:20		
220	0:40	AIR									52	53:20	0.5	Z
		AIR/O_2									**12**	13:20		
230	0:40	AIR									64	65:20	1	Z
		AIR/O_2									**16**	17:20		
240	0:40	AIR									75	76:20	1	Z
		AIR/O_2									**19**	20:20		
Exceptional Exposure: In-Water Air Decompression ------------ In-Water Air/O_2 Decompression or SurDO_2 Required														
270	0:40	AIR									101	102:20	1	Z
		AIR/O_2									**26**	27:20		
300	0:40	AIR									128	129:20	1.5	
		AIR/O_2									**33**	34:20		
330	0:40	AIR									160	161:20	1.5	
		AIR/O_2									**38**	44:20		
360	0:40	AIR									184	185:20	2	
		AIR/O_2									**44**	50:20		
420	0:40	AIR									248	249:20	2.5	
		AIR/O_2									**56**	62:20		
480	0:40	AIR									321	322:20	2.5	
		AIR/O_2									**68**	79:20		
Exceptional Exposure: In-Water Air/0_2 Decompression ------------ SurDO_2 Required														
540	0:40	AIR									372	373:20	3	
		AIR/O_2									**80**	91:20		
600	0:40	AIR									410	411:20	3.5	
		AIR/O_2									**93**	104:20		
660	0:40	AIR									439	440:20	4	
		AIR/O_2									**103**	119:20		
Exceptional Exposure: SurDO_2														
720	0:40	AIR									461	462:20	4.5	
		AIR/O_2									**112**	128:20		

Table 9-9. *Air Decompression Table (Continued).*
(DESCENT RATE 75 FPM—ASCENT RATE 30 FPM)

Bottom Time (min)	Time to First Stop (M:S)	Gas Mix	DECOMPRESSION STOPS (FSW) Stop times (min) include travel time, except first air and first O_2 stop 100	90	80	70	60	50	40	30	20	Total Ascent Time (M:S)	Chamber O_2 Periods	Repet Group
45 FSW														
125	1:30	AIR									0	1:30	0	N
		AIR/O_2									**0**	1:30		
130	0:50	AIR									2	3:30	0.5	O
		AIR/O_2									**1**	2:30		
140	0:50	AIR									14	15:30	0.5	O
		AIR/O_2									**5**	6:30		
In-Water Air/O_2 Decompression or SurDO_2 Recommended ----------														
150	0:50	AIR									25	26:30	0.5	Z
		AIR/O_2									**8**	9:30		
160	0:50	AIR									34	35:30	0.5	Z
		AIR/O_2									**11**	12:30		
170	0:50	AIR									41	42:30	1	Z
		AIR/O_2									**14**	15:30		
180	0:50	AIR									59	60:30	1	Z
		AIR/O_2									**17**	18:30		
190	0:50	AIR									75	76:30	1	Z
		AIR/O_2									**19**	20:30		
Exceptional Exposure: In-Water Air Decompression ------------ In-Water Air/O_2 Decompression or SurDO_2 Required ----------														
200	0:50	AIR									89	90:30	1	Z
		AIR/O_2									**23**	24:30		
210	0:50	AIR									101	102:30	1	Z
		AIR/O_2									**27**	28:30		
220	0:50	AIR									112	113:30	1.5	Z
		AIR/O_2									**30**	31:30		
230	0:50	AIR									121	122:30	1.5	Z
		AIR/O_2									**33**	34:30		
240	0:50	AIR									130	131:30	1.5	Z
		AIR/O_2									**37**	43:30		
270	0:50	AIR									173	174:30	2	
		AIR/O_2									**45**	51:30		
300	0:50	AIR									206	207:30	2	
		AIR/O_2									**51**	57:30		
330	0:50	AIR									243	244:30	2.5	
		AIR/O_2									**61**	67:30		
360	0:50	AIR									288	289:30	3	
		AIR/O_2									**69**	80:30		
Exceptional Exposure: In-Water Air/0_2 Decompression ------------ SurDO_2 Required----------														
420	0:50	AIR									373	374:30	3.5	
		AIR/O_2									**84**	95:30		
480	0:50	AIR									431	432:30	4	
		AIR/O_2									**101**	117:30		
Exceptional Exposure: SurDO_2 ----------														
540	0:50	AIR									473	474:30	4.5	
		AIR/O_2									**117**	133:30		

Table 9-9. *Air Decompression Table (Continued).*
(DESCENT RATE 75 FPM—ASCENT RATE 30 FPM)

Bottom Time (min)	Time to First Stop (M:S)	Gas Mix	DECOMPRESSION STOPS (FSW) Stop times (min) include travel time, except first air and first O_2 stop 100	90	80	70	60	50	40	30	20	Total Ascent Time (M:S)	Chamber O_2 Periods	Repet Group
50 FSW														
92	1:40	AIR									0	1:40	0	M
		AIR/O_2									**0**	1:40		
95	1:00	AIR									2	3:40	0.5	M
		AIR/O_2									**1**	2:40		
100	1:00	AIR									4	5:40	0.5	N
		AIR/O_2									**2**	3:40		
110	1:00	AIR									8	9:40	0.5	O
		AIR/O_2									**4**	5:40		
In-Water Air/O_2 Decompression or SurDO_2 Recommended														
120	1:00	AIR									21	22:40	0.5	O
		AIR/O_2									**7**	8:40		
130	1:00	AIR									34	35:40	0.5	Z
		AIR/O_2									**12**	13:40		
140	1:00	AIR									45	46:40	1	Z
		AIR/O_2									**16**	17:40		
150	1:00	AIR									56	57:40	1	Z
		AIR/O_2									**19**	20:40		
160	1:00	AIR									78	79:40	1	Z
		AIR/O_2									**23**	24:40		
Exceptional Exposure: In-Water Air Decompression ------------ In-Water Air/O_2 Decompression or SurDO_2 Required														
170	1:00	AIR									96	97:40	1	Z
		AIR/O_2									**26**	27:40		
180	1:00	AIR									111	112:40	1.5	Z
		AIR/O_2									**30**	31:40		
190	1:00	AIR									125	126:40	1.5	Z
		AIR/O_2									**35**	36:40		
200	1:00	AIR									136	137:40	1.5	Z
		AIR/O_2									**39**	45:40		
210	1:00	AIR									147	148:40	2	
		AIR/O_2									**43**	49:40		
220	1:00	AIR									166	167:40	2	
		AIR/O_2									**47**	53:40		
230	1:00	AIR									183	184:40	2	
		AIR/O_2									**50**	56:40		
240	1:00	AIR									198	199:40	2	
		AIR/O_2									**53**	59:40		
270	1:00	AIR									236	237:40	2.5	
		AIR/O_2									**62**	68:40		
300	1:00	AIR									285	286:40	3	
		AIR/O_2									**74**	85:40		
Exceptional Exposure: In-Water Air/O_2 Decompression ------------ SurDO_2 Required														
330	1:00	AIR									345	346:40	3.5	
		AIR/O_2									**83**	94:40		
360	1:00	AIR									393	394:40	3.5	
		AIR/O_2									**92**	103:40		
Exceptional Exposure: SurDO_2														
420	1:00	AIR									464	465:40	4.5	
		AIR/O_2									**113**	129:40		

***Table 9-9.** Air Decompression Table (Continued).*
(DESCENT RATE 75 FPM—ASCENT RATE 30 FPM)

Bottom Time (min)	Time to First Stop (M:S)	Gas Mix	DECOMPRESSION STOPS (FSW) Stop times (min) include travel time, except first air and first O_2 stop 100	90	80	70	60	50	40	30	20	Total Ascent Time (M:S)	Chamber O_2 Periods	Repet Group
55 FSW														
74	1:50	AIR									0	1:50	0	L
		AIR/O_2									**0**	1:50		
75	1:10	AIR									1	2:50	0.5	L
		AIR/O_2									**1**	2:50		
80	1:10	AIR									4	5:50	0.5	M
		AIR/O_2									**2**	3:50		
90	1:10	AIR									10	11:50	0.5	N
		AIR/O_2									**5**	6:50		
In-Water Air/O_2 Decompression or SurDO_2 Recommended														
100	1:10	AIR									17	18:50	0.5	O
		AIR/O_2									**8**	9:50		
110	1:10	AIR									34	35:50	0.5	O
		AIR/O_2									**12**	13:50		
120	1:10	AIR									48	49:50	1	Z
		AIR/O_2									**17**	18:50		
130	1:10	AIR									59	60:50	1	Z
		AIR/O_2									**22**	23:50		
140	1:10	AIR									84	85:50	1	Z
		AIR/O_2									**26**	27:50		
Exceptional Exposure: In-Water Air Decompression ------------ In-Water Air/O_2 Decompression or SurDO_2 Required														
150	1:10	AIR									105	106:50	1.5	Z
		AIR/O_2									**30**	31:50		
160	1:10	AIR									123	124:50	1.5	Z
		AIR/O_2									**34**	35:50		
170	1:10	AIR									138	139:50	1.5	Z
		AIR/O_2									**40**	46:50		
180	1:10	AIR									151	152:50	2	Z
		AIR/O_2									**45**	51:50		
190	1:10	AIR									169	170:50	2	
		AIR/O_2									**50**	56:50		
200	1:10	AIR									190	191:50	2	
		AIR/O_2									**54**	60:50		
210	1:10	AIR									208	209:50	2.5	
		AIR/O_2									**58**	64:50		
220	1:10	AIR									224	225:50	2.5	
		AIR/O_2									**62**	68:50		
230	1:10	AIR									239	240:50	2.5	
		AIR/O_2									**66**	77:50		
240	1:10	AIR									254	255:50	3	
		AIR/O_2									**69**	80:50		
Exceptional Exposure: In-Water Air/O_2 Decompression ------------ SurDO_2 Required														
270	1:10	AIR									313	314:50	3.5	
		AIR/O_2									**83**	94:50		
300	1:10	AIR									380	381:50	3.5	
		AIR/O_2									**94**	105:50		
330	1:10	AIR									432	433:50	4	
		AIR/O_2									**106**	122:50		
Exceptional Exposure: SurDO_2														
360	1:10	AIR									474	475:50	4.5	
		AIR/O_2									**118**	134:50		

Table 9-9. Air Decompression Table (Continued).
(DESCENT RATE 75 FPM—ASCENT RATE 30 FPM)

Bottom Time (min)	Time to First Stop (M:S)	Gas Mix	DECOMPRESSION STOPS (FSW) Stop times (min) include travel time, except first air and first O_2 stop									Total Ascent Time (M:S)	Chamber O_2 Periods	Repet Group
			100	90	80	70	60	50	40	30	20			
60 FSW														
60	2:00	AIR									0	2:00	0	K
		AIR/O_2									**0**	2:00		
65	1:20	AIR									2	4:00	0.5	L
		AIR/O_2									**1**	3:00		
70	1:20	AIR									7	9:00	0.5	L
		AIR/O_2									**4**	6:00		
80	1:20	AIR									14	16:00	0.5	N
		AIR/O_2									**7**	9:00		
In-Water Air/O_2 Decompression or SurDO_2 Recommended														
90	1:20	AIR									23	25:00	0.5	O
		AIR/O_2									**10**	12:00		
100	1:20	AIR									42	44:00	1	Z
		AIR/O_2									**15**	17:00		
110	1:20	AIR									57	59:00	1	Z
		AIR/O_2									**21**	23:00		
120	1:20	AIR									75	77:00	1	Z
		AIR/O_2									**26**	28:00		
Exceptional Exposure: In-Water Air Decompression ------------ In-Water Air/O_2 Decompression or SurDO_2 Required														
130	1:20	AIR									102	104:00	1.5	Z
		AIR/O_2									**31**	33:00		
140	1:20	AIR									124	126:00	1.5	Z
		AIR/O_2									**35**	37:00		
150	1:20	AIR									143	145:00	2	Z
		AIR/O_2									**41**	48:00		
160	1:20	AIR									158	160:00	2	Z
		AIR/O_2									**48**	55:00		
170	1:20	AIR									178	180:00	2	
		AIR/O_2									**53**	60:00		
180	1:20	AIR									201	203:00	2.5	
		AIR/O_2									**59**	66:00		
190	1:20	AIR									222	224:00	2.5	
		AIR/O_2									**64**	71:00		
200	1:20	AIR									240	242:00	2.5	
		AIR/O_2									**68**	80:00		
210	1:20	AIR									256	258:00	3	
		AIR/O_2									**73**	85:00		
220	1:20	AIR									278	280:00	3	
		AIR/O_2									**77**	89:00		
Exceptional Exposure: In-Water Air/0_2 Decompression ------------ SurDO_2 Required														
230	1:20	AIR									300	302:00	3.5	
		AIR/O_2									**82**	94:00		
240	1:20	AIR									321	323:00	3.5	
		AIR/O_2									**88**	100:00		
270	1:20	AIR									398	400:00	4	
		AIR/O_2									**102**	119:00		
Exceptional Exposure: SurDO_2														
300	1:20	AIR									456	458:00	4.5	
		AIR/O_2									**115**	132:00		

Table 9-9. *Air Decompression Table (Continued).*
(DESCENT RATE 75 FPM—ASCENT RATE 30 FPM)

Bottom Time (min)	Time to First Stop (M:S)	Gas Mix	DECOMPRESSION STOPS (FSW) Stop times (min) include travel time, except first air and first O_2 stop 100	90	80	70	60	50	40	30	20	Total Ascent Time (M:S)	Chamber O_2 Periods	Repet Group
70 FSW														
48	2:20	AIR									0	2:20	0	K
		AIR/O_2									**0**	2:20		
50	1:40	AIR									2	4:20	0.5	K
		AIR/O_2									**1**	3:20		
55	1:40	AIR									9	11:20	0.5	L
		AIR/O_2									**5**	7:20		
60	1:40	AIR									14	16:20	0.5	M
		AIR/O_2									**8**	10:20		
In-Water Air/O_2 Decompression or SurDO_2 Recommended														
70	1:40	AIR									24	26:20	0.5	N
		AIR/O_2									**13**	15:20		
80	1:40	AIR									44	46:20	1	O
		AIR/O_2									**17**	19:20		
90	1:40	AIR									64	66:20	1	Z
		AIR/O_2									**24**	26:20		
100	1:40	AIR									88	90:20	1.5	Z
		AIR/O_2									**31**	33:20		
Exceptional Exposure: In-Water Air Decompression ------------ In-Water Air/O_2 Decompression or SurDO_2 Required														
110	1:40	AIR									120	122:20	1.5	Z
		AIR/O_2									**38**	45:20		
120	1:40	AIR									145	147:20	2	Z
		AIR/O_2									**44**	51:20		
130	1:40	AIR									167	169:20	2	Z
		AIR/O_2									**51**	58:20		
140	1:40	AIR									189	191:20	2.5	
		AIR/O_2									**59**	66:20		
150	1:40	AIR									219	221:20	2.5	
		AIR/O_2									**66**	78:20		
160	1:20	AIR								1	244	247:00	3	
		AIR/O_2								**1**	**72**	85:00		
Exceptional Exposure: In-Water Air/O_2 Decompression ------------ SurDO_2 Required														
170	1:20	AIR								2	265	269:00	3	
		AIR/O_2								**1**	**78**	91:00		
180	1:20	AIR								4	289	295:00	3.5	
		AIR/O_2								**2**	**83**	97:00		
190	1:20	AIR								5	316	323:00	3.5	
		AIR/O_2								**3**	**88**	103:00		
200	1:20	AIR								9	345	356:00	4	
		AIR/O_2								**5**	**93**	115:00		
210	1:20	AIR								13	378	393:00	4	
		AIR/O_2								**7**	**98**	122:00		
Exceptional Exposure: SurDO_2														
240	1:20	AIR								25	454	481:00	5	
		AIR/O_2								**13**	**110**	140:00		

Table 9-9. *Air Decompression Table (Continued).*
(DESCENT RATE 75 FPM—ASCENT RATE 30 FPM)

Bottom Time (min)	Time to First Stop (M:S)	Gas Mix	DECOMPRESSION STOPS (FSW) Stop times (min) include travel time, except first air and first O_2 stop									Total Ascent Time (M:S)	Chamber O_2 Periods	Repet Group
			100	90	80	70	60	50	40	30	20			
80 FSW														
39	2:40	AIR									0	2:40	0	J
		AIR/O_2									**0**	2:40		
40	2:00	AIR									1	3:40	0.5	J
		AIR/O_2									**1**	3:40		
45	2:00	AIR									10	12:40	0.5	K
		AIR/O_2									**5**	7:40		
In-Water Air/O_2 Decompression or SurDO_2 Recommended ----------														
50	2:00	AIR									17	19:40	0.5	M
		AIR/O_2									**9**	11:40		
55	2:00	AIR									24	26:40	0.5	M
		AIR/O_2									**13**	15:40		
60	2:00	AIR									30	32:40	1	N
		AIR/O_2									**16**	18:40		
70	2:00	AIR									54	56:40	1	O
		AIR/O_2									**22**	24:40		
80	2:00	AIR									77	79:40	1.5	Z
		AIR/O_2									**30**	32:40		
Exceptional Exposure: In-Water Air Decompression ------------ In-Water Air/O_2 Decompression or SurDO_2 Required ----------														
90	2:00	AIR									114	116:40	1.5	Z
		AIR/O_2									**39**	46:40		
100	1:40	AIR								1	147	150:20	2	Z
		AIR/O_2								**1**	**46**	54:20		
110	1:40	AIR								6	171	179:20	2	Z
		AIR/O_2								**3**	**51**	61:20		
120	1:40	AIR								10	200	212:20	2.5	
		AIR/O_2								**5**	**59**	71:20		
130	1:40	AIR								14	232	248:20	3	
		AIR/O_2								**7**	**67**	86:20		
Exceptional Exposure: In-Water Air/0_2 Decompression ------------ SurDO_2 Required----------														
140	1:40	AIR								17	258	277:20	3.5	
		AIR/O_2								**9**	**73**	94:20		
150	1:40	AIR								19	285	306:20	3.5	
		AIR/O_2								**10**	**80**	102:20		
160	1:40	AIR								21	318	341:20	4	
		AIR/O_2								**11**	**86**	114:20		
170	1:40	AIR								27	354	383:20	4	
		AIR/O_2								**14**	**90**	121:20		
Exceptional Exposure: SurDO_2 ----------														
180	1:40	AIR								33	391	426:20	4.5	
		AIR/O_2								**17**	**96**	130:20		
210	1:40	AIR								50	474	526:20	5	
		AIR/O_2								**26**	**110**	158:20		

Table 9-9. *Air Decompression Table (Continued).*
(DESCENT RATE 75 FPM—ASCENT RATE 30 FPM)

Bottom Time (min)	Time to First Stop (M:S)	Gas Mix	DECOMPRESSION STOPS (FSW) Stop times (min) include travel time, except first air and first O_2 stop									Total Ascent Time (M:S)	Chamber O_2 Periods	Repet Group
			100	90	80	70	60	50	40	30	20			
90 FSW														
30	3:00	AIR									0	3:00	0	I
		AIR/O_2									**0**	3:00		
35	2:20	AIR									4	7:00	0.5	J
		AIR/O_2									**2**	5:00		
40	2:20	AIR									14	17:00	0.5	L
		AIR/O_2									**7**	10:00		
In-Water Air/O_2 Decompression or SurDO_2 Recommended														
45	2:20	AIR									23	26:00	0.5	M
		AIR/O_2									**12**	15:00		
50	2:20	AIR									31	34:00	1	N
		AIR/O_2									**17**	20:00		
55	2:20	AIR									39	42:00	1	O
		AIR/O_2									**21**	24:00		
60	2:20	AIR									56	59:00	1	O
		AIR/O_2									**24**	27:00		
70	2:20	AIR									83	86:00	1.5	Z
		AIR/O_2									**32**	35:00		
Exceptional Exposure: In-Water Air Decompression ------------ In-Water Air/O_2 Decompression or SurDO_2 Required														
80	2:00	AIR								5	125	132:40	2	Z
		AIR/O_2								**3**	**40**	50:40		
90	2:00	AIR								13	158	173:40	2	Z
		AIR/O_2								**7**	**46**	60:40		
100	2:00	AIR								19	185	206:40	2.5	
		AIR/O_2								**10**	**53**	70:40		
110	2:00	AIR								25	224	251:40	3	
		AIR/O_2								**13**	**61**	86:40		
Exceptional Exposure: In-Water Air/O_2 Decompression ------------ SurDO_2 Required														
120	1:40	AIR							1	29	256	288:20	3.5	
		AIR/O_2							1	**15**	**70**	98:40		
130	1:40	AIR							5	28	291	326:20	3.5	
		AIR/O_2							5	**15**	**78**	110:40		
140	1:40	AIR							8	28	330	368:20	4	
		AIR/O_2							8	**15**	**86**	126:40		
Exceptional Exposure: SurDO_2														
150	1:40	AIR							11	34	378	425:20	4.5	
		AIR/O_2							11	**17**	**94**	139:40		
160	1:40	AIR							13	40	418	473:20	4.5	
		AIR/O_2							13	**21**	**100**	151:40		
170	1:40	AIR							15	45	451	513:20	5	
		AIR/O_2							15	**23**	**106**	166:40		
180	1:40	AIR							16	51	479	548:20	5.5	
		AIR/O_2							16	**26**	**112**	176:40		
240	1:40	AIR							42	68	592	704:20	7.5	
		AIR/O_2							42	**34**	**159**	267:00		

Table 9-9. *Air Decompression Table (Continued).*
(DESCENT RATE 75 FPM—ASCENT RATE 30 FPM)

Bottom Time (min)	Time to First Stop (M:S)	Gas Mix	DECOMPRESSION STOPS (FSW) Stop times (min) include travel time, except first air and first O_2 stop									Total Ascent Time (M:S)	Chamber O_2 Periods	Repet Group
			100	90	80	70	60	50	40	30	20			
100 FSW														
25	3:20	AIR									0	3:20	0	H
		AIR/O_2									**0**	3:20		
30	2:40	AIR									3	6:20	0.5	J
		AIR/O_2									**2**	5:20		
35	2:40	AIR									15	18:20	0.5	L
		AIR/O_2									**8**	11:20		
In-Water Air/O_2 Decompression or SurDO_2 Recommended														
40	2:40	AIR									26	29:20	1	M
		AIR/O_2									**14**	17:20		
45	2:40	AIR									36	39:20	1	N
		AIR/O_2									**19**	22:20		
50	2:40	AIR									47	50:20	1	O
		AIR/O_2									**24**	27:20		
55	2:40	AIR									65	68:20	1.5	Z
		AIR/O_2									**28**	31:20		
60	2:40	AIR									81	84:20	1.5	Z
		AIR/O_2									**33**	35:20		
Exceptional Exposure: In-Water Air Decompression ------------ In-Water Air/O_2 Decompression or SurDO_2 Required														
70	2:20	AIR								11	124	138:00	2	Z
		AIR/O_2								**6**	**39**	53:00		
80	2:20	AIR								21	160	184:00	2.5	Z
		AIR/O_2								**11**	**45**	64:00		
90	2:00	AIR							2	28	196	228:40	2.5	
		AIR/O_2							2	**15**	**52**	82:00		
Exceptional Exposure: In-Water Air/O_2 Decompression ------------ SurDO_2 Required														
100	2:00	AIR							9	28	241	280:40	3	
		AIR/O_2							9	**14**	**66**	102:00		
110	2:00	AIR							14	28	278	322:40	3.5	
		AIR/O_2							14	**15**	**75**	117:00		
120	2:00	AIR							19	28	324	373:40	4	
		AIR/O_2							19	**15**	**84**	136:00		
Exceptional Exposure: SurDO_2														
150	1:40	AIR						3	26	46	461	538:20	5	
		AIR/O_2						3	26	**24**	**108**	183:40		

Table 9-9. *Air Decompression Table (Continued).*

(DESCENT RATE 75 FPM—ASCENT RATE 30 FPM)

Bottom Time (min)	Time to First Stop (M:S)	Gas Mix	DECOMPRESSION STOPS (FSW) Stop times (min) include travel time, except first air and first O_2 stop									Total Ascent Time (M:S)	Chamber O_2 Periods	Repet Group
			100	90	80	70	60	50	40	30	20			
110 FSW														
20	3:40	AIR									0	3:40	0	H
		AIR/O_2									**0**	3:40		
25	3:00	AIR									3	6:40	0.5	I
		AIR/O_2									**2**	5:40		
30	3:00	AIR									14	17:40	0.5	K
		AIR/O_2									**7**	10:40		
In-Water Air/O_2 Decompression or SurDO_2 Recommended ----------														
35	3:00	AIR									27	30:40	1	M
		AIR/O_2									**14**	17:40		
40	3:00	AIR									39	42:40	1	N
		AIR/O_2									**20**	23:40		
45	3:00	AIR									50	53:40	1	O
		AIR/O_2									**26**	29:40		
50	3:00	AIR									71	74:40	1.5	Z
		AIR/O_2									**31**	34:40		
Exceptional Exposure: In-Water Air Decompression ------------ In-Water Air/O_2 Decompression or SurDO_2 Required ----------														
55	2:40	AIR								5	85	93:20	1.5	Z
		AIR/O_2								**3**	**33**	44:20		
60	2:40	AIR								13	111	127:20	2	Z
		AIR/O_2								**7**	**36**	51:20		
70	2:40	AIR								26	155	184:20	2.5	Z
		AIR/O_2								**13**	**43**	64:20		
80	2:20	AIR							9	28	200	240:00	2.5	
		AIR/O_2							9	**15**	**53**	90:20		
Exceptional Exposure: In-Water Air/0$_2$ Decompression ------------ SurDO_2 Required----------														
90	2:20	AIR							17	29	248	297:00	3.5	
		AIR/O_2							17	**15**	**67**	112:20		
100	2:20	AIR							25	28	295	351:00	3.5	
		AIR/O_2							25	**15**	**78**	131:20		
110	2:00	AIR						5	26	28	353	414:40	4	
		AIR/O_2						5	26	**15**	**90**	154:00		
Exceptional Exposure: SurDO_2 ----------														
120	2:00	AIR						10	26	35	413	486:40	4.5	
		AIR/O_2						10	26	**18**	**101**	173:00		
180	1:40	AIR					3	23	47	68	593	736:20	7.5	
		AIR/O_2					3	23	47	**34**	**159**	298:00		

***Table 9-9.** Air Decompression Table (Continued).*
(DESCENT RATE 75 FPM—ASCENT RATE 30 FPM)

Bottom Time (min)	Time to First Stop (M:S)	Gas Mix	DECOMPRESSION STOPS (FSW) Stop times (min) include travel time, except first air and first O_2 stop									Total Ascent Time (M:S)	Chamber O_2 Periods	Repet Group
			100	90	80	70	60	50	40	30	20			
120 FSW														
15	4:00	AIR									0	4:00	0	F
		AIR/O_2									**0**	4:00		
20	3:20	AIR									2	6:00	0.5	H
		AIR/O_2									**1**	5:00		
25	3:20	AIR									8	12:00	0.5	J
		AIR/O_2									**4**	8:00		
In-Water Air/O_2 Decompression or SurDO_2 Recommended														
30	3:20	AIR									24	28:00	0.5	L
		AIR/O_2									**13**	17:00		
35	3:20	AIR									38	42:00	1	N
		AIR/O_2									**20**	24:00		
40	3:20	AIR									51	55:00	1	O
		AIR/O_2									**27**	31:00		
45	3:20	AIR									72	76:00	1.5	Z
		AIR/O_2									**33**	37:00		
Exceptional Exposure: In-Water Air Decompression ------------ In-Water Air/O_2 Decompression or SurDO_2 Required														
50	3:00	AIR								9	86	98:40	1.5	Z
		AIR/O_2								**5**	**33**	46:40		
55	3:00	AIR								19	116	138:40	2	Z
		AIR/O_2								**10**	**35**	53:40		
60	3:00	AIR								27	142	172:40	2	Z
		AIR/O_2								**14**	**39**	61:40		
70	2:40	AIR							12	29	189	233:20	2.5	
		AIR/O_2							12	**15**	**50**	85:40		
Exceptional Exposure: In-Water Air/O_2 Decompression ------------ SurDO_2 Required														
80	2:40	AIR							24	28	246	301:20	3	
		AIR/O_2							24	**14**	**67**	118:40		
90	2:20	AIR						7	26	28	303	367:00	3.5	
		AIR/O_2						7	26	**15**	**79**	140:20		
100	2:20	AIR						14	26	28	372	443:00	4	
		AIR/O_2						14	26	**15**	**94**	167:20		
Exceptional Exposure: SurDO_2														
110	2:20	AIR						21	25	38	433	520:00	5	
		AIR/O_2						21	25	**20**	**104**	188:20		
120	2:00	AIR					3	23	25	47	480	580:40	5.5	
		AIR/O_2					3	23	25	**24**	**113**	211:00		

Table 9-9. *Air Decompression Table (Continued).*
(DESCENT RATE 75 FPM—ASCENT RATE 30 FPM)

Bottom Time (min)	Time to First Stop (M:S)	Gas Mix	DECOMPRESSION STOPS (FSW) Stop times (min) include travel time, except first air and first O_2 stop									Total Ascent Time (M:S)	Chamber O_2 Periods	Repet Group
			100	90	80	70	60	50	40	30	20			
130 FSW														
10	4:20	AIR									0	4:20	0	E
		AIR/O_2									**0**	4:20		
15	3:40	AIR									1	5:20	0.5	G
		AIR/O_2									**1**	5:20		
20	3:40	AIR									4	8:20	0.5	I
		AIR/O_2									**2**	6:20		
In-Water Air/O_2 Decompression or SurDO_2 Recommended														
25	3:40	AIR									17	21:20	0.5	K
		AIR/O_2									**9**	13:20		
30	3:40	AIR									34	38:20	1	M
		AIR/O_2									**18**	22:20		
35	3:40	AIR									49	53:20	1	N
		AIR/O_2									**26**	30:20		
40	3:20	AIR								3	67	74:00	1.5	Z
		AIR/O_2								**2**	**31**	37:00		
Exceptional Exposure: In-Water Air Decompression ------------ In-Water Air/O_2 Decompression or SurDO_2 Required														
45	3:20	AIR								12	84	100:00	1.5	Z
		AIR/O_2								**6**	**33**	48:00		
50	3:20	AIR								22	116	142:00	2	Z
		AIR/O_2								**11**	**35**	55:00		
55	3:00	AIR							4	28	145	180:40	2	Z
		AIR/O_2							4	**15**	**39**	67:00		
60	3:00	AIR							12	28	170	213:40	2.5	Z
		AIR/O_2							12	**15**	**45**	81:00		
Exceptional Exposure: In-Water Air/0_2 Decompression ------------ SurDO_2 Required														
70	2:40	AIR						1	26	28	235	293:20	3	
		AIR/O_2						1	26	**14**	**63**	117:40		
80	2:40	AIR						12	26	28	297	366:20	3.5	
		AIR/O_2						12	26	**15**	**78**	144:40		
90	2:40	AIR						21	26	28	374	452:20	4	
		AIR/O_2						21	26	**15**	**94**	174:40		
Exceptional Exposure: SurDO_2														
100	2:20	AIR					6	23	26	38	444	540:00	5	
		AIR/O_2					6	23	26	**20**	**106**	204:20		
120	2:20	AIR					17	23	28	57	533	661:00	6	
		AIR/O_2					17	23	28	**29**	**130**	255:20		
180	2:00	AIR				13	21	45	57	94	658	890:40	9	
		AIR/O_2				13	21	45	57	**46**	**198**	417:20		

Table 9-9. *Air Decompression Table (Continued).*
(DESCENT RATE 75 FPM—ASCENT RATE 30 FPM)

Bottom Time (min)	Time to First Stop (M:S)	Gas Mix	DECOMPRESSION STOPS (FSW) Stop times (min) include travel time, except first air and first O_2 stop 100	90	80	70	60	50	40	30	20	Total Ascent Time (M:S)	Chamber O_2 Periods	Repet Group
140 FSW														
10	4:40	AIR									0	4:40	0	E
		AIR/O_2									**0**	4:40		
15	4:00	AIR									2	6:40	0.5	H
		AIR/O_2									**1**	5:40		
20	4:00	AIR									7	11:40	0.5	J
		AIR/O_2									**4**	8:40		
In-Water Air/O_2 Decompression or SurDO_2 Recommended														
25	4:00	AIR									26	30:40	1	L
		AIR/O_2									**14**	18:40		
30	4:00	AIR									**44**	48:40	1	N
		AIR/O_2									**23**	27:40		
35	3:40	AIR								4	59	67:20	1.5	O
		AIR/O_2								**2**	**30**	36:20		
Exceptional Exposure: In-Water Air Decompression ------------ In-Water Air/O_2 Decompression or SurDO_2 Required														
40	3:40	AIR								11	80	95:20	1.5	Z
		AIR/O_2								**6**	**33**	48:20		
45	3:20	AIR							3	21	113	141:00	2	Z
		AIR/O_2							3	**11**	**34**	57:20		
50	3:20	AIR							7	28	145	184:00	2	Z
		AIR/O_2							7	**14**	**40**	70:20		
55	3:20	AIR							16	28	171	219:00	2.5	Z
		AIR/O_2							16	**15**	**45**	85:20		
Exceptional Exposure: In-Water Air/0_2 Decompression ------------ SurDO_2 Required														
60	3:00	AIR						2	23	28	209	265:40	3	
		AIR/O_2						2	23	**15**	**55**	109:00		
70	3:00	AIR						14	25	28	276	346:40	3.5	
		AIR/O_2						14	25	**15**	**74**	142:00		
80	2:40	AIR					2	24	25	29	362	445:20	4	
		AIR/O_2					2	24	25	**15**	**91**	175:40		
Exceptional Exposure: SurDO_2														
90	2:40	AIR					12	23	26	38	443	545:20	5	
		AIR/O_2					12	23	26	**19**	**107**	210:40		

Table 9-9. *Air Decompression Table (Continued).*
(DESCENT RATE 75 FPM—ASCENT RATE 30 FPM)

Bottom Time (min)	Time to First Stop (M:S)	Gas Mix	DECOMPRESSION STOPS (FSW) Stop times (min) include travel time, except first air and first O_2 stop 100	90	80	70	60	50	40	30	20	Total Ascent Time (M:S)	Chamber O_2 Periods	Repet Group
150 FSW														
5	5:00	AIR									0	5:00	0	C
		AIR/O_2									**0**	5:00		
10	4:20	AIR									1	6:00	0.5	F
		AIR/O_2									**1**	6:00		
15	4:20	AIR									3	8:00	0.5	H
		AIR/O_2									**2**	7:00		
20	4:20	AIR									14	19:00	0.5	K
		AIR/O_2									**8**	13:00		
In-Water Air/O_2 Decompression or SurDO_2 Recommended														
25	4:20	AIR									35	40:00	1	M
		AIR/O_2									**19**	24:00		
30	4:00	AIR								3	51	58:40	1.5	O
		AIR/O_2								**2**	**26**	32:40		
35	4:00	AIR								11	72	87:40	1.5	Z
		AIR/O_2								**6**	**31**	46:40		
Exceptional Exposure: In-Water Air Decompression ------------ In-Water Air/O_2 Decompression or SurDO_2 Required														
40	3:40	AIR							4	18	102	128:20	2	Z
		AIR/O_2							4	**9**	**34**	56:40		
45	3:40	AIR							10	25	140	179:20	2	Z
		AIR/O_2							10	**13**	**39**	71:40		
50	3:20	AIR						3	15	28	170	220:00	2.5	Z
		AIR/O_2						3	15	**15**	**45**	87:20		
Exceptional Exposure: In-Water Air/O_2 Decompression ------------ SurDO_2 Required														
55	3:20	AIR						6	22	28	211	271:00	3	
		AIR/O_2						6	22	**15**	**56**	113:20		
60	3:20	AIR						11	26	28	248	317:00	3	
		AIR/O_2						11	26	**15**	**66**	132:20		
70	3:00	AIR					3	24	25	28	330	413:40	4	
		AIR/O_2					3	24	25	**15**	**84**	170:00		
Exceptional Exposure: SurDO_2														
80	3:00	AIR					15	23	26	35	430	532:40	4.5	
		AIR/O_2					15	23	26	**18**	**104**	205:00		
90	2:40	AIR				3	22	23	26	47	496	620:20	5.5	
		AIR/O_2				3	22	23	26	**24**	**118**	239:40		
120	2:20	AIR			3	20	22	23	50	75	608	804:00	8	
		AIR/O_2			3	20	22	23	50	**37**	**168**	355:40		
180	2:00	AIR		2	19	20	42	48	79	121	694	1027:40	10.5	
		AIR/O_2		2	19	20	42	48	79	**58**	**222**	537:20		

Table 9-9. *Air Decompression Table (Continued).*
(DESCENT RATE 75 FPM—ASCENT RATE 30 FPM)

Bottom Time (min)	Time to First Stop (M:S)	Gas Mix	DECOMPRESSION STOPS (FSW) Stop times (min) include travel time, except first air and first O_2 stop									Total Ascent Time (M:S)	Chamber O_2 Periods	Repet Group
			100	90	80	70	60	50	40	30	20			
160 FSW														
5	5:20	AIR									0	5:20	0	C
		AIR/O_2									**0**	5:20		
10	4:40	AIR									1	6:20	0.5	F
		AIR/O_2									**1**	6:20		
15	4:40	AIR									5	10:20	0.5	I
		AIR/O_2									**3**	8:00		
In-Water Air/O_2 Decompression or SurDO_2 Recommended ----------														
20	4:40	AIR									22	27:20	0.5	L
		AIR/O_2									**12**	17:20		
25	4:20	AIR								3	41	49:00	1	N
		AIR/O_2								**2**	**21**	28:00		
30	4:00	AIR							1	8	60	73:40	1.5	O
		AIR/O_2							1	**5**	**28**	39:00		
Exceptional Exposure: In-Water Air Decompression ------------ In-Water Air/O_2 Decompression or SurDO_2 Required ----------														
35	4:00	AIR							4	14	84	106:40	1.5	Z
		AIR/O_2							4	**8**	**32**	54:00		
40	4:00	AIR							12	20	130	166:40	2	Z
		AIR/O_2							12	**11**	**37**	70:00		
45	3:40	AIR						5	13	28	164	214:20	2.5	Z
		AIR/O_2						5	13	**14**	**44**	85:40		
Exceptional Exposure: In-Water Air/0_2 Decompression ------------ SurDO_2 Required----------														
50	3:40	AIR						10	19	28	207	268:20	3	
		AIR/O_2						10	19	**15**	**54**	112:40		
55	3:20	AIR					2	12	26	28	248	320:00	3	
		AIR/O_2					2	12	26	**14**	**67**	135:20		
60	3:20	AIR					5	18	25	29	290	371:00	3.5	
		AIR/O_2					5	18	25	**15**	**77**	154:20		
Exceptional Exposure: SurDO_2 ----------														
70	3:20	AIR					15	23	26	29	399	496:00	4.5	
		AIR/O_2					15	23	26	**15**	**99**	197:20		
80	3:00	AIR				6	21	24	25	44	482	605:40	5.5	
		AIR/O_2				6	21	24	25	**23**	**114**	237:00		

Table 9-9. Air Decompression Table (Continued).
(DESCENT RATE 75 FPM—ASCENT RATE 30 FPM)

Bottom Time (min)	Time to First Stop (M:S)	Gas Mix	DECOMPRESSION STOPS (FSW) Stop times (min) include travel time, except first air and first O_2 stop									Total Ascent Time (M:S)	Chamber O_2 Periods	Repet Group
			100	90	80	70	60	50	40	30	20			
170 FSW														
5	5:40	AIR									0	5:40	0	D
		AIR/O_2									**0**	5:40		
10	5:00	AIR									2	7:40	0.5	G
		AIR/O_2									**1**	6:40		
15	5:00	AIR									7	12:40	0.5	J
		AIR/O_2									**4**	9:40		
In-Water Air/O_2 Decompression or SurDO_2 Recommended														
20	4:40	AIR								1	29	35:20	1	L
		AIR/O_2								**1**	**15**	21:20		
25	4:20	AIR							1	6	46	58:00	1	N
		AIR/O_2							1	**4**	**23**	33:20		
Exceptional Exposure: In-Water Air Decompression ------------ In-Water Air/O_2 Decompression or SurDO_2 Required														
30	4:20	AIR							5	11	72	93:00	1.5	Z
		AIR/O_2							5	**6**	**29**	45:20		
35	4:00	AIR						2	9	17	113	145:40	2	Z
		AIR/O_2						2	9	**9**	**35**	65:00		
40	4:00	AIR						6	13	23	155	201:40	2.5	Z
		AIR/O_2						6	13	**12**	**43**	84:00		
Exceptional Exposure: In-Water Air/O_2 Decompression ------------ SurDO_2 Required														
45	4:00	AIR						12	16	28	194	254:40	2.5	
		AIR/O_2						12	16	**15**	**51**	109:00		
50	3:40	AIR					5	12	23	28	243	315:20	3	
		AIR/O_2					5	12	23	**15**	**65**	134:40		
55	3:40	AIR					9	16	25	28	287	369:20	3.5	
		AIR/O_2					9	16	25	**15**	**76**	155:40		
60	3:20	AIR				2	11	21	26	28	344	436:00	4	
		AIR/O_2				2	11	21	26	**15**	**87**	181:20		
Exceptional Exposure: SurDO_2														
70	3:20	AIR				7	19	24	25	39	454	572:00	5	
		AIR/O_2				7	19	24	25	**20**	**109**	228:20		
80	3:20	AIR				17	22	23	26	53	525	670:00	6	
		AIR/O_2				17	22	23	26	**27**	**128**	267:20		
90	3:00	AIR			7	20	22	23	37	66	574	752:40	7	
		AIR/O_2			7	20	22	23	37	**33**	**148**	318:20		
120	2:40	AIR		9	19	20	22	42	60	94	659	928:20	9	
		AIR/O_2		9	19	20	22	42	60	**46**	**198**	454:00		
180	2:20	AIR	10	18	19	40	43	70	97	156	703	1159:00	11.5	
		AIR/O_2	10	18	19	40	43	70	97	**75**	**228**	648:00		

Table 9-9. Air Decompression Table (Continued).

(DESCENT RATE 75 FPM—ASCENT RATE 30 FPM)

Bottom Time (min)	Time to First Stop (M:S)	Gas Mix	DECOMPRESSION STOPS (FSW) Stop times (min) include travel time, except first air and first O_2 stop 100	90	80	70	60	50	40	30	20	Total Ascent Time (M:S)	Chamber O_2 Periods	Repet Group
180 FSW														
5	6:00	AIR									0	6:00	0	D
		AIR/O_2									**0**	6:00		
10	5:20	AIR									3	9:00	0.5	G
		AIR/O_2									**2**	8:00		
15	5:20	AIR									11	17:00	0.5	J
		AIR/O_2									**6**	12:00		
In-Water Air/O_2 Decompression or SurDO_2 Recommended														
20	5:00	AIR								4	34	43:40	1	M
		AIR/O_2								**2**	**18**	25:40		
25	4:40	AIR							**4**	7	54	70:20	1.5	O
		AIR/O_2							**4**	**4**	**26**	39:40		
Exceptional Exposure: In-Water Air Decompression ------------ In-Water Air/O_2 Decompression or SurDO_2 Required														
30	4:20	AIR						2	7	14	83	111:00	1.5	Z
		AIR/O_2						2	7	**7**	**31**	57:20		
35	4:20	AIR						5	13	19	138	180:00	2	Z
		AIR/O_2						5	13	**10**	**40**	78:20		
Exceptional Exposure: In-Water Air/O_2 Decompression ------------ SurDO_2 Required														
40	4:00	AIR					2	11	12	28	175	232:40	2.5	Z
		AIR/O_2					2	11	12	**14**	**47**	96:00		
45	4:00	AIR					7	11	20	28	231	301:40	3	
		AIR/O_2					7	11	20	**15**	**61**	129:00		
50	3:40	AIR				1	11	13	25	28	276	358:20	3.5	
		AIR/O_2				1	11	13	25	**15**	**74**	153:40		
55	3:40	AIR				5	11	19	26	28	336	429:20	4	
		AIR/O_2				5	11	19	26	**14**	**87**	181:40		
Exceptional Exposure: SurDO_2														
60	3:40	AIR				8	13	24	25	31	405	510:20	4.5	
		AIR/O_2				8	13	24	25	**16**	**100**	205:40		
70	3:20	AIR			3	13	21	24	25	48	498	636:00	5.5	
		AIR/O_2			3	13	21	24	25	**25**	**118**	253:20		

Table 9-9. Air Decompression Table (Continued).
(DESCENT RATE 75 FPM—ASCENT RATE 30 FPM)

Bottom Time (min)	Time to First Stop (M:S)	Gas Mix	DECOMPRESSION STOPS (FSW) Stop times (min) include travel time, except first air and first O_2 stop									Total Ascent Time (M:S)	Chamber O_2 Periods	Repet Group
			100	90	80	70	60	50	40	30	20			
190 FSW														
5	6:20	AIR									0	6:20	0	D
		AIR/O_2									**0**	6:20		
10	5:40	AIR									4	10:20	0.5	H
		AIR/O_2									**2**	8:20		
In-Water Air/O_2 Decompression or SurDO_2 Recommended														
15	5:40	AIR									17	23:20	0.5	K
		AIR/O_2									**9**	15:20		
20	5:00	AIR							1	7	37	50:40	1	N
		AIR/O_2							1	**4**	**19**	30:00		
25	4:40	AIR						2	6	9	67	89:20	1.5	Z
		AIR/O_2						2	6	**5**	**28**	46:40		
Exceptional Exposure: In-Water Air Decompression ------------ In-Water Air/O_2 Decompression or SurDO_2 Required														
30	4:40	AIR						6	8	14	111	144:20	2	Z
		AIR/O_2						6	8	**8**	**35**	67:40		
35	4:20	AIR					3	8	13	22	160	211:00	2.5	Z
		AIR/O_2					3	8	13	**12**	**44**	90:20		
Exceptional Exposure: In-Water Air/0_2 Decompression ------------ SurDO_2 Required														
40	4:20	AIR					7	12	14	29	210	277:00	3	
		AIR/O_2					7	12	14	**15**	**56**	119:20		
45	4:00	AIR				2	11	12	23	28	262	342:40	3.5	
		AIR/O_2				2	11	12	23	**15**	**70**	148:00		
50	4:00	AIR				7	11	16	26	28	321	413:40	4	
		AIR/O_2				7	11	16	26	**15**	**83**	178:00		
Exceptional Exposure: SurDO2														
55	3:40	AIR			2	10	10	24	25	30	396	501:20	4.5	
		AIR/O_2			2	10	10	24	25	**16**	**98**	204:40		
60	3:40	AIR			5	10	16	24	25	40	454	578:20	5	
		AIR/O_2			5	10	16	24	25	**21**	**108**	233:40		
90	3:20	AIR		11	19	20	21	28	51	83	626	863:00	8.5	
		AIR/O_2		11	19	20	21	28	51	**42**	**177**	408:40		
120	3:00	AIR	15	17	19	20	37	46	79	113	691	1040:40	10.5	
		AIR/O_2	15	17	19	20	37	46	79	**55**	**219**	550:20		

Table 9-9. ***Table 9-9.*** *Air Decompression Table (Continued).*

(DESCENT RATE 75 FPM—ASCENT RATE 30 FPM)

Bottom Time (min)	Time to First Stop (M:S)	Gas Mix	DECOMPRESSION STOPS (FSW) Stop times (min) include travel time, except first air and first O_2 stop 100	90	80	70	60	50	40	30	20	Total Ascent Time (M:S)	Chamber O_2 Periods	Repet Group
200 FSW														
Exceptional Exposure														
5	6:00	AIR									1	7:40	0.5	
		AIR/O_2									**1**	7:40		
10	6:00	AIR									2	8:40	0.5	
		AIR/O_2									**1**	7:40		
15	5:40	AIR								2	22	30:20	0.5	
		AIR/O_2								**1**	**11**	18:20		
20	5:20	AIR							5	6	43	60:00	1	
		AIR/O_2							5	**4**	**21**	36:20		
25	5:00	AIR						5	6	11	78	105:40	1.5	
		AIR/O_2						5	6	**6**	**29**	52:00		
30	4:40	AIR					4	5	11	18	136	179:20	2	
		AIR/O_2					4	5	11	**9**	**40**	79:40		
35	4:20	AIR				1	6	10	13	26	179	240:00	2.5	
		AIR/O_2				1	6	10	13	**13**	**49**	102:20		
40	4:20	AIR				3	10	12	18	28	243	319:00	3	
		AIR/O_2				3	10	12	18	**15**	**65**	138:20		
45	4:20	AIR				8	11	12	26	28	300	390:00	3.5	
		AIR/O_2				8	11	12	26	**15**	**79**	166:20		
50	4:00	AIR			3	10	11	20	26	28	377	479:40	4.5	
		AIR/O_2			3	10	11	20	26	**15**	**95**	200:00		
210 FSW														
Exceptional Exposure														
5	6:20	AIR									1	8:00	0.5	
		AIR/O_2									**1**	8:00		
10	6:20	AIR									5	12:00	0.5	
		AIR/O_2									**3**	10:00		
15	6:00	AIR								5	26	37:40	1	
		AIR/O_2								**3**	**13**	22:40		
20	5:20	AIR						2	6	7	50	71:00	1.5	
		AIR/O_2						2	6	**4**	**24**	42:20		
25	5:00	AIR					2	6	7	13	94	127:40	1.5	
		AIR/O_2					2	6	7	**7**	**32**	65:00		
30	4:40	AIR				2	5	6	13	21	156	208:20	2	
		AIR/O_2				2	5	6	13	**11**	**43**	90:40		
35	4:40	AIR				5	6	12	14	28	214	284:20	3	
		AIR/O_2				5	6	12	14	**14**	**58**	124:40		
40	4:20	AIR			2	6	11	12	22	28	271	357:00	3.5	
		AIR/O_2			2	6	11	12	22	**15**	**74**	157:20		
45	4:20	AIR			4	10	11	16	25	29	347	447:00	4	
		AIR/O_2			4	10	11	16	25	**15**	**89**	190:20		
50	4:20	AIR			9	10	11	23	26	35	426	545:00	4.5	
		AIR/O_2			9	10	11	23	26	**18**	**104**	221:20		

Table 9-9. *Air Decompression Table (Continued).*
(DESCENT RATE 75 FPM—ASCENT RATE 30 FPM)

Bottom Time (min)	Time to First Stop (M:S)	Gas Mix	DECOMPRESSION STOPS (FSW) Stop times (min) include travel time, except first air and first O_2 stop									Total Ascent Time (M:S)	Chamber O_2 Periods	Repet Group
			100	90	80	70	60	50	40	30	20			
220 FSW														
Exceptional Exposure														
5	6:40	AIR									2	9:20	0.5	
		AIR/O_2									**1**	8:20		
10	6:40	AIR									8	15:20	0.5	
		AIR/O_2									**4**	11:20		
15	6:00	AIR							1	7	30	44:40	1	
		AIR/O_2							1	**4**	**15**	27:00		
20	5:40	AIR						5	6	7	63	87:20	1.5	
		AIR/O_2						5	6	**4**	**27**	48:40		
25	5:20	AIR					5	6	8	14	119	158:00	2	
		AIR/O_2					5	6	8	**7**	**38**	75:20		
30	5:00	AIR				5	5	8	13	24	174	234:40	2.5	
		AIR/O_2				5	5	8	13	**13**	**47**	102:00		
35	4:40	AIR			3	5	9	11	18	28	244	323:20	3	
		AIR/O_2			3	5	9	11	18	**15**	**66**	142:40		
40	4:20	AIR		1	4	9	11	11	26	28	312	407:00	4	
		AIR/O_2		1	4	9	11	11	26	**15**	**82**	179:20		
250 FSW														
Exceptional Exposure														
5	7:40	AIR									3	11:20	0.5	
		AIR/O_2									**2**	10:20		
10	7:20	AIR								2	15	25:00	0.5	
		AIR/O_2								**1**	**8**	17:00		
15	6:40	AIR						3	7	7	41	65:20	1	
		AIR/O_2						3	7	**4**	**21**	42:40		
20	6:00	AIR				2	6	5	7	12	106	144:40	2	
		AIR/O_2				2	6	5	7	**6**	**35**	73:00		
25	5:40	AIR			4	5	5	7	13	24	175	239:20	2.5	
		AIR/O_2			4	5	5	7	13	**13**	**47**	105:40		
30	5:20	AIR		4	4	5	9	11	20	28	257	344:00	3.5	
		AIR/O_2		4	4	5	9	11	20	**14**	**70**	153:20		
35	5:00	AIR	2	5	4	10	11	14	25	29	347	452:40	4	
		AIR/O_2	2	5	4	10	11	14	25	**15**	**89**	196:00		
300 FSW														
Exceptional Exposure														
5	9:20	AIR									6	16:00	0.5	
		AIR/O_2									**3**	13:00		
10	8:20	AIR						2	5	7	32	55:00	1	
		AIR/O_2						2	5	**4**	**16**	36:20		
15	7:20	AIR			1	4	5	6	6	10	102	142:00	1.5	
		AIR/O_2			1	4	5	6	6	**5**	**35**	75:20		
20	6:40	AIR	1	4	5	5	5	6	14	28	196	271:20	2.5	
		AIR/O_2	1	4	5	5	5	6	14	**15**	**52**	124:40		
25	6:40	AIR	7	4	5	5	10	12	25	29	305	409:00	3.5	
		AIR/O_2	7	4	5	5	10	12	25	**15**	**80**	180:20		

CHAPTER 10

Nitrogen-Oxygen Diving Operations

10-1 INTRODUCTION

Nitrogen-oxygen (NITROX) diving is a unique type of diving using nitrogen-oxygen breathing gas mixtures ranging from 75 percent nitrogen/25 percent oxygen to 60 percent nitrogen/40 percent oxygen. Using NITROX significantly increases the amount of time a diver can spend at depth without decompressing. It also decreases the required decompression time compared to a similar dive made to the same depth using air. NITROX may be used in all diving operations suitable for air, but its use is limited to a normal depth of 140 fsw.

NITROX breathing gas mixtures are normally used for shallow dives. The most benefit is gained when NITROX is used shallower than 50 fsw, but it can be advantageous when used to a depth of 140 fsw.

10-1.1 **Advantages and Disadvantages of NITROX Diving.** The advantages of using NITROX rather than air for diving include:

- Extended bottom times for no-decompression diving.
- Reduced decompression time.
- Reduced residual nitrogen in the body after a dive.
- Reduced possibility of decompression sickness.
- Reduced Nitrogen Narcosis

The disadvantages of using NITROX include:

- Increased risk of CNS oxygen toxicity.
- Producing NITROX mixtures requires special equipment.
- NITROX equipment requires special cleaning techniques.
- Long-duration NITROX dives can result in pulmonary oxygen toxicity.
- Working with NITROX systems requires special training.
- NITROX is expensive to purchase.

10-2 EQUIVALENT AIR DEPTH

The partial pressure of nitrogen in a NITROX mixture is the key factor determining the diver's decompression obligation. Oxygen plays no role. The decompression obligation for a NITROX dive therefore can be determined using the Standard Air Tables simply by selecting the depth on air that has the same partial pressure of nitrogen as the NITROX mixture. This depth is called the Equivalent Air Depth (EAD). For example, the nitrogen partial pressure in a 68% nitrogen 32% oxygen mixture at 63 fsw is 2.0 ata. This is the same partial pressure of nitrogen found in air at 50 fsw. 50 fsw is the Equivalent Air Depth.

10-2.1 Equivalent Air Depth Calculation.

The Equivalent Air Depth can be computed from the following formula:

$$\text{EAD} = \frac{(1 - \text{O}_2\%)(\text{D} + 33)}{0.79} - 33$$

Where:

EAD = equivalent depth on air (fsw)
D = diving depth on mixture (fsw)
$O_2\%$ = oxygen concentration in breathing medium (percentage decimal)

For example, while breathing a mixture containing 40 percent oxygen ($O_2\% = 0.40$) at 70 fsw (D = 70), the equivalent air depth would be:

$$\begin{aligned} \text{EAD} &= \frac{(1 - 0.40)(70 + 33)}{0.79} - 33 \\ &= \frac{(0.60)(103)}{0.79} - 33 \\ &= \frac{61.8}{0.79} - 33 \\ &= 78.22 - 33 \\ &= \mathbf{45.2\ fsw} \end{aligned}$$

Note that with NITROX, the Equivalent Air Depth is always shallower than the diver's actual depth. This is the reason that NITROX offers a decompression advantage over air.

10-3 OXYGEN TOXICITY

Although the use of NITROX can increase the diver's bottom time and reduce the risk of nitrogen narcosis, using a NITROX mixture raises the concern for oxygen toxicity. For example, using air as the breathing medium, an oxygen partial pressure (ppO_2) of 1.6 ata is reached at a depth of 218 fsw. In contrast, when using the NITROX mixture containing 60 percent nitrogen and 40 percent oxygen, a ppO_2 of 1.6 ata is reached at 99 fsw. Therefore, oxygen toxicity must be considered when diving a NITROX mixture and is a limiting factor when considering depth and duration of a NITROX dive.

Generally speaking, there are two types of oxygen toxicity—central nervous system (CNS) oxygen and pulmonary oxygen toxicity. CNS oxygen toxicity is usually not encountered unless the partial pressure of oxygen approaches or exceeds 1.6 ata, but it can result in serious symptoms including potentially life-threatening convulsions. Pulmonary oxygen toxicity may result from conducting long-duration dives at oxygen partial pressures in excess of 1.0 ata. For example, a dive longer than 240 minutes at 1.3 ata or a dive longer than 320 minutes at 1.1 ata may place the diver at risk if the exposure is on a daily basis. Pulmonary oxygen toxicity under these conditions can result in decrements of pulmonary function, but is not life threatening.

The NITROX Equivalent Air Depth (EAD) Decompression Selection Table (Table 10-1) was developed considering both CNS and pulmonary oxygen toxicity. Normal working dives that exceed a ppO_2 of 1.4 ata are not permitted, principally to avoid the risk of CNS oxygen toxicity. Dives with a ppO_2 less than 1.4 ata, however, can be conducted using the full range of bottom times allowed by the air tables without concern for CNS or pulmonary oxygen toxicity.

Supervisors must keep in mind that pulmonary oxygen toxicity may become an issue with frequent, repetitive diving. The effects of pulmonary oxygen toxicity can be cumulative and can reduce the underwater work performance of susceptible individuals after a long series of repetitive daily exposures. Fatigue, headache, flu-like symptoms, and numbness of the fingers and toes may also be experienced with repetitive exposures. Table 10-1 takes these repetitive exposures into account, and therefore problems with oxygen toxicity should not be encountered with its use. If symptoms are experienced, the diver should stop diving NITROX until they resolve.

10-3.1 Selecting the Proper NITROX Mixture. Considerable caution must be used when selecting the proper NITROX mixture for a dive. The maximum depth of the dive must be known as well as the planned bottom time. Once the maximum depth is known, the various NITROX mixtures can be evaluated to determine which one will provide the least amount of decompression while also allowing for a maximum bottom time. If a diver's depth exceeds that allowed for a certain NITROX mixture, the diver is at great risk of life-threatening oxygen toxicity.

10-4 NITROX DIVING PROCEDURES

10-4.1 NITROX Diving Using Equivalent Air Depths. NITROX diving is based upon the current Air Decompression Tables. The actual schedule used is adjusted for the oxygen percentage in the breathing gas. To use the EAD Decompression Selection Table (Table 10-1), find the actual oxygen percentage of the breathing gas in the heading and the diver's actual depth in the left column to determine the appropriate schedule to be used from the Air Decompression Tables. The EAD decompression schedule is where the column and row intersect. When using Table 10-1, round all gas mixtures using the standard rounding rule where gas mixes at or above 0.5% round up to the next whole percent and mixes of 0.1% to 0.4% round down to the next whole percent. Once an EAD is determined and an air table is selected, follow the rules of the air table using the EAD for the remainder of the dive.

Table 10-1. *Equivalent Air Depth Table.*

Diver's Actual Depth (fsw)	EAD Feet															
	25% O_2	26% O_2	27% O_2	28% O_2	29% O_2	30% O_2	31% O_2	32% O_2	33% O_2	34% O_2	35% O_2	36% O_2	37% O_2	38% O_2	39% O_2	40% O_2
20	20	20	20	20	20	20	20	15	15	15	15	15	10	10	10	10
30	30	30	30	30	30	30	30	25	25	25	20	20	20	20	20	20
40	40	40	40	40	40	40	40	35	30	30	30	30	30	30	25	25
50	50	50	50	50	50	50	50	40	40	40	40	40	35	35	35	35
60	60	60	60	60	60	60	50	50	50	50	50	50	50	50	40	40
70	70	70	70	70	70	60	60	60	60	60	60	60	50	50	50	50
80	80	80	80	80	70	70	70	70	70	70	70	60	60	60	60	60
90	90	90	90	90	80	80	80	80	80	80	70	70	70 (:107)	70 (:80)	70 (:61)	70 (:47)
100	100	100	100	90	90	90	90	90	90	80 (:113)	80 (:82)	80 (:61)	80 (:46)	80 (:36)	80 (:29)	70 (:23)
110	110	110	110	100	100	100	100	100 (:96)	100 (:69)	90 (:51)	90 (:39)	90 (:30)				
120	120	120	120	110	110	110 (:91)	110 (:64)	110 (:47)	100 (:35)	100 (:27)						
130	130	130	120	120 (:95)	120 (:65)	120 (:47)	120 (:35)	110 (:26)								
140	140	140 (:109)	130 (:73)	130 (:50)	130 (:36)											
150	150 (:89)	150 (:59)	140 (:41)													
160	160 (:50)	160 (:35)														

EAD = Equivalent Air Depth - For Decompression Table Selection Only Rounded to Next Greater Depth

▬ = 1.4 ata Normal working limit.

■ = Depth exceeds the normal working limit, requires the Commanding Officer's authorization and surface-supplied equipment. Repetitive dives are not authorized. Times listed in parentheses indicate maximum allowable exposure.

Note[1]: Depths not listed are considered beyond the safe limits of NITROX diving.

Note[2]: The EAD, 1.4 ata Normal Working Limit Line and Maximum Allowable Exposure Time for dives deeper than the Normal Working Limit Line are calculated assuming the diver rounds the oxygen percentage in the gas mixture using the standard rounding rule discussed in paragraph 10-4.1. The calculations also take into account the allowable ± 0.5 percent error in gas analysis.

10-4.2 **SCUBA Operations.** For SCUBA operations, analyze the nitrox mix in each bottle to be used prior to every dive.

10-4.3 **Special Procedures.** In the event there is a switch to air during the NITROX dive, using the diver's maximum depth and bottom time follow the Air Decompression Table for the actual depth of the dive.

10-4.4 **Omitted Decompression.** In the event that the loss of gas required a direct ascent to the surface, any decompression requirements must be addressed using the standard protocols for "omitted decompression." For omitted decompression dives that exceed the maximum depth listed on Table 10-1, the diving supervisor must rapidly calculate the diver's EAD and follow the omitted decompression procedures based on the diver's EAD, not his or her actual depth. If time will not permit this, the diving supervisor can elect to use the diver's actual depth and follow the omitted decompression procedures.

10-4.5 **Dives Exceeding the Normal Working Limit.** The EAD Table has been developed to restrict dives with a ppO_2 greater than 1.4 ata and limits dive duration based on CNS oxygen toxicity. Dives exceeding the normal working limits of Table 10-1 require the Commanding Officer's authorization and are restricted to surface-supplied diving equipment only. All Equivalent Air Depths provided below the normal working limit line have the maximum allowable exposure time listed alongside. This is the maximum time a diver can safely spend at that depth and avoid CNS oxygen toxicity. Repetitive dives are not authorized when exceeding the normal working limits of Table 10-1.

10-5 NITROX REPETITIVE DIVING

Repetitive diving is possible when using NITROX or combinations of air and NITROX. Once the EAD is determined for a specific dive, the Air Decompression Tables are used throughout the dive using the EAD from Table 10-1.

The Residual Nitrogen Timetable for Repetitive Air Dives will be used when applying the EAD for NITROX dives. Determine the Repetitive Group Designator for the dive just completed using either Table 9-7, Unlimited/No-Decompression Limits and Repetitive Group Designation Table for Unlimited/No-Decompression Air Dives or Table 9-8, Air Decompression Table.

Enter Table 9-7, Residual Nitrogen Timetable for Repetitive Air Dives, using the repetitive group designator. If the repetitive dive is an air dive, use Table 9-7 as is. If the repetitive dive is a NITROX dive, determine the EAD of the repetitive dive from Table 10-1 and use that depth as the repetitive dive depth.

10-6 NITROX DIVE CHARTING

The NITROX Diving Chart (Figure 10-1) should be used for NITROX diving and filled out as described in Chapter 9. The NITROX chart has an additional block for EAD with the percentage of gas written in the bottom mix block.

Date:	Type of Dive: N_2O_2	
Diver 1:	Diver 2:	Standby:
Rig: PSIG: O_2%:	Rig: PSIG: O_2%:	Rig: PSIG: O_2%:
Diving Supervisor:	Chartman:	Bottom Mix:

EVENT	STOP TIME	CLOCK TIME	EVENT	TIME/DEPTH
LS or 20 fsw			Descent Time (Water)	
RB			Stage Depth (fsw)	
LB			Maximum Depth (fsw)	
R 1st Stop			EAD (NITROX)	
190 fsw			Total Bottom Time	
180 fsw			Table/Schedule	
170 fsw			Time to 1st Stop (Planned)	
160 fsw			Time to 1st Stop (Actual)	
150 fsw			Delay to 1st Stop	
140 fsw			Travel/Shift/Vent Time	
130 fsw			Ascent Time-Water/SurD (Actual)	
120 fsw			Undress Time-SurD (Actual)	
110 fsw			Descent Chamber-SurD (Actual)	
100 fsw			Total SurD Surface Interval	
90 fsw			Ascent Time–Chamber (Actual)	
80 fsw			**HOLDS ON DESCENT**	
70 fsw			**DEPTH**	**PROBLEM**
60 fsw				
50 fsw				
40 fsw				
30 fsw			**DELAYS ON ASCENT**	
20 fsw			**DEPTH**	**PROBLEM**
RS				
RB CHAMBER				
50 fsw chamber			**DECOMPRESSION PROCEDURES USED**	
40 fsw chamber			N_2O_2 ☐ In-water N_2O_2 decompression ☐ In-water N_2O_2/O_2 decompression ☐ SurDO_2	
30 fsw chamber				
RS CHAMBER				
TDT	**TTD**			
			REPETITIVE GROUP:	

Remarks:

Figure 10-1. NITROX Diving Chart.

10-7 FLEET TRAINING FOR NITROX

A Master Diver shall conduct training for NITROX diving prior to conducting NITROX diving operations. Actual NITROX dives are not required for this training. The following are the minimum training topics to be covered:

- Pulmonary and CNS oxygen toxicity associated with NITROX diving.
- EAD tables and their association with the air tables.
- Safe handling of NITROX mixtures.

NITROX Charging and Mixing Technicians must be trained on the following topics:

- Oxygen handling safety.
- Oxygen analysis equipment.
- NITROX mixing techniques.
- NITROX cleaning requirements (MIL-STD-1330 Series).

10-8 NITROX DIVING EQUIPMENT

NITROX diving can be performed using a variety of equipment that can be broken down into two general categories: surface-supplied or closed- and open-circuit SCUBA. Closed-circuit SCUBA apparatus is discussed in Chapter 17.

10-8.1 **Open-Circuit SCUBA Systems.** Open-circuit SCUBA systems for NITROX diving are identical to air SCUBA systems with one exception: the SCUBA bottles are filled with NITROX (nitrogen-oxygen) rather than air. There are specific regulators authorized for NITROX diving, which are identified on the ANU list. These regulators have been tested to confirm their compatibility with the higher oxygen percentages encountered with NITROX diving.

10-8.1.1 **Regulators.** SCUBA regulators designated for NITROX use should be cleaned to the standards of MIL-STD-1330. Once designated for NITROX use and cleaned, the regulators should be maintained to the level of cleanliness outlined in MIL-STD-1330.

10-8.1.2 **Bottles.** SCUBA bottles designated for use with NITROX should be oxygen cleaned and maintained to that level. The bottles should have a NITROX label in large yellow letters on a green background. Once a bottle is cleaned and designated for NITROX diving, it should not be used for any other type of diving (Figure 10-2).

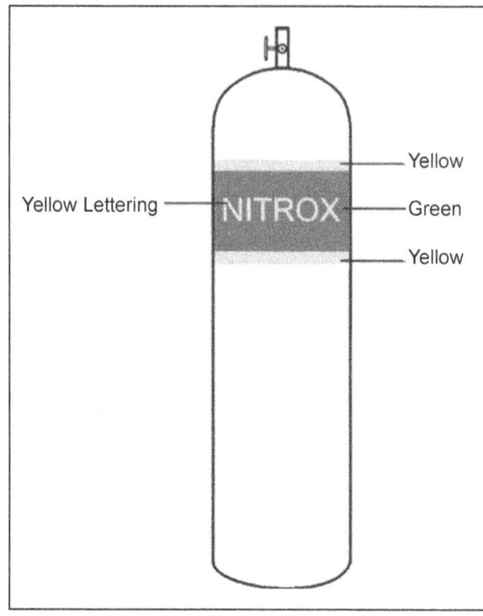

Figure 10-2. NITROX SCUBA Bottle Markings.

10-8.2 **General.** All high-pressure flasks, SCUBA cylinders, and all high-pressure NITROX charging equipment that comes in contact with 100 percent oxygen during NITROX diving, mixing, or charging evolutions must be cleaned and maintained for NITROX service in accordance with the current MIL-STD-1330 series.

10-8.3 **Surface-Supplied NITROX Diving.** Surface-supplied NITROX diving systems must be modified to make them compatible with the higher percentage of oxygen found in NITROX mixtures. A request to convert the system to NITROX must be forwarded to NAVSEA 00C for review and approval. The request must be accompanied by the proposed changes to the Pre-survey Outline Booklet (PSOB) permitting system use with NITROX. Once the system is designated for NITROX, it shall be labeled NITROX with large yellow letters on a green background. MIL-STD-1330D outlines the cleanliness requirements to which a surface-supplied NITROX system must be maintained.

Once a system has been cleaned and designated for NITROX use, only air meeting the requirements of Table 10-2 shall be used to charge the system gas flasks. Air diving, using a NITROX designated system, is authorized if the air meets the purity requirements of Table 10-2.

The EGS used in surface-supplied NITROX diving shall be filled with the same mixture that is being supplied to the diver ± 0.5 percent.

10-9 EQUIPMENT CLEANLINESS

Cleanliness and the procedures used to obtain cleanliness are a concern with NITROX systems. MIL-STD-1330 is applicable to anything with an oxygen level higher than 25 percent by volume. Therefore, MIL-STD-1330 must be followed when dealing with NITROX systems. Personnel involved in the maintenance and repair of NITROX equipment shall complete an oxygen clean worker course, as described in MIL-STD-1330. Even with oxygen levels of 25 to 40 percent, there is still a greater risk of fire than with compressed air. Materials that would not normally burn in air may burn at these higher O_2 levels. Normally combustible materials require less energy to ignite and will burn faster. The energy required for ignition can come from different sources, for example adiabatic compression or particle impact/spark. Another concern is that if improper cleaning agents or processes are used, the agents themselves can become fire or toxic hazards. It is therefore important to adhere to MIL-STD-1330 to reduce the risk of damage or loss of equipment and injury or death of personnel.

10-10 BREATHING GAS PURITY

It is essential that all gases used in producing a NITROX mixture meet the breathing gas purity standards for oxygen (Table 4-3) and nitrogen (Table 4-5). If air is to be used to produce a mixture, it must be compressed using an oil free NITROX approved compressor or meet the purity requirements of oil free air (Table 10-2). Prior to diving, all NITROX gases shall be analyzed using an ANU approved O_2 analyzer accurate to within ± 0.5 percent.

10-11 NITROX MIXING

NITROX mixing can be accomplished by a variety of techniques to produce a final predetermined nitrogen-oxygen mixture. The techniques for mixing NITROX are listed as follows:

1. **Continuous Flow Mixing.** There are two techniques for continuous flow mixing:

   a. **Mix-maker.** A mix-maker uses a precalibrated mixing system that proportions the amount of each gas in the mixture as it is delivered to a common mixing chamber. A mix-maker performs a series of functions that ensures accurate mixtures. The gases are regulated to the same temperature and pressure before they are sent through precision metering valves. The valves are precalibrated to provide the desired mixing pressure. The final mixture can be provided directly to the divers or be compressed using an oil-free compressor into storage banks.

   b. **Oxygen Induction.** Oxygen induction uses a system where low pressure oxygen is delivered to the intake header of an oil-free compressor, where it is mixed with the air being drawn into the compressor. Oxygen flow is adjusted and the compressor output is monitored for oxygen content. When the desired NITROX mixture is attained the gas is diverted to the storage banks for diver use while being continually monitored for oxygen content (Figure 10-3).

2. **Mixing by Partial Pressure.** Partial pressure mixing techniques are similar to those used in helium-oxygen mixed gas diving and are discussed in Chapter 16.

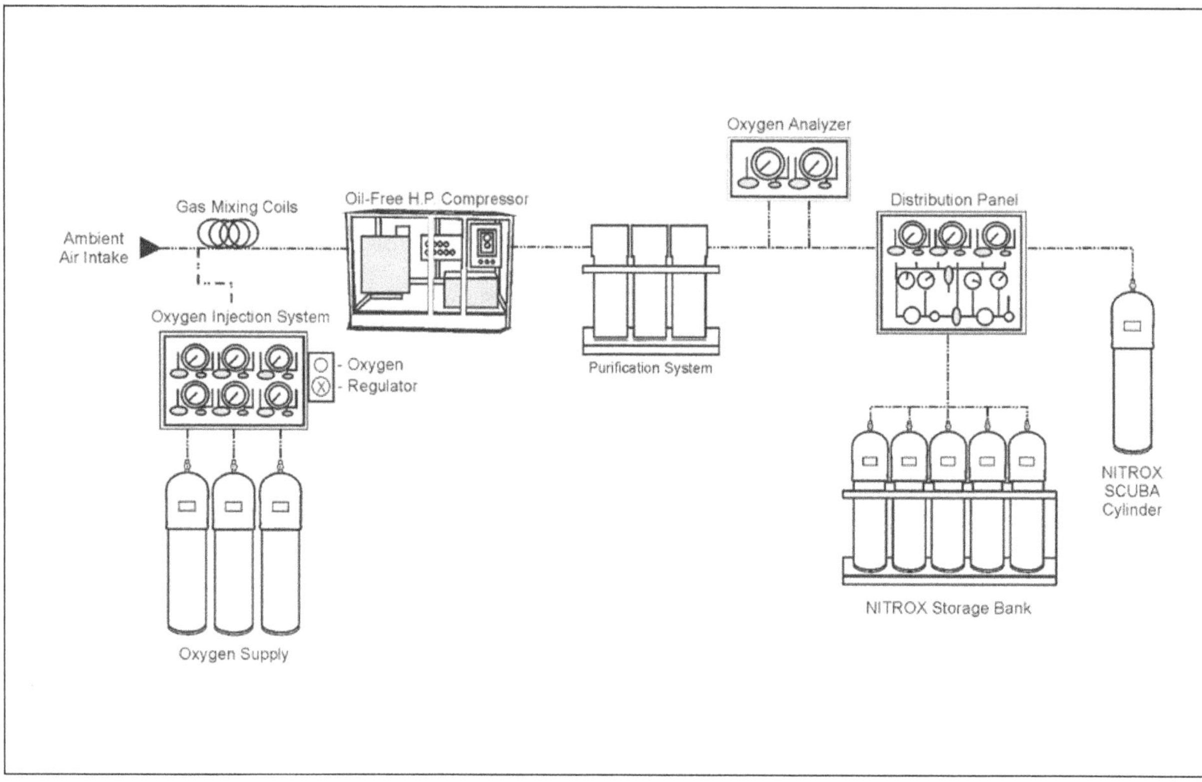

Figure 10-3. NITROX O_2 Injection System.

a. **Partial Pressure Mixing with Air.** Oil-free air can be used as a Nitrogen source for the partial pressure mixing of NITROX using the following procedures:

- Prior to charging air into a NITROX bottle, the NITROX mixing technician shall smell, taste, and feel the oil-free air coming from the compressor for signs of oil, mist, or particulates, or for any unusual smell. If any signs of compressor malfunction are found, the system must not be used until a satisfactory air sample has been completed.

- Prior to charging with oxygen, to produce a NITROX mix, the NITROX-charging technician shall charge the bottle to at least 100 psi with oil-free air. This will reduce the risk of adiabatic compression temperature increase. Once 100 psi of oil-free air has been added to the charging vessel, the required amount of oxygen should then be added. The remaining necessary amount of oil-free air can then be safely charged into the bottle. The charging rate for NITROX mixing shall not exceed 200 psi per minute.

WARNING **Mixing contaminated or non-oil free air with 100% oxygen can result in a catastrophic fire and explosion.**

- Compressed air for NITROX mixing shall meet the purity standards for "Oil Free Air," (Table 10-2). All compressors producing air for NITROX mixing shall have a filtration system designed to produce oil-free air that has been approved by NAVSEA 00C3. In addition, all compressors producing oil-free air for NITROX charging shall have an air sample taken within 90 days prior to use.

Table 10-2. *Oil Free Air.*

Constituent	Specification
Oxygen (percent by volume)	20-22%
Carbon dioxide (by volume)	500 ppm (max)
Carbon monoxide (by volume)	2 ppm (max)
Total hydrocarbons [as Methane (CH_4) by volume]	25 ppm (max)
Odor	Not objectionable
Oil, mist, particulates	0.1 mg/m^3 (max)
Separated Water	None
Total Water	0.02 mg/l (max)
Halogenated Compounds (by volume):	
Solvents	0.2 ppm (max)

3. **Mixing Using a Membrane System.** Membrane systems selectively separate gas molecules of different sizes such as nitrogen or oxygen from the air. By removing the nitrogen from the air in a NITROX membrane system the oxygen percent is increased. The resulting mixture is NITROX. Air is fed into an in-line filter canister system that removes hydrocarbons and other contaminants. It is then passed into the membrane canister containing thousands of hollow membrane fibers. Oxygen permeates across the membrane at a controlled rate. The amount of nitrogen removed is determined by a needle valve. Once the desired nitrogen-oxygen ratio is achieved, the gas is diverted through a NITROX approved compressor and sent to the storage banks (see Figure 10-4 and Figure 10-5). Membrane systems can also concentrate CO_2 and argon.

4. **Mixing Using Molecular Sieves.** Molecular sieves are columns of solid, highly selective chemical absorbent which perform a similar function to membrane systems, and are used in a similar fashion. Molecular sieves have the added advantage of absorbing CO_2 and moisture from the feed gas.

5. **Purchasing Premixed NITROX.** Purchasing premixed NITROX is an acceptable way of obtaining a NITROX mixture. When purchasing premixed NITROX it is requisite that the gases used in the mixture meet the minimum purity standards for oxygen (Table 4-3) and nitrogen (Table 4-5).

10-12 NITROX MIXING, BLENDING, AND STORAGE SYSTEMS

NITROX mixing, blending, and storage systems shall be designed for oxygen service and constructed using oxygen-compatible material following accepted military and commercial practices in accordance with either ASTM G-88, G-63, G-94, or MIL-STD-438 and -777. Commands should contact NAVSEA 00C for specific guidance on developing NITROX mixing, blending, or storage systems. Commands are not authorized to build or use a NITROX system without prior NAVSEA 00C review and approval.

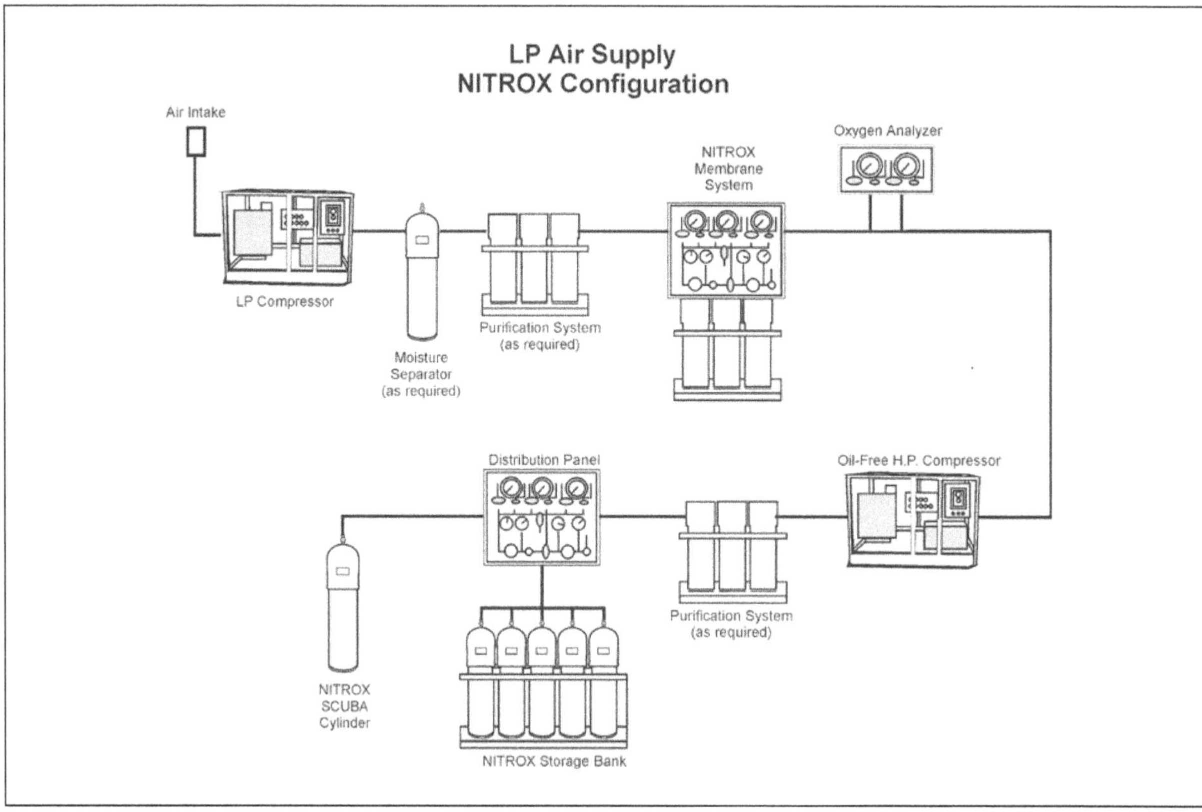

Figure 10-4. LP Air Supply NITROX Membrane Configuration.

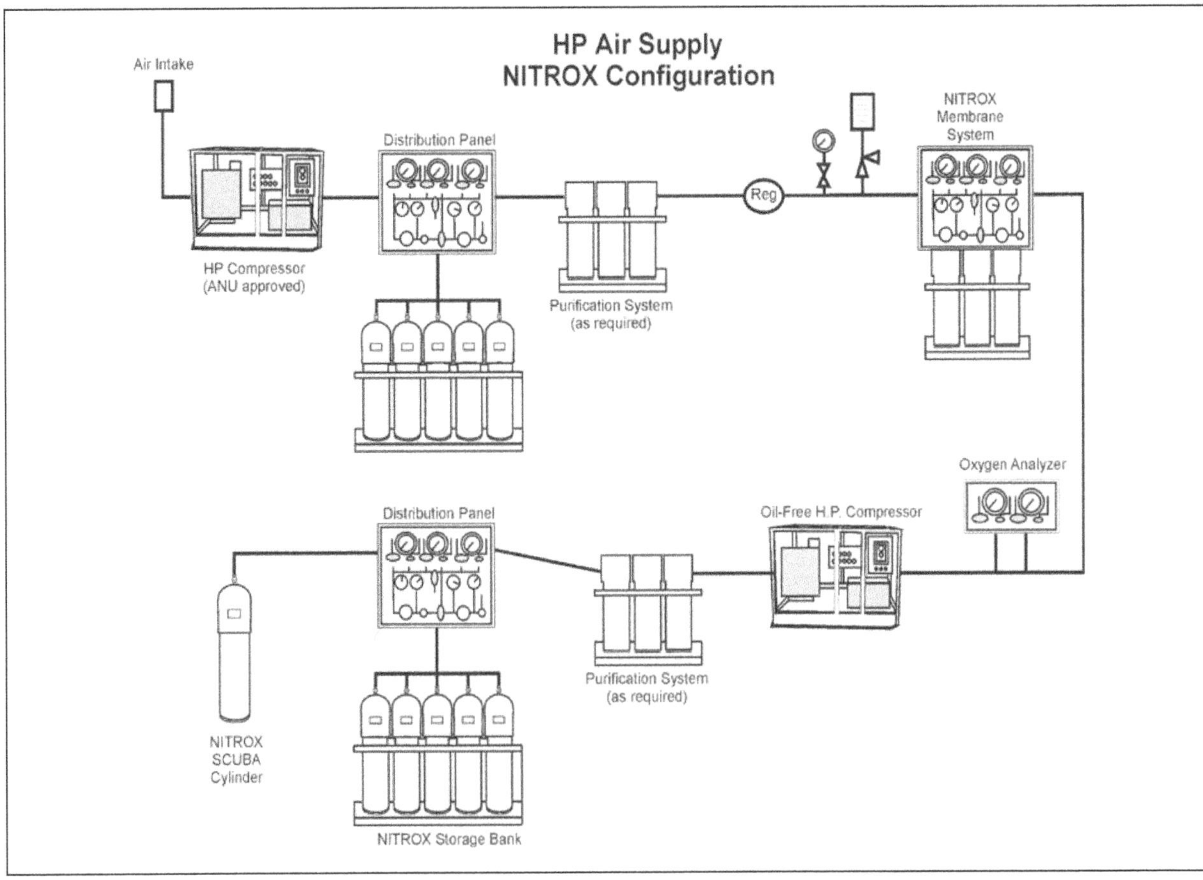

Figure 10-5. HP Air Supply NITROX Membrane Configuration.

PAGE LEFT BLANK INTENTIONALLY

CHAPTER 11

Ice and Cold Water Diving Operations

11-1 INTRODUCTION

11-1.1 **Purpose.** This chapter explains the special requirements for ice and cold water diving.

11-1.2 **Scope.** Polar regions and other cold weather environments are uniquely hostile to divers, topside support personnel, and equipment. Diving where ice cover is present can be extremely hazardous and requires special equipment as well as appropriate operating and support procedures. Awareness of environmental conditions, personnel and equipment selection, and adequate logistical support are vital to mission success and dive team safety.

11-2 OPERATIONS PLANNING

Normal diving procedures generally apply to diving in extremely cold environments. However, there are a number of significant equipment and procedural differences that enhance the diver's safety.

11-2.1 **Planning Guidelines.** The following special planning considerations relate to diving under/near ice cover or in water at or below a temperature of 37°F:

- The task and requirement for ice diving should be reviewed to ascertain that it is operationally essential.
- Environmental conditions such as ice thickness, water depth, temperature, wind velocity, current, visibility, and light conditions should be determined. Ideally, a reconnaissance of the proposed dive site is performed by the Diving Supervisor or a person with ice-covered or cold water diving experience.
- The type of dive equipment chosen must be suited for the operation.
- Logistical planning must include transportation, ancillary equipment, provisioning, fuel, tools, clothing and bedding, medical evacuation procedures, communications, etc.

NOTE **The water temperature of 37°F was set as a limit as a result of Naval Experimental Diving Unit's regulator freeze-up testing. For planning purposes, the guidance above may also be used for diving where the water temperature is above 37°F.**

11-2.2 **Navigational Considerations.** Conditions in cold and ice-covered water affect diver underwater navigation in the following ways:

- The proximity of the magnetic pole in polar regions makes the magnetic compass useless.

- The life of batteries in homing beacons, strobes, and communication equipment is shortened when used in cold water.

- Surface light is so diffused by ice cover that it is nearly impossible to determine its source.

- Direct ascent to the surface is impossible when under the ice and determining return direction is often hindered.

- In shallow ice-covered waters, detours are often required to circumvent keels or pressure ridges beneath the ice.

- With an ice cover, there are no waves and therefore no ripple patterns on the bottom to use for general orientation.

11-2.3 **SCUBA Considerations.** SCUBA equipment has advantages and disadvantages that should be considered when planning a cold water dive.

The advantages of using SCUBA are:

- Portability
- Quick deployment
- Minimal surface-support requirements

The disadvantages of using SCUBA are:

- Susceptibility of regulator to freezing
- Depth limitations
- Limited communications
- Severely limited ability to employ decompression diving techniques
- Duration limitations of CO_2 removal systems in closed-circuit UBA

11-2.4 **SCUBA Regulators.** Refer to the ANU for selection of proper regulator. The single-hose regulator is susceptible to freezing. The first and/or second stage of the single-hose regulator may freeze in the free-flow position after a few minutes of exposure in cold water. The single-hose regulator should be kept in a warm place before diving. It is important that the diver test the regulator in a warm place, then refrain from breathing it until submerging. When returning to the surface, the regulator should remain submerged and the diver should refrain from breathing from the regulator until resubmerging. The diver's time on the surface should be kept to a minimum. Once under the water, chances of a freeze-up are reduced. However, if a regulator is allowed to free-flow at depth for as little as five seconds, freeze-up may occur. The diver should therefore avoid purging the second stage of the regulator when diving in cold water. If water needs to be purged from the mouthpiece, the diver should do so by exhaling into it (Figure 11-1).

Figure 11-1. Ice Diving with SCUBA. Divers in Typhoon dry suits and Aga/Divator FFM SCUBA with approved cold-water regulators.

11-2.4.1 **Special Precautions.** Single-hose regulators should be equipped with an antifreeze cap, which is a special first-stage cap that can be filled with liquid silicone available from the manufacturer. Correct maintenance and application of an approved lubricant to the appropriate points are also essential. Extra precautions must also be taken to make sure that SCUBA cylinders are completely dry inside, that moisture-free air is used, and that the regulator is thoroughly dried prior to use.

11-2.4.2 **Octopus and Redundant Regulators.** Where water temperature is at or below 37°F, a redundant SCUBA system (twin SCUBA bottles, each having a K-valve and an approved cold water regulator) or twin SCUBA bottles with one common manifold and an approved cold water regulator (with octopus) shall be used.

11-2.5 **Life Preserver.** The use of life preservers with CO_2 actuation is prohibited only when diving under ice. The accidental inflation of a life preserver will force the diver upward and may cause a collision with the undersurface of the ice. Should the diver be caught behind a pressure ridge or other subsurface ice structure, recovery may be difficult even with tending lines. Also, the exhaust and inlet valves of the variable volume dry suit will be covered if a life preserver is worn. In the event of a dry suit blow-up, the inability to reach the exhaust dump valve could cause rapid ascent and collision with the surface ice.

11-2.6 **Face Mask.** The diver's mask may show an increased tendency to fog in cold water. An antifog solution should be used to prevent this from occurring. Saliva will not prevent cold water fogging.

11-2.7 **SCUBA Equipment.** The minimum equipment required by every Navy SCUBA diver for under-ice operations consists of:

- Wet suit/variable volume dry suit
- Approved cold water open-circuit SCUBA or closed-circuit UBA, see ANU
- Face mask or approved Full Faced Mask, see ANU
- Weight belt and weights as required
- Knife and scabbard
- Swim fins
- Wrist watch
- Depth gauge
- Submersible SCUBA bottle pressure gauge
- Harness such as an Integrated Divers Vest (IDV), MK 12 jocking harness, etc.
- Lifelines
- Stainless Steel Ice Screws

A variety of special equipment, such as underwater cameras and lift bags, is available to divers [see the NAVSEA/00C Authorized for Navy Use (ANU) list for specific identification of authorized equipment]. However, the effect of extreme cold on the operation of special equipment must be ascertained prior to use.

11-2.8 **Surface-Supplied Diving System (SSDS) Considerations.** Using SSDS in ice-covered or cold water requires detailed operations planning and extensive logistical support. This includes thermal protection for an elaborate dive station and recompression chamber and hot water heating equipment. In addition, dive equipment may require cold climate modification. Because of logistical considerations, SCUBA is used in most ice diving situations. However, SSDS may be required because of prolonged bottom times, depth requirements, and complex communications between topside and diver. When diving in cold water that is not ice covered, logistic and equipment support requirements are reduced; however, very cold water poses many of the same dangers to the surface-supplied diver as ice diving.

11-2.8.1 **Advantages and Disadvantages of SSDS.**

The advantages of using SSDS are:

- Configuration supports bottom-oriented work.
- Hot water suit and variable volume dry suit offer diver maximum thermal and environmental protection.
- Communications cable offers audio communications.
- Gas supply allows maximum duration to the maximum depth limits of diving.

The disadvantages of using SSDS are:

- Air console may freeze up.
- Low-pressure compressors do not efficiently remove moisture from the air which may freeze and clog filters or fracture equipment. This is more likely when the water is very cold and the air is warm. Banks of high-pressure cylinders may have to be used.
- Buildup of air or gas under the ice cover could weaken and fracture thin ice, endangering tenders, other topside personnel, and equipment.
- Movement of ice could foul or drag diver's umbilical.
- Battery life of electronic gear is severely reduced.
- Carbon dioxide removal recirculator components may have to be heated.
- Decompression under extreme cold conditions may be dangerous due to water temperature, ice movement, etc.
- Umbilicals are rigid and difficult to maneuver.
- Failure of hot water heater during in-water decompression must be considered during operational planning.

11-2.8.2 **Effect of Ice Conditions on SSDS.** Ice conditions can prevent or severely affect surface-supplied diving. In general, the ice field must be stationary and thick enough to support the dive station and support equipment. If the dive must be accomplished through an ice floe, the floe must be firmly attached to land or a stable ice field. Severe ice conditions seriously restrict or prohibit surface-supplied diving through the ice (i.e., moving, unstable ice or pack ice and bergs, and deep or jagged pressure ridges could obstruct or trap the diver). In cases where a diver is deployed from a boat in a fixed mooring, the boat, divers, and divers' umbilicals must not be threatened by moving ice floes.

11-2.9 **Suit Selection.** Custom wet suits designed for cold water diving, variable volume dry suits, and hot water suits have all been used effectively for diving in extremely cold water. Each has advantages and disadvantages that must be considered when planning a particular dive mission. All suits must be inspected before use to ensure they are in good condition with no seam separations or fabric cuts.

11-2.9.1 **Wet Suits.** Custom wet suits have the advantages of wide availability, simplicity and less danger of catastrophic failure than dry suits. Although the wet suit is not the equipment of choice, if used the following should be considered:

- The wet suit should be maintained in the best possible condition to reduce water flushing in and out of the suit.

- Wearing heavy insulating socks under the boots in a wet suit will help keep feet warm.

CAUTION **In very cold water, the wet suit is only a marginally effective thermal protective measure, and its use exposes the diver to hypothermia and restricts available bottom time. The use of alternative thermal protective equipment should be considered in these circumstances.**

11-2.9.2 **Variable Volume Dry Suits.** Variable volume dry suits provide superior thermal protection to the surface-supplied or SCUBA diver in the water and on the surface. They are constructed so the entry zipper or seal and all wrist and neck seals are waterproof, keeping the interior dry. They can be inflated orally or from a low-pressure air source via an inlet valve. Air can be exhausted from the suit via a second valve, allowing excellent buoyancy control. The level of thermal protection can be varied through careful selection of the type and thickness of long underwear. However, too much underwear is bulky and can cause overheating, sweating, and subsequent chilling of the standby diver. Dry suit disadvantages are increased swimmer fatigue due to suit bulk, possible malfunction of inlet and exhaust valves, and the need for additional weights for neutral buoyancy. Furthermore, if the diver is horizontal or deployed with the head below the rest of the body, air can migrate into the suit lower extremities, causing overinflation and loss of fins and buoyancy control. A parting seam or zipper could result in a dramatic loss of buoyancy control and thermal shock. Nevertheless, because of its superior thermal protection, the dry suit is an essential component of extremely cold water diving.

CAUTION **Prior to the use of variable volume dry suits and hot water suits in cold and ice-covered waters, divers must be trained in their use and be thoroughly familiar with the operation of these suits.**

11-2.9.3 **Extreme Exposure Suits/Hot Water Suits.** Hot water suits provide excellent thermal protection. If their use can be supported logistically, they are an excellent choice whenever bottom times are lengthy. They are impractical for use by standby divers exposed on the surface.

A hot water system failure can be catastrophic for a diver in very cold water since the hot water is a life support system under such conditions. Hot water temperature must be carefully monitored to ensure that the water is delivered at the proper temperature. When using the hot water suit, wet suit liners must be worn. The hose on the surface must be monitored to ensure it does not melt into the ice. When not in use, the heater and hoses must be thoroughly drained and dried to prevent freezing and rupture.

11-2.10 **Clothing.** Proper planning must include protecting tenders and topside support personnel from the environment. However, bulky clothing and heavy mittens make even routine tasks difficult for topside personnel. Waterproof outer gloves and boots may also be considered. Regardless of the type of clothing selected, the clothing must be properly fitted (loosely worn), and kept clean and dry to maximize insulation. In planning operations for such conditions, reduced efficiency resulting in longer on-site time must be considered. Refer to the *Polar Operations Manual* for complete information on thermal protection of support personnel and equipment.

11-2.11 Ancillary Equipment. A detailed reconnaissance of the dive site will provide the planner with information that is helpful in deciding what ancillary equipment is required. Diving under ice will require special accessory equipment such as a line with lights for underwater navigation, ice-cutting tools, platforms, engine protection kits, and stainless steel ice screws.

The method of cutting the hole through the ice depends on ice thickness and availability of equipment. Normally, two or more of the following tools are used: hand ice chipper, ice handsaw, ice auger, chain saw, thermal ice cutter or blasting equipment. In addition, equipment to lift the ice block, remove the slush, and mark the hole is required. Sandbags, burlap bags, or pallets for the tenders to stand on are also needed. Ladders should be in place in case a tender falls into the hole.

If there is a possibility of surface support personnel falling through the ice, floatable work platforms, such as an inflated Zodiac boat, should be used. With such flotation equipment, the operation could be continued or safely concluded if the ice breaks up.

Gasoline and diesel engines must be cold-weather modified to prevent engine freeze-up. Vibrations of engines running on the ice can be a problem and vibration dampening platforms may be required.

11-2.12 Dive Site Shelter. Tent equipment including framing and flooring material may be required to construct a dive site shelter and a windbreak. Depending on the severity of the climate, remoteness of the site, and duration of the mission, shelters can range from small tents to steel sea-land vans and elaborate insulated huts transported to the site and erected from kits. Dive site shelters should have storage areas for dry items and a place for drying equipment. Benches should be provided for dressing divers, flooring should be installed for insulation, and heating and lighting should be adequate. In an extremely cold and dry climate, fire and inadequate ventilation are ever-present dangers. A carbon monoxide detection kit should be available and periodic checks made of all living and working spaces. Fire extinguishers shall be available in each shelter.

11-3 PREDIVE PROCEDURES

11-3.1 Personnel Considerations. The supervisor of the dive must ensure that all personnel required to make the dive have been properly trained in ice diving techniques and are physically fit. No diver may be allowed to make the dive if, in the opinion of the Diving Supervisor, the diver is suffering from the psychological stress of an ice dive (anxiety, claustrophobia, or recklessness).

11-3.2 Dive Site Selection Considerations. The selection of the dive site will depend upon the purpose of the dive and the geographical environment of the area (ice thickness, ice surface conditions, etc.). Additionally, the diving method chosen, safe access routes, shelter location, emergency holes, and exposure of divers and required support personnel will also have a bearing on site selection.

11-3.3 **Shelter.** When ice diving is conducted, a shelter must be erected as close as possible to the diving site to reduce the probability of frostbite and equipment freeze-up. Normally, tents are not placed over the dive hole because they would restrict the movement of tenders and light available to the diver. However, a windbreak should be constructed. A shelter of modular tents and space heaters is ideal; although precautions must be taken to ensure that the ice beneath the shelter is not weakened. Extreme caution must be used when diving for objects, such as downed aircraft, that have fallen through the ice; the area around the original hole may be dangerously weakened.

11-3.4 **Entry Hole.** Proper equipment should be used to cut a suitable hole or holes through the ice in order to leave a clean edge around the hole. Using a sledgehammer to break through the ice is not recommended as it will weaken the surrounding ice. The hole should be a rectangle 6 feet by 3 feet, or a triangle with six-foot sides as shown in Figure 11-2. The triangular hole is easier to cut and is large enough to allow simultaneous exit by two divers. Slush and ice must be removed from the hole, not pushed under the ice surface, as it could slip back and block the hole. To assist exiting divers and improve footing for other team members on the ice surface, sand, wooden pallets, or burlap bags should be placed on the ice around the hole. Upon completing the dive, the hole must be clearly marked to prevent anyone from falling in accidentally. When possible, the pieces cut from the ice should be replaced to speed up the refreezing process.

11-3.5 **Escape Holes.** Escape holes provide alternative exit points and aid in searching for a lost diver. Downstream escape holes or emergency exit holes must be cut in the ice when diving in a river or bay where there is a current or tidal stream.

11-3.6 **Navigation Lines.** A weighted line should be hung through the hole to aid the diver in retaining his bearing and sense of direction. Suspending a light at the end of the line may be helpful, as well as attaching a series of strobe lights to indicate depth. After locating the work site, a distance line should be laid from the weighted line to the work site. Another method of aiding the diver in keeping his bearings in clear water is to shovel off the snow cover on the ice around the dive site in the form of a spoked wheel (see Figure 11-2). When the ice and snow cover is less than 2 feet thick, the diver should be able to see the spokes leading to the dive hole located at the center of the wheel. The wheel should have a minimum diameter of 60 feet.

11-3.7 **Lifelines.** Diver tending lines are mandatory when diving under ice to help the diver relocate the entrance hole. A polypropylene braided or twisted line has proven to be the best lifeline. It has the advantage of floating up and away from the diver and is available in yellow, white, and orange for high visibility. A bowline or a D-ring and snap hook spliced into the lifeline is the easiest method of attaching the lifeline to the diver. The attachment of the lifeline on both ends must be absolutely secure. Do not tie the line to a vehicle, shovel, first-aid box, or other portable

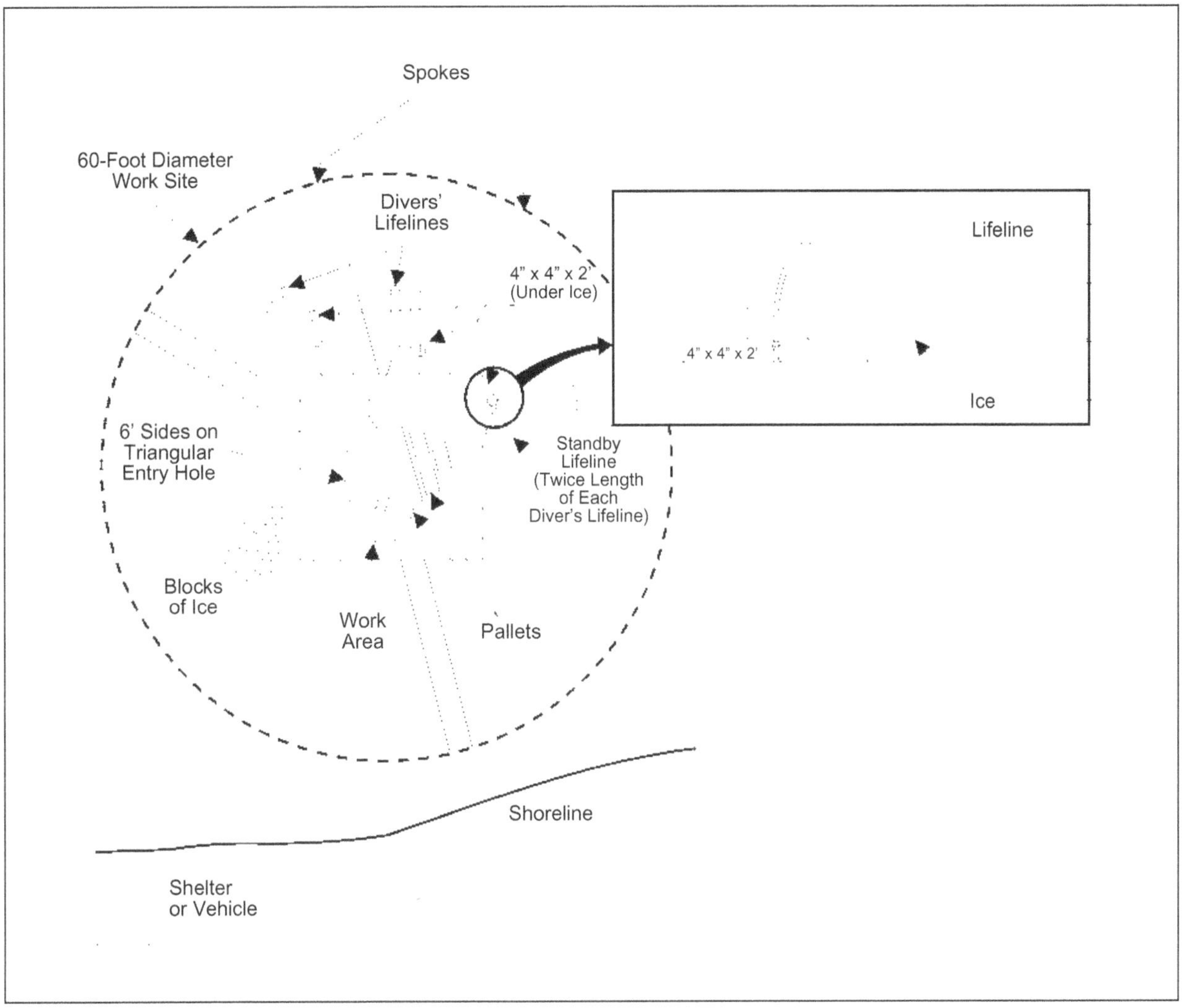

Figure 11-2. Typical Ice Diving Worksite.

equipment. The preferred method to secure the bitter end of the life-line is with a stainless steel ice screw threaded into the ice. Alternatively, a 4-inch by 4-inch by 2-foot board placed under the ice several yards away from the dive hole can be used to secure the bitter end of the lifeline (see Figure 11-2). The D-ring and snap hook allow the quickest transfer of the lifeline from diver to diver on the surface, provided the snap hooks are not frozen shut. The snap hooks should be checked for corrosion at frequent intervals. A wet lifeline must be kept off the bare ice to prevent it from freezing to the surface.

11-3.8 **Equipment Preparation.** The diver must wear a distress light that should be turned on upon entering the water. Divers should not be encumbered with unnecessary equipment during cold water dives. Snorkels should be removed and knives worn on the inside of the leg to help prevent the lifeline from snagging on the diver's equipment. Personnel, divers, and tenders must handle rubber accessories such as masks and fins carefully; extreme cold causes them to become brittle.

11-4 UNDERWATER PROCEDURES

11-4.1 **Buddy Diving.** Diving under the ice or in extremely cold waters requires the use of paired dive partners. When diving through the ice, the pair shall always be surface tended. The life-threatening consequences of suit failure, regulator freeze-up or other equipment problems make a solitary tended SCUBA diver particularly vulnerable. Divers must practice buddy breathing prior to the operation because of the increased possibility that buddy breathing will be required. Proficiency in the process will minimize loss of valuable time during an emergency. Using approved cold water SCUBA equipment will minimize or eliminate freeze-up problems (see paragraph 11-2.3).

11-4.2 **Tending the Diver.** The lifeline is to be held by the tender at all times. As an additional safety measure during ice diving, the end of the lifeline must be secured to a stationary object to prevent it from falling into the entry hole should it be dropped by the tender (see Figure 11-2). It is recommended that the lifeline be marked at 10-foot intervals to allow the tender and Diving Supervisor to estimate the diver's position. However, the diver's radial position can only be roughly estimated. The dive team must be thoroughly familiar with the procedures for lifeline tending in Chapter 8.

Tending line sensitivity and awareness of the diver's position by tenders may be difficult with the added factors of lifeline drag on subsurface ice formations, line drag over the lip of the under-ice hole, tending through heavy mittens, and the lack of surface bubbles.

11-4.3 **Standby Diver.** The standby diver and tender must be immediately available. The standby diver should be kept warm until the Diving Supervisor determines that the standby diver is needed. If possible a shelter or windbreak at the hole should be used. The lifeline of the standby diver should be twice the length of the diver's lifeline in order to perform a thorough circular search. The standby diver must be dressed with the exception of fins, mask, and tanks. These will be ready to don immediately.

11-5 OPERATING PRECAUTIONS

Normal procedures generally apply to diving in extremely cold environments. However, the increased likelihood of regulator freeze-up calls for total familiarity with the buddy breathing procedures described in Chapter 7. This section outlines some of the precautions for operating in cold and ice-covered water.

11-5.1 **General Precautions.** General precautions for ice and cold water diving operations include:

- Divers should be well rested, have a meal high in carbohydrates and protein, and should not consume any alcohol. Alcohol dilates the blood vessels in the skin, thus increasing body heat loss.

- Bathing is an important health measure to prevent infectious diseases prevalent in cold environments. If necessary, the body can be sponge-bathed under clothing.

- After bathing, a soothing ointment or lotion should be applied to the skin to keep it soft and protect it against evaporation caused by the dry air.

- Shaving and washing the face should be done in the evening because shaving removes protective oils from the skin. Shaving too close can also remove some of the protective layer of the skin, promoting frostbite.

11-5.2 **Ice Conditions.** The inconsistency and dynamics of ice conditions in any particular area can make diving operations extremely hazardous. The movement of ice floes can be very significant over a relatively short period of time, requiring frequent relocation of dive sites and the opening of new access holes in order to work a fixed site on the sea floor. Diving from drifting ice or in the midst of broken free ice is dangerous and should be conducted only if absolutely necessary.

Differential movement of surface and subsurface pressure ridges or icebergs could close an access hole, sever a diving umbilical, and isolate or crush a diver. The opening of a rift in the ice near a dive site could result in loss of support facilities on the ice, as well as diver casualties.

11-5.3 **Dressing Precautions.** With a properly fitting suit and all seals in place, the diver can usually be kept warm and dry for short periods in even the coldest water. When dressing for an ice or cold water dive:

- Thermal protection suits should be checked carefully for fabric cuts and separations. Thermal protection suits should expose only a minimum of facial area.

- Mittens, boots, and seals should prevent water entry, while causing no restriction of circulation. Wearing a knitted watchcap under the hood of a dry suit is effective in conserving body heat. With the cap pushed back far enough to permit the suit's face seal to seat properly, the head will be relatively dry and comfortable.

11-5.4 **On-Surface Precautions.** While on the surface:

- Suited divers should be protected from overheating and associated perspiring before entering the water. Overheating easily occurs when operating from a heated hut, especially if diver exertion is required to get to the dive site. The divers' comfort can be improved and sweating delayed before entering the water by cooling the divers face with a damp cloth and fanning every few minutes. Perspiration will dampen undergarments, greatly reducing their thermal insulating capabilities.

- While waiting to enter the water, divers should avoid sitting on or resting their feet on the ice or cold floor of a hut. Even in an insulated hut, the temperature at the floor may be near freezing.

- Time on the surface with the diver suited, but relatively inactive, should be minimized to prevent chilling of the diver. Surface time can also cool metal components of the diving gear, such as suit valves and SCUBA regulators, below the freezing point and cause the parts to ice up when the diver enters the water. Dressing rehearsals prior to diving will help minimize surface delays.

- When operating from an open boat, heavy parkas or windbreakers should be worn over the exposure suits.

- When operating at the surface in newly formed ice, care should be taken to avoid cutting exposed facial skin. Such wounds occur easily and, although painless because of the numbness of the skin, usually bleed profusely.

- Diving from a beach and without a support vessel should be limited to a distance that allows the divers to return to the beach if the suit floods.

- Extreme caution must be exercised when diving near ice keels in polar regions as they will often move with tidal action, wind, or current. In doing so, they can foul umbilicals and jeopardize the divers' safety.

11-5.5 In-Water Precautions.

- Because severe chilling can result in impaired judgment, the tasks to be performed under water must be clearly identified, practiced, and kept simple.

- A dive should be terminated upon the onset of involuntary shivering or severe impairment of manual dexterity.

- If the exposure suit tears or floods, the diver should surface immediately, regardless of the degree of flooding. The extreme chilling effect of frigid water can cause thermal shock within minutes, depending on the extent of flooding.

- Divers and Diving Supervisors must be aware of the cumulative thermal effect of repetitive diving. A thermal debt can accumulate over successive diving days, resulting in increased fatigue and reduced performance. The progressive hypothermia associated with long, slow cooling of the body appears to cause significant core temperature drop before shivering and heat production begins.

11-5.6 Postdive Precautions. Upon exiting cold water, a diver will probably be fatigued and greatly susceptible to additional chilling:

- If a wet suit was worn, immediate flushing with warm water upon surfacing will have a comforting, heat-replacing effect.

- Facilities must be provided to allow the diver to dry off in a comfortable, dry and relatively warm environment to regain lost body heat.

- The diver should remove any wet dress, dry off, and don warm protective clothing as soon as possible. Personnel should have warm, dry clothing, blankets, and hot non-alcoholic beverages available to them.

11-6 EMERGENCY PROCEDURES

11-6.1 **Lost Diver.** A diver who becomes detached from the lifeline and cannot locate the entrance hole should:

1. Ascend to the underside of the ice.
2. Remove weight belt and allow it to drop.
3. Thread an ice screw onto underside of the ice to maintain position.
4. Remain in a vertical position, to maximize vertical profile and thereby snag the searching standby diver's lifeline.
5. Watch for lifeline and the lifeline of the standby diver and wait for the standby diver to arrive. The lost diver MUST NOT attempt to relocate the hole. The diver must remain calm and watch for the standby diver.

11-6.2 **Searching for a Lost Diver.** As soon as the tender fails to get a response from the diver, the tender must notify the Diving Supervisor immediately. These procedures are to be implemented at once:

1. The Diving Supervisor shall immediately recall all other divers.
2. The Diving Supervisor must estimate the probable location of the lost diver by assessing the diver's speed and direction of travel.
3. As directed by the Diving Supervisor, the standby diver enters the water and swims in the indicated direction, a distance equal to twice that believed to be covered by the lost diver. The distance may be the full extent of the standby diver's lifeline since it is twice as long as the lost diver's lifeline.
4. The tender must keep the standby diver's lifeline taut.
5. The standby diver conducts a circular sweep.
6. When the lifeline snags on the lost diver, the standby diver swims toward the diver signaling the tender to take up slack.
7. Upon locating the lost diver, the standby diver assists the diver back to the hole.
8. If the first sweep fails, it should be repeated only once before moving the search to the most likely emergency hole.

11-6.3 **Hypothermia.** When diving in cold water, hypothermia may predispose the diver to decompression sickness. Hypothermia is easily diagnosed. The hypothermic diver loses muscle strength, the ability to concentrate and may become irrational or confused. The victim may shiver violently, or, with severe hypothermia, shivering may be replaced by muscle rigidity. Profound hypothermia may so depress the heartbeat and respiration that the victim appears dead. However, a diver should not be considered dead until the diver has been rewarmed and all resuscitation attempts have been proven to be unsuccessful.

Hypothermia demands immediate treatment and prompt evacuation to a medical facility. A hypothermic diver must not be allowed to walk; the diver should be transported in a horizontal position. Improper handling of the diver can cause dangerous rhythms of the heart and a drop in the body core temperature, known as after drop.

11-7 ADDITIONAL REFERENCES

For information on extreme cold weather conditions and the polar environment, refer to:

- *A Guide to Extreme Cold Weather Operations* (Naval Safety Center, July 1986)
- *Polar Operations Manual* S0300-A5-MAN-010 (Naval Coastal Systems Center) (NCSC)
- *Guide to Polar Diving* (Office of Naval Research, June 1976)
- *UCT Arctic Operation Manual* NAVFAC P-992
  (To obtain a copy of this manual, contact NAVFAC Ocean Facilities Programs.)

APPENDIX 2A

Optional Shallow Water Diving Tables

2-A1.1 **Introduction.** At the shallow depths typical of ship husbandry diving, a small change in the diver's maximum depth can make a significant difference in the allowable no-decompression time. For example, at 35 fsw the no-decompression time on air is 232 minutes; at 40 fsw it is only 163 minutes, more than an hour less. When the diver's maximum depth is accurately known at the beginning of the dive, e.g., in ballast tank dives, or when continuous electronic depth recording is available, e.g., with a decompression computer, use of a decompression table with depth listed in one-foot increments rather than five-foot increments may result in a significant gain in no-decompression time.

Shallow Water Diving Tables covering the depth range of 30–50 fsw in one-foot increments are given in Tables 2A-1 and 2A-2. These tables are simply an expansion of Tables 9-7 and 9-8 and the rules for using Tables 2A-1 and 2A-2 are identical to the rules for using Tables 9-7 and 9-8. These Shallow Water Diving Tables are optional. They may be used instead of Tables 9-7 and 9-8 if they offer a gain in no-decompression time. The Optional Shallow Water Diving Tables are most suited to ship husbandry diving, but can be used in other shallow air diving applications as well.

Table 2A-1. *No-Decompression Limits and Repetitive Group Designators for Shallow Water Air No-Decompression Dives.*

Depth (fsw)	No-Stop Limit (min)	Repetitive Group Designation															
		A	B	C	D	E	F	G	H	I	J	K	L	M	N	O	Z
30	371	17	27	38	50	62	76	91	107	125	145	167	193	223	260	307	371
31	334	16	26	37	48	60	73	87	102	119	138	158	182	209	242	282	334
32	304	15	25	35	46	58	70	83	98	114	131	150	172	197	226	261	304
33	281	15	24	34	45	56	67	80	94	109	125	143	163	186	212	243	281
34	256	14	23	33	43	54	65	77	90	104	120	137	155	176	200	228	256
35	232	14	23	32	42	52	63	74	87	100	115	131	148	168	190	215	232
36	212	14	22	31	40	50	61	72	84	97	110	125	142	160	180	204	212
37	197	13	21	30	39	49	59	69	81	93	106	120	136	153	172	193	197
38	184	13	21	29	38	47	57	67	78	90	102	116	131	147	164	184	
39	173	12	20	28	37	46	55	65	76	87	99	112	126	141	157	173	
40	163	12	20	27	36	44	53	63	73	84	95	108	121	135	151	163	
41	155	12	19	27	35	43	52	61	71	81	92	104	117	130	145	155	
42	147	11	19	26	34	42	50	59	69	79	89	101	113	126	140	147	
43	140	11	18	25	33	41	49	58	67	76	87	98	109	122	135	140	
44	134	11	18	25	32	40	48	56	65	74	84	95	106	118	130	134	
45	125	11	17	24	31	39	46	55	63	72	82	92	102	114	125		
46	116	10	17	23	30	38	45	53	61	70	79	89	99	110	116		
47	109	10	16	23	30	37	44	52	60	68	77	87	97	107	109		
48	102	10	16	22	29	36	43	51	58	67	75	84	94	102			
49	97	10	16	22	28	35	42	49	57	65	73	82	91	97			
50	92	9	15	21	28	34	41	48	56	63	71	80	89	92			

Table 2A-2. *Residual Nitrogen Time Table for Repetitive Shallow Water Air Dives.*

Locate the diver's repetitive group designation from his previous dive along the diagonal line above the table. Read horizontally to the interval in which the diver's surface interval lies.

Next, read vertically downward to the new repetitive group designation. Continue downward in this same column to the row that represents the depth of the repetitive dive. The time given at the intersection is residual nitrogen time, in minutes, to be applied to the repetitive dive.

* Dives following surface intervals longer than this are not repetitive dives. Use actual bottom times in the Air Decompression Tables to compute decompression for such dives.

Repetitive Group at Beginning of Surface Interval

Group at Beginning	Z	O	N	M	L	K	J	I	H	G	F	E	D	C	B	A
A																:10 2:20 *
B															:10 1:16	1:17 3:36 *
C														:10 :55	:56 2:11	2:12 4:31 *
D													:10 :52	:53 1:47	1:48 3:03	3:04 5:23 *
E												:10 :52	:53 1:44	1:45 2:39	2:40 3:55	3:56 6:15 *
F											:10 :52	:53 1:44	1:45 2:37	2:38 3:31	3:32 4:48	4:49 7:08 *
G										:10 :52	:53 1:44	1:45 2:37	2:38 3:29	3:30 4:23	4:24 5:40	5:41 8:00 *
H									:10 :52	:53 1:44	1:45 2:37	2:38 3:29	3:30 4:21	4:22 5:16	5:17 6:32	6:33 8:52 *
I								:10 :52	:53 1:44	1:45 2:37	2:38 3:29	3:30 4:21	4:22 5:13	5:14 6:08	6:09 7:24	7:25 9:44 *
J							:10 :52	:53 1:44	1:45 2:37	2:38 3:29	3:30 4:21	4:22 5:13	5:14 6:06	6:07 7:00	7:01 8:16	8:17 10:36 *
K						:10 :52	:53 1:44	1:45 2:37	2:38 3:29	3:30 4:21	4:22 5:13	5:14 6:06	6:07 6:58	6:59 7:52	7:53 9:09	9:10 11:29 *
L					:10 :52	:53 1:44	1:45 2:37	2:38 3:29	3:30 4:21	4:22 5:13	5:14 6:06	6:07 6:58	6:59 7:50	7:51 8:44	8:45 10:01	10:02 12:21 *
M				:10 :52	:53 1:44	1:45 2:37	2:38 3:29	3:30 4:21	4:22 5:13	5:14 6:06	6:07 6:58	6:59 7:50	7:51 8:42	8:43 9:37	9:38 10:53	10:54 13:13 *
N			:10 :52	:53 1:44	1:45 2:37	2:38 3:29	3:30 4:21	4:22 5:13	5:14 6:06	6:07 6:58	6:59 7:50	7:51 8:42	8:43 9:34	9:35 10:29	10:30 11:45	11:46 14:05 *
O		:10 :52	:53 1:44	1:45 2:37	2:38 3:29	3:30 4:21	4:22 5:13	5:14 6:06	6:07 6:58	6:59 7:50	7:51 8:42	8:43 9:34	9:35 10:27	10:28 11:21	11:22 12:37	12:38 14:58 *
Z	:10 :52	:53 1:44	1:45 2:37	2:38 3:29	3:30 4:21	4:22 5:13	5:14 6:06	6:07 6:58	6:59 7:50	7:51 8:42	8:43 9:34	9:35 10:27	10:28 11:19	11:20 12:13	12:14 13:30	13:31 15:50 *

Repetitive Group at the End of the Surface Interval

Dive Depth	Z	O	N	M	L	K	J	I	H	G	F	E	D	C	B	A
30	372	308	261	224	194	168	146	126	108	92	77	63	51	39	28	18
31	334	282	243	210	183	159	139	120	103	88	74	61	49	38	27	17
32	305	262	227	198	173	151	132	115	99	85	71	59	47	36	26	17
33	282	244	213	187	164	144	126	110	95	81	69	57	46	35	25	16
34	262	229	201	177	156	138	121	105	91	78	66	55	44	34	25	16
35	245	216	191	169	149	132	116	101	88	75	64	53	43	33	24	15
36	231	204	181	161	143	126	111	98	85	73	62	51	41	32	23	15
37	218	194	173	154	137	122	107	94	82	70	60	50	40	31	23	14
38	207	185	165	148	132	117	103	91	79	68	58	48	39	30	22	14
39	197	177	158	142	127	113	100	88	77	66	56	47	38	29	21	14
40	188	169	152	136	122	109	97	85	74	64	55	45	37	29	21	13
41	180	163	146	132	118	105	93	82	72	62	53	44	36	28	20	13
42	173	156	141	127	114	102	91	80	70	61	52	43	35	27	20	13
43	166	150	136	123	110	99	88	78	68	59	50	42	34	26	19	12
44	160	145	131	119	107	96	85	75	66	57	49	41	33	26	19	12
45	154	140	127	115	104	93	83	73	64	56	48	40	32	25	18	12
46	149	136	123	111	101	90	81	71	63	54	46	39	32	25	18	12
47	144	131	119	108	98	88	78	70	61	53	45	38	31	24	18	11
48	139	127	116	105	95	85	76	68	60	52	44	37	30	24	17	11
49	135	123	112	102	92	83	74	66	58	51	43	36	30	23	17	11
50	131	120	109	99	90	81	73	65	57	49	42	35	29	23	17	11

Residual Nitrogen Times (Minutes)

PAGE LEFT BLANK INTENTIONALLY

VOLUME 3

Mixed Gas Surface Supplied Diving Operations

U.S. NAVY DIVING MANUAL

Volume 3 - Table of Contents

Chap/Para Page

Chap/Para	Page

Chap/Para		Page

Chap/Para | Page

PAGE LEFT BLANK INTENTIONALLY

Volume 3 - List of Illustrations

PAGE LEFT BLANK INTENTIONALLY

Volume 3 - List of Tables

PAGE LEFT BLANK INTENTIONALLY

CHAPTER 12

Mixed-Gas Diving Theory

12-1 INTRODUCTION

12-1.1 **Purpose.** The fundamental laws and concepts of underwater physics presented in Chapter 2 (Volume 1) are basic to a proper understanding of mixed-gas diving techniques. In mixed-gas diving, calculations requiring the use of the various gas laws are vital to safe diving. A thorough working knowledge of the application of the gas laws is mandatory for the mixed-gas diver. This chapter reviews the gas laws.

12-1.2 **Scope.** This chapter discusses the theory and techniques used in mixed-gas diving.

12-2 BOYLE'S LAW

Boyle's law states that at constant temperature, the absolute pressure and the volume of gas are inversely proportional. As pressure increases, the gas volume is reduced; as the pressure is reduced, the gas volume increases.

The formula for expressing Boyle's law is:

$$C = P \times V$$

Where:

C *is* constant
P *is* absolute pressure
V *is* volume

Boyle's law can also be expressed as:

$$P_1V_1 = P_2V_2$$

Where:

P_1 = initial pressure
V_1 = initial volume
P_2 = final pressure
V_2 = final volume

When working with Boyle's law, absolute pressure may be measured in atmospheres absolute. To calculate absolute pressure using atmospheres absolute:

$$P_{ata} = \frac{\text{Depth fsw} + 33 \text{ fsw}}{33 \text{ fsw}} \quad \text{or} \quad P_{ata} = \frac{\text{psig} + 14.7\text{psi}}{14.7 \text{ psi}}$$

Sample Problem 1. The average gas flow requirements of a diver using a MK 21 MOD 1 UBA doing moderate work is 1.4 acfm when measured at the depth of the diver. Determine the gas requirement, expressed in volume per minute at surface conditions, for a diver working at 132 fsw.

1. Rearrange the formula for Boyle's law to find the initial volume (V_1):

$$V_1 = \frac{P_2 V_2}{P_1}$$

2. Calculate the final pressure (P_2):

$$P_2 = \frac{132\text{ fsw} + 33\text{ fsw}}{33\text{ fsw}}$$
$$= 5\text{ ata}$$

3. Substitute known values to find the initial volume (V_1):

$$V_1 = \frac{5\text{ ata} \times 1.4\text{ acfm}}{1\text{ ata}}$$
$$= 7.0\text{ acfm}$$

4. The gas requirement for a diver working at 132 fsw is 7.0 acfm.

Sample Problem 2. Determine the gas requirement, expressed in volume per minute at surface conditions, for a diver working at 231 fsw.

1. Rearrange the formula for Boyle's law to find the initial volume (V_1):

$$V_1 = \frac{P_2 V_2}{P_1}$$

2. Calculate the final pressure (P_2):

$$P_2 = \frac{231\text{ fsw} + 33\text{ fsw}}{33\text{ fsw}}$$
$$= 8\text{ ata}$$

3. Substitute the known values to find the initial volume (V_1):

$$V_1 = \frac{8\text{ ata} \times 1.4\text{ acfm}}{1\text{ ata}}$$
$$= 11.2\text{ acfm}$$

The gas requirement for a diver working at 231 fsw is 11.2 surface acfm.

Sample Problem 3. Determine the gas requirement, expressed in volume per minute at surface conditions, for a diver working at 297 fsw.

1. Rearrange the formula for Boyle's law to find the initial volume (V_1):

$$V_1 = \frac{P_2 V_2}{P_1}$$

2. Calculate the final pressure (P_2):

$$\begin{aligned} P_2 &= \frac{297 \text{ fsw} + 33 \text{ fsw}}{33 \text{ fsw}} \\ &= 10 \text{ ata} \end{aligned}$$

3. Substitute the known values to find the initial volume (V_1):

$$\begin{aligned} V_1 &= \frac{10 \text{ ata} \times 1.4 \text{ acfm}}{1 \text{ ata}} \\ &= 14.0 \text{ acfm} \end{aligned}$$

The gas requirement for a diver working at 297 fsw is 14.0 surface acfm.

Sample Problem 4. An open diving bell of 100-cubic-foot internal volume is to be used to support a diver at 198 fsw. Determine the pressure and total surface equivalent volume of the helium-oxygen gas that must be in the bell to balance the ambient water pressure at depth.

1. Calculate final pressure (P_2):

$$\begin{aligned} P_2 &= \frac{198 \text{ fsw} + 33 \text{ fsw}}{33 \text{ fsw}} \\ &= 7 \text{ ata} \end{aligned}$$

2. Rearrange the formula to solve for the initial volume (V_1):

$$V_1 = \frac{P_2 V_2}{P_1}$$

3. Substitute the known values to find the initial volume (V_1):

$$\begin{aligned} V_1 &= \frac{7 \text{ ata} \times 100 \text{ ft}^3}{1 \text{ ata}} \\ &= 700 \text{ ft}^3 \end{aligned}$$

There must be 700 ft^3 of helium-oxygen gas in the bell to balance the water pressure at depth.

Sample Problem 5. The open bell described in Sample Problem 4 is lowered to 297 fsw after pressurization to 198 fsw and no more gas is added. Determine the gas volume in the bell at 297 fsw.

1. Calculate the final pressure (P_2):

$$P_2 = \frac{297 \text{ fsw} + 33 \text{ fsw}}{33 \text{ fsw}}$$
$$= 10 \text{ ata}$$

2. Rearrange the formula to solve for the final volume (V_2):

$$V_2 = \frac{P_1 V_1}{P_2}$$

3. Substitute the known values to find the final volume (V_2):

$$V_2 = \frac{7 \text{ ata} \times 100 \text{ ft}^3}{10 \text{ ata}}$$
$$= 70 \text{ ft}^3$$

The gas volume in the bell at 297 fsw is 70 ft^3.

12-3 CHARLES'/GAY-LUSSAC'S LAW

Charles' and Gay-Lussac's laws state that at a constant pressure, the volume of a gas is directly proportional to the change in the absolute temperature. If the pressure is kept constant and the absolute temperature is doubled, the volume will double. If temperature decreases, volume decreases. If volume instead of pressure is kept constant (i.e., heating gas in a rigid container), then the absolute pressure will change in proportion to the absolute temperature.

The formula for expressing Charles'/Gay-Lussac's law when the pressure is constant is:

$$V_2 = \frac{V_1 T_2}{T_1}$$

Where:

V_1 = initial volume
V_2 = final volume
T_1 = initial absolute temperature
T_2 = final absolute temperature

The formula for expressing Charles'/Gay-Lussac's law when the volume is constant is:

$$P_2 = \frac{P_1 T_2}{T_1}$$

Where:

P_1 = initial absolute pressure
P_2 = final absolute pressure
T_1 = initial absolute temperature
T_2 = final absolute temperature

Sample Problem 1. The on-board gas supply of a PTC is charged on deck to 3,000 psig at an ambient temperature of 32°C. The capsule is deployed to a depth of 850 fsw where the water temperature is 7°C. Determine the pressure in the gas supply at the new temperature. Note that in this example the volume is constant; only pressure and temperature change.

1. Transpose the formula for Charles'/Gay-Lussac's law to solve for the final pressure:

$$P_2 = \frac{P_1 T_2}{T_1}$$

2. Convert Celsius temperatures to absolute temperature values (Kelvin):

$°K = °C + 273$

$T_1 = 32°C + 273 = 305°K$

$T_2 = 7°C + 273 = 280°K$

3. Convert initial pressure to absolute pressure:

$$P_1 = \frac{3,000 \text{ psig} + 14.7 \text{ psi}}{14.7 \text{ psi}}$$
$$= 205 \text{ ata}$$

4. Substitute known values to find the final pressure:

$$P_2 = \frac{205 \text{ ata} \times 280°K}{305°K}$$
$$= 188.19 \text{ ata}$$

5. Convert the final pressure to gauge pressure:

$$P_2 = (188.19 \text{ ata} - 1 \text{ ata}) \times (14.7 \text{ psi})$$

$$= 2{,}751.79 \text{ psig}$$

The pressure in the gas supply at the new temperature is 2,751 psig.

Sample Problem 2. A habitat is deployed to a depth of 627 fsw at which the water temperature is 40°F. It is pressurized from the surface to bottom pressure, and because of the heat of compression, the internal temperature rises to 110°F. The entrance hatch is opened at depth and the divers begin their work routine. During the next few hours, the habitat atmosphere cools down to the surrounding sea water temperature because of a malfunction in the internal heating system. Determine the percentage of the internal volume that would be flooded by sea water assuming no additional gas was added to the habitat. Note that in this example pressure is constant; only volume and temperature change.

1. Convert Fahrenheit temperatures to absolute temperature values (Rankine):

$$°R = °F + 460$$

$$T_1 = 110°F + 460$$

$$= 570°R$$

$$T_2 = 40°F + 460$$

$$= 500°R$$

2. Substitute known values to solve for the final volume:

$$V_2 = \frac{V_1 T_2}{T_1}$$

$$= V_1 \times \frac{500°R}{570°R}$$

$$= 0.88 V_1$$

3. Change the value to a percentage:

$$V_2 = (0.88 \times 100\%) V_1$$

$$= 88\% V_1$$

4. Calculate the flooded volume:

$$\text{Flooded volume} = 100\% - 88\%$$

$$= 12\%$$

Sample Problem 3. A 6-cubic-foot flask is charged to 3,000 psig and the temperature in the flask room is 72°F. A fire in an adjoining space causes the temperature in the flask room to reach 170°F. What will happen to the pressure in the flask?

1. Convert gauge pressure unit to absolute pressure unit:

$$P_1 = 3{,}000 \text{ psig} + 14.7$$
$$= 3{,}014.7 \text{ psia}$$

2. Convert Fahrenheit temperatures to absolute temperatures (Rankine):

$$°R = °F + 460$$
$$T_1 = 72°F + 460$$
$$= 532°R$$
$$T_2 = 170°F + 460$$
$$= 630°R$$

3. Transpose the formula for Charles's/Gay-Lussac's law to solve for the final pressure (P_2):

$$P_2 = \frac{P_1 T_2}{T_1}$$

4. Substitute known values and solve for the final pressure (P_2):

$$P_2 = \frac{3{,}014.7 \text{ psia} \times 630°R}{532°R}$$
$$= \frac{1{,}899{,}261}{532°R}$$
$$= 3{,}570.03 \text{ psia}$$

The pressure in the flask increased from 3,000 psig to 3,570.03 psia. Note that the pressure increased even though the flask's volume and the volume of the gas remained the same.

12-4 THE GENERAL GAS LAW

The general gas law is a combination of Boyle's law, Charles' law, and Gay-Lussac's law, and is used to predict the behavior of a given quantity of gas when pressure, volume, or temperature changes.

The formula for expressing the general gas law is:

$$\frac{P_1V_1}{T_1} = \frac{P_2V_2}{T_2}$$

Where:

P_1	=	initial absolute pressure
V_1	=	initial volume
T_1	=	initial absolute temperature
P_2	=	final absolute pressure
V_2	=	final volume
T_2	=	final absolute temperature

The following points should be noted when using the general gas law:

- There can be only one unknown value.

- If it is known that a value remains unchanged (such as the volume of a tank) or that the change in one of the variables will be of little consequence, cancel the value out of both sides of the equation to simplify the computations.

Sample Problem 1. A bank of cylinders having an internal volume of 20 cubic feet is to be charged with helium and oxygen to a final pressure of 2,200 psig to provide mixed gas for a dive. The cylinders are rapidly charged from a large premixed supply, and the gas temperature in the cylinders rises to 160°F by the time final pressure is reached. The temperature in the cylinder bank compartment is 75°F. Determine the final cylinder pressure when the gas has cooled.

1. Simplify the equation by eliminating the variables that will not change. The volume of the tank will not change, so V_1 and V_2 can be eliminated from the formula in this problem:

$$\frac{P_1}{T_1} = \frac{P_2}{T_2}$$

2. Multiply each side of the equation by T_2, then rearrange the equation to solve for the final pressure (P2):

$$P_2 = \frac{P_1T_2}{T_1}$$

3. Calculate the initial pressure by converting the gauge pressure unit to the atmospheric pressure unit:

$$P_1 = 2{,}200 \text{ psig} + 14.7 \text{ psi}$$
$$= 2{,}214.7 \text{ psia}$$

4. Convert Fahrenheit temperatures to absolute temperature values (Rankine):

$$°R = °F + 460$$

$$T_1 = 160°F + 460$$

$$= 620°R$$

$$T_2 = 75°F + 460$$

$$= 535°R$$

5. Fill in known values to find the final pressure (P_2):

$$P_2 = \frac{2,214.7 \text{ psia} \times 535°R}{620°R}$$

$$= 1,911.07 \text{ psia}$$

6. Convert final pressure (P_2) to gauge pressure:

$$P_2 = 1,191.07 \text{ psig}$$

$$= 1,896.3 \text{ psig}$$

The pressure when the cylinder cools will be 1896.3 psig.

Sample Problem 2. Using the same scenario as in Sample Problem 1, determine the volume of gas at standard temperature and pressure (STP = 70°F @ 14.7 psia) resulting from rapid charging.

1. Rearrange the formula to solve for the final volume (V_2):

$$V_2 = \frac{P_1 V_1 T_2}{P_2 T_1}$$

2. Convert Fahrenheit temperatures to absolute temperature values (Rankine):

$$°R = °F + 460$$

$$T_1 = 160°F + 460$$

$$= 620°R$$

$$T_2 = 70°F + 460$$

$$= 530°R$$

3. Fill in known values to find the final volume (V_2):

$$V_2 = \frac{2,214.7 \text{ psia} \times 20\text{ft}^3 \times 530°\text{R}}{14.7 \text{ psia} \times 620°\text{R}}$$
$$= 2,575.79 \text{ ft}^3\text{STP}$$

Sample Problem 3. Determine the volume of the gas at STP resulting from slow charging (maintaining 70°F temperature to 2,200 psig).

1. Rearrange the formula to solve for the final volume (V_2):

$$V_2 = \frac{P_1V_1T_2}{P_2T_1}$$

2. Convert Fahrenheit temperatures to absolute temperature values (Rankine):

$$T_1 = 75°\text{F} + 460$$
$$= 535°\text{R}$$
$$T_2 = 70°\text{F} + 460$$
$$= 530°\text{R}$$

3. Substitute known values to find the final volume (V_2):

$$V_2 = \frac{2,214.7 \text{ psia} \times 20\text{ft}^3 \times 530°\text{R}}{14.7 \text{ psia} \times 535°\text{R}}$$
$$= 2,985.03 \text{ ft}^3\text{STP}$$

Sample Problem 4. A 100-cubic-foot salvage bag is to be used to lift a 3,200-pound torpedo from the sea floor at a depth of 231 fsw. An air compressor with a suction of 120 cfm at 60°F and a discharge temperature of 140°F is to be used to inflate the bag. Water temperature at depth is 55°F. To calculate the amount of time required before the torpedo starts to rise (neglecting torpedo displacement, breakout forces, compressor efficiency and the weight of the salvage bag), the displacement of the bag required to lift the torpedo is computed as follows:

1. Calculate the final volume (V_2):

$$V_2 = \frac{3200 \text{ lbs}}{64 \text{ lb/ft}^3}$$
$$= 50 \text{ ft}^3$$

2. Calculate the final pressure (P_2):

$$P_2 = \frac{231\text{ fsw} + 33\text{ fsw}}{33\text{ fsw}}$$
$$= 8\text{ ata}$$

3. Convert Fahrenheit temperatures to absolute temperature values (Rankine):

$$°R = °F + 460$$
$$T_1 = 60°F + 460$$
$$= 520°R$$
$$T_2 = 55°F + 460$$
$$= 515°R$$

4. Rearrange the formula to solve for the initial volume (V_1):

$$V_1 = \frac{P_2 \times V_2 \times T_1}{P_1 \times T_2}$$

5. Substitute known values to find the initial volume (V_1):

$$V_1 = \frac{8\text{ ata} \times 50\text{ft}^3 \times 520°R}{1\text{ ata} \times 515°R}$$
$$= 403.8\text{ ft}^3$$

6. Compute the time:

$$\text{Time} = \frac{\text{Volume Required}}{\text{Compressor Displacement}}$$
$$= \frac{403.8\text{ ft}^3}{120\text{ ft}^3/\text{min}}$$
$$= :03::22$$

(Note that the 140°F compressor discharge temperature is an intermediate temperature and does not enter into the problem.)

12-5 DALTON'S LAW

Dalton's law states that the total pressure exerted by a mixture of gases is equal to the sum of the pressures of the different gases making up the mixture, with each gas acting as if it alone occupied the total volume. The pressure contributed by any gas in the mixture is proportional to the number of molecules of that gas in the total volume. The pressure of that gas is called its partial pressure (pp), meaning its part of the whole.

The formula for expressing Dalton's law is:

$$P_{Total} = pp_A + pp_B + pp_C + \ldots$$

Where: A, B, and C are gases and

$$pp_A = \frac{P_{Total} \times \%Vol_A}{100\%}$$

Sample Problem 1. A helium-oxygen mixture is to be prepared which will provide an oxygen partial pressure of 1.2 ata at a depth of 231 fsw. Compute the oxygen percentage in the mix.

1. Convert depth to pressure in atmospheres absolute:

$$P_{Total} = \frac{231 \text{ fsw} + 33 \text{ fsw}}{33 \text{ fsw}}$$
$$= 8 \text{ ata}$$

2. Calculate the oxygen percentage of the mix.

Since:

$$pp_A = P_{Total} \times \frac{\%Vol_A}{100\%}$$

Then:

$$\%Vol_A = \frac{pp_A}{P_{Total}} \times 100\%$$
$$= \frac{1.2 \text{ ata}}{8 \text{ ata}} \times 100\%$$
$$= 15\% \text{ oxygen}$$

The oxygen percentage of the mix is 15 percent.

Sample Problem 2. A 30-minute bottom time dive is to be conducted at 264 fsw. The maximum safe oxygen partial pressure for a dive should never exceed 1.3 ata while on the bottom. Two premixed supplies of HeO_2 are available: 84/16 percent and 86/14 percent. Which of these mixtures is safe for the intended dive?

1. Convert depth to pressure in atmospheres absolute:

$$P_{Total} = \frac{264\text{ fsw} + 33\text{ fsw}}{33\text{ fsw}}$$
$$= 9\text{ ata}$$

2. Calculate the maximum allowable O_2 percentage:

$$\%Vol_A = \frac{pp_A}{P_{Total}} \times 100\%$$
$$= \frac{1.3\text{ ata}}{9\text{ ata}} \times 100\%$$
$$= 14.4\%\text{ oxygen}$$

Result: The 14 percent O_2 mix is safe to use; the 16 percent O_2 mix is unsafe.

$$\text{The pp of the 14\% mix} = 9\text{ ata} \times \frac{14\%}{100\%}$$
$$= 1.26\text{ ata}O_2$$

1.26 ata O_2 is less than the maximum allowable.

$$\text{The pp of the 16\% mix} = 9\text{ ata} \times \frac{16\%}{100\%}$$
$$= 1.44\text{ ata}O_2$$

Use of this mixture will result in a greater risk of oxygen toxicity.

Sample Problem 3. Gas cylinders aboard a PTC are to be charged with an HeO_2 mixture. The mixture should provide a ppO_2 of 0.9 ata to the diver using a MK 21 MOD 0 helmet at a saturation depth of 660 fsw. Determine the oxygen percentage in the charging gas, then compute the oxygen partial pressure of the breathing gas if the diver makes an excursion from saturation depth to 726 fsw.

1. Convert depth to pressure in atmospheres absolute:

$$P_{Total} = \frac{660\text{ fsw} + 33\text{ fsw}}{33\text{ fsw}}$$
$$= 21\text{ ata}$$

2. Calculate the O_2 content of the charging mix:

$$\begin{aligned} \%VolO_2 &= \frac{0.9 \text{ ata}}{21 \text{ ata}} \times 100\% \\ &= 4.3\% \ O_2 \end{aligned}$$

3. Convert excursion depth to pressure in atmospheres absolute:

$$\begin{aligned} P_{Total} &= \frac{726 \text{ fsw} + 33 \text{ fsw}}{33 \text{ fsw}} \\ &= 23 \text{ ata} \end{aligned}$$

4. Calculate the O_2 partial pressure at excursion depth:

$$\begin{aligned} ppO_2 &= 23 \text{ ata} \times \frac{4.3\% \ O_2}{100\%} \\ &= 0.99 \text{ ata} \end{aligned}$$

12-6 HENRY'S LAW

Henry's law states that the amount of gas that will dissolve in a liquid at a given temperature is almost directly proportional to the partial pressure of that gas. If one unit of gas is dissolved at one atmosphere partial pressure, then two units will be dissolved at two atmospheres, and so on.

CHAPTER 13

Mixed Gas Operational Planning

13-1 INTRODUCTION

13-1.1 **Purpose.** This chapter discusses the planning associated with mixed gas diving operations. Most of the provisions in Chapter 6, Operational Planning and Risk Management, also apply to mixed gas operations and should be reviewed for planning. In planning any mixed gas operation, the principles and techniques presented in this chapter shall be followed.

13-1.2 **Scope.** This chapter outlines a comprehensive planning process that may be used in whole or in part to effectively plan and execute diving operations in support of military operations.

13-1.3 **Additional Sources of Information.** This chapter is not the only source of information available to the diving team when planning mixed gas diving operations. Operation and maintenance manuals for the diving equipment, intelligence reports, and oceanographic studies all contain valuable planning information. The nature of the operation will dictate the procedures to be employed and the planning and preparations required for each. While it is unlikely that even the best planned operation can ever anticipate all possible contingencies, attention to detail in planning will minimize complications that could threaten the success of a mission.

13-1.4 **Complexity of Mixed Gas Diving.** Mixed gas diving operations are complex, requiring constant support and close coordination among all personnel. Due to extended decompression obligations, mixed gas diving can be hazardous if not properly planned and executed. Seemingly minor problems can quickly escalate into emergency situations, leaving limited time to research dive protocols or operational orders to resolve the situation. Each member of the diving team must be qualified on his watch station and be thoroughly competent in executing applicable operating and emergency procedures. Safety is important in any diving operation and must become an integral part of all operations planning.

13-1.5 **Medical Considerations.** The Diving Officer, Master Diver, and Diving Supervisor must plan the operation to safeguard the physical and mental well being of each diver. All members of the team must thoroughly understand the medical aspects of mixed gas, oxygen, and saturation diving. A valuable source of guidance in operations planning is the Diving Medical Officer (DMO), a physician trained specifically in diving medicine and physiology.

Mixed gas diving entails additional risks and procedural requirements for the diver and the support team. At the surface, breathing a medium other than air causes physiological changes in the body. When a diver breathes an unusual medium under increased pressure, additional alterations in the functioning of the mind and body may occur. Each diver must be aware of the changes that can occur and how they may affect his performance and safety. Mixed gas diving procedures that minimize the effects of these changes are described in this and the following chapters. Every mixed gas diver must be thoroughly familiar with these procedures.

Typical medical problems in mixed gas and oxygen diving include decompression sickness, oxygen toxicity, thermal stress, and carbon dioxide retention. Deep saturation diving presents additional concerns, including high pressure nervous syndrome (HPNS), dyspnea, compression arthralgia, skin infections, and performance decrements. These factors directly affect the safety of the diver and the outcome of the mission and must be addressed during the planning stages of an operation. Specific information concerning medical problems particular to various mixed gas diving modes are contained in Volume 5.

13-2 ESTABLISH OPERATIONAL TASKS

Preparing a basic outline and schedule of events for the entire operation ensures that all phases will be properly coordinated. This chapter gives specific guidelines that should be considered when analyzing the operational tasks. Mixed gas diving requires additional considerations in the areas of gas requirements, decompression, and medical support.

Mixed gas diving requires a predetermined supply of breathing gases and carbon dioxide absorbent material. Operations must be planned thoroughly to determine usage requirements in order to effectively obtain required supplies in port or at sea prior to the start of the mission. See paragraph 13-3.10 and Table 13-1 for specific gas/material requirements. Logistic requirements may include planning for on-site resupply of mixed gases and other supplies and for relief of diving teams from Fleet units. Consult unit standing operating procedures for resupply guidance and personnel procurement (refer to OPNAVINST 3120.32 [series]).

Table 13-1. *Average Breathing Gas Consumption Rates.*

Diving Equipment	Gas Consumption (Normal)	Gas Consumption (Heavy Work)
MK 21 MOD 1 UBA EXO BR MS UBA KM-37	1.4 acfm (demand)	6.0 acfm (free flow)

13-3 SELECT DIVING METHOD AND EQUIPMENT

Selecting the appropriate diving method is essential to any diving operations planning. The method will dictate many aspects of an operation including personnel and equipment.

13-3.1 **Mixed Gas Diving Methods.** Mixed gas diving methods are defined by the type of mixed gas diving equipment that will be used. The three types of mixed gas diving equipment are:

- Surface supplied gear (MK 21 MOD 1, EXO BR MS, KM-37)
- Semiclosed circuit and closed circuit UBAs
- Saturation deep dive systems

For deep dives (190-300 fsw) of short duration, or for shallower dives where nitrogen narcosis reduces mental acuity and physical dexterity, helium-oxygen diving methods should be employed.

Because of the unusual hazards incurred by long exposures to extreme environmental conditions, extended excursions away from topside support, and great decompression obligations, semiclosed circuit and closed circuit diving should only be undertaken by specially trained divers. Semiclosed circuit and closed circuit diving operations are covered in depth in Volume 4.

Saturation diving is the preferred method for dives deeper than 300 fsw or for shallow dives where extensive in-water times are required. Disadvantages of saturation diving include the requirement for extensive logistic support and the inability of the support ship to easily shift position once the mooring is set. For this reason, it is very important that the ship be moored as closely over the work site as possible. Using side-scan sonar, remotely operated vehicles (ROVs) or precision navigation systems will greatly aid in the successful completion of the operation. Saturation diving is discussed in Chapter 15.

13-3.2 **Method Considerations.** In mixed gas diving, the principle factors influencing the choice of a particular method are:

- Depth and planned duration of the dive
- Equipment availability
- Quantities of gas mixtures available
- Qualifications and number of personnel available
- Type of work and degree of mobility required
- Environmental considerations such as temperature, visibility, type of bottom, current, and pollution levels
- Communication requirements
- Need for special operations procedures

13-3.3 **Depth.** Equipment depth limitations are contained in Table 13-2. The limitations are based on a number of interrelated factors such as decompression obligations, duration of gas supply and carbon dioxide absorbent material, oxygen tolerance, and the possibility of nitrogen narcosis when using emergency gas (air). Divers must be prepared to work at low temperatures and for long periods of time.

Table 13-2. *Equipment Operational Characteristics.*

Diving Equipment	Normal Working Limit (fsw)	Maximum Working Limit (fsw)	Chamber Requirement	Minimum Personnel
MK 21 MOD 1 UBA EXO BR MS UBA KM-37	300 (HeO_2) (Note 1)	380 (HeO_2) (Note 1)	On station (Note 2)	12

Notes:
1. Depth limits are based on considerations of working time, decompression obligation, oxygen tolerance and nitrogen narcosis.
2. An on-station chamber is defined as a certified and ready chamber at the dive site.

Operations deeper than 300 fsw usually require Deep Diving Systems (DDSs). The decompression obligation upon the diver is of such length that in-water decompression is impractical. Using a personnel transfer capsule (PTC) to transport divers to a deck decompression chamber (DDC) increases the margin of diver safety and support-ship flexibility.

13-3.4 **Bottom Time Requirements.** The nature of the operation may influence the bottom time requirements of the diver. An underwater search may be best undertaken by using multiple divers with short bottom times or by conducting a single bounce dive simply to identify a submerged object. Other tasks, such as underwater construction work, may require numerous dives with long bottom times requiring surface supplied or saturation diving techniques. Although primarily intended to support deep diving operations, saturation diving systems may be ideal to support missions as shallow as 150 fsw where the nature of the work is best accomplished using several dives with extended bottom times. Under these conditions, time is saved by eliminating in-water decompression obligations for each diver and by reducing the number of dive team changes, thus compensating for the increased logistical complexity such operations entail.

13-3.5 **Environment.** Environmental conditions play an important role in planning mixed gas diving operations. Environmental factors, such as those addressed in Chapter 6, should be considered when planning such operations. Mixed gas diving operations often involve prolonged dives requiring lengthy decompression and travels that carry divers great distances from a safe haven. Special attention should therefore be given to preventing diver hypothermia. Mixed gas diving apparatus are designed to minimize thermal stress, but the deepest, longest helium-oxygen dives place the greatest stress on the diver. Exposure to extreme surface conditions prior to the dive may leave the diver in a thermally compromised state. A diver who has been exposed to adverse environmental conditions should not be considered for mixed gas diving until complete rewarming of the diver has taken place, as shown by sweating, normal pulse, and return of normal core temperature. Subjective thermal comfort does not accurately indicate adequate rewarming.

13-3.6 **Mobility.** Some diving operations may dictate the use of a diving method that is selected as a result of special mobility requirements in addition to depth, bottom time and logistical requirements. The MK 21 MOD 1/KM-37 is the preferred method when operations require mobility in the water column (see Figure 13-1).

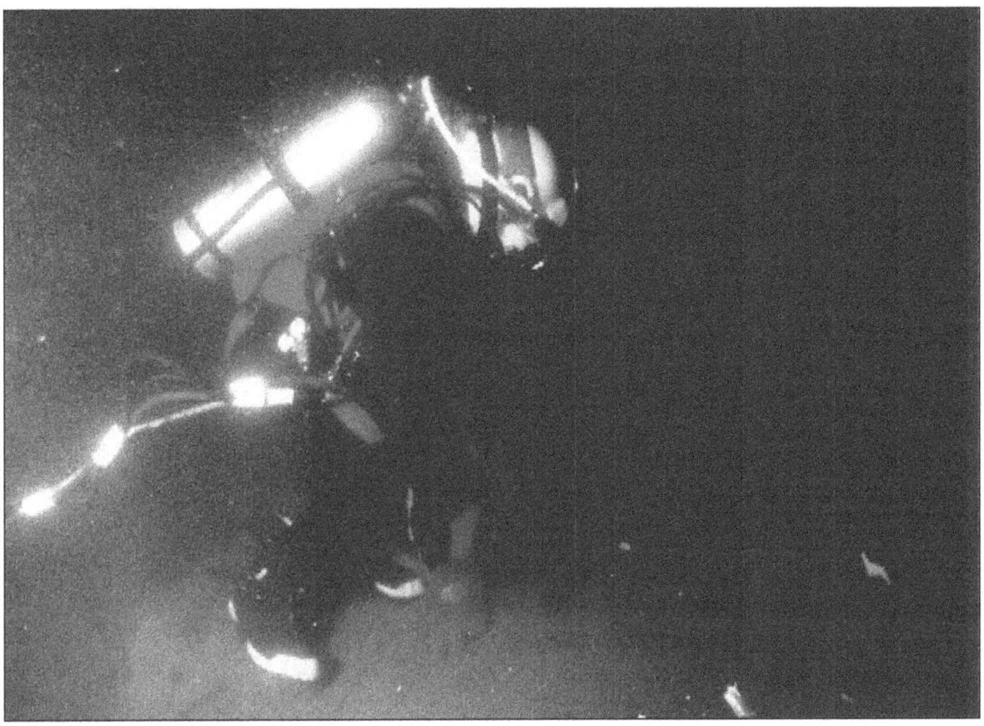

Figure 13-1. Searching Through Aircraft Debris on the Ocean Floor.

For missions where mobility is an essential operating element and depth and bottom time requirements are great, closed circuit diving may be the only available option. Such diving is frequently required by special warfare and/or explosive ordnance disposal (EOD) personnel.

13-3.7 **Equipment Selection.** Equipment and supplies available for mixed gas diving operations by U.S. Navy personnel have been tested under stringent conditions to ensure that they will perform according to design specifications under the most difficult conditions that may be encountered. Several types of equipment are available for mixed gas operations. Equipment selection is based upon the chosen diving method, depth of the dive and the operation to be performed. Table 13-3 outlines the differences between equipment configurations.

Table 13-3. *Mixed Gas Diving Equipment.*

Type	Principal Applications	Minimum Personnel	Advantages	Disadvantages	Restrictions and Depth Limits
MK 21 MOD 1 EXO BR MS KM-37 (Notes 1 & 2)	Deep search, inspection and repair.	12 (Note 2)	Horizontal mobility. Voice communications.	Support craft required. High rate of gas consumption.	Normal 300 fsw. Maximum: 380 fsw with CNO authorization.

Notes:
1. Surface supplied deep-sea
2. Minimum personnel consists of topside support and one diver in the water

13-3.8 **Operational Characteristics.** Equipment operational characteristics are reviewed in Table 13-2 and specific equipment information can be found in paragraph 13-8.

All diving equipment must be certified or authorized for Navy use. Authorized equipment is listed in the NAVSEA/00C Authorized for Navy Use (ANU) list. For proper operation and maintenance of U.S. Navy approved diving equipment, refer to the appropriate equipment operation and maintenance manual.

13-3.9 **Support Equipment and ROVs.** In addition to the UBA, support equipment must not be overlooked. Items commonly used include tools, underwater lighting, power sources, and communications systems. The Coordinated Shipboard Allowance List (COSAL) for the diving platform is a reliable source of support equipment. Commercial resources may also be available.

Occasionally, a mission is best undertaken with the aid of a remotely operated vehicle (ROV). ROVs offer greater depth capabilities with less risk to personnel but at the expense of the mobility, maneuverability, and versatility that only manned operations can incorporate.

13-3.9.1 **Types of ROV.** There are two types of ROVs, tethered and untethered. Tethered ROVs receive power, control signals, and data through an umbilical. Untethered ROVs can travel three to five times faster than tethered ROVs, but because their energy source must be contained in the vehicle their endurance is limited. ROVs used in support of diving operations must have ground fault interrupter (GFI) systems installed to protect the divers.

13-3.9.2 **ROV Capabilities.** Currently, much of the Fleet's requirements for observation diving are being met by using ROVs. They have been used for search and salvage since 1966. State-of-the-art ROVs combine short-range search, inspection, and recovery capabilities in a single system. A typical ROV system includes a control and display console, a power source, a launch and retrieval system, and the vehicle itself. Tethered systems are connected to surface support by an umbilical that supplies power, control signals and data. Untethered search systems that will greatly increase current search rates with extended endurance rates of 24 hours or more are currently under development. Figure 13-2 shows a typical NAVSEA ROV.

Figure 13-2. Remotely Operated Vehicle (ROV) Deep Drone.

13-3.10 **Diver's Breathing Gas Requirements.** In air diving, the breathing mixture is readily available, although pump and compressor capacities and the availability of back-up systems may impose operational limitations. The primary requirement for mixed gas diving is that there be adequate quantities of the appropriate gases on hand, as well as a substantial reserve, for all phases of the operation. The initial determinations become critical if the nearest point of resupply is far removed from the operation site.

13-3.10.1 **Gas Consumption Rates.** The gas consumption rates and carbon dioxide absorbent durations for various types of underwater breathing apparatus are shown in Table 13-1. Refer to Chapter 4 for required purity standards.

13-3.10.2 **Surface Supplied Diving Requirements.** For surface supplied diving, the diver gas supply system is designed so that helium-oxygen, oxygen, or air can be supplied to the divers as required. All surface supplied mixed gas diving systems require a primary and secondary source of breathing medium consisting of helium-oxygen and oxygen in cylinder banks and an emergency supply of air from compressors or high-pressure flasks. Each system must be able to support the gas flow and pressure requirements of the specified equipment. The gas capacity of the primary system must meet the consumption rate of the designated number of divers for the duration of the dive. The secondary system must be able to support recovery operations of all divers and equipment if the primary system fails. This may occur immediately prior to completing the planned bottom time at maximum depth when decompression obligations are the greatest. Emergency air supply is provided in the event all mixed gas supplies are lost.

13-4 SELECTING AND ASSEMBLING THE DIVE TEAM

Selecting a properly trained team for a particular diving mission is critical. Refer to Chapter 6 for an expanded discussion on dive team selection, as well as the criteria for selecting qualified personnel for various tasks. It is critical to ensure that only formally qualified personnel are assigned. The Diving Officer, Master Diver, and Diving Supervisor must verify the qualification level of each team member. The size and complexity of deep dive systems reinforces the need for a detailed and comprehensive watch station qualification program.

13-4.1 **Diver Training.** Training must be given the highest command priority. The command that dives infrequently, or with insufficient training and few work-up dives between operations, will be ill prepared in the event of an emergency. The dive team must be exercised on a regular diving schedule using both routine and nonroutine drills to remain proficient not only in the water but on topside support tasks as well. Cross-training ensures that divers are qualified to substitute for one another when circumstances warrant.

13-4.2 **Personnel Requirements.** To ensure a sufficient number of properly trained and qualified individuals are assigned to the most critical positions on a surface supplied mixed gas dive station, the following minimum stations shall be manned by formally trained (NDSTC) mixed gas divers:

- Diving Officer
- Master Diver
- Diving Supervisor

All other assignments to a surface supplied mixed gas dive station shall be filled in accordance with Table 13-4.

13-4.3 **Diver Fatigue.** Fatigue will predispose a diver to decompression sickness. A tired diver is not mentally alert. Mixed gas dives shall not be conducted using a fatigued diver. The command must ensure that all divers making a mixed gas dive are well rested prior to the dive. All divers making mixed gas dives must have at least 8 hours of sleep within the last 24 hours before diving.

Table 13-4. *Surface Supplied Mixed Gas Dive Team.*

Designation	Deep-Sea (MK 21, EXO BR MS, KM-37) One Diver	Two Divers
Diving Officer	1 (Note 1)	1 (Note 1)
Diving Medical Officer	1 (Notes 1 and 4)	1 (Notes 1 and 4)
Diving Supervisor/Master Diver	1 (Notes 1 and 5)	1 (Notes 1 and 5)
Diving Medical Technician	1 (Notes 1 and 6)	1 (Notes 1 and 6)
Diver	1 (Note 2)	2 (Note 2)
Standby Diver	1 (Note 2)	1 (Note 2)
Tender	3 (Note 2)	5 (Note 2)
Timekeeper/Recorder	1 (Note 2)	1 (Note 2)
Rack Operator	1 (Note 2)	1 (Note 2)
Winch Operator	1 (Note 3)	1 (Note 3)
Console Operator	1 (Note 2)	1 (Note 2)
Total Personnel Required	12	15

Notes:

1. To ensure sufficient properly trained and qualified individuals are assigned to the most critical positions on a surface supplied mixed gas dive station, the following minimum stations shall be manned by formally trained (NDSTC) mixed gas divers:

    Diving Officer
    Master Diver
    Diving Supervisor

2. The following stations shall be manned by formally trained (NDSTC) surface supplied divers:

    Diver
    Standby Diver
    Rack Operator
    Console Operator
    Timekeeper/Recorder

3. The following stations should be a qualified diver. When circumstances require the use of a non-diver, the Diving Officer, Master Diver, and Diving Supervisor must ensure that the required personnel has been thoroughly instructed in the required duties. These stations include:

    Tender
    Standby Tender
    Winch Operator

4. A Diving Medical Officer is required on dive station for all exceptional exposure dives and dives exceeding the equipment normal working limits.
5. Master Diver may serve as the Diving Officer if so designated in writing by the Commanding Officer.
6. A Diving Medical Technician is required on-site when a DMO is not available.

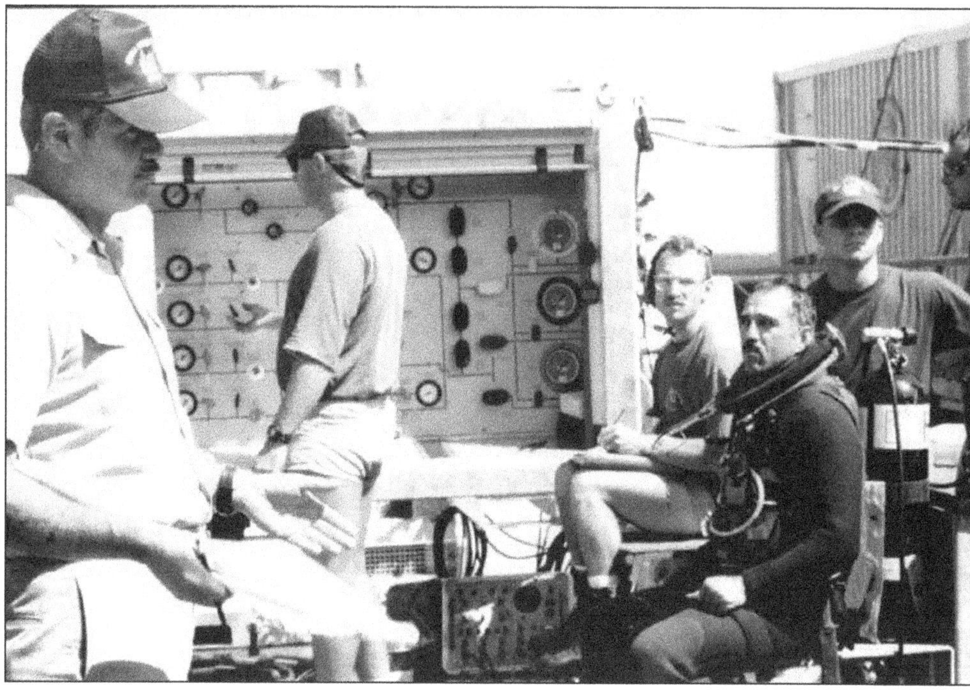

Figure 13-3. Dive Team Brief for Divers.

13-5 BRIEFING THE DIVE TEAM

Large personnel requirements and the increased complexities of mixed gas diving operations make comprehensive briefings of all personnel extremely important. For mixed gas surface supplied operations, briefings of each day's schedule are appropriate. In addition, during saturation diving operations, a dive protocol is required to be read and signed in accordance with the unit's instructions. The briefing should cover all aspects of the operation including communications, equipment, gas supply, and emergencies such as fouling and entrapment. Each diving member should understand his own role as well as that of his diving companions and the support crew (Figure 13-3).

While the operation is in progress, divers returning to the surface or to the PTC should be promptly debriefed. This ensures that topside personnel are kept advised of the progress of the dive and have the information necessary to modify the dive plan or protocol as appropriate.

13-6 FINAL PREPARATIONS AND SAFETY PRECAUTIONS

Prior to the start of a mixed gas diving operation, it is important to check that all necessary preparations have been made and that all safety precautions have been checked. This ensures that the diving team is properly supported in its mission and that all possible contingencies have been evaluated in case an unexpected circumstance should arise.

13-7 RECORD KEEPING

Chapter 5 describes the objectives and importance of maintaining accurate records. The Diving Officer, Master Diver, and Diving Supervisor should identify the records required for their respective systems and tailor them to suit their needs. The purpose of any record is to provide an accurate and detailed account of every facet of the diving operation and a tabulation of supplies expended to support the operation (e.g., gases, carbon dioxide absorbent, etc.). Any unusual circumstances regarding dive conduct (i.e., treatments, operational/emergency procedures, or deviation from procedures) established in the U.S. Navy Diving Manual shall be brought to the attention of the Commanding Officer and logged in the Command Smooth Diving Log.

13-8 MIXED GAS DIVING EQUIPMENT

There are several modes of diving that are characterized by the diving equipment used. The following descriptions outline capabilities and logistical requirements for various mixed gas diving systems.

13-8.1 Minimum Required Equipment.

Minimum required equipment for the pool phase of dive training conducted at Navy diving schools may be modified as necessary. Any modifications to the minimum required equipment listed herein must be noted in approved lesson training guides.

Minimum Equipment:

1. MK 21 MOD 1 helmet, KM-37, or EXO BR MS full face mask with tethered umbilical
2. Thermal protection garment
3. Weight belt
4. Dive knife
5. Swim fins or shoes/booties
6. EGS bottle with submersible tank pressure gauge
7. Integrated diver's vest/harness

Figure 13-4. MK 21 MOD 1 UBA.

13-8.2 Operational Considerations:

1. Adequate mixed gas supply
2. Master Diver required on dive station for mixed gas operations
3. Diving Medical Officer required on dive station for all exceptional exposure dives and dives exceeding the equipment normal working limits
4. Recompression chamber required on dive station

5. Planned exceptional exposure dives or dives exceeding normal working limits require CNO approval
6. Breathing gas heater
7. Hot water suit

13-8.3 **Flyaway Dive System III Mixed Gas System (FMGS).** The FADS III Mixed Gas System (FMGS) is a portable, self contained, surface supplied diver life support system designed to support mixed gas dive missions to 300 fsw (Figure 13-5 and Figure 13-6). The FMGS consists of five gas rack assemblies, one air supply rack assembly (ASRA), one oxygen supply rack assembly (OSRA), and three helium-oxygen supply rack assemblies (HOSRA). Each rack consists of nine 3.15 cu ft floodable volume composite flasks vertically mounted in rack assembly. The ASRA will hold 9600 scf of compressed air at 5000 psi. Compressed air is provided by a 5000 psi air compressor assembly, which includes an air purification system. Oxygen is stored at 3000 psi. The FMGS also includes a mixed gas control console assembly (MGCCA) and two gas booster assemblies for use in charging the OSRA and HOSRA. Three banks of two, three, and four flasks allow the ASRA to provide air to the divers as well as air to support chamber operations. Set-up and operating procedures for the FMGS are found in the Operating and Maintenance Technical Manual for Fly Away Dive System (FADS) III Mixed Gas System, S9592-B2-OMI-010.

Figure 13-5. FADS III Mixed Gas System (FMGS).

Figure 13-6. FMGS Control Console Assembly.

PAGE LEFT BLANK INTENTIONALLY

CHAPTER 14

Surface-Supplied Mixed Gas Diving Procedures

14-1 INTRODUCTION

14-1.1 **Purpose.** The purpose of this chapter is to familiarize divers with the U.S. Navy surface-supplied mixed gas diving procedures.

14-1.2 **Scope.** Surface-supplied, open-circuit mixed gas diving is conducted with helium-oxygen mixtures supplied from the surface by a flexible hose. Surface-supplied mixed gas diving is particularly suited for operations beyond the depth limits of air diving, yet short of the depths and times requiring the use of a saturation diving system. Surface-supplied mixed gas diving is also useful in the air diving range when freedom from nitrogen narcosis is required.

14-2 PLANNING THE OPERATION

Planning surface-supplied mixed gas dives involves many of the same considerations used when planning an air dive. Planning aspects that are unique to surface-supplied mixed gas diving include the logistics of providing several different gas mixtures to the diver and repetitive diving limitations discussed below.

14-2.1 **Depth and Exposure Limits.** The normal operational limit for surface-supplied mixed gas diving is 300 fsw for 30 minutes.

Within each decompression table (Table 14-3), an exceptional exposure limit line separates normal working dives from dives that are considered exceptional exposure. Exceptional exposure dives require lengthy decompression and are associated with an increased risk of decompression sickness and exposure to the elements. Exceptional exposures should be undertaken only at the Commanding Officer's discretion in an emergency. Planned exceptional exposure dives require prior CNO approval.

Repetitive diving is not allowed in surface-supplied helium-oxygen diving, except as outlined in paragraph 14-3.6. Following a "no-decompression dive" the diver must wait 12 hours before making a second dive. Following a decompression dive, the diver must wait 18 hours. To minimize pulmonary oxygen toxicity effects, a diver should take a one day break after four consecutive days of diving.

14-2.2 **Ascent to Altitude.** Following a no-decompression dive, the diver must wait 12 hours before ascent to altitude. Following a decompression dive, the diver must wait 24 hours.

14-2.3 **Water Temperature.** Loss of body temperature (hypothermia) can be a major problem during long, deep dives. A hot water suit is preferred for surface-supplied dives in cold water.

14-2.4 **Gas Mixtures.** Four gas mixtures are required to dive the surface-supplied mixed gas tables over their full range:

- Bottom Mixture - The bottom mixture may vary from 90% helium 10% oxygen to 60% helium 40% oxygen depending on the diver's depth. The allowable range of bottom mixtures for each depth is shown in Table 14-3.

- 50% Helium 50% Oxygen - This mixture is used from 90 fsw to 40 fsw during decompression. Oxygen concentration in the mixture may range from 49 to 51 percent.

- 100% Oxygen - Oxygen is used at the 30- and 20-fsw water stops during in-water decompression and at 50, 40 and 30 fsw in the chamber during surface decompression.

- Air - Air is used as an emergency backup gas throughout the dive and to provide air breaks during oxygen breathing.

Helium-oxygen mixtures must be analyzed for oxygen content with an instrument having an accuracy of ±0.5 percent.

14-2.5 **Emergency Gas Supply.** All divers are equipped with an emergency gas supply (EGS). The EGS gas mixture shall be the same as the bottom mixture unless the bottom mixture contains less than 16 percent oxygen, in which case the EGS gas mixture may range from 15 to 17 percent oxygen. The EGS shall be an adequately charged ANU approved SCUBA cylinder. An adequately charged SCUBA cylinder is defined as: the pressure that provides sufficient gas to bring the diver to his first decompression stop or the surface for no-decompression dives. It is assumed that this will give topside personnel enough time to perform required emergency procedures to restore surface-supplied gas.

14-3 SURFACE-SUPPLIED HELIUM-OXYGEN DESCENT AND ASCENT PROCEDURES

The Surface-Supplied Helium-Oxygen Decompression Table (Table 14-3) is used to decompress divers from surface-supplied helium-oxygen dives. The table is in a depth time format similar to the U.S. Navy Air Decompression Table and is used in a similar fashion.

14-3.1 **Selecting the Bottom Mix.** The Surface-Supplied Helium-Oxygen Decompression Table (Table 14-3) specifies maximum and minimum concentrations of oxygen allowable in the helium-oxygen mixture at depth. The maximum oxygen concentration has been selected so that the diver never exceeds an oxygen partial pressure of 1.3 ata while on the bottom. The minimum oxygen percentage allowed in the mixture is 14 percent for depths to 200 fsw and 10 percent for depths in excess of 200 fsw. Diving with a mixture near maximum oxygen percentage is encouraged as it offers a decompression advantage to the diver. For operational planning, the range of possible depths should be established and a mixture selected that will meet the maximum/minimum specification across the depth range.

14-3.2 **Selecting the Decompression Schedule.** To select a proper decompression table and schedule, measure the deepest depth reached by the diver and enter the table at the exact or next greater depth. When using a pneumofathometer to measure depth, correct the observed depth reading as shown in Table 14-1. Ensure the pneumofathometer is located at mid-chest level.

Table 14-1. *Pneumofathometer Correction Factors.*

Pneumofathometer Depth Reading	Correction Factor
0-100 fsw	+1 fsw
101-200 fsw	+2 fsw
201-300 fsw	+4 fsw
301-400 fsw	+7 fsw

Example: The diver's pneumofathometer reads 250 fsw. In the depth range of 201-300 fsw, the pneumofathometer underestimates the diver's depth by 4 fsw. To determine a diver's depth, add 4 fsw to the pneumofathometer reading giving the diver's depth as 254 fsw.

Bottom time is measured as the time from leaving the surface to leaving the bottom, rounded up to the next whole minute, except as noted in paragraph 14-3.5. Enter the table at the exact or next greater bottom time.

14-3.3 **Travel Rates and Stop Times.** The descent rate is not critical, but in general should not exceed 75 fsw/min. The ascent rate from the bottom to the first decompression stop, between decompression stops, and from the last decompression stop to the surface is 30 fsw/min. Minor variations in the rate of ascent between 20 and 40 fsw/min are acceptable. For surface decompression, the ascent rate from the 40 fsw water stop to the surface is 40 fsw/min.

The time at the first decompression stop begins when the diver arrives at the stop and ends when he leaves the stop. For all subsequent stops, the stop time begins when the diver leaves the previous stop and ends when he leaves the stop. In other words, ascent time between stops is included in the subsequent stop time. The single exception is the first oxygen stop at 30 fsw. The 30-fsw oxygen stop begins when the divers are confirmed to be on oxygen at 30 fsw and ends when the divers leave 30 fsw. The ascent time from the 30- to the 20-fsw oxygen stop is included in the 20-fsw oxygen stop time.

14-3.4 **Decompression Breathing Gases.** Decompress on bottom mixture to 90 fsw, then shift the diver to a 50% helium 50% oxygen mixture. Upon arrival at the 30 fsw stop, shift the diver to 100% oxygen.

For all dives, surface decompression may be used after completing the 40 fsw water stop as described in paragraph 14-3.11. During surface decompression, the diver surfaces while breathing 50% helium 50% oxygen.

14-3.5 Special Procedures for Descent with Less than 16 Percent Oxygen. To prevent hypoxia, a special descent procedure is required when the bottom mixture contains less than 16% oxygen:

1. Place the diver on air on the surface.

2. Make the appropriate predive checks.

3. Have the diver descend to 20 fsw.

4. At 20 fsw, shift the diver to the bottom mix and ventilate the diver for 20 seconds.

5. Confirm the diver is on bottom mix, then perform a final leak check. The diver is allowed 5 minutes to descend to 20 fsw, shift to the bottom mixture and perform equipment checks.

6. Have the diver begin descent.

7. Start bottom time.

   - If the diver spends 5 minutes or less performing above procedures, bottom time starts when the diver leaves 20 fsw.

   - If the diver spends more than 5 minutes performing above procedures, bottom time starts at the 5 minute mark.

8. If it is necessary to bring the diver back to the surface from 20 fsw to correct a problem:

   - Shift the diver from the bottom mixture back to air.
   - Ventilate the diver.
   - Confirm the diver is on air.
   - Have the diver begin ascent.
   - When the diver reenters the water, the 5 minute grace period begins again. No adjustment of bottom time is required for the previous exposure at 20 fsw.

14-3.6 Aborting Dive During Descent. Inability to equalize the ears or sinuses may force the dive to be aborted during descent.

1. If it is necessary to bring the diver back to the surface from depths of 100 fsw and shallower:

   - Ensure the diver is in a no-decompression status.
   - If the bottom mixture is 16% oxygen or greater, ascend directly to the surface at 30 fsw/min.
   - If the bottom mixture is less than 16% oxygen, ascend to 20 fsw at 30 fsw/min.

- Shift the diver from the bottom mixture back to air.
- Ventilate the diver.
- Confirm the diver is on air.
- Complete ascent to the surface on air.
- If desired, another dive may be performed following a dive aborted 100 fsw and shallower. Add the bottom time of all the dives to the bottom time of the new dive and use the deepest depth when calculating a table and schedule for the new dive.

2. If it is necessary to abort a dive deeper than 100 fsw:
   - Follow the normal decompression schedule to the surface.
   - Repetitive diving is not allowed following a dive aborted deeper than 100 fsw.

14-3.7 **Procedures for Shifting to 50 Percent Helium/50 Percent Oxygen at 90 fsw.** All dives except no-decompression dives require a shift from bottom mixture to 50% helium 50% oxygen at 90 fsw during decompression. Follow these steps:

1. Shift the console to 50% helium 50% oxygen when the diver reaches 90 fsw.
2. If there is a decompression stop at 90 fsw, ventilate each diver for 20 seconds at 90 fsw.
3. Confirm the divers are on 50% helium 50% oxygen.
4. If there is no decompression stop at 90 fsw, delay ventilation until arrival at the next shallower stop.

Gas shift time is included in the stop time.

14-3.8 **Procedures for Shifting to 100 Percent Oxygen at 30 fsw.** All in-water decompression dives require a shift to 100 percent oxygen at the 30-fsw stop. Upon arrival at the stop, ventilate each diver with oxygen following these steps:

1. Shift the console to 100% oxygen when the diver reaches 30 fsw.
2. Ventilate each diver for 20 seconds.
3. Verify the diver's voice change.

Time at the 30-fsw stop begins when the divers are confirmed to be on oxygen.

14-3.9 **Air Breaks at 30 and 20 fsw.** At the 30-fsw and 20-fsw water stops, the diver breathes oxygen for 30-minute periods separated by 5-minute air breaks. The air breaks do not count toward required decompression time. When an air break is required, shift the console to air for 5 minutes then back to 100% oxygen.

Ventilation of the divers is not required. For purposes of timing air breaks, begin clocking oxygen time when all divers are confirmed on oxygen. If the total oxygen stop time is 35 minutes or less, an air break is not required at 30 minutes. If the final oxygen period is 35 minutes or less, a final air break at the 30-minute mark is not required. In either case, surface the diver on 100% oxygen upon completion of the oxygen time.

Example

1. Divers are decompressing in the water on a 220 fsw for 20 minute decompression schedule. The schedule calls for 23 minutes on oxygen at 30 fsw and 41 minutes on oxygen at 20 fsw.

2. Divers start their 23 minute 30 fsw stop time when confirmed to be on oxygen at 30 fsw.

3. After 23 minute on oxygen at 30 fsw, divers travel to 20 fsw to complete their 41 minute 20 fsw stop. The 20 second travel time from 30 to 20 fsw on oxygen is included in the 41 minute stop time.

4. Seven minutes from the time the divers left their 30 fsw stop, the console is shifted to air. This is due to the divers having completed a total of 30 minutes on oxygen. No ventilation of the divers is required.

5. After five minutes on air, the console is shifted back to oxygen. No ventilation is required. The five-minute period is considered dead time from the decompression standpoint. A total of 34 minutes on oxygen remain to be completed at 20 fsw.

6. Since the remaining oxygen time is less than 35 minutes, the divers breathe oxygen for the last 34 minutes prior to ascent to the surface without taking an additional air break. Divers remain on oxygen for ascent to the surface.

14-3.10 **Ascent from the 20-fsw Water Stop.** For normal in-water decompression, the diver surfaces from 20 fsw on oxygen. Ascent rate is 30 fsw/min.

14-3.11 **Surface Decompression on Oxygen (SurDO_2).** Surface decompression on oxygen is preferred over in-water decompression on oxygen for routine operations. SurDO_2 procedures improve the diver's comfort and safety. A diver is eligible for surface decompression when he has completed the 40-fsw water stop. To initiate surface decompression:

WARNING **The interval from leaving 40-fsw in the water to arriving at 50-fsw in the chamber cannot exceed 5 minutes without incurring a penalty. (See paragraph 14-4.14.)**

1. Bring the diver to the surface at 40 fsw/min and undress him.

2. Place the diver in the recompression chamber. Use of an inside tender when two divers undergo surface decompression is at the discretion of the Dive

Supervisor. If an inside tender is not used, both divers will carefully monitor each other in addition to being closely observed by topside personnel.

3. Compress the diver on air to 50 fsw at a maximum compression rate of 100 fsw/min. The surface interval is the elapsed time from the time the diver leaves the 40-fsw water stop to the time the diver arrives at 50 fsw in the chamber. A normal surface interval should not exceed 5 minutes.

4. Upon arrival at 50 fsw, place the diver on 100 percent oxygen by mask. The mask will be strapped on both divers to ensure a good oxygen seal.

5. In the chamber, have the divers breathe oxygen for 30-minute periods separated by 5-minute air breaks. The number of oxygen periods required is indicated in Table 14-3. The first period consists of 15 minutes on oxygen at 50 fsw followed by 15 minutes on oxygen at 40 fsw. Periods 2, 3, and 4 are spent at 40 fsw. Periods 5, 6, 7 and 8 are spent at 30 fsw. Ascent from 50 to 40 and from 40 to 30 fsw is at 30 fsw/min. Ascent time is included in the oxygen/air time. Ascent from 40 to 30 fsw, if required, should take place during the air break.

6. When the last oxygen breathing period has been completed, return the diver to breathing chamber air.

7. Ascend to the surface at a rate of 30 fsw/min.

The diving supervisor can initiate surface decompression at any point during in-water oxygen decompression at 30 or 20 fsw, if desired. Surface decompression may become desirable if sea conditions are deteriorating, the diver feels ill, or some other contingency arises. Once in the chamber, the diver should receive the full number of chamber oxygen periods prescribed by the tables. Unlike in air diving, no credit is allowed for time already spent on oxygen in the water.

14-3.12 **Variation in Rate of Ascent.** The rate of ascent to the first stop and between subsequent stops is 30 fsw/minute. Minor variations in the rate of travel between 20 and 40 fsw/minute are acceptable.

14-3.12.1 **Early Arrival at the First Stop.** If the divers arrive early at the first stop:

1. Begin timing the first stop when the required travel time has been completed.

2. If the first stop requires a gas shift, initiate the gas shift and ventilation upon arrival at the stop, but begin the stop time only when the required travel time has been completed.

14-3.12.2 **Delays in Arriving at the First Stop.**

1. Delay less than 1 minute. Delays in arrival at the first stop of less than 1 minute may be ignored.

2. Delay greater than 1 minute. Round up the delay time to the next whole minute and add it to the bottom time. Recompute the decompression schedule. If no

change in schedule is required, continue on the planned decompression. If a change in schedule is required and the new schedule calls for a decompression stop deeper than the diver's current depth, perform any missed deeper stops at the diver's current depth. Do not go deeper.

Example: If the delay time to arrival at the first stop is 3 minutes and 25 seconds, round up to the next whole minute and add 4 minutes to the bottom time. Recheck the decompression table to see if the decompression stop depths or times have changed.

14-3.12.3 **Delays in Leaving a Stop or Arrival at the Next Stop.**

- Delays Deeper than 90 fsw.

  1. Delays less than 1 minute may be ignored.

  2. Greater than 1 minute. Add the delay to the bottom time and recalculate the required decompression. If a new schedule is required, pick up the new schedule at the present stop or subsequent stop if the delay occurs between stops. Ignore any missed stops or time deeper than the depth at which the delay occurred. If a delay occurs between stops, restart subsequent stop time at completion of the delay.

- Delays 90 fsw and shallower:

  1. Delays less than 1 minute may be ignored.

  2. Delays greater than 1 minute require no special action except as described below under special considerations when decompressing with high oxygen partial pressure. Resume the normal decompression schedule at the completion of the delay. If a delay occurs between stops, restart subsequent stop time at completion of the delay.

- Special considerations when decompressing with high oxygen partial pressure:

  1. Delays greater than 5 minutes between 90 and 70 fsw. Shift the diver to air to avoid the risk of CNS oxygen toxicity. At the completion of the delay, return the diver to 50% helium 50% oxygen. Add the time on air to the bottom time and recalculate the required decompression. If a new schedule is required, pick up the new schedule at the present stop or subsequent stop if delay occurs between stops. Ignore any missed stops or time deeper than the depth at which the delay occurred.

  2. Delays leaving the 30-fsw stop. Delays greater than 1 minute leaving the 30-fsw stop shall be subtracted from the 20-fsw stop time.

14-3.12.4 **Delays in Travel from 40 fsw to the Surface for Surface Decompression.** Disregard any delays in travel from 40 fsw to the surface during surface decompression unless the diver exceeds the 5-minute surface interval. If the diver exceeds the 5-minute surface interval, follow the guidance in paragraph 14-4.14.

14-4 SURFACE-SUPPLIED HELIUM-OXYGEN EMERGENCY PROCEDURES

In surface-supplied mixed gas diving, specific procedures are used in emergency situations. The following paragraphs detail these procedures. Other medical/physiological factors that surface-supplied mixed gas divers need to consider are covered in detail in Chapter 3. The U.S. Navy Treatment Tables are presented in Chapter 20.

14-4.1 Bottom Time in Excess of the Table.

In the rare instance of diver entrapment or umbilical fouling, bottom times may exceed 120 minutes, the longest value shown in the table. When it is foreseen that bottom time will exceed 120 minutes, immediately contact the Navy Experimental Diving Unit for advice on which decompression procedure to follow. If advice cannot be obtained in time:

1. Decompress the diver using the 120-minute schedule for the deepest depth attained.
2. Shift to 100 percent oxygen at 40 fsw.
3. Surface the diver after completing 30 minutes on oxygen at 40 fsw. Oxygen time at 40 fsw starts when divers are confirmed on oxygen.
4. Compress the diver to 60 fsw in the chamber as fast as possible not to exceed 100 fsw/min.
5. Treat the diver on an extended Treatment Table 6. Extend Treatment Table 6 for two oxygen breathing periods at 60 fsw (20 minutes on oxygen, then 5 minutes on air, then 20 minutes on oxygen) and two oxygen breathing periods at 30 fsw (60 minutes on oxygen, then 15 minutes on air, then 60 minutes on oxygen).

14-4.2 **Loss of Helium-Oxygen Supply on the Bottom.** Follow this procedure if the umbilical helium-oxygen supply is lost on the bottom:

1. Shift the diver to the emergency gas system (EGS).
2. Abort the dive.
3. Remain on the EGS until arrival at 90 fsw.
4. At 90 fsw, shift the diver to 50% helium 50% oxygen and complete the decompression as planned.
5. If the EGS becomes exhausted before 90 fsw is reached, shift the diver to air, complete decompression to 90 fsw, shift the diver to 50% helium 50% oxygen, and continue the decompression as planned.

14-4.3 **Loss of 50 Percent Oxygen Supply During In-Water Decompression.** If the diver cannot be shifted to 50% helium 50% oxygen at 90 fsw or the 50% helium 50% oxygen supply is lost during decompression:

1. Shift the diver to air and continue the decompression as planned while trying to correct the problem.

2. Shift the diver to 50% helium 50% oxygen once the problem is corrected. Time spent on air counts toward decompression.

3. If the problem cannot be corrected:

   - Continue the planned decompression on air.

   - Shift the diver from air to oxygen upon arrival at the 50-fsw stop.

   - Breathe oxygen at 50 and 40 fsw for the decompression times indicated in Table 14-3, but not to exceed 16 minutes at 50 fsw. Oxygen time at 50 fsw starts when divers are confirmed on oxygen. If the 50-fsw stop exceeds 16 minutes, travel divers to 40 fsw and add remaining 50-fsw stop time to the 40-fsw stop time on oxygen.

   - Surface decompress per paragraph 14-3.11 following completion of the 40-fsw stop.

14-4.4 **Loss of Oxygen Supply During In-Water Decompression.** If the diver cannot be shifted to oxygen at 30 fsw or the oxygen supply is lost during the 30- or 20-fsw water stops:

1. Switch back to 50% helium 50% oxygen. If a switch to 50% helium 50% oxygen is not possible, switch the diver to air.

2. If the problem can be quickly remedied, reventilate the diver with oxygen and resume the schedule at the point of interruption. Consider any time on helium-oxygen or air as dead time.

3. If the problem cannot be remedied, initiate surface decompression. Ignore any time already spent on oxygen at 30 or 20 fsw. The five minute surface interval requirement for surface decompression begins upon leaving the 30- or 20-fsw stop.

4. If the problem cannot be remedied and surface decompression is not feasible, complete the decompression on 50% helium 50% oxygen or air. For 50% helium 50% oxygen, double the remaining oxygen time at each water stop. For air, triple the remaining oxygen time.

Example: A diver loses oxygen 15 minutes into the 30-fsw water stop and is switched back to the 50% helium 50% oxygen decompression mixture. The problem cannot be corrected. The divers original schedule called for 32 minutes of oxygen at 30 fsw and 58 minutes of oxygen at 20 fsw.

Seventeen minutes of oxygen time (32 - 15) remain at 30 fsw. Fifty-eight minutes remain at 20 fsw. The diver should spend an additional 34 minutes (17 x 2) at 30 fsw on the 50/50 mixture, followed by 116 minutes (58 x 2) at 20 fsw. Surface the diver upon completion of the 20-fsw stop.

Example: A diver loses oxygen 10 minutes into the 30-fsw water stop and is switched to air. The problem cannot be corrected. The diver's original schedule called for 28 minutes of oxygen at 30 fsw and 50 minutes of oxygen at 20 fsw.

Eighteen minutes of oxygen time (28 - 10) remain at 30 fsw. Fifty minutes remain at 20 fsw. The diver should spend an additional 54 minutes (18 x 3) at 30 fsw on air followed by 150 minutes (50 x 3) on air at 20 fsw. Surface the diver upon completion of the 20-fsw stop.

14-4.5 **Loss of Oxygen Supply in the Chamber During Surface Decompression.** If the oxygen supply in the chamber is lost during surface decompression, have the diver breathe chamber air.

- Temporary Loss. Return the diver to oxygen breathing. Consider any time on air as dead time.
- Permanent Loss. Multiply the remaining oxygen time by three to obtain the equivalent chamber decompression time on air. If 50% helium 50% oxygen is available, multiply the remaining oxygen time by two to obtain the equivalent chamber decompression time on 50/50. If the loss occurred at 50 or 40 fsw, allocate 10% of the equivalent air or helium-oxygen time to the 40-fsw stop, 20% to the 30-fsw stop, and 70% to the 20-fsw stop. If the diver is at 50 fsw, ascend to 40 fsw to begin the stop time. If the loss occurred at 30 fsw, allocate 30% of the equivalent air or helium-oxygen time to the 30-fsw stop and 70% to the 20-fsw stop. Round the stop times to the nearest whole minute. Surface upon completion of the 20-fsw stop.

Example: The oxygen supply to the chamber is lost 10 minutes into the first 30-minute period on oxygen. Helium-oxygen is not available. The original surface decompression schedule called for three 30-min oxygen breathing periods (total of 90 minutes of oxygen). The diver is at 50 fsw.

The remaining oxygen time is 80 minutes (90-10). The equivalent chamber decompression time on air is 240 minutes (3 x 80). The 240 minutes of air stop time should be allocated as follows: Twenty-four minutes at 40 fsw (240 x 0.1), 48 minutes at 30 fsw (240 x 0.2), and 168 minutes at 20 fsw (240 x 0.7). As addressed above, the diver should ascend from 50 to 40 fsw and begin the 24 minute stop time at 40 fsw.

14-4.6 **Decompression Gas Supply Contamination.** If the decompression gas supply becomes contaminated with the bottom mixture, 50/50 mix, air, or oxygen:

1. Find the source of the contamination and correct the problem. Probable sources include:

- An improper valve line-up on the console. This can be verified by checking oxygen percentage on console oxygen analyzer.
- Accidental opening of the emergency gas supply (EGS) valve.

2. When the problem is corrected:
    - Ventilate each diver for 20 seconds and confirm divers are on decompression gas.
    - Continue decompression as planned. Do not lengthen stop times to compensate for the time spent correcting the problem.

14-4.7 **CNS Oxygen Toxicity Symptoms (Nonconvulsive) at the 90-60 fsw Water Stops.**

CNS oxygen toxicity symptoms are unlikely but possible while the diver is breathing 50% helium 50% oxygen in the water at depths 60 fsw and deeper. If symptoms of oxygen toxicity do appear, take the following actions:

1. Bring the divers up 10 feet and shift to air to reduce the partial pressure of oxygen. Shift the console as the divers are traveling.
2. Ventilate both divers upon arrival at the shallower stop. Ventilate the stricken diver first.
3. Remain at the shallower stop until the missed time at the previous stop is made up.
4. Resume the planned decompression breathing air.
5. Upon arrival at the next shallower stop, return the divers to the 50% helium 50% oxygen mixture. Ignore any missed time on the 50/50 mixture. A recurrence of symptoms is highly unlikely because of the reduced oxygen partial pressure at the shallower depth.

Example: Red Diver has an oxygen toxicity symptom 5 minutes into his scheduled 9-minute 80-fsw stop. The stage with both divers travels to 70 fsw and the console is shifted to air. Upon arrival at 70 fsw, Red diver is ventilated for 20 seconds followed by Green diver. The divers remain at 70 fsw for the remaining 4 minutes left from their 80-fsw stop and then start their 10 minute scheduled 70-fsw stop time at the completion of the 4 minutes. Upon reaching 60 fsw, the console is shifted back to their 50/50 mixture and both divers are ventilated. The normal decompression schedule is resumed at 60 fsw.

14-4.8 **Oxygen Convulsion at the 90-60 fsw Water Stop.** If symptoms of oxygen toxicity progress to an oxygen convulsion at 90-60 fsw despite the measures taken above, a serious emergency has developed. Only general management guidelines can be presented here. Topside supervisory personnel must take whatever action they deem necessary to bring the casualty under control.

Follow these procedures when the diver is convulsing at the 90-60 fsw water stops:

1. Shift both divers to air if this action has not already been taken.

2. Have the unaffected diver ventilate himself and then ventilate the stricken diver.

3. If only one diver is in the water, launch the standby diver immediately and have him ventilate the stricken diver.

4. Hold the divers at depth until the tonic-clonic phase of the convulsion has subsided. The tonic-clonic phase of a convulsion generally lasts 1 to 2 minutes.

5. At the end of the tonic-clonic phase, have the dive partner or standby diver ascertain whether the diver is breathing. The presence or absence of breath sounds will also be audible over the intercom.

6. If the diver appears not to be breathing, have the dive partner or standby diver attempt to reposition the head to open the airway. Airway obstruction will be the most common reason why an unconscious diver fails to breathe.

7. If the affected diver is breathing, have the dive partner or standby diver tend the stricken diver and decompress both divers on air following the original schedule. Shift the divers to 50% helium 50% oxygen upon arrival at 50 fsw. Surface decompress upon completion of the 40-fsw water stop.

8. If it is not possible to verify that the affected diver is breathing, leave the unaffected diver at the stop to complete decompression, and surface the affected diver and the standby diver at 30 fsw/min. Shift the unaffected diver back to his 50/50 mixture for completion of decompression. The standby diver should maintain an open airway on the stricken diver during ascent. On the surface the affected diver should receive any necessary airway support and be immediately recompressed and treated for arterial gas embolism and missed decompression in accordance with Figure 20-1.

14-4.9 **CNS Toxicity Symptoms (Nonconvulsive) at 50- and 40-fsw Water Stops.** It is very unlikely that a diver will develop symptoms of CNS oxygen toxicity while breathing 50% helium 50% oxygen at the 50- and 40-fsw water stops. Symptoms are much more likely if the diver is breathing 100% oxygen in accordance with paragraph 14-4.3. If the diver does experience symptoms of CNS oxygen toxicity at 50 or 40 fsw while breathing either 50% helium 50% oxygen or 100% oxygen, take the following actions:

1. Bring the divers up 10 feet and shift to air to reduce the partial pressure of oxygen. Shift the console as the divers are traveling to the shallower stop.

2. Ventilate both divers upon arrival at the shallower stop. Ventilate the stricken diver first.

3. Remain on air at the shallower depth for double the missed time from 50- and 40-fsw water stops, then surface decompress the diver in accordance with paragraph 14-3.11. If the diver was on 100% oxygen in accordance with paragraph 14-4.3, triple the missed time from the 50- and 40-fsw water stops, then surface decompress.

   Example: A diver on 50% helium 50% oxygen experiences an oxygen symptom five minutes into his 10 min stop at 50 fsw. He immediately ascends to 40 fsw and begins breathing air. The decompression schedule calls for a 10 min stop at 40 fsw. The diver missed 5 min of helium-oxygen at 50 fsw and will miss 10 min more at 40 fsw by virtue of the fact that he is on air. The total missed helium-oxygen time is 15 min. The diver should remain at 40 fsw for 30 min, then surface decompress.

   Example: A diver on 100% oxygen experiences an oxygen symptom five minutes into his 10 min stop at 40 fsw. He immediately ascends to 30 fsw and begins breathing air. The missed oxygen time at 40 fsw is 5 min. The diver should remain on air at 30 fsw for 15 min, then surface decompress.

4. If surface decompression is not feasible, continue decompression in the water on either air or oxygen depending on the diver's condition:

   - To continue on oxygen, ascend to 30 fsw (or remain at 30 fsw if already there). Take a 10 min period on air (Time on air does not count toward decompression). Then shift the diver to oxygen and complete decompression in the water according to the schedule.

   - To continue on air, ascend to 30 fsw (or remain at 30 fsw if already there). Compute the remaining 30- and 20-fsw air stop times by tripling the oxygen time given in the original schedule. Surface upon completion of the 20-fsw stop.

   - Alternatively, the diver may complete the 30-fsw stop on air by tripling the oxygen stop time, then switch to oxygen upon arrival at 20 fsw. Remain at 20 fsw for the oxygen time indicated in the original schedule. Surface upon completion of the 20-fsw stop.

14-4.10 **Oxygen Convulsion at the 50-40 fsw Water Stop.** If oxygen symptoms progress to an oxygen convulsion despite the measures described above or if a convulsion occurs suddenly without warning at 50 or 40 fsw, take the following actions:

1. Shift both divers to air if this action has not already been taken. Have the unaffected diver ventilate himself then ventilate the stricken diver.

2. Follow the guidance given in paragraph 14-4.8 for stabilizing the stricken diver and determining whether he is breathing. If the diver is breathing, hold him at his current depth until he is stable, then take one of the following actions:

   - If the diver missed helium-oxygen or oxygen decompression time at 50 fsw, hold the diver at depth until the total elapsed time on air is at least double the missed time on helium-oxygen, then surface decompress in accordance with paragraph 14-3.11. If the diver was on 100% oxygen in accordance with paragraph 14-4.3, remain at depth until the total elapsed time on air is at least triple the missed time on oxygen, then surface decompress. In either case, add the 40-fsw water stop time to the 50-fsw chamber oxygen stop time.

- If the diver did not miss any helium-oxygen or oxygen decompression time at 50 fsw, surface decompress in accordance with paragraph 14-3.11. Add any missed oxygen or helium-oxygen time at 40 fsw to the 50-fsw chamber oxygen stop time.

3. If surface decompression is not feasible, complete decompression in the water on air. Compute the remaining stop times on air by doubling the remaining helium-oxygen time, or tripling the remaining oxygen time at each stop.

4. If the diver is not breathing, surface the diver at 30 fsw/min while maintaining an open airway. Treat the diver for arterial gas embolism (Figure 20-1).

14-4.11 **CNS Oxygen Toxicity Symptoms (Nonconvulsive) at 30- and 20-fsw Water Stops.** If the diver develops symptoms of CNS toxicity at the 30- or 20-fsw water stops, take the following action:

1. If a recompression chamber is available on the dive station, initiate surface decompression. Shift the console to air during travel to the surface. Once in the chamber, take the full number of chamber oxygen periods prescribed by the tables. Unlike in air diving, no credit is allowed for time already spent on oxygen in the water.

2. If a recompression chamber is not available on the dive station and the event occurs at 30 fsw, bring the divers up 10 fsw and shift to air to reduce the partial pressure of oxygen. Shift the console as the divers are traveling to 20 fsw. Ventilate both divers with air upon arrival at 20 fsw. Ventilate the affected diver first. Complete the decompression on air in the water at 20 fsw. Compute the required air time at 20 fsw by tripling the sum of the missed oxygen time at 30 and 20 fsw.

3. If a recompression chamber is not available on the dive station and the event occurs at 20 fsw, shift the console to air, ventilate both divers, affected diver first, and complete the decompression in the water at 20 fsw on air. Compute the required air time at 20 fsw by tripling the missed oxygen time at 20 fsw.

14-4.12 **Oxygen Convulsion at the 30- and 20-fsw Water Stop.** If symptoms progress to an oxygen convulsion despite the above measures, a serious emergency has developed and the following actions must be taken.

1. Shift both divers to air and follow the guidance given in paragraph 14-4.8 for stabilizing the diver and determining whether he is breathing.

2. If the diver is breathing, hold him at depth until he is stable, then surface decompress.

3. If surface decompression is not feasible, ventilate both divers with air and complete decompression in the water on air. Compute the remaining stop times on air by tripling the remaining oxygen time at each stop. See paragraph 14-4.4 for example.

4. If the diver is not breathing, surface the diver at 30 fsw/min while maintaining an open airway and treat the diver for arterial gas embolism (Figure 20-1).

14-4.13 **Oxygen Toxicity Symptoms in the Chamber.** At the first sign of CNS oxygen toxicity, the patient should be removed from oxygen and allowed to breathe chamber air. Fifteen minutes after all symptoms have completely subsided, resume oxygen breathing at the point of interruption. If symptoms of CNS oxygen toxicity develop again or if the first symptom is a convulsion, take the following action:

1. Remove the mask.

2. After all symptoms have completely subsided, decompress 10 feet at a rate of 1 fsw/min. For a convulsion, begin travel when the patient is fully relaxed and breathing normally.

3. Resume oxygen breathing at the shallower depth at the point of interruption.

4. If another oxygen symptom occurs, complete decompression on chamber air. Follow the guidance given in paragraph 14-4.5 for permanent loss of chamber oxygen supply to compute the air decompression schedule.

14-4.14 **Surface Interval Greater than 5 Minutes.** If the time from leaving 40 fsw in the water to the time of arrival at 50 fsw in the chamber during surface decompression exceeds 5 minutes, take the following action:

1. If the surface interval is less than or equal to 7 minutes, add one-half oxygen period to the total number of chamber periods required by increasing the time on oxygen at 50 fsw from 15 to 30 minutes. Ascend to 40 fsw during the subsequent air break. The 15-min penalty is considered a part of the normal surface decompression procedure, not an emergency procedure.

2. If the surface interval is greater than 7 minutes, continue compression to a depth of 60 fsw. Treat the divers on Treatment Table 5 if the original schedule required 2 or fewer oxygen periods in the chamber. Treat the divers on Treatment Table 6 if the original schedule required 3 or more oxygen periods in the chamber.

3. On rare occasions a diver may not be able to reach 50 fsw in the chamber because of difficulty equalizing middle ear pressure. In this situation, an alternative procedure for surface decompression on oxygen may be used. Compress the diver to the deepest depth he can attain initially. This will usually be less than 20 fsw. Begin oxygen breathing at that depth. Once oxygen breathing has begun,

attempt to gradually compress the diver to 30 fsw. If surface decompression was initiated while the diver was decompressing on oxygen in the water at 20 fsw, attempt to gradually compress the diver to 20 fsw. In either case, double the number of chamber oxygen periods indicated in the table and have the diver take these periods at whatever depth he is able to attain. Oxygen time starts when the diver initially goes on oxygen. Interrupt oxygen breathing every 60 minutes with a 15-min air break. The air break does not count toward the total oxygen time. Upon completion of the oxygen breathing periods, surface the diver at 30 fsw/min. Carefully observe the diver post-dive for the onset of decompression sickness. This "safe way out" procedure is not intended to be used in place of normal surface decompression procedures. Repetitive diving is not allowed following a dive in which the "safe way out" procedure is used.

14-4.15 **Asymptomatic Omitted Decompression.** Certain emergencies may interrupt or prevent required decompression. Unexpected surfacing, exhausted gas supply and bodily injury are examples of such emergencies. Table 14-2 shows the initial management steps to be taken when the diver has an uncontrolled ascent.

Table 14-2. *Management of Asymptomatic Omitted Decompression.*

Deepest Decompression Stop Omitted	Decompression Status	Surface Interval (Note 1)	ACTION
None	No-D	Any	Observe on surface for one hour
20 or 30 fsw	Stops Required	Less than 1 min	Return to depth of stop. Increase stop time by 1 min. Resume decompression according to original schedule
		1–7 min	Use Surface Decompression Procedure (Note 2)
		Greater than 7 min	Treatment Table 5 if 2 or fewer SurDO_2 periods Treatment Table 6 if 3 or more SurDO_2 periods
40 or 50 fsw	Stops Required	Any	Treatment Table 6
Deeper than 50 fsw	Stops Required: Less than 60 min missed	Any	Treatment Table 6A
	Stops Required: More than 60 min missed	Any	Compress to depth of dive not to exceed 225 fsw. Use Treatment Table 8 For saturation systems: Compress to depth of dive. Saturate two hours. Use saturation decompression without an initial upward excursion

Notes:
1. Surface interval is the time from leaving the stop to arriving at depth in the chamber.
2. For surface intervals greater than 5 minutes but less than or aequal to 7 minutes, increase the oxygen time at 50 fsw from 15 to 30 minutes.

14-4.15.1 **Omitted Decompression Stop Deeper Than 50 fsw.** An omitted decompression stop deeper than 50 fsw when more than 60 minutes of decompression are missed is an extreme emergency. The diver shall be returned as rapidly as possible to the full depth of the dive or the deepest depth of which the chamber is capable, whichever is shallower.

14-4.15.1.1 ***For Nonsaturation Systems.*** For nonsaturation systems, the diver shall be rapidly compressed on air to the depth of the dive or to 225 feet, whichever is shallower. For compressions deeper than 165 feet, remain at depth for 30 minutes. For compressions to 165 feet and shallower, remain at depth for a minimum of two hours. Decompress on USN Treatment Table 8. While deeper than 165 feet, a helium-oxygen mixture with 16 percent to 21 percent oxygen, if available, may be breathed by mask to reduce narcosis.

14-4.15.1.2 ***For Saturation Systems.*** For saturation systems, the diver should be rapidly compressed on air to 60 fsw, followed by compression on pure helium to the full depth of the dive or deeper if symptom onset warrants. The diver shall breathe 84% helium/16% oxygen by mask during the compression (if possible) to avoid the possibility of hypoxia as a result of gas pocketing in the chamber. Once at the saturation depth, the length of time spent can be dictated by the circumstances of the diver, but should not be less than 2 hours. During this 2 hours, treatment gas should be administered to the diver as outlined in Chapter 15, paragraph 15-23.8.2. The chamber oxygen partial pressure should be allowed to fall passively to 0.44-0.48 ata. Begin saturation decompression without an upward excursion.

14-4.16 **Symptomatic Omitted Decompression.** If the diver develops symptoms of decompression sickness or gas embolism before recompression for omitted decompression can be accomplished, immediate treatment using the appropriate oxygen or air recompression table is essential. Guidance for table selection and use is given in Chapter 20. If the depth of the deepest stop omitted was greater than 50 fsw and more than 60 minutes of decompression have been missed, use of Treatment Table 8 (Figure 20-10) or saturation treatment is indicated. See USN Treatment Table 4 and Treatment Table 7 (Chapter 20) for guidance on oxygen breathing.

In all cases of deep blowup, the services of a Diving Medical Officer shall be sought at the earliest possible moment.

14-4.17 **Light Headed or Dizzy Diver on the Bottom.** Dizziness is a common term used to describe a number of feelings, including lightheadedness, unsteadiness, vertigo (a sense of spinning), or the feeling that one might pass out. There are a number of potential causes of dizziness in surface-supplied diving, including hypoxia, a gas supply contaminated with toxic gases such as methylchloroform, and trauma to the inner ear caused by difficult clearing of the ear. At the low levels of oxygen percentage specified for surface-supplied diving, oxygen toxicity is an unlikely cause unless the wrong gas has been supplied to the diver.

14-4.17.1 **Initial Management.** The first step to take is to have the diver stop work and ventilate the rig while topside checks the oxygen content of the supply gas. These actions should eliminate hypoxia and hypercapnia as a cause. If ventilation does not improve symptoms, the cause may be a contaminated gas supply. Shift banks to the standby helium-oxygen supply and continue ventilation. If the condition clears, isolate the contaminated bank for future analysis and abort the dive on the standby gas supply. If the entire gas supply is suspect, place the diver on the EGS and abort the dive. Follow the guidance of paragraph 14-4.2 for ascents.

14-4.17.2 **Vertigo.** Vertigo due to inner ear problems will not respond to ventilation and in fact may worsen. One form of vertigo, however, alternobaric vertigo, may be so short lived that it will disappear during ventilation. Alternobaric vertigo will usually occur just as the diver arrives on the bottom and often can be related to a difficult clearing of the ear. It would be unusual for alternobaric vertigo to occur after the diver has been on the bottom for more than a few minutes. Longer lasting vertigo due to inner ear barotrauma will not respond to ventilation and will be accompanied by an intense sensation of spinning and marked nausea. Also, it is usually accompanied by a history of difficult clearing during the descent. These characteristic symptoms may allow the diagnosis to be made. A wide variety of ordinary medical conditions may also lead to dizziness. These conditions may occur while the diver is on the bottom. If symptoms of dizziness are not cleared by ventilation and/or shifting to alternate gas supplies, have the dive partner or standby diver assist the diver(s) and abort the dive.

14-4.18 Unconscious Diver on the Bottom. An unconscious diver on the bottom constitutes a serious emergency. Only general guidance can be given here. Management decisions must be made on site, taking into account all known factors. The advice of a Diving Medical Officer shall be obtained at the earliest possible moment.

If the diver becomes unconscious on the bottom:

1. Make sure that the breathing medium is adequate and that the diver is breathing. Verify manifold pressure and oxygen percentage.

2. Check the status of any other divers.

3. Have the dive partner or standby diver ventilate the afflicted diver to remove any accumulated carbon dioxide in the helmet and ensure the correct oxygen concentration.

4. If there is any reason to suspect gas contamination, shift to the standby helium-oxygen supply and ventilate both divers, ventilating the non-affected diver first.

5. When ventilation is complete, have the dive partner or standby diver ascertain whether the diver is breathing. The presence or absence of breath sounds will be audible over the intercom.

6. If the diver appears not to be breathing, the dive partner/standby diver should attempt to reposition the diver's head to open the airway. Airway obstruction will be the most common reason why an unconscious diver fails to breathe.

7. Check afflicted diver for signs of consciousness:

    - If the diver has regained consciousness, allow a short period for stabilization and then abort the dive.
    - If the diver remains unresponsive but is breathing, have the dive partner or standby diver move the afflicted diver to the stage. This action need not be rushed.
    - If the diver appears not to be breathing, maintain an open airway while moving the diver rapidly to the stage.

8. Once the diver is on the stage, observe again briefly for the return of consciousness.

    - If consciousness returns, allow a period for stabilization, then begin decompression.
    - If consciousness does not return, bring the diver to the first decompression stop at a rate of 30 fsw/min (or to the surface if the diver is in a no-decompression status).

9. At the first decompression stop:

    - If consciousness returns, decompress the diver on the standard decompression schedule using surface decompression.
    - If the diver remains unconscious but is breathing, decompress on the standard decompression schedule using surface decompression.
    - If the diver remains unconscious and breathing cannot be detected in spite of repeated attempts to position the head and open the airway, an extreme emergency exists. One must weigh the risk of catastrophic, even fatal, decompression sickness if the diver is brought to the surface, versus the risk of asphyxiation if the diver remains in the water. As a general rule, if there is any doubt about the diver's breathing status, assume he is breathing and continue normal decompression in the water. If it is absolutely certain that the diver is not breathing, leave the unaffected diver at his first decompression stop to complete decompression and surface the affected diver at 30 fsw/minute, deploying the standby diver as required. Recompress the diver immediately and treat for omitted decompression according to Table 14-2.

14-4.19 **Decompression Sickness in the Water.** Decompression sickness may develop in the water during surface-supplied diving. This possibility is one of the principal reasons for limiting dives to 300 fsw and allowing exceptional exposures only under emergency circumstances. The symptoms of decompression sickness may be joint pain or more serious manifestations such as numbness, loss of muscular function, or vertigo.

Management of decompression sickness in the water will be difficult under the best of circumstances. Only general guidance can be presented here. Management decisions must be made on site taking into account all known factors. The advice of a Diving Medical Officer shall be obtained at the earliest possible moment.

14-4.19.1 **Decompression Sickness Deeper than 30 fsw.** If symptoms of decompression sickness occur deeper than 30 fsw, recompress the diver 10 fsw. The diver may remain on 50% helium 50% oxygen during recompression from 90 to 100 fsw. Remain at the deeper stop for 1.5 times the stop time called for in the decompression table. If no stop time is indicated in the table, use the next shallower stop time to make the calculation. If symptoms resolve or stabilize at an acceptable level, decompress the diver to the 40-fsw water stop by multiplying each intervening stop time by 1.5 or more as needed to control the symptoms. Shift to 50% helium 50% oxygen at 90 fsw if the diver is not already on this mixture. Shift to 100 percent oxygen at 40 fsw and complete a 30 minute stop, then surface decompress and treat on Treatment Table 6. If during this scenario, symptoms worsen to the point that it is no longer practical for the diver to remain in the water, surface the diver and follow the guidelines for treatment of decompression sickness outlined in Chapter 20.

14-4.19.2 **Decompression Sickness at 30 fsw and Shallower.** If symptoms of decompression sickness occur at 30 fsw or shallower, remain on oxygen and recompress the diver 10 fsw. Remain at the deeper stop for 30 minutes. If symptoms resolve, surface decompress the diver at the end of the 30 minute period and treat on Treatment Table 6. If symptoms do not resolve, but stabilize at an acceptable level, decompress the diver to the surface on oxygen by multiplying each intervening stop time by 1.5 or more as needed to control symptoms. Treat on Treatment Table 6 upon reaching the surface. If during this scenario symptoms worsen to the point that it is no longer practical for the diver to remain in the water, surface the diver and follow guidelines for treatment of decompression sickness outlined in Chapter 20.

14-4.20 Decompression Sickness During the Surface Interval. If symptoms of Type I decompression sickness occur during travel from 40 fsw to the surface during surface decompression or during the surface undress phase, compress the diver to 50 fsw following normal surface decompression procedures. Delay neurological exam until the diver reaches the 50-fsw stop and is on oxygen. If Type I symptoms resolve during the 15-minute 50-fsw stop, the surface interval was 5 minutes or less, and no neurological signs are found, increase the oxygen time at 50 fsw from 15 to 30 minutes, then continue normal decompression for the schedule of the dive. Ascend from 50 to 40 fsw during the subsequent air break.

If Type I symptoms do not resolve during the 15-minute 50-fsw stop or symptoms resolve but the surface interval was greater than 5 minutes, compress the diver to 60 fsw on oxygen. Treat the diver on Treatment Table 5 if the original schedule required 2 or fewer oxygen periods in the chamber. Treat the diver on Treatment Table 6 if the original schedule required 3 or more oxygen periods in the chamber. Treatment table time starts upon arrival at 60 fsw. Follow the guidelines for treatment of decompression sickness given in Chapter 20.

If symptoms of Type II decompression sickness occur during travel from 40 fsw to the surface, during the surface undress phase, or the neurological examination at 50 fsw is abnormal, compress the diver to 60 fsw on oxygen. Treat the diver on Treatment Table 6. Treatment table time starts upon arrival at 60 fsw. Follow the guidelines for treatment of decompression sickness given in Chapter 20.

14-5 CHARTING SURFACE SUPPLIED HELIUM OXYGEN DIVES

Chapter 5 provides information for maintaining a Command Diving Log and personal diving log and for reporting individual dives to the Naval Safety Center. In addition to these records, every Navy HeO_2 dive shall be recorded on a diving chart similar to Figure 14-1. The diving chart is a convenient means of collecting the dive data, which in turn will be transcribed in the dive log. It is also useful in completing a mishap report for a diving related accident.

14-5.1 **Charting an HeO_2 Dive.** Figure 14-1 is a blank HeO_2 diving chart. Figure 14-2 is an example of a Surface Decompression dive. Figure 14-3 is an example of an In-water Decompression dive. Figure 14-4 is an example of a Surface Decompression dive with a hold on descent and delay on ascent.

When logging times on an HeO_2 diving chart, times will be recorded in a minute and second format. Clock time, however, will be logged in hours and minutes. All ascent times are rounded up to the next whole minute.

14-6 DIVING AT ALTITUDE

Surface-supplied helium-oxygen dives can be performed at altitude. The procedures for measuring water depth, obtaining the Sea Level Equivalent Depth and correcting in-water decompression stop depths are identical to the procedures for air diving (see Paragraph 9-13, Chapter 9). The procedures for performing surface decompression are also identical. The chamber stop depths during surface decompression are not adjusted for the altitude. Table 14-3 gives the maximum and minimum percentage of oxygen allowed in the bottom mixture at each depth. When diving at altitude, the maximum and minimum percentage of oxygen associated with the diver's actual depth rather than his Sea Level Equivalent Depth should be used. There are two important differences between diving helium-oxygen and diving air at altitude:

- Table 9-5 and Figure 9-15 cannot be used to correct the bottom time of a diver who is not fully equilibrated at altitude. The diver should wait 12 hours after arrival at altitude before making the first dive.

- Repetitive diving is not allowed during surface-supplied helium-oxygen diving at altitude. Following a no-decompression dive, the diver must wait 12 hours before making a another dive. Following a decompression dive, the diver must wait 18 hours before making another dive. A second dive is allowed following an abort during descent at depth of 100 fsw or less. Follow the guidance given in paragraph 14-3.6. Substitute the diver's maximum Sea Level Equivalent Depth for the diver's maximum depth when computing the table and schedule for the second dive.

Date:	Type of Dive: AIR HeO_2	
Diver 1:	Diver 2:	Standby:
Rig: PSIG: O_2%:	Rig: PSIG: O_2%:	Rig: PSIG: O_2%:
Diving Supervisor:	Chartman:	Bottom Mix:

EVENT	STOP TIME	CLOCK TIME
LS or 20 fsw		
RB		
LB		
R 1st Stop		
190 fsw		
180 fsw		
170 fsw		
160 fsw		
150 fsw		
140 fsw		
130 fsw		
120 fsw		
110 fsw		
100 fsw		
90 fsw		
80 fsw		
70 fsw		
60 fsw		
50 fsw		
40 fsw		
30 fsw		
20 fsw		
RS		
RB CHAMBER		
50 fsw chamber		
40 fsw chamber		
30 fsw chamber		
RS CHAMBER		

TDT	TTD

EVENT	TIME/DEPTH
Descent Time (Water)	
Stage Depth (fsw)	
Maximum Depth (fsw)	
Total Bottom Time	
Table/Schedule	
Time to 1st Stop (Actual)	
Time to 1st Stop (Planned)	
Delay to 1st Stop	
Travel/Shift/Vent Time	
Ascent Time-Water/SurD (Actual)	
Undress Time-SurD (Actual)	
Descent Chamber-SurD (Actual)	
Total SurD Surface Interval	
Ascent Time–Chamber (Actual)	

HOLDS ON DESCENT	
DEPTH	PROBLEM

DELAYS ON ASCENT	
DEPTH	PROBLEM

DECOMPRESSION PROCEDURES USED

AIR

- ☐ In-water Air decompression
- ☐ In-water Air/O_2 decompression
- ☐ $SurDO_2$

HeO_2

- ☐ In-water HeO_2/O_2 decompression
- ☐ $SurDO_2$

REPETITIVE GROUP:

Remarks:

Figure 14-1. Diving Chart.

1210		
Date: 4 Sept 07	Type of Dive: AIR (HeO$_2$) circled	
Diver 1: NDC Credle	Diver 2: ND1 Hopkins	Standby: NDC Fleming
Rig: MK 21 PSIG: 3000 O$_2$%: 16.2	Rig: MK 21 PSIG: 3000 O$_2$%: 16.2	Rig: MK 21 PSIG: 3000 O$_2$%:16.2
Diving Supervisor: NDCM Boyd	Chartman: EN2 Golden	Bottom Mix: 15.2

EVENT	STOP TIME	CLOCK TIME
LS or 20 fsw		0800
RB		0804
LB		0839
R 1st Stop		0843
190 fsw		
180 fsw		
170 fsw		
160 fsw		
150 fsw		
140 fsw		
130 fsw		
120 fsw		
110 fsw	:07	0850
100 fsw		
90 fsw	:03	0853
80 fsw	:07	0900
70 fsw	:09	0909
60 fsw	:13	0922
50 fsw	:13	0935
40 fsw	:13	0948
30 fsw		
20 fsw		
RS		0950
RB CHAMBER		0953
50 fsw chamber	:15	1008
40 fsw chamber	:15+:5+:30+:5+:30 :5+:30	1208
30 fsw chamber		
RS CHAMBER		1210

TDT	TTD
3:31	4:10

EVENT	TIME/DEPTH
Descent Time (Water)	:04
Stage Depth (fsw)	212
Maximum Depth (fsw)	222+4=226
Total Bottom Time	:39
Table/Schedule	230/40
Time to 1st Stop (Actual)	:03::49
Time to 1st Stop (Planned)	:03::24
Delay to 1st Stop	::25
Travel/Shift/Vent Time	
Ascent Time-Water/SurD (Actual)	:01::03
Undress Time-SurD (Actual)	:02::15
Descent Chamber-SurD (Actual)	::58
Total SurD Surface Interval	:04::16
Ascent Time–Chamber (Actual)	:01::20

HOLDS ON DESCENT

DEPTH	PROBLEM

DELAYS ON ASCENT

DEPTH	PROBLEM

DECOMPRESSION PROCEDURES USED

AIR

- ☐ In-water Air decompression
- ☐ In-water Air/O_2 decompression
- ☐ SurDO_2

HeO$_2$

- ☐ In-water HeO$_2$/O_2 decompression
- ☑ SurDO_2

REPETITIVE GROUP:

Remarks:

Figure 14-2. Completed HeO$_2$ Diving Chart: Surface Decompression Dive.

1139		
Date: 4 Sept 07	Type of Dive: AIR (HeO$_2$) circled	
Diver 1: NDC Allred	Diver 2: ND1 Wittman	Standby: ND1 Schlabach
Rig: KM-37 PSIG: 2950 O$_2$%:16.2	Rig: KM-37 PSIG: 2950 O$_2$%:16.2	Rig: KM-37 PSIG:2950 O$_2$%:16.2
Diving Supervisor: NDCM Van Horn	Chartman: NDC Parsons	Bottom Mix: 15.2

EVENT	STOP TIME	CLOCK TIME
LS or 20 fsw		0800
RB		0804
LB		0839
R 1st Stop		0843
190 fsw		
180 fsw		
170 fsw		
160 fsw		
150 fsw		
140 fsw		
130 fsw		
120 fsw		
110 fsw	:07	0850
100 fsw		
90 fsw	:03	0853
80 fsw	:07	0900
70 fsw	:09	0909
60 fsw	:13	0922
50 fsw	:13	0935
40 fsw	:13	0948
30 fsw	:2+:30+:5+:4	1029
20 fsw	:26+:5+:30+:5+:8	1143
RS		1144
RB CHAMBER		
50 fsw chamber		
40 fsw chamber		
30 fsw chamber		
RS CHAMBER		

TDT	TTD
3:05	3:44

EVENT	TIME/DEPTH
Descent Time (Water)	:04
Stage Depth (fsw)	212
Maximum Depth (fsw)	222+4=226
Total Bottom Time	:39
Table/Schedule	230/40
Time to 1st Stop (Actual)	:03::49
Time to 1st Stop (Planned)	:03::24
Delay to 1st Stop	::25
Travel/Shift/Vent Time	:02
Ascent Time-Water/SurD (Actual)	::40
Undress Time-SurD (Actual)	
Descent Chamber-SurD (Actual)	
Total SurD Surface Interval	
Ascent Time–Chamber (Actual)	

HOLDS ON DESCENT

DEPTH	PROBLEM

DELAYS ON ASCENT

DEPTH	PROBLEM

DECOMPRESSION PROCEDURES USED

AIR

- ☐ In-water Air decompression
- ☐ In-water Air/O$_2$ decompression
- ☐ SurDO$_2$

HeO$_2$

- ☑ In-water HeO$_2$/O$_2$ decompression
- ☐ SurDO$_2$

REPETITIVE GROUP:

Remarks:

Figure 14-3. Completed HeO$_2$ Diving Chart: In-water Decompression Dive.

1210		
Date: 4 Sept 07	Type of Dive: AIR (HeO₂ circled)	
Diver 1: ND2 Costin	Diver 2: ND1 Hatter	Standby: NDC Keller
Rig: KM-37 PSIG: 2950 O_2%:16.2	Rig: KM-37 PSIG: 2950 O_2%:16.2	Rig: KM-37 PSIG:2950 O_2%:16.2
Diving Supervisor: NDCM Pratschner	Chartman: ND2 Juarez	Bottom Mix: 15.2

EVENT	STOP TIME	CLOCK TIME	EVENT	TIME/DEPTH
LS or 20 fsw		0800	Descent Time (Water)	:07
RB		0807	Stage Depth (fsw)	212
LB		0838	Maximum Depth (fsw)	222+4=226
R 1st Stop		0843	Total Bottom Time	:38 + :02 = :40
190 fsw			Table/Schedule	230/40 Sur D
180 fsw			Time to 1st Stop (Actual)	:04::47
170 fsw			Time to 1st Stop (Planned)	:03::24
160 fsw			Delay to 1st Stop	:01::23
150 fsw			Travel/Shift/Vent Time	
140 fsw			Ascent Time-Water/SurD (Actual)	:01::03
130 fsw			Undress Time-SurD (Actual)	:02::05
120 fsw			Descent Chamber-SurD (Actual)	:01::25
110 fsw	:07	0850	Total SurD Surface Interval	:04::33
100 fsw			Ascent Time–Chamber (Actual)	:01::20
90 fsw	:03	0853	**HOLDS ON DESCENT**	
80 fsw	:07	0900	**DEPTH**	**PROBLEM**
70 fsw	:09	0909	32'	Red—right ear
60 fsw	:13	0922		
50 fsw	:13	0935		
40 fsw	:13	0948	**DELAYS ON ASCENT**	
30 fsw			**DEPTH**	**PROBLEM**
20 fsw			150'	Winch wire (fixed)
RS		0950		
RB CHAMBER		0953		
50 fsw chamber	:15	1008	**DECOMPRESSION PROCEDURES USED**	
40 fsw chamber	:15+:5+:30+:5+ :30+:5+:30	1208	AIR ☐ In-water Air decompression ☐ In-water Air/O_2 decompression ☐ SurDO_2 HeO_2 ☐ In-water HeO_2/O_2 decompression ☑ SurDO_2	
30 fsw chamber				
RS CHAMBER		1210		
TDT	**TTD**			
3:32	4:10		REPETITIVE GROUP:	

Remarks: 1. Delay on Ascent. Added :02 to bottom time. Did not change schedule.
2. Red diver had trouble clearing due to position of nose clearing device. DMT checked ears post dive. No barotrauma noted.

Figure 14-4. Completed HeO_2 Diving Chart: Surface Decompression Dive with Hold on Descent and Delay on Ascent.

Table 14-3. *Surface-Supplied Helium-Oxygen Decompression Table.*
(DESCENT RATE 75 FPM—ASCENT RATE 30 FPM)

Depth (fsw)	Bottom Time (min.)	Time to First Stop (min:sec)	**Decompression Stops (fsw)** Stop times (min) include travel time, except first HeO_2 and first O_2 stop																		Chamber O_2 Periods
			190	180	170	160	150	140	130	120	110	100	90	80	70	60	50	40	30	20	
			BOTTOM MIX										50% O_2						100% O_2		
60	10	2:00																		0	0
	20	2:00																		0	0
	30	2:00																		0	0
	40	2:00																		0	0
Max O_2=40.0%	60	0:40																10	11	16	1
Min O_2=14.0%	80	0:40																10	13	22	2
	100	0:40																10	16	27	2
	120	0:40																10	17	28	2
70	10	2:20																		0	0
	20	2:20																		0	0
	30	2:20																		0	0
	40	1:00																10	10	16	1
Max O_2=40.0%	60	1:00																10	14	24	2
Min O_2=14.0%	80	1:00																10	18	30	2
	100	1:00																10	19	34	2
	120	1:00																10	21	37	2
80	10	2:40																		0	0
	20	2:40																		0	0
	25	2:40																		0	0
	30	1:20																10	11	16	1
Max O_2=38.0%	40	1:20																10	13	21	2
Min O_2=14.0%	60	1:20																10	18	32	2
	80	1:20																10	21	38	2
	100	1:20																10	24	42	3
	120	1:20																10	25	45	3
90	10	3:00																		0	0
	20	3:00																		0	0
	30	1:40																10	13	21	2
	40	1:40																10	16	26	2
Max O_2=34.9%	60	1:40																10	21	38	2
Min O_2=14.0%	80	1:40																10	25	45	3
	100	1:40																10	28	50	3
	120	1:40																10	29	52	3

Table 14-3. Surface-Supplied Helium-Oxygen Decompression Table (Continued).
(DESCENT RATE 75 FPM—ASCENT RATE 30 FPM)

			Decompression Stops (fsw) Stop times (min) include travel time, except first HeO_2 and first O_2 stop																		
Depth (fsw)	Bottom Time (min.)	Time to First Stop (min:sec)	190	180	170	160	150	140	130	120	110	100	90	80	70	60	50	40	30	20	Chamber O_2 Periods
			BOTTOM MIX										50% O_2						100% O_2		
100	10	3:20																		0	0
	15	3:20																		0	0
	20	2:00																10	11	17	1
	30	2:00																10	15	24	2
Max O_2=*32.3%*	40	2:00																10	18	32	2
Min O_2=*14.0%*	60	2:00																10	25	44	3
	80	2:00																10	28	52	3
	100	2:00																10	31	56	3
	120	2:00																10	32	58	3
110	10	2:20																10	8	11	1
	20	2:20																10	12	20	1
	30	2:20																10	17	28	2
	40	2:20																10	20	36	2
Max O_2=*30.0%*	60	2:20																10	27	49	3
Min O_2=*14.0%*	80	2:20																10	31	58	3
	100	2:20																10	33	62	4
	Exceptional Exposure																				
	120	2:20																10	35	64	4
120	10	2:40																10	9	13	1
	20	2:40																10	14	23	2
	30	2:40																10	19	33	2
	40	2:40																10	23	42	3
Max O_2=*28.0%*	60	2:40																10	30	55	3
Min O_2=*14.0%*	80	2:40																10	34	63	4
	100	2:40																10	36	66	4
	Exceptional Exposure																				
	120	2:40															10	10	35	65	4
130	10	2:40															10	10	6	8	1
	20	2:40															10	10	12	19	1
	30	2:40															10	10	18	30	2
	40	2:20														7	10	10	22	40	3
Max O_2=*26.3%*	60	2:20														7	10	10	29	52	3
Min O_2=*14.0%*	80	2:20														7	10	10	33	60	3
	Exceptional Exposure																				
	100	2:20														7	10	10	35	64	4
	120	2:20														7	11	11	35	66	4

Table 14-3. *Surface-Supplied Helium-Oxygen Decompression Table (Continued).*
(DESCENT RATE 75 FPM—ASCENT RATE 30 FPM)

Decompression Stops (fsw)
Stop times (min) include travel time, except first HeO_2 and first O_2 stop

Depth (fsw)	Bottom Time (min.)	Time to First Stop (min:sec)	190	180	170	160	150	140	130	120	110	100	90	80	70	60	50	40	30	20	Chamber O_2 Periods
			BOTTOM MIX										50% O_2						100% O_2		
140	10	3:00															10	10	6	8	1
	20	3:00															10	10	12	19	1
	30	3:00															10	10	18	30	2
	40	2:40														7	10	10	22	40	2
Max O_2=*24.8%*	60	2:40														7	10	10	29	52	3
Min O_2=*14.0%*	80	2:40														7	10	10	33	60	3
	Exceptional Exposure																				
	100	2:40														7	10	10	35	64	4
	120	2:40														7	11	11	35	66	4
150	10	3:20															10	10	7	8	1
	20	3:00														7	10	10	14	22	2
	30	3:00														7	10	10	19	34	2
	40	3:00														7	10	10	24	44	3
Max O_2=*23.4%*	60	3:00														7	10	10	31	56	3
Min O_2=*14.0%*	80	3:00														7	10	10	35	64	4
	Exceptional Exposure																				
	100	3:00														7	13	13	36	66	4
	120	3:00														9	16	16	36	66	5
160	10	3:20														7	10	10	8	10	1
	20	3:20														7	10	10	15	24	2
	30	3:20														7	10	10	21	37	2
	40	3:20														7	10	10	26	47	3
Max O_2=*22.2%*	60	3:00													7	6	10	10	30	56	3
Min O_2=*14.0%*	**Exceptional Exposure**																				
	80	3:00													7	9	10	10	35	66	4
	100	3:00													7	13	14	14	35	66	5
	120	3:00													7	17	17	17	36	66	5
170	10	3:20													7	0	10	10	8	12	1
	20	3:20													7	0	10	10	16	28	2
	30	3:20													7	1	10	10	23	42	3
	40	3:20													7	4	10	10	28	52	3
Max O_2=*21.1%*	60	3:20													7	10	10	10	33	62	4
Min O_2=*14.0%*	**Exceptional Exposure**																				
	80	3:20													9	14	14	14	35	66	4
	100	3:00												7	5	18	18	18	36	66	5
	120	3:00												7	9	21	21	21	36	66	5

Table 14-3. Surface-Supplied Helium-Oxygen Decompression Table (Continued).
(DESCENT RATE 75 FPM—ASCENT RATE 30 FPM)

Depth (fsw)	Bottom Time (min.)	Time to First Stop (min:sec)	Decompression Stops (fsw) Stop times (min) include travel time, except first HeO_2 and first O_2 stop																		Chamber O_2 Periods
			190	180	170	160	150	140	130	120	110	100	90	80	70	60	50	40	30	20	
			BOTTOM MIX										50% O_2						100% O_2		
180	10	3:40													7	0	10	10	9	14	1
	20	3:40													7	0	10	10	17	30	2
	30	3:40													7	4	10	10	25	45	3
	40	3:20												7	0	8	10	10	30	54	3
Max O_2=*20.1%*	60	3:20												7	5	11	11	11	35	64	4
Min O_2=*14.0%*	**Exceptional Exposure**																				
	80	3:20												7	9	15	15	15	36	66	4
	100	3:20												7	13	19	19	19	36	66	5
	120	3:20												7	17	23	23	23	36	66	6
190	10	4:00													7	0	10	10	10	15	1
	20	3:40												7	0	2	10	10	19	34	2
	30	3:40												7	0	7	10	10	26	46	3
	40	3:40												7	4	9	10	10	31	56	3
	Exceptional Exposure																				
Max O_2=*19.2%*	60	3:40												7	9	13	13	13	34	62	4
Min O_2=*14.0%*	80	3:20											7	3	13	18	18	18	36	66	5
	100	3:20											7	6	16	21	21	21	36	66	6
	120	3:20											7	8	20	23	23	23	36	66	7
200	10	4:00												7	0	0	10	10	11	17	1
	20	4:00												7	0	4	10	10	20	36	2
	30	3:40											7	0	3	7	10	10	27	50	3
	40	3:40											7	0	7	10	10	10	31	58	3
	Exceptional Exposure																				
Max O_2=*18.4%*	60	3:40											7	4	10	14	14	14	35	66	4
Min O_2=*14.0%*	80	3:40											7	8	14	18	18	18	36	66	5
	100	3:40											7	12	17	23	23	23	36	66	6
	120	3:40											8	15	21	23	23	23	36	66	7
210	10	4:20												7	0	0	10	10	12	19	1
	20	4:00											7	0	1	6	10	10	22	38	2
	30	4:00											7	0	6	7	10	10	29	53	3
	40	4:00											7	3	9	10	10	10	33	60	3
	Exceptional Exposure																				
Max O_2=*17.7%*	60	3:40										7	0	9	11	17	17	17	35	66	5
Min O_2=*10.0%*	80	3:40										7	3	11	15	20	20	20	36	66	6
	100	3:40										7	6	14	19	23	23	23	36	66	7
	120	3:40										7	8	18	23	23	23	23	36	66	7

Table 14-3. *Surface-Supplied Helium-Oxygen Decompression Table (Continued).*
(DESCENT RATE 75 FPM—ASCENT RATE 30 FPM)

Depth (fsw)	Bottom Time (min.)	Time to First Stop (min:sec)	Decompression Stops (fsw) Stop times (min) include travel time, except first HeO_2 and first O_2 stop																		Chamber O_2 Periods
			190	180	170	160	150	140	130	120	110	100	90	80	70	60	50	40	30	20	
			BOTTOM MIX										50% O_2						100% O_2		
220	10	4:40												7	0	2	10	10	13	20	1
	20	4:20											7	0	3	7	10	10	23	41	3
	30	4:20											7	2	6	9	10	10	30	54	3
	40	4:00										7	0	6	9	11	11	11	34	62	4
	Exceptional Exposure																				
Max O_2=17.0% Min O_2=10.0%	60	4:00										7	4	9	12	18	18	18	36	66	5
	80	4:00										7	8	12	17	21	21	21	36	66	6
	100	4:00										7	12	15	20	23	23	23	36	66	7
	120	4:00										8	14	19	23	23	23	23	36	66	8
230	10	4:40											7	0	0	3	10	10	14	22	2
	20	4:20										7	0	3	4	7	10	10	24	44	3
	30	4:20										7	0	5	7	10	10	10	31	57	3
	40	4:00									7	0	3	7	9	13	13	13	34	64	4
	Exceptional Exposure																				
Max O_2=16.3% Min O_2=10.0%	60	4:00									7	0	8	10	14	18	18	18	36	66	6
	80	4:00									7	3	10	14	18	23	23	23	36	66	7
	100	4:00									7	6	12	17	23	23	23	23	36	66	8
	120	4:00									7	7	16	19	23	23	23	23	36	66	8
240	10	4:40										7	0	0	3	4	10	10	14	24	2
	20	4:40										7	0	3	5	7	10	10	25	46	3
	30	4:20									7	0	3	6	7	10	10	10	32	58	3
	40	4:20									7	0	5	8	9	14	14	14	35	64	4
	Exceptional Exposure																				
Max O_2=15.7% Min O_2=10.0%	60	4:20									7	4	8	11	14	19	19	19	36	66	6
	80	4:20									7	7	11	16	18	23	23	23	36	66	7
	100	4:20									7	10	14	19	23	23	23	23	36	66	8
	120	4:00								7	3	12	17	19	23	23	23	23	36	66	8
250	10	5:00										7	0	0	3	4	10	10	15	25	2
	20	4:40									7	0	0	3	7	7	10	10	26	47	3
	30	4:40									7	0	4	6	8	10	10	10	32	60	4
	40	4:40									7	2	5	9	9	14	14	14	35	64	4
	Exceptional Exposure																				
Max O_2=15.2% Min O_2=10.0%	60	4:20								7	0	7	9	12	16	21	21	21	36	66	6
	80	4:20								7	3	9	13	15	21	23	23	23	36	66	7
	100	4:20								7	6	11	14	19	23	23	23	23	36	66	8
	120	4:20								7	8	13	19	20	23	23	23	23	36	66	8

Table 14-3. *Surface-Supplied Helium-Oxygen Decompression Table (Continued).*
(DESCENT RATE 75 FPM—ASCENT RATE 30 FPM)

Depth (fsw)	Bottom Time (min.)	Time to First Stop (min:sec)	Decompression Stops (fsw) Stop times (min) include travel time, except first HeO_2 and first O_2 stop																		Chamber O_2 Periods
			190	180	170	160	150	140	130	120	110	100	90	80	70	60	50	40	30	20	
			BOTTOM MIX										50% O_2						100% O_2		
260	10	5:00									7	0	0	0	4	4	10	10	16	27	2
	20	5:00									7	0	3	4	6	7	10	10	27	50	3
	30	4:40								7	0	2	5	6	9	10	10	10	33	62	4
	40	4:40								7	0	3	8	9	10	15	15	15	35	64	5
	Exceptional Exposure																				
Max O_2=14.6%	60	4:40								7	3	7	10	14	16	21	21	21	36	66	6
Min O_2=10.0%	80	4:40								7	6	10	13	17	23	23	23	23	36	66	7
	100	4:20							7	2	9	13	16	20	23	23	23	23	36	66	8
	120	4:20							7	4	11	14	19	20	23	23	23	23	36	66	8
270	10	5:20									7	0	0	3	3	4	10	10	17	28	2
	20	5:00								7	0	0	3	6	6	8	10	10	29	52	3
	30	5:00								7	0	3	6	6	9	13	13	13	34	62	4
	Exceptional Exposure																				
	40	5:00							7	0	2	5	8	8	12	16	16	16	35	66	5
Max O_2=14.2%	60	4:40							7	0	6	8	10	14	19	23	23	23	36	66	6
Min O_2=10.0%	80	4:40							7	3	8	11	14	17	23	23	23	23	36	66	7
	100	4:40							7	5	11	13	16	20	23	23	23	23	36	66	8
	120	4:40							7	8	12	16	19	20	23	23	23	23	36	66	8
280	10	5:40									7	0	0	3	3	4	10	10	18	31	2
	20	5:20								7	0	0	4	6	7	7	10	10	30	54	3
	30	5:00							7	0	1	5	5	9	9	12	12	12	35	64	4
	Exceptional Exposure																				
	40	5:00							7	0	4	6	8	9	12	17	17	17	35	66	5
Max O_2=13.7%	60	5:00							7	4	6	8	12	15	18	23	23	23	36	66	7
Min O_2=10.0%	80	4:40						7	0	7	9	11	15	17	23	23	23	23	36	66	8
	100	4:40						7	2	9	11	15	17	20	23	23	23	23	36	66	8
	120	4:40						7	4	11	13	16	19	20	23	23	23	23	36	66	8
290	10	5:40								7	0	0	0	4	3	4	10	10	19	33	2
	20	5:20							7	0	0	2	6	6	6	9	10	10	30	56	3
	30	5:20							7	0	2	5	5	9	9	14	14	14	34	63	5
	Exceptional Exposure																				
	40	5:20							7	0	5	7	8	11	13	17	17	17	35	66	5
Max O_2=13.3%	60	5:00						7	0	6	7	9	12	15	20	23	23	23	36	66	7
Min O_2=10.0%	80	5:00						7	2	8	10	12	16	19	23	23	23	23	36	66	8
	100	5:00						7	5	10	12	15	19	20	23	23	23	23	36	66	8
	120	5:00						7	8	11	16	17	19	20	23	23	23	23	36	66	8

Table 14-3. *Surface-Supplied Helium-Oxygen Decompression Table (Continued).*
(DESCENT RATE 75 FPM—ASCENT RATE 30 FPM)

Depth (fsw)	Bottom Time (min.)	Time to First Stop (min:sec)	Decompression Stops (fsw) Stop times (min) include travel time, except first HeO_2 and first O_2 stop																		Chamber O_2 Periods
			190	180	170	160	150	140	130	120	110	100	90	80	70	60	50	40	30	20	
			BOTTOM MIX										50% O_2						100% O_2		
300	10	6:00								7	0	0	0	4	3	4	10	10	19	33	2
	20	5:40							7	0	0	2	6	6	6	9	10	10	30	56	3
	30	5:40							7	0	2	5	5	9	9	14	14	14	34	63	5
	Exceptional Exposure																				
	40	5:40							7	0	5	7	8	11	13	17	17	17	35	66	6
Max O_2=12.9% Min O_2=10.0%	60	5:20						7	0	6	7	9	12	15	20	23	23	23	36	66	7
	80	5:20						7	2	8	10	12	16	19	23	23	23	23	36	66	8
	100	5:20						7	5	10	12	15	19	20	23	23	23	23	36	66	8
	120	5:20						7	8	11	16	17	19	20	23	23	23	23	36	66	8
310	**Exceptional Exposure**																				
	10	6:00							7	0	0	0	3	3	3	7	10	10	21	36	2
	20	5:40						7	0	0	2	4	5	6	7	10	10	10	31	57	4
	30	5:40						7	0	2	4	5	7	8	11	15	15	15	35	66	5
Max O_2=12.5% Min O_2=10.0%	40	5:20					7	0	1	4	6	7	8	12	15	19	19	19	36	66	7
	60	5:20					7	0	5	6	9	11	13	17	20	23	23	23	36	66	8
	80	5:20					7	3	7	9	11	13	17	20	23	23	23	23	36	66	8
	100	5:20					7	5	9	11	13	17	19	20	23	23	23	23	36	66	8
	120	5:20					7	7	12	13	16	17	19	20	23	23	23	23	36	66	8
320	**Exceptional Exposure**																				
	10	6:20							7	0	0	0	4	3	3	7	10	10	21	38	2
	20	6:00						7	0	0	3	5	5	6	8	10	10	10	32	59	4
	30	5:40					7	0	0	4	4	6	7	9	11	17	17	17	35	66	5
	40	4:40					7	0	4	4	6	7	9	12	16	20	20	20	36	66	6
Max O_2=12.2% Min O_2=10.0%	60	5:20				7	0	2	6	8	9	11	14	17	23	23	23	23	36	66	8
	80	5:20				7	0	6	8	8	13	14	19	20	23	23	23	23	36	66	8
	100	5:20				7	2	7	10	13	16	17	19	20	23	23	23	23	36	66	8
	120	5:20				7	4	9	12	13	16	17	19	20	23	23	23	23	36	66	8
330	**Exceptional Exposure**																				
	10	6:20						7	0	0	0	2	3	3	4	7	10	10	22	40	2
	20	6:00					7	0	0	2	3	4	6	5	10	10	10	10	33	60	4
	30	6:00					7	0	1	4	5	6	8	8	13	17	17	17	35	66	6
Max O_2=11.8% Min O_2=10.0%	40	5:40				7	0	1	4	5	7	7	10	12	17	22	22	22	36	66	7
	60	5:40				7	0	5	6	8	9	11	15	20	23	23	23	23	36	66	8
	80	5:40				7	2	7	8	10	13	15	19	20	23	23	23	23	36	66	8
	100	5:40				7	5	9	9	13	16	17	19	20	23	23	23	23	36	66	8
	120	5:20			7	1	7	10	13	15	16	17	19	20	23	23	23	23	36	66	8

Table 14-3. *Surface-Supplied Helium-Oxygen Decompression Table (Continued).*
(DESCENT RATE 75 FPM—ASCENT RATE 30 FPM)

Depth (fsw)	Bottom Time (min.)	Time to First Stop (min:sec)	Decompression Stops (fsw) Stop times (min) include travel time, except first HeO_2 and first O_2 stop																		Chamber O_2 Periods
			190	180	170	160	150	140	130	120	110	100	90	80	70	60	50	40	30	20	
			BOTTOM MIX										50% O_2						100% O_2		
340	**Exceptional Exposure**																				
	10	6:40						7	0	0	0	3	3	3	4	7	10	10	23	41	3
	20	6:20					7	0	0	2	4	5	7	8	9	10	10	10	33	60	5
	30	6:00				7	0	0	3	5	5	6	8	9	13	18	18	18	35	66	6
	40	6:00				7	0	2	4	6	7	8	10	13	16	22	22	22	36	66	7
Max O_2=11.5%	60	5:40			7	0	3	5	6	9	10	13	16	18	21	23	23	23	36	66	8
Min O_2=10.0%	80	5:40			7	0	7	7	8	11	13	15	19	20	23	23	23	23	36	66	8
	100	5:40			7	2	8	8	12	13	16	17	19	20	23	23	23	23	36	66	8
	120	5:40			7	4	9	11	13	15	16	17	19	20	23	23	23	23	36	66	8
350	**Exceptional Exposure**																				
	10	6:40					7	0	0	0	2	2	3	3	5	7	10	10	24	43	3
	20	6:20				7	0	0	0	4	4	5	5	7	9	13	13	13	33	63	5
	30	6:20				7	0	1	4	4	5	7	8	11	13	18	18	18	36	66	6
	40	6:00			7	0	1	3	5	6	7	8	11	14	17	23	23	23	36	66	7
Max O_2=11.2%	60	6:00			7	0	5	5	8	8	11	12	16	19	23	23	23	23	36	66	8
Min O_2=10.0%	80	6:00			7	2	7	7	10	11	13	17	19	20	23	23	23	23	36	66	8
	100	5:40		7	0	6	8	9	11	15	16	17	19	20	23	23	23	23	36	66	8
	120	5:40		7	1	7	9	12	14	15	16	17	19	20	23	23	23	23	36	66	8
360	**Exceptional Exposure**																				
	10	7:00					7	0	0	0	2	2	3	3	7	7	10	10	25	44	3
	20	6:40				7	0	0	2	3	4	5	5	8	10	13	13	13	34	63	5
	30	6:20			7	0	0	3	3	5	6	7	8	11	13	19	19	19	36	66	7
	40	6:20			7	0	2	4	5	7	7	9	10	14	20	23	23	23	36	66	8
Max O_2=10.9%	60	6:20			7	2	5	6	7	9	11	14	16	19	23	23	23	23	36	66	8
Min O_2=10.0%	80	6:00		7	0	6	6	8	11	12	14	16	19	20	23	23	23	23	36	66	8
	100	6:00		7	2	7	8	11	13	13	16	17	19	20	23	23	23	23	36	66	8
	120	6:00		7	4	8	10	12	14	15	16	17	19	20	23	23	23	23	36	66	8

Table 14-3. *Surface-Supplied Helium-Oxygen Decompression Table (Continued).*
(DESCENT RATE 75 FPM—ASCENT RATE 30 FPM)

Depth (fsw)	Bottom Time (min.)	Time to First Stop (min:sec)	Decompression Stops (fsw) Stop times (min) include travel time, except first HeO_2 and first O_2 stop																		Chamber O_2 Periods
			190	180	170	160	150	140	130	120	110	100	90	80	70	60	50	40	30	20	
			BOTTOM MIX										50% O_2						100% O_2		
370	**Exceptional Exposure**																				
	10	7:00				7	0	0	0	0	3	3	3	3	7	7	10	10	25	46	3
	20	6:40			7	0	0	0	3	4	4	5	5	8	10	13	13	13	34	63	5
	30	6:20		7	0	0	2	3	4	4	7	7	8	11	16	19	19	19	36	66	7
	40	6:20		7	0	0	4	4	5	6	8	10	11	14	20	23	23	23	36	66	8
Max O_2=10.6%	60	6:20		7	0	4	5	7	8	9	11	13	17	20	23	23	23	23	36	66	8
Min O_2=10.0%	80	6:00	7	0	3	6	7	9	10	12	15	17	19	20	23	23	23	23	36	66	8
	100	6:00	7	0	6	7	9	10	14	15	16	17	19	20	23	23	23	23	36	66	8
	120	6:00	7	1	7	9	11	13	14	15	16	17	19	20	23	23	23	23	36	66	8
380	**Exceptional Exposure**																				
	10	7:20				7	0	0	0	0	3	3	3	3	7	7	10	10	25	46	3
	20	7:00			7	0	0	0	3	4	4	5	5	8	10	13	13	13	34	63	6
	30	6:40		7	0	0	2	3	4	4	7	7	8	11	16	19	19	19	36	66	7
	40	6:40		7	0	0	4	4	5	6	8	10	11	14	20	23	23	23	36	66	8
Max O_2=10.4%	60	6:20		7	0	4	5	7	8	9	11	13	17	20	23	23	23	23	36	66	8
Min O_2=10.0%	80	6:20	7	0	3	6	7	9	10	12	15	17	19	20	23	23	23	23	36	66	8
	100	6:20	7	0	6	7	9	10	14	15	16	17	19	20	23	23	23	23	36	66	8
	120	6:20	7	1	7	9	11	13	14	15	16	17	19	20	23	23	23	23	36	66	8

PAGE LEFT BLANK INTENTIONALLY

CHAPTER 15

Saturation Diving

15-1 INTRODUCTION

15-1.1 **Purpose.** The purpose of this chapter is to familiarize divers with U.S. Navy saturation diving systems and deep diving equipment.

15-1.2 **Scope.** Saturation diving is used for deep salvage or recovery using U.S. Navy deep diving systems or equipment. These systems and equipment are designed to support personnel at depths to 1000 fsw for extended periods of time.

SECTION ONE — DEEP DIVING SYSTEMS

15-2 APPLICATIONS

The Deep Diving System (DDS) is a versatile tool in diving and its application is extensive. Most of today's systems employ a multilock deck decompression chamber (DDC) and a personnel transfer capsule (PTC).

- **Non-Saturation Diving.** Non-saturation diving can be accomplished with the PTC pressurized to a planned depth. This mode of operation has limited real time application and therefore is seldom used in the U.S. Navy.

- **Saturation Diving.** Underwater projects that demand extensive bottom time (i.e., large construction projects, submarine rescue, and salvage) are best conducted with a DDS in the saturation mode.

- **Conventional Diving Support.** The DDC portion of a saturation system can be employed as a recompression chamber in support of conventional, surface-supplied diving operations.

15-3 BASIC COMPONENTS OF A SATURATION DIVE SYSTEM

The configuration and the specific equipment composing a deep diving system vary greatly based primarily on the type mission for which it is designed. Modern systems however, have similar major components that perform the same functions despite their actual complexity. Major components include a PTC, a PTC handling system, and a DDC.

15-3.1 **Personnel Transfer Capsule.** The PTC (Figure 15-1) is a spherical, submersible pressure vessel that can transfer divers in full diving dress, along with work tools and associated operating equipment, from the deck of the surface platform to their designated working depth.

15-3.1.1 **Gas Supplies.** During normal diving operations, the divers' breathing and PTC gas are supplied from the surface through a gas supply hose. In addition, all PTCs

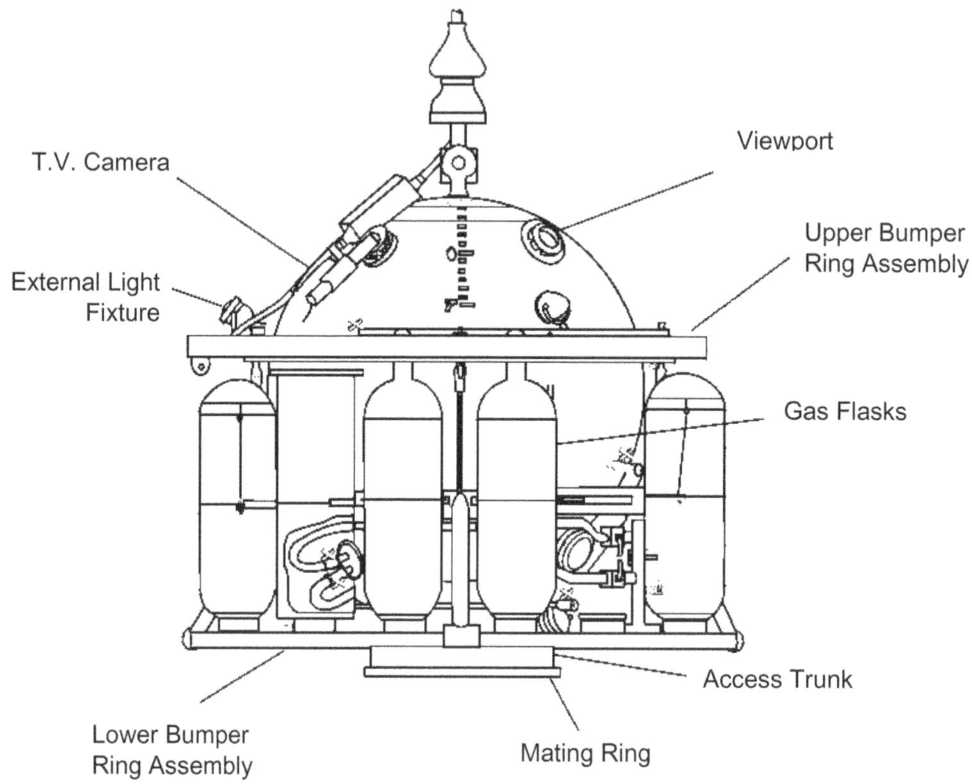

Figure 15-1. Typical Personnel Transfer Capsule Exterior.

carry emergency supplies of helium, helium-oxygen, and oxygen in externally mounted flasks. Internal PTC pressure, gas supply pressures, and water depth are continuously monitored from the PTC.

The typical helium system is designed to maintain PTC pressurization and purge oxygen from all PTC electrical units to alleviate any fire hazard.

The helium-oxygen mixed-gas system consists of an internal built-in breathing system (BIBS) with associated valves, piping, and fittings. The mixed-gas system supplies emergency breathing gas to the diver umbilicals when the topside supply is interrupted, and supplies the BIBS if the internal PTC atmosphere is contaminated.

15-3.1.2 **PTC Pressurization/Depressurization System.** The gas supply and exhaust system control and regulate internal PTC pressure. Relief valves and manual vent valves prevent overpressurization of the PTC in case a line rupture causes a full flask to discharge into the PTC. Needle valves are employed to control depressurization. Depth gauges, calibrated in feet of seawater, monitor internal and external PTC depth. Equalization and vent valves are also provided for the access trunk.

15-3.1.3 **PTC Life-Support System.** The life-support equipment for the PTC includes carbon dioxide scrubbers, a gas supply to provide metabolic oxygen, oxygen, and carbon dioxide analyzers.

15-3.1.4 **Electrical System.** The electrical system uses a multiple voltage distribution system that may be used for heating, internal and external lighting, instrumentation, and communications. Power for normal PTC operation is surface-supplied and is transmitted through power and communications cables. A battery supplies critical loads such as atmosphere monitoring, emergency CO_2 scrubber, and communications if the surface-supplied power is interrupted.

15-3.1.5 **Communications System.** A typical communications system is divided into four individual systems to ensure efficient operation under a variety of conditions.

- **Hardwire Intercom System.** The intercom system is an amplified voice system employing a helium speech unscrambler providing communications within the PTC and between the Main Control Console (MCC), divers, deck winch operator, Deck Officer, and the DDCs.
- **Underwater Mobile Sound Communications Set (UQC).** The UQC system is a wireless emergency system providing voice communications between the PTC and underwater telephone system of the attending ship. The UQC system is used if the power and communications cables fail or are disconnected.
- **Closed-Circuit Television (CCTV).** The CCTV consists of video channels from the PTC to the MCC. Cameras are usually mounted outside the PTC.
- **Sound-Powered Phones.** The PTC is equipped with a sound-powered phone system for communication with the MCC in case the normal system is lost.

15-3.1.6 **Strength, Power, and Communications Cables (SPCCs).** The strength, power, and communications cables typically provide electrical power, wired communications, instrumentation signals, a strength member, and coaxial transmission (CCTV signals) between the MCC and the PTC.

15-3.1.7 **PTC Main Umbilical.** The typical PTC main umbilical consists of a breathing-gas supply hose, a hot water hose, a pneumofathometer, and a strength member.

15-3.1.8 **Diver Hot Water System.** Hot water may be necessary when conducting saturation dives. The surface ship supplies hot water via the PTC main umbilical to the diver's suit and breathing gas heater. The PTC operator monitors the water temperature and ensures that the flow is adequate.

15-3.2 **Deck Decompression Chamber (DDC).** The DDC furnishes a dry environment for accomplishing decompression and, if necessary, recompression. The DDC is a multi-compartment, horizontal pressure vessel mounted on the surface-support platform. Each DDC is equipped with living, sanitary, and resting facilities for the dive team. A service lock provides for the passage of food, medical supplies, and other articles between the diving crew inside the chamber and topside support personnel.

15-3.2.1 **DDC Life-Support System (LSS).** The DDC Life Support-System maintains the chamber environment within acceptable limits for the comfort and safety of the divers. The typical system consists of temperature and humidity control, carbon dioxide removal, and equipment monitoring. Processing consists of filtering particulate matter, removing carbon dioxide and gaseous odors, and controlling heat and humidity.

15-3.2.2 **Sanitary System.** The sanitary system consists of hot and cold water supplies for operating the wash basin, shower, and head. Waste from the head discharges into a separate holding tank for proper disposal through the support platform's collection, holding, and transfer system.

15-3.2.3 **Fire Suppression System.** All DDCs have fire-fighting provisions ranging from portable fire extinguishers to installed, automatic systems. DDCs and recompression chambers have similar hyperbaric flammability hazards. Ignition sources and combustion materials should be minimized during critical fire zone times. (At the normal operating depth of PTCs, the oxygen concentration will not support combustion, so they have no built-in fire-fighting equipment.)

15-3.2.4 **Main Control Console (MCC).** The MCC is a central control and monitoring area. The MCC houses the controls for the gas supply and atmosphere analysis for the DDC, atmosphere monitoring for the PTC, pressure gauges for gas banks, clocks, communications systems controls, recorders, power supplies, and CCTV monitors and switches for the DDC and PTC.

15-3.2.5 **Gas Supply Mixing and Storage.** The DDC gas system provides oxygen, helium-oxygen mixtures, helium, and air for pressurization and diver life support. A BIBS is installed in every lock for emergency breathing in contaminated atmospheres, as well as for administering treatment gas during recompression treatment. Normal pressurizing or depressurizing of the DDC is done from the MCC. A means of sampling the internal atmosphere is provided for monitoring carbon dioxide and oxygen partial pressure. An oxygen-addition system maintains oxygen partial pressure at required levels. A pressure-relief system prevents overpressurization of the chamber.

A DDS should be outfitted with gas-mixing equipment, commonly referred to as a "Mixmaker," which provides additional flexibility when conducting deep saturation diving. The Mixmaker can provide mixed gas at precise percentages and quantities needed for any given dive. If necessary, the gas coming from the Mixmaker can be sent directly to the divers for consumption.

15-3.3 **PTC Handling Systems.** Of all the elements of DDS, none are more varied than PTC handling systems. Launch and retrieval of the PTC present significant hazards to the divers during heavy weather and are major factors in configuring and operating the handling system.

15-3.3.1 **Handling System Characteristics.** All handling systems have certain common characteristics. The system should:

- Be adequately designed and maintained to withstand the elements and dynamic loads imposed by heavy weather.
- Have the ability to control the PTC through the air-sea interface at sufficient speed to avoid excessive wave action.
- Keep the PTC clear of the superstructure of the surface-support platform to avoid impact damage.
- Have lifting capability of sufficient power to permit fast retrieval of the PTC, and controls and brakes that permit precision control for PTC mating and approach to the seafloor.
- Include a handling system to move the suspended PTC to and from the launch/retrieval position to the DDC.
- Have a method of restraining PTC movement during mating to the DDC.

15-3.4 **Saturation Mixed-Gas Diving Equipment.** The UBA MK 21 MOD 0 is an open circuit, demand-regulated diving helmet designed for saturation, mixed-gas diving at depths in excess of 300 fsw and as deep as 950 fsw (Figure 15-2). With the exception of the demand regulator, it is functionally identical to the UBA MK 21 MOD 1, which is used for air and mixed-gas diving. The regulator for the MK 21 MOD 0 helmet is the Ultraflow 500, which provides improved breathing resistance and gas flow over the MK 21 MOD 1.

The UBA MK 22 MOD 0 is an open circuit, demand-regulated, band-mask version of the UBA MK 21 MOD 0 (Figure 15-3). It is used for the standby diver for saturation, mixed-gas diving at depths in excess of 300 fsw and as deep as 950 fsw. It is provided with a hood and head harness instead of the helmet shell to present a smaller profile for storage.

15-4 U.S. NAVY SATURATION FACILITIES

15-4.1 **Navy Experimental Diving Unit (NEDU), Panama City, FL.** NEDU's mission is to test and evaluate diving, hyperbaric, and other life-support systems and procedures, and to conduct research and development in biomedical and environmental physiology. NEDU then provides technical recommendations to Commander, Naval Sea Systems Command to support operational requirements of our the U.S. Armed Forces.

NEDU houses the Ocean Simulation Facility (OSF), one of the world's largest man-rated hyperbaric facilities. The OSF consists of five chambers with a wet pot and transfer trunk. The wet pot holds 55,000 gallons of water. The OSF can simulate depths to 2,250 fsw and can accommodate a wide range of experiments in its dry and wet chambers (see Figure 15-4, Figure 15-5, and Figure 15-6).

Figure 15-2. MK 21 MOD 0 with Hot Water Suit, Hot Water Shroud, and Come-Home Bottle.

Figure 15-3. MK 22 MOD 0 with Hot Water Suit, Hot Water Shroud, and Come-Home Bottle.

15-4.2 **Naval Submarine Medical Research Laboratory (NSMRL), New London, CT.** The mission of the Naval Submarine Medical Research Laboratory is to conduct medical research and development in the fields of hyperbaric physiology, operational psychology and physiology, human factors engineering, and other allied sciences as they apply to biomedical programs in operational environments (Figure 15-7).

SECTION TWO — DIVER LIFE-SUPPORT SYSTEMS

15-5 INTRODUCTION

Saturation diver life-support systems must provide adequate respiratory and thermal protection to allow work in the water at extreme depths and temperatures. Because of the increased stresses placed upon the diver by deep saturation dives, this equipment must be carefully designed and tested in its operating environment. The diver life-support system consists of two components: an underwater breathing apparatus (UBA) and a thermal protection system. The actual in-water time a diver can work effectively depends on the adequacy of his life-support apparatus and his physical conditioning. Important considerations in the duration of effective in-water time are the rate of gas consumption for the system and the degree of thermal protection. Present U.S. Navy saturation diving UBAs are designed to operate effectively underwater for at least 4 hours. Although a given diving apparatus may be able to provide longer diver life support, experience has shown that cumulative dive time at deep depths will progressively reduce diver effectiveness after a 4-hour in-water exposure.

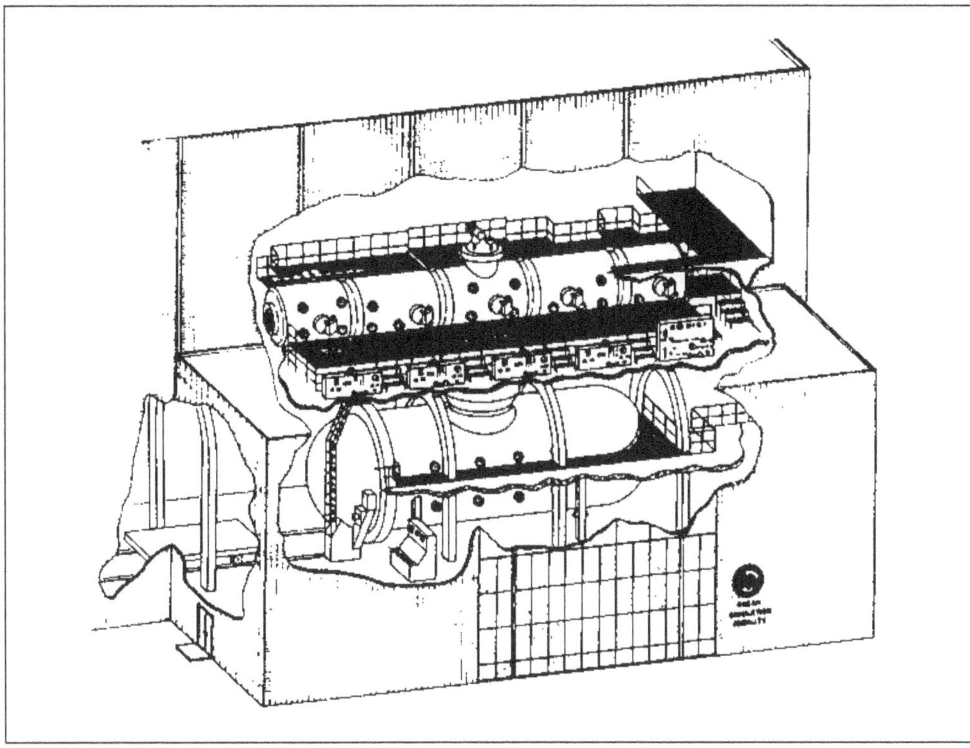

Figure 15-4. NEDU's Ocean Simulation Facility (OSF).

Figure 15-5. NEDU's Ocean Simulation Facility Saturation Diving Chamber Complex.

Figure 15-6. NEDU's Ocean Simulation Facility Control Room.

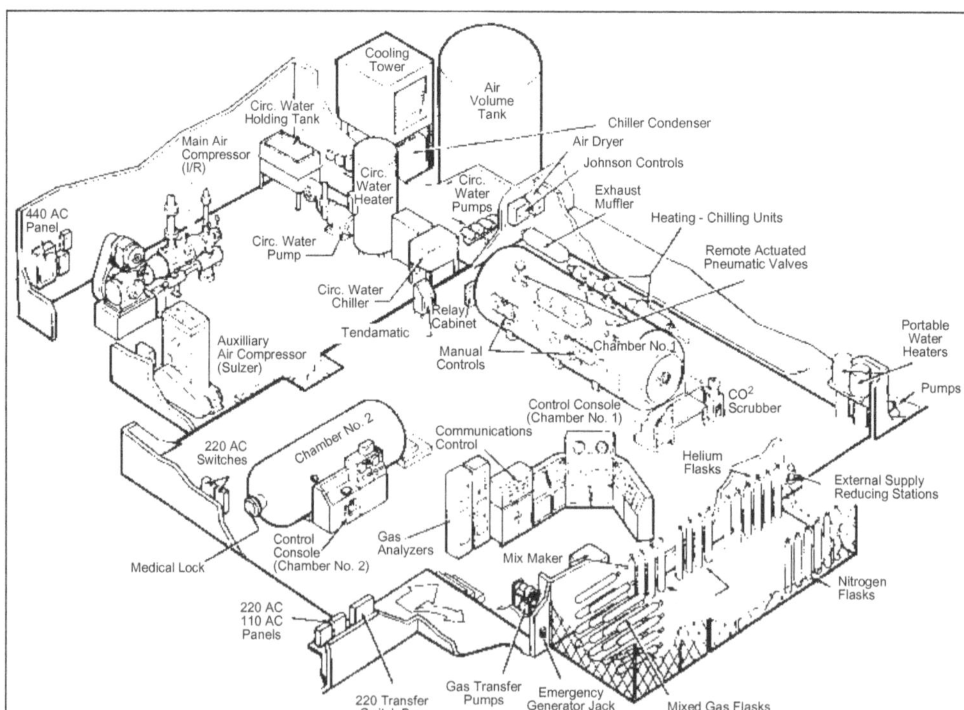

Figure 15-7. Naval Submarine Medical Research Laboratory (NSMRL).

15-6 THERMAL PROTECTION SYSTEM

All saturation diver life-support systems include diver thermal protection consisting of a hot water suit and a breathing gas heater. The thermal protection is designed to minimize the diver's heat loss caused by helium's high thermal conductivity. Helium conducts heat away from the body rapidly and causes a significant heat loss via the diver's breathing gas. The diver's metabolic rate may not be great enough to compensate for the heat loss when breathing cold gas, resulting in a drop in body temperature and increasing the chance of hypothermia.

15-6.1 **Diver Heating.** Because of the high thermal conductivity of helium and depths attained, most conventional diving suits (i.e., wet suits/dry suits) provide inadequate insulation in a helium environment. As a result, thermal protection garments for helium-oxygen saturation diving must employ active heating. The most successful thermal protection currently used is the non-return valve (NRV) hot water suit using circulating hot water as the heat source. The typical NRV hot water suit is constructed from closed-cell, pre-crushed neoprene with an outer layer of tough canvas-type nylon. The interior is lined with a softer nylon with perforated hot water hoses along the limbs, chest, and backbone. Divers are required to wear Polartec Diveskins or Neoprene liners under their NRV suits. The liners or Diveskins offer almost no protection from cold water. The liners or Diveskins keep the divers from getting burned by hot water discharge from the NRV suit and minimize chafing of skin.

The effectiveness of the hot water suit in keeping the divers warm is dependent upon maintaining an adequate flow of water at the proper temperature. A 4-gallon per minute (gpm) (3 gpm to the suit and 1 gpm to the breathing gas heater) hot water flow rate with the suit inlet temperature adjusted to diver's comfort generally provides adequate protection. During normal operation, hot water is distributed through the NRV hot water suit and is then discharged to the sea through the NRV. If there is a diver heating system failure, the diver shuts the NRV and opens the bypass valve, trapping sufficient hot water in the suit to allow him to return to the PTC. To prevent burn injury to the diver, the water temperature at the suit inlet should not exceed 110°F. Hot water thermal protection systems should be designed to provide individual control of water temperature and rate of flow supplied to each diver. All divers normally use umbilicals of similar length.

15-6.2 **Inspired Gas Heating.** The thermal protection system includes a breathing-gas heater to warm the gas to a temperature sufficient to minimize respiratory heat loss. A typical breathing-gas heater is a hot water heat exchanger that can raise the breathing-gas temperature by 30–50°F. Breathing cold helium-oxygen at deep saturation diving depths can cause incapacitating nasal and trachea-bronchial secretions, breathing difficulties, chest pain, headache, and severe shivering. These symptoms may begin within minutes of starting the dive excursion. Breathing apparently comfortable but low-temperature helium-oxygen at deep depths can rapidly lower body temperature through respiratory heat loss, even though the skin is kept warm by the hot water suit. The diver usually remains unaware of respiratory heat loss, has no symptoms, and will not begin to shiver until his core temperature has fallen. Metabolic heat production may not compensate for continuing respiratory heat loss. Table 15-1 contains guidelines for the minimum allowable temperatures for helium-oxygen breathing gas. These limits are based on a 4-hour excursion with a maximum core body temperature drop of 1.8°F (1.0°C) in a diver wearing a properly fitted and functioning NRV or hot water suit.

Table 15-1. *Guidelines for Minimum Inspired HeO_2 Temperatures for Saturation Depths Between 350 and 1,500 fsw.**

	Minimum Inspired Gas Temperature	
Depth (fsw)	**°C**	**°F**
350	-3.1	26.4
400	1.2	34.2
500	7.5	45.5
600	11.7	53.1
700	14.9	58.8
800	17.3	63.1
900	19.2	66.6
1000	20.7	69.3
1100	22.0	71.6
1200	23.0	73.4
1300	23.9	75.0
1400	24.7	76.5
1500	25.4	77.72

* Ref: C. A. Piantadosi, "Respiratory Heat Loss Limits in Helium Oxygen Saturation Diving," Navy Experimental Diving Unit Report NR 10-80 Revised 1982 (ADA 094132).

15-7 SATURATION DIVING UNDERWATER BREATHING APPARATUS

The rate of gas consumption and the composition of the gas supply depend in part upon the design of the UBA. Three types of underwater breathing apparatus have been used successfully to support saturation diving operations: demand open-circuit, semiclosed-circuit, and closed-circuit.

UBA systems should be designed to support saturation diving excursions of at least 4 hours duration in temperatures as low as 29°F. Specific information on U.S. Navy certified diving equipment can be found in the applicable system-specific technical manuals.

15-8 UBA GAS USAGE

Gas usage can be the controlling factor in the planning for a mission and determining appropriate excursions. However, gas usage is UBA- and platform-specific.

15-8.1 **Specific Dives.** For a specific dive, storage of gas to support the mission may be the controlling parameter. The following formulas may be used to calculate gas usage by divers:

$$\text{ata} = \frac{D + 33}{33}$$

scfm (for one diver at depth) = ata × acfm

total scfm = scfm × number of divers

scf required = scfm × minutes

D = depth of diver

ata = atmosphere absolute

acfm = actual cubic feet per minute required by specific UBA being used (refer to the tech manual)

number of divers = total number of divers making excursion

minutes = duration of excursion

scf required = standard cubic feet of gas required to support the divers

Example. Two divers and one standby diver using the MK 21 MOD 0 and MK 22 MOD 0 UBAs at 300 fsw are deployed for a 15-minute excursion. Determine the gas usage.

1. Convert the depth to atmospheres:

$$\frac{300 \text{ fsw} + 33 \text{ fsw}}{33 \text{ fsw}} = 10.09 \text{ ata}$$

2. Calculate gas usage for 1 diver:

$$\begin{array}{l} 10.09 \text{ ata} \\ \underline{\text{x } 1.4 \text{ acfm for MK21 MOD 0}} \\ 14.13 \text{ scfm for 1 diver at 300 fsw} \end{array}$$

3. Calculate gas usage for 3 divers:

14.13 scfm for 1 diver at 300 fsw
x 3 divers (2) and standby (1)

42.39 scfm for 3 divers at 300 fsw

4. Calculate the total gas usage requirement:

42.39 scfm
x 15 minutes excursion time

635.85 scf (round up to 636 scf)

A gas usage requirement of 636 Standard Cubic Feet of helium-oxygen can be expected for this two-diver excursion.

NOTE **Usage for three divers is computed even though the standby would not normally be using gas for the entire 15 minutes.**

15-8.2 **Emergency Gas Supply Duration.** The gas computation in paragraph 15-8.1 is used to determine excursion limits based on diver's gas storage. The diver's emergency gas supply (EGS) duration should also be calculated using the following formulas:

$$\text{mmp} = (D \times .445) + \text{psi (obp)}$$

$$\text{psi available for use} = \text{psi (cylinder)} - \text{mmp}$$

$$\text{scf gas available} = \frac{\text{psi (Available)} + 14.7}{14.7} \times \text{fv}$$

$$\text{scfm} = \text{acfm} \times \text{ata}$$

$$\text{duration in minutes} = \frac{\text{scf}}{\text{scfm}}$$

D = depth of diver

psi (obp) = over-bottom pressure required for specific UBA

mmp = minimum manifold pressure

fv = floodable volume of cylinder

acfm = actual cubic feet per minute at excursion depth required by specific UBA being used

scfm = standard cubic feet per minute required to deliver acfm

Example. Using an 80-cubic-foot aluminum cylinder (floodable volume = .399 cu. ft.) filled to 3,000 psig, calculate the diver's EGS duration at 300 fsw.

1. Calculate the psi available for use:

$$\begin{array}{r} 185.0 \text{ overbottom psi, MK 21 MOD 0} \\ +\ 133.5 \text{ psi (300 fsw converted to psi)} \\ \hline 318.5 \text{ psi (round up to 319 psi)} \end{array}$$

2. Calculate the psig available for use:

$3{,}000 - 319 \text{ psig} = 2{,}681 \text{ psig available for use}$

3. Calculate the scf of gas available:

$$\frac{2681 + 14.7}{14.7} \times 0.399 = 73.2 \text{ scf of gas available}$$

4. Calculate the standard cubic feet per minute required:

$1.4 \text{ acfm} \times 10.09 \text{ ata} = 14.13 \text{ scfm}$

5. Calculate the duration of the gas supply:

$$\frac{73.2 \text{ scf}}{14.13 \text{ scfm}} = 5.18 \text{ minutes}$$

The duration of the emergency gas supply is very short, especially at greater depths.

15-8.3 **Gas Composition.** The percentage of oxygen in the mix depends on diver depth and can be calculated as follows:

1. $\% \text{ decimal equivalent} = \dfrac{ppO_2 \text{ desired}}{\text{ata}}$

2. % decimal equivalent × 100 = % of O_2 required to maintain desired ppO_2

Example. Calculate the minimum and maximum percentage of O_2 required to sustain a .44 to 1.25 ppO_2 range at 300 fsw.

1. Calculate the minimum percentage of O_2 required to sustain the lower value of the range:

$$\frac{0.44 \text{ ata}}{10.09 \text{ ata}} = 0.0436 \times 100 = 4.36\%$$

4.36% O_2 in He provides the minimum ppO_2.

2. Calculate the maximum percentage of O_2 required to sustain the lower value of the range:

$$\frac{1.25 \text{ ata}}{10.09 \text{ ata}} = 0.1239 \times 100 = 12.39\%$$

12.39% O_2 in He provides the maximum ppO_2.

SECTION THREE — SATURATION DIVING OPERATIONS

15-9 INTRODUCTION

Saturation diving is the mode of choice for diving operations requiring long bottom times or diving operations deeper than surface-supplied tables permit. Saturation diving allows divers to remain at working depths without concern for decompression. The Unlimited Duration Excursion Tables (Table 15-7 and Table 15-8) allow a large vertical range of working depths without time limits.

15-10 OPERATIONAL CONSIDERATIONS

Saturation diving requires complex saturation diving systems designed to precisely control depth, atmosphere composition, and temperature. Commanding Officers, Diving Officers, and Master Divers must consider personnel and training requirements, the physiological stress imposed by depth and dive duration, logistics, and gas supply requirements. Refer to Table 15-2 for the personnel requirements for saturation diving.

15-10.1 **Dive Team Selection.** All candidates for a saturation dive shall be physically qualified to make the dive as determined by a Saturation Diving Medical Officer. With the exceptions of authorized research, testing of equipment, or training purposes, all divers shall be qualified and experienced with the UBA being used and in the particular dive system to which they are assigned. Depending on mission requirements, divers may need to have special skills that are required for the operation.

15-10.2 **Mission Training.** When the schedule permits, training in preparation for a specific saturation diving mission shall be conducted. This training provides an opportunity to ensure that all personnel are in optimal physical condition and facilitates the development of special skills required for the operation. Training also provides an opportunity for individuals to function as a team and to identify an individual with leadership skills necessary to fill the role of dive team leader. Alternate divers should be identified and trained with the team in the event of illness or injury to a primary diver.

15-11 SELECTION OF STORAGE DEPTH

The selection of the storage depth for the deck decompression chamber (DDC) is based on the approximate planned diver working depth. This can be achieved by comparing the storage depth and planned diver working depth with the descent

Table 15-2. Typical Saturation Diving Watch Stations.

Watch Station
Diving Officer
Diving Medical Officer (Note 2)
Master Diver
Diving Supervisor
Atmosphere Monitor
MCC Gas-Control Operator
Life-Support Operator
MCC Communications and Log Operator
Surface-Support Divers
Gas King
PTC Operators
PTC Divers
Main Deck Supervisors

Note:

1. A Diving Medical Officer is required on site for all saturation diving operations. ("On site" is defined as accessible within 30 minutes of the dive site by available transportation.)

and ascent limits of the Unlimited Duration Excursion Tables (Table 15-7 and Table 15-8). When the diver's working depth range is small, the DDC should be compressed to approximately the middle of the range. This minimizes the amount of gas used in pressurizing or depressurizing the personnel transfer capsule (PTC).

When the expected diver work range is large or multiple objectives at different depths are to be accomplished, several different storage depths will be required. The unlimited excursion procedures may be used at several progressively shallower storage depths to accomplish the objective.

15-12 RECORDS

This section covers the records required to be maintained during the conduct of a saturation dive.

15-12.1 **Command Diving Log.** An official diving log shall be maintained at all times throughout the dive. It shall contain a chronological record of the dive procedure in addition to any significant events. A narrative of significant events is to be recorded by the Diving Officer (or Diving Supervisor) and Saturation Diving Medical Officer (as necessary). This log shall be retained for 3 years.

15-12.2 **Master Protocol.** Each diving operation shall have a master protocol submitted by the Master Diver, reviewed by the Saturation Diving Medical Officer and Diving Officer, and approved by the Commanding Officer. This master protocol shall contain all the information needed to ensure that the dive follows a program consistent with the requirements for saturation diving as defined in this manual and shall include the necessary information to carry out these procedures on the specific operational platform.

A copy of the protocol shall be maintained as the master copy at the MCC. No alterations except those made by the Diving Officer and approved by the Commanding Officer are permitted. Any changes to this protocol shall be signed and dated.

15-12.2.1 **Modifications.** Because saturation dives generally follow a predictable pattern, only a few elements of protocol need to be modified from mission to mission. Consequently, once a complete and carefully written protocol is available, only minor modifications will be needed to support future missions.

15-12.2.2 **Elements.** The dive protocol shall include, but is not limited to, the following:

- A detailed gas-usage plan, including projected gas supply requirements (paragraph 15-15). The required mixtures for supplying emergency, treatment, and excursion gas shall be specified for the depth ranges expected with specific depths to shift mixes indicated.
- A compression schedule, including planned rate of travel with rest stops, if applicable.
- Manning requirements, including a watchbill.
- Predive and postdive procedures.

15-12.3 **Chamber Atmosphere Data Sheet.** Hourly readings of chamber pressure, temperature, humidity, oxygen, and carbon dioxide concentrations shall be recorded. In addition, time of operation of the carbon dioxide scrubbers and time of carbon dioxide absorbent replenishment shall be recorded.

15-12.4 **Service Lock.** The following information shall be recorded: date, depth, clock time upon leaving the surface or leaving the bottom, and items locked in or out of the chamber. This information is useful in controlling the spread of contaminants and in minimizing the combustibles in the chamber while in the fire zone.

15-12.5 **Machinery Log/Gas Status Report.** A record of the status of all gas banks, including their pressure and mixture, and of the status of all DDS gas delivery equipment, shall be maintained. This log shall be reviewed by each oncoming Diving Supervisor prior to assuming the watch and daily by the Diving Officer and Master Diver.

15-12.6 **Operational Procedures (OPs).** Currently approved operational procedure sheets are to be properly completed and signed by the operator and then reviewed and signed by the Diving Supervisor and Dive Watch Officer and logged in the Command Smooth Log.

15-12.7 **Emergency Procedures (EPs).** A set of approved emergency procedures with each individual watch station's responsibilities shall be separately bound and available at the main control console throughout a saturation dive. The convenience of having emergency procedures on station does not relieve any diver or any saturation diving watch team member from being sufficiently knowledgeable, thoroughly trained, and fully qualified to react efficiently and instantaneously to any emergency. Constant training in these emergency procedures is necessary to maintain watchstanding proficiency.

15-12.8 **Individual Dive Record.** Use the Dive Reporting System (DRS) to record and report dives, as outlined in paragraph 5-9.

15-13 LOGISTICS

In planning an extended diving operation, care must be taken to ensure that sufficient supplies and power to support a diving mission are available. When operating at remote sites, the Commanding Officer and Diving Officer must carefully evaluate the availability of shore-based support. Loss of steam and/or electrical power at sea is an emergency situation. The loss of either of these vital services to the saturation dive system with a dive team committed to lengthy decompression constitutes a major emergency that must be acted upon quickly. Accordingly, transit times and contingency plans must be made prior to commencing saturation diving operations at remote sites in case support services for the dive complex are threatened or lost.

15-14 DDC AND PTC ATMOSPHERE CONTROL

The hyperbaric atmosphere within the DDC and PTC is controlled to maintain the gaseous components as follows:

Oxygen Partial Pressure	.44 – .48 ata
Carbon Dioxide Partial Pressure	Less than 0.005 $ppCO_2$ (.5% SEV) (3.8 millimeters of mercury)
Helium and Nitrogen	Balance of total pressure

Oxygen levels and time limits are presented in Table 15-3.

These levels, particularly that of oxygen, are essential for safe decompression and the use of the Unlimited Duration Excursion Tables. Increases in the oxygen partial pressure above 0.6 ata for extended periods (greater than 24 hours) risk pulmonary oxygen toxicity and should only be used in emergency situations. A ppO_2 below 0.42 ata may result in inadequate decompression, and a ppO_2 below 0.16 ata will result in hypoxia. Once carbon dioxide concentration reaches 0.5 percent surface equivalent (3.8 millimeters of mercury) for 1 hour, the scrubber canister should be changed, because carbon dioxide levels tend to rise rapidly thereafter. An inspired carbon dioxide level of 2 percent surface equivalent (15.2 millimeters of mercury) can be tolerated for periods of up to 4 hours at depth. Nitrogen concentration tends to decrease with time at depth, due to purging by helium during service lock operation.

Table 15-3. Chamber Oxygen Exposure Time Limits.

	Oxygen Level (ata)	Time
Storage	.44 – .48	Unlimited
Excursion	.40 – .60	4 hours (6 hours)***
Excursion associated with decompression	.42 – .48*	Unlimited
Emergency	.60**	24 hours

Notes:
* This level may be exceeded prior to starting the upward excursion for decompression.
** If oxygen levels exceed this limit, switch to emergency gas.
*** Diver performance exponentially decreases between 4 and 6 hours of an in-water excursion.

NOTE Discharging UBA gas into the PTC during diving operations may make it difficult to control the oxygen level.

15-15 GAS SUPPLY REQUIREMENTS

The following gases shall be available for use in a UBA, for emergency supply, and for the treatment of decompression sickness.

15-15.1 **UBA Gas.** An adequate quantity of gas within an oxygen partial pressure range of 0.44–1.25 ata shall be available for use.

15-15.2 **Emergency Gas.** Emergency gas is used as a backup breathing supply in the event of DDC or PTC atmosphere contamination. An emergency gas with an oxygen partial pressure of 0.16 to 1.25 ata shall be immediately available to the built-in breathing system (BIBS). The volume of emergency breathing gas shall be sufficient to supply the divers for the time needed to correct the DDC atmosphere.

Upward excursions of the PTC or DDC or decompression shall not be started during emergency gas breathing unless the oxygen partial pressure of the diver's inspired gas is 0.42 ata or above.

Example. An emergency gas schedule for a dive to 850 fsw is:

Bank Mix	Allowable Depth Range (fsw)	Shift Depth (fsw)
#1 84/16 HeO_2	0–224	200
#2 96/4 HeO_2	99–998	

15-15.3 **Treatment Gases.** Treatment gases having an oxygen partial pressure range of 1.5 to 2.8 shall be available in the event of decompression sickness. The premixed gases shown in Table 15-4 may be used over the depth range of 0 – 1,600 fsw. A source of treatment gas shall be available as soon as treatment depth is reached. The source shall be able to supply a sufficient volume of breathing gas to treat each chamber occupant.

Table 15-4. Treatment Gases.

Depth (fsw)	Mix
0–60	100% O_2
60–100	40/60% HeO_2
100–200	64/36% HeO_2
200–350	79/21% HeO_2
350–600	87/13% HeO_2
600–1000	92/08% HeO_2
1000–1600	95/05% HeO_2

15-16 ENVIRONMENTAL CONTROL

Helium-oxygen gas mixtures conduct heat away from the diver very rapidly. As a result, temperatures higher than those required in an air environment are necessary to keep a diver comfortable. As depth increases, the temperature necessary to achieve comfort may increase to the 85–93°F range.

As a general guideline to achieve optimum comfort for all divers, the temperature should be kept low enough for the warmest diver to be comfortable. Cooler divers can add clothing as needed. All divers should be questioned frequently about their comfort.

The relative humidity should be maintained between 30 and 80 percent with 50 to 70 percent being the most desirable range for diver comfort, carbon dioxide scrubber performance, and fire protection.

15-17 FIRE ZONE CONSIDERATIONS

Every effort shall be made to eliminate any fire hazard within a chamber. When oxygen percentages are elevated as during the later stages of decompression, a fire will burn rapidly once started, perhaps uncontrollably. As a result, special precautions are necessary to protect the diver's safety when in the fire zone. The fire zone is where the oxygen concentration in the chamber is 6 percent or greater. Using standard saturation diving procedures (oxygen partial pressure between 0.44 and 0.48 ata), fire is possible at depths less than 231 fsw. Thus, during a saturation dive the divers will be in the fire zone during initial compression to depth and during the final stages of decompression.

Example. The chamber atmosphere is 0.48 ata ppO_2. The minimum oxygen percentage for combustion is 6 percent. Compute the fire zone depth.

The fire zone depth is computed as follows:

$$\text{Fire zone depth (fsw)} = \frac{ppO_2 \times 33}{O_2\%/100} - 33$$

$$= \frac{0.48 \times 33}{0.06} - 33$$

$$= 231 \text{ fsw}$$

Although the design of the DDS minimizes fire potential, personnel must remain vigilant at all times to prevent fires. Appropriate precautions for fire prevention include:

- Fire-suppression systems, if available, must be operational at all times when in the fire zone.
- Chamber clothing, bed linen, and towels shall be made of 100% cotton. Diver swim trunks made of a 65% polyester–35% cotton material is acceptable.
- Mattresses and pillows shall be made of fire-retardant material when in the fire zone.
- Limit combustible personal effects to essential items.
- Limit reading material, notebooks, etc., in the fire zone.
- All potential combustibles shall be locked in only with the permission of the Diving Supervisor.
- Whenever possible, stow all combustibles, including trash, in fire-retardant containers, and lock out trash as soon as possible.
- Being thoroughly familiar with all emergency procedures (EPs) regarding fire inside and outside the Deep Diving System.

15-18 HYGIENE

Once a saturation dive begins, any illness that develops is likely to affect the entire team, reducing their efficiency and perhaps requiring the dive to be aborted. To minimize this possibility, the Saturation Diving Medical Officer should conduct a brief review of the diver's physical condition within 24 hours of compression. If an infectious process or illness is suspected, it shall be carefully evaluated by the Saturation Diving Medical Officer for possible replacement of the diver with a previously designated alternate diver. Strict attention to personal hygiene, chamber cleanliness, and food-handling procedures should be maintained once the dive begins to minimize the development and spread of infection.

15-18.1 Personal Hygiene. Personal hygiene and cleanliness is the most important factor in preventing infections, especially skin and ear infections. All divers should wash at least daily, and as soon as possible after wet excursions. Fresh linens and clothing should be locked into the complex every day. To prevent foot injury, clean, dry footwear should be worn at all times except while showering, sleeping, or in diving dress. Feet must be thoroughly dry, especially between the toes, to minimize local infections. A personal toiletry bag shall be maintained by each chamber occupant. These bags shall be inspected by the Diving Supervisor or Master Diver prior to commencing the dive to prevent potential contaminants or fire hazards from being carried into the chamber.

15-18.2 Prevention of External Ear Infections. Severe ear infections can develop unless preventative measures are taken. An effective preventative regime includes irrigating each ear with 2 percent acetic acid in aluminum acetate solution (i.e., DOMEBORO) for 5 minutes at least twice daily. Irrigation shall be observed by the Diving Supervisor, timed by the clock, and logged.

After a week or so, even with the ear prophylaxis regimen, the ear canals may become occluded with debris. Once this happens, an ear infection may develop rapidly. In order to prevent this occurrence, all divers should be trained to detect and treat blockage. Before beginning a dive, all divers should be trained by qualified medical personnel to use an otoscope to view the ear drum. Also, they should be trained to use an ear syringe. At least weekly during a dive, divers should examine each other's ear canals. If the ear drum cannot be viewed because of a blockage, then the canal should be gently irrigated with the ear syringe until the canal is unplugged.

15-18.3 Chamber Cleanliness. Strict attention shall be paid to chamber cleanliness at all times, particularly in the area of the toilet, wash basin, shower, and service locks. Only approved compounds shall be used to clean the chamber, components, and clothing used in the pressurized environment. During wet excursions, close attention shall be paid to routine postdive cleaning of the diver-worn equipment to prevent rashes and skin infections.

Upon completing a saturation dive, the chamber should be well ventilated, emptied, and liberally washed down with non-ionic detergent (MIL-D-16791) and water and then closed. Additionally, all chamber bedding, linens, and clothing shall be washed.

15-18.4 Food Preparation and Handling. All food provided to the divers during a saturation diving evolution shall meet the standards prescribed in NAVMED P-5010. All food locked in shall be inspected by the Dive Watch Supervisor or Dive Watch Officer. The Saturation Diving Medical Officer should inspect food preparation areas daily.

15-19 ATMOSPHERE QUALITY CONTROL

Preventing chamber atmosphere contamination by toxic gases is extremely important to the health of the divers. Once introduced into the chambers, gaseous contaminants are difficult to remove and may result in prolonged diver exposure.

15-19.1 **Gaseous Contaminants.** Gaseous contaminants can be introduced into the chamber through a contaminated gas supply, through chamber piping and/or gas flasks containing residual lubricants or solvents, or by the divers or maintenance personnel.

The hazard of atmospheric contamination can be reduced by ensuring that only gases that meet the appropriate federal specifications are used and that appropriate gas transfer procedures are used. All gas flasks and chamber piping used with helium, oxygen, or mixed gases shall be cleaned using approved cleaning procedures to remove substances that may become chamber contaminants. Once cleaned, care shall be taken to prevent introduction of contaminants back into these systems during maintenance by marking and bagging openings into the piping system. Finally, inadvertent chamber contamination can be prevented by limiting the items that may be taken inside. Only approved paints, lubricants, solvents, glues, equipment, and other materials known not to off-gas potential toxic contaminants are allowed in the chamber. Strict control of all substances entering the chamber is an essential element in preventing chamber contamination.

15-19.2 **Initial Unmanned Screening Procedures.** To ensure that chamber systems are free of gaseous contaminants, the chamber atmosphere shall be screened for the presence of the common contaminants found in hyperbaric systems when contamination of the chamber and/or gas supply is suspected, or after any major chamber repair or overhaul has been completed. Only NAVFAC- or NAVSEA-approved procedures may be used to collect screening samples.

Table 15-5 lists a few selected contaminants that may be present in hyperbaric complexes, with their 90-day continuous exposure limits (or 7-day limits where a 90-day limit is not available). In the absence of specific guidelines for hyperbaric exposures, these limits shall be used as safe limits for saturation diving systems.

When any one of these contaminants is reported in chamber samples, the calculated Surface Equivalent Value (SEV) shall be compared to the limit on this list. If the calculated SEV exceeds this limit, the chamber shall be cleaned and retested. Assistance with any contamination identification and resolution can be obtained by contacting NEDU or the system certification authority for guidance.

15-20 COMPRESSION PHASE

The initial phase of the dive is the compression of the dive team to the selected storage depth. This phase includes establishing the chamber oxygen partial pressure at a value between 0.44 and 0.48 ata, instrument and systems checkouts, and the actual compression of the divers to storage depth.

Table 15-5. *Limits for Selected Gaseous Contaminants in Saturation Diving Systems.*

Contaminant	Limit
Acetone	200 ppm (Note 1) (Note 3: Same limit)
Benzene	1 ppm (Note 3)
Chloroform	1 ppm (Note 1)
Ethanol	100 ppm (Note 3)
Freon 113	100 ppm (Note 1)
Freon 11	100 ppm (Note 1)
Freon 12	100 ppm (Note 1) (Note 3: Same limit)
Freon 114	100 ppm (Note 1)
Isopropyl Alcohol	1 ppm (Note 1)
Methanol	10 ppm (Note 3)
Methyl Chloroform	30 ppm (Note 2) (Note 3: 90-day limit = 2.5 ppm, 24-hour limit = 10 ppm)
Methyl Ethyl Ketone	20 ppm (Note 2)
Methyl Isobutyl Ketone	20 ppm (Note 2)
Methylene Chloride	25 ppm (Note 2)
Toulene	20 ppm (Note 1) (Note 3: Same limit)
Trimethyl Benzenes	3 ppm (Note 2)
Xylenes	50 ppm (Note 1) (Note 3: Same limit)

Notes:

1. 90-day continuous exposure limit. *National Research Council Committee on Toxicology Emergency and Continuous Exposure Limits for Selected Airborne Contaminants,* Vols. 1-8, Washington, D.C., National Academy Press, 1984–1988.
2. 7-day maximum allowable concentration in manned spacecraft. National Aeronautics and Space Administration, Office of Space Transportation Systems. *Flammability, Odor, and Offgassing Requirements and Test Procedures for Materials in Environments that Support Combustion,* NHB 8060, 1B, Washington, D.C., U.S. Government Printing Office, 1981.
3. 90-day limit. *U.S. Naval Sea Systems Command Nuclear Powered Submarine Atmosphere Control Manual,* NAVSEA S9510-AB-ATM-010 (U), Vol. 1, Revision 2, 30 July 1992.

15-20.1 **Establishing Chamber Oxygen Partial Pressure.** Prior to compression to storage depth, the chamber oxygen partial pressure shall be raised from 0.21 ata to 0.44–0.48 ata. There are two methods of raising the oxygen partial pressure to the desired level.

- **Air Method.** Compress the chamber with air at a moderate rate to 36 fsw. This will raise the chamber ppO_2 to 0.44 ata. If desired, further elevation of the chamber ppO_2 to 0.48 ata can be undertaken by using the oxygen makeup system.

- **Helium-Oxygen Method.** Compress the chamber at a moderate rate with a helium-oxygen mixture containing less than 21 percent oxygen. The depth of the required compression can be calculated using the following formula:

$$\text{Compression Depth (fsw)} = 33 \times \frac{(ppO_2 - 0.21)}{O_2\%} \times 100$$

Example. If a 20 percent mixture of helium-oxygen is used and the desired ppO_2 is 0.44 ata, calculate the compression depth.

$$\text{Compression Depth (fsw)} = 33 \times \frac{(0.44 - 0.21)}{20} \times 100$$
$$= 37.95 \text{ fsw}$$

15-20.2 **Compression to Storage Depth.** Rapid compression to saturation storage depth may provoke symptoms of High-Pressure Nervous Syndrome (HPNS) and may intensify compression joint pains. To avoid these complications, the slowest rate of compression consistent with operational requirements should be used. Table 15-6 shows the range of allowable compression rates.

Table 15-6. *Saturation Diving Compression Rates.*

Depth Range	Compression Rate
0–60 fsw	0.5 – 30 fsw/min
60–250 fsw	0.5 – 10 fsw/min
250–750 fsw	0.5 – 3 fsw/min
750–1000 fsw	0.5 – 2 fsw/min

If operational necessity dictates, compression to storage depth of 400 fsw or shallower can be made at the maximum rates indicated in Table 15-6 with little risk of HPNS. Direct compression at maximum rates to deeper storage depths, however, may produce symptoms of HPNS in some divers. These divers may be unable to perform effectively for a period of 24 to 48 hours. Experience has shown that the appearance of such symptoms can be minimized by slowing compression rates or introducing holds during compression.

The depth and time duration of holds, if used, may be adjusted to suit operational requirements and diver comfort.

15-20.3 **Precautions During Compression.** During compression the chamber atmosphere shall be monitored carefully. The chamber atmosphere may not mix well during rapid compression, resulting in areas of low oxygen concentration.

15-20.4 Abort Procedures During Compression. The following abort procedure is authorized if a casualty occurs during compression. Consult with a Saturation Diving Medical Officer prior to committing to this procedure. This procedure is normally used for shallow aborts where the maximum depth and bottom time do not exceed the limits of the table.

Using the Surface Supplied HeO_2 Tables, the following procedure applies:

- **Depth.** Use the actual chamber depth.
- **Bottom Time.** If the initial compression uses air, time spent shallower than 40 fsw, up to a maximum of 60 minutes, is not counted as bottom time. If the initial compression uses helium, time starts when leaving the surface.
- **BIBS Gas.** Maintain BIBS between 1.5 – 2.8 ppO_2.
- **Stops.** Follow the scheduled stops of the Surface Supplied HeO_2 Tables.
- **O_2 Breaks.** For every 25 minutes of breathing BIBS gas, take a 5-minute break breathing a gas between 0.16 to 1.25 ata ppO_2. The 5-minute break counts as a stop time. The lower oxygen percentage shall not be less than 0.16 ata ppO_2.

Upon completing abort decompression, all divers shall be closely monitored and observed for a minimum of 24 hours. For deeper emergency aborts beyond the limits of the Surface-supplied HeO_2 Tables, refer to paragraph 15-23.7.2.

15-21 STORAGE DEPTH

The Unlimited Duration Excursion Tables (Table 15-7 and Table 15-8) allow multiple diver excursions to be conducted during the course of a saturation dive. When using these excursion procedures, the diving supervisor need only be concerned with the depth of the divers. To use these tables when planning the dive, select a chamber storage depth in a range that allows diver excursions shallower or deeper than the storage depth. The actual depth of the work site or PTC may be significantly different from the storage depth.

When using Table 15-8, enter the table at the deepest depth attained at any time within the last 48 hours. While the DDC may be at 400 fsw, if one diver had reached a depth of 460 fsw during an in-water excursion, the maximum upward excursion depth for the divers is 360 fsw instead of 307 fsw. After completing work at one depth and then compressing DDC to a deeper storage depth, unlimited downward or upward excursions are permitted immediately upon reaching the new storage depth. When decompressing the DDC from a deeper depth using standard saturation decompression procedures, unlimited downward excursions, as defined in Table 15-7, may begin immediately upon reaching the new chamber storage depth. A minimum of 48 hours shall elapse at the new storage depth before any upward excursions may be made.

Table 15-7. Unlimited Duration Downward Excursion Limits.

Storage Depth (fsw)	Deepest Excursion Distance (ft)	Deepest Excursion Depth (fsw)
0	29	29
10	33	43
20	37	57
30	40	70
40	43	83
50	46	96
60	48	108
70	51	121
80	53	133
90	56	146
100	58	158
110	60	170
120	62	182
130	64	194
140	66	206
150	68	218
160	70	230
170	72	242
180	73	253
190	75	265
200	77	277
210	78	288
220	80	300
230	82	312
240	83	323
250	85	335
260	86	346
270	88	358
280	89	369
290	90	380
300	92	392
310	93	403
320	95	415
330	96	426
340	97	437
350	98	448
360	100	460
370	101	471
380	102	482
390	103	493
400	105	505
410	106	516
420	107	527
430	108	538
440	109	549
450	111	561
460	112	572
470	113	583
480	114	594
490	115	605
500	116	616
510	117	627
520	118	638
530	119	649
540	120	660
550	122	672
560	123	683
570	124	694
580	125	705
590	126	716
600	127	727
610	128	738
620	129	749
630	130	760
640	131	771
650	132	782
660	133	793
670	133	803
680	134	814
690	135	825
700	136	836
710	137	847
720	138	858
730	139	869
740	140	880
750	141	891
760	142	902
770	143	913
780	144	924
790	144	934
800	145	945
810	146	956
820	147	967
830	148	978
840	149	989
850	150	1000

Table 15-8. *Unlimited Duration Upward Excursion Limits.*

Storage Depth (fsw)	Shallowest Excursion Distance (ft)	Shallowest Excursion Depth (fsw)
29	29	0
30	29	1
40	32	8
50	35	15
60	37	23
70	40	30
80	42	38
90	44	46
100	47	53
110	49	61
120	51	69
130	53	77
140	55	85
150	56	94
160	58	102
170	60	110
180	62	118
190	63	127
200	65	135
210	67	143
220	68	152
230	70	160
240	71	169
250	73	177
260	74	186
270	76	194
280	77	203
290	79	211
300	80	220
310	81	229
320	83	237
330	84	246
340	85	255
350	87	263
360	88	272
370	89	281
380	90	290
390	92	298
400	93	307
410	94	316
420	95	325
430	96	334
440	97	343
450	99	351
460	100	360
470	101	369
480	102	378
490	103	387
500	104	396
510	105	405
520	106	414
530	107	423
540	108	432
550	110	440
560	111	449
570	112	458
580	113	467
590	114	476
600	115	485
610	116	494
620	117	503
630	118	512
640	119	521
650	119	531
660	120	540
670	121	549
680	122	558
690	123	567
700	124	576
710	125	585
720	126	594
730	127	603
740	128	612
750	129	621
760	130	630
770	131	639
780	131	649
790	132	658
800	133	667
810	134	676
820	135	685
830	136	694
840	137	703
850	137	713
860	138	722
870	139	731
880	140	740
890	141	749
900	142	758
910	142	768
920	143	777
730	144	786
940	145	795
950	146	804
960	146	814
970	147	823
980	148	832
990	149	841
1000	150	850

Example. After decompression from 1,000 fsw to 400 fsw, the maximum downward excursion is 105 fsw. After 48 hours have elapsed at 400 fsw, a full upward excursion of 93 fsw to 307 fsw is permitted.

If less than 48 hours is spent at the new storage depth, the maximum upward excursion is based on the deepest depth attained in the preceding 48 hours.

Example. Decompression from a 1,000 fsw dive has been conducted to the 400 fsw depth. Twenty-four hours have been spent at 400 fsw. The dive log shows that the deepest depth attained in the preceding 48 hours is 496 fsw. The maximum upward excursion from Table 15-8, based on a 496 fsw depth, is to 396 fsw (500 – 104) allowing a maximum of a 4 fsw upward excursion. After 36 hours have elapsed at 400 fsw, the dive log shows that the deepest depth attained in the preceding 48 hours was 448 fsw. From Table 15-8, the shallowest excursion depth is now 351 fsw.

The ascent rate should not exceed 60 fsw/min during an excursion. When it is detected that a diver is ascending faster than 60 fsw/min, the diver shall immediately stop and wait until enough time has elapsed to return to the 60 fsw/min schedule. The diver may then resume ascent at a rate not to exceed 60 fsw/min from that depth.

If storage depth falls between the depths listed in Table 15-7, use the next shallower depth (e.g., if the storage depth is 295 fsw, enter Table 15-7 at 290 fsw). If storage depth falls between the depths listed in Table 15-8, use the next deeper depth (e.g., if the storage depth is 295 fsw, enter Table 15-8 at 300 fsw).

15-21.1 Excursion Table Examples.

Example 1. The chamber was compressed to 400 fsw from the surface. The initial depth in Table 15-7 is 400 fsw. The maximum downward excursion for an unlimited period not requiring decompression is 105 fsw, allowing a maximum diver depth of 505 fsw. If the diver descends to 450 fsw, the maximum depth achieved from the 400 fsw storage depth will be 450 fsw. Table 15-8 at 450 fsw allows a 99 fsw upward excursion to a depth of 351 fsw. Thus, these divers may move freely between the depths of 351 and 450 fsw while at a storage depth of 400 fsw.

Example 2. At a storage depth of 600 fsw, during which dives were made to 650 fsw, the maximum upward excursion that may by made to begin saturation decompression is:

- If less than 48 hours have elapsed since the 650 fsw excursion, Table 15-8 allows a maximum upward excursion of 119 fsw from a deepest depth of 650 fsw to a depth of 531 fsw.

- If more than 48 hours have elapsed since the excursion, the maximum upward excursion allowed is 115 fsw from 600 fsw to 485 fsw.

Example 3. At the new shallower storage depth of 350 fsw, divers conduct an excursion to 400 fsw. Using the deepest depth of 400 fsw achieved during storage at 350 fsw, a maximum upward ascent from Table 15-8 of 93 fsw to a depth of 307 fsw is allowed, provided the chamber and the divers have been at the storage depth of 350 fsw for at least 48 hours. Otherwise, no upward excursion is permitted.

15-21.2 **PTC Diving Procedures.** Actual PTC diving operations are dictated by the Unit's operating instructions. In conducting these operations, experience indicates that a maximum in-water time of 4 hours is optimal for diver efficiency. Longer dive times result in a loss of diver effectiveness because of fatigue and exposure, while shorter dives will significantly increase the time at depth for the completion of operations. Standard practice is to rotate in-water divers with the PTC operators, allowing two 4-hour dives to be conducted during a single PTC excursion to the work site. Proper positioning of the PTC near the objective is important in ensuring that the diver does not exceed the maximum permitted excursion limits (Figure 15-8).

15-21.2.1 **PTC Deployment Procedures.** A brief overview of PTC deployment procedures follows:

1. For initial pressurization, the PTC, with internal hatch open, is usually mated to the DDC. Divers enter the DDC and secure the hatches.

2. The DDC and PTC are pressurized to bottom depth. The divers transfer to the PTC and secure the DDC and PTC hatches after them.

3. The trunk space is vented to the atmosphere and then the PTC is deployed and lowered to working depth. The hatch is opened when seawater and internal PTC pressures are equal. The divers don diving equipment and deploy from the PTC.

4. Divers return to the PTC and secure the hatch. The PTC is raised and mated to the DDC, and the divers transfer to the DDC. Until they are decompressed in the DDC, the divers rotate between periods of living in the DDC and working on the bottom. Deep underwater projects requiring moderate bottom time or diver activities involving work at various depths are conducted in the saturation mode with excursion dives. The PTC and DDC are pressurized to a storage depth within the ascent and descent limits of the Unlimited Duration Excursion Tables (Table 15-7 and Table 15-8), maximizing diving efficiency for deep, long dives. Once tissue saturation is reached, decompression requirements no longer increase.

15-22 DEEP DIVING SYSTEM (DDS) EMERGENCY PROCEDURES

Major DDS emergencies include loss of atmosphere control, loss of depth control and fire in the DDC. Emergencies will be covered by locally prepared and NAVSEA- or NAVFAC-approved emergency procedures. The following are guidelines for establishing these procedures.

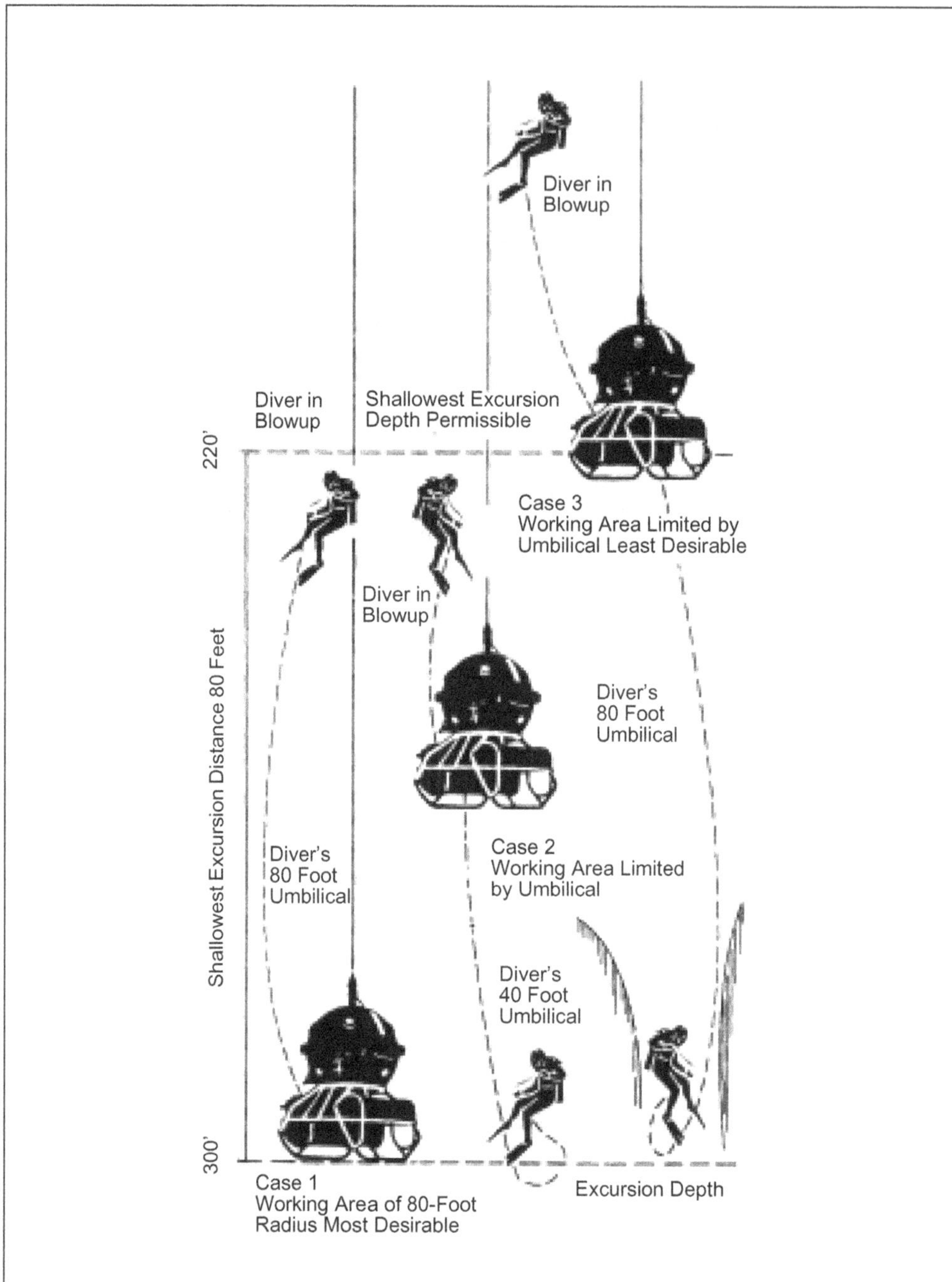

Figure 15-8. PTC Placement Relative to Excursion Limits.

15-22.1 Loss of Chamber Atmosphere Control. Loss of chamber atmosphere control includes loss of oxygen control, high carbon dioxide level, chamber atmosphere contamination and loss of temperature control.

15-22.1.1 **Loss of Oxygen Control.** Divers can be safely exposed to chamber oxygen partial pressures between 0.16 and 1.25 ata; however, efforts should be implemented immediately to correct the problem and reestablish normal oxygen levels. For an oxygen partial pressure from 0.16 to 0.48 ata, the normal oxygen addition system can be used to increase the oxygen level slowly over time. For an oxygen partial pressure above 0.48, it may be necessary to secure the oxygen addition system and allow the divers to breathe down the chamber oxygen to a normal level. Table 15-3 lists the chamber oxygen exposure time limits. If these limits are exceeded, the divers should be placed on BIBS and the chamber ventilated to reduce the oxygen level.

15-22.1.2 **Loss of Carbon Dioxide Control.** When the DDC's life-support system loses its ability to absorb carbon dioxide, the level of carbon dioxide within the chamber will rise at a rate depending on the chamber size and the combined carbon dioxide production rate of the divers. An increasing carbon dioxide level may be the result of exhaustion of the carbon dioxide absorbent or inadequate gas flow through the carbon dioxide absorbent canister. If, after the carbon dioxide absorbent canister is changed, chamber carbon dioxide level still cannot be brought under 0.005 ata (3.8 mmhg), the flow through the canister may be inadequate. Divers shall don BIBS when the chamber carbon dioxide level exceeds 0.06 ata (45.6 mmhg).

15-22.1.3 **Atmosphere Contamination.** If an abnormal odor is detected or if several divers report symptoms of eye or lung irritation, coughing, headache, or impaired performance, contamination of the chamber atmosphere should be suspected. The divers shall be placed on BIBS and emergency procedures executed. The divers should be isolated in the part of the complex thought to be least contaminated. Test the chamber atmosphere using chemical detector tubes or by collecting a gas sample for analysis on the surface, as described in paragraph 15-19.2. If atmosphere contamination is found, the divers should be moved to the chamber or PTC with the least level of contamination and this chamber isolated from the rest of the complex.

15-22.1.4 **Interpretation of the Analysis.** The allowable contaminant limits within a diving system are based upon the Threshold Limit Values (TLV) for Chemical Substances and Physical Agents guidelines published by the American Conference of Governmental Industrial Hygienists (ACGIH). TLVs are the time-weighted average concentration for an 8-hour work day and a 40-hour work week, to which nearly all workers can be repeatedly exposed day after day without adverse effect. These guidelines are published yearly and should be used to determine acceptability. Because the partial pressure of a gas generally causes its physiological effects, the published limits must be corrected for the expected maximum operating depth (ata) of the diving system.

The solution to an atmosphere contamination problem centers around identifying the source of contamination and correcting it. Gas samples from suspected sources must be checked for contaminants. Special attention should be given to recently changed and cleaned piping sections, gas hoses, and diver umbilicals, any of which may contain residual cleaning solvents. Surfaced chambers should be thoroughly ventilated with air or a breathable helium-oxygen mixture (to prevent hypoxia in maintenance personnel), inspected, and thoroughly scrubbed down to remove residual contaminants. These chambers can then be compressed to depth using a gas bank that is free of contaminants, the divers can be transferred to this chamber, and the surface cleaning process can be repeated on the remaining chamber(s). After cleaning and compression to depth, the chamber should be checked periodically for recurrence of the contamination.

15-22.1.5 **Loss of Temperature Control.** Loss of temperature control of more than 2–3°F above or below the comfort level may lead to severe thermal stress in the divers. Studies have shown that heat loss by perspiring is less effective in a hyperbaric atmosphere. Heating a chamber to warm up cold divers may result in the divers rapidly becoming overheated. Heat stroke may then become a possibility. The potential for uncontrolled chamber heating occurs when chambers and PTCs are exposed to direct sunlight.

When the chamber temperature falls, the divers begin intense shivering and hypothermia develops unless rapid and aggressive measures are taken to correct the problem. Divers may be provided with insulated clothing, blankets, and sleeping bags. The best of these insulators are of limited effectiveness within the helium-oxygen environment and will provide marginal protection until the problem can be corrected. Special thermal protection systems have been designed for the use within DDCs. These systems include thermal protection garments, insulating deck pads or hammocks, and combination carbon dioxide absorbent and respiratory-heat regenerator systems.

15-22.2 **Loss of Depth Control.** Loss of depth control is defined as a pressure loss or gain that cannot be controlled within the normal capabilities of the system. When loss of depth control is encountered, all deployed divers shall be recovered immediately and all divers placed on BIBS. Attempt to control depth by exhausting excess gas or adding helium to minimize depth loss until the cause can be found and corrected. If the depth change is in excess of that allowed by the Unlimited Duration Excursion Tables, the divers should be returned to the original storage depth immediately and the Diving Medical Officer notified.

15-22.3 **Fire in the DDC.** Because fire within a DDC may progress rapidly, the divers and watchstanders must immediately activate the fire suppression system and secure the oxygen system as soon as a fire is suspected. When the fire suppression system is activated, all divers shall immediately go on BIBS. Watchstanders should monitor depth carefully because an extensive fire will cause an increase in depth. If the fire suppression system fails to extinguish the fire, rapid compression of the chamber with helium may extinguish the fire, in that helium lowers the oxygen concentration and promotes heat transfer. After the fire is extinguished, chamber atmosphere contaminant emergency procedures shall be followed.

15-22.4 **PTC Emergencies.** PTC emergencies, like DDC emergencies, require specific, timely, and uniform responses in order to prevent injury or casualty to divers, watchstanders, and equipment.

15-23 SATURATION DECOMPRESSION

Saturation decompression may be initiated by an upward excursion as long as the excursion remains within the limits permitted by the Unlimited Duration Excursion Tables. The alternative is to begin travel at the appropriate decompression rate without the upward excursion. Decompression travel rates are found on Table 15-9.

Table 15-9. *Saturation Decompression Rates.*

Depth	Rate
1,600 – 200 fsw	6 feet per hour
200 – 100 fsw	5 feet per hour
100 – 50 fsw	4 feet per hour
50 – 0 fsw	3 feet per hour

15-23.1 **Upward Excursion Depth.** The minimum depth to which the upward excursion may be made is found by entering Table 15-8 with the deepest depth attained by any diver in the preceding 48 hours. The total upward excursion actually chosen is determined by the Diving Officer and Master Diver, and approved by the Commanding Officer, taking into consideration environmental factors, the diver's workload, and the diver's physical condition.

15-23.2 **Travel Rate.** The travel rate for the upward excursion is 2 fsw/min. Beginning decompression with an upward excursion will save considerable time and may be used whenever practical.

15-23.3 **Post-Excursion Hold.** Due to the increased risk of decompression sickness following an upward excursion for dives with a storage depth of 200 fsw or less, a 2-hour post-excursion hold should be utilized. The 2-hour hold begins upon arrival at upward excursion depth.

15-23.4 **Rest Stops.** During decompression, traveling stops for a total of 8 hours out of every 24 hours. The 8 hours should be divided into at least two periods known as "Rest Stops." At what hours these rest stops occur are determined by the daily routine and operations schedule. The 2-hour post-excursion hold may be considered as one of the rest stops.

15-23.5 **Saturation Decompression Rates.** Table 15-9 shows saturation decompression rates. Saturation decompression is executed by decompressing the DDC in 1-foot increments not to exceed 1 fsw per minute. For example, using a travel rate of 6 feet per hour will decompress the chamber 1 foot every 10 minutes. The last decompression stop before surfacing may be taken at 4 fsw to ensure early surfacing does not occur and that gas flow to atmosphere monitoring instruments remains adequate. This last stop would be 80 minutes, followed by direct ascent to the surface at 1 fsw/min.

Traveling is conducted for 16 hours in each 24-hour period. A 16-hour daily travel/rest outline example consistent with a normal day/night cycle is:

Daily Routine Schedule

2400–0600	Rest Stop
0600–1400	Travel
1400–1600	Rest Stop
1600–2400	Travel

This schedule minimizes travel when the divers are normally sleeping. Such a daily routine is not, however, mandatory. Other 16-hour periods of travel per 24-hour routines are acceptable, although they shall include at least two stop periods dispersed throughout the 24-hour period and travel may continue while the divers sleep. An example of an alternate schedule is:

Alternate Sample Schedule

2300–0500	Travel
0500–0700	Rest Stop
0700–0900	Travel
0900–1500	Rest Stop
1500–2300	Travel

The timing of the stop is dependent upon operational requirements.

15-23.6 **Atmosphere Control at Shallow Depths.** As previously stated, the partial pressure of oxygen in the chamber shall be maintained between 0.44 and 0.48 ata, with two exceptions. The first is just before making the initial Upward Excursion and the second during the terminal portion of saturation decompression. Approximately 1 hour before beginning an Upward Excursion, the chamber ppO_2 may be increased up to a maximum of 0.6 ata to ensure that the ppO_2 after excursion does not fall excessively. The ppO_2 should be raised just enough so the post-excursion ppO_2 does not exceed 0.48 ata. However, when excursions begin from depths of 200 fsw or shallower, a pre-excursion ppO_2 of 0.6 ata will result in a post-excursion ppO_2 of less than 0.44 ata. In these cases, the pre-excursion ppO_2 should not exceed 0.6 ata, but the post-excursion ppO_2 should be increased as rapidly as possible.

The second exception is at shallow chamber depth. As chamber depth decreases, the fractional concentration of oxygen necessary to maintain a given partial pressure increases. If the chamber ppO_2 were maintained at 0.44–0.48 ata all the way to the surface, the chamber oxygen percentage would rise to 44–48 percent. Accordingly, for the terminal portion of saturation decompression, the allowable oxygen percentage is between 19 and 23 percent. The maximum oxygen percentage for the terminal portion of the decompression shall not exceed 23 percent, based upon fire-risk considerations.

15-23.7 Saturation Dive Mission Abort. If it is necessary to terminate a saturation dive after exceeding the abort limits (see paragraph 15-20.4), standard saturation decompression procedures shall be followed.

15-23.7.1 Emergency Cases. In exceptional cases it could be necessary to execute a mission abort and not be able to adhere to standard saturation decompression procedures. The emergency abort procedures should only be conducted for grave, unforeseen casualties that require deviation from the standard decompression procedures such as:

- An unrepairable failure of key primary and related backup equipment in the dive system that would prevent following standard decompression procedures.
- Unrepairable damage to the diving support vessel or diving support facility.
- A life-threatening medical emergency where the risk of not getting the patient to a more specialized medical care facility outweighs the increased risk of pulmonary oxygen toxicity and increased risk of decompression sickness imposed upon the patient by not following standard saturation decompression procedures.

An Emergency Abort Procedure was developed and has received limited testing. It enables the divers to surface earlier than would be allowed normally. However, the time saved may be insignificant to the total decompression time still required, especially if the divers have been under pressure for 12 hours or more. In addition, executing the Emergency Abort Procedure increases the diver's risk for decompression sickness and complications from pulmonary oxygen toxicity.

Before executing a mission abort procedure that does not follow standard decompression procedures or the abort procedures contained in paragraph 15-20.4, the Commanding Officer must carefully weigh the risk of the action, relying on the advice and recommendations of the Master Diver, Diving Officer, and Saturation Diving Medical Officer. Specifically, it must be determined if the time saved will benefit the diver's life despite the increased risks, and whether the Emergency Abort Procedure can be supported logistically.

NOTE **USN dive system design incorporates separate primary, secondary, and treatment gas supplies and redundancy of key equipment. It is neither the intent of this section nor a requirement that saturation dive systems be configured with additional gas stores specifically dedicated to execution of an emergency abort procedure. Augmentation gas supplies if required will be gained by returning to port or receiving additional supplies on site.**

Except in situations where the nature or time sensitivity of the emergency does not allow, technical and medical assistance should be sought from the Navy Experimental Diving Unit prior to deviating from standard saturation decompression procedures.

15-23.7.2 **Emergency Abort Procedure.** Emergency Abort Procedures should only be conducted for grave casualties that are time critical. Decompression times and chamber oxygen partial pressures for emergency aborts from helium-oxygen saturation are shown in Table 15-10.

Table 15-10. *Emergency Abort Decompression Times and Oxygen Partial Pressures.*

Post Excursion Depth (fsw)	ppO_2 (ata)	One-Foot Stop Time (min) 1000–200 fsw	One-Foot Stop Time (min) 200–0 fsw
0–203	0.8	11	18
204–272	0.7	11	19
273–1000	0.6	12	21

Emergency Abort decompression is begun by making the maximum Upward Excursion allowed by Table 15-8. Rate of travel should not exceed 2 fsw/min. The upward excursion includes a 2-hour hold at the upward excursion limit. Travel time is included as part of the 2-hour hold. Following the Upward Excursion, the chamber oxygen partial pressure is raised to the value shown in Table 15-10. Decompression is begun in 1-foot increments using the times indicated in Table 15-10. Rate of travel between stops is not to exceed 1 fsw/min. Travel time is included in the next stop time. The partial pressure of oxygen is controlled at the value indicated until the chamber oxygen concentration reaches 23 percent. The oxygen concentration is then controlled between 19 and 23 percent for the remainder of the decompression. Stop travel at 4 fsw until total decompression time has elapsed and then travel to the surface at 1 fsw/min.

For example, the maximum depth of the diver in the last 48 hours was 400 fsw, and the Commanding Officer approves using the Emergency Abort Procedure. From the Upward Excursion Table, the complex travels to 307 fsw at a rate not to exceed 2 fsw/min. It takes 46.5 minutes to travel. This time is part of a 2-hour hold requirement as part of the upward excursion for emergency aborts.

Because the post-excursion depth is between 273–1,000 fsw, the chamber oxygen partial pressure is raised to 0.6 ata. Once the atmosphere is established and the remainder of the 2-hour hold completed, begin decompression in 1-foot increments with stop times of 12 minutes from 307 to 200 fsw. The travel rate between stops should not exceed 1 fsw/min. Travel time is included in the stop time. It will take 21.4 hours to arrive at 200 fsw.

At 200 fsw the 1-foot stop time changes to 21 minutes. It will take 70 hours to reach the surface. The total decompression time is 93.4 hours (3 days, 21 hours, 21 minutes, 36 seconds). By contrast, standard saturation decompression would take approximately 4 days and 3 hours to complete.

During and following the dive, the divers should be monitored closely for signs of decompression sickness and for signs of pulmonary oxygen toxicity. The latter includes burning chest pain and coughing. The divers should be kept under close observation for at least 24 hours following the dive.

If the emergency ceases to exist during the decompression, hold for a minimum of 2 hours, revert to standard decompression rates, and allow the oxygen partial pressure to fall to normal control values as divers consume the oxygen. Venting to reduce the oxygen level is not necessary.

15-23.8 **Decompression Sickness (DCS).** Decompression sickness may occur during a saturation dive as a result of an Upward Excursion or as a result of standard saturation decompression. The decompression sickness may manifest itself as musculoskeletal pain (Type I) or as involvement of the central nervous system and organs of special sense (Type II). Due to the subtleness of decompression sickness pain, all divers should be questioned about symptoms when it is determined that one diver is suffering from decompression sickness. For treatment, refer to Figure 15-9.

15-23.8.1 **Type I Decompression Sickness.** Type I Decompression Sickness may result from an Upward Excursion or as the result of standard saturation decompression. It is usually manifested as the gradual onset of musculoskeletal pain most often involving the knee. Divers report that it begins as knee stiffness that is relieved by motion but which increases to pain over a period of several hours. Care must be taken to distinguish knee pain arising from compression arthralgia or injury incurred during the dive from pain due to decompression sickness. This can usually be done by obtaining a clear history of the onset of symptoms and their progression. Pain or soreness present prior to decompression and unchanged after ascent is unlikely to be decompression sickness. Type I Decompression Sickness that occurs during an Upward Excursion or within 60 minutes immediately after an Upward Excursion shall be treated in the same manner as Type II Decompression Sickness, as it may herald the onset of more severe symptoms. Type I Decompression Sickness occurring more than 60 minutes after an Upward Excursion or during saturation decompression should be treated by recompressing in increments of 5 fsw at 5 fsw/min until distinct improvement of symptoms is indicated. Recompression of more than 30 fsw is usually unnecessary. Once treatment depth is reached, the stricken diver is given a treatment gas, by BIBS mask, with an oxygen partial pressure between 1.5 and 2.8 ata. Interrupt treatment gas breathing every 25 minutes with 5 minutes of breathing chamber atmosphere. Divers should remain at treatment depth for at least 2 hours on treatment gas following resolution of symptoms. Decompression can then be resumed using standard saturation decompression rates. Further Upward Excursions are not permitted.

15-23.8.2 **Type II Decompression Sickness.** Type II Decompression Sickness in saturation diving most often occurs as a result of an Upward Excursion. The onset of symptoms is usually rapid, occurring during the Upward Excursion or within the first hour following an excursion ascent. Inner ear decompression sickness manifests itself as nausea and vomiting, vertigo, loss of equilibrium, ringing in the ears and hearing loss. Central nervous system (CNS) decompression sickness may present itself as weakness, muscular paralysis, or loss of mental alertness and

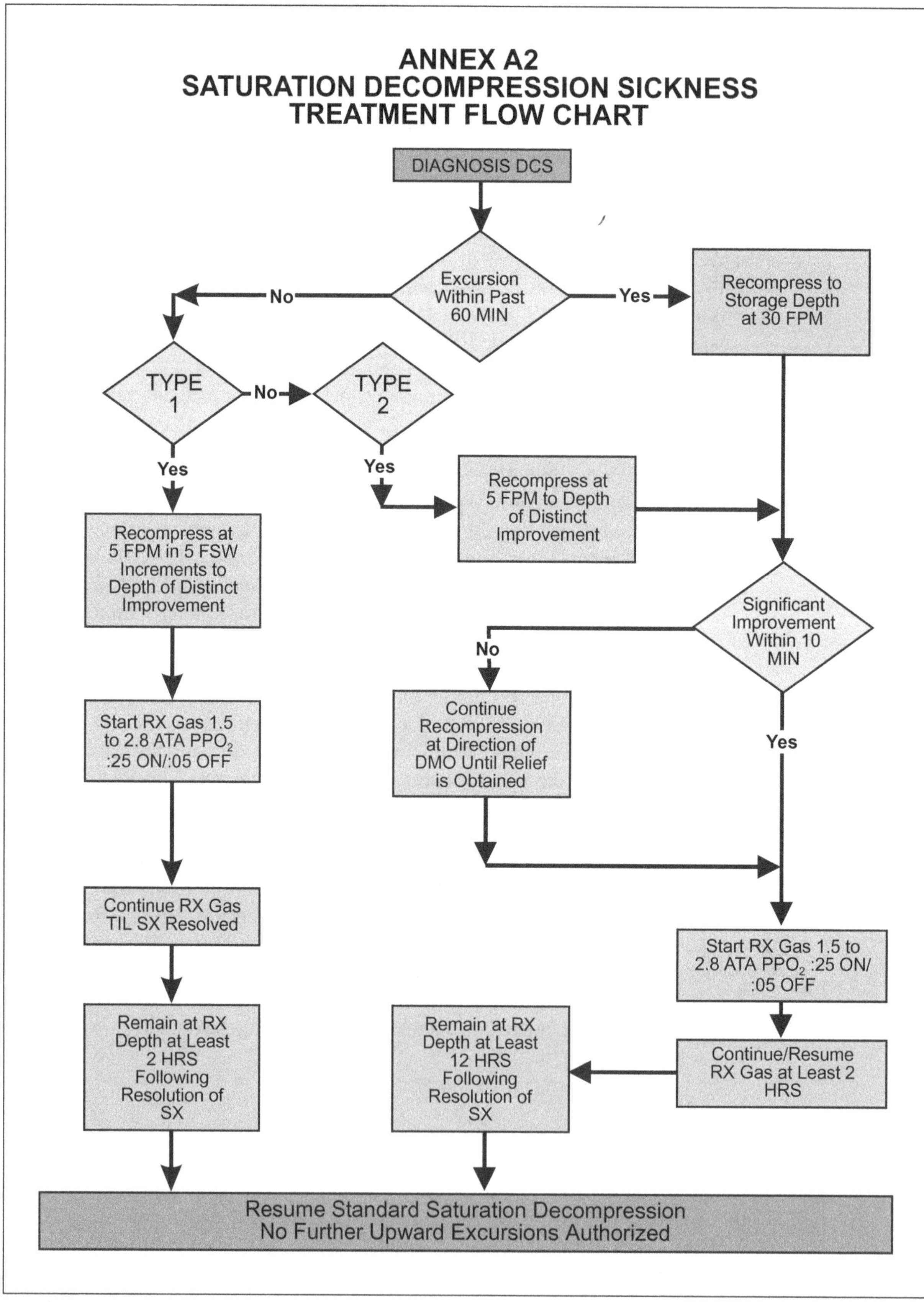

Figure 15-9. Saturation Decompression Sickness Treatment Flow Chart.

memory. Type II Decompression Sickness resulting from an Upward Excursion is a medical emergency and shall be treated by immediate recompression at 30 fsw/min to the depth from which the Upward Excursion originated. When Type II Decompression Sickness symptoms do not occur in association with an Upward Excursion, compression at 5 fsw/min to the depth where distinct improvement is noted should take place. Upon reaching treatment depth, symptoms usually begin to abate rapidly. If symptoms are not significantly improved within 5 to 10 minutes at the initial treatment depth, deeper recompression at the recommendation of a Saturation Diving Medical Officer should be started until significant relief is obtained. After reaching the final treatment depth, treatment gas having an oxygen partial pressure of 1.5 to 2.8 ata shall be administered to the stricken diver for 25-minute periods interspersed with 5 minutes of breathing chamber atmosphere. Treatment gas shall be administered for at least 2 hours and the divers shall remain at the final treatment depth for at least 12 hours following resolution of symptoms. Decompression can then be resumed using standard saturation decompression using rates shown in Table 15-9. Further Upward Excursions are not permitted.

15-24 POSTDIVE PROCEDURES

After surfacing from the dive, the divers are still at risk from decompression sickness. Divers shall remain in the immediate vicinity of a chamber for 2 hours and within 30 minutes travel of a chamber for 48 hours after the dive. Divers shall not fly for 72 hours after the dive surfaces.

PAGE LEFT BLANK INTENTIONALLY

CHAPTER 16

Breathing Gas Mixing Procedures

16-1 INTRODUCTION

16-1.1 **Purpose.** The purpose of this chapter is to familiarize divers with the techniques used to mix divers' breathing gas.

16-1.2 **Scope.** This chapter outlines the procedures used in mixing divers' breathing and treatment gas.

16-2 MIXING PROCEDURES

Two or more pure gases, or gas mixtures, may be combined by a variety of techniques to form a final mixture of predetermined composition. This section discusses the techniques for mixing gases. Aboard ships, where space is limited and motion can affect the accuracy of precision scales, gases are normally mixed by partial pressure or by continuous-flow mixing systems. The methods of mixing by volume or weight are most suitable for use in shore-based facilities because the procedure requires large, gas-tight holding tanks and precision scales.

16-2.1 **Mixing by Partial Pressure.** Mixing gases in proportion to their partial pressures in the final mixture is the method commonly used at most Navy facilities. The basic principle behind this method is Dalton's Law of Partial Pressures, which states that the total pressure of a mixture is equal to the sum of the partial pressures of all the gases in the mixture.

The partial pressure of a gas in a mixture can be calculated using the ideal-gas (perfect-gas) method or the real-gas method. The ideal-gas method assumes that pressure is directly proportional to the temperature and density of a gas. The real-gas method additionally accounts for the fact that some gases will compress more or less than other gases.

Compressibility is a physical property of every gas. Helium does not compress as much as oxygen.

If two cylinders with the same internal volume are filled to the same pressure, one with oxygen and the other with helium, the oxygen cylinder will hold more cubic feet of gas than the helium cylinder. As pressure is increased, and/or as temperature is decreased in both cylinders, the relative difference in the amount of gas in each cylinder increases accordingly. The same phenomenon results when two gases are mixed in one cylinder. If an empty cylinder is filled to 1,000 psia with oxygen and topped off to 2,000 psia with helium, the resulting mixture contains more oxygen than helium.

Being aware of the differences in the compressibility of various gases is usually sufficient to avoid the problems that are often encountered when mixing gases.

When using the ideal-gas procedures, a diver should add less oxygen than is called for, analyze the resulting mixture, and compensate as required. These procedures take into consideration the compressibility of the gases being mixed. Regardless of the basis of the calculations used to determine the final partial pressures of the constituent gases, the mixture shall always be analyzed for oxygen content prior to use.

16-2.2 **Ideal-Gas Method Mixing Procedure.** Gas mixing may be prepared one cylinder at a time or to and from multiple cylinders. The required equipment is inert gas, oxygen, mix cylinders or flasks, an oxygen analyzer, and a mixing manifold. A gas transfer system may or may not be used. Typical mixing arrangements are shown in Figure 16-1 and Figure 16-2. To mix gas using the idea-gas method:

1. Measure the pressure in the inert-gas cylinder(s) P_I.

2. Calculate the pressure in the mixed-gas cylinder(s) after mixing, using the following equation:

$$P_F = \frac{P_I + 14.7}{A} - 14.7$$

Where:

P_F = Final mix cylinder pressure, psig*
P_I = Inert gas cylinder pressure, psig
A = Decimal percent of inert gas in the final mixture

* PF cannot exceed the working pressure of the inert gas cylinder.

3. Measure the pressure in the oxygen cylinder(s), P_O.

4. Determine if there is sufficient pressure in the oxygen cylinder(s) to accomplish mixing with or without an oxygen transfer pump.

$$P_O \geq (2P_F 0 - P_I) + 5$$

Where:

P_O = Pressure in the oxygen cylinder, psig
50 = Required minimum over pressure, psi
≥ *means* greater than or equal to

5. Connect the inert-gas and oxygen cylinder(s) using an arrangement shown in Figure 16-1 or Figure 16-2.

6. Open the mix gas cylinders valve(s).

7. Open the oxygen cylinders valve. Bleed oxygen into the mix gas cylinders at a maximum rate of 70 psi minute until the desired P_F is reached.

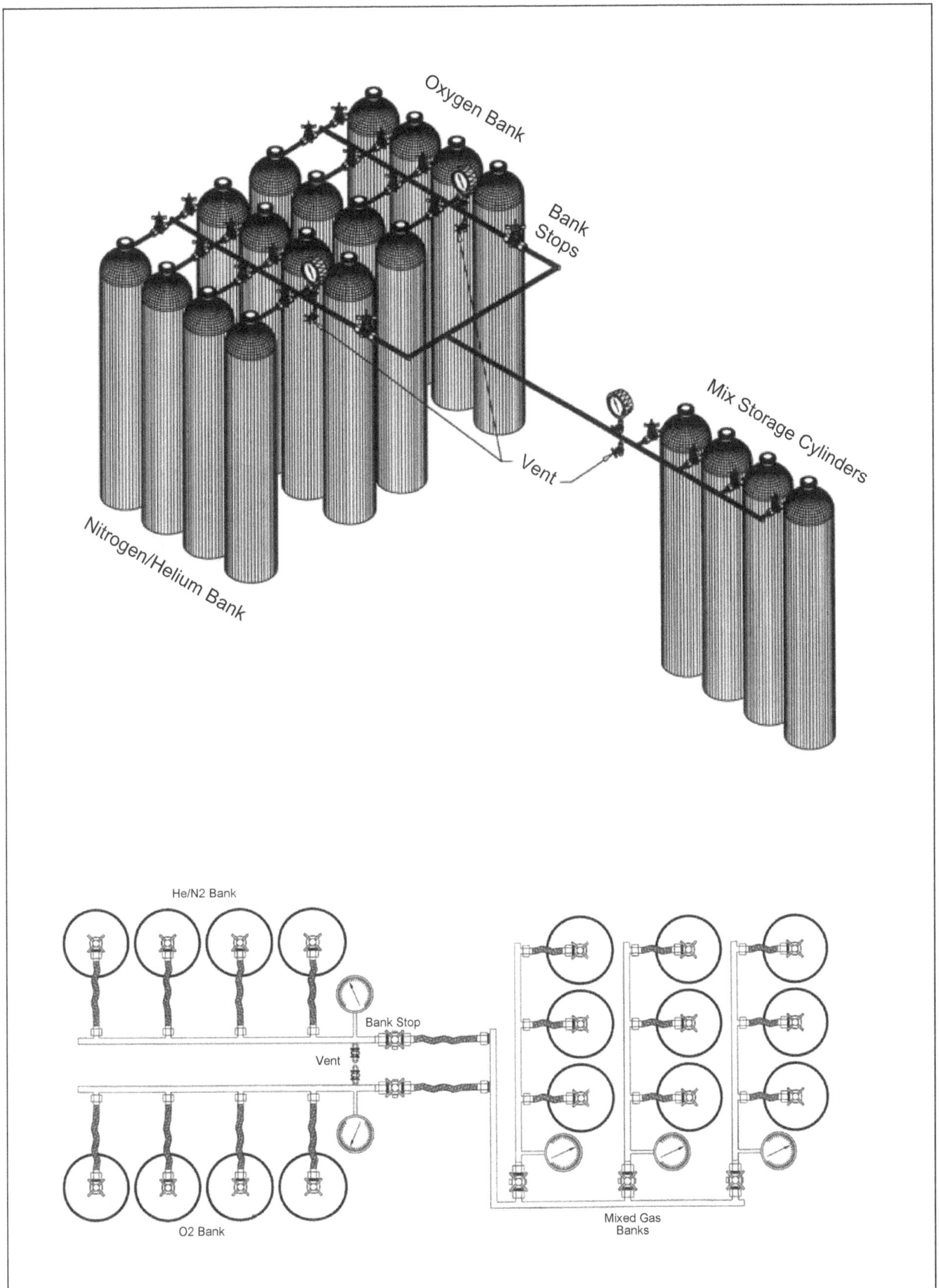

Figure 16-1. Mixing by Cascading.

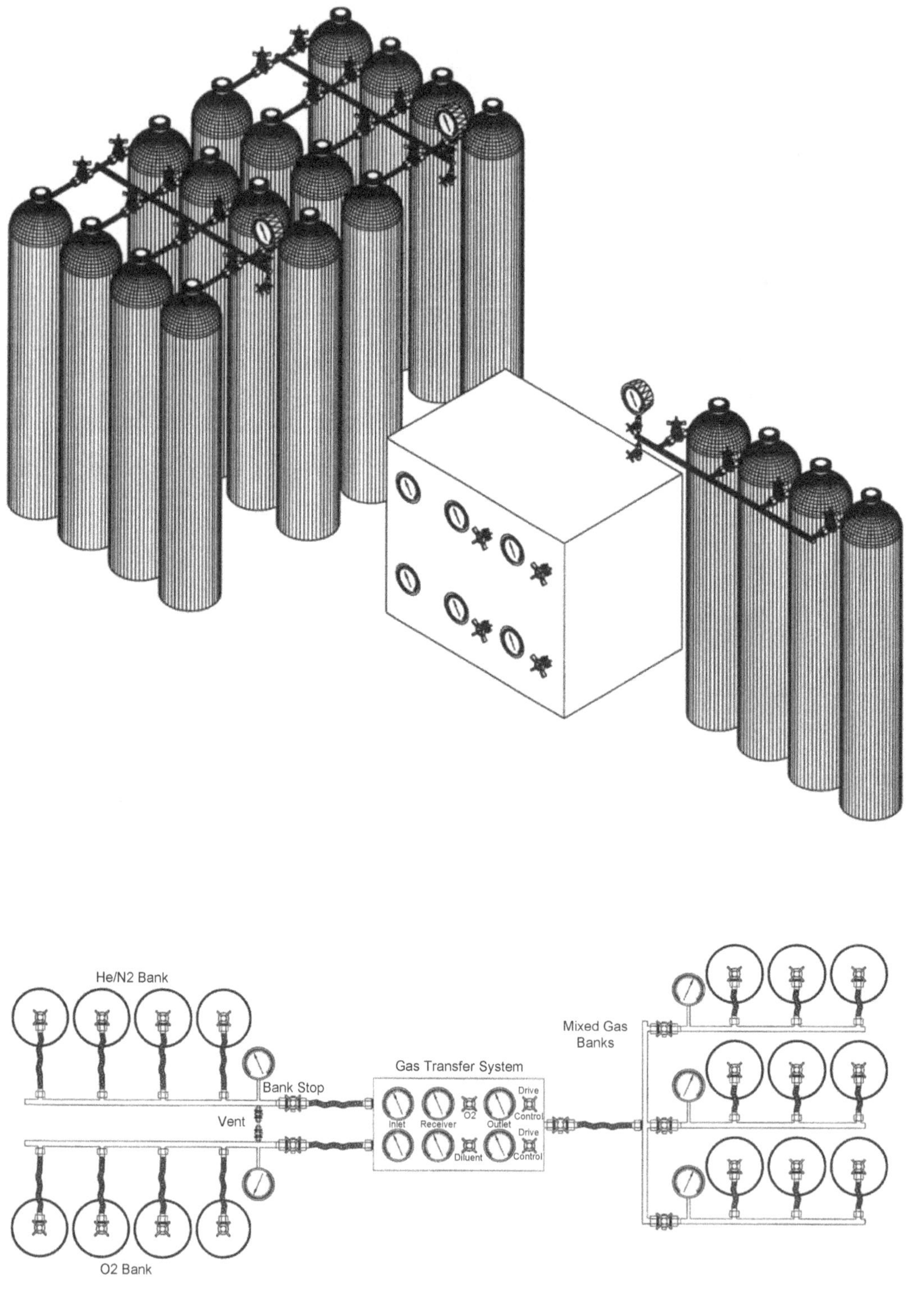

Figure 16-2. Mixing with Gas Transfer System.

8. Close the oxygen and mixed-gas cylinder valves. The heat of compression will have increased the temperature of the mixed-gas cylinders and will give a false indication of the pressure in the cylinder. The calculation requires the P_F to be taken at the same temperature as P_I. However, because of the compressibility effects, more oxygen will normally have to be bled into the mixed-gas cylinders than expected. Therefore, allow the cylinders to stand for at least six hours to permit the gases to mix homogeneously, or if equipment is available, roll the cylinder for at least one hour. Analyze the gas mixture to determine its oxygen percentage. The percentage of oxygen should be near or slightly below the desired percentage.

9. Add oxygen as necessary and reanalyze the mixture. Repeat this step until the desired mixture is attained.

16-2.3 **Adjustment of Oxygen Percentage.** After filling a mixed-gas cylinder, it may be necessary to increase or decrease the percentage of oxygen in the cylinder.

16-2.3.1 **Increasing the Oxygen Percentage.** To increase the oxygen percentage:

1. Subtract the known percentage of oxygen from 100 to obtain the existing percentage of helium.

2. Multiply the helium percentage by the cylinder pressure to obtain the pressure of helium in the cylinder.

3. Subtract the desired oxygen percentage from 100 to obtain the desired percentage of helium.

4. Divide the existing helium pressure (Step 2) by the desired helium percentage (Step 3) in decimal form. (This step gives the cylinder pressure that will exist when enough oxygen has been added to yield the desired percentage.)

5. Add oxygen until this pressure is reached.

6. Allow temperature and pressure to stabilize and add more oxygen, if necessary.

   The following formula sums up the computation:

$$F = \frac{P \times (1.00 - O_O)}{(1.00 - O_f)}$$

Where:

F = Final cylinder pressure
P = Original Cylinder pressure
O_o = Original oxygen % (decimal form)
O_f = Final oxygen % (decimal form)

Sample Problem. An oxygen cylinder contains 1,000 psi of a 16 percent oxygen mixture, and a 20 percent oxygen mixture is desired.

$$F = \frac{1{,}000 \times (1.00 - 0.16)}{1.00 - 0.20}$$
$$= \frac{1{,}000 \times 84}{0.80}$$
$$= \frac{840}{0.80}$$
$$= 1{,}050 \text{ psi}$$

Add 50 psi of oxygen to obtain a cylinder pressure of 1,050 psi.

16-2.3.2 **Reducing the Oxygen Percentage.** To reduce the oxygen percentage, use the following procedure:

1. Multiply oxygen percentage (decimal form) by the cylinder pressure to obtain the psi of oxygen pressure.

2. Divide this figure by the desired oxygen percentage (decimal form). This yields the final pressure to be obtained by adding helium.

3. Add helium until this pressure is reached.

4. Allow temperature and pressure to stabilize and add more helium, if necessary.

The following formula sums up the computation:

$$F = \frac{P \times O_O}{O_f}$$

Where:

F = Final cylinder pressure
P = Original Cylinder pressure
O_o = Original oxygen % (decimal form)
O_f = Final oxygen % (decimal form)

Sample Problem. For a cylinder containing 1,000 psi of a 20 percent oxygen mixture and a 16 percent oxygen mixture is desired.

$$\begin{aligned} F &= \frac{1{,}000 \times 0.20}{0.16} \\ &= \frac{200}{0.16} \\ &= 1{,}250 \text{ psi} \end{aligned}$$

Add 250 psi of helium to obtain a cylinder pressure of 1,250 psi.

These mixing procedures also apply to mixing by means of an oxygen-transfer pump. Instead of being bled directly from an oxygen cylinder into a helium cylinder, oxygen may be drawn from a cylinder at low pressure by the oxygen-transfer pump until the proper cylinder pressure is reached. This allows most of the oxygen in the cylinder to be used, and it also conserves gas.

16-2.4 **Continuous-Flow Mixing.** Continuous-flow mixing is a precalibrated mixing system that proportions the amounts of each gas in a mixture by controlling the flow of each gas as it is delivered to a common mixing chamber. Continuous-flow gas mixing systems perform a series of functions that ensure extremely accurate mixtures. Constituent gases are regulated to the same pressure and temperature before they are metered through precision micro-metering valves. The valve settings are precalibrated and displayed on curves that are provided with every system and relate final mixture percentages with valve settings. After mixing, the mixture is analyzed on-line to provide a continuous history of the oxygen percentage. Many systems have feedback controls that automatically adjust the valve settings when the oxygen percentage of the mixture varies from preset tolerance limits. The final mixture may be supplied directly to a diver or a chamber or be compressed into storage tanks for later use.

16-2.5 **Mixing by Volume.** Mixing by volume is a technique where known volumes of each gas are delivered to a constant-pressure gas holder at near-atmospheric pressure. The final mixture is subsequently compressed into high-pressure cylinders. Mixing by volume requires accurate gas meters for measuring the volume of each gas added to the mixture. When preparing mixtures with this technique, the gases being mixed shall be at the same temperature unless the gas meters are temperature compensated.

The volumes of each of the constituent gases are calculated based on their desired percentages in the final mixture. For example, if 1,000 scf of a 90 percent helium/10 percent oxygen mixture is needed, 900 scf of helium will be added to 100 scf of oxygen. Normally, an inflatable bag large enough to contain the required volume of gas at near-atmospheric pressure is used as the mixing chamber. The pure gases, which are initially contained in high-pressure cylinders, are regulated at atmospheric pressure, metered, and then piped into the mixing chamber. Finally, the mixture is compressed and stored in high-pressure flasks or cylinders.

Provided that the temperatures of the constituent gases are essentially the same, extremely accurate mixtures are possible by using the volume technique of mixing. Additionally, care must be taken to ensure that the mixing chamber is either completely empty or has been filled with a known mixture of uncontaminated gas before mixing.

16-2.6 **Mixing by Weight.** Mixing by weight is most often employed where small, portable cylinders are used. This proportions the gases in the final mixture by the weight that each gas adds to the initial weight of the container. When mixing by weight, the empty weight of the container must be known as well as the weight of any gases already inside the container. Although the accuracy of the mixture when using this technique is not affected by variations in gas temperature, it is directly dependent on the accuracy of the scale being used to weigh the gases. This accuracy shall be known and the operator must be aware of its effect on the accuracy of the composition of the final mixture. As a safeguard, the final mixture must be analyzed for composition using an accurate method of analysis.

16-3 GAS ANALYSIS

The precise determination of the type and concentration of the constituents of breathing gas is of vital importance in many diving operations. Adverse physiological reactions can occur when exposure time and concentrations of various components in the breathing atmosphere vary from prescribed limits. Analysis of oxygen content of helium-oxygen mixtures shall be accurate to within ± 0.5 percent.

The quality of the breathing gas is important in both air and mixed-gas diving. In air diving, the basic gas composition is fixed, and the primary consideration is directed toward determining if gaseous impurities are present in the air supply (i.e. carbon monoxide, hydrocarbons) and the effects of inadequate ventilation (carbon dioxide). Using analytical equipment in air diving is not routine practice. Analytical equipment is generally employed only when it is suspected that the air supply is not functioning properly or when evaluating new equipment.

Gas analysis is essential in mixed-gas diving. Because of the potential hazards presented by anoxia and by CNS and pulmonary oxygen toxicity, it is mandatory that the oxygen content of the gas supply be determined before a dive. Oxygen analysis is the most common, but not the only type of analytical measurement that is performed in mixed-gas diving. In deep diving systems, scrubbing equipment performance must be monitored by carbon dioxide analysis of the atmosphere. Long-term maintenance of personnel under hyperbaric conditions often necessitates the use of a range of analytical procedures. Analyses are required to determine the presence and concentration of minor quantities of potentially toxic impurities resulting from the off-gassing of materials, metabolic processes, and other sources.

16-3.1 **Instrument Selection.**

Selecting an instrument for analyzing hyperbaric atmospheric constituents shall be determined on an individual command basis. Two important characteristics are accuracy and response time. Accuracy within the range of expected concentration must be adequate to determine the true value of the constituent being studied. This characteristic is of particular importance when a sample must be taken at elevated pressure and expanded to permit analysis. The instrument's response time to changes in concentration is important when measuring constituents that may rapidly change and result in quick development of toxic conditions.

Response times of up to 10 seconds are adequate for monitoring gas concentrations such as oxygen and carbon dioxide in a diving apparatus. When monitoring hyperbaric chamber atmospheres, response times of up to 30 seconds are acceptable. The instruments used should accurately measure concentrations to within 1/10 of the maximum allowable concentration. Thus, to analyze for carbon dioxide with a maximum permissible concentration of 5,000 ppm (SEV), an instrument with an accuracy of at least 500 ppm (SEV) must be used.

In addition to accuracy and response time, portability is a factor in choosing the correct instrument. While large, permanently-mounted instruments are acceptable for installation on fixed-chamber facilities, small hand-carried instruments are better suited for emergency use inside a chamber or at remote dive sites.

16-3.2 **Techniques for Analyzing Constituents of a Gas.** The constituents of a gas may be analyzed both qualitatively (type determination) and quantitatively (type and amount) using many different techniques and instruments. Guidance regarding instrument selection can be obtained from NAVSEA, NEDU, or from instrument manufacturer technical representatives. Although each technique is not discussed, the major types are listed below as a reference for those who desire to study them in detail.

- Mass spectrometry
- Colorimetric detection
- Ultraviolet spectrophotometry
- Infrared spectrophotometry
- Gas chromatography
- Electrolysis
- Paramagnetism

PAGE LEFT BLANK INTENTIONALLY

VOLUME 4

Closed Circuit and Semiclosed Circuit Diving Operations

U.S. NAVY DIVING MANUAL

PAGE LEFT BLANK INTENTIONALLY

Volume 4 - Table of Contents

Chap/Para Page

Chap/Para **Page**

Chap/Para **Page**

PAGE LEFT BLANK INTENTIONALLY

Volume 4 - List of Illustrations

PAGE LEFT BLANK INTENTIONALLY

Volume 4 - List of Tables

CHAPTER 17

MK 16 MOD 0 Closed-Circuit Mixed-Gas UBA Diving

17-1 INTRODUCTION

The MK16 MOD 0 is a 0.75 ata constant partial pressure of oxygen (ppO_2) closed-circuit mixed-gas underwater breathing apparatus (UBA) primarily employed by Naval Special Warfare (SPECWAR) forces. The U.S. Navy's use of mixed-gas closed circuit UBAs was developed to satisfy the operational requirements of SPECWAR combat swimmers and EOD divers. This equipment combines the mobility of a free-swimming diver with the depth advantages of mixed gas. The term closed circuit refers to the recirculation of 100 percent of the mixed-gas breathing medium and results in bubble-free operation, except during ascent or inadvertent gas release. This capability makes closed circuit UBAs well suited for special warfare operations. The maximum working limits for the MK16 MOD 0 UBA are 150 feet of seawater (fsw) when N_2O_2 (air) is used as a diluent or 200 fsw when 84/16 HeO_2 mix is used as a diluent.

17-1.1 **Purpose.** This chapter provides general guidelines for MK 16 MOD 0 UBA diving, operations and procedures (Figure 17-1). For detailed operation and maintenance instructions, see technical manual SS600-AH-MMA-010 (MK 16 MOD 0).

17-1.2 **Scope.** This chapter covers MK 16 MOD 0 UBA principles of operations, operational planning, dive procedures, and medical aspects of mixed-gas closed-circuit diving. Refer to Chapter 16 for procedures for mixing divers' breathing gas.

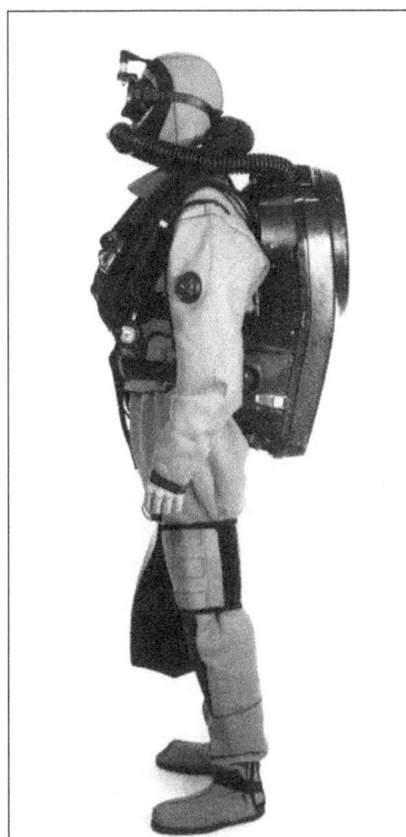

Figure 17-1. MK 16 MOD 0 Closed-Circuit Mixed-Gas UBA.

17-2 PRINCIPLES OF OPERATION

Closed circuit UBAs efficiently use the available gas supply to extend underwater duration by recirculating the breathing gas. To do this efficiently a closed circuit UBA must be able to:

- Remove carbon dioxide produced by metabolic action of the body.
- Monitor the ppO_2 and add oxygen in order to replace the oxygen consumed by metabolic action of the body.

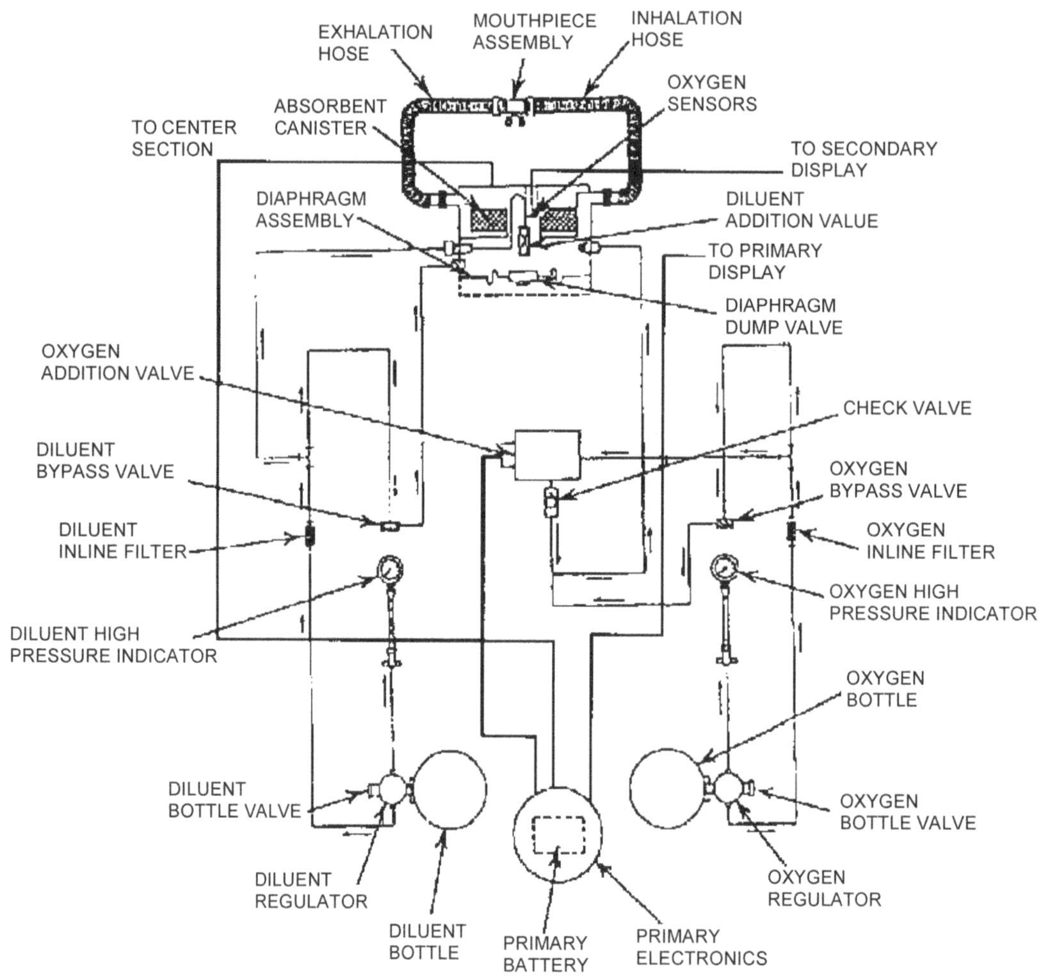

Figure 17-2. MK 16 MOD 0 UBA Functional Block Diagram.

17-2.1 **Diving Safety.** Closed-circuit mixed-gas UBAs are more complex than open-circuit SCUBA and require a high level of diver training and situational awareness. Careful dive planning is essential. Diving safety is achieved only when:

- The diver has been thoroughly trained and qualified in the proper use of the UBA.
- All equipment has been prepared for the specific diving conditions expected.
- The dive is conducted within specified depth and duration limits.
- The diver strictly adheres to and immediately implements all operational and emergency procedures.

17-2.2 **Advantages of Closed-Circuit Mixed-Gas UBA.** While functionally simpler in principle, the closed-circuit mixed-gas UBA tends to be more complex than the semi-closed UBA because of the oxygen analysis and control circuits required. Offsetting this complexity, however, are several inherent advantages:

- Aside from mixed or diluent gas addition during descent, the only gas required at depth is oxygen to make up for metabolic consumption.
- The partial pressure of oxygen in the system is automatically controlled throughout the dive to a preset value. No adjustment is required during a dive for variations in depth and work rate.
- No inert gas leaves the system except by accident or during ascent, making the closed circuit UBA relatively bubble-free and well suited for SPECWAR operations.

17-2.3 **Recirculation and Carbon Dioxide Removal.** The diver's breathing medium is recirculated in a closed circuit UBA to remove carbon dioxide and permit reuse of the inert diluent and unused oxygen in the mixture. The basic recirculation system consists of a closed loop that incorporates inhalation and exhalation hoses and associated check valves, a mouthpiece or full face mask (FFM), a carbon dioxide removal unit, and a diaphragm assembly.

17-2.3.1 **Recirculating Gas.** Recirculating gas is normally moved through the circuit by the natural inhalation-exhalation action of the diver's lungs. Because the lungs can produce only small pressure differences, the entire circuit must be designed for minimum flow restriction.

17-2.3.2 **Full Face Mask.** The FFM uses an integral oral-nasal mask or T-bit to reduce dead space and the possibility of rebreathing carbon dioxide-rich gas. Similarly, check valves used to ensure one-way flow of gas through the circuit must be close to the diver's mouth and nose to minimize dead space. All breathing hoses in the system must be of relatively large diameter to minimize breathing resistance.

17-2.3.3 **Carbon Dioxide Scrubber.** Carbon dioxide is removed from the breathing circuit in a watertight canister filled with an approved carbon dioxide-absorbent material. The bed of carbon dioxide-absorbent material chemically combines with the diver's exhaled carbon dioxide, while allowing the unused oxygen and diluent to pass through it. If the canister is improperly filled, channels may form in the absorbent granules permitting gas to bypass the absorbent material and allow the build up of carbon dioxide in the UBA. The canister design must also provide a low flow resistance for the gas while ensuring maximum contact between the gas and the absorbent. Flow resistance is minimized in the MK 16 MOD 0 UBA by employing a radially designed canister to reduce gas flow distance. Since inadvertent wetting of the absorbent material may produce a caustic solution, water absorbent pads are usually placed above and below the canister to collect water produced from both the reaction between carbon dioxide and the carbon dioxide absorbent and by the diver himself. The amount of CO_2 absorbent capacity is one of the major limiting factors for any closed circuit UBA. Absorbent duration is also directly affected by the environmental operating temperature and depth. Absorbent duration decreases as temperature decreases and as depth increases.

17-2.3.4 **Diaphragm Assembly.** A diaphragm assembly or counter lung is used in all closed circuit UBAs to permit free breathing in the circuit. The need for such devices can be readily demonstrated by attempting to exhale and inhale into an empty bottle. The bottle, similar to the recirculation system without a bag, is unyielding and presents extreme back pressure. In order to compensate, flexible diaphragms or a breathing bag must be placed in the UBA circuit with a maximum displacement equal to the combined volume of both lungs.

Constant buoyancy is inherent in the system because the gas reservoir acts counter to normal lung action. In open-circuit scuba, diver buoyancy decreases during exhalation due to a decrease in lung volume. In closed-circuit scuba, expansion of the breathing bag keeps buoyancy constant. On inhalation, the process is reversed. This cycle is shown in Figure 17-3.

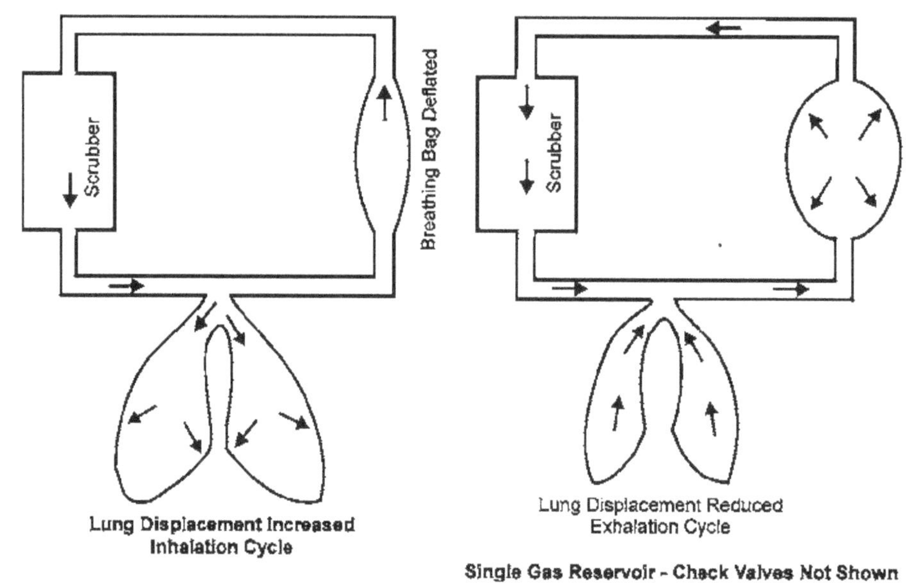

Figure 17-3. UBA Breathing Bag Acts to Maintain the Diver's Constant Buoyancy by Responding Counter to Lung Displacement.

The flexible gas reservoir must be located as close to the diver's chest as possible to minimize hydrostatic pressure differences between the lungs and the reservoir as the diver changes attitude in the water.

The MK 16 MOD 0 UBA uses a single reservoir built into a streamlined backpack assembly. Using a single reservoir located within the backpack affords minimum encumbrance to the diver and maximum protection for the reservoir.

17-2.3.5 **Recirculation System.** Optimal performance of the recirculation system depends on proper maintenance of equipment, proper filling with fresh absorbent, and accurate metering of oxygen input. To ensure efficient carbon dioxide removal throughout the dive, personnel must carefully limit dive time to the specified canister duration. Any factor that reduces the efficiency of carbon dioxide removal increases the risk of carbon dioxide poisoning.

WARNING **The MK 16 MOD 0 UBA provides no visual warning of excess CO_2 problems. The diver should be aware of CO_2 toxicity symptoms.**

17-2.3.6 Gas Addition, Exhaust, and Monitoring.

In addition to the danger of carbon dioxide toxicity, the closed circuit UBA diver encounters the potential hazards of hypoxia and central nervous system (CNS) oxygen toxicity. The UBA must control the partial pressure of oxygen (ppO_2) in the breathing medium within narrow limits for safe operation and be monitored frequently by the diver.

Hypoxia can occur when there is insufficient oxygen in the recirculation circuit to meet metabolic requirements. If oxygen is not added to the breathing circuit, the oxygen in the loop will be gradually consumed over a period of 2-5 minutes, at which point the oxygen in the mixture is incapable of sustaining life.

CNS oxygen toxicity can occur whenever the oxygen partial pressure in the diver's breathing medium exceeds specified concentration and exposure time limits. Consequently, the UBA must function to limit the ppO_2 level to the appropriate value.

The closed-circuit mixed-gas UBA uses a direct control method of maintaining oxygen concentration in the system, rather than the indirect method of a preset mass flow, common to semi-closed apparatus.

17-3 MK16 MOD 0 Closed Circuit UBA

The MK 16 MOD 0 UBA is broken down into four basic systems (housing, recirculation, pneumatics, and electronics) and their subassemblies as described in the following paragraphs. These systems provide a controlled ppO_2 breathing gas to the diver.

17-3.1 **Housing System.** Major components of the MK 16 MOD 0 UBA are housed in a reinforced ABS or fiberglass molded case. The equipment case is a contoured backpack assembly designed for minimum interference while swimming, and is equipped with an integral harness assembly. A streamlined, readily detachable outer cover minimizes the danger of underwater entanglement. External to the housing are components such as the mouthpiece, pressure indicators, hoses, and primary and secondary displays.

17-3.2 **Recirculation System.** The recirculation system consists of a closed loop incorporating inhalation and exhalation hoses, a mouthpiece or FFM, a carbon dioxide-absorbent canister, and a flexible breathing diaphragm. The diver's breathing gases are recirculated to remove carbon dioxide and permit reuse of the inert component of the diluent and residual oxygen in the breathing mixture. Inhalation and exhalation check valves in the mouthpiece assembly (or manifold of the FFM) ensure the unidirectional flow of gas through the system.

17-3.2.1 **Closed-Circuit Subassembly.** The closed-circuit subassembly has a removable cover, a center section attached to the fiberglass equipment case, a flexible rubber breathing diaphragm, and a CO_2 scrubber assembly. Moisture-absorbent pads inside the scrubber assembly absorb any condensation formed on the cover walls. The space between the scrubber canister and the cover serves as a gas plenum, insulating the canister from the ambient cold water.

17-3.2.2 **Scrubber Functions.** The scrubber has two functions:

- **Carbon Dioxide Removal.** Before the diver's exhaled breath reaches the breathing diaphragm, it passes through the scrubber canister. The scrubber canister is filled with an approved, high efficiency, granular carbon dioxide-absorbent material. Two filter discs in the scrubber canister serve as gas distributors to minimize effects of any channeling in the absorbent. After passing through the filters, the exhaled gas passes through the carbon dioxide-absorbent bed, chemically combining with the carbon dioxide created by metabolic use of the diver's breathing oxygen but allowing the diluent and unused oxygen to pass through it.

- **Water Removal.** Moisture produced by diver exhalation and the reaction between carbon dioxide and carbon dioxide-absorbent is absorbed by moisture-absorbent pads located outside the canister.

17-3.3 **Pneumatics System.** The pneumatics system comprises:

- High-pressure bottles for storing oxygen and diluent gases.

- Indicators to permit monitoring of the remaining gas supply.

- Regulators, fittings, tubing, filters and valves to regulate and deliver oxygen and diluent gases to the recirculation system.

17-3.4 **Electronics System.** The electronics system maintains a constant partial pressure of oxygen in the closed-circuit UBA by processing and conditioning signal outputs from the oxygen sensors located in the breathing loop, stimulating the oxygen-addition valve, and controlling the output of the primary display.

17-3.4.1 **Oxygen Sensing.** The partial pressure of oxygen within the recirculation system is monitored by three sensors. Each sensor's output is evaluated by the primary electronics package through a voting logic circuit negating the output from a faulty sensor. Sensor averages are shown by the primary display. Backup reading of each individual sensor can be read on the secondary display which requires no outside power source.

17-3.4.2 **Oxygen Control.** Oxygen concentration in the recirculation system is measured by sensors. The sensors send signals to the primary electronics assembly and the secondary display. The primary electronics assembly compares these sensor signals with the setpoint value, providing output to the primary display and controlling the oxygen-addition valve. An actual ppO_2 value less than the setpoint automatically actuates the oxygen-addition valve to admit oxygen to the breathing loop.

Oxygen control involves several factors:

- **System Redundancy.** The primary electronics assembly in the MK 16 MOD 0 UBA treats each of the sensor signals as a vote. The sensor vote is either above or below the predetermined setpoint. If a simple majority of the sensors is below the predetermined setpoint, a drive signal is sent to the oxygen-addition valve; when a majority of the sensors is above the predetermined setpoint, the signal is terminated. In effect, the electronics circuit ignores the highest and lowest sensor signals and controls the oxygen-addition valve with the middle sensor. Similarly, the electronics circuit displays a high-oxygen alarm (flashing green) if a majority of the sensors' signals indicates a high oxygen level and displays a low-oxygen alarm (flashing red) if a majority of the sensors' signals indicates a low oxygen level. If only one sensor indicates a high oxygen level and/or only one sensor indicates a low oxygen level, the electronics circuit output alternates between the two alarm states (alternating red/green).

- **Setpoint Calibration.** The normal operational ppO_2 setpoint for the MK 16 Mod 0 UBA is 0.75 ata. Appropriate calibration procedures are used to preset the specific ppO_2 setting.

- **Oxygen Addition.** In response to the sensor outputs, the oxygen-addition valve admits oxygen to the breathing loop in the recirculation system. The control circuits continuously monitor the average ppO_2 level. If the oxygen partial pressure in the recirculation system is lower than the setpoint level, the oxygen-addition valve is energized to admit oxygen. When the ppO_2 reaches the required level, the automatic control system maintains the oxygen-addition valve in the SHUT position. Should the oxygen-addition valve fail in an OPEN position, the resulting free flow of oxygen in the MK 16 MOD 0 is restricted by the tubing diameter and the orifice size of the piezoelectric oxygen-addition valve.

17-3.4.3 **Displays.** The MK 16 MOD 0 UBA has two displays that provide continuous information to the diver about ppO_2, battery condition, and oxygen sensor malfunction.

17-3.4.3.1 ***Primary Display.*** The primary display consists of two light-emitting diodes (LEDs) that are contained within the primary display housing. This display is normally mounted on the face mask, within the peripheral vision of the diver. The two LEDs (one red and one green) powered by the primary electronics assembly battery indicate the general overall condition of various electronic components and the ppO_2 in the breathing loop as follows:

- **Steady green:** Normal oxygen range, 0.60 to 0.90 ata ppO_2 (using a set point of 0.75 ata).
- **Steady red or simultaneously illuminated steady red and green:** Primary electronics failure.

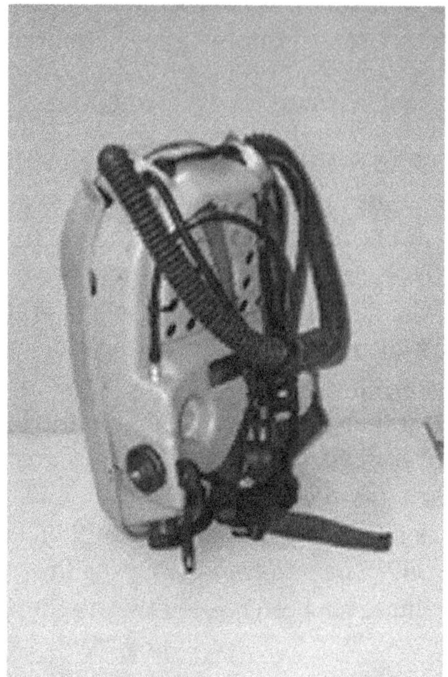

Figure 17-4. Underwater Breathing Apparatus MK 16 MOD 0.

- **Flashing green:** High oxygen content, greater than 0.90 ata ppO_2.
- **Flashing red:** Low oxygen content, less than 0.60 ata ppO_2.
- **Alternating red/green:** Normal transition period (ppO_2 is transitioning from normal to low, from low to normal, from normal to high, or from high to normal), one sensor out of limits, low primary battery power (displayed on secondary display) or primary electronics failure.
- **No display (display blanked):** Electronics assembly or primary battery failure.

17-3.4.3.2 ***Secondary Display.*** The MK 16 MOD 0 secondary display is designed to provide quantitative information to the diver on the condition of the breathing medium, the primary battery voltage and the condition of the secondary batteries. It also serves as a backup for the primary display in the event of a failure or malfunction to the primary electronics assembly, the primary display, or the primary battery. The secondary display functions concurrently with, but independently of, the primary display and displays the O_2 sensor readings and primary battery information in digital form. The secondary display is powered by four 1.5-volt batteries for illumination of the LED display only. It does not rely on the primary electronics subassembly, but receives signals directly from the oxygen sensors and the primary battery. It will continue to function in the event of a primary electronics assembly failure.

17-4 OPERATIONAL PLANNING

Chapter 6 provides general guidelines for operational planning. The information provided in this section is supplemental to Chapter 6 and the MK 16 MOD 0 UBA

O&M manual. Units should allow frequent opportunity for training, ensuring diver familiarity with equipment and procedures. Workup dives are strongly recommended prior to diving at depths greater than 130 fsw. MK 16 MOD 0 diver qualifications may be obtained by completion of the Naval Special Warfare Center MK 16 MOD 0 qualifications course. Qualifications remain in effect as long as diver qualifications are maintained in accordance with Military Personnel Manual Article 1220. However, a diver who has not made a MK 16 MOD 0 dive in the previous six months must refamiliarize himself with the MK 16 MOD 0 EPs and OPs and must complete a training dive prior to making an operational dive. Prior to conducting a decompression dive, a diver who has not conducted a MK 16 MOD 0 decompression dive within the previous six months must complete open water decompression training dives. The minimum personnel requirements for MK 16 MOD 0 diving operations are the same as open circuit SCUBA, see Figure 6-16.

17-4.1 **Operating Limitations.** The dive depth and time limits are based on considerations of working time, decompression obligation, oxygen tolerance, and nitrogen narcosis. The expected duration of the gas supply, the expected duration of the carbon dioxide absorbent, the adequacy of thermal protection, or other factors may also limit both the depth and the duration of the dive. Diving Supervisors must consider these limiting factors when planning closed-circuit UBA operations.

17-4.1.1 **Oxygen Flask Endurance.** In calculating the endurance of the MK 16 MOD 0, only the oxygen flask is considered. The endurance of the oxygen flask is dependent upon the following:

- Flask floodable volume
- Initial predive pressure
- Required reserve pressure
- Oxygen consumption by the diver
- Effect of cold water immersion on flask pressure

17-4.1.1.1 ***Flask Floodable Volume.*** The oxygen flask floodable volume (fv) is 0.1 cubic foot (2.9 liters).

17-4.1.1.2 ***Initial Predive Pressure.*** The initial pressure is the pressure of the oxygen flask at ambient temperature when it has cooled following charging. A reserve pressure of 500 psig is required to drive the reducer. Calculation of initial pressure must also account for gas loss resulting from UBA predive calibration. Oxygen consumption by the diver is computed as 0.049 scfm (1.4 lpm). This is a conservative value for a diver swimming at 0.85 knots. Refer to Table 17-1 for information on the average breathing gas consumption rates and CO_2 absorbent usage.

17-4.1.1.3 ***Effect of Cold Water Immersion on Flask Pressure.*** Immersion in cold water will reduce the flask pressure and actual cubic feet (acf) of gas available for the diver, in accordance with Charles'/Gay-Lussac's gas law. Based upon direct measurement, available data, or experience, the coldest temperature expected during the dive is used.

Table 17-1. *Average Breathing Gas Consumption Rates and CO_2 Absorbent Usage.*

Diving Equipment	Gas Consumption (Normal)	Gas Consumption (Heavy Work)	CO_2 Absorbent Capacity (lbs.)	Duration 40°F (Note 1)	Duration 70°F (Note 1)
MK 16 MOD 0 UBA	12-15 psi/min	15-17 psi/min	7.75-8.0	5h	6h 40m

Note:

1. CO_2 absorbent duration is based upon a comfortable work rate (0.8-knot swimming speed).

17-4.1.1.4 ***Calculating Gas Endurance.*** Combining these factors produces the formula for MK 16 MOD 0 gas endurance:

MK 16 MOD 0 gas endurance =

$$F_V \times \frac{\left[\left(P_I \times \frac{T_2}{T_1}\right) - P_R\right]}{VO_2 \times 14.7 \text{ psi}} \times \frac{492}{T_2}$$

Where:

F_V = Floodable volume of flask in cubic feet
P_I = Initial Pressure in psia
P_R = Reserve Pressure in psia
VO_2 = Oxygen consumption in medical scfm (32°F)
T_1 = Ambient air temperature in °R
T_2 = Coldest water temperature expected in °R

Rankine conversion factor:
°R = °F + 460

All pressure and temperature units must be absolute.

17-4.1.1.5 ***Example.*** The endurance of a MK 16 MOD 0 UBA charged to 2,500 psig for a dive in 50° F water when the ambient air temperature is 65° F would be computed as follows:

$$\text{MK 16 MOD 0 gas endurance} = 0.1 \times \frac{[(2{,}514.7 \times 510/525) - 514.7]}{0.049 \times 14.7} \times \frac{492}{510}$$

$$= 258 \text{ minutes}$$

This duration assumes no gas loss from the UBA during the dive and only considers metabolic consumption of oxygen by the diver. Divers must be trained to minimize gas loss by avoiding leaks and unnecessary depth changes. Clearing a flooded face mask is a common cause of gas loss from the UBA. When a full face mask (FFM) is used, gas can pass from the UBA breathing loop into the FFM and escape into the surrounding seawater due to a poor face seal. Leaks that continue unchecked can deplete UBA gas supply rapidly. Additionally, during diver ascent, the dump valve opens to discharge breathing gas into the surrounding water, thereby preventing overinflation of the breathing diaphragm. Depth changes should be avoided as much as possible to minimize this gas loss.

17-4.1.2 **Diluent Flask Endurance.** Under normal conditions the anticipated duration of the MK 16 MOD 0 diluent flask will exceed that of the oxygen flask. The MK 16 MOD 0 diluent bottle holds approximately 21 standard cubic feet (595 liters) of gas at a stored pressure of 3,000 psig. Diluent gas is used to maintain the required gas volume in the breathing loop and is not depleted by metabolic consumption. As the diver descends, diluent is added to maintain the total pressure within the recirculation system at ambient water pressure. Loss of UBA gas due to off gassing at depth requires the addition of diluent gas to the breathing loop either automatically through the diluent add valve or manually through the diluent bypass valve to make up lost volume. Excessive gas loss caused by face mask leaks, frequent depth changes, or improper UBA assembly will deplete the diluent gas supply rapidly.

17-4.1.3 **Canister Duration.** Canister duration is estimated by using a working diver scenario. This allows an adequate safety margin for the diver in any situation. Table 17-2 shows the canister duration limits for the MK 16 MOD 0 UBA.

Table 17-2. *MK 16 MOD 0 Canister Duration Limits.*

	Canister Duration with HeO_2	
Water Temperature (°F)	**Depth (fsw)**	**Time (minutes)**
40 and above	0-300	300
29-39	0-100	300
35-39	101-300	240
29-34	101-300	120
	Canister Duration with N_2O_2	
Water Temperature (°F)	**Depth (fsw)**	**Time (minutes)**
29 and above	0-50	300
40 and above	51-150	200
29-39	51-150	100

NAVSEA-approved Sodalime CO_2 absorbents for the MK 16 MOD 0 are listed in the ANU list.

17-4.1.4 **Thermal Protection.** Divers must be equipped with adequate thermal protection to perform effectively and safely. A cold diver will either begin to shiver or increase his exercise rate, both of which will increase oxygen consumption and decrease oxygen supply duration and canister duration. Refer to Chapter 11 for guidance on thermal protection.

17-4.2 **Equipment Requirements.** Equipment requirements for MK16 MOD 0 training dives are provided in Table 17-3. Two equipment items merit special comment:

Table 17-3. *MK 16 MOD 0 UBA Diving Equipment Requirements.*

General	Diving Supervisor	Divers	Standby Diver
1. Motorized safety boat (Note 1) 2. Radio (communications with parent unit, chamber, communication between safety boats when feasible) 3. High-intensity, wide-beam light (night operations) 4. Dive flags and/or special operations lights as required 5. Sufficient (2 quarts) fresh water in case of chemical injury	1. Dive watch 2. Dive Bill list 3. Air Decompression Tables 4. Closed-Circuit Mixed-Gas UBA Decompression Tables using 0.7 ATA Constant Partial Pressure Oxygen in Nitrogen and in Helium. 5. Recall device	1. Dive watch (Note 2) 2. Face mask 3. Fins 4. Dive knife 5. Approved life preserver or buoyancy compensator (BC) 6. Appropriate thermal protection 7. Depth gauge (Note 2) 8. Buddy line (as appropriate for SPECWAR operations) (Note 1) 9. Tending line	1. Dive watch 2. Face mask 3. Fins 4. Dive knife 5. Approved life preserver or buoyancy compensator (BC) 6. Appropriate thermal protection 7. UBA with same depth capability 8. Depth gauge 9. Weight belt (if needed) 10. Tending line

Notes:

1. See paragraph 17-4.2
2. See paragraph 17-4.2.6

- **Safety Boat.** A minimum of one motorized safety boat must be present for all open-water dives. A safety boat is also recommended for tended pier dives or diving from shore. Safe diving practice in many situations, however, will require the presence of more than one safety boat. The Diving Supervisor must determine the number of boats required based on the diving area, medical evacuation plan, night operations, and the number of personnel participating in the dive operation.

- **Buddy Lines.** Buddy lines are considered important safety equipment for closed-circuit UBA dives. In special diving situations, such as certain combat swimmer operations or tended diving, the use of buddy lines may not be feasible. The Diving Supervisor shall conduct dives without buddy lines only in situations where their use is not feasible or where their use will pose a greater hazard to the divers than by diving without them.

17-4.2.1 **Distance Line.** Any buddy line over 10 feet (3 meters) in length is referred to as a distance line. The length of the distance line shall not exceed 81 feet (25 meters). Distance lines shall be securely attached to both divers.

17-4.2.2 **Standby Diver.** When appropriate during training and non-influence diving operations, open circuit SCUBA may be used to a maximum depth of 130 fsw.

17-4.2.3 **Lines.** Diver marker lines shall be manufactured from any light line that is buoyant and easily marked as directed in paragraph 17-4.2.4 (one-quarter inch polypropylene is quite suitable).

17-4.2.4 **Marking of Lines.** Lines used for controlling the depth of the diver(s) for decompression diving shall be marked. This includes tending lines, marker lines, and lazy-shot lines. Lines shall be marked with red and yellow or black bands starting at the diver(s) or clump end. Red bands will indicate 50 feet and yellow or black bands will mark every 10 feet.

17-4.2.5 **Diver Marker Buoy.** Diver marker buoys will be constructed to provide adequate visual reference to monitor the divers location. Additionally, the amount of line will be of sufficient length for the planned dive profile.

17-4.2.6 **Depth Gauge/Wrist Watch.** A single depth gauge and wrist watch may be used when diving with a partner and using a distance line.

17-4.3 **Recompression Chamber Considerations.** A recompression chamber and a Diving Medical Officer are not required on the dive station (on the dive station is defined as at the dive location) as prerequisites for closed-circuit UBA diving operations, unless the dive(s) will exceed the maximum working limit. However, the following items should be determined prior to beginning diving operations:

- Location of the nearest functional recompression chamber. Positive confirmation of the chamber's availability in case of emergency should be obtained.
- Location of the nearest available Diving Medical Officer if not at the nearest recompression chamber.
- Location of the nearest medical facility for treatment of injuries and medical problems not requiring recompression therapy.
- The optimal method of transportation to the treatment chamber or medical facility. If coordination with other units for aircraft/boat/vehicle support is necessary, the Diving Supervisor shall know the telephone numbers and points of contact necessary to make these facilities available as quickly as possible in case of emergency. A medical evacuation plan should be included in the Diving Supervisor brief. Preparing an emergency assistance checklist similar to that in Chapter 6 is recommended.

17-4.4 **Ship Safety.** When operations are to be conducted in the vicinity of ships, the guidelines provided in the Ship Repair Safety Checklist (see Chapter 6) must be followed.

17-4.5 **Operational Area Clearance.** Notification of intent to conduct diving operations should be coordinated in accordance with local directives.

17-5 PREDIVE PROCEDURES

17-5.1 **Diving Supervisor Brief.** A thorough, well-prepared dive briefing reinforces the confidence level of the divers and increases safety, and is an important factor in successful mission accomplishment. It should normally be given by the Diving Supervisor, who will be in charge of all diving operations on the scene. The briefing shall be given separately from the overall mission briefing and shall focus on the diving portion of the operation, with special attention to the items shown in Table 17-4. For MK 16 MOD 0 UBA diving, use the appropriate checklist provided in the MK16 MOD 0 UBA O&M Manual.

17-5.2 **Diving Supervisor Check.** As the divers set up their UBAs prior to the dive, the Diving Supervisor must ensure that each diver checks his own equipment, that setup is completed properly by checking the UBA, and that each diver completes a UBA predive checklist from the appropriate UBA operation and maintenance manual. The second phase of the Diving Supervisor check is a predive inspection conducted after the divers are dressed. The Diving Supervisor ensures that the UBA and related gear (life preserver, weight belt, etc.) are properly donned, that mission-related equipment (compass, depth gauge, dive watch, buddy lines, tactical equipment, etc.) are available, and that the UBA functions properly before allowing the divers to enter the water. Appropriate check lists to confirm proper functioning of the UBA are provided in the MK 16 MOD 0 O&M manual.

17-6 WATER ENTRY AND DESCENT

The maximum descent rate is 60 feet per minute. During descent, the UBA will automatically compensate for increased water pressure and provide an adequate volume of gas for breathing. During descent the oxygen partial pressure may increase as oxygen is added to the breathing mixture as a portion of the diluent. Depending on rate and depth of descent, the primary display on the MK 16 MOD 0 UBA may illuminate flashing green. It may take from 2 to 15 minutes to consume the additional oxygen added by the diluent during descent. While breathing down the ppO_2, the diver should continuously monitor the primary and secondary display until the ppO_2 returns to setpoint level.

Table 17-4. *MK 16 MOD 0 UBA Dive Briefing.*

A. **Dive Plan**
   1. Operating depth
   2. Dive times
   3. CSMD tables or decompression tables
   4. Distance, bearing, and transit times
   5. All known obstacles or hazards

B. **Environment**
   1. Weather conditions
   2. Water/air temperatures
   3. Water visibility
   4. Tides/currents
   5. Depth of water
   6. Bottom type
   7. Geographic location

C. **Personnel Assignments**
   1. Dive pairs
   2. Diving Supervisor
   3. Diving Officer
   4. Standby diver
   5. Diving medical personnel
   6. Base of operations support personnel

D. **Special Equipment for:**
   1. Divers (include thermal garments)
   2. Diving Supervisor
   3. Standby diver
   4. Medical personnel

E. **Review of Dive Signals**
   1. Hand signals

F. **Communications**
   1. Frequencies, primary/secondary
   2. Call signs

G. **Emergency Procedures**
   1. Symptoms of CO_2 buildup
   2. Review of management of CO_2 toxicity, hypoxia, chemical injury, unconscious diver
   3. UBA malfunction (refer to maintenance manual for detailed discussion)
      - Oxygen sensor failure
      - Low partial pressure of oxygen
      - High partial pressure of oxygen
      - Electronics failure
      - Low battery
      - Diluent free flow
      - Diluent addition valve failure
      - System flooding
   4. Lost swim pair procedures
   5. Omitted decompression plan
   6. Medical evacuation plan
      - Nearest available chamber
      - Nearest Diving Medical Officer
      - Transportation plan
      - Recovery of other swim pairs

H. **Times for Operations**

I. **Time Check**

17-7 UNDERWATER PROCEDURES

17-7.1 **General Guidelines.** The divers shall adhere to the following guidelines as the dive is conducted.

WARNING **Failure to adhere to these guidelines could result in serious injury or death.**

- Monitor primary and secondary display frequently (every 2-3 minutes).
- Wear adequate thermal protection.
- Know and use the proper amount of weights for the thermal protection worn and the equipment carried.
- Check each other's equipment carefully for leaks at the start of the dive.
- Do not exceed the UBA canister duration and depth limitations for the dive (paragraph 17-4.1.3).

- Minimize gas loss from the UBA (avoid mask leaks and frequent depth changes, if possible).
- Maintain frequent visual or touch checks with buddy.
- Be alert for symptoms suggestive of a medical disorder (paragraph 17-11).
- Use tides and currents to maximum advantage.

17-7.2 **At Depth.** If the UBA is performing normally at depth, no adjustments will be required. The ppO_2 control system will add oxygen from time to time. Monitor UBA primary and secondary displays and high pressure gauges in strict accordance with the MK 16 MOD 0 O&M manual. Items to monitor include:

- **Primary Display.** Primary Display. Check the primary display frequently to ensure that the oxygen level remains at the setpoint during normal activity at a constant depth.
- **Secondary Display.** Secondary Display. Check the secondary display frequently (every 2-3 minutes) to ensure that all sensors are consistent with the primary display and that plus and minus battery voltages are properly indicating.
- **High-Pressure Indicators.** Check the oxygen and diluent pressure indicators frequently to ensure that the gas supply is adequate to complete the dive.

17-8 ASCENT PROCEDURES

The maximum ascent rate for the MK 16 MOD 0 is 30 feet per minute. During ascent, when water pressure decreases, the diaphragm dump valve compensates for increased gas volume by discharging the excess gas into the water. As a result, oxygen in the breathing gas mixture may be vented faster than O_2 is replaced by the addition valve. In this case, the primary display may alternate red/green before the low ppO_2 signal (blinking red) appears. This is a normal transition period and shall not cause concern. Monitor the secondary display frequently on ascent and add oxygen by depressing the bypass valve during this instance.

17-9 POSTDIVE PROCEDURES

Postdive procedures shall be completed in accordance with the appropriate post-dive checklists in the MK 16 MOD 0 UBA O&M manual.

17-10 DECOMPRESSION PROCEDURES

When diving with an open-circuit UBA, ppO_2 increases with depth. With a closed circuit UBA, ppO_2 remains constant at a preset level regardless of depth. Therefore, standard U.S. Navy decompression tables cannot be used. The three methods to determine a MK16 MOD 0 diver's decompression obligation are listed below.

- **Navy Dive Computer.** The Navy Dive Computer (NDC) is a diver worn decompression computer that calculates the divers decompression obligation in real time. It is authorized for use with the MK16 MOD 0 UBA when air is used as a diluent. The NDC assumes the diver is breathing air at depths shallower than 78 fsw and is using a MK16 MOD 0 at deeper depths.

- **Combat Swimmer Multilevel Dive Tables.** Combat Swimmer Multilevel Dive (CSMD) procedures provide SPECWAR divers with the option of conducting multiple-depth diving with the MK 16 MOD 0 UBA to a depth of 70 fsw. However, the CSMD procedures may be used for dives between 70 and 110 fsw by adding 10 fsw to the depth when entering the table.

- **Constant 0.7 ata ppO_2 Decompression Tables.** The constant 0.7 ata ppO_2 decompression tables Oxygen in Nitrogen (Table 17-9) and Oxygen in Helium (Table 17-10) are discussed in paragraph 17-10.1 below. These tables were computed assuming an oxygen setpoint of 0.70 ata.

NOTE **Surface decompression is not authorized for MK 16 MOD 0 operations. Appropriate surface decompression tables have not been developed for constant 0.7 ata ppO_2 closed-circuit diving.**

17-10.1 Rules for Using 0.7 ata Constant ppO_2 in Nitrogen and in Helium Decompression Tables.

NOTE **The rules using the 0.7 ata ppO_2 tables are the same for nitrogen and helium; however, the tables are not interchangeable.**

- These tables are designed to be used with the MK 16 MOD 0 UBA (or any other constant ppO_2 closed-circuit UBA) with an oxygen setpoint of 0.70 ata or greater.

- When using helium as the inert gas, the amount of nitrogen must be minimized in the breathing loop. Flush the UBA well with helium-oxygen using the purge procedure given in the MK 16 MOD 0 UBA O&M manual.

- Tables are grouped by depth. Within each decompression table, exceptional exposure dives are separated by a dashed line. These tables are designed to be dived to the exceptional exposure line. Exceptional exposure schedules are provided in case of unforseen circumstances when a diver might experience an inadvertent downward excursion or for an unforeseen reason overstay the planned bottom time. Planned exceptional exposure dives require prior CNO approval.

- Tables/schedules are selected according to the maximum depth obtained during the dive and the bottom time (time from leaving the surface to leaving the bottom).

- Monitoring ppO_2. During decompression, it is very important to frequently monitor the secondary display and ensure a 0.75 ata ppO_2 is maintained as closely as possible. Always use the appropriate decompression table when surfacing, even if UBA malfunction has significantly altered the ppO_2.

- General rules for using these tables are the same as for the air decompression tables and include the use of the RNT exception rule when calculating the equivalent single dive time for repetitive dives.

1. Enter the table at the listed depth that is exactly equal to or is next greater than the maximum depth attained during the dive.

2. Select the bottom time from those listed for the selected depth that is exactly equal to or is next greater than the bottom time of the dive.

3. Never attempt to interpolate between decompression schedules.

4. Use the decompression stops listed for the selected bottom time.

5. Ensure that the diver's chest is maintained as close as possible to each decompression depth for the number of minutes listed.

6. Maximum ascent rate is 30 feet per minute. The rules for compensating for variations in the rate of ascent are identical to those for air diving (see Chapter 9, paragraph 9-11).

7. Begin timing the first stop when the diver arrives at the stop. For all subsequent stops, begin timing the stop when the diver leaves the previous stop. Ascent time between stops is included in the subsequent stop time.

8. The last stop may be taken at 20 fsw if desired. After completing the prescribed 20 fsw stop, remain at any depth between 10 fsw and 20 fsw inclusive for the 10 fsw stop time as noted in the appropriate decompression table.

9. Use the appropriate decompression table for the selected decompression method unless an emergency or equipment malfunction has occurred. Interpolating between different methods of decompression in order to shorten the decompression obligation is not authorized.

10. When selecting the proper decompression table, all dives within the past 18 hours must be considered. Repetitive dives are allowed. Repetitive diving decompression procedures vary depending on the breathing medium(s) selected for past dives and for the current dive. If a dive resulted in breathing from the an alternate air supply then no repetitive dives shall be made within the next 18 hours. Refer to the following tables and figures for repetitive diving.

    - Table 17-5 for Repetitive Dive Procedures for Various Gas Mediums.

    - Figure 17-5 for the Dive Worksheet for Repetitive 0.7 ata Constant Partial Pressure Oxygen in Nitrogen Dives.

    - Table 17-6 for the No-Decompression Limits and Repetitive Group Designation Table for No-Decompression 0.7 ata Constant Partial Pressure Oxygen in Nitrogen Dives.

    - Table 17-7 for the Residual Nitrogen Timetable for Repetitive 0.7 ata Constant Partial Pressure Oxygen in Nitrogen Dives.

Table 17-5. *Repetitive Dive Procedures for Various Gas Mediums.*

WARNING
No repetitive dives are authorized after an emergency procedure requiring a shift to the EBS.

Selection of Repetitive Procedures for Various Gas Mediums

Previous Breathing Medium (Refer to Notes 1, 2, and 3)	Repetitive Dive Breathing Medium	Note
N_2O_2	N_2O_2	A
Air	N_2O_2	B
N_2O_2	Air	C
HeO_2	HeO_2	D
HeO_2	Air	E
Air	HeO_2	F
HeO_2	N_2O_2	G
N_2O_2	HeO_2	H

Notes:

1. If a breathing medium containing helium was breathed at any time during the 18-hour period immediately preceding a dive, use HeO_2 as the previous breathing medium.

2. If 100 percent oxygen rebreathers are used on a dive in conjunction with other breathing gases, treat that portion of the dive as if 0.7 ATA O_2 in N_2 was breathed.

3. If both air and 0.7 ATA O_2 in N_2 are breathed during a dive, treat the entire dive as an air dive. If the 0.7 ata O_2 in N_2 is breathed at depths 80 fsw or deeper, add the following correction factors to the maximum depth when selecting the appropriate air table.

Maximum Depth on N_2O_2	Correction Factor
Not exceeding 80 FSW	0
81-99	Plus 5
100-119	Plus 10
120-139	Plus 15
140-150	Plus 20

Notes:

A. (1) If the surface interval is less than 10 minutes, determine the table and schedule for the repetitive dive by adding the bottom times and taking the deepest depth of all the dives in the series, including the planned repetitive dive.

Table 17-5. *Repetitive Dive Procedures for Various Gas Mediums. (Continued)*

Notes continued:

A. (2) If the surface interval is longer than 10 minutes, use the repetitive dive worksheet (Figure 17-5) to determine the table and schedule:

a) Determine the repetitive group letter for the depth and time of dive conducted from Table 17-6 for no-decompression dives or from Table 17-9 for decompression dives. If the exact time or depth is not found, go to the longer time or the next deeper depth.

b) Locate the repetitive group letter in Table 17-7. Move across the table to the correct surface interval time. Move down to the bottom of the column for the new group designation.

c) Move down the column of the new group designation to the depth of the planned dive. This is the residual Nitrogen time (RNT). Add this to the planned bottom time of the next dive to find the decompression schedule and the new group designation.

(3) The RNT exception rule applies to repetitive MK 16 MOD 0 diving. Determine the table and schedule for the repetitive dive by adding the bottom times and taking the deepest depth of all the MOD 0 dives in the series, including the planned repetitive dive. If the resultant table and schedule requires less decompression than the table and schedule obtained using the repetitive dive worksheet, it may be used instead of the worksheet table and schedule.

B. Use the repetitive group designation from the air decompression table or the no-decompression limits and repetitive group designation table for no-decompression air dives to enter Table 17-7. Compute the RNT as in Note A. Do not use the residual nitrogen timetable for repetitive air dives to find the RNT. The RNT exception rule applies to repetitive air/MK 16 MOD 0 diving. In order to apply the RNT exception rule, convert the depth of any air dive in the series to its equivalent MK 16 MOD 0 depth before taking the deepest depth in the series. Equivalent MOD 0 Depth = (0.79 x Depth on Air) + 18 fsw.

C. (1) Determine the repetitive group designation for depth and time of dive conducted from Table 17-6 or Table 17-9. If the exact time or depth is not found, go to the next longer time or the next deeper depth.

(2) Using the repetitive group designator, enter Table 9-8 on the diagonal. Move across the table to the correct surface interval time. Move down to the bottom of the column for the new repetitive group designation.

(3) Continue to read down the column of Table 9-8 to the depth that is exactly equal to or greater than the depth of the repetitive dive to find the RNT.

D. Add the bottom time of the planned repetitive dive to the sum of the bottom times for all dives within the past 18 hours to get the adjusted bottom time. Use the maximum depth attained within the past 18 hours and the adjusted bottom time to select the appropriate schedule from Table 17-10.

E. Add the bottom times of all dives within the past 18 hours to get an adjusted bottom time. Using the air decompression table, find the maximum depth attained during the past 18 hours and the adjusted bottom time. The repetitive group from this air table may then be used as the surfacing repetitive group from the last dive. The residual nitrogen timetable for repetitive air dives is used to find the repetitive group at the end of the current surface interval and the appropriate residual nitrogen time for the repetitive air dive.

F. Compute the RNT from the residual nitrogen timetable for repetitive air dives using the depth of the planned dive. Add the RNT to the planned bottom time to get the adjusted bottom time. Use Table 17-10 for the adjusted bottom time at the planned depth.

G. Add the bottom times of all dives within the past 18 hours to get an adjusted bottom time. Using Table 17-9, find the maximum depth attained during the past 18 hours and the adjusted bottom time. The repetitive group from the table may then be used as the surfacing repetitive group from the last dive. Table 17-7 is used to find the repetitive group at the end of the current surface interval and the appropriate RNT for the current dive.

H. Compute the RNT from Table 17-7 using the depth of the planned dive. Add the RNT to the planned bottom time to get the adjusted bottom time. Use Table 17-10 for the adjusted bottom time at the planned depth.

REPETITIVE DIVE WORKSHEET
FOR
0.7 ATA N_2O_2 DIVES

Part 1. Previous Dive:

______________ minutes

______________ feet

______________ repetitive group designator from Table 17-6 or 17-9

Part 2. Surface Interval:

______________ hours ______________ minutes on the surface

______________ final repetitive group from Table 17-7

Part 3. Equivalent Single Dive Time:

Enter Table 17-10 at the depth row for the new dive and the column of the final repetitive group to find the corresponding Residual Nitrogen Time (RNT).

______________ minutes RNT

\+ ______________ minutes planned bottom time

= ______________ minutes equivalent single dive time

Part 4. Decompression Schedule for the Repetitive Dive:

______________ minutes equivalent single dive time from Part 3

______________ feet, depth of the repetitive dive.

Ensure RNT exception rule does not apply.

Figure 17-5. Dive Worksheet for Repetitive 0.7 ata Constant Partial Pressure Oxygen in Nitrogen Dives.

Table 17-6. *No-Decompression Limits and Repetitive Group Designation Table for 0.7 ata Constant ppO_2 in Nitrogen Dives.*

Depth (fsw)	No-Stop Limit	Repetitive Group Designator															
		A	B	C	D	E	F	G	H	I	J	K	L	M	N	O	Z
10	Unlimited	–	–	–	–	–	–	–	–	–	–	–	–	–	–	–	–
15	Unlimited	–	–	–	–	–	–	–	–	–	–	–	–	–	–	–	–
20	Unlimited	154	425	*													
30	Unlimited	31	50	73	98	128	165	211	274	375	643	*					
40	369	17	27	38	50	63	76	91	107	125	144	167	192	223	259	305	369
50	143	12	19	26	33	41	50	59	68	78	88	99	111	123	137	143	
60	74	9	14	19	25	31	37	43	50	56	63	71	74				
70	51	7	11	15	20	25	29	34	39	44	50	51					
80	40	6	9	13	16	20	24	28	32	36	40						
90	32	5	8	11	14	17	20	24	27	31	32						
100	27	4	7	9	12	15	18	21	24	27							
110	23	3	6	8	11	13	16	18	21	23							
120	20	3	5	7	9	12	14	16	18	20							
130	16		4	6	8	10	12	14	16								
140	14		4	6	7	9	11	13	14								
150	11		3	5	7	8	10	11									
Exceptional Exposure																	
160	10		3	4	6	8	9	10									
170	9		3	4	5	7	8	9									

– Diver does not acquire a repetitive group designator during dives to these depths.

* Highest repetitive group that can be achieved at this depth regardless of bottom time.

Table 17-7. *Residual Nitrogen Timetable for Repetitive 0.7 ata Constant ppO_2 in Nitrogen Dives.*

Locate the diver's repetitive group designation from his previous dive along the diagonal line above the table. Read horizontally to the interval in which the diver's surface interval lies.

Next, read vertically downward to the new repetitive group designation. Continue downward in this same column to the row that represents the depth of the repetitive dive. The time given at the intersection is residual nitrogen time, in minutes, to be applied to the repetitive dive.

* Dives following surface intervals longer than this are not repetitive dives. Use actual bottom times in the Table 17-9 to compute decompression for such dives.

Repetitive Group at Beginning of Surface Interval (rows) / **Repetitive Group at the End of the Surface Interval** (columns)

	Z	O	N	M	L	K	J	I	H	G	F	E	D	C	B	A
A																0:10 2:20 *
B															0:10 1:16	1:17 3:36 *
C														0:10 0:55	0:56 2:11	2:12 4:31 *
D													0:10 0:52	0:53 1:47	1:48 3:03	3:04 5:23 *
E												0:10 0:52	0:53 1:44	1:45 2:39	2:40 3:55	3:56 6:15 *
F											0:10 0:52	0:53 1:44	1:45 2:37	2:38 3:31	3:32 4:48	4:49 7:08 *
G										0:10 0:52	0:53 1:44	1:45 2:37	2:38 3:29	3:30 4:23	4:24 5:40	5:41 8:00 *
H									0:10 0:52	0:53 1:44	1:45 2:37	2:38 3:29	3:30 4:21	4:22 5:16	5:17 6:32	6:33 8:52 *
I								0:10 0:52	0:53 1:44	1:45 2:37	2:38 3:29	3:30 4:21	4:22 5:13	5:14 6:08	6:09 7:24	7:25 9:44 *
J							0:10 0:52	0:53 1:44	1:45 2:37	2:38 3:29	3:30 4:21	4:22 5:13	5:14 6:06	6:07 7:00	7:01 8:16	8:17 10:36 *
K						0:10 0:52	0:53 1:44	1:45 2:37	2:38 3:29	3:30 4:21	4:22 5:13	5:14 6:06	6:07 6:58	6:59 7:52	7:53 9:09	9:10 11:29 *
L					0:10 0:52	0:53 1:44	1:45 2:37	2:38 3:29	3:30 4:21	4:22 5:13	5:14 6:06	6:07 6:58	6:59 7:50	7:51 8:44	8:45 10:01	10:02 12:21 *
M				0:10 0:52	0:53 1:44	1:45 2:37	2:38 3:29	3:30 4:21	4:22 5:13	5:14 6:06	6:07 6:58	6:59 7:50	7:51 8:42	8:43 9:37	9:38 10:53	10:54 13:13 *
N			0:10 0:52	0:53 1:44	1:45 2:37	2:38 3:29	3:30 4:21	4:22 5:13	5:14 6:06	6:07 6:58	6:59 7:50	7:51 8:42	8:43 9:34	9:35 10:29	10:30 11:45	11:46 14:05 *
O		0:10 0:52	0:53 1:44	1:45 2:37	2:38 3:29	3:30 4:21	4:22 5:13	5:14 6:06	6:07 6:58	6:59 7:50	7:51 8:42	8:43 9:34	9:35 10:27	10:28 11:21	11:22 12:37	12:38 14:58 *
Z	0:10 0:52	0:53 1:44	1:45 2:37	2:38 3:29	3:30 4:21	4:22 5:13	5:14 6:06	6:07 6:58	6:59 7:50	7:51 8:42	8:43 9:34	9:35 10:27	10:28 11:19	11:20 12:13	12:14 13:30	13:31 15:50 *

Dive Depth	Z	O	N	M	L	K	J	I	H	G	F	E	D	C	B	A
10	–	–	–	–	–	–	–	–	–	–	–	–	–	–	–	–
15	–	–	–	–	–	–	–	–	–	–	–	–	–	–	–	–
20	**	**	**	**	**	**	**	**	**	**	**	**	**	**	420	153
30	**	**	**	**	**	**	626	372	273	211	165	129	99	73	51	31
40	365	303	258	222	192	167	144	125	107	91	77	63	51	39	28	18
50	167	151	137	123	111	99	88	78	68	59	50	42	34	27	19	12
60	113	104	95	87	79	71	64	57	50	44	38	32	26	20	15	10
70	86	79	73	67	61	56	50	45	40	35	30	25	21	16	12	8
80	69	64	60	55	50	46	41	37	33	29	25	21	18	14	10	7
90	58	54	50	46	43	39	35	32	28	25	22	18	15	12	9	6
100	50	47	44	40	37	34	31	28	25	22	19	16	13	11	8	5
110	44	41	38	36	33	30	27	25	22	19	17	14	12	9	7	5
120	39	37	34	32	29	27	25	22	20	18	15	13	11	9	6	4
130	36	33	31	29	27	24	22	20	18	16	14	12	10	8	6	4
140	33	30	28	26	24	22	20	18	17	15	13	11	9	7	5	4
150	30	28	26	24	22	21	19	17	15	14	12	10	8	7	5	3
160	28	26	24	23	21	19	18	16	14	13	11	9	8	6	5	3
170	26	24	23	21	19	18	16	15	13	12	10	9	7	6	4	3

Residual Nitrogen Times (Minutes)

– Repetitive dives to these depths are equivalent to remaining on the surface. Add the bottom time of the dive to the preceding surface interval. Use the Surface Interval Credit Table (SICT) to determine the repetitive group at the end of the dive.

** Residual Nitrogen Time cannot be determined using this table (see paragraph 9-9.1 for instructions).

11. The partial pressure of nitrogen in the MK 16 MOD 0 UBA at depths up to 15 fsw is lower than the partial pressure of nitrogen in air at the surface. A diver diving to these depths, therefore, loses rather than gains body nitrogen during the dive. Accordingly, the diver does not acquire a repetitive group designator when making these shallow dives. If the dive is a repetitive dive up to 15 fsw, the diver will lose more nitrogen during the repetitive dive than if he remained on the surface. The dive can be considered the equivalent of remaining on the surface for the duration of the dive. The repetitive group designator at the end of the repetitive dive can be determined by adding the bottom time of the repetitive dive to the preceding surface interval, then using the surface interval credit table to determine the ending repetitive group.

17-10.2 **PPO_2 Variances.** The ppO_2 in the MK 16 UBAs is expected to vary slightly from 0.6 - 0.9 ata for irregular brief intervals. This does not constitute a rig malfunction.

When addition of oxygen to the UBA is manually controlled, ppO_2 should be maintained in accordance with techniques and emergency procedures listed in the MK 16 MOD 0 O&M manual.

The Diving Supervisor and medical personnel should recognize that a diver who has been breathing a mixture with ppO_2 lower than 0.6 ata for any length of time may have a greater risk of developing decompression sickness. Once the diver reaches surface he will be given a neurological exam and observed for an hour. The diver will not require recompression treatment unless symptoms of decompression sickness occur.

17-10.3 **Emergency Breathing System (EBS).** When planning a MK16 MOD 0 decompression dive, the Diving Supervisor must ensure an alternate air source is available to the diver in the event of a MK 16 failure. The air source must be sufficient to allow the diver to complete his decompression obligation as determined below. See Chapter 7 for procedures to calculate the volume of air required.

17-10.3.1 **Emergency Decompression on Air.** In emergency situations (e.g., UBA flood-out or failure), the diver should immediately ascend to the first decompression stop according to the original decompression schedule and shift to the alternate air supply. An alternate air supply can be any ANU approved SCUBA bottle(s) and regulator. The subsequent decompression is modified according to the diluent gas originally breathed.

- Helium-Oxygen Diluent. Follow the original HeO_2 decompression schedule without modification while breathing air.

- Nitrogen-Oxygen (Air) Diluent. Double all remaining decompression stops while breathing air. If the switch to emergency air is made while at a decompression stop, then double the remaining time at that stop and all shallower stops. If the dive falls within the no-decompression limit and a switch to an alternate air supply has occurred, a mandatory 10-minute stop at 20 fsw is required.

If either of these procedures is used, the diver will be given a neurological exam and observed on the surface for one hour. The diver will not require recompression treatment unless symptoms of decompression sickness occur.

17-10.4 Asymptomatic Omitted Decompression. Certain emergencies may interrupt or prevent specified decompression. UBA failure, exhausted diluent or oxygen gas supply, and bodily injury are examples that constitute such emergencies. The omitted decompression procedures for an asymptomatic MK 16 MOD 0 diver are contained in Table 17-8. If the diver switches from the MK 16 MOD 0 to an alternate air source then the decompression obligations must also be modified in accordance with paragraph 17-10.3.1.

Table 17-8. *Management of Asymptomatic Omitted Decompression MK 16 MOD 0 Diver.*

Deepest Decompression Stop Omitted	Decompression Status	Surface Interval	Action	
			Chamber Available	**No Chamber Available**
None	No decompression stops required	NA	Observe on surface for 1 hour	Observe on surface for 1 hour
20 fsw or shallower	Decompression stops required	<1 min	Return to depth of stop. Increase stop time by 1 minute. Resume decompression according to original schedule.	Return to depth of stop. Increase stop time by 1 minute. Resume decompression according to original schedule.
		>1 min	Return to depth of stop. Multiply 20-fsw and/or 10-fsw stop times by 1.5. Resume decompression. Or: Treatment Table 5 for surface interval <5 min Or: Treatment Table 6 for surface interval >5 min	Return to depth of stop. Multiply 20-fsw and/or 10-fsw stop times by 1.5. Resume decompression.
Deeper than 20 fsw	Decompression stops required (<30 min missed)	<5 min	Treatment Table 5	Descend to the deepest stop omitted. Multiply all stops 40 fsw and shallower by 1.5. Resume decompression.
		>5 min	Treatment Table 6	Descend to the deepest stop omitted. Multiply all stops 40 fsw and shallower by 1.5. Resume decompression.
	Decompression stops required (>30 min missed)	Any	Treatment Table 6	Descend to the deepest stop omitted. Multiply all stops 40 fsw and shallower by 1.5. Resume decompression.

17-10.5 Symptomatic Omitted Decompression. If the diver shows evidence of decompression sickness or arterial gas embolism before recompression for omitted decompression can be carried out, immediate treatment using the appropriate oxygen or air treatment table is essential. Guidance for table selection and use is given in Chapter 20.

17-11 MEDICAL ASPECTS OF CLOSED-CIRCUIT MIXED-GAS UBA

When using a closed-circuit mixed-gas UBA, the diver is susceptible to the usual diving-related illnesses (i.e., decompression sickness, arterial gas embolism, barotraumas, etc.). Only the diving disorders that merit special attention for closed-circuit mixed gas divers are addressed in this chapter. Refer to Chapter 3 for a detailed discussion of diving related physiology and related disorders.

17-11.1 **Central Nervous System (CNS) Oxygen Toxicity.** High pressure oxygen poisoning is known as CNS oxygen toxicity. High partial pressures of oxygen are associated with many biochemical changes in the brain, but which specific changes are responsible for the signs and symptoms of CNS oxygen toxicity is presently unknown. CNS oxygen toxicity is not likely to occur at oxygen partial pressures below 1.3 ata, though relatively brief exposure to partial pressures above this, when it occurs at depth or in a pressurized chamber, can result in CNS oxygen toxicity causing CNS-related symptoms.

17-11.1.1 **Causes of CNS Oxygen Toxicity.** Factors that increase the likelihood of CNS oxygen toxicity are:

- Increased partial pressure of oxygen.
- Increased time of exposure.
- Prolonged immersion.
- Stress from strenuous physical exercise.
- Carbon dioxide buildup. The increased risk for CNS oxygen toxicity may occur even before the diver is aware of any symptoms of carbon dioxide buildup.
- Cold stress resulting from shivering or an increased exercise rate as the diver attempts to keep warm.
- Systemic diseases that increase oxygen consumption. Conditions associated with increased metabolic rates (such as certain thyroid or adrenal disorders) tend to cause an increase in oxygen sensitivity. Divers with these diseases should be excluded from mixed gas diving.

17-11.1.2 **Symptoms of CNS Oxygen Toxicity.** The symptoms of CNS oxygen toxicity may not always appear and most are not exclusively symptoms of oxygen toxicity. The most serious symptom of CNS oxygen toxicity is convulsion, which may occur suddenly without any previous symptoms, and may result in drowning or arterial gas embolism. Twitching is perhaps the clearest warning of oxygen toxicity, but it may occur late if at all. The mnemonic device VENTID-C is a helpful reminder of the most common symptoms of CNS oxygen toxicity. The appearance of any one of these symptoms usually represents a bodily signal of distress of some kind and should be heeded.

V: Visual symptoms. Tunnel vision, a decrease in the diver's peripheral vision, and other symptoms, such as blurred vision, may occur.

E: Ear symptoms. Tinnitus is any sound perceived by the ears but not resulting from an external stimulus. The sound may resemble bells ringing, roaring, or a machinery-like pulsing sound.

N: Nausea or spasmodic vomiting. These symptoms may be intermittent.

T: Twitching and tingling symptoms. Any of the small facial muscles, lips, or muscles of the extremities may be affected. These are the most frequent and clearest symptoms.

I: Irritability. Any change in the diver's mental status; including confusion, agitation, and anxiety.

D: Dizziness. Symptoms include clumsiness, incoordination, and unusual fatigue.

C: Convulsions.

The following additional factors should be noted regarding an oxygen convulsion:

- The diver is unable to carry on any effective breathing during the convulsion.
- After the diver is brought to the surface, there will be a period of unconsciousness or neurologic impairment following the convulsion; these symptoms are indistinguishable from those of arterial gas embolism.
- No attempt should be made to insert any object between the clenched teeth of a convulsing diver. Although a convulsive diver may suffer a lacerated tongue, this trauma is preferable to the trauma that may be caused during the insertion of a foreign object. In addition, the person providing first aid may incur significant hand injury if bitten by the convulsing diver.
- There may be no warning of an impending convulsion to provide the diver the opportunity to return to the surface. Therefore, buddy lines are essential to safe closed-circuit mixed gas diving.

17-11.1.3 **Treatment of Non-Convulsive Symptoms.** If non-convulsive symptoms of CNS oxygen toxicity occur, action must be taken immediately to lower the oxygen partial pressure. Such actions include:

- Ascend. Dalton's law will lower the oxygen partial pressure.
- Add diluent to the breathing loop.
- Secure the oxygen cylinder if oxygen addition is uncontrolled.

Though an ascent from depth will lower the partial pressure of oxygen, the diver may still suffer other or worsening symptoms. The divers should notify the Diving Supervisor and terminate the dive.

17-11.1.4 **Treatment of Underwater Convulsion.** The following steps should be taken when treating a convulsing diver:

1. Assume a position behind the convulsing diver. Release the victim's weight belt only if progress to the surface is significantly impeded.

2. Do not ascend in the water until the convulsion subsides.

3. Open the victim's airway and leave the mouthpiece in his mouth. If it is not in his mouth, do not attempt to replace it; however, ensure that the mouthpiece is switched to the SURFACE POSITION to prevent unnecessary negative buoyancy from a flooded UBA.

4. Grasp the victim around his chest above the UBA or between the UBA and his body. If difficulty is encountered in gaining control of the victim in this manner, the rescuer should use the best method possible to obtain control.

5. Ventilate the UBA with diluent to lower the ppO_2 and maintain depth until the convulsion subsides.

6. Make a controlled ascent to the first decompression stop, maintaining a slight pressure on the diver's chest to assist exhalation.

- If the diver regains control, continue with appropriate decompression.
- If the diver remains incapacitated, surface at a moderate rate, establish an airway, and treat for symptomatic omitted decompression as outlined in paragraph 17-10.5.
- Frequent monitoring of the primary and secondary displays as well as the oxygen- and diluent-bottle pressure gauges will keep the diver well informed of his breathing gas and rig status.

7. If additional buoyancy is required, activate the victim's life jacket. The rescuer should not release his own weight belt or inflate his life jacket.

8. Upon reaching the surface, inflate the victim's life jacket if not previously done.

9. Remove the victim's mouthpiece and switch the valve to SURFACE to prevent the possibility of the rig flooding and weighing down the victim.

10. Signal for emergency pickup.

11. Ensure the victim is breathing. Mouth-to-mouth breathing may be initiated if necessary.

12. If an upward excursion occurred during the actual convulsion, transport to the nearest chamber and have the victim evaluated by an individual trained to recognize and treat diving-related illness.

17-11.1.5 **Prevention of CNS Oxygen Toxicity.** All predive checks must be performed to ensure proper functioning of the oxygen sensors and the oxygen-addition valve.

Frequent monitoring of both the primary and secondary displays will help ensure that the proper ppO_2 is maintained.

17-11.1.6 **Off-Effect.** The off-effect, a hazard associated with CNS oxygen toxicity, may occur several minutes after the diver comes off gas or experiences a reduction of oxygen partial pressure. The off-effect is manifested by the onset or worsening of CNS oxygen toxicity symptoms. Whether this paradoxical effect is truly caused by the reduction in partial pressure or whether the association is coincidental is unknown.

17-11.2 **Pulmonary Oxygen Toxicity.** Pulmonary oxygen toxicity can result from prolonged exposure to elevated partial pressures of oxygen. This form of oxygen toxicity produces lung irritation with symptoms of chest pain, cough, and pain on inspiration that develop slowly and become increasingly worse as long as the elevated level of oxygen is breathed. Although hyperbaric oxygen may cause serious lung damage, if the oxygen exposure is discontinued before the symptoms become too severe, the symptoms will slowly abate. This form of oxygen toxicity is generally seen during oxygen recompression treatment and saturation diving, and on long, shallow, in-water oxygen exposures.

17-11.3 **Oxygen Deficiency (Hypoxia).** Hypoxia is an abnormal deficiency of oxygen in the arterial blood in which the partial pressure of oxygen is too low to meet the metabolic needs of the body. Chapter 3 contains an in-depth description of this disorder. Although all cells in the body need oxygen, the initial symptoms of hypoxia are a manifestation of central nervous system dysfunction.

17-11.3.1 **Causes of Hypoxia.** The primary cause of hypoxia for a MK16 diver is failure of the oxygen addition valve or primary electronics. However, during a rapid ascent Dalton's law may cause the ppO_2 to fall faster than can be compensated for by the oxygen-addition system. If, during ascent, low levels of oxygen are displayed, slow the ascent and add oxygen if necessary. Depletion of the oxygen supply or malfunctioning oxygen sensors can also lead to a hypoxic gas mixture.

17-11.3.2 **Symptoms of Hypoxia.** Hypoxia may have no warning symptoms prior to loss of consciousness. Other symptoms that may appear include confusion, loss of coordination, dizziness, and convulsion. It is important to note that if symptoms of unconsciousness or convulsion occur at the beginning of a closed-circuit dive, hypoxia, not oxygen toxicity, is the most likely cause.

17-11.3.3 **Treating Hypoxia.** If symptoms of hypoxia develop, the diver must take immediate action to raise the oxygen partial pressure. If unconsciousness occurs, the buddy diver should add oxygen to the rig while monitoring the secondary display. If the diver does not require decompression, the buddy diver should bring the afflicted diver to the surface at a moderate rate, remove the mouthpiece or mask, and have him breathe air. If the event was clearly related to hypoxia and the diver recovers fully with normal neurological function shortly after breathing surface air, the diver does not require treatment for arterial gas embolism.

17-11.3.4 **Treatment of Hypoxic Divers Requiring Decompression.** If the divers require decompression, the buddy diver should bring the afflicted diver to the first decompression stop.

- If consciousness is regained, continue with normal decompression.
- If consciousness is not regained, ascend to the surface at a moderate rate (not to exceed 30 fpm), establish an airway, administer 100-percent oxygen, and treat for symptomatic omitted decompression as outlined in paragraph 17-10.5. If possible, immediate assistance from the standby diver should be obtained and the unaffected diver should continue normal decompression.

17-11.4 Carbon Dioxide Toxicity (Hypercapnia). Carbon dioxide toxicity, or hypercapnia, is an abnormally high level of carbon dioxide in the blood and body tissues.

17-11.4.1 **Causes of Hypercapnia.** Hypercapnia is generally a result of the failure of the carbon dioxide-absorbent material. The failure may be a result of channeling, flooding or saturation of the absorbent material. Skip breathing or controlled ventilation by the diver, which results in an insufficient removal of CO_2 from the diver's body, may also cause hypercapnia.

17-11.4.2 **Symptoms of Hypercapnia.** Symptoms of hypercapnia are:

- Increased breathing rate
- Shortness of breath, sensation of difficult breathing or suffocation (dyspnea)
- Confusion or feeling of euphoria
- Inability to concentrate
- Increased sweating
- Drowsiness
- Headache
- Unconsciousness.

WARNING **Hypoxia and hypercapnia may give the diver little or no warning prior to onset of unconsciousness.**

Symptoms are dependent on the partial pressure of carbon dioxide, which is a function of both the fraction of carbon dioxide and the absolute pressure. Thus, symptoms would be expected to increase as depth increases. The presence of a high partial pressure of oxygen may also reduce the early symptoms of hypercapnia. Elevated levels of carbon dioxide may result in an episode of CNS oxygen toxicity on a normally safe dive profile.

17-11.4.3 **Treating Hypercapnia.** If symptoms of hypercapnia develop, the diver should:

- Immediately stop work and take several deep breaths.
- Increase ventilation if skip-breathing is a possible cause.
- Ascend. This will reduce the partial pressure of carbon dioxide both in the rig and the lungs.
- If symptoms do not rapidly abate, the diver should abort the dive.
- During ascent, while maintaining a vertical position, the diver should activate his bypass valve, adding fresh gas to his UBA. If the symptoms are a result of canister floodout, an upright position decreases the likelihood that the diver will sustain chemical injury.
- If unconsciousness occurs at depth, the same principles of management for underwater convulsion as described in paragraph 17-11.1.4 apply.

17-11.4.4 **Prevention of Hypercapnia.** To minimize the risk of hypercapnia:

- Use only an approved carbon dioxide absorbent in the UBA canister.
- Follow the prescribed canister-filling procedure to ensure that the canister is correctly packed with carbon dioxide absorbent.
- Dip test the UBA carefully before the dive. Watch for leaks that may result in canister floodout.
- Do not exceed canister duration limits for the water temperature.
- Ensure that the one-way valves in the supply and exhaust hoses are installed and working properly.
- Swim at a relaxed, comfortable pace.
- Avoid skip-breathing. There is no advantage to this type of breathing in a closed-circuit rig and it may cause elevated blood carbon dioxide levels even with a properly functioning canister.

17-11.5 **Chemical Injury.** The term chemical injury refers to the introduction of a caustic solution from the carbon dioxide scrubber of the UBA into the upper airway of a diver.

17-11.5.1 **Causes of Chemical Injury.** A caustic alkaline solution results when water leaking into the canister comes in contact with the carbon dioxide absorbent. When the diver is in a horizontal or head down position, this solution may travel through the inhalation hose and irritate or injure the upper airway.

17-11.5.2 **Symptoms of Chemical Injury.** Before actually inhaling the caustic solution, the diver may experience labored breathing or headache, which are symptoms of carbon dioxide buildup in the breathing gas. This occurs because an accumulation of the caustic solution in the canister may be impairing carbon dioxide absorption. If the problem is not corrected promptly, the alkaline solution may travel into the breathing hoses and consequently be inhaled or swallowed. Choking, gagging, foul taste, and burning of the mouth and throat may begin immediately. This condition is sometimes referred to as a "caustic cocktail." The extent of the injury depends on the amount and distribution of the solution.

17-11.5.3 **Management of a Chemical Incident.** If the caustic solution enters the mouth, nose, or face mask, the diver must take the following steps:

- Immediately assume an upright position in the water.
- Depress the manual diluent bypass valve continuously.
- If the dive is a no-decompression dive, make a controlled ascent to the surface, exhaling through the nose to prevent overpressurization.
- If the dive requires decompression, shift to the EBS or another alternative breathing supply. If it is not possible to complete the planned decompression, surface the diver and treat for omitted decompression as outlined in paragraph 17-10.4.

Using fresh water, rinse the mouth several times. Several mouthfuls should then be swallowed. If only sea water is available, rinse the mouth but do not swallow. Other fluids may be substituted if available, but the use of weak acid solutions (vinegar or lemon juice) is not recommended. Do not attempt to induce vomiting.

A chemical injury may cause the diver to have difficulty breathing properly on ascent. He should be observed for signs of an arterial gas embolism and should be treated if necessary. A victim of a chemical injury should be evaluated by a physician or corpsman as soon as possible. Respiratory distress which may result from the chemical trauma to the air passages requires immediate hospitalization.

17-11.5.4 **Prevention of Chemical Injury.** Chemical injuries are best prevented by the performance of a careful dip test during predive set-up to detect any system leaks. Special attention should also be paid to the position of the mouthpiece rotary valve upon water entry and exit to prevent the entry of water into the breathing loop. Additionally, dive buddies should perform a careful leak check on each other before leaving the surface at the start of a dive.

17-11.6 Decompression Sickness in the Water. Decompression sickness may develop in the water during MK 16 MOD 0 diving. The symptoms of decompression sickness may be joint pain or may be more serious manifestations such as numbness, loss of muscular function, or vertigo.

Managing decompression sickness in the water will be difficult in the best of circumstances. Only general guidance can be presented here. Management decisions must be made on site, taking into account all known factors. The advice of a Diving Medical Officer should be sought whenever possible.

17-11.6.1 **Diver Remaining in Water.** If the diver signals that he has decompression sickness but feels that he can remain in the water:

1. Dispatch the standby diver to assist.
2. Have the diver descend to the depth of relief of symptoms in 10-fsw increments, but no deeper than two increments (i.e., 20 fsw).
3. Compute a new decompression profile by multiplying all stops by 1.5. If recompression went deeper than the depth of the first stop on the original decompression schedule, use a stop time equal to 1.5 times the first stop in the original decompression schedule for the one or two stops deeper than the original first stop.
4. Ascend on the new profile.
5. Lengthen stops as needed to control symptoms.
6. Upon surfacing, transport the diver to the nearest chamber. If he is asymptomatic, treat on Treatment Table 5. If he is symptomatic, treat in accordance with the guidance given in Chapter 20.

17-11.6.2 **Diver Leaving the Water.** If the diver signals that he has decompression sickness but feels that he cannot remain in the water:

1. Surface the diver at a moderate rate (not to exceed 30 fpm).
2. If a recompression chamber is on site (i.e., within 30 minutes), recompress the diver immediately. Guidance for treatment table selection and use is given in Chapter 20.
3. If a recompression chamber is not on site, follow the management guidance given in Volume 5.

17-11.7. Altitude Diving Procedures and Flying After Diving

Ascent to altitude following a MK 16 MOD 0 dive at sea level will increase the risk of decompression sickness if the interval on the surface before ascent is not long enough to permit excess nitrogen or helium to be eliminated from the body. To determine the safe surface interval before ascent, take the following steps:

- **Nitrogen-Oxygen Dives**
  1. Determine the highest repetitive group designator obtained in the previous 24-hour period using either Table 17-6 or Table 17-9.

2. Using the highest repetitive group designator, enter Table 9-6 in Chapter 9. Read across the row to the altitude that is exactly equal to or next higher than the planned change in altitude. The required surface interval is given at the intersection of the row and the column.

- **Helium-Oxygen Dives**

1. For no-decompression dives with bottom times less than 2 hours, wait 12 hours on the surface before ascending to altitude.

2. For no-decompression dives with bottom times greater than 2 hours or for decompression dives, wait 24 hours on the surface before ascending to altitude.

The MK 16 MOD 0 decompression procedures may be used for diving at altitudes up to 1000 feet without modification. Contact NAVSEA 00C for guidance for any planned dives at altitudes greater than 1000 feet.

17-12 MK 16 MOD 0 DIVING EQUIPMENT REFERENCE DATA

Figure 17-6 outlines the capabilities and logistical requirements of the MK 16 MOD 0 UBA mixed-gas diving system. Minimum required equipment for the pool phase of diving conducted at Navy diving schools and MK 16 MOD 0 RDT&E commands may be modified as necessary. Any modification to the minimum required equipment listed herein must be noted in approved lesson training guides or SOPs.

MK 16 MOD 0 UBA General Characteristics

Principle of Operation:

Self-contained closed-circuit constant ppO_2 system

Minimum Equipment:

1. An approved Life Preserver or Buoyancy Compensator (BC). When using an approved BC, a Full Face Mask is required.
2. Dive knife
3. Swim fins
4. Face mask or full face mask (FFM)
5. Weight belt (as required)
6. Dive watch or Dive Timer/Depth Gauge (DT/DG) (as required)
7. Depth gauge or DT/DG (as required)

Principal Applications:

1. Special warfare
2. Search and inspection
3. Light repair and recovery

Advantages:

1. Minimal surface bubbles
2. Optimum efficiency of gas supply
3. Portability
4. Excellent mobility
5. Communications (when used with an approved FFM)
6. Modularized assembly
7. Low acoustic signature

Disadvantages:

1. Extended decompression requirement for long bottom times or deep dives.
2. Limited physical and thermal protection
3. No voice communications (unless FFM used)
4. Extensive predive/postdive procedures

Restrictions:

Working limit 150 feet, air diluent; 200 fsw, HeO_2 diluent

Operational Considerations:

1. Dive team
2. Safety boat(s) required
3. MK 16 MOD 0 decompression schedule must be used (unless using NDC, CSMD procedure 110 fsw and shallower, or air decompression procedures 70 fsw and shallower)

Figure 17-6. MK 16 MOD 0 General Characteristics.

Table 17-9. *Closed-Circuit Mixed-Gas UBA Decompression Table Using 0.7 ata Constant Partial Pressure Oxygen in Nitrogen.*
(DESCENT RATE 60 FPM—ASCENT RATE 30 FPM)

Bottom Time (min)	Time to First Stop (M:S)	DECOMPRESSION STOPS (FSW) Stop times (min) include travel time, except first stop								Total Ascent Time (M:S)	Repet Group
		80	70	60	50	40	30	20	10		
40 FSW											
369	1:20								0	1:20	Z
370	1:00								1	2:20	Z
380	1:00								2	3:20	Z
390	1:00								3	4:20	Z
50 FSW											
143	1:40								0	1:40	O
150	1:20								3	4:40	O
160	1:20								8	9:40	O
170	1:20								12	13:40	O
180	1:20								15	16:40	Z
190	1:20								19	20:40	Z
200	1:20								22	23:40	Z
210	1:20								25	26:40	Z
220	1:20								29	30:40	Z
230	1:20								33	34:40	Z
240	1:20								37	38:40	Z
250	1:20								42	43:40	Z
260	1:20								45	46:40	Z
270	1:20								49	50:40	Z
280	1:20								52	53:40	Z
290	1:20								56	57:40	Z
300	1:20								59	60:40	Z
310	1:20								61	62:40	Z
320	1:20								64	65:40	Z
330	1:20								67	68:40	Z
Exceptional Exposure ----------											
340	1:20								69	70:40	
350	1:20								73	74:40	
360	1:20								77	78:40	
370	1:20								80	81:40	
380	1:20								83	84:40	
390	1:20								87	88:40	

Table 17-9. *Closed-Circuit Mixed-Gas UBA Decompression Table Using 0.7 ata Constant Partial Pressure Oxygen in Nitrogen (Continued).*
(DESCENT RATE 60 FPM—ASCENT RATE 30 FPM)

Bottom Time (min)	Time to First Stop (M:S)	DECOMPRESSION STOPS (FSW) Stop times (min) include travel time, except first stop 80	70	60	50	40	30	20	10	Total Ascent Time (M:S)	Repet Group
60 FSW											
74	2:00								0	2:00	L
75	1:40								1	3:00	L
80	1:40								3	5:00	L
90	1:40								8	10:00	M
100	1:40								12	14:00	N
110	1:40								16	18:00	O
120	1:40								24	26:00	O
130	1:40								32	34:00	O
140	1:40								38	40:00	Z
150	1:40								44	46:00	Z
160	1:40								50	52:00	Z
170	1:40								55	57:00	Z
180	1:20							3	60	64:40	Z
190	1:20							8	62	71:40	Z
200	1:20							12	65	78:40	Z
210	1:20							15	69	85:40	Z
220	1:20							19	71	91:40	Z
230	1:20							22	74	97:40	Z
240	1:20							25	76	102:40	Z
250	1:20							27	80	108:40	Z
Exceptional Exposure ----------											
260	1:20							30	82	113:40	
270	1:20							32	85	118:40	
280	1:20							35	88	124:40	
290	1:20							40	90	131:40	
300	1:20							43	93	137:40	
310	1:20							47	94	142:40	
320	1:20							51	96	148:40	
330	1:20							54	98	153:40	
340	1:20							57	100	158:40	
350	1:20							60	102	163:40	
360	1:20							63	105	169:40	
370	1:20							65	109	175:40	
380	1:20							68	112	181:40	
390	1:20							70	115	186:40	

Table 17-9. *Closed-Circuit Mixed-Gas UBA Decompression Table Using 0.7 ata Constant Partial Pressure Oxygen in Nitrogen (Continued).*
(DESCENT RATE 60 FPM—ASCENT RATE 30 FPM)

Bottom Time (min)	Time to First Stop (M:S)	DECOMPRESSION STOPS (FSW) Stop times (min) include travel time, except first stop								Total Ascent Time (M:S)	Repet Group
		80	70	60	50	40	30	20	10		
70 FSW											
51	2:20								0	2:20	K
55	2:00								4	6:20	K
60	2:00								9	11:20	K
70	2:00								17	19:20	L
80	2:00								24	26:20	M
90	1:40							2	29	33:00	N
100	1:40							7	34	43:00	O
110	1:40							12	39	53:00	O
120	1:40							15	46	63:00	O
130	1:40							18	52	72:00	Z
140	1:40							21	57	80:00	Z
150	1:40							29	58	89:00	Z
160	1:40							36	62	100:00	Z
170	1:40							42	66	110:00	Z
180	1:40							48	70	120:00	Z
Exceptional Exposure ----------											
190	1:20						1	53	73	128:40	
200	1:20						2	57	77	137:40	
210	1:20						6	57	81	145:40	
220	1:20						10	57	84	152:40	
230	1:20						14	59	87	161:40	
240	1:20						18	62	89	170:40	
250	1:20						21	66	91	179:40	
260	1:20						24	69	94	188:40	
270	1:20						26	72	97	196:40	
280	1:20						29	75	99	204:40	
290	1:20						31	78	102	212:40	
300	1:20						33	81	105	220:40	
310	1:20						35	83	110	229:40	
320	1:20						37	86	113	237:40	
330	1:20						41	86	118	246:40	
340	1:20						45	86	124	256:40	
350	1:20						49	88	127	265:40	

Table 17-9. *Closed-Circuit Mixed-Gas UBA Decompression Table Using 0.7 ata Constant Partial Pressure Oxygen in Nitrogen (Continued).*
(DESCENT RATE 60 FPM—ASCENT RATE 30 FPM)

Bottom Time (min)	**Time to First Stop (M:S)**	**DECOMPRESSION STOPS (FSW)** Stop times (min) include travel time, except first stop								**Total Ascent Time (M:S)**	**Repet Group**
		80	**70**	**60**	**50**	**40**	**30**	**20**	**10**		
80 FSW											
40	2:40								0	2:40	J
45	2:20								8	10:40	K
50	2:20								15	17:40	K
55	2:20								21	23:40	L
60	2:20								27	29:40	L
70	2:00							9	28	39:20	M
80	2:00							17	29	48:20	N
90	2:00							24	36	62:20	O
100	1:40						2	29	43	76:00	O
110	1:40						7	29	50	88:00	Z
120	1:40						12	29	57	100:00	Z
Exceptional Exposure --											
130	1:40						15	37	58	112:00	
140	1:40						18	43	62	125:00	
150	1:40						21	49	67	139:00	
160	1:40						23	56	70	151:00	
170	1:40						29	57	75	163:00	
180	1:40						36	57	80	175:00	
190	1:40						42	57	85	186:00	
200	1:20					1	48	60	86	196:40	
210	1:20					2	52	64	90	209:40	
220	1:20					2	57	68	93	221:40	
230	1:20					6	57	73	96	233:40	
240	1:20					10	57	77	100	245:40	
250	1:20					14	57	81	104	257:40	
260	1:20					18	56	85	110	270:40	
270	1:20					21	59	86	116	283:40	
280	1:20					24	63	85	124	297:40	
290	1:20					26	67	86	129	309:40	
300	1:20					29	70	88	134	322:40	
310	1:20					31	73	92	137	334:40	
320	1:20					33	76	95	141	346:40	

Table 17-9. Closed-Circuit Mixed-Gas UBA Decompression Table Using 0.7 ata Constant Partial Pressure Oxygen in Nitrogen (Continued).
(DESCENT RATE 60 FPM—ASCENT RATE 30 FPM)

Bottom Time (min)	Time to First Stop (M:S)	DECOMPRESSION STOPS (FSW) Stop times (min) include travel time, except first stop								Total Ascent Time (M:S)	Repet Group
		80	70	60	50	40	30	20	10		
90 FSW											
32	3:00								0	3:00	J
35	2:40								5	8:00	J
40	2:40								14	17:00	K
45	2:40								23	26:00	K
50	2:20							3	28	33:40	L
55	2:20							10	28	40:40	L
60	2:20							17	28	47:40	M
70	2:20							28	29	59:40	N
80	2:00						10	29	34	75:20	O
90	2:00						18	29	44	93:20	Z
Exceptional Exposure ----------											
100	2:00						25	29	52	108:20	
110	1:40					3	29	33	56	123:00	
120	1:40					8	29	41	62	142:00	
130	1:40					12	29	49	67	159:00	
140	1:40					16	29	56	73	176:00	
150	1:40					19	36	57	76	190:00	
160	1:40					21	43	57	81	204:00	
170	1:40					23	50	57	89	221:00	
180	1:40					25	56	62	91	236:00	
190	1:40					31	57	67	95	252:00	
100 FSW											
27	3:20								0	3:20	I
30	3:00								6	9:20	J
35	3:00								18	21:20	J
40	3:00								28	31:20	K
45	2:40							10	28	41:00	L
50	2:40							19	28	50:00	M
55	2:40							27	29	59:00	M
60	2:20						7	28	28	65:40	N
65	2:20						14	28	28	72:40	O
Exceptional Exposure ----------											
70	2:20						20	28	32	82:40	
75	2:20						26	28	37	93:40	
80	2:00					3	28	29	42	104:20	
90	2:00					12	29	28	53	124:20	
100	2:00					20	29	34	61	146:20	
110	2:00					27	28	44	66	167:20	

Table 17-9. *Closed-Circuit Mixed-Gas UBA Decompression Table Using 0.7 ata Constant Partial Pressure Oxygen in Nitrogen (Continued).*
(DESCENT RATE 60 FPM—ASCENT RATE 30 FPM)

Bottom Time (min)	Time to First Stop (M:S)	DECOMPRESSION STOPS (FSW) Stop times (min) include travel time, except first stop								Total Ascent Time (M:S)	Repet Group
		80	70	60	50	40	30	20	10		
110 FSW											
23	3:40								0	3:40	I
25	3:20								4	7:40	J
30	3:20								18	21:40	J
35	3:00							3	28	34:20	K
40	3:00							14	29	46:20	L
45	3:00							25	29	57:20	L
50	2:40						7	29	28	67:00	M
55	2:40						16	29	28	76:00	N
Exceptional Exposure ----------											
60	2:40						25	28	29	85:00	
65	2:20					4	29	28	33	96:40	
70	2:20					11	29	28	40	110:40	
80	2:20					24	28	29	52	135:40	
90	2:00				6	29	28	34	65	164:20	
120 FSW											
20	4:00								0	4:00	I
25	3:40								14	18:00	J
30	3:20							3	27	33:40	J
35	3:20							15	29	47:40	K
40	3:00						4	25	28	60:20	L
45	3:00						12	29	28	72:20	M
Exceptional Exposure ----------											
50	2:40					1	23	28	28	83:00	
55	2:40					5	29	28	29	94:00	
60	2:40					15	28	28	35	109:00	
70	2:20				3	28	29	28	50	140:40	
80	2:20				17	28	29	31	68	175:40	

Table 17-9. Closed-Circuit Mixed-Gas UBA Decompression Table Using 0.7 ata Constant Partial Pressure Oxygen in Nitrogen (Continued).

(DESCENT RATE 60 FPM—ASCENT RATE 30 FPM)

130 FSW

Bottom Time (min)	Time to First Stop (M:S)	DECOMPRESSION STOPS (FSW) Stop times (min) include travel time, except first stop 80	70	60	50	40	30	20	10	Total Ascent Time (M:S)	Repet Group
16	4:20								0	4:20	H
20	4:00								5	9:20	I
25	3:40							4	20	28:00	J
30	3:20						2	11	28	44:40	K
35	3:20						7	21	29	60:40	L
Exceptional Exposure ----------											
40	3:00					1	14	28	28	74:20	
45	3:00					7	21	28	29	88:20	
50	3:00					12	28	28	29	100:20	
55	2:40				3	20	28	29	34	117:00	
60	2:40				7	26	28	29	43	136:00	
70	2:40				23	28	28	29	67	178:00	

140 FSW

Bottom Time (min)	Time to First Stop (M:S)	80	70	60	50	40	30	20	10	Total Ascent Time (M:S)	Repet Group
14	4:40								0	4:40	H
15	4:20								1	5:40	H
20	4:00							3	11	18:20	J
25	3:40						3	7	24	38:00	K
30	3:20					1	7	17	28	56:40	L
Exceptional Exposure ----------											
35	3:20					4	13	24	29	73:40	
40	3:20					11	18	28	28	88:40	
45	3:00				4	14	25	29	28	103:20	
50	3:00				10	18	28	29	35	123:20	
60	2:40			5	18	28	29	28	61	172:00	
70	2:40			14	28	29	28	36	80	218:00	

Table 17-9. *Closed-Circuit Mixed-Gas UBA Decompression Table Using 0.7 ata Constant Partial Pressure Oxygen in Nitrogen (Continued).*
(DESCENT RATE 60 FPM—ASCENT RATE 30 FPM)

Bottom Time (min)	Time to First Stop (M:S)	DECOMPRESSION STOPS (FSW) Stop times (min) include travel time, except first stop								Total Ascent Time (M:S)	Repet Group
		80	70	60	50	40	30	20	10		

150 FSW

Bottom Time (min)	Time to First Stop (M:S)	80	70	60	50	40	30	20	10	Total Ascent Time (M:S)	Repet Group
11	5:00								0	5:00	G
15	4:40								6	11:00	H
20	4:00						2	7	14	27:20	J
25	3:40					2	7	9	27	49:00	K
30	3:40					7	9	20	28	68:00	M
Exceptional Exposure ----------											
35	3:20				3	10	14	28	28	86:40	
40	3:20				7	14	22	28	29	103:40	
45	3:00			1	14	15	29	28	35	125:20	
50	3:00			7	14	23	29	28	49	153:20	
60	2:40		3	14	24	29	28	32	76	209:00	
70	2:40		10	24	28	29	28	52	91	265:00	

160 FSW

Bottom Time (min)	Time to First Stop (M:S)	80	70	60	50	40	30	20	10	Total Ascent Time (M:S)	Repet Group
Exceptional Exposure ----------											
10	5:20								0	5:20	
15	4:40							3	7	15:00	
20	4:20						6	8	17	35:40	
25	4:00					7	7	12	29	59:20	
30	3:40				6	7	12	23	28	80:00	
35	3:20			3	7	12	17	29	28	99:40	
40	3:20			5	13	14	25	29	35	124:40	
45	3:20			12	14	19	29	28	49	154:40	
50	3:00		4	15	14	28	28	29	65	186:20	

170 FSW

Bottom Time (min)	Time to First Stop (M:S)	80	70	60	50	40	30	20	10	Total Ascent Time (M:S)	Repet Group
Exceptional Exposure ----------											
9	5:40								0	5:40	
10	5:20								2	7:40	
15	4:40						2	6	7	20:00	
20	4:20					5	7	7	21	44:40	
25	4:00				6	7	7	17	28	69:20	
30	3:40			5	7	8	14	26	29	93:00	
35	3:20		2	7	9	14	21	28	35	119:40	
40	3:20		5	9	14	15	28	29	46	149:40	
45	3:20		8	15	14	24	28	29	65	186:40	
50	3:00	2	14	14	19	28	29	36	76	221:20	

Table 17-10. *Closed-Circuit Mixed-Gas UBA Decompression Table Using 0.7 ata Constant Partial Pressure Oxygen in Helium.*
(DESCENT RATE 60 FPM—ASCENT RATE 30 FPM)

Bottom Time (min)	Time to First Stop (M:S)	DECOMPRESSION STOPS (fsw) Stop times (min) include travel time, except first stop																			Total Ascent Time (M:S)
		190	180	170	160	150	140	130	120	110	100	90	80	70	60	50	40	30	20	10	
40 FSW																					
390	1:20																			0	1:20
50 FSW																					
205	1:40																			0	1:40
210	1:20																			3	4:40
220	1:20																			9	10:40
230	1:20																			14	15:40
240	1:20																			20	21:40
250	1:20																			24	25:40
Exceptional Exposure ----------																					
260	1:20																			29	30:40
270	1:20																			33	34:40
280	1:20																			37	38:40
290	1:20																			41	42:40
300	1:20																			45	46:40
310	1:20																			48	49:40
320	1:20																			52	53:40
330	1:20																			55	56:40
340	1:20																			58	59:40
350	1:20																			60	61:40
360	1:20																			63	64:40
370	1:20																			65	66:40
380	1:20																			68	69:40
390	1:20																			70	71:40
60 FSW																					
133	2:00																			0	2:00
140	1:40																			8	10:00
150	1:40																			20	22:00
160	1:40																			30	32:00
170	1:40																			40	42:00
Exceptional Exposure ----------																					
180	1:40																			50	52:00
190	1:40																			59	61:00
200	1:40																			67	69:00
210	1:40																			75	77:00
220	1:40																			82	84:00
230	1:40																			90	92:00
240	1:40																			96	98:00
250	1:40																			103	105:00
260	1:40																			109	111:00
270	1:20																		1	113	115:40

Table 17-10. Closed-Circuit Mixed-Gas UBA Decompression Table Using 0.7 ata Constant Partial Pressure Oxygen in Helium (Continued).
(DESCENT RATE 60 FPM—ASCENT RATE 30 FPM)

Bottom Time (min)	Time to First Stop (M:S)	DECOMPRESSION STOPS (fsw) Stop times (min) include travel time, except first stop																			Total Ascent Time (M:S)
		190	180	170	160	150	140	130	120	110	100	90	80	70	60	50	40	30	20	10	
60 FSW Continued																					
280	1:20																		7	113	121:40
290	1:20																		12	113	126:40
300	1:20																		16	114	131:40
310	1:20																		21	113	135:40
320	1:20																		25	113	139:40
330	1:20																		29	113	143:40
340	1:20																		33	113	147:40
350	1:20																		36	113	150:40
360	1:20																		40	113	154:40
370	1:20																		43	113	157:40
380	1:20																		46	113	160:40
390	1:20																		49	113	163:40
70 FSW																					
82	2:20																			0	2:20
85	2:00																			2	4:20
90	2:00																			6	8:20
95	2:00																			9	11:20
100	2:00																			12	14:20
110	2:00																			19	21:20
120	2:00																			35	37:20
130	2:00																			51	53:20
140	2:00																			65	67:20
Exceptional Exposure																					
150	2:00																			79	81:20
160	2:00																			92	94:20
170	2:00																			104	106:20
180	1:40																		7	109	118:00
190	1:40																		14	113	129:00
200	1:40																		24	113	139:00
210	1:40																		34	113	149:00
220	1:40																		43	113	158:00
230	1:40																		52	113	167:00
240	1:40																		60	113	175:00
250	1:40																		68	113	183:00
260	1:40																		75	113	190:00
270	1:40																		82	113	197:00

Table 17-10. *Closed-Circuit Mixed-Gas UBA Decompression Table Using 0.7 ata Constant Partial Pressure Oxygen in Helium (Continued).*

(DESCENT RATE 60 FPM—ASCENT RATE 30 FPM)

Bottom Time (min)	Time to First Stop (M:S)	DECOMPRESSION STOPS (fsw) Stop times (min) include travel time, except first stop																			Total Ascent Time (M:S)
		190	180	170	160	150	140	130	120	110	100	90	80	70	60	50	40	30	20	10	
80 FSW																					
52	2:40																			0	2:40
55	2:20																			2	4:40
60	2:20																			5	7:40
65	2:20																			8	10:40
70	2:20																			14	16:40
75	2:20																			19	21:40
80	2:20																			24	26:40
85	2:20																			29	31:40
90	2:20																			33	35:40
95	2:20																			36	38:40
100	2:00																		3	44	49:20
110	2:00																		9	58	69:20
120	2:00																		14	73	89:20
Exceptional Exposure																					
130	2:00																		18	87	107:20
140	2:00																		22	100	124:20
150	2:00																		33	105	140:20
160	2:00																		43	111	156:20
170	2:00																		55	113	170:20
180	2:00																		69	113	184:20
190	2:00																		82	113	197:20
90 FSW																					
37	3:00																			0	3:00
40	2:40																			4	7:00
45	2:40																			10	13:00
50	2:40																			15	18:00
55	2:40																			19	22:00
60	2:20																		1	23	26:40
65	2:20																		4	27	33:40
70	2:20																		6	32	40:40
75	2:20																		8	36	46:40
80	2:20																		12	38	52:40
85	2:20																		17	38	57:40
90	2:20																		22	44	68:40
95	2:20																		26	53	81:40
100	2:20																		30	61	93:40
110	2:20																		38	77	117:40
120	2:00																	6	38	94	140:20
Exceptional Exposure																					
130	2:00																	11	46	102	161:20
140	2:00																	15	55	109	181:20
150	2:00																	19	66	113	200:20
160	2:00																	22	81	113	218:20

Table 17-10. *Closed-Circuit Mixed-Gas UBA Decompression Table Using 0.7 ata Constant Partial Pressure Oxygen in Helium (Continued).*
(DESCENT RATE 60 FPM—ASCENT RATE 30 FPM)

Bottom Time (min)	Time to First Stop (M:S)	DECOMPRESSION STOPS (fsw) Stop times (min) include travel time, except first stop																			Total Ascent Time (M:S)
		190	180	170	160	150	140	130	120	110	100	90	80	70	60	50	40	30	20	10	
100 FSW																					
29	3:20																			0	3:20
30	3:00																			1	4:20
35	3:00																			11	14:20
40	3:00																			19	22:20
50	2:40																		9	22	34:00
60	2:40																		18	27	48:00
70	2:20																	2	22	38	64:40
80	2:20																	7	31	41	81:40
90	2:20																	11	38	59	110:40
100	2:20																	21	38	78	139:40
Exceptional Exposure																					
110	2:20																	29	39	96	166:40
120	2:20																	36	50	103	191:40
130	2:00																4	38	61	111	216:20
140	2:00																9	38	76	113	238:20
110 FSW																					
23	3:40																			0	3:40
25	3:20																			2	5:40
30	3:20																			14	17:40
35	3:00																		3	22	28:20
40	3:00																		11	22	36:20
50	2:40																	3	22	22	50:00
60	2:40																	13	22	33	71:00
70	2:40																	20	28	37	88:00
80	2:20																3	23	37	55	120:40
90	2:20																7	31	38	76	154:40
100	2:20																11	38	39	96	186:40
Exceptional Exposure																					
110	2:20																20	38	52	103	215:40
120	2:20																28	38	64	111	243:40
130	2:20																34	40	80	113	269:40
140	2:00															2	38	51	89	113	295:20

Table 17-10. *Closed-Circuit Mixed-Gas UBA Decompression Table Using 0.7 ata Constant Partial Pressure Oxygen in Helium (Continued).*
(DESCENT RATE 60 FPM—ASCENT RATE 30 FPM)

Bottom Time (min)	Time to First Stop (M:S)	DECOMPRESSION STOPS (fsw) Stop times (min) include travel time, except first stop																			Total Ascent Time (M:S)
		190	180	170	160	150	140	130	120	110	100	90	80	70	60	50	40	30	20	10	
120 FSW																					
18	4:00																			0	4:00
20	3:40																			2	6:00
25	3:40																			13	17:00
30	3:20																		5	22	30:40
35	3:20																		16	22	41:40
40	3:00																	4	22	22	51:20
50	3:00																	19	23	24	69:20
60	2:40																9	22	22	37	93:00
70	2:40																16	22	34	52	127:00
80	2:40																22	29	38	72	164:00
Exceptional Exposure ----------																					
90	2:20															4	24	37	38	95	200:40
100	2:20															7	32	38	50	104	233:40
110	2:20															12	37	38	65	112	266:40
120	2:20															20	38	41	83	113	297:40
130 FSW																					
15	4:20																			0	4:20
20	4:00																			8	12:20
25	3:40																		6	18	28:00
30	3:20																	2	16	22	43:40
35	3:20																	8	22	22	55:40
40	3:20																	19	22	22	66:40
50	3:00																14	22	22	28	89:20
60	2:40															4	22	22	26	48	125:00
70	2:40															12	22	24	38	70	169:00
Exceptional Exposure ----------																					
80	2:40															18	22	36	38	93	210:00
90	2:20														1	22	32	37	46	107	247:40
100	2:20														4	26	38	37	64	113	284:40
110	2:20														6	35	38	40	84	113	318:40
120	2:20														12	38	38	55	93	113	351:40

Table 17-10. *Closed-Circuit Mixed-Gas UBA Decompression Table Using 0.7 ata Constant Partial Pressure Oxygen in Helium (Continued).*
(DESCENT RATE 60 FPM—ASCENT RATE 30 FPM)

Bottom Time (min)	Time to First Stop (M:S)	DECOMPRESSION STOPS (fsw) Stop times (min) include travel time, except first stop																			Total Ascent Time (M:S)
		190	180	170	160	150	140	130	120	110	100	90	80	70	60	50	40	30	20	10	
140 FSW																					
12	4:40																			0	4:40
15	4:20																			4	8:40
20	4:00																		5	12	21:20
25	3:40																	4	10	22	40:00
30	3:40																	10	20	22	56:00
35	3:20																4	18	22	22	69:40
40	3:20																12	22	22	22	81:40
50	3:00															8	22	22	22	35	112:20
60	3:00															21	22	22	31	66	165:20
70	2:40														9	22	22	29	38	93	216:00
Exceptional Exposure ----------																					
80	2:40														15	22	27	38	40	113	258:00
90	2:40														20	23	38	38	63	113	298:00
100	2:20													1	22	35	38	37	88	113	336:40
150 FSW																					
10	5:00																			0	5:00
15	4:20																		2	7	13:40
20	4:00																	2	10	15	31:20
25	3:40																2	9	15	22	52:00
30	3:40																7	14	22	22	69:00
35	3:20															3	11	22	22	22	83:40
40	3:20															6	21	22	22	22	96:40
45	3:20															15	22	22	22	33	117:40
50	3:00														2	23	22	22	22	56	150:20
55	3:00														10	22	22	22	27	74	180:20
60	3:00														16	22	23	22	35	88	209:20
Exceptional Exposure ----------																					
70	2:40													5	22	22	22	35	40	113	262:00
80	2:40													12	22	22	34	38	65	113	309:00
90	2:40													17	22	31	38	38	90	113	352:00

Table 17-10. *Closed-Circuit Mixed-Gas UBA Decompression Table Using 0.7 ata Constant Partial Pressure Oxygen in Helium (Continued).*

(DESCENT RATE 60 FPM—ASCENT RATE 30 FPM)

Bottom Time (min)	Time to First Stop (M:S)	DECOMPRESSION STOPS (fsw) Stop times (min) include travel time, except first stop																			Total Ascent Time (M:S)
		190	180	170	160	150	140	130	120	110	100	90	80	70	60	50	40	30	20	10	
155 FSW																					
9	5:10																			0	5:10
10	4:50																			1	6:10
15	4:30																		3	9	16:50
20	4:10																	5	10	17	36:30
25	3:50																5	9	17	22	57:10
30	3:30															2	9	17	22	22	75:50
35	3:30															6	15	22	22	22	90:50
40	3:30															12	22	22	22	22	103:50
45	3:10														3	20	22	22	22	44	136:30
50	3:10														10	23	22	22	22	68	170:30
55	3:10														18	22	22	22	30	84	201:30
60	2:50													3	22	22	22	22	38	100	232:10
Exceptional Exposure --------																					
70	2:50													14	22	22	22	38	52	113	286:10
80	2:50													21	22	22	38	37	77	113	333:10
90	2:30												5	22	22	35	38	37	103	113	377:50
160 FSW																					
9	5:20																			0	5:20
10	5:00																			2	7:20
15	4:20																	1	4	10	19:40
20	4:00																1	8	9	19	41:20
25	4:00																8	10	19	22	63:20
30	3:40															5	10	19	22	22	82:00
35	3:20														1	9	18	22	22	22	97:40
40	3:20														4	15	22	22	23	27	116:40
45	3:20														9	22	22	22	22	55	155:40
50	3:20														18	22	23	22	22	79	189:40
Exceptional Exposure --------																					
55	3:00													5	22	22	22	22	31	97	224:20
60	3:00													12	22	22	22	24	38	113	256:20
70	2:40												1	22	22	22	25	38	64	113	310:00
80	2:40												8	22	23	25	37	38	91	113	360:00
90	2:40												14	22	24	37	38	43	111	113	405:00

Table 17-10. *Closed-Circuit Mixed-Gas UBA Decompression Table Using 0.7 ata Constant Partial Pressure Oxygen in Helium (Continued).*
(DESCENT RATE 60 FPM—ASCENT RATE 30 FPM)

Bottom Time (min)	Time to First Stop (M:S)	DECOMPRESSION STOPS (fsw) Stop times (min) include travel time, except first stop																			Total Ascent Time (M:S)
		190	180	170	160	150	140	130	120	110	100	90	80	70	60	50	40	30	20	10	
165 FSW																					
8	5:30																			0	5:30
10	5:10																			3	8:30
15	4:30																	2	6	9	21:50
20	4:10																2	10	9	21	46:30
25	3:50															2	10	9	22	22	69:10
30	3:50															9	9	22	22	22	88:10
35	3:30														5	9	21	22	22	22	104:50
40	3:30														8	19	22	22	22	39	135:50
45	3:10													1	16	22	22	22	22	66	174:30
50	3:10													5	22	22	22	22	24	92	212:30
Exceptional Exposure ----------																					
55	3:10													13	22	22	22	22	34	108	246:30
60	3:10													20	22	22	22	27	48	113	277:30
70	2:50												10	22	22	22	28	38	79	113	337:10
80	2:50												18	22	22	28	38	38	105	113	387:10
170 FSW																					
8	5:40																			0	5:40
10	5:00																		1	3	9:20
15	4:40																	4	7	9	25:00
20	4:20																5	10	10	22	51:40
25	4:00															6	9	11	22	22	74:20
30	3:40														3	10	12	22	22	22	95:00
35	3:40														8	12	22	22	22	22	112:00
40	3:20													3	9	22	22	22	22	50	153:40
45	3:20													5	19	22	23	22	22	78	194:40
50	3:20													13	22	22	22	22	26	104	234:40
Exceptional Exposure ----------																					
55	3:20													21	23	22	22	22	42	113	268:40
60	3:00												7	22	22	22	22	29	62	113	302:20
70	3:00												19	22	22	22	31	38	92	113	362:20
80	2:40											5	22	22	22	32	38	43	113	113	413:00

Table 17-10. *Closed-Circuit Mixed-Gas UBA Decompression Table Using 0.7 ata Constant Partial Pressure Oxygen in Helium (Continued).*
(DESCENT RATE 60 FPM—ASCENT RATE 30 FPM)

Bottom Time (min)	Time to First Stop (M:S)	DECOMPRESSION STOPS (fsw) Stop times (min) include travel time, except first stop																			Total Ascent Time (M:S)
		190	180	170	160	150	140	130	120	110	100	90	80	70	60	50	40	30	20	10	
175 FSW																					
7	5:50																			0	5:50
10	5:10																		2	4	11:30
15	4:30																1	4	8	10	27:50
20	4:10															1	7	10	12	22	56:30
25	4:10															9	9	14	22	22	80:30
30	3:50														7	9	15	22	22	22	101:10
35	3:30													3	9	15	22	22	22	31	127:50
40	3:30													7	13	22	22	22	22	62	173:50
Exceptional Exposure																					
45	3:30													10	22	22	22	22	22	91	214:50
50	3:10												2	19	22	22	22	22	30	113	255:30
55	3:10												8	22	22	22	22	22	58	113	292:30
60	3:10												16	22	22	22	22	31	76	113	327:30
65	3:10												22	22	22	22	25	38	90	113	357:30
70	2:50											6	22	22	22	22	34	38	106	113	388:10
75	2:50											10	22	22	23	27	37	45	113	113	415:10
80	2:50											14	22	22	22	36	38	58	113	113	441:10
180 FSW																					
7	6:00																			0	6:00
10	5:20																		3	4	12:40
15	4:40																3	4	9	11	32:00
20	4:20															3	8	10	14	22	61:40
25	4:00														3	9	10	16	22	22	86:20
30	3:40													1	10	9	17	22	22	23	108:00
35	3:40													7	9	17	22	23	22	41	145:00
40	3:20												1	10	16	22	22	22	22	73	191:40
Exceptional Exposure																					
45	3:20												4	14	22	22	22	22	22	105	236:40
50	3:20												7	22	22	22	22	22	44	113	277:40
55	3:20												16	22	22	22	22	24	70	113	314:40
60	3:00											3	22	22	22	22	22	33	90	113	352:20
65	3:00											9	22	22	22	22	28	38	105	113	384:20
70	3:00											15	22	22	22	22	37	45	113	113	414:20

Table 17-10. *Closed-Circuit Mixed-Gas UBA Decompression Table Using 0.7 ata Constant Partial Pressure Oxygen in Helium (Continued).*
(DESCENT RATE 60 FPM—ASCENT RATE 30 FPM)

Bottom Time (min)	Time to First Stop (M:S)	DECOMPRESSION STOPS (fsw) Stop times (min) include travel time, except first stop 190	180	170	160	150	140	130	120	110	100	90	80	70	60	50	40	30	20	10	Total Ascent Time (M:S)
185 FSW																					
6	6:10																			0	6:10
10	5:30																		4	4	13:50
15	4:50																4	5	10	12	36:10
20	4:10														1	4	10	9	16	22	66:30
25	4:10														6	10	9	19	22	22	92:30
30	3:50													5	9	10	20	22	22	22	114:10
35	3:30												1	10	9	21	22	22	22	52	162:50
40	3:30												5	10	19	22	22	22	22	86	211:50
Exceptional Exposure ----------																					
45	3:30												8	18	22	22	22	22	28	113	258:50
50	3:10											1	14	22	22	22	22	22	58	113	299:30
55	3:10											3	22	22	22	22	22	26	84	113	339:30
60	3:10											11	22	22	22	22	22	36	103	113	376:30
65	3:10											18	22	22	22	22	30	44	113	113	409:30
70	2:50										2	22	22	22	22	24	38	60	113	113	441:10
190 FSW																					
6	6:20																			0	6:20
10	5:20																	1	4	5	15:40
15	4:40															2	4	6	9	15	41:00
20	4:20														2	6	10	9	18	22	71:40
25	4:00													1	9	9	10	20	23	22	98:20
30	4:00													8	10	10	22	22	22	27	125:20
35	3:40												5	9	11	22	22	22	22	63	180:00
40	3:40												9	11	22	22	22	22	22	99	233:00
Exceptional Exposure ----------																					
45	3:20											3	9	22	22	22	22	22	41	113	279:40
50	3:20											5	18	22	22	22	22	22	73	113	322:40
55	3:20											11	22	22	22	22	22	28	99	113	364:40
60	3:20											20	22	22	22	22	22	42	114	113	402:40
65	3:00										5	22	22	22	22	22	33	59	113	113	436:20
70	3:00										11	22	22	22	22	27	38	76	113	113	469:20

Table 17-10. *Closed-Circuit Mixed-Gas UBA Decompression Table Using 0.7 ata Constant Partial Pressure Oxygen in Helium (Continued).*

(DESCENT RATE 60 FPM—ASCENT RATE 30 FPM)

Bottom Time (min)	Time to First Stop (M:S)	DECOMPRESSION STOPS (fsw) Stop times (min) include travel time, except first stop																			Total Ascent Time (M:S)
		190	180	170	160	150	140	130	120	110	100	90	80	70	60	50	40	30	20	10	
195 FSW																					
6	6:30																			0	6:30
10	5:30																	3	3	6	17:50
15	4:50															3	4	8	9	16	45:10
20	4:30														4	7	10	9	20	22	76:50
25	4:10													4	9	10	10	22	22	22	103:30
30	3:50												3	9	10	12	22	23	22	37	142:10
35	3:50												9	9	14	22	22	22	22	75	199:10
40	3:30											4	9	14	22	22	22	22	22	112	252:50
Exceptional Exposure																					
45	3:30											7	12	22	22	22	22	22	55	113	300:50
50	3:30											9	22	22	22	22	22	22	88	113	345:50
55	3:10										1	19	22	22	22	22	22	30	113	113	389:30
60	3:10										6	22	22	22	22	22	26	55	113	113	426:30
200 FSW																					
6	6:40																			0	6:40
10	5:40																	4	4	6	20:00
15	4:40														1	4	4	8	10	17	49:00
20	4:20													2	4	9	9	9	22	22	81:40
25	4:20													7	10	9	13	22	22	22	109:40
30	4:00												6	10	9	16	22	22	22	48	159:20
35	3:40											3	10	9	17	22	22	22	22	87	218:00
Exceptional Exposure																					
40	3:40											7	10	17	22	22	22	22	34	113	273:00
45	3:20										1	10	16	22	22	22	22	22	70	113	323:40
50	3:20										4	14	22	22	22	22	22	22	106	113	372:40
55	3:20										6	22	22	22	22	22	22	46	113	113	413:40
60	3:20										15	22	22	22	22	22	27	72	113	114	454:40
205 FSW																					
5	6:50																			0	6:50
10	5:30																1	4	4	8	22:50
15	4:50														2	4	5	9	9	19	53:10
20	4:30													3	5	9	10	11	22	22	86:50
25	4:10												2	9	9	10	15	22	22	22	115:30
30	3:50											1	9	10	9	18	22	22	22	59	176:10
35	3:50											7	9	10	20	22	22	22	22	100	238:10
Exceptional Exposure																					
40	3:30										2	10	9	21	22	22	22	22	48	113	294:50
45	3:30										5	10	20	22	22	22	22	22	85	113	346:50
50	3:30										8	18	22	22	22	22	22	30	113	113	395:50
55	3:30										14	22	22	22	22	22	22	62	113	113	437:50
60	3:10									2	22	22	22	22	22	22	30	87	113	113	480:30

Table 17-10. Closed-Circuit Mixed-Gas UBA Decompression Table Using 0.7 ata Constant Partial Pressure Oxygen in Helium (Continued).
(DESCENT RATE 60 FPM—ASCENT RATE 30 FPM)

Bottom Time (min)	Time to First Stop (M:S)	DECOMPRESSION STOPS (fsw) Stop times (min) include travel time, except first stop																			Total Ascent Time (M:S)
		190	180	170	160	150	140	130	120	110	100	90	80	70	60	50	40	30	20	10	
210 FSW																					
5	7:00																			0	7:00
10	5:40																2	4	4	9	25:00
15	5:00														4	3	6	10	9	20	57:20
20	4:20												1	4	6	10	9	13	22	22	91:40
25	4:20												5	9	9	10	17	22	22	26	124:40
30	4:00											4	10	9	9	21	22	23	22	68	192:20
35	3:40										1	10	9	11	22	22	22	22	22	112	257:00
Exceptional Exposure																					
40	3:40										6	9	12	22	22	22	22	22	61	113	315:00
45	3:40										9	11	22	23	22	22	22	22	100	113	370:00
50	3:20									2	10	22	22	22	22	22	22	45	113	113	418:40
55	3:20									4	19	22	22	22	22	22	22	81	113	113	465:40
60	3:20									10	22	22	22	22	22	22	32	103	113	113	506:40
215 FSW																					
5	7:10																			0	7:10
10	5:50																3	4	4	10	27:10
15	4:50													1	4	4	7	9	10	22	62:10
20	4:30												2	4	8	10	9	15	22	22	96:50
25	4:10											1	7	10	9	9	20	22	22	36	140:30
30	4:10											8	9	10	11	22	22	22	22	81	211:30
Exceptional Exposure																					
35	3:50										5	10	9	14	22	22	22	22	35	113	278:10
40	3:30									1	9	10	15	22	22	22	22	22	77	113	338:50
45	3:30									4	9	15	22	22	22	23	22	24	113	113	392:50
50	3:30									6	14	22	22	22	22	22	22	62	113	114	444:50
55	3:30									9	22	22	22	22	22	22	23	97	113	113	490:50
60	3:30									19	22	22	22	22	22	22	41	112	113	113	533:50
220 FSW																					
5	7:20																			0	7:20
10	5:40															1	4	4	5	9	29:00
15	5:00													3	3	4	9	9	11	22	66:20
20	4:40												4	4	9	10	9	17	22	22	102:00
25	4:20											3	8	10	9	10	22	22	22	45	155:40
30	4:00										2	10	9	9	14	22	22	22	22	93	229:20
Exceptional Exposure																					
35	4:00										9	9	10	17	22	22	22	22	48	113	298:20
40	3:40									5	9	9	19	22	22	22	22	22	92	113	361:00
45	3:40									8	9	19	22	22	22	22	22	41	113	113	417:00
50	3:20								1	10	17	22	22	22	22	22	22	80	113	113	469:40
55	3:20								3	15	22	22	22	22	22	22	30	108	113	113	517:40

Table 17-10. *Closed-Circuit Mixed-Gas UBA Decompression Table Using 0.7 ata Constant Partial Pressure Oxygen in Helium (Continued).*
(DESCENT RATE 60 FPM—ASCENT RATE 30 FPM)

Bottom Time (min)	Time to First Stop (M:S)	DECOMPRESSION STOPS (fsw) Stop times (min) include travel time, except first stop																			Total Ascent Time (M:S)
		190	180	170	160	150	140	130	120	110	100	90	80	70	60	50	40	30	20	10	
225 FSW																					
4	7:30																			0	7:30
5	7:10																			1	8:30
10	5:50															2	4	4	6	9	31:10
15	5:10													4	4	4	9	10	12	22	70:30
20	4:30											2	4	5	10	9	9	19	22	22	106:50
25	4:10										1	5	9	9	10	12	22	22	22	56	172:30
30	4:10										6	9	9	10	16	22	22	23	22	104	247:30
Exceptional Exposure																					
35	3:50									3	10	9	10	20	22	22	22	22	61	113	318:10
40	3:50									8	10	9	22	22	22	22	22	22	106	113	382:10
45	3:30								3	9	10	22	22	22	22	22	22	56	113	113	439:50
50	3:30								5	10	21	22	22	22	22	22	22	97	113	113	494:50
55	3:30								7	19	22	22	22	22	22	22	42	113	113	114	543:50
230 FSW																					
4	7:40																			0	7:40
5	7:20																			2	9:40
10	6:00															3	4	4	7	9	33:20
15	5:00												2	4	3	6	9	9	14	22	74:20
20	4:40											3	4	7	9	10	9	21	22	22	112:00
25	4:20										2	7	9	10	9	14	22	22	22	66	187:40
30	4:20										9	10	9	9	20	22	22	22	26	113	266:40
Exceptional Exposure																					
35	4:00									7	9	10	10	22	22	22	22	22	74	113	337:20
40	3:40								3	9	10	13	22	22	22	22	22	31	113	113	406:00
45	3:40								7	9	14	22	22	22	22	22	22	74	113	113	466:00
50	3:40								9	13	22	22	22	22	22	22	27	109	113	113	520:00
55	3:20							2	10	22	22	22	23	22	22	22	60	113	113	113	569:40
235 FSW																					
4	7:50																			0	7:50
5	7:30																			3	10:50
10	5:50														1	4	3	4	8	10	36:10
15	5:10												3	4	4	6	10	9	15	22	78:30
20	4:30										1	4	4	8	10	9	10	22	22	22	116:50
25	4:30										4	8	9	10	9	17	22	22	22	76	203:50
30	4:10									4	9	9	10	9	22	22	22	22	38	113	284:30
Exceptional Exposure																					
35	3:50								2	9	9	10	13	22	22	23	22	22	88	113	359:10
40	3:50								7	9	10	16	22	22	22	22	22	46	113	113	428:10
45	3:30							1	10	9	17	23	22	22	22	22	22	90	113	113	489:50
50	3:30							4	9	17	22	22	22	22	22	22	40	113	113	113	544:50

Table 17-10. *Closed-Circuit Mixed-Gas UBA Decompression Table Using 0.7 ata Constant Partial Pressure Oxygen in Helium (Continued).*
(DESCENT RATE 60 FPM—ASCENT RATE 30 FPM)

Bottom Time (min)	Time to First Stop (M:S)	DECOMPRESSION STOPS (fsw) Stop times (min) include travel time, except first stop																			Total Ascent Time (M:S)
		190	180	170	160	150	140	130	120	110	100	90	80	70	60	50	40	30	20	10	
240 FSW																					
4	8:00																			0	8:00
5	7:40																			3	11:00
10	6:00														2	4	4	3	9	10	38:20
15	5:00											1	4	4	3	8	9	10	17	22	83:20
20	4:40										3	3	5	9	10	9	12	22	22	32	132:00
25	4:20									2	4	10	9	9	10	19	22	22	22	87	220:40
Exceptional Exposure ----------																					
30	4:20									7	9	10	9	12	22	22	22	22	51	113	303:40
35	4:00								5	10	9	10	16	22	22	22	22	22	104	113	381:20
40	3:40							1	10	9	10	19	22	22	22	22	22	60	113	113	449:00
45	3:40							5	10	9	21	22	22	22	22	22	22	107	113	113	514:00
50	3:40							8	9	21	22	22	22	22	22	22	58	113	113	113	571:00
245 FSW																					
5	7:30																		1	4	12:50
10	6:10														3	4	4	4	9	11	41:30
15	5:10											2	4	4	4	9	9	9	19	22	87:30
20	4:50										4	4	6	9	10	9	14	22	22	41	146:10
25	4:30									3	6	10	9	10	9	21	22	22	22	98	236:50
Exceptional Exposure ----------																					
30	4:10								1	10	9	10	9	15	22	22	22	22	64	113	323:30
35	4:10								9	9	10	9	20	22	22	22	22	27	113	113	402:30
40	3:50							5	10	9	11	22	22	22	22	22	22	77	114	113	475:10
45	3:50							9	10	12	22	22	22	22	22	22	33	113	113	113	539:10
50	3:30						3	9	12	22	22	22	22	22	23	22	75	113	114	113	597:50
250 FSW																					
5	7:40																		1	4	13:00
10	6:20														4	4	4	5	9	12	44:40
15	5:20											3	4	4	5	9	9	10	20	22	91:40
20	4:40									2	4	4	7	9	10	9	16	22	22	50	160:00
25	4:20								1	4	8	9	10	9	11	22	22	22	22	110	254:40
Exceptional Exposure ----------																					
30	4:20								5	9	10	9	10	17	22	22	22	22	78	113	343:40
35	4:00							4	9	9	10	10	22	22	22	22	22	41	113	114	424:20
40	4:00							9	9	10	14	22	22	22	22	22	22	94	113	113	498:20
45	3:40						4	9	10	16	22	22	22	22	22	22	51	113	113	113	565:00
50	3:40						7	9	16	22	22	22	22	22	22	22	95	113	113	113	624:00

Table 17-10. *Closed-Circuit Mixed-Gas UBA Decompression Table Using 0.7 ata Constant Partial Pressure Oxygen in Helium (Continued).*
(DESCENT RATE 60 FPM—ASCENT RATE 30 FPM)

255 FSW

Bottom Time (min)	Time to First Stop (M:S)	DECOMPRESSION STOPS (fsw) Stop times (min) include travel time, except first stop 190	180	170	160	150	140	130	120	110	100	90	80	70	60	50	40	30	20	10	Total Ascent Time (M:S)
5	7:50																		2	4	14:10
10	6:10													1	4	4	4	6	10	12	47:30
15	5:10										1	4	4	4	5	10	9	10	22	22	96:30
20	4:50									3	4	4	9	9	10	9	18	22	22	59	174:10
25	4:30								3	4	9	10	9	10	13	22	22	22	31	113	272:50
Exceptional Exposure																					
30	4:10							1	8	9	10	9	9	21	22	22	22	22	91	113	363:30
35	4:10							7	10	9	9	14	22	22	22	22	22	56	113	113	445:30
40	3:50						4	9	10	9	17	22	22	22	22	22	25	107	113	113	521:10
45	3:50						8	9	10	19	22	22	22	22	22	22	68	113	113	113	589:10
50	3:30					2	9	10	20	22	22	22	22	22	22	32	104	113	113	113	651:50

260 FSW

Bottom Time (min)	Time to First Stop (M:S)	190	180	170	160	150	140	130	120	110	100	90	80	70	60	50	40	30	20	10	Total Ascent Time (M:S)
5	8:00																		3	4	15:20
10	6:20													2	4	4	4	7	10	14	51:40
15	5:20										2	4	4	4	7	9	10	11	22	22	100:40
20	4:40								1	4	4	5	9	10	9	9	20	22	22	69	189:00
25	4:20							1	4	5	10	9	10	9	16	22	22	22	43	113	290:40
Exceptional Exposure																					
30	4:20							3	9	10	9	9	11	22	22	22	22	22	105	113	383:40
35	4:00						2	9	10	9	9	17	22	22	22	22	22	72	113	113	468:20
40	4:00						8	9	9	10	20	22	22	23	22	22	34	113	113	113	544:20
45	3:40					3	9	9	11	22	22	22	22	22	22	22	86	113	113	113	615:00

265 FSW

Bottom Time (min)	Time to First Stop (M:S)	190	180	170	160	150	140	130	120	110	100	90	80	70	60	50	40	30	20	10	Total Ascent Time (M:S)
5	8:10																		4	4	16:30
10	6:30													4	3	4	4	8	10	15	54:50
15	5:30										4	4	3	4	9	9	9	13	22	22	104:50
20	4:50								3	4	3	7	9	10	9	9	22	22	22	78	203:10
25	4:30							2	4	8	9	10	9	9	18	22	22	22	55	113	307:50
Exceptional Exposure																					
30	4:30							6	10	9	9	10	13	22	22	22	22	27	113	113	402:50
35	4:10						5	10	9	10	9	19	22	23	22	22	22	87	113	113	490:30
40	3:50					2	10	9	10	11	22	22	22	22	22	22	52	113	113	113	569:10
45	3:50					7	9	9	15	22	22	22	22	22	22	26	100	113	113	113	641:10

Table 17-10. *Closed-Circuit Mixed-Gas UBA Decompression Table Using 0.7 ata Constant Partial Pressure Oxygen in Helium (Continued).*
(DESCENT RATE 60 FPM—ASCENT RATE 30 FPM)

Bottom Time (min)	Time to First Stop (M:S)	DECOMPRESSION STOPS (fsw) Stop times (min) include travel time, except first stop																			Total Ascent Time (M:S)
		190	180	170	160	150	140	130	120	110	100	90	80	70	60	50	40	30	20	10	
270 FSW																					
5	8:00																	1	4	4	17:20
10	6:20												1	4	4	4	4	9	9	16	57:40
15	5:20									1	4	4	4	4	9	10	9	15	22	22	109:40
20	5:00								4	4	4	8	9	9	10	11	22	22	22	88	218:20
25	4:40							4	4	9	9	10	9	10	20	22	22	22	66	113	325:00
Exceptional Exposure																					
30	4:20						2	8	9	10	9	10	16	22	22	22	22	41	113	113	423:40
35	4:20						9	9	10	9	10	22	22	22	22	22	22	102	113	113	511:40
40	4:00					6	9	10	9	15	22	22	22	22	22	22	69	113	113	113	593:20
45	3:40				1	10	9	10	18	22	22	22	22	22	22	37	107	113	113	113	667:00
275 FSW																					
5	8:10																	2	4	4	18:30
10	6:30												2	4	4	4	4	10	9	18	61:50
15	5:30									3	4	3	4	5	10	9	10	16	22	24	115:50
20	4:50							2	4	4	4	9	9	10	9	14	22	22	22	99	235:10
Exceptional Exposure																					
25	4:30						2	4	5	9	10	9	10	10	22	22	22	22	79	113	343:50
30	4:30						4	9	10	9	10	9	19	22	22	22	22	55	113	113	443:50
35	4:10					4	9	9	10	9	13	22	22	22	22	22	32	108	113	113	534:30
40	3:50				1	9	10	9	9	19	22	22	22	22	22	22	86	113	113	114	619:10
45	3:50				5	10	9	9	22	22	22	22	22	22	22	48	113	113	113	113	691:10
280 FSW																					
5	8:20																	3	4	3	18:40
10	6:40												3	4	4	4	5	10	9	19	65:00
15	5:40									4	4	4	4	6	9	10	9	18	22	32	128:00
20	5:00							3	4	4	5	10	9	10	9	15	23	22	22	109	250:20
Exceptional Exposure																					
25	4:40						3	4	7	10	9	10	9	12	22	22	22	22	92	113	362:00
30	4:20					2	6	9	10	9	10	9	21	22	22	22	22	70	113	113	464:40
35	4:20					7	10	9	9	10	16	22	22	22	22	22	43	113	113	113	557:40
40	4:00				4	10	9	10	9	22	22	22	22	22	22	26	99	113	113	113	642:20
45	4:00				9	9	10	13	22	22	22	22	22	22	22	68	113	113	113	113	719:20

Table 17-10. *Closed-Circuit Mixed-Gas UBA Decompression Table Using 0.7 ata Constant Partial Pressure Oxygen in Helium (Continued).*
(DESCENT RATE 60 FPM—ASCENT RATE 30 FPM)

285 FSW

Bottom Time (min)	Time to First Stop (M:S)	190	180	170	160	150	140	130	120	110	100	90	80	70	60	50	40	30	20	10	Total Ascent Time (M:S)
		DECOMPRESSION STOPS (fsw) Stop times (min) include travel time, except first stop																			
5	8:30																	3	4	4	19:50
10	6:30											1	4	4	3	4	7	9	10	20	68:50
15	5:30								2	4	4	3	4	8	9	10	9	21	22	40	141:50
20	4:50						1	4	4	4	7	9	9	10	9	18	22	22	29	113	266:10
Exceptional Exposure																					
25	4:30					1	4	4	9	9	10	9	10	14	22	22	22	23	104	113	380:50
30	4:30					3	8	10	9	10	9	11	22	22	22	22	22	84	113	113	484:50
35	4:10				2	9	10	9	9	10	19	22	22	22	22	22	59	113	113	113	580:30
40	4:10				8	10	9	10	12	22	22	22	22	22	22	38	104	113	113	113	666:30
45	3:50			4	9	10	9	17	22	22	22	22	22	22	22	87	113	113	113	113	746:10

290 FSW

Bottom Time (min)	Time to First Stop (M:S)	190	180	170	160	150	140	130	120	110	100	90	80	70	60	50	40	30	20	10	Total Ascent Time (M:S)
5	8:20																1	4	3	5	21:40
10	6:40											2	4	4	4	3	8	9	10	22	73:00
15	5:40								3	4	4	4	4	8	10	9	10	22	22	48	154:00
20	5:00						3	4	3	4	9	9	9	10	9	20	22	22	40	113	282:20
Exceptional Exposure																					
25	4:40					3	4	5	9	9	10	9	10	17	22	22	22	31	109	113	400:00
30	4:20				1	5	9	10	9	9	10	14	22	22	22	22	23	99	113	113	507:40
35	4:20				5	10	9	10	9	10	22	22	22	22	22	22	76	113	113	113	604:40
40	4:00			3	9	10	9	10	15	22	23	22	22	22	22	49	111	113	113	113	692:20
45	4:00			8	9	10	9	20	22	22	22	22	22	22	31	95	113	113	113	113	770:20

295 FSW

Bottom Time (min)	Time to First Stop (M:S)	190	180	170	160	150	140	130	120	110	100	90	80	70	60	50	40	30	20	10	Total Ascent Time (M:S)
5	8:30																1	4	4	5	22:50
10	6:50											3	4	4	4	3	9	9	11	22	76:10
15	5:30							1	4	4	3	4	5	9	10	9	12	22	22	56	166:50
20	4:50					1	3	4	4	4	10	9	10	9	10	22	22	22	50	113	298:10
Exceptional Exposure																					
25	4:30				1	4	4	6	10	9	9	10	9	20	22	22	22	41	112	113	418:50
30	4:30				3	6	10	9	9	10	9	17	22	22	22	22	33	103	113	113	527:50
35	4:30				9	9	10	9	10	12	22	22	22	22	22	23	91	113	113	113	626:50
40	4:10			7	9	10	9	9	20	22	22	22	22	22	22	66	113	113	113	113	718:30
45	3:50		2	10	9	10	11	22	22	22	22	22	22	22	43	102	113	113	113	113	797:10

Table 17-10. Closed-Circuit Mixed-Gas UBA Decompression Table Using 0.7 ata Constant Partial Pressure Oxygen in Helium (Continued).

(DESCENT RATE 60 FPM—ASCENT RATE 30 FPM)

Bottom Time (min)	Time to First Stop (M:S)	DECOMPRESSION STOPS (fsw) Stop times (min) include travel time, except first stop																			Total Ascent Time (M:S)
		190	180	170	160	150	140	130	120	110	100	90	80	70	60	50	40	30	20	10	
300 FSW																					
5	8:40																2	4	4	6	25:00
10	7:00											4	4	4	4	4	9	9	12	22	79:20
15	5:40							2	4	4	4	4	5	10	9	10	14	22	22	64	180:00
20	5:00					2	4	4	4	5	10	9	10	9	12	22	22	22	62	113	315:20
Exceptional Exposure																					
25	4:40				2	4	4	8	10	9	10	9	9	22	22	23	22	51	113	113	436:00
30	4:20			1	4	8	9	10	9	10	9	20	22	22	22	22	43	108	113	113	549:40
35	4:20			4	9	9	10	9	10	15	22	22	22	22	23	32	97	113	113	113	649:40
40	4:00		1	10	9	10	9	10	22	22	22	22	22	22	22	83	113	113	113	113	742:20
310 FSW																					
Exceptional Exposure																					
6	8:20														1	4	4	4	6	9	36:40
10	7:00										2	4	4	4	4	6	9	10	15	22	87:20
15	5:40						1	4	4	4	4	4	8	9	9	10	18	22	22	81	206:00
20	5:00				1	4	4	4	4	8	10	9	10	9	17	22	22	22	85	113	349:20
25	4:40			2	4	3	7	9	10	9	9	10	14	22	22	22	22	81	113	113	477:00
30	4:40			4	6	10	9	10	9	10	12	22	22	22	22	22	69	113	113	113	593:00
35	4:20		2	9	10	9	9	10	9	22	22	22	22	22	22	54	109	113	113	113	696:40
40	4:20		9	9	10	9	10	16	22	22	22	22	23	22	41	98	113	113	113	113	791:40
320 FSW																					
Exceptional Exposure																					
6	8:40														3	4	4	4	7	10	41:00
10	7:00									1	4	4	4	4	4	7	10	9	19	22	95:20
15	6:00						4	3	4	4	4	5	10	9	9	10	22	22	22	98	232:20
20	5:20				4	4	4	4	6	9	10	9	9	10	22	22	22	28	102	113	383:40
25	4:40		1	4	4	4	9	10	9	10	9	10	19	22	22	22	34	96	113	113	516:00
30	4:40		3	5	10	9	9	10	9	10	18	22	22	22	22	31	91	113	113	113	637:00
35	4:20	1	8	10	9	10	9	9	16	22	22	22	22	22	24	84	113	113	113	113	746:40
40	4:20	7	10	9	10	9	11	22	22	22	22	22	22	22	66	112	113	113	113	113	844:40

PAGE LEFT BLANK INTENTIONALLY

CHAPTER 18

MK 16 MOD 1 Closed-Circuit Mixed-Gas UBA Diving

18-1 INTRODUCTION

Chapter 18 is intended for use by Explosive Ordnance Disposal (EOD) divers using the MK 16 MOD 1 Closed-Circuit Mixed-Gas Underwater Breathing Apparatus (UBA), Figure 18-1. This equipment combines the mobility of a free-swimming diver with the advantages of mixed-gas diving. The term closed-circuit refers to the recirculation of 100 percent of the mixed-gas breathing medium. This results in bubble-free operation, except during ascent or inadvertent gas release. This capability makes closed-circuit UBA's well-suited for EOD operations and for operations requiring a low acoustic signature. Improvements in gas usage, dive duration, and depth capabilities provided by the UBA greatly increase the effectiveness of the divers. Dives to 190 feet of seawater (fsw) can be made when N_2O_2 (air) is used as a diluent and 300 fsw when using HeO_2 (88/12) as a diluent, see Table 18-1. Due to the increased breathing resistance, and concerns about carbon dioxide retention and CNS O_2 toxicity, planned N_2O_2 dives deeper than 150 fsw are considered exceptional exposure dives and require prior CNO approval.

Figure 18-1. MK 16 MOD 1 Closed-Circuit Mixed-Gas UBA.

18-1.1 **Purpose.** This chapter provides general guidelines for MK 16 MOD 1 UBA diving, operations and procedures. For detailed operation and maintenance instructions, see NAVSEA SS600-AQ-MMO-010 Underwater Breathing Apparatus MK 16 MOD 1 Technical Manual.

18-1.2 **Scope.** This chapter covers MK 16 MOD 1 UBA operational planning, dive procedures, and medical aspects of mixed-gas closed-circuit diving. Refer to Chapter 16 for procedures for mixing divers' breathing gas.

18-2 OPERATIONAL PLANNING

Table 18-1 lists the operational characteristics of the MK 16 MOD 1. Because the MK 16 MOD 1 UBA maintains a constant partial pressure of oxygen (ppO_2) and only adds oxygen or diluent gas as needed, dives of long duration are possible. Mission capabilities, dive procedures, and decompression procedures are radically different from other diving methods. This requires a high level of diver training and awareness and necessitates careful dive planning. Chapter 6 provides general guidelines for operational planning. The information provided in this section is supplemental to the MK 16 MOD 1 UBA O&M manual and provides specific

Table 18-1. *MK 16 MOD 1 Operational Characteristics.*

MK 16 MOD 1 UBA	Normal Working Limit (fsw)	Maximum Working Limit (fsw)
N_2O_2 Diluent (Notes 1,2,4,5)	150	190
HeO_2 Diluent (Notes 1,3,4)	300	300

Notes:

1. Within each decompression table, exceptional exposure dives are separated by a dashed line. These tables are designed to be dived to the exceptional exposure line. Exceptional exposure schedules are provided in case of unforeseen circumstances. The expected duration of the gas supply, the expected duration of the carbon dioxide absorbent, the adequacy of thermal protection, or other factors may also limit both the depth and duration of the dive. The breathing resistance and risks of CO_2 retention, oxygen toxicity, decompression sickness and nitrogen narcosis are increased when using N_2O_2 as a diluent on dives below the exceptional exposure line. Planned exceptional exposure dives require prior CNO approval.

2. Permissible 1.3 ata N_2O_2 dive profiles:
   - Multiple repetitive dives from 0–150 fsw provided all the dives are no-decompression dives.
   - One decompression dive from 0–150 fsw plus up to three additional no-decompression dives from 0–150 fsw. The three no-decompression dives may precede, follow, or bracket the decompression dive.
   - Two decompression dives (initial decompression dive plus one repetitive decompression dive) from 0–150 fsw. Additional no-decompression dives are not allowed before or after the decompression dives.
   - Repetitive dives from 151–190 fsw under the rules above providing CNO grants a waiver for exceptional exposure diving.

3. Permissible 1.3 ata HeO_2 dive profiles:
   - Multiple repetitive dives from 0–200 fsw provided all the dives are no-decompression dives.
   - One decompression dive from 0–200 fsw plus up to three additional no-decompression dives from 0–200 fsw. The three no-decompression dives may precede, follow, or bracket the decompression dive.
   - Two decompression dives (initial decompression dive plus one repetitive decompression dive) from 0–200 fsw. Additional no-decompression dives are not allowed before or after the decompression dives.
   - One no-decompression dive or one decompression dive from 201–300 fsw. Repetitive diving is not allowed deeper than 200 fsw.

4. Switching diluents between dives is NOT authorized in the MK 16 MOD 1. There are no procedures for performing a repetitive dive on helium following a dive on nitrogen or for performing a repetitive dive on nitrogen following a dive on helium.

5. A repetitive dive on air can be performed following a MK 16 MOD 1 dive on nitrogen. See paragraph 9-9.3 for guidance. A repetitive MK 16 MOD 1 dive on nitrogen can be performed following a dive on air. See paragraph 18-7.2, subparagraph 14 for guidance.

guidelines for MK 16 MOD 1 UBA dive planning. In addition to any other requirements, at least half of all dive training should be at night or in conditions of restricted visibility. Units should allow frequent opportunity for training, ensuring diver familiarity with equipment and procedures. Workup dives are strongly recommended prior to diving at depths greater than 130 fsw. MK 16 MOD 1 diver qualifications may be obtained only by completion of the MK 16 MOD 1 Basic Course *(A-431-0075)*. MK 16 MOD 1 qualifications remain in effect as long as diver qualifications are maintained in accordance with Military Personnel Manual article *1220-100*. However, a diver who has not made a MK 16 MOD 1 dive in the previous six months must refamiliarize himself with MK 16 MOD 1 EPs and OPs and must complete a MK 16 MOD 1 training dive prior to making a MK 16 MOD 1 operational dive. Prior to conducting MK 16 MOD 1 decompression diving, a diver who has not conducted a MK 16 MOD 1 decompression dive within the previous six months must complete open water decompression training dives. Refer to Table 18-2 for the personnel requirements for MK 16 MOD 1 diving.

Table 18-2. *Personnel Requirements Chart for MK 16 MOD 1 Diving.*

MK16 MOD 1 UBA Dive Team
Minimum Manning Requirements

Designation	**One Diver**		**Two Divers**	
Diving Supervisor	1	(Note 1)	1	
Diver	1		2	
Standby Diver	1	(Note 2)	1	(Note 2)
Diver Tender	1	(Note 3)	1	(Note 3)
Standby Diver Tender		(Note 1)		(Note 1)
Timekeeper/Recorder		(Note 1)		(Note 1)
EBS Operator		(Note 4)		(Note 4)
Total Personnel Required	4		5	

Notes:

1. Diving Supervisor may act as time keeper/recorder, standby tender.
2. At the Diving Supervisor's discretion, the standby diver shall be fully dressed with the exception of SCUBA or MK 16 MOD 1, mask, and fins. These items shall be ready to don.
3. One tender per diver when divers are surface tended. If using a buddy line, one tender is required for each buddy pair.
4. EBS Operator is required for MK 16 MOD 1 decompression dives.

18-2.1 **Operating Limitations.** Diving Supervisors must also consider the limiting factors presented in the following paragraphs when planning closed-circuit UBA operations.

18-2.1.1 **Oxygen Flask Endurance.** Table 18-3(a) thru (e) were calculated using the formula for Oxygen Flask Endurance that is presented in the MK 16 MOD 1 Operations and Maintenance Manual and should be used when planning maximum dive times for 1.3 ata dives.

Table 18-3a. *Flask Endurance for 29°F Water Temperature.*

29° F Water Temp											
Air Temp	**2000 PSI**	**2100 PSI**	**2200 PSI**	**2300 PSI**	**2400 PSI**	**2500 PSI**	**2600 PSI**	**2700 PSI**	**2800 PSI**	**2900 PSI**	**3000 PSI**
20	215	229	243	257	272	286	300	314	329	343	357
30	209	223	237	251	265	279	293	307	320	334	348
40	203	217	231	244	258	272	285	299	313	326	340
50	198	211	225	238	252	265	278	292	305	318	332
60	193	206	219	232	245	258	272	285	298	311	324
70	188	201	214	226	239	252	265	278	291	304	317
80	183	196	208	221	234	246	259	271	284	297	309
90	178	191	203	216	228	240	253	265	278	290	303
100	174	186	198	210	223	235	247	259	271	284	296
110	170	182	193	205	217	229	241	253	265	277	289

Table 18-3b. *Flask Endurance for 40°F Water Temperature.*

40° F Water Temp											
Air Temp	**2000 PSI**	**2100 PSI**	**2200 PSI**	**2300 PSI**	**2400 PSI**	**2500 PSI**	**2600 PSI**	**2700 PSI**	**2800 PSI**	**2900 PSI**	**3000 PSI**
20	216	231	245	259	273	288	302	316	330	344	359
30	211	224	238	252	266	280	294	308	322	336	350
40	205	219	232	246	260	273	287	301	314	328	342
50	200	213	226	240	253	266	280	293	307	320	333
60	194	207	221	234	247	260	273	286	299	313	326
70	189	202	215	228	241	254	267	280	292	305	318
80	185	197	210	222	235	248	260	273	286	298	311
90	180	192	205	217	230	242	254	267	279	292	304
100	175	188	200	212	224	236	249	261	273	285	297
110	171	183	195	207	219	231	243	255	267	279	291

Table 18-3c. *Flask Endurance for 60°F Water Temperature.*

60° F Water Temp											
Air Temp	**2000 PSI**	**2100 PSI**	**2200 PSI**	**2300 PSI**	**2400 PSI**	**2500 PSI**	**2600 PSI**	**2700 PSI**	**2800 PSI**	**2900 PSI**	**3000 PSI**
20	219	233	248	262	276	290	304	319	333	347	361
30	213	227	241	255	269	283	297	311	325	339	353
40	208	221	235	249	262	276	290	303	317	331	344
50	202	216	229	242	256	269	283	296	309	323	336
60	197	210	223	236	250	263	276	289	302	315	328
70	192	205	218	231	244	256	269	282	295	308	321
80	187	200	213	225	238	250	263	276	288	301	314
90	183	195	207	220	232	245	257	270	282	294	307
100	178	190	203	215	227	239	251	264	276	288	300
110	174	186	198	210	222	234	246	258	270	282	294

Table 18-3d. *Flask Endurance for 80°F Water Temperature.*

80° F Water Temp											
Air Temp	**2000 PSI**	**2100 PSI**	**2200 PSI**	**2300 PSI**	**2400 PSI**	**2500 PSI**	**2600 PSI**	**2700 PSI**	**2800 PSI**	**2900 PSI**	**3000 PSI**
20	222	236	250	264	279	293	307	321	335	350	364
30	216	230	244	258	271	285	299	313	327	341	355
40	210	224	237	251	265	278	292	306	319	333	347
50	205	218	232	245	258	272	285	298	312	325	339
60	200	213	226	239	252	265	278	291	305	318	331
70	195	207	220	233	246	259	272	285	298	311	323
80	190	202	215	228	240	253	266	278	291	304	316
90	185	198	210	222	235	247	260	272	284	297	309
100	181	193	205	217	229	242	254	266	278	290	303
110	176	188	200	212	224	236	248	260	272	284	296

Table 18-3e. *Flask Endurance for 104°F Water Temperature.*

104° F Water Temp											
Air Temp	**2000 PSI**	**2100 PSI**	**2200 PSI**	**2300 PSI**	**2400 PSI**	**2500 PSI**	**2600 PSI**	**2700 PSI**	**2800 PSI**	**2900 PSI**	**3000 PSI**
20	224	239	253	267	281	296	310	324	338	352	367
30	219	232	246	260	274	288	302	316	330	344	358
40	213	227	240	254	268	281	295	309	322	336	350
50	207	221	234	248	261	274	288	301	315	328	341
60	202	215	229	242	255	268	281	294	307	321	334
70	197	210	223	236	249	262	275	288	300	313	326
80	193	205	218	230	243	256	268	281	294	306	319
90	188	200	213	225	238	250	262	275	287	300	312
100	183	196	208	220	232	244	257	269	281	293	305
110	179	191	203	215	227	239	251	263	275	287	299

18-2.1.2 **Effect of Cold Water Immersion on Flask Pressure.** Immersion in cold water will reduce the flask pressure and actual cubic feet (acf) of gas available for the diver, in accordance with Charles'/Gay-Lussac's gas law. Based upon direct measurement, available data, or experience, the coldest temperature expected during the dive is used.

18-2.1.3 **Diluent Flask Endurance.** Under normal conditions the anticipated duration of the MK 16 MOD 1 diluent flask will exceed that of the oxygen flask. The MK 16 MOD 1 diluent bottle holds approximately 21 standard cubic feet (595 liters) of gas at a stored pressure of 3,000 psig. Diluent gas is used to maintain the required gas volume in the breathing loop and is not depleted by metabolic consumption. As the diver descends, diluent is added to maintain the total pressure within the recirculation system at ambient water pressure. Loss of UBA gas due to off gassing at depth requires the addition of diluent gas to the breathing loop either automatically through the diluent addition valve or manually through the diluent bypass valve to make up lost volume. Excessive gas loss caused by face mask leaks, frequent depth changes, or improper UBA assembly will deplete the diluent gas supply rapidly.

18-2.1.4 **Canister Duration.** Table 18-4 shows the canister duration limits for the MK 16 MOD 1 UBA. Above 94°F, the primary concern limiting the dive duration is diver physiological considerations vice the canister duration.

Table 18-4. *MK 16 MOD 1 Canister Duration Limits.*

Canister Duration with N_2O_2		
Water Temperature (°F)	**Depth (fsw)**	**Time (minutes)**
97-99	0-190	60 (Note 1)
95-97	0-190	180 (Note 1)
40-94	51-190	200
29-39	51-190	100
29-94	0-50	300

Canister Duration with HeO_2		
Water Temperature (°F)	**Depth (fsw)**	**Time (minutes)**
97-99	0-300	60 (Note 1)
95-97	0-300	180 (Note 1)
40-94	0-300	300
35-39	101-300	300
29-34	101-300	240
29-39	0-100	120

Notes:

(1) Based on physiological limits. Refer to Chapter 6 Para 6-5.3.

NAVSEA-approved Sodalime CO_2 absorbents are listed in the ANU list.

18-2.2 **Equipment Requirements.** The minimum equipment requirements for MK 16 MOD 1 UBA dives are provided in Table 18-5 and explained in the following paragraphs.

18-2.2.1 **Safety Boat.** A minimum of one motorized safety boat must be present for all open-water dives. A safety boat is also recommended for tended pier dives or diving from shore. Safe diving practice in many situations, however, will require the presence of more than one safety boat. The Diving Supervisor must determine the number of boats required based on the diving area, medical evacuation plan, night operations, and the number of personnel participating in the dive operation.

18-2.2.2 **Buddy Lines.** Buddy lines are considered important safety equipment for closed-circuit UBA dives. In special diving situations, such as tended diving, the use of buddy lines may not be feasible. The Diving Supervisor shall conduct dives without buddy lines only in situations where their use is not feasible or where their use will pose a greater hazard to the divers than diving without them.

18-2.2.3 **Distance Line.** Any buddy line over 10 feet (3 meters) in length is referred to as a distance line. The length of the distance line shall not exceed 81 feet (25 meters). Distance lines shall be securely attached to both divers.

18-2.2.4 **Standby Diver.** When appropriate during training and non-influence diving operations open circuit scuba may be used. Refer to Chapter 6 Figure 6-23 for guidance.

Table 18-5. MK 16 MOD 1 UBA Diving Equipment Requirements.

General	Diving Supervisor	Divers	Standby Diver
1. Motorized safety boat	1. Dive watch	1. Dive watch	1. Dive watch
2. Radio (communications with parent unit, chamber, communications between safety boats when feasible)	2. Dive Bill List	2. Face Mask	2. Face Mask
3. High intensity, wide beam light (night operations)	3. Appropriate Decom-pression Tables	3. Fins	3. Fins
4. Dive flags and/or special operations lights as required	4. Recall device	4. Dive Knife	4. Dive Knife
5. Sufficient (2 quarts) fresh water in case of chemical injury.		5. Approved life preserver or Buoyancy Control Device (BCD)	5. Approved life preserver or Buoyancy Control Device (BCD)
6. Emergency Breathing System for planned decompression dives.		6. Appropriate Thermal Protection	6. Appropriate Thermal Protection
		7. MK 16 MOD 1 UBA	7. UBA with same depth capability. For non-in-fluence ordnance diving and training dives, standby diver may use SCUBA.
		8. Depth Gauge	8. Depth Gauge
		9. Weight Belt (as needed)	9. Weight Belt (as needed)
		10. Buddy lines as appropriate for EOD diving operations.	10. Tending line
		11. Tending line as appropriate for EOD diving operations	

18-2.2.5 **Tending Lines.** Diver tending lines should be manufactured from any light line that is buoyant and easily marked as directed in paragraph 18-2.2.6 (one-quarter inch polypropylene is quite suitable).

18-2.2.6 **Marking of Lines.** Lines used for controlling the depth of the diver(s) for decompression diving shall be marked. This includes tending lines, marker lines, and lazy-shot lines. Lines shall be marked with red and yellow or black bands starting at the diver(s) or clump end. Red bands will indicate 50 feet and yellow or black bands will mark every 10 feet.

18-2.2.7 **Diver Marker Buoy.** Diver marker buoys will be constructed to provide adequate visual reference to monitor the diver's location. Additionally, the amount of line will be of sufficient length for the planned dive profile.

18-2.2.8 **Depth Gauge/Wrist Watch.** A single depth gauge and wrist watch may be used when diving with a partner and using a buddy line.

18-2.2.9 **Thermal Protection.** Divers must be equipped with adequate thermal protection to perform effectively and safely. A cold diver will either begin to shiver or increase his exercise rate, both of which will increase oxygen consumption and decrease oxygen supply duration and canister duration. Refer to Chapter 6 paragraph 6-5.3 for guidance on warm water diving and Chapter 11 for guidance on cold water diving.

18-2.2.10 **Approved Life Preserver or Buoyancy Compensator (BC).** An approved life preserver or optional BC may be used at the discretion of the diving supervisor. The MK 4 life preserver is authorized to 200 fsw and shall be fitted with 4 (30-34 gram) MIL-C-16385 Type II (non-magnetic) cartridges for EOD operations. If the diver is wearing an approved dry suit, the use of a life preserver or BC is not required.

18-2.2.11 **Full Face Mask (FFM).** An authorized full face mask shall be used when deploying a single untended diver, single marked diver, paired marked diver, and when using an approved BC.

18-2.2.12 **Emergency Breathing System (EBS).** The Emergency Breathing System provides an alternate breathing source for decompressing diver(s) in the event of a MK 16 MOD 1 failure. The EBS consists of a MK 16 MOD 1 UBA mounted on the EBS frame assembly and charged with the same diluent gas as for the planned dive.

18-2.3 **Recompression Chamber Considerations.** A recompression chamber is required on-site for all MK 16 MOD 1 UBA decompression dives deeper than 200 fsw, regardless of the mission.

The following items should be determined prior to beginning any diving operation:

- Location of the nearest functional recompression chamber. Positive confirmation of the chamber's availability should be obtained.
- The optimal method of transportation to the treatment chamber or medical facility. If coordination with other units for aircraft/boat/vehicle support is necessary, the Diving Supervisor shall know the telephone numbers and points of contact necessary to make these facilities available as quickly as possible. A medical evacuation plan should be included in the Diving Supervisor brief. Preparing an emergency assistance checklist similar to that in Chapter 6 is recommended.

A recompression chamber, CNO waiver, and a Diving Medical Officer are required prior to any planned dive which exceeds the maximum working limits.

18-2.4 Diving Procedures for MK 16 MOD 1.

18-2.4.1 **EOD Standard Safety Procedures.** The following standard safety procedures shall be observed during EOD diving operations:

- An EOD Diving Officer is required to be in Tactical Control of all EOD diving operations that involve Render Safe Procedures (RSP). Tactical Control is defined as either on-station or in continuous, full time tactical voice communications with the dive team conducting the RSP.

- When diving on unknown or influence ordnance, the standby diver's equipment shall be the same type as the diver performing the actual procedure.

18-2.4.2 **Diving Methods.** MK 16 MOD 1 Diving methods include:

18-2.4.2.1 ***Deploying a Single, Untended EOD Diver.*** Generally, it is safer for divers to work in pairs rather than singly. However, to do so when diving on underwater influence ordnance doubles the diver bottom time expended, increases the risk to life from live ordnance detonation, and increases the risk of detonation caused by the additional influence signature of the second diver. The EOD Diving Officer may authorize the employment of a single, untended diver when it is deemed that the ordnance hazard is greater than the hazard presented by diving alone. All single, untended divers shall use a full face mask (FFM). The EOD Diving Officer or Diving Supervisor shall consider the following factors when deciding whether to operate singly or in pairs:

- Experience of the diver
- Confidence of the team
- Type and condition of ordnance suspected
- Environmental conditions
- Degree of operational urgency required

18-2.4.2.2 ***Single Marked Diving.*** Consists of a single diver with FFM marked with a lightweight buoyant line attached to a surface float. Upon completion of a dive requiring decompression, the diver will signal the diving supervisor that he is ready to surface. The diving boat will then approach the surface float and recover the diver.

18-2.4.2.3 ***Paired Marked Diving.*** Procedures for paired marked diving are identical to the procedures for a single marked diver, but with the addition of the second diver connected by a buddy/distance line.

18-2.4.2.4 ***Tended Diving.*** Tended diving consists of a single surface-tended diver or a pair of divers using a buddy/distance line, with one diver wearing a depth-marked line that is continuously tended at the surface.

18-2.4.2.5 ***Diver Training Scenarios.*** Simulated ordnance training scenarios do not constitute a real threat, therefore single untended divers shall not be used in training operations. The diver(s) shall be surface tended or marked by a buoy.

18-2.5 **Ship Safety.** When operations are to be conducted in the vicinity of ships, the guidelines provided in the Ship Repair Safety Checklist (see Figure 6-20) must be followed.

18-2.6 **Operational Area Clearance.** Notification of intent to conduct diving operations should be coordinated in accordance with local directives.

18-3 PREDIVE PROCEDURES

18-3.1 **Diving Supervisor Brief.** A thorough, well-prepared dive briefing reinforces the confidence level of the divers and increases safety, and is an important factor in successful mission accomplishment. It should normally be given by the Diving Supervisor, who will be in charge of all diving operations on the scene. The briefing shall be given separately from the overall mission briefing and shall focus on the diving portion of the operation, with special attention to the items shown in Table 18-6. MK 16 MOD 1 UBA line-pull dive signals are listed in Table 18-7. For MK 16 MOD 1 UBA diving, use the appropriate checklist provided in the MK 16 MOD 1 UBA O&M Manual. It is recommended that the Dive Record Sheet shown in Figure 18-2 be used by Diving Supervisors for MK 16 MOD 1 diving.

18-3.2 **Diving Supervisor Check.** Prior to the dive, the Diving Supervisor must ensure each UBA is setup properly and a predive checklist is completed. The second phase of the Diving Supervisor check is a predive inspection conducted after the divers are dressed (refer to Figure 3-3 of the MK 16 MOD 1 O & M manual). The Diving Supervisor ensures that the UBA and related gear (life preserver, weight belt, etc.) are properly donned, that mission-related equipment (compass, depth gauge, dive watch, buddy lines, tactical equipment, etc.) are available, and that the UBA functions properly before allowing the divers to enter the water. Appropriate check lists to confirm proper functioning of the UBA are provided in the MK 16 MOD 1 O&M manual.

Table 18-6. *MK 16 MOD 1 UBA Dive Briefing.*

A. **Dive Plan**
   1. Operating depth
   2. Dive times
   3. Decompression tables
   4. Distance, bearing, and transit times
   5. All known obstacles or hazards

B. **Environment**
   1. Weather conditions
   2. Water/air temperatures
   3. Water visibility
   4. Tides/currents
   5. Depth of water
   6. Bottom type
   7. Geographic location

C. **Personnel Assignments**
   1. Dive pairs
   2. Diving Supervisor
   3. Diving Officer
   4. Standby diver
   5. Diving medical personnel
   6. Base of operations support personnel

D. **Special Equipment for:**
   1. Divers (include thermal garments)
   2. Diving Supervisor
   3. Standby diver
   4. Medical personnel

E. **Review of Dive Signals**
   1. Hand signals
   2. MK 16 MOD 1 UBA Line-Pull Dive Signals (Table 18-7)

F. **Communications**
   1. Frequencies, primary/secondary
   2. Call signs

G. **Emergency Procedures**
   1. Symptoms of CNS O_2 toxicity and CO_2 buildup
   2. Review of management of CNS O_2 toxicity, CO_2 toxicity, hypoxia, chemical injury, unconscious diver
   3. UBA malfunction (refer to maintenance manual for detailed discussion)
      - Oxygen sensor failure
      - Low partial pressure of oxygen
      - High partial pressure of oxygen
      - Electronics failure
      - Flashing Red Primary display on ascent
      - (non-emergency situation)
      - Low battery
      - Diluent free flow
      - Diluent addition valve failure
      - System flooding
      - Failure of Primary Electronics to switch over to 1.3 ATA ppO_2 on descent
   4. Lost swim pair procedures
   5. Omitted decompression plan
   6. Medical evacuation plan
      - Nearest available chamber
      - Nearest Diving Medical Officer
      - Transportation plan
      - Recovery of other swim pairs

H. **Times for Operations**

I. **Time Check**

Table 18-7. *MK 16 MOD 1 UBA Line-Pull Signals.*

Signal	From	To	Meaning
1 Pull	Diver	Tender	Arrived at lazy shot (given on lazy shot)
7 Pulls	Diver	Tender	I have started, found, or completed work
2-3 Pulls	Diver	Tender	I have decompression symptoms.
3-2 Pulls	Diver	Tender	Breathing from EBS (EBS UBA is functioning properly)
4-2 Pulls	Diver	Tender	Rig malfunction
2-1 Pulls	Diver Tender	Tender Diver	Unshackle from the lazy shot
5 Pulls	Diver	Tender	I have exceeded the planned depth of the dive. (This is followed by 1 pull for every 5 fsw of depth the planned depth was exceeded)

MK 16 MOD 1 DIVE RECORD SHEET		
Diving Supervisor		Date
Water Temp	Air Temp	Depth (FSW)
Table	Schedule	Planned Bottom Time
EBS Oxygen Bottle Pressure	EBS Diluent Bottle Pressure	

	Name	Repet Group	Rig No.	O_2 Pressure	Diluent Pressure	BATT Percent	LS	LB	RS	TBT
Diver 1										
Diver 2										
Standby Diver										

Descent Rate	Schedule Time at Stop		Stop Depth	Actual time at Stop		Travel Time	Remarks
	Diver	Standby		Diver	Standby		
			20				
			30				
			40				
			50				
			60				
			70				
			80				
			90				
			100				
			110				
			120				
			130				
			140				
			150				

Figure 18-2. MK 16 MOD 1 Dive Record Sheet.

18-4 DESCENT

The maximum descent rate is 60 feet per minute. During descent, the UBA will automatically compensate for increased water pressure and provide an adequate volume of gas for breathing. During descent the oxygen partial pressure will increase as oxygen is added to the breathing mixture as a portion of the diluent. Depending on rate and depth of descent, the primary display on the MK 16 MOD 1 UBA may illuminate flashing green. It may take from 2 to 5 minutes to consume the additional oxygen added by the diluent during descent. While breathing down the ppO_2, the diver should continuously monitor the primary and secondary displays until the ppO_2 returns to the control setpoint level of 1.3 ata.

CAUTION **There is an increased risk of CNS oxygen toxicity when diving the MK 16 MOD 1 compared to diving the MK 16 MOD 0, especially during the descent phase of the dive. Diving supervisors and divers should be aware that oxygen partial pressures of 1.6 ata or higher may be temporarily experienced during descent on N_2O_2 dives deeper than 120 fsw (21% oxygen diluent) and on HeO_2 dives deeper than 200 fsw (12% oxygen diluent). Refer to** paragraph 18-11.1.1 **for information on recognizing and preventing CNS oxygen toxicity.**

The MK 16 MOD 1 UBA primary display should indicate a transition from 0.75 to 1.3 ata at 33 fsw. The diver should verify this transition by monitoring his secondary display. If there is no indication of this transition with continued descent past 40 fsw, the dive should be terminated and the diver should ascend to the surface in accordance with the appropriate decompression schedule.

18-5 UNDERWATER PROCEDURES

18-5.1 General Guidelines. The divers should adhere to the following guidelines as the dive is conducted:

WARNING **Failure to adhere to these guidelines could result in serious injury or death.**

- Monitor primary and secondary display frequently.
- The diver should not add oxygen on descent, except as part of an emergency procedure, or at any time while on the bottom due to the increased risk of CNS oxygen toxicity.
- Wear adequate thermal protection.
- Know and use the proper amount of weights for the thermal protection worn and the equipment carried.
- Check each other's equipment carefully for leaks.
- Do not exceed the UBA canister duration and depth limitations for the dive, see paragraph 18-2.1.4.

- Minimize gas loss from the UBA (avoid mask leaks and frequent depth changes, if possible).
- Maintain frequent visual or touch checks with buddy.
- Be alert for symptoms suggestive of a medical disorder, see paragraph 18-11.
- Use tides and currents to maximum advantage

18-5.2 **At Depth.** If the UBA is operating properly at depth, no adjustments will be required. The ppO_2 control system will add oxygen as necessary to ensure the oxygen level remains at the set-point. Monitor the following displays in accordance with the MK 16 MOD 1 O&M manual:

- **Primary Display.** Check the primary display frequently to ensure that the oxygen level remains at the set-point during normal activity at a constant depth (the oxygen-addition valve operation on the MK 16 MOD 1 cannot be heard).
- **Secondary Display.** Check the secondary display every 2-3 minutes to ensure that all sensors are consistent with the primary display and the battery voltages are properly indicating.
- **High-Pressure Indicators.** Check the oxygen and diluent pressure indicators frequently to ensure the gas supply is adequate to complete the dive.

18-6 ASCENT PROCEDURES

The maximum ascent rate for the MK 16 MOD 1 is 30 feet per minute (fpm). During ascent, when water pressure decreases, the ppO_2 in the breathing gas mixture may decrease faster than O_2 can be added via the O_2 addition valve. Under these circumstances, the primary display may show alternate red/green, then flashing red for low ppO_2. This is a normal reaction to the decrease in partial pressure and is an indication that the UBA is functioning correctly. Even with strict adherence to an ascent rate of 30 fpm, the diver may experience flashing red on the primary display. This may also be an indication of a rig malfunction and manually adding oxygen to the UBA may be necessary. Adding O_2 while observing the secondary display will help the diver discriminate between a normal decrease in oxygen partial pressure due to ascent and a UBA malfunction. Other actions the diver may take are:

- Ensure the ascent rate of 30 fpm is not exceeded.
- Upon arrival at the first decompression stop allow the UBA to stabilize. If after four minutes of arrival at the first stop a flashing red persists on the primary display the diver should initiate the appropriate emergency procedure for low ppO_2.

18-7 DECOMPRESSION PROCEDURES

Standard U.S. Navy decompression tables cannot be used with a closed-circuit UBA since the ppO_2 remains constant at a preset level regardless of depth. Therefore the decompression tables 18-9 through 18-14 have been specifically developed and tested for the MK 16 MOD 1.

18-7.1 **Monitoring ppO_2.** During decompression, it is very important to constantly monitor the secondary display and ensure a 1.3 ppO_2 is maintained as closely as possible. Always use the appropriate decompression table when surfacing, even if UBA malfunction has significantly altered the ppO_2.

NOTE Surface decompression is not authorized for MK 16 MOD 1 operations. Appropriate surface decompression tables have not been developed for constant 1.3 ata ppO_2 closed-circuit diving.

18-7.2 Rules for Using MK 16 MOD 1 Decompression Tables.

NOTE The rules for using the decompression tables are the same for nitrogen and helium; however, the tables are NOT interchangeable.

- These tables are designed to be used with MK 16 MOD 1 UBA.
- For HeO_2 dives, flush the UBA well with helium-oxygen using the purge procedure given in the MK 16 MOD 1 O&M manual.
- Tables are grouped by depth and within each depth group is an exceptional exposure line. These tables are designed to be dived to the exceptional exposure line. Schedules below the exceptional exposure line are provided for unforeseen circumstances when a diver might experience an inadvertent downward excursion or for an unforeseen reason overstay the planned bottom time.
- Tables/schedules are selected according to the maximum depth obtained during the dive and the bottom time (time from leaving the surface to leaving the bottom).
- General rules for using these tables are the same as for air decompression tables, and include the use of the Residual Nitrogen Time (RNT) and Residual Helium Time (RHT) exception rules when calculating the table and schedule for repetitive dives.

NOTE The repetitive group designators are not interchangeable between the nitrogen and helium decompression tables. There are no procedures for performing repetitive dives when the inert gas in the diluent mixture changes between dives.

1. Enter the table at the listed depth that is exactly equal to or is next greater than the maximum depth attained during the dive.
2. Select the bottom time from those listed for the selected depth that is exactly equal to or is next greater than the bottom time of the dive.
3. Never attempt to interpolate between decompression schedules.
4. Use the decompression stops listed for the selected bottom time.

5. Ensure that the diver's chest is maintained as close as possible to each decompression depth for the number of minutes listed.

6. Maximum ascent rate is 30 feet per minute. The rules for compensating for variations in the rate of ascent are identical to those for air diving (see Chapter 9, paragraph 9-11).

7. Begin timing the first stop when the diver arrives at the stop. For all subsequent stops, begin timing the stop when the diver leaves the previous stop. Ascent time between stops is included in the subsequent stop time.

8. The last stop taken will be at 20 fsw. There are no stops shallower than 20 fsw allowed for 1.3 ata ppO_2 diving as the primary electronics will switch from 1.3 ata ppO_2 to 0.75 ata ppO_2 upon ascent above 13 fsw.

9. Always use the appropriate decompression table when surfacing even if UBA malfunction has significantly altered ppO_2.

10. In emergency situations (e.g., UBA flood-out or failure), immediately ascend to the first decompression stop according to the original decompression schedule if deeper than the first stop, and shift to the Emergency Breathing System (EBS).

11. When selecting the proper decompression table, all dives within the past 18 hours must be considered. Repetitive dives are allowed provided the same diluent is used as the previous dives. Refer to the following tables and figures.

    - Figure 18-5 for the Repetitive Dive Worksheet for MK 16 MOD 1 N_2O_2 Dives.
    - Table 18-9 for the No-Decompression Limits and Repetitive Group Designators for MK 16 MOD 1 N_2O_2 Dives.
    - Table 18-10 for the Residual Nitrogen Timetable for MK 16 MOD 1 N_2O_2 Dives.
    - Figure 18-6 for the Repetitive Dive Worksheet for MK 16 MOD 1 HeO_2 Dives.
    - Table 18-11 for the MK 16 MOD 1 N_2O_2 Decompression Tables.
    - Table 18-12 for the No-Decompression Limits and Repetitive Group Designators for MK 16 MOD 1 HeO_2 Dives.
    - Table 18-13 for the Residual Helium Timetable for MK 16 MOD 1 HeO_2 Dives.
    - Table 18-14 MK 16 MOD 1 HeO_2 Decompression Tables

12. The partial pressure of inert gas (nitrogen or helium) in the MK 16 MOD 1 UBA at depths down to 15 fsw is lower than the partial pressure of nitrogen in air at the surface. A diver diving to these depths, therefore, will lose rather than gain inert gas during the dive. Accordingly, the diver does not acquire a repetitive group designator when making these shallow dives. If the dive is a repetitive dive to 15 fsw or shallower, the diver will lose more inert gas during the repetitive dive

than if he remained on the surface. The dive can be considered the equivalent of remaining on the surface for the duration of the dive. The repetitive group designator at the end of the repetitive dive can be determined by adding the bottom time of the repetitive dive to the preceding surface interval, then using the surface interval credit table to determine the ending repetitive group.

13. The RNT and RHT exception rules apply to repetitive MK 16 MOD 1 diving. The RNT and RHT exception rules read identically. The only difference is the table to which the rule refers: Table 18-10 for the RNT exception rule and Table 18-13 for the RHT exception rule. Determine the table and schedule for the repetitive dive by adding the bottom times and taking the deepest depth of all the MOD 1 dives in the series, including the planned repetitive dive. If the resultant table and schedule require less decompression than the table and schedule obtained using the repetitive dive worksheet, it may be used instead of the worksheet table and schedule.

During descent, the MK 16 MOD 1 switches from the 0.75 ata mode to the 1.3 ata mode at 33 fsw ± 2 fsw. The decompression tables assume that the diver is in the 0.75 ata mode up to a depth of 35 fsw. The RNT and RHT exception rules can be used as written above when all the dives in the series are to 35 fsw and shallower (PO_2 = 0.75 ata) or when all the dives in the series are deeper than 35 fsw (PO_2 = 1.3 ata). However, when some dives are shallower than 35 fsw and others are deeper, the shallow 0.75 ata dives must first be converted to their 1.3 ata equivalent depth before the deepest depth in the series is determined. The equivalent depth on 1.3 ata can be obtained by adding 20 fsw to the depth of the dive on 0.75 ata.

14. To perform a repetitive MK 16 MOD 1 nitrogen-oxygen dive following an air dive, take the following steps.

- Obtain the repetitive group designator following the air dive from either Table 9-7 or Table 9-9.
- Using that repetitive group designator, enter Table 18-10 on the diagonal. Read across the row to the surface interval between the air and MK 16 dive, then down to the depth of the MK 16 dive to obtain the residual nitrogen time.
- Add the residual nitrogen time to the bottom time of the repetitive MK 16 dive to obtain the Equivalent Single Dive Time.
- Enter Table 18-9 or Table 18-11 at the depth that is exactly equal to or next deeper than the actual depth of the repetitive dive. Select the schedule that is exactly equal to or next longer than the Equivalent Single Dive Time. Follow the prescribed decompression to the surface.
- The RNT exception rule does apply to Air-MK 16 MOD 1 N_20_2 repetitive diving. If *all* the MK 16 MOD 1 dives in the series, including the repetitive dive, are to 35 fsw or shallower, convert the depth(s) of the air dive(s) in the series to the equivalent depth on 0.75 ata before taking the deepest depth in the series. If any of the MK 16 MOD 1 dives in the series, including the repetitive dive, are to a depth greater than 35 fsw, convert the depth(s) of

the air dive(s) in the series to the equivalent depth on 1.3 ata before taking the deepest depth in the series. Equivalent Depth on 0.75 ata = (0.79 x Depth on Air) + 18 fsw. Equivalent Depth on 1.3 ata = (0.79 x Depth on Air) + 36 fsw.

18-7.3 **PPO_2 Variances.** The ppO_2 in the MK 16 MOD 1 UBA is expected to vary slightly from 1.15 – 1.45 ata for irregular brief intervals. This does not constitute a malfunction. When addition of oxygen to the UBA is manually controlled, ppO_2 should be maintained in accordance with techniques and emergency procedures listed in the MK 16 MOD 1 O&M manual.

The Diving Supervisor and medical personnel should recognize that a diver who has been breathing a mixture with ppO_2 lower than 1.15 ata for any length of time may have a greater risk of developing decompression sickness. Such a diver requires observation after surfacing, but need not be treated unless symptoms of decompression sickness occur.

18-7.4 **Emergency Breathing System (EBS).** The Emergency Breathing System (Figure 18-3) provides an alternate breathing source for decompressing diver(s) in the event of a MK 16 MOD 1 failure. The EBS consists of a MK 16 MOD 1 UBA mounted on the Emergency Breathing System frame assembly and charged with the same diluent gas as for the planned dive.

18-7.4.1 **EBS Deployment Procedures.** Regardless of the depth of the first decompression stop the EBS must be lowered to at least 40 fsw to allow the hydrostatic switch in the primary electronics to switch from 0.75 ata ppO_2 to 1.3 ata ppO_2. The EBS can then be raised or lowered to ten feet below the first decompression stop. Refer to Chapter 3 of the MK 16 MOD 1 Operations and Maintenance Manual for detailed EBS procedures.

It is recommended to lower EBS to 50 fsw if the first decompression stop is shallower than 40 fsw. This allows for topside personnel to track delays in ascent deeper than 50 fsw.

When conducting decompression dives it is highly recommended to utilize through water communications.

When diving in excessive currents, attach a tag line to the EBS frame on the side facing current, lower EBS maintaining slight tension. Once EBS is at correct depth, ensure tag line tends into the current and secure at opposite end of dive boat from EBS winch. Tag line will prevent EBS from spinning and maintain a straight up and down position. The tag line may also be used for line-pull signals.

18-7.4.2 **EBS Ascent Procedures.** As a diver prepares to leave bottom, diver movement should be in the direction of the tend. When tend is straight up and down, topside will marry EBS and tending line with weighted carabineer or other appropriate attachment device and allow to drop to EBS. When diver is directed to leave bottom, he will be able to follow the tend directly to EBS.

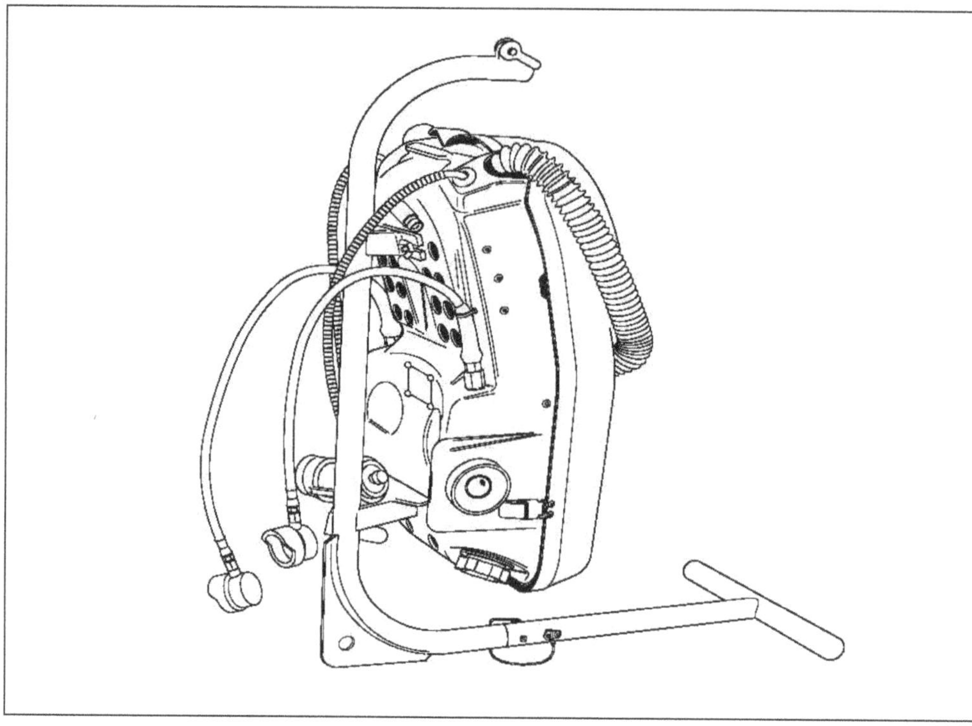

Figure 18-3. Emergency Breathing System.

18-8 MULTI-DAY DIVING

Repetitive exposure to an oxygen partial pressure of 1.3 ata over a multi-day period may result in the gradual development of pulmonary oxygen toxicity. The diver may experience burning substernal pain on inspiration, chest tightness, cough, and/or shortness of breath. The diver may also experience a temporary change in visual acuity. Distant objects may appear out of focus. To minimize the possibility of pulmonary or visual oxygen toxicity during multi-day diving with the MK 16 MOD 1, the diver shall adhere to the following rules:

- Limit total dive time on the MK 16 MOD 1 to a maximum of 4 hours per day.
- Limit total dive time on the MK 16 MOD 1 to a maximum of 16 hours per week.
- Shallow dives in which the rig remains in the 0.75 ata mode throughout are excluded from the above totals.
- Commanding Officer's permission is required to exceed these limits.
- If more dive time is required to accomplish a specific mission, contact Navy Experimental Diving Unit for additional guidance.
- If symptoms of pulmonary or visual oxygen toxicity develop at any time during a multi-day mission, stop diving until all symptoms have resolved and the diver remains symptom-free for a minimum of 24 hours.

18-9 ALTITUDE DIVING PROCEDURES AND FLYING AFTER DIVING

Ascent to altitude following a MK 16 MOD 1 dive at sea level will increase the risk of decompression sickness if the interval on the surface before ascent is not long enough to permit excess nitrogen or helium to be eliminated from the body. To determine the safe surface interval before ascent, take the following steps:

- Nitrogen-Oxygen Dives

  1. Determine the highest repetitive group designator obtained in the previous 24-hour period using either Table 18-9 or Table 18-11.

  2. Using the highest repetitive group designator, enter Table 9-6 in Chapter 9. Read across the row to the altitude that is exactly equal to or next higher than the planned change in altitude to obtain the safe surface interval.

- Helium-Oxygen Dives

  1. For no-decompression dives with total bottom times, including repetitive dives, less than 2 hours, wait 12 hours on the surface before ascending to altitude.

  2. For no-decompression dives with bottom times, including repetitive dives, greater than 2 hours or for decompression dives, wait 24 hours on the surface before ascending to altitude.

The MK 16 MOD 1 decompression procedures may be used for diving at altitudes up to 1000 feet without modification. For diving at altitudes above 1000 feet, the diving supervisor must contact NAVSEA 00C for guidance.

18-10 POSTDIVE PROCEDURES

Postdive procedures shall be completed in accordance with the appropriate postdive checklists in the MK 16 MOD 1 UBA O&M manual.

18-11 MEDICAL ASPECTS OF CLOSED-CIRCUIT MIXED-GAS UBA

When using a closed-circuit mixed-gas UBA, the diver is susceptible to the usual diving-related illnesses (i.e., decompression sickness, arterial gas embolism, barotraumas, etc.). Only the diving disorders that merit special attention for closed-circuit mixed gas divers are addressed in this chapter. Refer to Chapter 3 for a detailed discussion of diving related physiology and related disorders.

18-11.1 **Central Nervous System (CNS) Oxygen Toxicity.** High pressure oxygen poisoning is known as CNS oxygen toxicity. High partial pressures of oxygen are associated with many biochemical changes in the brain, but which specific changes are responsible for the signs and symptoms of CNS oxygen toxicity is presently unknown. CNS oxygen toxicity is not likely to occur at oxygen partial pressures below 1.3 ata, though relatively brief exposure to partial pressures above this, when it occurs at depth or in a pressurized chamber, can result in CNS oxygen toxicity causing CNS-related symptoms.

18-11.1.1 **Causes of CNS Oxygen Toxicity.** Factors that increase the likelihood of CNS oxygen toxicity are:

- Increased partial pressure of oxygen
- Increased time of exposure
- Prolonged immersion
- Stress from strenuous physical exercise
- Carbon dioxide buildup. The increased risk for CNS oxygen toxicity may occur even before the diver is aware of any symptoms of carbon dioxide buildup.
- Cold stress resulting from shivering or an increased exercise rate as the diver attempts to keep warm.
- Systemic diseases that increase oxygen consumption. Conditions associated with increased metabolic rates (such as certain thyroid or adrenal disorders) tend to cause an increase in oxygen sensitivity. Divers with these diseases should be excluded from mixed gas diving.

Though the MK 16 MOD 1 UBA maintains a ppO_2 of approximately 1.3 ata, a rapid descent may not allow the oxygen already in the circuit to be consumed fast enough resulting in a high ppO_2. ***When high levels of oxygen are displayed, the descent must be slowed. If the diver is in less than 20 fsw, little danger of oxygen toxicity exists. If the diver is deeper than 20 fsw, and an*** ppO_2 ***of 1.45 ata or higher persists within the UBA for a period of 15 consecutive minutes this condition should be treated as a malfunction of the UBA and the appropriate emergency procedures should be followed.***

18-11.1.2 **Symptoms of CNS Oxygen Toxicity.** The symptoms of CNS oxygen toxicity may not always appear and most are not exclusively symptoms of oxygen toxicity. The most serious symptom of CNS oxygen toxicity is convulsion, which may occur suddenly without any previous symptoms, and may result in drowning or arterial gas embolism. Twitching is perhaps the clearest warning of oxygen toxicity, but it may occur late if at all. The mnemonic VENTID-C is a helpful reminder of the most common symptoms of CNS oxygen toxicity. The appearance of any one of these symptoms usually represents a bodily signal of distress of some kind and should be heeded.

V: Visual symptoms. Tunnel vision, a decrease in the diver's peripheral vision, and other symptoms, such as blurred vision, may occur.

E: Ear symptoms. Tinnitus is any sound perceived by the ears but not resulting from an external stimulus. The sound may resemble bells ringing, roaring, or a machinery-like pulsing sound.

N: Nausea or spasmodic vomiting. These symptoms may be intermittent.

T: Twitching and tingling symptoms. Any of the small facial muscles, lips, or muscles of the extremities may be affected. These are the most frequent and clearest symptoms.

I: Irritability. Any change in the diver's mental status including confusion, agitation, and anxiety.

D: Dizziness. Symptoms include clumsiness, incoordination, and unusual fatigue.

C: Convulsions.

The following additional factors should be noted regarding an oxygen convulsion:

- The diver is unable to carry on any effective breathing during the convulsion.
- After the diver is brought to the surface, there will be a period of unconsciousness or neurologic impairment following the convulsion; these symptoms are indistinguishable from those of arterial gas embolism.
- No attempt should be made to insert any object between the clenched teeth of a convulsing diver. Although a convulsive diver may suffer a lacerated tongue, this trauma is preferable to the trauma that may be caused during the insertion of a foreign object. In addition, the person providing first aid may incur significant hand injury if bitten by the convulsing diver.
- There may be no warning of an impending convulsion to provide the diver the opportunity to return to the surface. Therefore, buddy lines are essential to safe closed-circuit mixed gas diving.

18-11.1.3 **Treatment of Nonconvulsive Symptoms.** If non-convulsive symptoms of CNS oxygen toxicity occur, action must be taken immediately to lower the oxygen partial pressure. Such actions include:

- Ascend. Dalton's law will lower the oxygen partial pressure.
- Add diluent to the breathing loop.
- Secure the oxygen cylinder if oxygen addition is uncontrolled.

Though an ascent from depth will lower the partial pressure of oxygen, the diver may still suffer other or worsening symptoms. The divers should notify the Diving Supervisor and terminate the dive.

18-11.1.4 **Treatment of Underwater Convulsion.** The following steps should be taken when treating a convulsing diver:

1. Assume a position behind the convulsing diver. Release the victim's weight belt only if progress to the surface is significantly impeded.
2. Leave the victim's mouthpiece in his mouth. If it is not in his mouth, do not attempt to replace it; however, if time permits, ensure that the mouthpiece is

switched to the SURFACE position to prevent unnecessary negative buoyancy from a flooded UBA.

3. Grasp the victim around his chest above the UBA or between the UBA and his body. If difficulty is encountered in gaining control of the victim in this manner, the rescuer should use the best method possible to obtain control.

4. Ventilate the UBA with diluent to lower the ppO_2 and maintain depth until the convulsion subsides.

5. Make a controlled ascent to the first decompression stop, maintaining a slight pressure on the diver's chest to assist exhalation.

   - If the diver regains control, continue with appropriate decompression.

   - If the diver remains incapacitated, surface at a moderate rate, establish an airway, and treat for symptomatic omitted decompression as outlined in paragraph 18-11.6.

   - Frequent monitoring of the primary and secondary displays as well as the oxygen- and diluent-bottle pressure gauges will keep the diver well informed of his breathing gas and rig status.

6. If additional buoyancy is required, activate the victim's life jacket. The rescuer should not release his own weight belt or inflate his life jacket.

7. Upon reaching the surface, inflate the victim's life jacket if not previously done.

8. Remove the victim's mouthpiece and switch the valve to SURFACE to prevent the possibility of the rig flooding and weighing down the victim.

9. Signal for emergency pickup.

10. Ensure the victim is breathing. Mouth-to-mouth breathing may be initiated if necessary.

11. If an upward excursion occurred during the actual convulsion, transport to the nearest chamber and have the victim evaluated by an individual trained to recognize and treat diving-related illness.

18-11.1.5 **Prevention of CNS Oxygen Toxicity.** All predive checks must be performed to ensure proper functioning of the oxygen sensors and the oxygen-addition valve. Frequent monitoring of both the primary and secondary displays will help ensure that the proper ppO_2 is maintained.

18-11.1.6 **Off-Effect.** The off-effect, a hazard associated with CNS oxygen toxicity, may occur several minutes after the diver comes off gas or experiences a reduction of oxygen partial pressure. The off-effect is manifested by the onset or worsening of CNS oxygen toxicity symptoms. Whether this paradoxical effect is truly caused by the reduction in partial pressure or whether the association is coincidental is unknown.

18-11.2 **Pulmonary Oxygen Toxicity.** Pulmonary oxygen toxicity can result from prolonged exposure to elevated partial pressures of oxygen. This form of oxygen toxicity produces lung irritation with symptoms of chest pain, cough, and pain on inspiration that develop slowly and become increasingly worse as long as the elevated level of oxygen is breathed. Although hyperbaric oxygen may cause serious lung damage, if the oxygen exposure is discontinued before the symptoms become too severe, the symptoms will slowly abate. Follow the guidance in paragraph 18-8 for multi-day exposures to oxygen.

18-11.3 **Oxygen Deficiency (Hypoxia).** Hypoxia is an abnormal deficiency of oxygen in the arterial blood in which the partial pressure of oxygen is too low to meet the metabolic needs of the body. Chapter 3 contains an in-depth description of this disorder. Although all cells in the body need oxygen, the initial symptoms of hypoxia are a manifestation of central nervous system dysfunction.

18-11.3.1 **Causes of Hypoxia.** The primary cause of hypoxia for a MK16 diver is failure of the oxygen addition valve or primary electronics. However, during a rapid ascent Dalton's law may cause the ppO_2 to fall faster than can be compensated for by the oxygen-addition system. If, during ascent, low levels of oxygen are displayed, slow the ascent and add oxygen if necessary. Depletion of the oxygen supply or malfunctioning oxygen sensors can also lead to a hypoxic gas mixture.

18-11.3.2 **Symptoms of Hypoxia.** Hypoxia may have no warning symptoms prior to loss of consciousness. Other symptoms that may appear include confusion, loss of coordination, dizziness, and convulsion. It is important to note that if symptoms of unconsciousness or convulsion occur at the beginning of a closed-circuit dive, hypoxia, not oxygen toxicity, is the most likely cause.

18-11.3.3 **Treating Hypoxia.** If symptoms of hypoxia develop, the diver must take immediate action to raise the oxygen partial pressure. If unconsciousness occurs, the buddy diver should add oxygen to the rig while monitoring the secondary display. If the diver does not require decompression, the buddy diver should bring the afflicted diver to the surface at a moderate rate, remove the mouthpiece or mask, and have him breathe air. If the event was clearly related to hypoxia and the diver recovers fully with normal neurological function shortly after breathing surface air, the diver does not require treatment for arterial gas embolism.

18-11.3.4 **Treatment of Hypoxic Divers Requiring Decompression.** If the divers require decompression, the buddy diver should bring the afflicted diver to the first decompression stop.

- If consciousness is regained, continue with normal decompression.
- If consciousness is not regained, ascend to the surface at a moderate rate (not to exceed 30 fpm), establish an airway, administer 100-percent oxygen, and treat for symptomatic omitted decompression as outlined in paragraph 18-11.6.4. If possible, immediate assistance from the standby diver should be obtained and the unaffected diver should continue normal decompression.

18-11.4 **Carbon Dioxide Toxicity (Hypercapnia).** Carbon dioxide toxicity, or hypercapnia, is an abnormally high level of carbon dioxide in the blood and body tissues.

18-11.4.1 **Causes of Hypercapnia.** Hypercapnia is generally a result of the failure of the carbon dioxide-absorbent material. The failure may be a result of channeling, flooding or saturation of the absorbent material. Skip breathing or controlled ventilation by the diver, which results in an insufficient removal of CO_2 from the diver's body, may also cause hypercapnia.

18-11.4.2 **Symptoms of Hypercapnia.** Symptoms of hypercapnia are:

- Increased rate and depth of breathing
- Labored breathing (similar to that seen with heavy exercise)
- Headache
- Confusion
- Unconsciousness

WARNING **Hypoxia and hypercapnia may give the diver little or no warning prior to onset of unconsciousness.**

Symptoms are dependent on the partial pressure of carbon dioxide, which is a function of both the fraction of carbon dioxide and the absolute pressure. Thus, symptoms would be expected to increase as depth increases. The presence of a high partial pressure of oxygen may also reduce the early symptoms of hypercapnia. Elevated levels of carbon dioxide may result in an episode of CNS oxygen toxicity on a normally safe dive profile.

18-11.4.3 **Treating Hypercapnia.** If symptoms of hypercapnia develop, the diver should:

- Immediately stop work and take several deep breaths.
- Increase ventilation if skip-breathing is a possible cause.
- Ascend. This will reduce the partial pressure of carbon dioxide both in the rig and the lungs.
- If symptoms do not rapidly abate, the diver should abort the dive.
- During ascent, while maintaining a vertical position, the diver should activate his bypass valve, adding fresh gas to his UBA. If the symptoms are a result of canister floodout, an upright position decreases the likelihood that the diver will sustain chemical injury.
- If unconsciousness occurs at depth, the same principles of management for underwater convulsion as described in paragraph 18-11.1.4 apply.

18-11.4.4 **Prevention of Hypercapnia.** To minimize the risk of hypercapnia:

- Use only an approved carbon dioxide absorbent in the UBA canister.
- Follow the prescribed canister-filling procedure to ensure that the canister is correctly packed with carbon dioxide absorbent.

- Dip test the UBA carefully before the dive. Watch for leaks that may result in canister floodout.
- Do not exceed canister duration limits for the water temperature.
- Ensure that the one-way valves in the supply and exhaust hoses are installed and working properly.
- Swim at a relaxed, comfortable pace.
- Avoid skip-breathing. There is no advantage to this type of breathing in a closed-circuit rig and it may cause elevated blood carbon dioxide levels even with a properly functioning canister.

18-11.5 Chemical Injury. The term chemical injury refers to the introduction of a caustic solution from the carbon dioxide scrubber of the UBA into the upper airway of a diver.

18-11.5.1 **Causes of Chemical Injury.** A caustic alkaline solution results when water leaking into the canister comes in contact with the carbon dioxide absorbent. The water may enter through a leak in the breathing loop or incorrect position of the mouthpiece rotary valve during a leak check. When the diver is in a horizontal or head down position, this solution may travel through the inhalation hose and irritate or injure the upper airway.

18-11.5.2 **Symptoms of Chemical Injury.** Before actually inhaling the caustic solution, the diver may experience labored breathing or headache, which are symptoms of carbon dioxide buildup in the breathing gas. This occurs because an accumulation of the caustic solution in the canister may be impairing carbon dioxide absorption. If the problem is not corrected promptly, the alkaline solution may travel into the breathing hoses and consequently be inhaled or swallowed. Choking, gagging, foul taste, and burning of the mouth and throat may begin immediately. This condition is sometimes referred to as a "caustic cocktail." The extent of the injury depends on the amount and distribution of the solution.

18-11.5.3 **Management of a Chemical Incident.** If the caustic solution enters the mouth, nose, or face mask, the diver must take the following steps:

- Immediately assume an upright position in the water.
- Depress the manual diluent bypass valve continuously.
- If the dive is a no-decompression dive, make a controlled ascent to the surface, exhaling through the nose to prevent over pressurization.
- If the dive requires decompression, shift to the EBS or another alternative breathing supply. If it is not possible to complete the planned decompression, surface the diver and treat for omitted decompression as outlined in paragraph 18-11.6.

Using fresh water, rinse the mouth several times. Several mouthfuls should then be swallowed. If only sea water is available, rinse the mouth but do not swallow. Other fluids may be substituted if available, but the use of weak acid solutions (vinegar or lemon juice) is not recommended. Do not attempt to induce vomiting.

A chemical injury may cause the diver to have difficulty breathing properly on ascent. He should be observed for signs of an arterial gas embolism and should be treated if necessary. A victim of a chemical injury should be evaluated by a physician or corpsman as soon as possible. Respiratory distress which may result from the chemical trauma to the air passages requires immediate hospitalization.

18-11.5.4 **Prevention of Chemical Injury.** Chemical injuries are best prevented by the performance of a careful dip test during predive set-up to detect any system leaks. Special attention should also be paid to the position of the mouthpiece rotary valve upon water entry and exit to prevent the entry of water into the breathing loop. Additionally, dive buddies should perform a careful leak check on each other before leaving the surface at the start of a dive.

18-11.6 Omitted Decompression. Certain emergencies may interrupt or prevent specified decompression. UBA failure, exhausted diluent or oxygen gas supply, and bodily injury are examples that constitute such emergencies. Omitted decompression must be made up to avoid later difficulty. Table 18-8 contains specific guidance for the initial management of omitted decompression in an asymptomatic MK 16 MOD 1 diver.

18-11.6.1 **At 20 fsw.** If the deepest decompression stop omitted is 20 fsw, the diver may be returned to the water stop at which the omission occurred.

- If the surface interval was less than 1 minute, add 1 minute to the stop time and resume the planned decompression at the point of interruption.
- If the surface interval was greater than 1 minute, compute a new decompression schedule by multiplying the 20-foot stop time by 1.5.
- Ascend on the new decompression schedule. Alternatively, the diver may be removed from the water and treated on Treatment Table 5 if the surface interval is less than 5 minutes, or Treatment Table 6 if the surface interval is greater than 5 minutes.

18-11.6.2 **Deeper than 20 fsw.** If the deepest decompression stop omitted is deeper than 20 fsw, a more serious situation exists. The use of a recompression chamber when immediately available is mandatory.

- If less than 30 minutes of decompression were missed and the surface interval is less than 5 minutes, treat the diver on Treatment Table 5.
- If less than 30 minutes of decompression were missed but the surface interval exceeds 5 minutes, treat the diver on Treatment Table 6.
- If more than 30 minutes of decompression were missed, treat the diver on Treatment Table 6 regardless of the length of the surface interval.

18-11.6.3 **Deeper than 20 fsw/No Recompression Chamber Available.** If the deepest decompression stop omitted is deeper than 20 fsw and a recompression chamber

Table 18-8. *Initial Management of Asymptomatic Omitted Decompression MK 16 MOD 1 Diver.*

Deepest Decompression Stop Omitted	Decompression Status	Surface Interval	Action	
			Chamber Available	No Chamber Available
None	No Decompression stops required	NA	Observe on surface for 1 hour	Observe on surface for 1 hour
20 fsw (Note 1)	Decompression stop required	<1 min	Return to depth of stop. Increase stop time by 1 minute. Resume decompression according to original schedule.	Return to depth of stop. Increase stop time by 1 minute. Resume decompression according to original schedule.
		>1 min	Return to depth of stop. Multiply stop time by 1.5. Resume decompression. Or treatment table 5 for surface interval <5 min. or treatment table 6 for surface interval >5 min.	Return to depth of stop. Multiply stop time by 1.5. Resume decompression.
Deeper than 20 fsw (Note 1)	Decompression stop required (<30 min missed)	<5 min	Treatment Table 5	Descend to the deepest stop omitted. Multiply all stops 40 fsw and shallower by 1.5. Resume decompression
		>5 min	Treatment Table 6	Descend to the deepest stop omitted. Multiply all stops 40 fsw and shallower by 1.5. Resume decompression
	Decompression stop required (>30 min missed)	Any	Treatment Table 6	Descend to the deepest stop omitted. Multiply all stops 40 fsw and shallower by 1.5. Resume decompression

Note 1 If the diver is returned to an omitted decompression stop that is shallower than 33 feet, then the diver must manually add oxygen to his UBA to maintain 1.3 ata ppO_2.

is not immediately available, recompression in the water is required. Recompress the diver in the water using the appropriate 1.3 ata constant ppO_2 decompression table. Descend to the deepest decompression stop omitted and repeat this stop in its entirety. Complete decompression on the original schedule, lengthening all stops 40 fsw and shallower by multiplying the stop time by 1.5. If the deepest stop was 40 fsw or shallower, this stop should also be multiplied by 1.5. After arrival at 40 fsw or shallower, the oxygen partial pressure may be manually adjusted to 1.3 ata (increased-rate oxygen supply depletion shall be taken into consideration). When recompression in the water is required, keep the surface interval as short as possible. The diver's UBA must be checked to ensure that it will sustain the diver for the additional decompression obligation. Switching to a standby UBA may be necessary so that the decompression time will not be compromised by depletion of gas supplies or carbon dioxide-absorbent failure. Maintain depth control, keep the diver at rest, and provide a buddy diver.

18-11.6.4 **Evidence of Decompression Sickness or Arterial Gas Embolism.** If the diver shows evidence of decompression sickness or arterial gas embolism before recompression for omitted decompression can be carried out, immediate treatment using the appropriate oxygen or air treatment table is essential. Guidance for table selection and use is given in Chapter 20. Symptoms that develop during treatment of omitted decompression should be managed in the same manner as recurrences during treatment.

18-11.7 **Decompression Sickness in the Water.** Decompression sickness may develop in the water during MK 16 MOD 1 diving. The symptoms of decompression sickness may be joint pain or may be more serious manifestations such as numbness, loss of muscular function, or vertigo.

Managing decompression sickness in the water will be difficult in the best of circumstances. Only general guidance can be presented here. Management decisions must be made on site, taking into account all known factors. The advice of a Diving Medical Officer should be sought whenever possible.

18-11.7.1 **Diver Remaining in Water.** If prior to surfacing the diver signals that he has decompression sickness but feels that he can remain in the water:

1. Dispatch the standby diver to assist.

2. Have the diver descend to the depth of relief of symptoms in 10-fsw increments, but no deeper than two increments (i.e., 20 fsw).

3. Compute a new decompression profile by multiplying all stops by 1.5. If recompression went deeper than the depth of the first stop on the original decompression schedule, use a stop time equal to 1.5 times the first stop in the original decompression schedule for the one or two stops deeper than the original first stop.

4. Ascend on the new profile.

5. Lengthen stops as needed to control symptoms.

6. Upon surfacing, transport the diver to the nearest chamber. If he is asymptomatic, treat on Treatment Table 5. If he is symptomatic, treat in accordance with the guidance given in Volume 5, Chapter 20.

18-11.7.2 **Diver Leaving the Water.**

If prior to surfacing the diver signals that he has decompression sickness but feels that he cannot remain in the water:

1. Surface the diver at a moderate rate (not to exceed 30 fpm).

2. If a recompression chamber is on site (i.e., within 30 minutes), recompress the diver immediately. Guidance for treatment table selection and use is given in Chapter 20.

3. If a recompression chamber is not on site, follow the management guidance given in Volume 5.

18-12 MK 16 MOD 1 Diving Equipment Reference Data

Figure 18-4 outlines the capabilities and logistical requirements of the MK 16 MOD 1 UBA mixed-gas diving system. Minimum required equipment for the pool phase of diving conducted at Navy diving schools and MK 16 MOD 1 RDT&E commands may be modified as necessary. Any modification to the minimum required equipment listed herein must be noted in approved lesson training guides or SOPs.

MK 16 MOD 1 UBA General Characteristics

Principle of Operation:

Self-contained closed-circuit constant ppO_2 system

Minimum Equipment:

1. An approved Life Preserver or Buoyancy Compensator (BC). When using an approved BC, a Full Face Mask is required.
2. Dive knife
3. Swim fins
4. Face mask or full face mask (FFM)
5. Weight belt (as required)
6. Dive watch or Dive Timer/Depth Gauge (DT/DG) (as required)
7. Depth gauge or DT/DG (as required)

Principal Applications:

1. EOD operations
2. Search and inspection
3. Light repair and recovery

Advantages:

1. Minimal surface bubbles
2. Optimum efficiency of gas supply
3. Portability
4. Excellent mobility
5. Communications (when used with FFM)
6. Modularized assembly
7. Low magnetic signature (lo-mu)
8. Low acoustic signature

Disadvantages:

1. Extended decompression requirement for long bottom times or deep dives.
2. Limited physical and thermal protection
3. No voice communications (unless FFM used)
4. Extensive predive/postdive procedures

Restrictions:

Working limit 150 feet, N_2O_2 (air) diluent; 300 fsw, HeO_2 diluent

Operational Considerations:

1. Dive team (Table 18-2)
2. Safety boat(s) required
3. MK 16 MOD 1 decompression schedule must be used.

Figure 18-4. MK 16 MOD 1 UBA General Characteristics.

Table 18-9. *No Decompression Limits and Repetitive Group Designators for MK 16 MOD 1 N_2O_2 Dives.*

Depth (fsw)	No-Stop Limit	Repetitive Group Designator															
		A	B	C	D	E	F	G	H	I	J	K	L	M	N	O	Z
10	Unlimited	–	–	–	–	–	–	–	–	–	–	–	–	–	–	–	–
15	Unlimited	–	–	–	–	–	–	–	–	–	–	–	–	–	–	–	–
20	Unlimited	153	420	*													
25	Unlimited	51	87	133	196	296	557	*									
30	Unlimited	31	50	72	98	128	164	210	273	372	629	*					
35	Unlimited	22	35	50	66	84	103	126	151	181	217	263	326	425	680	*	
40	Unlimited	89	168	318	*												
50	Unlimited	27	44	63	84	108	136	169	210	265	344	496	*				
60	297	16	25	36	46	58	70	83	97	113	130	149	170	194	222	255	297
70	130	11	18	25	32	39	47	55	64	73	83	93	103	115	127	130	
80	70	9	14	19	24	30	36	42	48	54	61	68	70				
90	50	7	11	15	20	24	29	33	38	43	48	50					
100	39	6	9	13	16	20	24	28	32	36	39						
110	32	5	8	11	14	17	20	24	27	30	32						
120	27	4	7	9	12	15	18	20	23	26	27						
130	23	3	6	8	11	13	16	18	21	23							
140	21	3	5	7	9	12	14	16	18	21							
150	17	3	5	6	8	10	12	15	17								
Exceptional Exposure ----------																	
160	15	3	4	6	8	9	11	13	15								
170	13	4	5	7	9	10	12	13									
180	12		3	5	6	8	9	11	12								
190	10			4	6	7	9	10									

– Diver does not acquire a repetitive group designator during dives to these depths.

* Highest repetitive group that can be achieved at this depth regardless of bottom time.

Table 18-10. Residual Nitrogen Timetable for MK 16 MOD 1 N_2O_2 Dives.

Locate the diver's repetitive group designation from his previous dive along the diagonal line above the table. Read horizontally to the interval in which the diver's surface interval lies.

Next, read vertically downward to the new repetitive group designation. Continue downward in this same column to the row that represents the depth of the repetitive dive. The time given at the intersection is residual nitrogen time, in minutes, to be applied to the repetitive dive.

* Dives following surface intervals longer than this are not repetitive dives. Use actual bottom times in the Table 18-11 to compute decompression for such dives.

Repetitive Group at Beginning of Surface Interval (rows) / **Repetitive Group at the End of the Surface Interval** (columns)

Group at Beginning	Z	O	N	M	L	K	J	I	H	G	F	E	D	C	B	A
A																0:10 2:20 *
B															0:10 1:16	1:17 3:36 *
C														0:10 0:55	0:56 2:11	2:12 4:31 *
D													0:10 0:52	0:53 1:47	1:48 3:03	3:04 5:23 *
E												0:10 0:52	0:53 1:44	1:45 2:39	2:40 3:55	3:56 6:15 *
F											0:10 0:52	0:53 1:44	1:45 2:37	2:38 3:31	3:32 4:48	4:49 7:08 *
G										0:10 0:52	0:53 1:44	1:45 2:37	2:38 3:29	3:30 4:23	4:24 5:40	5:41 8:00 *
H									0:10 0:52	0:53 1:44	1:45 2:37	2:38 3:29	3:30 4:21	4:22 5:16	5:17 6:32	6:33 8:52 *
I								0:10 0:52	0:53 1:44	1:45 2:37	2:38 3:29	3:30 4:21	4:22 5:13	5:14 6:08	6:09 7:24	7:25 9:44 *
J							0:10 0:52	0:53 1:44	1:45 2:37	2:38 3:29	3:30 4:21	4:22 5:13	5:14 6:06	6:07 7:00	7:01 8:16	8:17 10:36 *
K						0:10 0:52	0:53 1:44	1:45 2:37	2:38 3:29	3:30 4:21	4:22 5:13	5:14 6:06	6:07 6:58	6:59 7:52	7:53 9:09	9:10 11:29 *
L					0:10 0:52	0:53 1:44	1:45 2:37	2:38 3:29	3:30 4:21	4:22 5:13	5:14 6:06	6:07 6:58	6:59 7:50	7:51 8:44	8:45 10:01	10:02 12:21 *
M				0:10 0:52	0:53 1:44	1:45 2:37	2:38 3:29	3:30 4:21	4:22 5:13	5:14 6:06	6:07 6:58	6:59 7:50	7:51 8:42	8:43 9:37	9:38 10:53	10:54 13:13 *
N			0:10 0:52	0:53 1:44	1:45 2:37	2:38 3:29	3:30 4:21	4:22 5:13	5:14 6:06	6:07 6:58	6:59 7:50	7:51 8:42	8:43 9:34	9:35 10:29	10:30 11:45	11:46 14:05 *
O		0:10 0:52	0:53 1:44	1:45 2:37	2:38 3:29	3:30 4:21	4:22 5:13	5:14 6:06	6:07 6:58	6:59 7:50	7:51 8:42	8:43 9:34	9:35 10:27	10:28 11:21	11:22 12:37	12:38 14:58 *
Z	0:10 0:52	0:53 1:44	1:45 2:37	2:38 3:29	3:30 4:21	4:22 5:13	5:14 6:06	6:07 6:58	6:59 7:50	7:51 8:42	8:43 9:34	9:35 10:27	10:28 11:19	11:20 12:13	12:14 13:30	13:31 15:50 *

Repetitive Group at the End of the Surface Interval

Dive Depth	Z	O	N	M	L	K	J	I	H	G	F	E	D	C	B	A
10	–	–	–	–	–	–	–	–	–	–	–	–	–	–	–	–
15	–	–	–	–	–	–	–	–	–	–	–	–	–	–	–	–
20	**	**	**	**	**	**	**	**	**	**	**	**	**	**	420	153
25	**	**	**	**	**	**	**	**	**	**	556	296	196	134	88	51
30	**	**	**	**	**	**	626	372	273	211	165	129	99	73	51	31
35	**	**	671	423	325	263	218	181	152	126	104	84	67	51	36	22
40	**	**	**	**	**	**	**	**	**	**	**	**	**	311	166	88
50	**	**	**	**	**	481	339	262	209	168	135	107	84	63	44	27
60	293	252	220	192	168	148	129	112	97	83	70	58	46	36	26	16
70	153	139	126	114	103	92	82	73	64	56	47	40	32	25	18	12
80	107	98	90	82	75	68	61	54	48	42	36	30	25	19	14	9
90	82	76	70	64	59	54	48	43	38	34	29	25	20	16	12	8
100	67	62	58	53	49	44	40	36	32	28	24	21	17	13	10	7
110	57	53	49	45	41	38	34	31	28	24	21	18	15	12	9	6
120	49	46	42	39	36	33	30	27	24	21	19	16	13	10	8	5
130	43	40	38	35	32	29	27	24	22	19	17	14	12	9	7	5
140	39	36	34	31	29	26	24	22	19	17	15	13	11	8	6	4
150	35	33	31	28	26	24	22	20	18	16	14	12	10	8	6	4
160	32	30	28	26	24	22	20	18	16	14	13	11	9	7	5	4
170	30	28	26	24	22	20	19	17	15	13	12	10	8	7	5	3
180	27	26	24	22	21	19	17	16	14	12	11	9	8	6	5	3
190	25	24	22	21	19	18	16	15	13	12	10	9	7	6	4	3

Residual Nitrogen Time (Minutes)

– Repetitive dives to these depths are equivalent to remaining on the surface. Add the bottom time of the dive to the preceding surface interval. Use the Surface Interval Credit Table (SICT) to determine the repetitive group at the end of the dive.

** Residual Nitrogen Time cannot be determined using this table (see paragraph 9-9.1 for instructions).

REPETITIVE DIVE WORKSHEET FOR MK 16 MOD 1 N_2O_2 DIVES

Part 1. Previous Dive

______________ *minutes*

______________ *feet*

______________ *repetitive group designator from Table 18-9 if the dive was a no-decompression dive, or Table 18-11 if the dive was a decompression dive.*

Part 2. Surface Interval:

Enter the top section of Table 18-10 at the row for the repetitive group designator from Part 1 and move horizontally to the column in which the actual or planned surface interval time lies. Read the final repetitive group designator from the bottom of this column.

_________ *hours* ______ *minutes on the surface*

_________ *final repetitive group from Table 18-10*

Part 3. Equivalent Single Dive Time for the Repetitive Dive:

Enter the bottom section of Table 18-10 at the row for the maximum depth of the planned repetitive dive. Move horizontally to the column of the final repetitive group designator from Part 2 to find the Residual Nitrogen Time (RNT). Add this RNT to the planned bottom time for the repetitive dive to obtain the equivalent single dive time.

_____ *minutes: RNT*

+_____ *minutes: planned bottom time*

=_____ *minutes: equivalent single dive time*

Part 4. Decompression Schedule for the Repetitive Dive:

Locate the row for the depth of the planned repetitive dive in Table 18-9. Move horizontally to the column with bottom time equal to or just greater than the equivalent single dive time and read the surfacing repetitive group for the repetitive dive from the top of the column. If the equivalent single dive time exceeds the no-decompression limit, locate the row for the depth and equivalent single dive time in Table 18-11. Read the required decompression stops and surfacing repetitive group from the columns to the right along this row.

_____ *minutes: equivalent single dive time from Part 3*

_____ *feet: depth of the repetitive dive.*

_____ *Schedule (depth/bottom time) from Table 18-9 or Table 18-11.*

Ensure RNT Exception Rule does not apply.

Figure 18-5. Repetitive Dive Worksheet for MK 16 MOD 1 N_20_2.

Table 18-11. *MK 16 MOD 1 N_2O_2 Decompression Tables.*

(DESCENT RATE 60 FPM—ASCENT RATE 30 FPM)

Bottom Time (min)	Time to First Stop (M:S)	DECOMPRESSION STOPS (fsw) Stop times (min) include travel time, except first stop 80	70	60	50	40	30	20	Total Ascent Time (M:S)	Repet Group
60 FSW										
297	2:00							0	2:00	Z
300	1:20							1	3:00	Z
310	1:20							2	4:00	Z
320	1:20							3	5:00	Z
330	1:20							4	6:00	Z
Exceptional Exposure										
340	1:20							5	7:00	
350	1:20							6	8:00	
360	1:20							7	9:00	
370	1:20							8	10:00	
380	1:20							9	11:00	
390	1:20							10	12:00	
70 FSW										
130	2:20							0	2:20	O
140	1:40							3	5:20	O
150	1:40							6	8:20	O
160	1:40							8	10:20	Z
170	1:40							10	12:20	Z
180	1:40							12	14:20	Z
190	1:40							14	16:20	Z
200	1:40							16	18:20	Z
210	1:40							19	21:20	Z
220	1:40							22	24:20	Z
230	1:40							24	26:20	Z
Exceptional Exposure										
240	1:40							26	28:20	
250	1:40							29	31:20	
260	1:40							31	33:20	
270	1:40							33	35:20	
280	1:40							35	37:20	
290	1:40							37	39:20	
300	1:40							38	40:20	
310	1:40							40	42:20	
320	1:40							42	44:20	
340	1:40							47	49:20	
350	1:40							49	51:20	

Table 18-11. *MK 16 MOD 1 N_2O_2 Decompression Tables (Continued).*

(DESCENT RATE 60 FPM—ASCENT RATE 30 FPM)

Bottom Time (min)	Time to First Stop (M:S)	DECOMPRESSION STOPS (fsw) Stop times (min) include travel time, except first stop							Total Ascent Time (M:S)	Repet Group
		80	70	60	50	40	30	20		
80 FSW										
70	2:40							0	2:40	L
75	2:00							2	4:40	L
80	2:00							4	6:40	M
85	2:00							5	7:40	M
90	2:00							6	8:40	N
95	2:00							7	9:40	N
100	2:00							9	11:40	N
110	2:00							12	14:40	O
120	2:00							16	18:40	O
130	2:00							20	22:40	Z
140	2:00							24	26:40	Z
150	2:00							27	29:40	Z
160	2:00							30	32:40	Z
170	2:00							34	36:40	Z
Exceptional Exposure ----------										
180	2:00							39	41:40	
190	2:00							43	45:40	
200	2:00							47	49:40	
210	2:00							50	52:40	
220	2:00							54	56:40	
230	2:00							57	59:40	
240	2:00							60	62:40	
250	2:00							63	65:40	
260	2:00							67	69:40	
270	2:00							70	72:40	
280	2:00							74	76:40	
290	2:00							77	79:40	
300	2:00							81	83:40	
310	2:00							84	86:40	
320	2:00							87	89:40	

Table 18-11. *MK 16 MOD 1 N_2O_2 Decompression Tables (Continued).*

(DESCENT RATE 60 FPM—ASCENT RATE 30 FPM)

Bottom Time (min)	Time to First Stop (M:S)	DECOMPRESSION STOPS (fsw) Stop times (min) include travel time, except first stop							Total Ascent Time (M:S)	Repet Group
		80	**70**	**60**	**50**	**40**	**30**	**20**		
90 FSW										
50	3:00							0	3:00	K
55	2:20							3	6:00	K
60	2:20							6	9:00	L
65	2:20							8	11:00	L
70	2:20							11	14:00	M
75	2:20							13	16:00	M
80	2:20							14	17:00	N
85	2:20							16	19:00	N
90	2:20							18	21:00	O
95	2:20							21	24:00	O
100	2:20							24	27:00	O
110	2:20							30	33:00	O
120	2:20							35	38:00	Z
130	2:20							40	43:00	Z
Exceptional Exposure --										
140	2:20							45	48:00	
150	2:20							51	54:00	
160	2:20							57	60:00	
170	2:00						1	62	65:40	
180	2:00						2	66	70:40	
190	2:00						2	71	75:40	
100 FSW										
39	3:20							0	3:20	J
40	2:40							1	4:20	J
45	2:40							5	8:20	K
50	2:40							9	12:20	L
55	2:40							12	15:20	L
60	2:40							15	18:20	M
65	2:40							18	21:20	M
70	2:40							21	24:20	N
75	2:40							23	26:20	N
80	2:40							26	29:20	O
85	2:40							30	33:20	O
90	2:40							34	37:20	O
95	2:20						1	37	41:00	O
100	2:20						3	39	45:00	O
Exceptional Exposure --										
110	2:20						6	43	52:00	
120	2:20						8	47	58:00	

Table 18-11. *MK 16 MOD 1 N_2O_2 Decompression Tables (Continued).*

(DESCENT RATE 60 FPM—ASCENT RATE 30 FPM)

Bottom Time (min)	Time to First Stop (M:S)	DECOMPRESSION STOPS (fsw) Stop times (min) include travel time, except first stop							Total Ascent Time (M:S)	Repet Group
		80	70	60	50	40	30	20		
110 FSW										
32	3:40							0	3:40	J
35	3:00							3	6:40	J
40	3:00							8	11:40	K
45	3:00							13	16:40	L
50	3:00							17	20:40	L
55	3:00							21	24:40	M
60	3:00							25	28:40	M
65	3:00							28	31:40	N
70	2:40						1	30	34:20	O
75	2:40						4	32	39:20	O
80	2:40						7	34	44:20	O
Exceptional Exposure --										
85	2:40						9	37	49:20	
90	2:40						11	39	53:20	
95	2:40						13	42	58:20	
100	2:40						15	44	62:20	
110	2:20					3	15	49	70:00	
120	2:20					6	15	56	80:00	
120 FSW										
27	4:00							0	4:00	J
30	3:20							4	8:00	J
35	3:20							10	14:00	K
40	3:20							16	20:00	L
45	3:20							21	25:00	L
50	3:20							26	30:00	M
55	3:20							30	34:00	M
60	3:00						4	31	38:40	N
65	3:00						8	30	41:40	O
Exceptional Exposure --										
70	3:00						12	32	47:40	
75	3:00						15	35	53:40	
80	2:40					3	15	38	59:20	
85	2:40					6	15	41	65:20	
90	2:40					8	15	44	70:20	
95	2:40					10	15	47	75:20	
100	2:40					12	15	51	81:20	

Table 18-11. *MK 16 MOD 1 N_2O_2 Decompression Tables (Continued).*

(DESCENT RATE 60 FPM—ASCENT RATE 30 FPM)

Bottom Time (min)	Time to First Stop (M:S)	DECOMPRESSION STOPS (fsw) Stop times (min) include travel time, except first stop							Total Ascent Time (M:S)	Repet Group
		80	70	60	50	40	30	20		
130 FSW										
23	4:20							0	4:20	I
25	3:40							2	6:20	J
30	3:40							10	14:20	K
35	3:40							17	21:20	K
40	3:40							23	27:20	L
45	3:40							29	33:20	M
50	3:20						4	30	38:00	N
55	3:20						9	30	43:00	N
Exceptional Exposure ----------										
60	3:20						14	30	48:00	
65	3:00					3	15	33	54:40	
70	3:00					7	15	36	61:40	
75	3:00					11	15	39	68:40	
80	3:00					14	15	42	74:40	
140 FSW										
21	4:40							0	4:40	I
25	4:00							7	11:40	J
30	4:00							16	20:40	K
35	4:00							23	27:40	L
40	3:40						2	29	35:20	L
45	3:40						7	30	41:20	M
Exceptional Exposure ----------										
50	3:20					1	12	30	47:00	
55	3:20					4	15	30	53:00	
60	3:20					9	15	33	61:00	
65	3:20					13	15	36	68:00	
70	3:00				3	15	15	40	76:40	
75	3:00				7	15	15	44	84:40	
80	3:00				10	15	15	50	93:40	

Table 18-11. *MK 16 MOD 1 N_2O_2 Decompression Tables (Continued).*

(DESCENT RATE 60 FPM—ASCENT RATE 30 FPM)

Bottom Time (min)	Time to First Stop (M:S)	DECOMPRESSION STOPS (fsw) Stop times (min) include travel time, except first stop							Total Ascent Time (M:S)	Repet Group
		80	70	60	50	40	30	20		
150 FSW										
17	5:00							0	5:00	H
20	4:20							3	8:00	I
25	4:20							13	18:00	J
30	4:20							22	27:00	K
35	4:00						3	27	34:40	L
40	4:00						8	30	42:40	M
Exceptional Exposure --										
45	3:40					4	11	30	49:20	
50	3:40					7	15	30	56:20	
55	3:20				2	11	15	33	65:00	
60	3:20				4	14	15	37	74:00	
65	3:20				8	15	15	40	82:00	
70	3:20				13	15	15	46	93:00	
75	3:00			2	15	15	15	52	102:40	
80	3:00			6	15	15	15	59	113:40	
160 FSW										
Exceptional Exposure --										
15	5:20							0	5:20	H
20	4:40							7	12:20	J
25	4:20						1	17	23:00	K
30	4:20						3	25	33:00	L
35	4:00					1	8	28	41:40	M
40	4:00					5	10	30	49:40	
45	3:40				2	7	14	30	57:20	
50	3:40				5	10	15	33	67:20	
55	3:40				8	14	15	36	77:20	
60	3:20			3	10	15	15	41	88:00	
65	3:20			5	13	15	15	48	100:00	
70	3:20			8	15	15	15	55	112:00	
75	3:20			13	15	15	15	61	123:00	
80	3:00		3	15	15	15	15	68	134:40	

Table 18-11. *MK 16 MOD 1 N_2O_2 Decompression Tables (Continued).*

(DESCENT RATE 60 FPM—ASCENT RATE 30 FPM)

Bottom Time (min)	Time to First Stop (M:S)	DECOMPRESSION STOPS (fsw) Stop times (min) include travel time, except first stop							Total Ascent Time (M:S)	Repet Group
		80	70	60	50	40	30	20		

170 FSW

Bottom Time (min)	Time to First Stop (M:S)	80	70	60	50	40	30	20	Total Ascent Time (M:S)	Repet Group
Exceptional Exposure ----------										
13	5:40							0	5:40	H
15	5:00							2	7:40	I
20	5:00							12	17:40	J
25	4:40						3	20	28:20	K
30	4:20					3	5	26	39:00	L
35	4:00				1	5	8	30	48:40	
40	4:00				4	7	12	30	57:40	
45	4:00				8	8	15	32	67:40	
50	3:40			4	7	13	15	36	79:20	
55	3:40			7	9	15	15	41	91:20	
60	3:20		2	7	14	15	15	48	105:00	

180 FSW

Bottom Time (min)	Time to First Stop (M:S)	80	70	60	50	40	30	20	Total Ascent Time (M:S)	Repet Group
Exceptional Exposure ----------										
12	6:00							0	6:00	H
15	5:20							4	10:00	I
20	5:00						2	14	21:40	K
25	4:40					3	3	23	34:20	L
30	4:20				2	4	7	27	45:00	
35	4:00			1	3	8	9	30	55:40	
40	4:00			2	7	8	14	30	65:40	
45	4:00			6	7	11	15	35	78:40	
50	3:40		2	8	8	15	15	40	92:20	
55	3:40		5	8	12	15	15	49	108:20	
60	3:20	1	7	9	15	15	15	57	123:00	

190 FSW

Bottom Time (min)	Time to First Stop (M:S)	80	70	60	50	40	30	20	Total Ascent Time (M:S)	Repet Group
Exceptional Exposure ----------										
10	6:20							0	6:20	G
15	5:40							6	12:20	J
20	5:00					1	4	16	26:40	K
25	4:40				2	4	4	24	39:20	L
30	4:20			2	3	5	8	29	52:00	
35	4:20			4	5	8	11	30	63:00	
40	4:00		2	5	8	8	15	34	76:40	
45	4:00		4	8	7	14	15	39	91:40	
50	3:40	1	7	8	11	15	15	47	108:20	
55	3:40	4	8	8	15	15	15	56	125:20	
60	3:40	7	7	13	15	15	15	65	141:20	

Table 18-12. *No Decompression Limits and Repetitive Group Designators for MK 16 MOD 1 HeO_2 Dives.*

Depth (fsw)	No-Stop Limit	Repetitive Group Designator															
		A	B	C	D	E	F	G	H	I	J	K	L	M	N	O	Z
10	Unlimited	–	–	–	–	–	–	–	–	–	–	–	–	–	–	–	–
15	Unlimited	–	–	–	–	–	–	–	–	–	–	–	–	–	–	–	–
20	Unlimited	129	269	*													
25	Unlimited	45	72	106	146	200	278	425	*								
30	332	27	43	60	78	100	124	152	185	227	281	332					
35	190	19	30	41	54	67	81	97	114	133	154	178	190				
40	Unlimited	122	246	*													
50	325	27	43	59	78	99	123	150	183	223	276	325					
60	134	15	23	32	41	51	61	71	83	95	108	123	134				
70	86	11	16	22	28	34	41	47	54	61	69	77	85	86			
80	63	8	12	17	21	26	30	35	40	45	51	56	62	63			
90	44	6	10	13	17	20	24	28	32	36	40	44					
100	31	5	8	11	14	17	20	23	26	30	31						
110	24	4	7	9	12	14	17	20	22	24							
120	20	4	6	8	10	13	15	17	19	20							
130	17	3	5	7	9	11	13	15	17								
140	15	3	4	6	8	10	12	13	15								
150	13	3	4	6	7	9	10	12	13								
160	12		3	5	6	8	9	11	12								
170	11		3	4	6	7	9	10	11								
180	10		3	4	5	6	8	9	10								
190	9			4	5	6	7	8	9								
200	8				4	5	7	8									

– Diver does not acquire a repetitive group designator during dives to these depths.

* Highest repetitive group that can be achieved at this depth regardless of bottom time.

Table 18-13. *Residual Helium Timetable for MK 16 MOD 1 HeO_2 Dives.*

Locate the diver's repetitive group designation from his previous dive along the diagonal line above the table. Read horizontally to the interval in which the diver's surface interval lies.

Next, read vertically downward to the new repetitive group designation. Continue downward in this same column to the row that represents the depth of the repetitive dive. The time given at the intersection is residual helium time, in minutes, to be applied to the repetitive dive.

* Dives following surface intervals longer than this are not repetitive dives. Use actual bottom times in the Table 18-14 to compute decompression for such dives.

Repetitive Group at Beginning of Surface Interval

Group at beginning	Z	O	N	M	L	K	J	I	H	G	F	E	D	C	B	A
A																0:10 2:01 *
B															0:10 1:10	1:11 3:11 *
C														0:10 0:50	0:51 2:00	2:01 4:01 *
D													0:10 0:42	0:43 1:32	1:33 2:43	2:44 4:44 *
E												0:10 0:42	0:43 1:25	1:26 2:15	2:16 3:25	3:26 5:26 *
F											0:10 0:42	0:43 1:25	1:26 2:07	2:08 2:57	2:58 4:08	4:09 6:08 *
G										0:10 0:42	0:43 1:25	1:26 2:07	2:08 2:49	2:50 3:39	3:40 4:50	4:51 6:51 *
H									0:10 0:42	0:43 1:25	1:26 2:07	2:08 2:49	2:50 3:32	3:33 4:22	4:23 5:32	5:33 7:33 *
I								0:10 0:42	0:43 1:25	1:26 2:07	2:08 2:49	2:50 3:32	3:33 4:14	4:15 5:04	5:05 6:15	6:16 8:15 *
J							0:10 0:42	0:43 1:25	1:26 2:07	2:08 2:49	2:50 3:32	3:33 4:14	4:15 4:56	4:57 5:46	5:47 6:57	6:58 8:58 *
K						0:10 0:42	0:43 1:25	1:26 2:07	2:08 2:49	2:50 3:32	3:33 4:14	4:15 4:56	4:57 5:39	5:40 6:29	6:30 7:39	7:40 9:40 *
L					0:10 0:42	0:43 1:25	1:26 2:07	2:08 2:49	2:50 3:32	3:33 4:14	4:15 4:56	4:57 5:39	5:40 6:21	6:22 7:11	7:12 8:22	8:23 10:22 *
M				0:10 0:42	0:43 1:25	1:26 2:07	2:08 2:49	2:50 3:32	3:33 4:14	4:15 4:56	4:57 5:39	5:40 6:21	6:22 7:03	7:04 7:53	7:54 9:04	9:05 11:05 *
N			0:10 0:42	0:43 1:25	1:26 2:07	2:08 2:49	2:50 3:32	3:33 4:14	4:15 4:56	4:57 5:39	5:40 6:21	6:22 7:03	7:04 7:46	7:47 8:36	8:37 9:46	9:47 11:47 *
O		0:10 0:42	0:43 1:25	1:26 2:07	2:08 2:49	2:50 3:32	3:33 4:14	4:15 4:56	4:57 5:39	5:40 6:21	6:22 7:03	7:04 7:46	7:47 8:28	8:29 9:18	9:19 10:29	10:30 12:29 *
Z	0:10 0:42	0:43 1:25	1:26 2:07	2:08 2:49	2:50 3:32	3:33 4:14	4:15 4:56	4:57 5:39	5:40 6:21	6:22 7:03	7:04 7:46	7:47 8:28	8:29 9:10	9:11 10:00	10:01 11:11	11:12 13:12 *

Repetitive Group at the End of the Surface Interval

Dive Depth	Z	O	N	M	L	K	J	I	H	G	F	E	D	C	B	A
10	–	–	–	–	–	–	–	–	–	–	–	–	–	–	–	–
15	–	–	–	–	–	–	–	–	–	–	–	–	–	–	–	–
20	**	**	**	**	**	**	**	**	**	**	**	**	**	**	269	129
25	**	**	**	**	**	**	**	**	**	425	279	201	147	106	73	45
30	†	†	†	†	515	361	281	227	186	152	124	100	79	60	43	28
35	420	338	283	241	207	179	155	133	114	97	82	68	54	42	31	20
40	**	**	**	**	**	**	**	**	**	**	**	**	**	**	240	120
50	†	†	†	†	474	345	272	220	181	149	122	98	78	59	42	27
60	217	194	173	154	137	122	108	95	83	71	61	51	41	32	24	16
70	122	112	102	93	85	77	69	61	54	47	41	34	28	22	17	11
80	86	80	73	68	62	56	51	46	40	36	31	26	22	17	13	9
90	67	62	57	53	49	44	40	36	32	29	25	21	17	14	10	7
100	55	51	47	44	40	37	33	30	27	24	21	18	15	12	9	6
110	46	43	40	37	34	31	29	26	23	20	18	15	13	10	8	5
120	40	37	35	32	30	27	25	23	20	18	16	13	11	9	7	5
130	35	33	31	29	27	24	22	20	18	16	14	12	10	8	6	4
140	32	30	28	26	24	22	20	18	16	14	13	11	9	7	6	4
150	29	27	25	23	22	20	18	17	15	13	12	10	8	7	5	4
160	26	25	23	21	20	18	17	15	14	12	11	9	8	6	5	3
170	24	23	21	20	18	17	15	14	13	11	10	8	7	6	4	3
180	22	21	20	18	17	16	14	13	12	10	9	8	7	5	4	3
190	21	20	18	17	16	15	13	12	11	10	9	7	6	5	4	3
200	20	18	17	16	15	14	13	11	10	9	8	7	6	5	4	3

Residual Helium Time (Minutes)

– Repetitive dives to these depths are equivalent to remaining on the surface. Add the bottom time of the dive to the preceding surface interval. Use the Surface Interval Credit Table (SICT) to determine the repetitive group at the end of the dive.

** Residual Helium Time cannot be determined using this table (see paragraph 9-9.1 for instructions).

† Read vertically down to the 35 or 60 fsw repetitive dive depth to obtain the RHT. Decompress on the 35 or 60 fsw table.

REPETITIVE DIVE WORKSHEET FOR MK 16 MOD 1 HeO_2 DIVES

Part 1. Previous Dive:

_____________ *minutes*

_____________ *feet*

_____________ *repetitive group designator from Table 18-12 if the dive was a no-decompression dive, or from Table 18-14 if the dive was a decompression dive.*

Part 2. Surface Interval:

Enter the top section of Table 18-13 at the row for the repetitive group designator from Part 1 and move horizontally to the right to the column in which the time equal to or just greater than the actual or planned surface interval time lies. Read the final repetitive group designator from the bottom of this column.

_________ *hours* ______ *minutes on the surface*

_________ *final repetitive group from Table 18-13*

Part 3. Equivalent Single Dive Time for the Repetitive Dive:

Enter the bottom section of Table 18-13 at the row for the maximum depth of the planned repetitive dive. Move horizontally to the right to the column of the final repetitive group designator from Part 2 to find the Residual Helium Time (RHT). Add this RHT to the planned bottom time for the repetitive dive to obtain the equivalent single dive time.

_____ *minutes: RHT*

+_____ *minutes: planned bottom time*

=_____ *minutes: equivalent single dive time*

Part 4. Decompression Schedule for the Repetitive Dive:

Locate the row for the depth of the planned repetitive dive in Table 18-12. Move horizontally to the right to the column with bottom time equal to or just greater than the equivalent single dive time and read the surfacing repetitive group for the repetitive dive from the top of the column. If the equivalent single dive time exceeds the no-decompression limit, locate the row for the depth and equivalent single dive time in Table 18-14. Read the required decompression stops and surfacing repetitive group from the columns to the right along this row.

_____ *minutes: equivalent single dive time from Part 3*

_____ *feet: depth of the repetitive dive*

_____ *Schedule (depth/bottom time) from Table 18-12 or Table 18-14*

Ensure RHT Exception Rule does not apply.

Figure 18-6. Repetitive Dive Worksheet for MK 16 MOD 1 HeO_2 Dives.

Table 18-14. *MK 16 MOD 1 HeO_2 Decompression Tables.*

(DESCENT RATE 60 FPM—ASCENT RATE 30 FPM)

Bottom Time (min)	Time to First Stop (M:S)	DECOMPRESSION STOPS (fsw) Stop times (min) include travel time, except first stop																Total Ascent Time (M:S)	Repet Group
		170	160	150	140	130	120	110	100	90	80	70	60	50	40	30	20		
30 FSW																			
332	1:00																0	1:00	
340	0:20																4	5:00	
360	0:20																13	14:00	
420	0:20																34	35:00	
480	0:20																48	49:00	
540	0:20																59	60:00	
600	0:20																70	71:00	
660	0:20																87	88:00	
720	0:20																101	102:00	
35 FSW																			
190	1:10																0	1:10	L
200	0:30																12	13:10	L
210	0:30																23	24:10	
220	0:30																33	34:10	
230	0:30																42	43:10	
240	0:30																50	51:10	
270	0:30																71	72:10	
300	0:30																89	90:10	
330	0:30																103	104:10	
360	0:30																115	116:10	
390	0:30																126	127:10	
420	0:30																145	146:10	
450	0:30																162	163:10	
480	0:30																177	178:10	
50 FSW																			
325	1:40																0	1:40	K
330	1:00																1	2:40	K
340	1:00																2	3:40	K
350	1:00																3	4:40	K
360	1:00																5	6:40	K
420	1:00																11	12:40	
480	1:00																15	16:40	
540	1:00																18	19:40	
600	1:00																21	22:40	
660	1:00																25	26:40	
720	1:00																29	30:40	

Table 18-14. *MK 16 MOD 1 HeO_2 Decompression Tables (Continued).*

(DESCENT RATE 60 FPM—ASCENT RATE 30 FPM)

Bottom Time (min)	Time to First Stop (M:S)	DECOMPRESSION STOPS (fsw) Stop times (min) include travel time, except first stop																Total Ascent Time (M:S)	Repet Group
		170	160	150	140	130	120	110	100	90	80	70	60	50	40	30	20		
60 FSW																			
134	2:00																0	2:00	L
140	1:20																3	5:00	L
150	1:20																8	10:00	L
160	1:20																12	14:00	L
170	1:20																16	18:00	L
180	1:20																20	22:00	
190	1:20																24	26:00	
200	1:20																27	29:00	
210	1:20																31	33:00	
220	1:20																34	36:00	
230	1:20																37	39:00	
240	1:20																40	42:00	
250	1:20																42	44:00	
260	1:20																45	47:00	
270	1:20																47	49:00	
280	1:20																49	51:00	
290	1:20																51	53:00	
300	1:20																53	55:00	
310	1:20																55	57:00	
320	1:20																57	59:00	
330	1:20																59	61:00	
340	1:20																61	63:00	
350	1:20																64	66:00	
360	1:20																66	68:00	
70 FSW																			
86	2:20																0	2:20	M
90	1:40																3	5:20	M
95	1:40																8	10:20	
100	1:40																12	14:20	
110	1:40																19	21:20	
120	1:40																26	28:20	
130	1:40																33	35:20	
140	1:40																39	41:20	
150	1:40																45	47:20	
160	1:40																50	52:20	
170	1:40																55	57:20	
180	1:40																60	62:20	
190	1:40																64	66:20	
200	1:40																68	70:20	
210	1:40																72	74:20	
220	1:40																76	78:20	

Table 18-14. *MK 16 MOD 1 HeO_2 Decompression Tables (Continued).*

(DESCENT RATE 60 FPM—ASCENT RATE 30 FPM)

80 FSW

Bottom Time (min)	Time to First Stop (M:S)	DECOMPRESSION STOPS (fsw) Stop times (min) include travel time, except first stop 170	160	150	140	130	120	110	100	90	80	70	60	50	40	30	20	Total Ascent Time (M:S)	Repet Group
63	2:40																0	2:40	M
65	2:00																2	4:40	M
70	2:00																8	10:40	
75	2:00																14	16:40	
80	2:00																19	21:40	
85	2:00																24	26:40	
90	2:00																29	31:40	
95	2:00																34	36:40	
100	2:00																39	41:40	
110	2:00																48	50:40	
120	2:00																56	58:40	
130	2:00																63	65:40	
140	2:00																70	72:40	
150	2:00																76	78:40	
160	2:00																82	84:40	
170	2:00																88	90:40	
180	2:00																93	95:40	
190	2:00																98	100:40	

90 FSW

Bottom Time (min)	Time to First Stop (M:S)	170	160	150	140	130	120	110	100	90	80	70	60	50	40	30	20	Total Ascent Time (M:S)	Repet Group
44	3:00																0	3:00	K
45	2:20																1	4:00	K
50	2:20																2	5:00	L
55	2:20																7	10:00	M
60	2:20																15	18:00	
65	2:20																22	25:00	
70	2:20																29	32:00	
75	2:20																35	38:00	
80	2:20																41	44:00	
85	2:20																47	50:00	
90	2:20																53	56:00	
95	2:20																58	61:00	
100	2:20																63	66:00	
110	2:20																73	76:00	
120	2:20																82	85:00	
130	2:20																90	93:00	
140	2:20																97	100:00	
150	2:20																105	108:00	
160	2:20																112	115:00	

Table 18-14. *MK 16 MOD 1 HeO_2 Decompression Tables (Continued).*

(DESCENT RATE 60 FPM—ASCENT RATE 30 FPM)

Bottom Time (min)	Time to First Stop (M:S)	DECOMPRESSION STOPS (fsw) Stop times (min) include travel time, except first stop																Total Ascent Time (M:S)	Repet Group
		170	160	150	140	130	120	110	100	90	80	70	60	50	40	30	20		
100 FSW																			
31	3:20																0	3:20	J
35	2:40																2	5:20	K
40	2:40																4	7:20	L
45	2:40																6	9:20	M
50	2:40																16	19:20	
55	2:40																24	27:20	
60	2:40																33	36:20	
65	2:40																41	44:20	
70	2:40																48	51:20	
75	2:40																55	58:20	
80	2:40																62	65:20	
85	2:40																68	71:20	
90	2:40																74	77:20	
95	2:40																80	83:20	
100	2:40																85	88:20	
110	2:40																96	99:20	
120	2:40																105	108:20	
130	2:20															1	114	118:00	
140	2:20															1	124	128:00	
110 FSW																			
24	3:40																0	3:40	I
25	3:00																1	4:40	I
30	3:00																4	7:40	J
35	3:00																7	10:40	L
40	3:00																10	13:40	M
45	3:00																21	24:40	
50	3:00																31	34:40	
55	3:00																40	43:40	
60	2:40															1	49	53:20	
65	2:40															2	57	62:20	
70	2:40															3	64	70:20	
75	2:40															4	71	78:20	
80	2:40															5	77	85:20	
85	2:40															5	84	92:20	
90	2:40															6	89	98:20	
95	2:40															6	95	104:20	
100	2:40															6	101	110:20	
110	2:40															7	112	122:20	
EXCEPTIONAL EXPOSURE ----------																			
120	2:40															7	123	133:20	
130	2:40															7	136	146:20	
140	2:20														1	7	149	160:00	

Table 18-14. MK 16 MOD 1 HeO_2 Decompression Tables (Continued).

(DESCENT RATE 60 FPM—ASCENT RATE 30 FPM)

Bottom Time (min)	Time to First Stop (M:S)	DECOMPRESSION STOPS (fsw) Stop times (min) include travel time, except first stop 170	160	150	140	130	120	110	100	90	80	70	60	50	40	30	20	Total Ascent Time (M:S)	Repet Group
120 FSW																			
20	4:00																0	4:00	I
25	3:20																4	8:00	J
30	3:20																8	12:00	K
35	3:20																12	16:00	M
40	3:20																23	27:00	
45	3:00															2	34	39:40	
50	3:00															4	43	50:40	
55	3:00															6	52	61:40	
60	3:00															7	60	70:40	
65	2:40														2	7	68	80:20	
70	2:40														3	7	76	89:20	
75	2:40														3	8	83	97:20	
80	2:40														4	7	91	105:20	
85	2:40														5	7	97	112:20	
90	2:40														5	8	103	119:20	
95	2:40														6	7	110	126:20	
EXCEPTIONAL EXPOSURE ----------																			
100	2:40														6	7	117	133:20	
110	2:40														7	7	131	148:20	
120	2:40														7	7	145	162:20	
130 FSW																			
17	4:20																0	4:20	H
20	3:40																3	7:20	I
25	3:40																8	12:20	K
30	3:40																13	17:20	L
35	3:20															2	21	27:00	L
40	3:20															5	32	41:00	L
45	3:00														1	7	43	54:40	L
50	3:00														3	7	53	66:40	
55	3:00														5	7	63	78:40	
60	3:00														6	8	71	88:40	
65	2:40													1	7	7	81	99:20	
70	2:40													2	7	7	89	108:20	
75	2:40													3	7	7	97	117:20	
80	2:40													3	8	7	104	125:20	
85	2:40													4	8	7	111	133:20	
EXCEPTIONAL EXPOSURE ----------																			
90	2:40													5	7	7	119	141:20	
95	2:40													5	8	7	127	150:20	
100	2:40													6	7	7	136	159:20	
110	2:40													6	8	7	152	176:20	
120	2:40													7	7	18	159	194:20	

Table 18-14. *MK 16 MOD 1 HeO_2 Decompression Tables (Continued).*

(DESCENT RATE 60 FPM—ASCENT RATE 30 FPM)

Bottom Time (min)	Time to First Stop (M:S)	DECOMPRESSION STOPS (fsw) Stop times (min) include travel time, except first stop 170	160	150	140	130	120	110	100	90	80	70	60	50	40	30	20	Total Ascent Time (M:S)	Repet Group
140 FSW																			
15	4:40																0	4:40	H
20	4:00																7	11:40	J
25	4:00																12	16:40	K
30	3:40															3	16	23:20	M
35	3:40															7	29	40:20	
40	3:20														3	7	42	56:00	
45	3:20														6	7	53	70:00	
50	3:00													1	8	7	64	83:40	
55	3:00													3	8	7	74	95:40	
60	3:00													5	8	7	84	107:40	
65	3:00													7	7	7	93	117:40	
70	2:40												1	7	8	7	101	127:20	
75	2:40												2	7	8	7	110	137:20	
EXCEPTIONAL EXPOSURE ----------																			
80	2:40												3	7	8	7	118	146:20	
85	2:40												4	7	7	8	127	156:20	
90	2:40												4	8	7	7	137	166:20	
95	2:40												5	7	7	8	146	176:20	
100	2:40												5	8	7	8	155	186:20	
150 FSW																			
13	5:00																0	5:00	H
15	4:20																3	8:00	H
20	4:20																10	15:00	J
25	4:00															2	14	20:40	L
30	4:00															7	24	35:40	L
35	3:40														4	8	37	53:20	L
40	3:20													1	7	8	50	70:00	
45	3:20													4	8	7	63	86:00	
50	3:20													7	7	8	74	100:00	
55	3:00												2	8	7	7	86	113:40	
60	3:00												4	8	7	7	96	125:40	
65	3:00												6	7	7	8	105	136:40	
70	3:00												7	7	8	7	114	146:40	
EXCEPTIONAL EXPOSURE ----------																			
75	2:40											1	8	7	7	8	124	158:20	
80	2:40											2	8	7	7	8	135	170:20	
85	2:40											3	7	8	7	7	146	181:20	
90	2:40											4	7	7	8	9	155	193:20	

Table 18-14. MK 16 MOD 1 HeO_2 Decompression Tables (Continued).

(DESCENT RATE 60 FPM—ASCENT RATE 30 FPM)

Bottom Time (min)	Time to First Stop (M:S)	DECOMPRESSION STOPS (fsw) Stop times (min) include travel time, except first stop 170	160	150	140	130	120	110	100	90	80	70	60	50	40	30	20	Total Ascent Time (M:S)	Repet Group
160 FSW																			
12	5:20																0	5:20	H
15	4:40																5	10:20	I
20	4:40																13	18:20	K
25	4:20															6	16	27:00	M
30	4:00														4	8	31	47:40	
35	3:40													2	7	8	46	67:20	
40	3:40													6	8	7	60	85:20	
45	3:20												3	7	7	8	73	102:00	
50	3:20												6	7	7	8	85	117:00	
55	3:00											1	7	8	7	7	97	130:40	
60	3:00											3	7	8	7	8	107	143:40	
EXCEPTIONAL EXPOSURE --------																			
65	3:00											5	7	8	7	7	118	155:40	
70	3:00											6	8	7	7	8	130	169:40	
75	3:00											8	7	7	8	7	142	182:40	
80	2:40										2	7	7	8	7	7	154	195:20	
85	2:40										2	8	7	8	7	16	158	209:20	
90	2:40										3	8	7	7	8	25	161	222:20	
170 FSW																			
11	5:40																0	5:40	H
15	5:00																8	13:40	I
20	4:40															2	15	22:20	K
25	4:20														2	8	22	37:00	L
30	4:00													2	7	7	39	59:40	L
35	4:00													7	7	8	55	81:40	
40	3:40												4	8	7	7	70	100:20	
45	3:20											1	7	8	7	7	84	118:00	
50	3:20											4	7	8	7	8	96	134:00	
55	3:20											7	7	7	8	7	108	148:00	
EXCEPTIONAL EXPOSURE --------																			
60	3:00										2	7	8	7	7	8	120	162:40	
65	3:00										4	7	8	7	7	8	134	178:40	
70	3:00										5	8	7	8	7	7	148	193:40	
75	3:00										7	7	8	7	7	12	157	208:40	
80	2:40									1	7	8	7	7	8	22	160	223:20	

Table 18-14. *MK 16 MOD 1 HeO_2 Decompression Tables (Continued).*

(DESCENT RATE 60 FPM—ASCENT RATE 30 FPM)

Bottom Time (min)	Time to First Stop (M:S)	DECOMPRESSION STOPS (fsw) Stop times (min) include travel time, except first stop																Total Ascent Time (M:S)	Repet Group
		170	160	150	140	130	120	110	100	90	80	70	60	50	40	30	20		
180 FSW																			
10	6:00																0	6:00	H
15	5:20																11	17:00	J
20	5:00															6	14	25:40	L
25	4:40														6	8	29	48:20	L
30	4:20													6	7	8	47	73:00	
35	4:00												4	8	7	8	64	95:40	
40	3:40											2	8	7	7	8	80	116:20	
45	3:40											6	8	7	7	8	94	134:20	
50	3:20										3	7	7	8	7	7	108	151:00	
EXCEPTIONAL EXPOSURE ----------																			
55	3:20										5	8	7	8	7	7	121	167:00	
60	3:00									1	7	8	7	7	8	7	136	184:40	
65	3:00									3	7	8	7	7	8	7	151	201:40	
70	3:00									5	7	7	8	7	7	16	158	218:40	
190 FSW																			
9	6:20																0	6:20	H
10	5:40																2	8:20	H
15	5:40																14	20:20	J
20	4:40													1	1	8	16	31:20	M
25	3:20									1	0	0	0	4	7	7	38	61:00	
30	3:00								1	0	0	2	2	7	7	8	57	87:40	
35	2:40							1	0	0	2	0	8	7	8	7	75	111:20	
40	2:20						1	0	0	0	2	6	8	7	7	8	91	133:00	
45	2:20						1	0	0	0	5	7	8	7	7	8	105	151:00	
50	2:20						1	0	0	0	8	8	7	8	7	7	120	169:00	
EXCEPTIONAL EXPOSURE ----------																			
55	2:20						1	0	0	4	8	7	7	8	7	7	138	190:00	
60	2:20						1	0	0	7	7	8	7	7	8	7	153	208:00	
65	2:20						1	0	2	7	7	8	7	7	8	19	159	228:00	
70	2:20						1	0	3	8	7	8	7	7	8	31	164	247:00	

Table 18-14. *MK 16 MOD 1 HeO_2 Decompression Tables (Continued).*

(DESCENT RATE 60 FPM—ASCENT RATE 30 FPM)

Bottom Time (min)	Time to First Stop (M:S)	DECOMPRESSION STOPS (fsw) Stop times (min) include travel time, except first stop																Total Ascent Time (M:S)	Repet Group
		170	160	150	140	130	120	110	100	90	80	70	60	50	40	30	20		
200 FSW																			
8	6:40																0	6:40	G
10	6:00																5	11:40	H
15	5:20														1	1	15	23:00	K
20	3:20								1	0	0	2	0	0	5	7	25	44:00	L
25	2:00				1	0	0	0	2	0	1	0	1	7	7	7	47	75:40	L
30	1:20		1	0	0	2	0	0	0	2	0	1	7	7	8	7	69	106:00	
35	1:20		1	0	1	1	0	0	2	0	0	7	7	7	8	7	87	130:00	
40	1:00	1	0	1	1	0	0	2	0	0	5	8	7	7	8	7	104	152:40	
45	1:00	1	0	1	1	0	0	2	0	2	7	8	7	8	7	7	120	172:40	
EXCEPTIONAL EXPOSURE ----------																			
50	1:00	1	0	1	1	0	1	0	1	6	7	7	8	7	8	7	139	195:40	
55	1:00	1	0	1	1	0	1	0	2	8	7	7	8	7	8	8	155	215:40	
60	1:00	1	0	1	1	0	1	0	5	7	8	7	7	8	7	22	161	237:40	
210 FSW																			
5	7:00																0	7:00	
10	6:20																5	12:00	
15	6:00															7	5	18:40	
20	5:00												5	3	2	2	28	45:40	
25	4:20										3	3	3	2	3	3	57	79:00	
30	4:20										6	3	2	2	6	12	76	112:00	
35	3:40								3	3	3	2	3	5	12	12	95	142:20	
40	3:20							3	2	3	2	3	5	12	11	12	113	170:00	
EXCEPTIONAL EXPOSURE ----------																			
45	3:20							4	2	3	2	4	11	12	12	11	131	196:00	
50	3:20							4	3	2	3	10	11	12	12	11	149	221:00	
55	3:00						3	2	3	2	7	11	11	12	11	12	165	242:40	
60	3:20							5	3	2	11	12	11	11	12	21	173	265:00	

Table 18-14. *MK 16 MOD 1 HeO_2 Decompression Tables (Continued).*

(DESCENT RATE 60 FPM—ASCENT RATE 30 FPM)

Bottom Time (min)	Time to First Stop (M:S)	DECOMPRESSION STOPS (fsw) Stop times (min) include travel time, except first stop																Total Ascent Time (M:S)	Repet Group
		170	160	150	140	130	120	110	100	90	80	70	60	50	40	30	20		
220 FSW																			
5	7:20																0	7:20	
10	6:40																5	12:20	
15	5:40													4	3	2	6	21:20	
20	5:00											4	3	2	3	2	37	56:40	
25	5:00											7	3	3	2	8	65	93:40	
30	4:00								3	3	2	3	3	3	10	12	84	127:40	
35	4:20									8	2	3	2	12	12	11	106	161:00	
40	4:20									9	3	2	12	11	12	11	126	191:00	
EXCEPTIONAL EXPOSURE																			
45	3:40							6	2	3	2	10	12	11	12	11	144	217:20	
50	4:00								8	3	8	11	12	11	11	12	164	244:40	
55	4:00								9	4	12	11	12	11	11	18	177	269:40	
230 FSW																			
5	7:40																0	7:40	
10	7:00																6	13:40	
15	6:00													5	3	2	9	25:40	
20	5:00										3	3	2	3	3	2	46	67:40	
25	4:40									5	2	3	3	2	3	12	71	106:20	
30	4:00							3	3	2	3	2	3	6	12	12	93	143:40	
35	4:00							5	3	2	3	2	8	12	12	11	116	178:40	
EXCEPTIONAL EXPOSURE																			
40	3:20					2	3	2	3	2	3	8	12	11	12	11	137	210:00	
45	4:00							8	2	3	7	12	11	11	12	11	159	240:40	
50	3:20					4	3	2	3	5	11	13	11	11	11	16	174	268:00	
55	3:00				2	3	2	4	2	12	11	11	11	11	11	38	172	293:40	
240 FSW																			
5	8:00																0	8:00	
10	7:20																8	16:00	
15	6:00												4	3	2	4	15	34:40	
20	5:20										5	2	3	2	3	3	54	78:00	
25	5:20										9	3	2	2	8	12	80	122:00	
30	4:20							5	3	2	2	3	3	11	12	12	103	161:00	
35	4:20							7	3	2	3	4	12	11	12	12	127	198:00	
EXCEPTIONAL EXPOSURE																			
40	4:20							8	3	3	4	12	12	11	12	12	150	232:00	
45	4:20							10	2	4	12	12	11	12	11	12	173	264:00	
50	3:40					6	3	2	3	12	11	11	12	11	11	32	174	292:20	

Table 18-14. *MK 16 MOD 1 HeO_2 Decompression Tables (Continued).*

(DESCENT RATE 60 FPM—ASCENT RATE 30 FPM)

250 FSW

Bottom Time (min)	Time to First Stop (M:S)	DECOMPRESSION STOPS (fsw) Stop times (min) include travel time, except first stop 170	160	150	140	130	120	110	100	90	80	70	60	50	40	30	20	Total Ascent Time (M:S)	Repet Group
5	8:20																0	8:20	
10	7:40																9	17:20	
15	6:20												5	3	3	2	24	44:00	
20	5:40										6	3	2	3	3	6	61	90:20	
25	5:00								6	3	2	2	3	3	12	12	87	135:40	
30	4:20						4	3	3	2	3	2	8	11	12	12	112	177:00	
EXCEPTIONAL EXPOSURE ----------																			
35	4:40							9	2	3	2	10	12	12	11	12	139	217:20	
40	4:20						8	3	2	3	11	12	11	11	12	11	164	253:00	
45	4:00					7	3	3	2	11	11	12	11	11	12	25	175	287:40	
50	3:40				6	2	3	3	9	12	11	11	12	11	11	49	175	319:20	

260 FSW

Bottom Time (min)	Time to First Stop (M:S)	170	160	150	140	130	120	110	100	90	80	70	60	50	40	30	20	Total Ascent Time (M:S)	Repet Group
5	8:40																0	8:40	
10	8:00																11	19:40	
15	6:20											4	3	3	2	3	31	53:00	
20	5:40									5	3	3	2	3	3	10	67	102:20	
25	5:20								8	3	2	2	3	7	13	12	96	152:00	
30	4:40						6	3	2	3	2	3	12	12	13	11	123	195:20	
EXCEPTIONAL EXPOSURE ----------																			
35	4:40						8	3	3	2	6	12	12	11	12	11	151	236:20	
40	4:20					8	3	2	3	7	12	12	11	11	12	14	175	275:00	
45	4:00				7	3	2	3	8	12	11	11	11	12	11	42	173	310:40	

270 FSW

Bottom Time (min)	Time to First Stop (M:S)	170	160	150	140	130	120	110	100	90	80	70	60	50	40	30	20	Total Ascent Time (M:S)	Repet Group
5	8:20																5	14:00	
10	8:20																13	22:00	
15	6:20										3	3	3	2	3	3	39	63:00	
20	6:20										9	3	2	3	5	12	75	116:00	
25	5:40								9	3	2	3	3	12	11	12	105	166:20	
EXCEPTIONAL EXPOSURE ----------																			
30	5:00						8	3	2	3	2	9	11	12	11	12	134	212:40	
35	4:40					8	3	2	3	3	11	12	12	11	11	12	163	256:20	
40	4:20				8	3	3	1	5	12	12	11	11	11	12	30	174	298:00	
45	4:20				9	3	2	5	12	13	10	11	11	12	11	56	176	336:00	

Table 18-14. *MK 16 MOD 1 HeO_2 Decompression Tables (Continued).*

(DESCENT RATE 60 FPM—ASCENT RATE 30 FPM)

280 FSW

Bottom Time (min)	Time to First Stop (M:S)	DECOMPRESSION STOPS (fsw) Stop times (min) include travel time, except first stop 170	160	150	140	130	120	110	100	90	80	70	60	50	40	30	20	Total Ascent Time (M:S)	Repet Group
5	8:40																5	14:20	
10	8:40																14	23:20	
15	7:00											7	3	2	3	3	47	72:40	
20	6:20									9	2	3	2	3	9	12	82	129:00	
25	5:20						6	3	3	2	3	2	7	12	12	12	114	182:00	
EXCEPTIONAL EXPOSURE ----------																			
30	5:20						10	3	2	3	3	12	12	11	12	12	145	231:00	
35	4:40				8	2	3	2	3	8	12	12	11	11	11	13	176	277:20	
40	4:40				10	2	3	2	11	12	11	12	12	10	12	45	174	321:20	
45	4:40				11	3	3	11	11	12	11	11	11	12	11	72	178	362:20	

290 FSW

Bottom Time (min)	Time to First Stop (M:S)	170	160	150	140	130	120	110	100	90	80	70	60	50	40	30	20	Total Ascent Time (M:S)	Repet Group
5	9:00																5	14:40	
10	8:00													4	4	2	6	24:40	
15	7:00										6	3	2	3	3	2	55	81:40	
20	6:20								8	2	3	2	3	4	12	12	88	141:00	
25	5:40						8	3	2	3	3	2	12	12	11	12	122	196:20	
EXCEPTIONAL EXPOSURE ----------																			
30	5:00				7	3	2	3	3	2	9	12	12	11	11	12	156	248:40	
35	5:00				10	2	3	2	5	12	11	12	11	11	12	28	176	300:40	
40	5:00				12	2	3	7	12	11	12	11	11	11	12	59	177	345:40	
45	5:00				13	3	9	11	12	11	11	11	11	11	18	82	180	388:40	

300 FSW

Bottom Time (min)	Time to First Stop (M:S)	170	160	150	140	130	120	110	100	90	80	70	60	50	40	30	20	Total Ascent Time (M:S)	Repet Group
5	9:20																5	15:00	
10	8:20													6	3	2	9	29:00	
15	7:00									5	3	2	3	2	3	5	61	91:40	
20	6:20							7	3	2	3	2	4	6	12	12	96	154:00	
25	5:20				5	3	2	3	3	2	3	7	12	11	12	11	132	212:00	
EXCEPTIONAL EXPOSURE ----------																			
30	5:20				9	3	2	3	2	5	12	12	11	11	12	12	169	269:00	
35	5:20				12	2	3	2	10	12	11	12	11	11	12	41	176	321:00	
40	5:20				14	2	4	12	12	11	11	12	11	11	11	74	180	371:00	

Table 18-14. MK 16 MOD 1 HeO_2 Decompression Tables (Continued).

(DESCENT RATE 60 FPM—ASCENT RATE 30 FPM)

Bottom Time (min)	Time to First Stop (M:S)	DECOMPRESSION STOPS (fsw) Stop times (min) include travel time, except first stop																Total Ascent Time (M:S)	Repet Group
		170	160	150	140	130	120	110	100	90	80	70	60	50	40	30	20		
310 FSW																			
EXCEPTIONAL EXPOSURE																			
10	8:20												5	2	3	3	14	36:00	
15	7:20									6	3	3	2	3	2	9	66	102:00	
20	6:20						6	3	2	3	2	3	3	12	11	12	103	167:00	
25	6:00					9	3	2	3	3	2	12	11	12	12	11	142	228:40	
30	5:40				11	3	2	2	3	10	12	11	11	12	12	17	176	288:20	
35	5:40				14	2	3	6	12	11	12	11	11	11	12	55	178	344:20	
40	5:40				16	2	10	12	11	12	11	11	11	11	19	83	182	397:20	
320 FSW																			
EXCEPTIONAL EXPOSURE																			
10	8:20											4	2	3	3	2	21	44:00	
15	7:40									8	3	2	3	2	3	12	71	112:20	
20	6:20					6	2	3	2	3	2	4	5	12	12	12	111	181:00	
25	6:20					11	3	2	2	3	7	12	11	12	11	12	153	246:00	
30	6:00				13	2	3	2	6	12	11	12	11	11	12	30	177	308:40	
35	6:00				15	3	3	11	12	11	12	11	11	11	12	68	182	368:40	
40	6:00				18	7	11	12	11	11	11	12	11	11	35	83	185	424:40	

PAGE LEFT BLANK INTENTIONALLY

CHAPTER 19

Closed-Circuit Oxygen UBA Diving

19-1 INTRODUCTION

Figure 19-1. Diver in MK-25 UBA.

The term *closed-circuit oxygen rebreather* describes a specialized underwater breathing apparatus (UBA) in which the diver breathes 100% oxygen and all gases are kept within the UBA. The use of 100% oxygen prevents inert gas buildup in the diver and allows all of the gas carried by the diver to be used for metabolic needs. The exhaled gas is carried via the exhalation hose to a carbon dioxide-absorbent bed, which removes the carbon dioxide produced by the diver through a chemical reaction. Metabolically consumed oxygen is then replaced through an oxygen addition system. The gas then travels to the breathing bag where it is available again to the diver. Closed-circuit oxygen UBAs offer advantages valuable to special warfare, including stealth (no escaping bubbles), extended operating duration, and less weight than open-circuit air SCUBA. Weighed against these advantages are the disadvantages of increased hazards to the diver, greater training requirements, and greater expense. However, when compared to a closed-circuit mixed-gas UBA, an oxygen UBA offers the advantages of reduced training and maintenance requirements, lower cost, and reduction in weight and size.

19-1.1 **Purpose.** This chapter provides general guidance for MK 25 diving operations and procedures. For detailed operation and maintenance instructions see the MK 25 MOD 2 Operation and Maintenance Manual, SS600-A3-MMA-010/53833 (Marine Corps TM 09603B-14 & P/1).

19-1.2 **Scope.** This chapter covers MK 25 UBA principles of operations, operational planning, dive procedures, and medical aspects of closed-circuit oxygen diving.

19-2 MEDICAL ASPECTS OF CLOSED-CIRCUIT OXYGEN DIVING

Closed-circuit oxygen divers are subject to many of the same medical problems as other divers. Volume I, Chapter 3 provides in-depth coverage of all medical considerations of diving. Only the diving disorders that merit special attention for closed-circuit oxygen divers are addressed in this chapter.

19-2.1 **Central Nervous System (CNS) Oxygen Toxicity.** High pressure oxygen poisoning is known as CNS oxygen toxicity. High partial pressures of oxygen are associated with many biochemical changes in the brain, but which specific changes are responsible for the signs and symptoms of CNS oxygen toxicity is presently unknown. CNS oxygen toxicity is not likely to occur at oxygen partial pressures below 1.3 ata, though relatively brief exposure to partial pressures above this, when it occurs at depth or in a pressurized chamber, can result in CNS oxygen toxicity causing CNS-related symptoms.

19-2.1.1 **Causes of CNS Oxygen Toxicity.** Factors that increase the likelihood of CNS oxygen toxicity are:

- Increased partial pressure of oxygen. At depths less than 25 fsw, a change in depth of five fsw increases the risk of oxygen toxicity only slightly, but a similar depth increase in the 30-fsw to 50-fsw range may significantly increase the likelihood of a toxicity episode.
- Increased time of exposure
- Prolonged immersion
- Stress from strenuous physical exercise
- Carbon dioxide buildup. The increased risk for CNS oxygen toxicity may occur even before the diver is aware of any symptoms of carbon dioxide buildup.
- Cold stress resulting from shivering or an increased exercise rate as the diver attempts to keep warm.
- Systemic diseases that increase oxygen consumption. Conditions associated with increased metabolic rates (such as certain thyroid or adrenal disorders) tend to cause an increase in oxygen sensitivity. Divers with these diseases should be excluded from oxygen diving.

19-2.1.2 **Symptoms of CNS Oxygen Toxicity.** In diving, the most serious effects of oxygen toxicity are CNS symptoms. These symptoms may not always appear and most are not exclusively symptoms of oxygen toxicity. The appearance of any one of these symptoms usually represents a bodily signal of distress of some kind and should be heeded. Twitching is perhaps the clearest warning of oxygen toxicity, but it may occur late if at all. The most serious symptom of CNS oxygen toxicity is convulsion, which may occur suddenly without any previous symptoms, and may result in drowning or arterial gas embolism. The mnemonic device VENTID-C is a helpful reminder of the most common symptoms of CNS oxygen toxicity"

V: Visual symptoms. Tunnel vision, a decrease in the diver's peripheral vision, and other symptoms, such as blurred vision, may occur.

E: Ear symptoms. Tinnitus is any sound perceived by the ears but not resulting from an external stimulus. The sound may resemble bells ringing, roaring, or a machinery-like pulsing sound.

N: Nausea or spasmodic vomiting. These symptoms may be intermittent.

T: Twitching and tingling symptoms. Any of the small facial muscles, lips, or muscles of the extremities may be affected. These are the most frequent and clearest symptoms.

I: Irritability. Any change in the diver's mental status; including confusion, agitation, and anxiety.

D: Dizziness. Symptoms include clumsiness, incoordination, and unusual fatigue.

C: Convulsions.

The following additional factors should be noted regarding an oxygen convulsion:

- The diver is unable to carry on any effective breathing during the convulsion.

- After the diver is brought to the surface, there will be a period of unconsciousness or neurologic impairment following the convulsion; these symptoms are indistinguishable from those of arterial gas embolism.

- No attempt should be made to insert any object between the clenched teeth of a convulsing diver. Although a convulsing diver may suffer a lacerated tongue, this trauma is preferable to the trauma that may be caused during the insertion of a foreign object. In addition, the person providing first aid may incur significant hand injury if bitten by the convulsing diver.

- There may be no warning of an impending convulsion to provide the diver the opportunity to return to the surface. Therefore, buddy lines are essential to safe closed-circuit oxygen diving.

19-2.1.3 **Treatment of Nonconvulsive Symptoms.** The stricken diver should alert his dive buddy and make a controlled ascent to the surface. The victim's life preserver should be inflated (if necessary) with the dive buddy watching him closely for progression of symptoms. Though an ascent from depth will lower the partial pressure of oxygen, the diver may still suffer other or worsening symptoms. The divers should notify the Diving Supervisor and terminate the dive.

19-2.1.4 **Treatment of Underwater Convulsion.** The following steps should be taken when treating a convulsing diver:

1. Assume a position behind the convulsing diver. The weight belt should be left in place to prevent the diver from assuming a face down position on the surface. Release the victim's weight belt only if progress to the surface is significantly impeded.

2. Leave the victim's mouthpiece in his mouth. If it is not in his mouth, do not attempt to replace it; however, if time permits, ensure that the mouthpiece is switched to the SURFACE position.

3. Grasp the victim around his chest above the UBA or between the UBA and his body. If difficulty is encountered in gaining control of the victim in this manner, the rescuer should use the best method possible to obtain control. The UBA harnesses may be grasped if necessary.

4. Make a controlled ascent to the surface, maintaining a slight pressure on the diver's chest to assist exhalation.

5. If additional buoyancy is required, activate the victim's life jacket. The rescuer should not release his own weight belt or inflate his own life jacket.

6. Upon reaching the surface, inflate the victim's life jacket if not previously done.

7. Remove the victim's mouthpiece and switch the valve to SURFACE to prevent the possibility of the rig flooding and weighing down the victim.

8. Signal for emergency pickup.

9. Once the convulsion has subsided, open the victim's airway by tilting his head back slightly.

10. Ensure the victim is breathing. Mouth-to-mouth breathing may be initiated if necessary.

11. If an upward excursion occurred during the actual convulsion, transport to the nearest chamber and have the victim evaluated by an individual trained to recognize and treat diving-related illness.

19-2.1.5 **Off-Effect.** The off-effect, a hazard associated with CNS oxygen toxicity, may occur several minutes after the diver comes off gas or experiences a reduction of oxygen partial pressure. The off-effect is manifested by the onset or worsening of CNS oxygen toxicity symptoms. Whether this paradoxical effect is truly caused by the reduction in partial pressure or whether the association is coincidental is unknown.

19-2.2 **Pulmonary Oxygen Toxicity.** Pulmonary oxygen toxicity can result from prolonged exposure to elevated partial pressures of oxygen. This form of oxygen toxicity produces lung irritation with symptoms of chest pain, cough, and pain on inspiration that develop slowly and become increasingly worse as long as the elevated level of oxygen is breathed. Although hyperbaric oxygen may cause serious lung damage, if the oxygen exposure is discontinued before the symptoms become too severe, the symptoms will slowly abate. This form of oxygen toxicity is generally seen during oxygen recompression treatment and saturation diving, and on long, shallow, in-water oxygen exposures.

19-2.3 **Oxygen Deficiency (Hypoxia).** Hypoxia is an abnormal deficiency of oxygen in the arterial blood in which the partial pressure of oxygen is too low to meet the metabolic needs of the body. Chapter 3 contains an in-depth description of this disorder. Although all cells in the body need oxygen, the initial symptoms of hypoxia are a manifestation of central nervous system dysfunction.

19-2.3.1 **Causes of Hypoxia with the MK 25 UBA.** The primary cause of hypoxia in the MK25 is inadequate/incorrect purge of the UBA. The risk of hypoxia is greatest when the diver is breathing the UBA on the surface. In the MK25, oxygen is only added on a demand basis as the breathing bag is emptied on inhalation. On the surface as the diver consumes oxygen, the oxygen fraction in the breathing loop will begin to decrease, as will the gas volume in the breathing bag. If there is sufficient nitrogen in the breathing loop to prevent the breathing bag from being emptied no oxygen will be added and the oxygen fraction may drop to ten percent or lower. Since there is sufficient gas volume in the breathing bag for normal inhalation, hypoxia can occur without warning. Hypoxia on descent or while diving is less likely, because as the diver descends pure oxygen is added to the breathing loop to maintain volume which increases both the oxygen fraction in the breathing loop and the oxygen partial pressure.

19-2.3.2 **MK 25 UBA Purge Procedures.** The detailed purge procedures in the MK 25 Operation and Maintenance Manual are designed to remove as much of the inert gas (nitrogen) from a diver's lungs as possible prior to the start of a dive and have been thoroughly tested. They ensure the oxygen fraction in the breathing loop is sufficiently high to prevent the occurrence of hypoxia. The purge procedures should be strictly followed.

19-2.3.3 **Underwater Purge.** If the diver conducts an underwater purge or purge under pressure, the increase in oxygen fraction caused by volume make up described above may not occur and the diver may be more susceptible to hypoxia. Therefore, strict adherence to the under pressure purge procedures prescribed in the operations and maintenance manual is extremely important.

19-2.3.4 **Symptoms of Hypoxia.** Hypoxia may have no warning symptoms prior to loss of consciousness. Other symptoms that may appear include confusion, loss of coordination, dizziness, and convulsion. It is important to note that if symptoms of unconsciousness or convulsion occur at the beginning of a closed-circuit oxygen dive, hypoxia, not oxygen toxicity, is the most likely cause.

19-2.3.5 **Treatment of Hypoxia.** Treatment for a suspected case of hypoxia consists of the following:

- If the diver becomes unconscious or incoherent at depth, the dive buddy should add oxygen to the stricken diver's UBA.
- The diver must be brought to the surface. Remove the mouthpiece and allow the diver to breathe fresh air. If unconscious, check breathing and circulation, maintain an open airway and administer 100-percent oxygen. Switch mouthpiece valve to the SURFACE position.

- If the diver surfaces in an unconscious state, transport to the nearest chamber and have the victim evaluated by an individual trained to recognize and treat diving-related illness. If the diver recovers fully with normal neurological function, he does not require immediate treatment for arterial gas embolism.

19-2.4 **Carbon Dioxide Toxicity (Hypercapnia).** Carbon dioxide toxicity, or hypercapnia, is an abnormally high level of carbon dioxide in the blood and body tissues. Hypercapnia is generally the result of a buildup of carbon dioxide in the breathing supply or in the body. Inadequate ventilation (breathing volume) by the diver or failure of the carbon dioxide-absorbent canister to remove carbon dioxide from the exhaled gas will cause a buildup to occur.

19-2.4.1 **Symptoms of Hypercapnia.** Symptoms of hypercapnia are:

- Increased rate and depth of breathing
- Labored breathing (similar to that seen with heavy exercise)
- Headache
- Confusion
- Unconsciousness

It is important to note that the presence of a high partial pressure of oxygen may reduce the early symptoms of hypercapnia. As previously mentioned, elevated levels of carbon dioxide may result in an episode of CNS oxygen toxicity on a normally safe dive profile.

19-2.4.2 **Treating Hypercapnia.** To treat hypercapnia:

- Increase ventilation if skip-breathing is a possible cause.
- Decrease exertion level.
- Abort the dive. Return to the surface and breathe air.
- During ascent, while maintaining a vertical position, the diver should activate his bypass valve, adding fresh oxygen to his UBA. If the symptoms are a result of canister floodout, an upright position decreases the likelihood that the diver will sustain chemical injury (paragraph 19-2.5).
- If unconsciousness occurs at depth, the same principles of management for underwater convulsion as described in paragraph 19-2.1.4 apply.

19-2.4.3 **Prevention of Hypercapnia.** To minimize the risk of hypercapnia:

- Use only an approved carbon dioxide absorbent in the UBA canister.
- Follow the prescribed canister-filling procedure to ensure that the canister is correctly packed with carbon dioxide absorbent.
- Dip test the UBA carefully before the dive. Watch for leaks that may result in canister floodout.
- Do not exceed canister duration limits for the water temperature.
- Ensure that the one-way valves in the supply and exhaust hoses are installed and working properly.
- Swim at a relaxed, comfortable pace.
- Avoid skip-breathing. There is no advantage to this type of breathing in a closed-circuit rig and it may cause elevated blood carbon dioxide levels even with a properly functioning canister.

19-2.5 **Chemical Injury.** The term "chemical injury" refers to the introduction of a caustic solution from the carbon dioxide scrubber of the UBA into the upper airway of a diver.

19-2.5.1 **Causes of Chemical Injury.** The caustic alkaline solution results from water leaking into the canister and coming in contact with the carbon dioxide absorbent. When the diver is in a horizontal or head-down position, this solution may travel through the inhalation hose and irritate or injure his upper airway.

19-2.5.2 **Symptoms of Chemical Injury.** The diver may experience rapid breathing or headache, which are symptoms of carbon dioxide buildup in the breathing gas. This occurs because an accumulation of the caustic solution in the canister may be impairing carbon dioxide absorption. If the problem is not corrected promptly, the alkaline solution may travel into the breathing hoses and consequently be inhaled or swallowed. Choking, gagging, foul taste, and burning of the mouth and throat may begin immediately. This condition is sometimes referred to as a "caustic cocktail." The extent of the injury depends on the amount and distribution of the solution.

19-2.5.3 **Treatment of a Chemical Incident.** If the caustic solution enters the mouth, nose, or face mask, the diver must take the following steps:

- Immediately assume an upright position in the water.
- Depress the manual bypass valve continuously and make a controlled ascent to the surface, exhaling through the nose to prevent overpressurization.

- Should signs of system flooding occur during underwater purging, abort the dive and return to open-circuit or mixed-gas UBA if possible.

Using fresh water, rinse the mouth several times. Several mouthfuls should then be swallowed. If only sea water is available, rinse the mouth, but do not swallow. Other fluids may be substituted if available, but the use of weak acid solutions (vinegar or lemon juice) is not recommended. Do not attempt to induce vomiting.

As a result of the chemical injury, the diver may have difficulty breathing properly on ascent. He should be observed for signs of an arterial gas embolism and treated if necessary. A Diving Medical Officer or a Diving Medical Technician/Special Operations Technician should evaluate a victim of a chemical injury as soon as possible. Respiratory distress, which may result from the chemical trauma to the air passages, requires immediate hospitalization.

19-2.5.4 **Prevention of Chemical Injury.** Chemical injuries are best prevented by the performance of a careful dip test during predive set up to detect any system leaks. Special attention should also be paid to the position of the mouthpiece rotary valve upon water entry and exit to prevent the entry of water into the breathing loop. Additionally, dive buddies should perform a careful leak check on each other before leaving the surface at the start of a dive.

19-2.6 Middle Ear Oxygen Absorption Syndrome. Middle ear oxygen absorption syndrome refers to the negative pressure that may develop in the middle ear following a long oxygen dive.

19-2.6.1 **Causes of Middle Ear Oxygen Absorption Syndrome.** Gas with a very high percentage of oxygen enters the middle ear cavity during the course of an oxygen dive. Following the dive, the oxygen is slowly absorbed by the tissues of the middle ear. If the Eustachian tube does not open spontaneously, a negative pressure relative to ambient may result in the middle ear cavity. There may also be fluid (serous otitis media) present in the middle ear as a result of the differential pressure.

19-2.6.2 **Symptoms of Middle Ear Oxygen Absorption Syndrome.** Symptoms are often noted the morning after a long oxygen dive and include:

- A sense of pressure or mild discomfort in one or both ears.
- Muffled hearing in one or both ears.
- A moist, crackling sensation in one or both ears as a result of fluid accumulation in the middle ear.

19-2.6.3 **Treating Middle Ear Oxygen Absorption Syndrome.** Equalizing the pressure in the middle ear using a normal Valsalva maneuver or the diver's procedure of choice (e.g., swallowing, yawning) will usually relieve the symptoms. Discomfort and hearing loss resolve quickly, but the middle ear fluid is absorbed more slowly.

If symptoms persist, a Diving Medical Technician or Diving Medical Officer shall be consulted.

19-2.6.4 **Prevention of Middle Ear Oxygen Absorption Syndrome.** Middle ear oxygen absorption syndrome is difficult to avoid but usually does not pose a significant problem because symptoms are generally mild and easily eliminated. To prevent Middle Ear Oxygen Absorption Syndrome the diver should perform several gentle Valsalva maneuvers throughout the day after a long oxygen dive to ensure the Eustachian tube remains open.

19-3 MK-25

The closed-circuit oxygen UBA currently used by U.S. Navy combat swimmers is the MK 25 MOD 2. Figure 19-2 lists the operational characteristics of the MK 25 MOD 2.

MK 25 MOD 2 Characteristics

Principle of Operation:

Special Warfare only, closed-circuit system

Minimum Equipment:

1. MK 25 MOD 2 UBA
2. Approved life jacket
3. Face mask
4. Weight belt
5. Dive knife
6. Swim fins
7. Dive watch
8. Appropriate thermal protection
9. Whistle
10. Buddy line (one per pair)
11. Depth gauge (large face; accurate at shallow depths; one per pair)
12. Compass (one per pair if on compass course)

Principal Applications:

1. Special Warfare
2. Search
3. Inspection

Advantages:

1. No Surface Bubbles
2. Minimum Support
3. Long Duration
4. Fast deployment
5. Good Horizontal Mobility

Disadvantages:

1. Limited to shallow depths
2. CNS Oxygen toxicity hazards
3. Limited physical and thermal protection

Restrictions:

1. Normal working limit-25 fsw for 240 minutes
2. Maximum working limit-50 fsw for 10 minutes
3. No excursion allowed when using Single Depth Diving Limits

Operational Considerations:

1. Minimum personnel-5
2. Buddy diver required
3. Chase boat

Figure 19-2. MK 25 MOD 2 Operational Characteristics.

19-3.1 **Gas Flow Path.** The gas flow path of the MK 25 UBA is shown in Figure 19-3. The gas is exhaled by the diver into the mouthpiece. One-way valves in the breathing hoses direct the flow of gas through the exhalation hose and into the carbon dioxide-absorbent canister, which is packed with an approved carbon

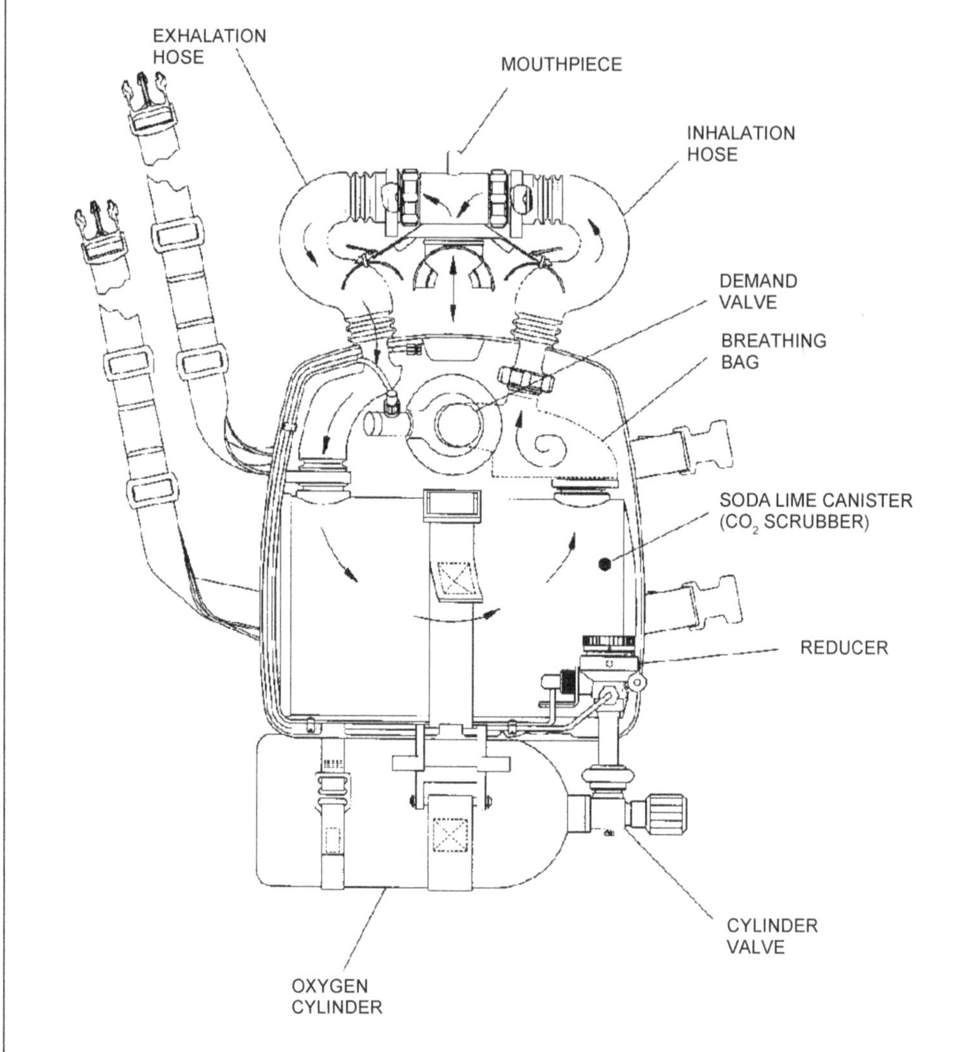

Figure 19-3. Gas Flow Path of the MK 25.

dioxide absorbent material. The carbon dioxide is removed by passing through the CO_2-absorbent bed and chemically combining with the CO_2-absorbent material in the canister. Upon leaving the canister the used oxygen enters the breathing bag. When the diver inhales, the gas is drawn from the breathing bag through the inhalation hose and mouthpiece and back into the diver's lungs. The gas flow described is entirely breath activated. As the diver exhales, the gas in the UBA is pushed forward by the exhaled gas, and upon inhalation the one-way valves in the hoses allow fresh gas to be pulled into the diver's lungs from the breathing bag.

19-3.1.1 **Breathing Loop.** The demand valve adds oxygen to the breathing bag of the UBA from the oxygen cylinder only when the diver empties the bag on inhalation. The demand valve also contains a manual bypass knob to allow for manual filling of the breathing bag during rig setup and as required. There is no constant flow of fresh oxygen to the diver. This feature of the MK 25 UBA makes it essential that nitrogen be purged from the apparatus prior to the dive. If too much nitrogen is present in the breathing loop, the breathing bag may not be emptied and the demand valve may not add oxygen even when metabolic consumption by the diver has reduced the oxygen in the UBA to dangerously low levels (see paragraph 19-2.3).

19-3.2 **Operational Duration of the MK 25 UBA.** The operational duration of the MK 25 UBA may be limited by either the oxygen supply or the canister duration. Refer to Table 19-1 for the breathing gas consumption rates for the MK 25 UBA.

19-3.2.1 **Oxygen Supply.** The MK 25 oxygen bottle is charged to 3,000 psig (207 BAR). The oxygen supply may be depleted in two ways: by the diver's metabolic consumption or by the loss of gas from the UBA. A key factor in maximizing the duration of the oxygen supply is for the diver to swim at a relaxed, comfortable pace. A diver swimming at a high exercise rate may have an oxygen consumption of two liters per minute (oxygen supply duration = 150 minutes) while one swimming at a relaxed pace may have an oxygen consumption of one liter per minute (oxygen supply duration = 300 minutes).

Oxygen pressure is monitored during the dive by the UBA oxygen pressure gauge, displayed in bars.

Table 19-1. *Average Breathing Gas Consumption.*

Diving Equipment	Overbottom Pressure Minimum	Gas Consumption (Normal)	Gas Consumption (Heavy Work)
MK 25 UBA (100% 0_2)	50 psi (3.4 BAR)	15-17 psi/min	(See Note)

NOTE: Heavy work is not recommended for the MK 25.

19-3.2.2 **Canister Duration.** The canister duration is dependent on water temperature, exercise rate, and the mesh size of the NAVSEA-approved carbon dioxide absorbent. (Table 19-2 lists NAVSEA-approved absorbents.) The canister will function adequately as long as the UBA has been set up properly. Factors that may cause the canister to fail early are discussed under carbon dioxide buildup in paragraph 19-2.4. Dives should be planned so as not to exceed the canister duration limits.

The duration of the oxygen supply will be dependent on the factors discussed in paragraph 19-5.2 and must be estimated using the anticipated swim speed and the expertise of the divers in avoiding gas loss.

Table 19-2. *NAVSEA-Approved CO_2 Absorbents.*

Name	Vendor
High Performance Sodasorb, Regular	W.R. Grace
Sofnolime 4-8 Mesh NI, L Grade	O.C. Lugo
Sofnolime 8-12 Mesh NI, D Grade **WARNING: Sofnolime 8-12 is only approved for use in the MK25 Mod 2 (urethane) canister. See the technical manual for further guidance.**	O.C. Lugo
Divesorb Pro 5-8 Mesh	Drager

19-3.3 **Packing Precautions.** Caution should be used when packing the carbon dioxide canister to ensure the canister is completely filled with carbon dioxide-absorbent material to minimize the possibility of channeling. Channeling allows the diver's exhaled carbon dioxide to pass through channels in the absorbent material without being absorbed, resulting in an ever-increasing concentration of carbon dioxide in the breathing bag, leading to hypercapnia. Channeling can be avoided by following the canister-packing instructions provided by the specific MK 25 Operation and Maintenance Manual. Basic precautions include orienting the canister vertically and filling the canister to approximately 1/3 full with the approved absorbent material and tapping the sides of the canister with the hand or a rubber mallet. This process should be repeated by thirds until the canister is filled to the fill line scribed on the inside of the absorbent canister. Mashing the material with a balled fist is not recommended as it may cause the approved absorbent material to fracture, thereby producing dust which would then be transported through the breathing loop to the diver's lungs while breathing the UBA.

19-3.4 **Preventing Caustic Solutions in the Canister.** Additional concerns include ensuring water is not inadvertently introduced into the canister by leaving the mouthpiece in the "DIVE" position when on the surface or through system leaks. The importance of performing the tightness and dip test while performing predive setup procedures cannot be overemphasized. When water combines with the absorbent material, it creates strong caustic solution commonly referred to as "caustic cocktail," which is capable of producing chemical burns in the diver's mouth and airway. In the event of a "caustic cocktail," the diver should immediately maintain a head-up attitude in the water column, depress the manual bypass knob on the demand valve, and terminate the dive.

19-4 CLOSED-CIRCUIT OXYGEN EXPOSURE LIMITS

The U.S. Navy closed-circuit oxygen exposure limits have been extended and revised to allow greater flexibility in closed-circuit oxygen diving operations. The revised limits are divided into two categories: Transit with Excursion Limits and Single Depth Limits.

19-4.1 **Transit with Excursion Limits Table.** The Transit with Excursion Limits (Table 19-3) call for a maximum dive depth of 20 fsw or shallower for the majority of the dive, but allow the diver to make a brief excursion to depths as great as 50 fsw. The Transit with Excursion Limits is normally the preferred mode of operation because maintaining a depth of 20 fsw or shallower minimizes the possibility of CNS oxygen toxicity during the majority of the dive, yet allows a brief downward excursion if needed (see Figure 19-4). Only a single excursion is allowed.

Table 19-3. *Excursion Limits.*

Depth	Maximum Time
21-40 fsw	15 minutes
41-50 fsw	5 minutes

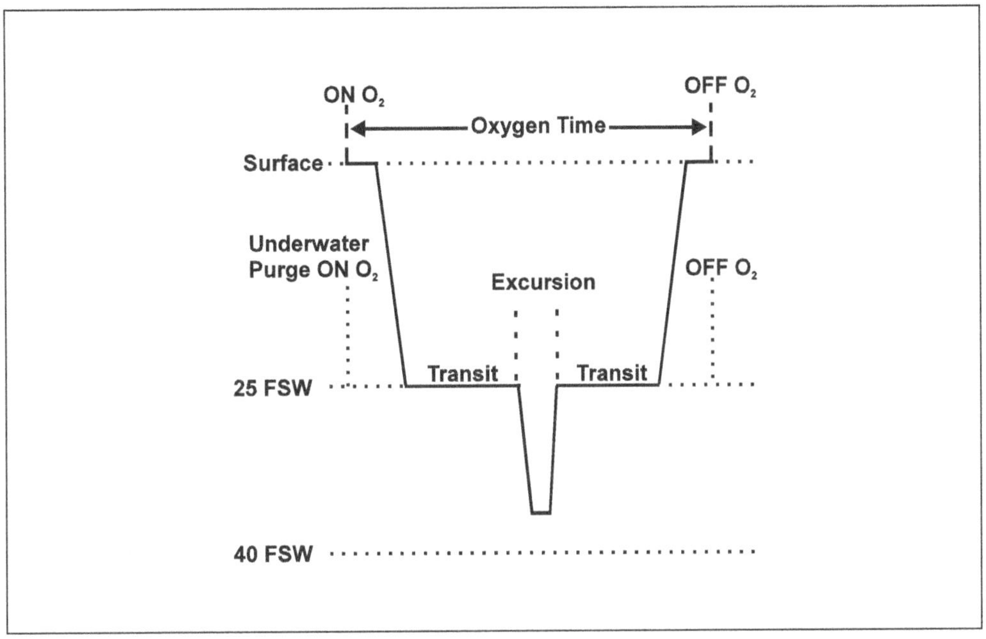

Figure 19-4. Example of Transit with Excursion.

19-4.2 **Single-Depth Oxygen Exposure Limits Table.** The Single-Depth Limits (Table 19-4) allow maximum exposure at the greatest depth, but have a shorter overall exposure time and do not allow for excursions. Single-depth limits may, however, be useful when maximum bottom time is needed deeper than 20 fsw.

19-4.3 **Oxygen Exposure Limit Testing.** The Transit with Excursion Limits and Single-Depth Limits have been tested extensively over the entire depth range and are acceptable for routine diving operations. They are not considered exceptional exposure. It must be noted that the limits shown in this section apply only to closed-circuit 100-percent oxygen diving and are not applicable to deep mixed-gas diving. Separate oxygen exposure limits have been established for deep, helium-oxygen mixed-gas diving.

Table 19-4. Single-Depth Oxygen Exposure Limits.

Depth	Maximum Oxygen Time
25 fsw	240 minutes
30 fsw	80 minutes
35 fsw	25 minutes
40 fsw	15 minutes
50 fsw	10 minutes

19-4.4 **Individual Oxygen Susceptibility Precautions.** Although the limits described in this section have been thoroughly tested and are safe for the vast majority of individuals, occasional episodes of CNS oxygen toxicity may occur. This is the basis for requiring buddy lines on closed-circuit oxygen diving operations.

19-4.5 **Transit with Excursion Limits.** A 20 foot maximum depth for transit with one excursion, if necessary, will be the preferred option in most combat swimmer operations. When operational considerations necessitate a descent to deeper than 20 fsw for longer than allowed by the excursion limits, the appropriate single-depth limit should be used (paragraph 19-4.6).

19-4.5.1 **Transit with Excursion Limits Definitions.** The following definitions are illustrated in Figure 19-4:

- Transit is the portion of the dive spent at 20 fsw or shallower.
- Excursion is the portion of the dive deeper than 20 fsw.
- Excursion time is the time between the diver's initial descent below 20 fsw and his return to 20 fsw or shallower at the end of the excursion.
- Oxygen time is calculated as the time interval between when the diver begins breathing from the closed-circuit oxygen UBA (on-oxygen time) and the time when he discontinues breathing from the closed-circuit oxygen UBA (off-oxygen time).

19-4.5.2 **Transit with Excursion Rules.** A diver who has maintained a transit depth of 20 fsw or shallower may make one brief downward excursion as long as he observes these rules:

- Maximum total time of dive (oxygen time) may not exceed 240 minutes.
- A single excursion may be taken at any time during the dive.
- The diver must have returned to 20 fsw or shallower by the end of the prescribed excursion limit.

- The time limit for the excursion is determined by the maximum depth attained during the excursion (Table 19-3). Note that the Excursion Limits are different from the Single-Depth Limits.

***Example:* Dive Profile Using Transit with Excursion Limits.** A dive mission calls for a swim pair to transit at 15 fsw for 45 minutes, descend to 36 fsw, and complete their objective. As long as the divers do not exceed a maximum depth of 40 fsw, they may use the 40-fsw excursion limit of 15 minutes. The time at which they initially descend below 20 fsw to the time at which they finish the excursion must be 15 minutes or less.

19-4.5.3 **Inadvertent Excursions.** If an inadvertent excursion should occur, one of the following situations will apply:

- If the depth and/or time of the excursion exceeds the limits in Table 19-3 or if an excursion has been taken previously, the dive must be aborted and the diver must return to the surface.

- If the excursion was within the allowed excursion limits, the dive may be continued to the maximum allowed oxygen dive time, but no additional excursions deeper than 20 fsw may be taken.

- The dive may be treated as a single-depth dive applying the maximum depth and the total oxygen time to the Single-Depth Limits shown in Table 19-4.

Example 1. A dive pair is having difficulty with a malfunctioning compass. They have been on oxygen (oxygen time) for 35 minutes when they notice that their depth gauge reads 55 fsw. Because this exceeds the maximum allowed oxygen exposure depth, the dive must be aborted and the divers must return to the surface.

Example 2. A diver on a compass swim notes that his depth gauge reads 32 fsw. He recalls checking his watch 5 minutes earlier and at that time his depth gauge read 18 fsw. As his excursion time is less than 15 minutes, he has not exceeded the excursion limit for 40 fsw. He may continue the dive, but he must maintain his depth at 20 fsw or less and make no additional excursions.

NOTE **If the diver is unsure how long he was below 20 fsw, the dive must be aborted.**

19-4.6 **Single-Depth Limits.** The term Single-Depth Limits does not mean that the entire dive must be spent at one depth, but refers to the time limit applied to the dive based on the maximum depth attained during the dive.

19-4.6.1 **Single-Depth Limits Definitions.** The following definitions apply when using the Single-Depth Limits:

- **Oxygen time** is calculated as the time interval between when the diver begins breathing from the closed-circuit oxygen UBA (on-oxygen time) and the time when he discontinues breathing from the closed-circuit oxygen UBA (off-oxygen time).

- **The depth** of the dive used for determining the allowable exposure time is determined by the maximum depth attained during the dive. For intermediate depth, the next deeper depth limit will be used.

19-4.6.2 **Depth/Time Limits.** The Single-Depth Limits are provided in Table 19-4. No excursions are allowed when using these limits.

Example. Twenty-two minutes (oxygen time) into a compass swim, a dive pair descends to 28 fsw to avoid the propeller of a passing boat. They remain at this depth for 8 minutes. They now have two choices for calculating their allowed oxygen time: (1) they may return to 20 fsw or shallower and use the time below 25 fsw as an excursion, allowing them to continue their dive on the Transit with Excursion Limits to a maximum time of 240 minutes; or (2) they may elect to remain at 28 fsw and use the 30-fsw Single-Depth Limits to a maximum dive time of 80 minutes.

19-4.7 **Exposure Limits for Successive Oxygen Dives.** If an oxygen dive is conducted after a previous closed-circuit oxygen exposure, the effect of the previous dive on the exposure limit for the subsequent dive is dependent on the Off-Oxygen Interval.

19-4.7.1 **Definitions for Successive Oxygen Dives.** The following definitions apply when using oxygen exposure limits for successive oxygen dives.

- **Off-Oxygen Interval.** The interval between off-oxygen time and on-oxygen time is defined as the time from when the diver discontinues breathing from his closed-circuit oxygen UBA on one dive until he begins breathing from the UBA on the next dive.

- **Successive Oxygen Dive.** A successive oxygen dive is one that follows a previous oxygen dive after an Off-Oxygen Interval of less than 2 hours.

19-4.7.2 **Off-Oxygen Exposure Limit Adjustments.** If an oxygen dive is a successive oxygen dive, the oxygen exposure limit for the dive must be adjusted as shown in Table 19-5. If the Off-Oxygen Interval is 2 hours or greater, no adjustment is required for the subsequent dive. An oxygen dive undertaken after an Off-Oxygen Interval of more than 2 hours is considered to be the same as an initial oxygen exposure. If a negative number is obtained when adjusting the single-depth exposure limits as shown in Table 19-5, a 2-hour Off-Oxygen Interval must be taken before the next oxygen dive.

NOTE **A maximum of 4 hours oxygen time is permitted within a 24-hour period.**

Table 19-5. *Adjusted Oxygen Exposure Limits for Successive Oxygen Dives.*

	Adjusted Maximum Oxygen Time	Excursion
Transit with Excursion Limits	Subtract oxygen time on previous dives from 240 minutes	Allowed if none taken on previous dives
Single-Depth Limits	1. Determine maximum oxygen time for deepest exposure. 2. Subtract oxygen time on previous dives from maximum oxygen time in Step 1 (above)	No excursion allowed when using Single-Depth Limits to compute remaining oxygen time

Example. Ninety minutes after completing a previous oxygen dive with an oxygen time of 75 minutes (maximum dive depth 19 fsw), a dive pair will be making a second dive using the Transit with Excursion Limits. Calculate the amount of oxygen time for the second dive, and determine whether an excursion is allowed.

Solution. The second dive is considered a successive oxygen dive because the Off-Oxygen Interval was less than 2 hours. The allowed exposure time must be adjusted as shown in Table 19-5. The adjusted maximum oxygen time is 165 minutes (240 minutes minus 75 minutes previous oxygen time). A single excursion may be taken because the maximum depth of the previous dive was 19 fsw.

Example. Seventy minutes after completing a previous oxygen dive (maximum depth 28 fsw) with an oxygen time of 60 minutes, a dive pair will be making a second oxygen dive. The maximum depth of the second dive is expected to be 25 fsw. Calculate the amount of oxygen time for the second dive, and determine whether an excursion is allowed.

Solution. First compute the adjusted maximum oxygen time. This is determined by the Single-Depth Limits for the deeper of the two exposures (30 fsw for 80 minutes), minus the oxygen time from the previous dive. The adjusted maximum oxygen time for the second dive is 20 minutes (80 minutes minus 60 minutes previous oxygen time). No excursion is permitted using the Single-Depth Limits.

19-4.8 Exposure Limits for Oxygen Dives Following Mixed-Gas or Air Dives. When a subsequent dive must be conducted and if the previous exposure was an air or MK 16 MOD 0 dive, the exposure limits for the subsequent oxygen dive require no adjustment.

19-4.8.1 **Mixed-Gas to Oxygen Rule.** If the previous dive used a mixed-gas breathing mix having an oxygen partial pressure of 1.0 ata or greater, the previous exposure must be treated as a closed-circuit oxygen dive as described in paragraph 19-4.7. In this case, the Off-Oxygen Interval is calculated from the time the diver discontinued breathing the previous breathing mix until he begins breathing from the closed-circuit oxygen rig.

19-4.8.2 **Oxygen to Mixed-Gas Rule.** If a diver employs the MK 25 UBA for a portion of the dive and another UBA that uses a breathing gas other than oxygen for another portion of the dive, only the portion of the dive during which the diver was breathing oxygen is counted as oxygen time. The use of multiple UBAs is generally restricted to special operations. Decompression procedures for multiple-UBA diving must be in accordance with approved procedures.

Example. A dive scenario calls for three swim pairs to be inserted near a harbor using a SEAL Delivery Vehicle (SDV). The divers will be breathing compressed air for a total of 3 hours prior to leaving the SDV. No decompression is required as determined by the Combat Swimmer Multilevel Dive (CSMD) procedures. The SDV will surface and the divers will purge their oxygen rigs on the surface, take a compass bearing and begin the oxygen dive. The Transit with Excursion Limits rules will be used. There would be no adjustment necessary for the oxygen time as a result of the 3 hour compressed air dive.

19-4.9 **Oxygen Diving at High Elevations.** The oxygen exposure limits and procedures as set forth in the preceding paragraphs may be used without adjustment for closed-circuit oxygen diving at altitudes above sea level.

19-4.10 **Flying After Oxygen Diving.** Flying is permitted immediately after oxygen diving unless the oxygen dive has been part of a multiple-UBA dive profile in which the diver was also breathing another breathing mixture (air, N_2O_2, or HeO_2). In this case, the rules found in the paragraph 9-14 apply.

19-4.11 **Combat Operations.** The oxygen exposure limits in this section are the only limits approved for use by the U.S. Navy and should not be exceeded in a training or exercise scenario. Should combat operations require a more severe oxygen exposure, an estimate of the increased risk of CNS oxygen toxicity may be obtained from a Diving Medical Officer or the Navy Experimental Diving Unit. The advice of a Diving Medical Officer is essential in such situations and should be obtained whenever possible.

19-5 OPERATIONS PLANNING

Certain factors must be taken into consideration in the planning of the oxygen dive operation. The following gives detailed information on specific areas of planning.

19-5.1 **Operating Limitations.** Diving Officers and Diving Supervisors must consider the following potential limiting factors when planning closed-circuit oxygen combat swimmer operations:

- UBA oxygen supply (paragraph 19-3.2)
- UBA canister duration (NAVSEA 00C3 ltr 3151 ser 00C34/3160, 27 Sep 01)
- Oxygen exposure limits (paragraph 19-4)
- Thermal factors (Chapter 11 and Chapter 3)

19-5.2 **Maximizing Operational Range.** The operational range of the UBA may be maximized by adhering to these guidelines:

- Whenever possible, plan the operation using the turtleback technique, in which the diver swims on the surface part of the time, breathing air where feasible.
- Use tides and currents to maximum advantage. Avoid swimming against a current when possible.
- Ensure that oxygen bottles are charged to a full 3,000 psig (207 bar) before the dive.
- Minimize gas loss from the UBA by avoiding leaks and unnecessary depth changes.
- Maintain a comfortable, relaxed swim pace during the operation. For most divers, this is a swim speed of approximately 0.8 knot. At high exercise rates, the faster swim speed is offset by a disproportionately higher oxygen consumption, resulting in a net decrease in operating range. High exercise rates may reduce the oxygen supply duration below the canister carbon dioxide scrubbing duration and become the limiting factor for the operation (paragraph 19-3.2).
- Ensure divers wear adequate thermal protection. A cold diver will begin shivering or increase his exercise rate, either of which will increase oxygen consumption and decrease the operating duration of the oxygen supply.

WARNING **The MK 25 does not have a carbon dioxide-monitoring capability. Failure to adhere to canister duration operations planning could lead to unconsciousness and/or death.**

19-5.3 **Training.** Training and requalification dives shall be performed with the following considerations in mind:

- Training dives shall be conducted with equipment that reflects what the diver will be required to use on operations. This should include limpets, demolitions, and weapons as deemed appropriate.
- Periodic classroom refresher training shall be conducted in oxygen diving procedures, CNS oxygen toxicity and management of diving accidents.
- Develop a simple set of hand signals, including the following signals:

- Surface
- Emergency Surface
- Descend
- Ascend
- Speed Up
- Slow Down
- Okay
- Feel Strange
- Ear Squeeze
- Stop
- Caution
- Excursion

- Match swim pairs according to swim speed.

- If long duration oxygen swims are to be performed, work-up dives of gradually increasing length are recommended.

19-5.4 **Personnel Requirements.** The following topside personnel must be present on all training and exercise closed-circuit oxygen dives:

- Diving Supervisor/Boat Coxswain

- Standby diver/surface swimmer with air (not oxygen) SCUBA

- Diving Medical Technician or other individual specifically trained in diagnosis/ emergency treatment of diving injuries. Must have completed formal training at a DOD recognized course of instruction (COI).

19-5.5 **Equipment Requirements.** The operational characteristics of the MK 25 UBA are shown in Figure 19-2. Equipment requirements for training and exercise closed-circuit oxygen dives are shown in Table 19-6. Several equipment items merit special consideration as noted below:

- **Motorized Chase Boat.** A minimum of one motorized chase boat must be present for the dive. Safe diving practice in many situations, however, would require the presence of more than one chase boat (e.g., night operations). The Diving Supervisor must determine the number of boats required based on the diving area, medical evacuation plan and number of personnel participating in the dive. When more than one safety craft is used, communications between support craft should be available.

- **Buddy Lines.** Because the risk is greater that a diver will become unconscious or disabled during a closed-circuit oxygen dive than during other types of dives, buddy lines are required equipment for oxygen dives. In a few special diving scenarios, when their use may hinder or endanger the divers, buddy lines may not be feasible. The Diving Supervisor must carefully consider each situation and allow buddy lines to be disconnected only when their use will impede the performance of the mission.

- **Depth Gauge.** The importance of maintaining accurate depth control on oxygen swims mandates that a depth gauge be worn by each diver.

- **Witness Float.** During Combat Swimmer training operations divers do not have to be surface tended to swim under the hull of a vessel. However, they

Table 19-6. *Closed-Circuit Oxygen Diving Equipment.*

A. General

1. Motorized chase boat*
2. Radio (radio communications with parent unit, chamber, medevac units, and support craft when feasible)
3. High-intensity, wide-beam light (night operations)
4. Dive flags and/or dive lights as required

B. Diving Supervisor

1. Dive watch
2. Dive pair list
3. Recall devices
4. Copy of Oxygen Exposure Limits
5. Copy of Air Tables

C. Standby Diver

1. Compressed-air SCUBA
2. Weight belt (if needed)
3. Approved life jacket
4. Face mask
5. Fins
6. Appropriate thermal protection
7. Dive knife
8. Flare
9. Tending line
10. Depth gauge
11. Dive watch

D. Diving Medical Technician

1. Self-inflating bag-mask ventilator with medium adult mask
2. Oro-pharyngeal airway, adaptable to mask used
3. First aid kit/portable O_2
4. Two canteens of fresh water for treating chemical injury

E. Divers

Required:

1. Approved life jacket
2. Weight belt (Jettisonable)
3. Face mask
4. Fins
5. Dive knife
6. Flare or Strobe
7. Dive watch
8. Appropriate thermal protection
9. Whistle
10. Buddy line (one per pair)*
11. Depth gauge (large face; accurate at shallow depths; one per diver)*
12. Compass (one per pair if on compass course)

Optional:

1. Gloves
2. Buoy (one per pair)
3. Slate with writing device

* See paragraph 19-5.5

must be marked by a witness float which must be visible on the surface at all times. After sunset, the float must be illuminated to be readily visible to topside personnel e.g. CHEMLITEs. The Diving Supervisor must consider the draft of the vessel and the appropriate environmental factors, e.g. current and sea state, to determine the required length of the witness float line.

19-5.6 **Predive Precautions.** The following items shall be determined prior to the diving operation:

- Means of communicating with the nearest available Diving Medical Officer.
- Location of the nearest functional recompression chamber. Positive confirmation of the chamber's availability must be obtained prior to diving.
- Nearest medical facility for treatment of injuries or medical problems not requiring recompression therapy.

- Optimal method of transportation to recompression chamber or medical facility. If coordination with other units for aircraft/boat/vehicle support is necessary, the Diving Supervisor must know the frequencies, call signs and contact personnel needed to make transportation available in case of emergency. A medical evacuation plan must be included in the Diving Supervisor brief.

- The preparation of a checklist similar to that found in Chapter 6 is recommended.

- When operations are to be conducted in the vicinity of ships, the guidelines provided in the Ship Repair Safety Checklist (Chapter 6) and appropriate Naval Special Warfare Group instructions shall be followed.

- Notification of intent to conduct diving operations must be sent to the appropriate authority in accordance with local directives.

19-6 PREDIVE PROCEDURES

This section provides the predive procedures for closed-circuit oxygen dives.

19-6.1 Equipment Preparation. The predive set up of the MK 25 UBA is performed using the appropriate checklist from the appropriate MK 25 UBA Operation and Maintenance Manual.

19-6.2 Diving Supervisor Brief. The Diving Supervisor brief shall be given separately from the overall mission brief and shall focus on the diving portion of the operation with special attention to the items shown in Table 19-7.

19-6.3 Diving Supervisor Check

19-6.3.1 **First Phase.** The Diving Supervisor check is accomplished in two stages. As the divers set up their rigs prior to the dive, the Diving Supervisor must ensure that the steps in the set up procedure are accomplished properly in accordance with the MK25 UBA Operation and Maintenance Manual. The Diving Supervisor signs the UBA predive checklist, verifying that the procedures were completed correctly.

19-6.3.2 **Second Phase.** The second phase of the Diving Supervisor check is done after the divers are dressed. At this point, the Diving Supervisor must check for the following items:

- Adequate oxygen pressure

- Proper functioning of hose one-way valves

- UBA harness for proper donning and fit.

Table 19-7. *Diving Supervisor Brief.*

A. **Dive Plan**
   1. Operating depth
   2. Distance, bearings, transit lines
   3. Dive time
   4. Known obstacles or hazards

B. **Environmental**
   1. Weather conditions
   2. Water/air temperatures
   3. Water/air visibility
   4. Current/Tides

C. **Special Equipment for:**
   1. Divers (include thermal garment)
   2. Diving supervisor
   3. Standby Diver
   4. Diving medical technician

D. **Review of Hand Signals**

E. **Communications**
   1. Frequencies
   2. Call signs

F. **Emergency Procedures**
   1. Symptoms of O_2 Toxicity - review in detail
   2. Symptoms of CO_2 buildup - review in detail
   3. Review management of underwater convulsion, nonconvulsive O_2 hit, CO_2 buildup, hypoxia, chemical injury, unconscious diver
   4. UBA malfunction
   5. Lost swim-pair procedures
   6. Medical evacuation plan
      - nearest available chamber
      - nearest Diving Medical Officer (DMO)
      - transportation plan
      - recovery of other swim pairs

G. **Review of Purge Procedure**

H. **Times for Operations**

- Proper donning of UBA, life jacket and weight belt. The weight belt is worn so it may be easily released
- Presence of required items such as compasses, depth gauges, dive watches, buddy lines, and tactical equipment

19-7 WATER ENTRY AND DESCENT

The diver is required to perform a purge procedure prior to or during any dive in which closed-circuit oxygen UBA is to be used. The purge procedure is designed to eliminate the nitrogen from the UBA and the diver's lungs as soon as he begins breathing from the rig. This procedure prevents the possibility of hypoxia as a result of excessive nitrogen in the breathing loop. The gas volume from which this excess nitrogen must be eliminated is comprised of more than just the UBA breathing bag. The carbon dioxide-absorbent canister, inhalation/exhalation hoses, and diver's lungs must also be purged of nitrogen.

19-7.1 **Purge Procedure.** Immediately prior to entering the water, the divers shall carry out the appropriate purge procedure. It is both difficult and unnecessary to eliminate nitrogen completely from the breathing loop. The purge procedure need only raise the fraction of oxygen in the breathing loop to a level high enough to prevent the diver from becoming hypoxic.

If the dive is part of a tactical scenario that requires a turtleback phase, the purge must be done in the water after the surface swim, prior to submerging. If the tactical scenario requires an underwater purge procedure, this will be completed while submerged after an initial subsurface transit on open-circuit SCUBA or other UBA. When the purge is done in either manner, the diver must be thoroughly familiar with the purge procedure and execute it carefully with attention to detail so that it may be accomplished correctly in this less favorable environment.

19-7.2 **Avoiding Purge Procedure Errors.** The following errors may result in a dangerously low percentage of oxygen in the UBA and should be avoided:

- Exhaling back into the bag with the last breath rather than to the atmosphere while emptying the breathing bag.
- Underinflating the bag during the fill segment of the fill/empty cycle.
- Adjusting the UBA harnesses or adjustment straps of the life jacket too tightly. Lack of room for bag expansion may result in underinflation of the bag and inadequate purging.
- Breathing gas volume deficiency caused by failure to turn on the oxygen-supply valve prior to underwater purge procedures.

19-8 UNDERWATER PROCEDURES

19-8.1 **General Guidelines.** During the dive, the divers shall adhere to the following guidelines:

- Know and observe the oxygen exposure limits.
- Observe the UBA canister limit for the expected water temperature, see NAVSEA 00C3 ltr 3151ser 00C34/3160, 27 Sep 01.
- Wear the appropriate thermal protection.
- Use the proper weights for the thermal protection worn and for equipment carried.
- Wear a depth gauge to allow precise depth control. The depth for the pair of divers is the greatest depth attained by either diver.
- Dive partners check each other carefully for leaks at the onset of the dive. This should be done in the water after purging, but before descending to transit depth.
- Swim at a relaxed, comfortable pace as established by the slower swimmer of the pair.
- Maintain frequent visual or touch checks with buddy.

- Be alert for any symptoms suggestive of a medical disorder (CNS oxygen toxicity, carbon dioxide buildup, etc.).
- Use tides and currents to maximum advantage.
- Swim at 20 fsw or shallower unless operational requirements dictate otherwise.
- Use the minimum surface checks consistent with operational necessity.
- Minimize gas loss from the UBA.
- Do not use the UBA breathing bag as a buoyancy compensation device.
- Do not perform additional purges during the dive unless the mouthpiece is removed and air is breathed.
- If an excursion is taken, the diver not using the compass will note carefully the starting and ending time of the excursion.

19-8.2 **UBA Malfunction Procedures.** The diver shall be thoroughly familiar with the malfunction procedures unique to his UBA. These procedures are described in the UBA MK 25 UBA Operational and Maintenance Manual.

19-9 ASCENT PROCEDURES.

The ascent rate shall never exceed 30 feet per minute.

19-10 POSTDIVE PROCEDURES AND DIVE DOCUMENTATION

UBA postdive procedures should be accomplished using the Postdive checklist from the MK 25 UBA Operation and Maintenance Manual.

All dives and mishap reporting shall be accomplished in accordance with guidance in Chapter 5.

PAGE LEFT BLANK INTENTIONALLY

VOLUME 5

Diving Medicine & Recompression Chamber Operations

U.S. NAVY DIVING MANUAL

PAGE LEFT BLANK INTENTIONALLY

Volume 5 - Table of Contents

Chap/Para Page

Chap/Para Page

Chap/Para Page

Chap/Para		Page

Chap/Para | Page

Volume 5 - List of Illustrations

Volume 5 - List of Tables

Table **Page**

PAGE LEFT BLANK INTENTIONALLY

CHAPTER 20

Diagnosis and Treatment of Decompression Sickness and Arterial Gas Embolism

20-1 INTRODUCTION

20-1.1 **Purpose.** This chapter describes the diagnosis and treatment of diving disorders with recompression therapy and/or hyperbaric oxygen therapy. Immediate recompression therapy is indicated for treating decompression sickness, arterial gas embolism and several other disorders. In those cases where diagnosis or treatment are not clear, contact the Diving Medical Officers at NEDU or NDSTC for clarification. The recompression procedures described in this chapter are designed to handle most situations that will be encountered operationally. They are applicable to both surface-supplied and open and closed circuit SCUBA diving as well as recompression chamber operations, whether on air, nitrogen-oxygen, helium-oxygen, or 100 percent oxygen. Treatment of decompression sickness during saturation dives is covered separately in Chapter 15 of this manual. Periodic evaluation of U.S. Navy recompression treatment procedures has shown they are effective in relieving symptoms over 90 percent of the time when used as published.

20-1.2 **Scope.** The procedures outlined in this chapter are to be performed only by trained personnel. Because these procedures cover disorders ranging from mild pain to life-threatening disorders, the degree of medical expertise necessary to carry out proper treatment will vary. Certain procedures, such as starting intravenous (IV) fluid lines and inserting chest tubes, require special training and must not be attempted by untrained individuals. Treatment tables can be initiated without consulting a Diving Medical Officer (DMO), however a DMO should always be contacted at the earliest possible opportunity. A DMO must be contacted prior to releasing the treated individual.

20-1.3 **Diving Supervisor's Responsibilities.** Experience has shown that symptoms of severe decompression sickness or arterial gas embolism may occur following seemingly uneventful dives within the prescribed limits. This fact, combined with the many operational scenarios under which diving is conducted, means that treatment of severely ill individuals will be required occasionally when qualified medical personnel are not immediately on scene. Therefore, it is the Diving Supervisor's responsibility to ensure that every member of the diving team:

- Is thoroughly familiar with all recompression procedures.
- Knows the location of the nearest, certified recompression facility.
- Knows how to contact a qualified Diving Medical Officer if one is not at the site.
- Has successfully completed Basic Life Support training.

20-1.4 **Prescribing and Modifying Treatments.** Because all possible outcomes cannot be anticipated, additional medical expertise should be sought immediately in all cases of decompression sickness or arterial gas embolism that do not show substantial improvement on standard treatment tables. Deviation from these protocols shall be made only with the recommendation of a Diving Medical Officer (DMO).

Not all Medical Officers are DMOs. The DMO shall be a graduate of the Diving Medical Officer course taught at the Naval Diving and Salvage Training Center (NDSTC) and have a subspecialty code of 16U0 (Basic Undersea Medical Officer) or 16U1 (Residency in Undersea Medicine trained Undersea Medical Officer). Medical Officers who complete only the nine-week diving medicine course at NDSTC do not receive DMO subspecialty codes, but are considered to have the same privileges as DMOs, with the exception that they are not granted the privilege of modifying treatment protocols. Only DMOs with subspecialty codes 16U0 or 16U1 may modify the treatment protocols as warranted by the patient's condition with the concurrence of the Commanding Officer or Officer in Charge. Other physicians may assist and advise treatment and care of diving casualties but may not modify recompression procedures.

20-1.5 **When Treatment is Not Necessary.** If the reason for postdive symptoms is firmly established to be due to causes other than decompression sickness or arterial gas embolism (e.g. injury, sprain, poorly fitting equipment), then recompression is not necessary. If the diving supervisor cannot rule out the need for recompression then commence treatment.

20-1.6 **Emergency Consultation.** Modern communications allow access to medical expertise from even the most remote areas. Emergency consultation is available 24 hours a day with:

Primary:
Navy Experimental Diving Unit (NEDU)
Commercial (850) 230-3100 or (850) 235-1668, DSN 436-4351

Secondary:
Navy Diving Salvage and Training Center (NDSTC)
Commercial (850) 234-4651, DSN 436-4651

20-2 ARTERIAL GAS EMBOLISM

Arterial gas embolism is caused by entry of gas bubbles into the arterial circulation as a result of pulmonary over inflation syndrome (POIS). Gas embolism can manifest during any dive where breaths are taken utilizing underwater breathing equipment, even a brief, shallow dive, or one made in a swimming pool. The onset of symptoms is usually sudden and dramatic, often occurring within minutes after arrival on the surface or even before reaching the surface. Because the supply of blood to the central nervous system is almost always compromised, arterial gas embolism may result in death or permanent neurological damage unless treated with immediate recompression.

20-2.1 **Diagnosis of Arterial Gas Embolism.** As a basic rule, any diver who has obtained a breath of compressed gas from any source at depth, whether from diving apparatus or from a diving bell, and who surfaces unconscious, loses consciousness, or has any obvious neurological symptoms within 10 minutes of reaching the surface, must be assumed to be suffering from arterial gas embolism. Recompression treatment shall be started immediately. A diver who surfaces unconscious and recovers when exposed to fresh air shall receive a neurological evaluation to rule out arterial gas embolism. Victims of near-drowning who have no neurological symptoms should be carefully evaluated by a DMO for pulmonary aspiration.

The symptoms of AGE may be masked by environmental factors or by other less significant symptoms. A chilled diver may not be concerned with numbness in an arm, which may actually be the sign of CNS involvement. Pain from any source may divert attention from other symptoms. The natural anxiety that accompanies an emergency situation, such as the failure of the diver's air supply, might mask a state of confusion caused by an arterial gas embolism to the brain.

If pain is the only symptom, arterial gas embolism is unlikely and decompression sickness or one of the other pulmonary overinflation syndromes should be considered.

20-2.1.1 **Symptoms of AGE.** The signs and symptoms of AGE may include near immediate onset of dizziness, paralysis or weakness in the extremities, large areas of abnormal sensation (paresthesias), vision abnormalities, convulsions or personality changes. During ascent, the diver may have noticed a sensation similar to that of a blow to the chest. The victim may become unconscious without warning and may stop breathing. Additional symptoms of AGE include:

- Extreme fatigue
- Difficulty in thinking
- Vertigo
- Nausea and/or vomiting
- Hearing abnormalities
- Bloody sputum
- Loss of control of bodily functions
- Tremors
- Loss of coordination
- Numbness

Symptoms of subcutaneous / mediastinal emphysema, pneumothorax and/or pneumopericardium may also be present (see paragraph 3-8). In all cases of arterial gas embolism, the possible presence of these associated conditions should not be overlooked.

20-2.2 **Treating Arterial Gas Embolism.** Arterial gas embolism is treated in accordance with Figure 20-1 with initial compression to 60 fsw. If symptoms are improved within the first oxygen breathing period, then treatment is continued using Treatment Table 6. If symptoms are unchanged or worsen, assess the patient upon descent and compress to depth of relief (or significant improvement), not to exceed 165 fsw and follow Figure 20-1.

20-2.3 **Resuscitation of a Pulseless Diver.** The following are intended as guidelines. For a diver with no pulse or respirations (cardiopulmonary arrest) immediate cardiopulmonary resuscitation (CPR) and use of the Automated External Defibrillator (AED) is a higher priority than recompression. Advanced cardiac life support (ACLS), which requires special medical training and equipment, is not always available. CPR, patient monitoring, and drug administration may be able to be performed at depth, but electrical therapy (defibrillation and cardioversion) **must** be performed on the surface.

CAUTION Defibrillation is not currently authorized at depth.

If a qualified provider with the necessary equipment (i.e., AED) can administer the potentially lifesaving therapies within 10 minutes, the stricken diver should be kept at the surface until a pulse is obtained. Unless defibrillation is administered within 10 minutes, the diver likely will die, even if adequate CPR is performed, with or without recompression. If defibrillation is not available and a Diving Medical Officer (DMO) is not present, the Diving Supervisor should compress the diver to 60 feet and continue CPR and attempt to contact a DMO.

If defibrillation becomes available within 20 minutes, the pulseless diver shall be brought to the surface at 30 fpm and defibrillated when appropriate on the surface. (Current data indicate that successful restoration of a perfusing rhythm after 20 minutes of cardiac arrest with only CPR is unlikely.) If the pulseless diver does not regain vital signs with defibrillation, continue CPR. Avoid **recompressing** a pulseless diver who has failed to regain vital signs after defibrillation. Resuscitation efforts shall continue until the diver recovers, the tenders are unable to continue CPR, or a physician pronounces the patient dead. If the pulseless diver does regain vital signs, proceed with recompression therapy if indicated.

CAUTION If the tender is outside of no-decompression limits, he should not be brought directly to the surface. Either take the decompression stops appropriate to the tender or lock in a new tender and decompress the patient and new tender to the surface in the outer lock, while maintaining the original tender at depth.

20-3 DECOMPRESSION SICKNESS

While a history of diving (or altitude exposure) is necessary for the diagnosis of decompression sickness to be made, the depth and duration of the dive are useful only in establishing if required decompression was missed. Decompression sickness can occur in divers well within no-decompression limits or in divers who have carefully followed decompression tables. Any decompression sickness that occurs must be treated by recompression.

For purposes of deciding the appropriate treatment, symptoms of decompression sickness are generally divided into two categories, Type I and Type II. Because the treatment of Type I and Type II symptoms may be different, it is important to distinguish between these two types of decompression sickness. The diver may exhibit certain signs that only trained observers will identify as decompression sickness. Some of the symptoms or signs will be so pronounced that there will be little doubt as to the cause. Others may be subtle and some of the more important signs could be overlooked in a cursory examination. Type I and Type II symptoms may or may not be present at the same time.

20-3.1 **Diagnosis of Decompression Sickness.** Decompression sickness symptoms usually occur shortly following the dive or other pressure exposure. If the controlled decompression during ascent has been shortened or omitted, the diver could be suffering from decompression sickness before reaching the surface. In analyzing several thousand air dives in a database set up by the U.S. Navy for developing decompression models, the time of onset of symptoms after surfacing was as follows:

- 42 percent occurred within 1 hour.
- 60 percent occurred within 3 hours.
- 83 percent occurred within 8 hours.
- 98 percent occurred within 24 hours.

Appendix 5A contains a set of guidelines for performing a neurological examination and an examination checklist to assist trained personnel in evaluating decompression sickness cases.

20-3.2 **Symptoms of Type I Decompression Sickness.** Type I decompression sickness includes joint pain (musculoskeletal or pain-only symptoms) and symptoms involving the skin (cutaneous symptoms), or swelling and pain in lymph nodes.

20-3.2.1 **Musculoskeletal Pain-Only Symptoms.** The most common symptom of decompression sickness is joint pain. Other types of pain may occur which do not involve joints. The pain may be mild or excruciating. The most common sites of joint pain are the shoulder, elbow, wrist, hand, knee, and ankle. The characteristic pain of Type I decompression sickness usually begins gradually, is slight when first noticed and may be difficult to localize. It may be located in a joint or muscle, may increase in intensity, and is usually described as a deep, dull ache. The pain may or may not be increased by movement of the affected joint, and the limb may be held preferentially in certain positions to reduce the intensity (so-called guarding). The hallmark of Type I pain is its dull, aching quality and confinement to particular areas. It is always present at rest and is usually unaffected by movement.

Any pain occurring in the abdominal and thoracic areas, including the hips, should be considered as symptoms arising from spinal cord involvement and treated as Type II decompression sickness. The following symptoms may indicate spinal cord involvement:

- Pain localized to joints between the ribs and spinal column or joints between the ribs and sternum.
- A shooting-type pain that radiates from the back around the body (radicular or girdle pain).
- A vague, aching pain in the chest or abdomen (visceral pain).

20-3.2.1.1 ***Differentiating Between Type I Pain and Injury.*** The most difficult differentiation is between the pain of Type I decompression sickness and the pain resulting from a muscle strain or bruise. If there is any doubt as to the cause of the pain, assume the diver is suffering from decompression sickness and treat accordingly. Frequently, pain may mask other more significant symptoms. Pain should not be treated with drugs in an effort to make the patient more comfortable. The pain may be the only way to localize the problem and monitor the progress of treatment.

20-3.2.2 **Cutaneous (Skin) Symptoms.** The most common skin manifestation of decompression sickness is itching. Itching by itself is generally transient and does not require recompression. Faint skin rashes may be present in conjunction with itching. These rashes also are transient and do not require recompression. Mottling or marbling of the skin, known as cutis marmorata (marbling), may precede a symptom of serious decompression sickness and shall be treated by recompression as Type II decompression sickness. This condition starts as intense itching, progresses to redness, and then gives way to a patchy, dark-bluish discoloration of the skin. The skin may feel thickened. In some cases the rash may be raised.

20-3.2.3 **Lymphatic Symptoms.** Lymphatic obstruction may occur, creating localized pain in involved lymph nodes and swelling of the tissues drained by these nodes. Recompression may provide prompt relief from pain. The swelling, however, may take longer to resolve completely and may still be present at the completion of treatment.

20-3.3 Treatment of Type I Decompression Sickness.

Type I Decompression Sickness is treated in accordance with Figure 20-2. If a full neurological exam is not completed before initial recompression, treat as a Type II symptom.

Symptoms of musculoskeletal pain that have shown absolutely no change after the second oxygen breathing period at 60 feet may be due to orthopedic injury rather than decompression sickness. If, after reviewing the patient's history, the Diving Medical Officer feels that the pain can be related to specific orthopedic trauma or injury, a Treatment Table 5 may be completed. If a Diving Medical Officer is not consulted, Treatment Table 6 shall be used.

20-3.4 Symptoms of Type II Decompression Sickness.

In the early stages, symptoms of Type II decompression sickness may not be obvious and the stricken diver may consider them inconsequential. The diver may feel fatigued or weak and attribute the condition to overexertion. Even as weakness becomes more severe the diver may not seek treatment until walking, hearing, or urinating becomes difficult. Initial denial of DCS is common. For this reason, symptoms must be anticipated during the postdive period and treated before they become too severe. Type II, or serious, symptoms are divided into three categories: neurological, inner ear (staggers), and cardiopulmonary (chokes). Type I symptoms may or may not be present at the same time.

20-3.4.1 **Neurological Symptoms.** These symptoms may be the result of involvement of any level of the nervous system. Numbness, paresthesias (a tingling, pricking, creeping, "pins and needles," or "electric" sensation on the skin), decreased sensation to touch, muscle weakness, paralysis, mental status changes, or motor performance alterations are the most common symptoms. Disturbances of higher brain function may result in personality changes, amnesia, bizarre behavior, lightheadedness, lack of coordination, and tremors. Lower spinal cord involvement can cause disruption of urinary function. Some of these signs may be subtle and can be overlooked or dismissed by the stricken diver as being of no consequence.

The occurrence of any neurological symptom after a dive is abnormal and should be considered a symptom of Type II decompression sickness or arterial gas embolism, unless another specific cause can be found. Normal fatigue is not uncommon after long dives and, by itself, is not usually treated as decompression sickness. If the fatigue is unusually severe, a complete neurological examination is indicated to ensure there is no other neurological involvement.

20-3.4.2 **Inner Ear Symptoms ("Staggers").** The symptoms of inner ear decompression sickness include: tinnitus (ringing in the ears), hearing loss, vertigo, dizziness, nausea, and vomiting. Inner ear decompression sickness has occurred most often in helium-oxygen diving and during decompression when the diver switched from breathing helium-oxygen to air. Inner ear decompression sickness should be differentiated from inner ear barotrauma, since the treatments are different. The "Staggers" has been used as another name for inner ear decompression sickness because of the afflicted diver's difficulty in walking due to vestibular system dysfunction. However, symptoms of imbalance may also be due to neurological decompression sickness involving the cerebellum. Typically, rapid involuntary eye movement (nystagmus) is not present in cerebellar decompression sickness.

20-3.4.3 **Cardiopulmonary Symptoms ("Chokes").** If profuse intravascular bubbling occurs, symptoms of chokes may develop due to congestion of the lung circulation. Chokes may start as chest pain aggravated by inspiration and/or as an irritating cough. Increased breathing rate is usually observed. Symptoms of increasing lung congestion may progress to complete circulatory collapse, loss of consciousness, and death if recompression is not instituted immediately. Careful examination for signs of pneumothorax should be performed on patients presenting with shortness of breath. Recompression is not indicated for pneumothorax if no other signs of DCS or AGE are present.

20-3.4.4 **Differentiating Between Type II DCS and AGE.** Many of the symptoms of Type II decompression sickness are the same as those of arterial gas embolism, although the time course is generally different. (AGE usually occurs within 10 minutes of surfacing.) Since the initial treatment of these two conditions is the same and since subsequent treatment conditions are based on the response of the patient to treatment, treatment should not be delayed unnecessarily in order to make the diagnosis.

20-3.5 **Treatment of Type II Decompression Sickness.** Type II Decompression Sickness is treated with initial compression to 60 fsw in accordance with Figure 20-1. If symptoms are improved within the first oxygen breathing period, then treatment is continued on a Treatment Table 6. If severe symptoms (e.g. paralysis, major weakness, memory loss) are unchanged or worsen within the first 20 minutes at 60 fsw, assess the patient during descent and compress to depth of relief (or significant improvement), not to exceed to 165 fsw. Treat on Treatment Table 6A. To limit recurrence, severe Type II symptoms warrant full extensions at 60 fsw even if symptoms resolve during the first oxygen breathing period.

20-3.6 **Decompression Sickness in the Water.** In rare instances, decompression sickness may develop in the water while the diver is undergoing decompression. The predominant symptom will usually be joint pain, but more serious manifestations such as numbness, weakness, hearing loss, and vertigo may also occur. Decompression sickness is most likely to appear at the shallow decompression stops just prior to surfacing. Some cases, however, have occurred during ascent to the first stop or shortly thereafter. Treatment of decompression sickness in the water will vary depending on the type of diving equipment in use. Specific guidelines are given in Chapter 9 for air dives, Chapter 14 for surface-supplied helium-oxygen dives, Chapter 17 for MK 16 MOD 0 dives, and Chapter 18 for MK 16 MOD 1 dives.

20-3.7 **Symptomatic Omitted Decompression.** If a diver has had an uncontrolled ascent and has any symptoms, he should be compressed immediately in a recompression chamber to 60 fsw. Conduct a rapid assessment of the patient and treat accordingly. Treatment Table 5 is not an appropriate treatment for symptomatic omitted decompression. If the diver surfaced from 50 fsw or shallower, compress to 60 fsw and begin Treatment Table 6. If the diver surfaced from a greater depth, compress to 60 fsw or the depth where the symptoms are significantly improved, not to exceed 165 fsw, and begin Treatment Table 6A. Consultation with a Diving Medical Officer should be obtained as soon as possible. For uncontrolled ascent deeper than 165 feet, the diving supervisor may elect to use Treatment Table 8 at the depth of relief, not to exceed 225 fsw.

Treatment of symptomatic divers who have surfaced unexpectedly is difficult when no recompression chamber is on site. Immediate transportation to a recompression facility is indicated; if this is impossible, the guidelines in paragraph 20-4.4 may be useful.

20-3.8 **Altitude Decompression Sickness.** Decompression sickness may also occur with exposure to subatmospheric pressures (altitude exposure), as in an altitude chamber or sudden loss of cabin pressure in an aircraft. Aviators exposed to altitude may experience symptoms of decompression sickness similar to those experienced by divers. The only major difference is that symptoms of spinal cord involvement are less common and symptoms of brain involvement are more frequent in altitude decompression sickness than hyperbaric decompression sickness. Simple pain, however, still accounts for the majority of symptoms.

20-3.8.1 **Joint Pain Treatment.** If only joint pain was present but resolved before reaching one ata from altitude, then the individual may be treated with two hours of 100 percent oxygen breathing at the surface followed by 24 hours of observation.

20-3.8.2 **Other Symptoms and Persistent Symptoms.** For other symptoms or if joint pain symptoms are present after return to one ata, the stricken individual should be transferred to a recompression facility and treated on the appropriate treatment table, even if the symptoms resolve while in transport. Individuals should be kept on 100 percent oxygen during transfer to the recompression facility.

20-4 RECOMPRESSION TREATMENT FOR DIVING DISORDERS

20-4.1 **Primary Objectives.** Table 20-1 gives the basic rules that shall be followed for all recompression treatments. The primary objectives of recompression treatment are:

- Compress gas bubbles to a small volume, thus relieving local pressure and restarting blood flow,
- Allow sufficient time for bubble resorption, and
- Increase blood oxygen content and thus oxygen delivery to injured tissues.

20-4.2 **Guidance on Recompression Treatment.** Certain facets of recompression treatment have been mentioned previously, but are so important that they cannot be stressed too strongly:

- Treat promptly and adequately.
- The effectiveness of treatment decreases as the length of time between the onset of symptoms and the treatment increases.
- Do not ignore seemingly minor symptoms. They can quickly become major symptoms.
- Follow the selected treatment table unless changes are recommended by a Diving Medical Officer.
- If multiple symptoms occur, treat for the most serious condition.

20-4.3 **Recompression Treatment When Chamber Is Available.** Oxygen treatment tables are significantly more effective than air treatment tables. Air treatment tables shall only be used after oxygen system failure or intolerable patient oxygen toxicity problems with DMO recommendation. Treatment Table 4 can be used with or without oxygen but should always be used with oxygen if it is available.

Table 20-1. *Rules for Recompression Treatment.*

ALWAYS:

1. Follow the treatment tables accurately, unless modified by a Diving Medical Officer with concurrence of the Commanding Officer.
2. Have a qualified tender in chamber at all times during treatment.
3. Maintain the normal descent and ascent rates as much as possible.
4. Examine the patient thoroughly at depth of relief or treatment depth.
5. Treat an unconscious patient for arterial gas embolism or serious decompression sickness unless the possibility of such a condition can be ruled out without question.
6. Use air treatment tables only if oxygen is unavailable.
7. Be alert for warning signs of oxygen toxicity if oxygen is used.
8. In the event of an oxygen convulsion, remove the oxygen mask and keep the patient from self-harm. Do not force the mouth open during a convulsion.
9. Maintain oxygen usage within the time and depth limitations prescribed by the treatment table.
10. Check the patient's condition and vital signs periodically. Check frequently if the patient's condition is changing rapidly or the vital signs are unstable.
11. Observe patient after treatment for recurrence of symptoms. Observe 2 hours for pain-only symptoms, 6 hours for serious symptoms. Do not release patient without consulting a DMO.
12. Maintain accurate timekeeping and recording.
13. Maintain a well-stocked Primary and Secondary Emergency Kit.

NEVER:

1. Permit any shortening or other alteration of the tables, except under the direction of a Diving Medical Officer.
2. Wait for a bag resuscitator. Use mouth-to-mouth resuscitation with a barrier device immediately if breathing ceases.
3. Interrupt chest compressions for longer than 10 seconds.
4. Permit the use of 100 percent oxygen below 60 feet in cases of DCS or AGE.
5. Fail to treat doubtful cases.
6. Allow personnel in the chamber to assume a cramped position that might interfere with complete blood circulation.

20-4.3.1 **Recompression Treatment With Oxygen.** Use Oxygen Treatment Table 5, 6, 6A, 4, or 7, according to the flowcharts in Figure 20-1, Figure 20-2 and Figure 20-3. The descent rate for all these tables is 20 feet per minute. Upon reaching a treatment depth of 60 fsw or shallower place the patient on oxygen. For treatment depths deeper than 60 fsw, use treatment gas if available.

20-4.3.2 **Recompression Treatments When Oxygen Is Not Available.** Air Treatment Tables 1A, 2A, and 3 (Figures 20-11, 20-12, and 20-13) are provided for use only as a last resort when oxygen is not available. Use Air Treatment Table 1A if pain is relieved at a depth less than 66 feet. If pain is relieved at a depth greater than 66 feet, use

Treatment Table 2A. Treatment Table 3 is used for treatment of serious symptoms where oxygen cannot be used. Use Treatment Table 3 if symptoms are relieved within 30 minutes at 165 feet. If symptoms are not relieved in less than 30 minutes at 165 feet, use Treatment Table 4.

20-4.4 **Recompression Treatment When No Recompression Chamber is Available.** The Diving Supervisor has two alternatives for recompression treatment when the diving facility is not equipped with a recompression chamber. If recompression of the patient is not immediately necessary, the diver may be transported to the nearest certified recompression chamber or the Diving Supervisor may elect to complete in-water recompression.

20-4.4.1 **Transporting the Patient.** In certain instances, some delay may be unavoidable while the patient is transported to a recompression chamber. While moving the patient to a recompression chamber, the patient should be kept supine (lying horizontally). Do not put the patient head-down. Additionally, the patient should be kept warm and monitored continuously for signs of obstructed (blocked) airway, cessation of breathing, cardiac arrest, or shock. Always keep in mind that a number of conditions may exist at the same time. For example, the victim may be suffering from both decompression sickness and hypothermia.

20-4.4.1.1 ***Medical Treatment During Transport.*** Always have the patient breathe 100 percent oxygen during transport, if available. If symptoms of decompression sickness or arterial gas embolism are relieved or improve after breathing 100 percent oxygen, the patient should still be recompressed as if the original symptom(s) were still present. Always ensure the patient is adequately hydrated. Give fluids by mouth if the patient is alert and able to tolerate them. Otherwise, an IV should be inserted and intravenous fluids should be started before transport. If the patient must be transported, initial arrangements should have been made well in advance of the actual diving operations. These arrangements, which would include an alert notification to the recompression chamber and determination of the most effective means of transportation, should be posted on the Job Site Emergency Assistant Checklist for instant referral.

20-4.4.1.2 ***Transport by Unpressurized Aircraft.*** If the patient is moved by helicopter or other unpressurized aircraft, the aircraft should be flown as low as safely possible, preferably less than 1,000 feet. Exposure to altitude results in an additional reduction in external pressure and possible additional symptom severity or other complications. If available, always use aircraft that can be pressurized to one atmosphere. If available, transport using the Emergency Evacuation Hyperbaric Stretcher should be considered.

20-4.4.1.3 ***Communications with Chamber.*** Call ahead to ensure that the chamber will be ready and that qualified medical personnel will be standing by. If two-way communications can be established, consult with the doctor as the patient is being transported.

20-4.4.2 **In-Water Recompression.** Recompression in the water should be considered an option of last resort, to be used only when no recompression facility is on site, symptoms are significant and there is no prospect of reaching a recompression facility within a reasonable timeframe (12–24 hours). In an emergency, an uncertified chamber may be used if, in the opinion of a qualified Chamber Supervisor (DSWS Watchstation 305), it is safe to operate. In divers with severe Type II symptoms, or symptoms of arterial gas embolism (e.g., unconsciousness, paralysis, vertigo, respiratory distress (chokes), shock, etc.), the risk of increased harm to the diver from in-water recompression probably outweighs any anticipated benefit. Generally, these individuals should not be recompressed in the water, but should be kept at the surface on 100 percent oxygen, if available, and evacuated to a recompression facility regardless of the delay. The stricken diver should begin breathing 100 percent oxygen immediately (if it is available). Continue breathing oxygen at the surface for 30 minutes before committing to recompress in the water. If symptoms stabilize, improve, or relief on 100 percent oxygen is noted, do not attempt in-water recompression unless symptoms reappear with their original intensity or worsen when oxygen is discontinued. Continue breathing 100 percent oxygen as long as supplies last, up to a maximum time of 12 hours. The patient may be given air breaks as necessary. If surface oxygen proves ineffective after 30 minutes, begin in-water recompression. To avoid hypothermia, it is important to consider water temperature when performing in-water recompression.

20-4.4.2.1 ***In-Water Recompression Using Air.*** In-water recompression using air is always less preferable than in-water recompression using oxygen.

- Follow Air Treatment Table 1A as closely as possible.
- Use either a full face mask or, preferably, a surface-supplied helmet UBA.
- Never recompress a diver in the water using a SCUBA with a mouth piece unless it is the only breathing source available.
- Maintain constant communication.
- Keep at least one diver with the patient at all times.
- Plan carefully for shifting UBAs or cylinders.
- Have an ample number of tenders topside.
- If the depth is too shallow for full treatment according to Air Treatment Table 1A:
  - Recompress the patient to the maximum available depth.
  - Remain at maximum depth for 30 minutes.
  - Decompress according to Air Treatment Table 1A. Do not use stops shorter than those of Air Treatment Table 1A.

20-4.4.2.2 ***In-Water Recompression Using Oxygen.*** If 100 percent oxygen is available to the diver using an oxygen rebreather, an ORCA, or other device, the following in-water recompression procedure should be used instead of Air Treatment Table 1A:

- Put the stricken diver on the UBA and have the diver purge the apparatus at least three times with oxygen.
- Descend to a depth of 30 feet with a standby diver.
- Remain at 30 feet, at rest, for 60 minutes for Type I symptoms and 90 minutes for Type II symptoms. Ascend to 20 feet even if symptoms are still present.
- Decompress to the surface by taking 60-minute stops at 20 feet and 10 feet.
- After surfacing, continue breathing 100 percent oxygen for an additional 3 hours.
- If symptoms persist or recur on the surface, arrange for transport to a recompression facility regardless of the delay.

20-4.4.2.3 ***Symptoms After In-Water Recompression.*** The occurrence of Type II symptoms after in-water recompression is an ominous sign and could progress to severe, debilitating decompression sickness. It should be considered life-threatening. Operational considerations and remoteness of the dive site will dictate the speed with which the diver can be evacuated to a recompression facility.

20-5 TREATMENT TABLES

20-5.1 **Air Treatment Tables.** Air Treatment Tables 1A, 2A, and 3 (Figures 20-11, 20-12, and 20-13) are provided for use only as a last resort when oxygen is not available. Oxygen treatment tables are significantly more effective than air treatment tables and shall be used whenever possible.

20-5.2 **Treatment Table 5.** Treatment Table 5, Figure 20-4, may be used for the following:

- Type I DCS (except for cutis marmorata) symptoms when a complete neurological examination has revealed no abnormality. After arrival at 60 fsw a neurological exam shall be performed to ensure that no overt neurological symptoms (e.g., weakness, numbness, loss of coordination) are present. If any abnormalities are found, the stricken diver should be treated using Treatment Table 6.
- Asymptomatic omitted decompression
- Treatment of resolved symptoms following in-water recompression
- Follow-up treatments for residual symptoms
- Carbon monoxide poisoning
- Gas gangrene

20-5.3 **Treatment Table 6.** Treatment Table 6, Figure 20-5, is used for the following:

- Arterial gas embolism
- Type II DCS symptoms

- Type I DCS symptoms where relief is not complete within 10 minutes at 60 feet or where pain is severe and immediate recompression must be instituted before a neurological examination can be performed
- Cutis marmorata
- Severe carbon monoxide poisoning, cyanide poisoning, or smoke inhalation
- Asymptomatic omitted decompression
- Symptomatic uncontrolled ascent
- Recurrence of symptoms shallower than 60 fsw

20-5.4 **Treatment Table 6A.** Treatment Table 6A, Figure 20-6, is used to treat arterial gas embolism or decompression symptoms when severe symptoms remain unchanged or worsen within the first 20 minutes at 60 fsw. The patient is compressed to depth of relief (or significant improvement), not to exceed 165 fsw. Once at the depth of relief, begin treatment gas (N_2O_2, HeO_2) if available. Consult with a Diving Medical Officer at the earliest opportunity. If the severity of the patient's condition warrants, the Diving Medical Officer may recommend conversion to a Treatment Table 4.

NOTE **If deterioration or recurrence of symptoms is noted during ascent to 60 feet, treat as a recurrence of symptoms (Figure 20-3).**

20-5.5 **Treatment Table 4.** Treatment Table 4, Figure 20-7, is used when it is determined that the patient would receive additional benefit at depth of significant relief, not to exceed 165 fsw. The time at depth shall be between 30 to 120 minutes, based on the patient's response. If a shift from Treatment Table 6A to Treatment Table 4 is contemplated, a Diving Medical Officer should be consulted before the shift is made.

If oxygen is available, the patient should begin oxygen breathing periods immediately upon arrival at the 60-foot stop. Breathing periods of 25 minutes on oxygen, interrupted by 5 minutes of air, are recommended because each cycle lasts 30 minutes. This simplifies timekeeping. Immediately upon arrival at 60 feet, a minimum of four oxygen breathing periods (for a total time of 2 hours) should be administered. After that, oxygen breathing should be administered to suit the patient's individual needs and operational conditions. Both the patient and tender must breathe oxygen for at least 4 hours (eight 25-minute oxygen, 5-minute air periods), beginning no later than 2 hours before ascent from 30 feet is begun. These oxygen-breathing periods may be divided up as convenient, but at least 2 hours' worth of oxygen breathing periods should be completed at 30 feet.

NOTE **If deterioration or recurrence of symptoms is noted during ascent to 60 feet, treat as a recurrence of symptoms (Figure 20-3).**

20-5.6 **Treatment Table 7.** Treatment Table 7, Figure 20-8, is an extension at 60 feet of Treatment Table 6, 6A, or 4 (or any other nonstandard treatment table). This means that considerable treatment has already been administered. Treatment Table 7 is considered a heroic measure for treating non-responding severe gas embolism or life-threatening decompression sickness and is not designed to treat all residual symptoms that do not improve at 60 feet and should never be used to treat residual pain. Treatment Table 7 should be used only when loss of life may result if the currently prescribed decompression from 60 feet is undertaken. Committing a patient to a Treatment Table 7 involves isolating the patient and having to minister to his medical needs in the recompression chamber for 48 hours or longer. Experienced diving medical personnel shall be on scene.

A Diving Medical Officer should be consulted before shifting to a Treatment Table 7 and careful consideration shall be given to life support capability of the recompression facility. Because it is difficult to judge whether a particular patient's condition warrants Treatment Table 7, additional consultation may be obtained from either NEDU or NDSTC.

When using Treatment Table 7, a minimum of 12 hours should be spent at 60 feet, including time spent at 60 feet from Treatment Table 4, 6, or 6A. Severe Type II decompression sickness and/or arterial gas embolism cases may continue to deteriorate significantly over the first several hours. This should not be cause for premature changes in depth. Do not begin decompression from 60 feet for at least 12 hours. At completion of the 12-hour stay, the decision must be made whether to decompress or spend additional time at 60 feet. If no improvement was noted during the first 12 hours, benefit from additional time at 60 feet is unlikely and decompression should be started. If the patient is improving but significant residual symptoms remain (e.g., limb paralysis, abnormal or absent respiration), additional time at 60 feet may be warranted. While the actual time that can be spent at 60 feet is unlimited, the actual additional amount of time beyond 12 hours that should be spent can only be determined by a Diving Medical Officer (in consultation with on-site supervisory personnel), based on the patient's response to therapy and operational factors. When the patient has progressed to the point of consciousness, can breathe independently, and can move all extremities, decompression can be started and maintained as long as improvement continues. Solid evidence of continued benefit should be established for stays longer than 18 hours at 60 feet. Regardless of the duration at the recompression deeper than 60 feet, at least 12 hours must be spent at 60 feet and then Treatment Table 7 followed to the surface. Additional recompression below 60 feet in these cases should not be undertaken unless adequate life support capability is available.

20-5.6.1 **Decompression.** Decompression on Treatment Table 7 is begun with an upward excursion at time zero from 60 to 58 feet. Subsequent 2-foot upward excursions are made at time intervals listed as appropriate to the rate of decompression:

Table 20-2. Decompression

Depth	Ascent Rate	Time Interval
58-40 feet	3 ft/hr	40 min
40-20 feet	2 ft/hr	60 min
20-4 feet	1 ft/hr	120 min

The travel time between stops is considered as part of the time interval for the next shallower stop. The time intervals shown above begin when ascent to the next shallower stop has begun.

20-5.6.2 **Tenders.** When using Treatment Table 7, tenders breathe chamber atmosphere throughout treatment and decompression.

20-5.6.3 **Preventing Inadvertent Early Surfacing.** Upon arrival at 4 feet, decompression should be stopped for 4 hours. At the end of 4 hours, decompress to the surface at 1 foot per minute. This procedure prevents inadvertent early surfacing.

20-5.6.4 **Oxygen Breathing.** On a Treatment Table 7, patients should begin oxygen breathing periods as soon as possible at 60 feet. Oxygen breathing periods of 25 minutes on 100 percent oxygen, followed by 5 minutes breathing chamber atmosphere, should be used. Normally, four oxygen breathing periods are alternated with 2 hours of continuous air breathing. In conscious patients, this cycle should be continued until a minimum of eight oxygen breathing periods have been administered (previous 100 percent oxygen breathing periods may be counted against these eight periods). Beyond that, oxygen breathing periods should be continued as recommended by the Diving Medical Officer, as long as improvement is noted and the oxygen is tolerated by the patient. If oxygen breathing causes significant pain on inspiration, it should be discontinued unless it is felt that significant benefit from oxygen breathing is being obtained. In unconscious patients, oxygen breathing should be stopped after a maximum of 24 oxygen breathing periods have been administered. The actual number and length of oxygen breathing periods should be adjusted by the Diving Medical Officer to suit the individual patient's clinical condition and response to pulmonary oxygen toxicity.

20-5.6.5 **Sleeping, Resting, and Eating.** At least two tenders should be available when using Treatment Table 7, and three may be necessary for severely ill patients. Not all tenders are required to be in the chamber, and they may be locked in and out as required following appropriate decompression tables. The patient may sleep anytime except when breathing oxygen deeper than 30 feet. While asleep, the patient's pulse, respiration, and blood pressure should be monitored and recorded at intervals appropriate to the patient's condition. Food may be taken at any time and fluid intake should be maintained.

20-5.6.6 **Ancillary Care.** Patients on Treatment Table 7 requiring intravenous and/or drug therapy should have these administered in accordance with paragraph 20-11 and associated subparagraphs.

20-5.6.7 **Life Support.** Before committing to a Treatment Table 7, the life-support considerations in paragraph 20-7 must be addressed. Do not commit to a Treatment Table 7 if the internal chamber temperature cannot be maintained at 85°F (29°C) or less.

20-5.7 **Treatment Table 8.** Treatment Table 8, Figure 20-9, is an adaptation of Royal Navy Treatment Table 65 mainly for treating deep uncontrolled ascents (see Chapter 14) when more than 60 minutes of decompression have been missed. Compress symptomatic patient to depth of relief not to exceed 225 fsw. Initiate Treatment Table 8 from depth of relief. The schedule for Treatment Table 8 from 60 fsw is the same as Treatment Table 7. The guidelines for sleeping and eating are the same as Treatment Table 7.

20-5.8 **Treatment Table 9.** Treatment Table 9, Figure 20-10, is a hyperbaric oxygen treatment table providing 90 minutes of oxygen breathing at 45 feet. This table is used only on the recommendation of a Diving Medical Officer cognizant of the patient's medical condition. Treatment Table 9 is used for the following:

1. Residual symptoms remaining after initial treatment of AGE/DCS
2. Selected cases of carbon monoxide or cyanide poisoning
3. Smoke inhalation

This table may also be recommended by the cognizant Diving Medical Officer when initially treating a severely injured patient whose medical condition precludes long absences from definitive medical care.

20-6 RECOMPRESSION TREATMENT FOR NON-DIVING DISORDERS

In addition to individuals suffering from diving disorders, U.S. Navy recompression chambers are also permitted to conduct emergent hyperbaric oxygen (HBO_2) therapy to treat individuals suffering from cyanide poisoning, carbon monoxide poisoning, gas gangrene, smoke inhalation, necrotizing soft-tissue infections, or arterial gas embolism arising from surgery, diagnostic procedures, or thoracic trauma. If the chamber is to be used for treatment of non-diving related medical conditions other than those listed above, authorization from BUMED Code M3B42 shall be obtained before treatment begins (BUMEDINST 6320.38 series.) Any treatment of a non-diving related medical condition shall be done under the cognizance of a Diving Medical Officer.

The guidelines given in Table 20-3 for conducting HBO_2 therapy are taken from the Undersea and Hyperbaric Medical Society's Hyperbaric Oxygen (HBO_2) Therapy Committee Report-2003: Approved Indications for Hyperbaric Oxygen Therapy. For each condition, the guidelines prescribe the recommended Treatment Table, the frequency of treatment, and the minimum and maximum number of treatments.

Table 20-3. *Guidelines for Conducting Hyperbaric Oxygen Therapy.*

Indication	Treatment Table	Minimum # Treatments	Maximum # Treatments
Carbon Monoxide Poisoning and Smoke Inhalation	Treatment Table 5 or Table 6 as recommended by the DMO	1	5
Gas Gangrene (Clostridial Myonecrosis)	Treatment Table 5 TID × 1 day then BID × 4-5 days	5	10
Crush Injury, Compartment Syndrome, and other Acute Traumatic Ischemia	Treatment Table 9 TID × 2 days BID × 2 days QD × 2 days	3	12
Enhancements of Healing in Selected Wounds	Treatment Table 9 QD or BID	10	60
Necrotizing Soft-Tissue Infections (subcutaneous tissue, muscle, fascia)	Treatment Table 9 BID initially, then QD	5	30
Osteomyelitis (refractory)	Treatment Table 9 QD	20	60
Radiation Tissue Damage (osteoradinecrosis)	Treatment Table 9 QD	20	60
Skin Grafts and Flaps (compromised)	Treatment Table 9 BID initially, then QD	6	40
Thermal Burns	Treatment Table 9 TID × 1 day, then BID	5	45

QD = 1 time in 24 hours BID = 2 times in 24 hours TID = 3 times in 24 hours
For further information, see Hyperbaric Oxygen Therapy: A Committee Report, 2003 Revision.

20-7 RECOMPRESSION CHAMBER LIFE-SUPPORT CONSIDERATIONS

The short treatment tables (Oxygen Treatment Tables 5, 6, 6A, 9; Air Treatment Tables 1A and 2A) can be accomplished easily without significant strain on either the recompression chamber facility or support crew. The long treatment tables (Tables 3, 4, 7, and 8) will require long periods of decompression and may tax both personnel and hardware severely.

20-7.1 **Minimum Manning Requirements.** The minimum team for conducting any recompression operation shall consist of three individuals. In case of emergency, the recompression chamber can be manned with two individuals.

- The Diving Supervisor is in complete charge at the scene of the operation, keeping individual and overall times on the operation, logging progress, and communicating with personnel inside the chamber.
- The Outside Tender is responsible for the operation of gas supplies, ventilation, pressurization, and exhaust of the chamber.
- The Inside Tender is familiar with the diagnosis and treatment of diving-related illnesses.

20-7.2 **Optimum Manning Requirements.** The optimum team for conducting recompression operations consists of four individuals:

- The Diving Supervisor is in complete charge at the scene of the operation.
- The Outside Tender #1 is responsible for the operation of the gas supplies, ventilation, pressurization, and exhaust of the chamber.
- The Outside Tender #2 is responsible for keeping individuals' and overall times on the operation, logging progress as directed by the Diving Supervisor, and communicating with personnel inside the chamber.
- The Inside Tender is familiar with the diagnosis and treatment of diving-related illnesses.

20-7.2.1 **Additional Personnel.** If the patient has symptoms of serious decompression sickness or arterial gas embolism, the team will require additional personnel. If the treatment is prolonged, a second team may have to relieve the first team. Patients with serious decompression sickness and gas embolism would initially be accompanied inside the chamber by a Diving Medical Technician or Diving Medical Officer, if possible. However, treatment should not be delayed to comply with this recommendation.

20-7.2.2 **Required Consultation by a Diving Medical Officer.** A Diving Medical Officer shall be consulted as early as possible in all recompression treatments, and, if at all possible, before committing the patient to a Treatment Table 4, 7, or 8. The Diving Medical Officer may be on scene or in communication with the Diving Supervisor. In all cases a DMO must be consulted prior to releasing a patient from treatment.

20-7.3 **Oxygen Control.** All treatment schedules listed in this chapter are usually performed with a chamber atmosphere of air. To accomplish safe decompression, the oxygen percentage should not be allowed to fall below 19 percent. Oxygen may be added to the chamber by ventilating with air or by bleeding in oxygen from an oxygen breathing system. If a portable oxygen analyzer is available, it can be used to determine the adequacy of ventilation and/or addition of oxygen. If no oxygen analyzer is available, ventilation of the chamber in accordance with paragraph 20-7.6 will ensure adequate oxygenation. Chamber oxygen percentages as high as 25 percent are permitted. If the chamber is equipped with a life-support system so that ventilation is not required and an oxygen analyzer is available, the oxygen level should be maintained between 19 percent and 25 percent. If chamber oxygen goes above 25 percent, ventilation with air should be used to bring the oxygen percentage down.

20-7.4 **Carbon Dioxide Control.** Ventilation of the chamber in accordance with paragraph 20-7.6 will ensure that carbon dioxide produced metabolically does not cause the chamber carbon dioxide level to exceed 1.5 percent SEV (11.4 mmHg).

20-7.4.1 **Carbon Dioxide Monitoring.** Chamber carbon dioxide should be monitored with electronic carbon dioxide monitors. Monitors generally read CO_2 percentage once chamber air has been exhausted to the surface. The CO_2 percent reading at the surface 1 ata must be corrected for depth. To keep chamber CO_2 below 1.5 percent SEV (11.4 mmHg), the surface CO_2 monitor values should remain below 0.78 percent with chamber depth at 30 feet, 0.53 percent with chamber depth at 60 feet, and 0.25 percent with the chamber at 165 feet. If the CO_2 analyzer is within the chamber, no correction to the CO_2 readings is necessary.

20-7.4.2 **Carbon Dioxide Scrubbing.** If the chamber is equipped with a carbon dioxide scrubber, the absorbent should be changed when the partial pressure of carbon dioxide in the chamber reaches 1.5 percent SEV (11.4 mmHg). If absorbent cannot be changed, supplemental chamber ventilation will be required to maintain chamber CO_2 at acceptable levels. With multiple or working chamber occupants, supplemental ventilation may be necessary to maintain chamber CO_2 at acceptable levels.

20-7.4.3 **Carbon Dioxide Absorbent.** CO_2 absorbent may be used beyond the expiration date when used in a recompression chamber equipped with a CO_2 monitor. When used in a recompression chamber that has no CO_2 monitor, CO_2 absorbent in an opened but resealed bucket may be used until the expiration date on the bucket is reached. Pre-packed, double-bagged canisters shall be labeled with the expiration date from the absorbent bucket for recompression chambers with no CO_2 monitor.

20-7.5 **Temperature Control.** Internal chamber temperature should be maintained at a level comfortable to the occupants whenever possible. Cooling can usually be accomplished by chamber ventilation. If the chamber is equipped with a heater/chiller unit, temperature control can usually be maintained for chamber occupant comfort under any external environmental conditions. Usually, recompression chambers will become hot and must be cooled continuously. Chambers should always be shaded from direct sunlight. The maximum durations for chamber occupants will depend on the internal chamber temperature as listed in Table 20-4. Never commit to a treatment table that will expose the chamber occupants to greater temperature/time combinations than listed in Table 20-4 unless qualified medical personnel who can evaluate the trade-off between the projected heat stress and the anticipated treatment benefit are consulted. A chamber temperature below 85°F (29°C) is always desirable, no matter which treatment table is used.

For patients with brain or spinal cord damage, the current evidence recommends aggressive treatment of elevated body temperature. When treating victims of AGE or severe neurological DCS, hot environments that elevate body temperature above normal should be avoided, whenever possible. As in DCS, patient temperature should be a routinely monitored vital sign.

Table 20-4. Maximum Permissible Recompression Chamber Exposure Times at Various Temperatures.

Internal Temperature	Maximum Tolerance Time	Permissible Treatment Tables
Over 104°F (40°C)	Intolerable	No treatments
95–104°F (34.4–40°C)	2 hours	Table 5, 9
85–94°F (29–34.4°C)	6 hours	Tables 5, 6, 6A, 1A, 9
Under 85°F (29°C)	Unlimited	All treatments

NOTE:
Internal chamber temperature can be kept considerably below ambient by venting or by using an installed chiller unit. Internal chamber temperature can be measured using electronic, bimetallic, alcohol, or liquid crystal thermometers. Never use a mercury thermometer in or around hyperbaric chambers. Since chamber ventilation will produce temperature swings during ventilation, the above limits should be used as averages when controlling temperature by ventilation. Always shade chamber from direct sunlight.

20-7.5.1 **Patient Hydration.** Always ensure patients are adequately hydrated. Fully conscious patients may be given fluid by mouth to maintain adequate hydration. One to two liters of water, juice, or non-carbonated drink, over the course of a Treatment Table 5 or 6, is usually sufficient. Patients with Type II symptoms, or symptoms of arterial gas embolism, should be considered for IV fluids. Stuporous or unconscious patients should always be given IV fluids, using large-gauge plastic catheters. If trained personnel are present, an IV should be started as soon as possible and kept dripping at a rate of 75 to 100 cc/hour, using isotonic fluids (Lactated Ringer's Solution, Normal Saline) until specific instructions regarding the rate and type of fluid administration are given by qualified medical personnel. Avoid solutions containing glucose (Dextrose) if brain or spinal cord injury is present. Intravenously administered glucose may worsen the outcome. In some cases, the bladder may be paralyzed. The victim's ability to void shall be assessed as soon as possible. If the patient cannot empty a full bladder, a urinary catheter shall be inserted as soon as possible by trained personnel. Always inflate catheter balloons with liquid, not air. Adequate fluid is being given when urine output is at least 0.5cc/kg/hr. Thirst is an unreliable indicator of the water intake to compensate for heavy sweating. A useful indicator of proper hydration is a clear colorless urine.

20-7.6 **Chamber Ventilation.** Ventilation is the usual means of controlling oxygen level, carbon dioxide level, and temperature. Ventilation using air is required for chambers without carbon dioxide scrubbers and atmospheric analysis. A ventilation rate of two acfm for each resting occupant, and four acfm for each active occupant, should be used. These procedures are designed to assure that the effective concentration of carbon dioxide will not exceed 1.5 percent sev (11.4 mmHg) and that, when oxygen is being used, the percentage of oxygen in the chamber will not exceed 25 percent.

20-7.7 **Access to Chamber Occupants.** Recompression treatments usually require access to occupants for passing in items such as food, water, and drugs and passing out such items as urine, excrement, and trash. Never attempt a treatment longer than a Treatment Table 6 unless there is access to inside occupants. When doing a Treatment Table 4, 7, or 8, a double-lock chamber is mandatory because additional personnel may have to be locked in and out during treatment.

20-7.8 **Inside Tenders.** When conducting a recompression treatment, at least one qualified tender shall be inside the chamber. The inside tender shall be familiar with all treatment procedures and the signs, symptoms, and treatment of all diving-related disorders. Medical personnel may have to be locked into the chamber as the patient's condition dictates.

20-7.8.1 **Inside Tender Responsibilities.** During the early phases of treatment, the inside tender must monitor the patient constantly for signs of relief. Drugs that mask signs of the illness should not be given. Observation of these signs is the principal method of diagnosing the patient's illness. Furthermore, the depth and time of their relief helps determine the treatment table to be used. The inside tender is also responsible for:

- Releasing the door latches (dogs) after a seal is made
- Communicating with outside personnel
- Providing first aid as required by the patient
- Administering treatment gas to the patient at treatment depth
- Providing normal assistance to the patient as required
- Ensuring that sound attenuators for ear protection are worn during compression and ventilation portions of recompression treatments.
- Ensuring that the patient is lying down and positioned to permit free blood circulation to all extremities.

20-7.8.2 **DMO or DMT Inside Tender.** If it is known before the treatment begins that adjunctive therapy or advanced medical support must be administered to the patient (examples include an IV, or airway maintenance), or if the patient is suspected of suffering from arterial gas embolism, a Diving Medical Technician or Diving Medical Officer should accompany the patient inside the chamber. However, recompression treatment must not be delayed while awaiting the arrival of the DMO or DMT.

20-7.8.3 **Use of Diving Medical Officer as Inside Tender.** If only one Diving Medical Officer is on site, the Medical Officer should lock in and out as the patient's condition dictates, but should not commit to the entire treatment unless absolutely necessary. Once committed to remain in the chamber, the Diving Medical Officer effectiveness in directing the treatment is greatly diminished and consultation with other medical personnel becomes more difficult. If periods in the chamber are necessary, visits should be kept within no-decompression limits if possible.

20-7.8.4 **Non-Diver Inside Tender - Medical.** Non-diving medical personnel may be qualified as Inside Tenders (examples would include U.S. Naval Reserve Corpsmen, and nursing personnel). Qualifications may be achieved through Navy Diver Inside Tender PQS. Prerequisites: Current diving physical exam, conformance to Navy physical standards, and diver candidate pressure test.

20-7.8.5 **Specialized Medical Care.** Emergency situations that require specialized medical care should always have the best qualified person provide it. The best qualified person may be a surgeon, respiratory therapist, IDC, etc. Since these are emergency exposures, no special medical or physical prerequisites exist. A qualified Inside Tender is required inside the chamber to handle any system related requirements.

20-7.8.6 **Inside Tender Oxygen Breathing.** During treatments, all chamber occupants may breathe 100 percent oxygen at depths of 45 feet or shallower without locking in additional personnel. Tenders should not fasten the oxygen masks to their heads, but should hold them on their faces. When deeper than 45 feet, at least one chamber occupant must breathe air. Tender oxygen breathing requirements are specified in the figure for each Treatment Table.

20-7.8.7 **Tending Frequency.** Normally, tenders should allow a surface interval of at least 18 hours between consecutive treatments on Treatment Tables 1A, 2A, 3, 5, 6, and 6A, and at least 48 hours between consecutive treatments on Tables 4, 7, and 8. If necessary, however, tenders may repeat Treatment Tables 5, 6, or 6A within this 18-hour surface interval if oxygen is breathed at 30 feet and shallower as outlined in Table 20-6. Minimum surface intervals for Treatment Tables 1A, 2A, 3, 4, 7, and 8 shall be strictly observed.

20-7.9 **Equalizing During Descent.** Descent rates may have to be decreased as necessary to allow the patient to equalize; however, it is vital to attain treatment depth in a timely manner for a suspected arterial gas embolism patient.

20-7.10 **Use of High Oxygen Mixes.** High oxygen N_2O_2/HeO_2 mixtures may be used to treat patients when recompression deeper than 60 fsw is required. These mixtures offer significant therapeutic advantages over air. Select a treatment gas that will produce a ppO_2 between 1.5 and 3.0 ata at the treatment depth. The standardized gas mixtures shown in Table 20-4 are suitable over the depth range of 61-225 fsw.

Decompression sickness following helium dives can be treated with either nitrogen or helium mixtures. For recompression deeper than 165 fsw, helium mixtures are preferred to avoid narcosis. The situation is less clear for treatment of decompression sickness following air or nitrogen-oxygen dives. Experimental studies have shown both benefit and harm with helium treatment. Until more experience is obtained, high oxygen mixtures with nitrogen as the diluent gas are preferred if available. High oxygen mixtures may also be substituted for 100% oxygen at 60 fsw and shallower on Treatment Tables 4, 7, and 8 if the patient is unable to tolerate 100% oxygen.

Table 20-5. *High Oxygen Treatment Gas Mixtures.*

Depth (fsw)	Mix (HeO_2 or N_2O_2)	ppO_2
0-60	100%	1.00-2.82
61-165	50/50	1.42-3.00
166-225	64/36 (HeO_2 only)	2.17-2.81

20-7.11 **Oxygen Toxicity During Treatment.** Acute CNS oxygen toxicity may develop on any oxygen treatment table.

During prolonged treatments on Treatment Tables 4, 7, or 8, and with repeated Treatment Table 6, pulmonary oxygen toxicity may also develop.

20-7.11.1 **Central Nervous System Oxygen Toxicity.** When employing the oxygen treatment tables, tenders must be particularly alert for the early symptoms of CNS oxygen toxicity. The symptoms can be remembered readily by using the mnemonic VENTID-C (Vision, Ears, Nausea, Twitching\Tingling, Irritability, Dizziness, Convulsions). Unfortunately, a convulsion may occur without early warning signs or before the patient can be taken off oxygen in response to the first sign of CNS oxygen toxicity. CNS oxygen toxicity is unlikely in resting individuals at chamber depths of 50 feet or shallower and very unlikely at 30 feet or shallower, regardless of the level of activity. However, patients with severe Type II decompression sickness or arterial gas embolism symptoms may be abnormally sensitive to CNS oxygen toxicity. Convulsions unrelated to oxygen toxicity may also occur and may be impossible to distinguish from oxygen seizures.

20-7.11.1.1 ***Procedures in the Event of CNS Oxygen Toxicity.*** At the first sign of CNS oxygen toxicity, the patient should be removed from oxygen and allowed to breathe chamber air. Fifteen minutes after all symptoms have subsided, resume oxygen breathing. For Treatment Tables 5, 6, 6A resume treatment at the point of interruption. For Treatment Tables 4, 7 and 8 no compensatory lengthening of the table is required. If symptoms of CNS oxygen toxicity develop again or if the first symptom is a convulsion, take the follow action:

CAUTION **Inserting an airway device or bite block is not recommended while the patient is convulsing; it is not only difficult, but may cause harm if attempted.**

For Treatment Tables 5, 6, and 6A:

- Remove the mask
- After all symptoms have completely subsided, decompress 10 feet at a rate of 1 fsw/min. For a convulsion, begin travel when the patient is fully relaxed and breathing normally.
- Resume oxygen breathing at the shallower depth at the point of interruption.

- If another oxygen symptom occurs after ascending 10 fsw, contact a Diving Medical Officer to recommend appropriate modifications to the treatment schedule.

For Treatment Tables 4, 7, and 8:

- Remove the mask.
- Consult with a Diving Medical Officer before administering further oxygen breathing. No compensatory lengthening of the table is required for interruption in oxygen breathing

20-7.11.2 **Pulmonary Oxygen Toxicity.** Pulmonary oxygen toxicity is unlikely to develop on single Treatment Tables 5, 6, or 6A. On Treatment Tables 4, 7, or 8 or with repeated Treatment Tables 5, 6, or 6A (especially with extensions) prolonged exposure to oxygen may result in end-inspiratory discomfort, progressing to substernal burning and severe pain on inspiration. If a patient who is responding well to treatment complains of substernal burning, discontinue use of oxygen and consult with a DMO. However, if a significant neurological deficit remains and improvement is continuing (or if deterioration occurs when oxygen breathing is interrupted), oxygen breathing should be continued as long as considered beneficial or until pain limits inspiration. If oxygen breathing must be continued beyond the period of substernal burning, or if the 2-hour air breaks on Treatment Tables 4, 7, or 8 cannot be used because of deterioration upon the discontinuance of oxygen, the oxygen breathing periods should be changed to 20 minutes on oxygen, followed by 10 minutes breathing chamber air or alternative treatment gas mixtures with a lower percentage of oxygen should be considered. The Diving Medical Officer may tailor the above guidelines to suit individual patient response to treatment.

20-7.12 Loss of Oxygen During Treatment. Loss of oxygen breathing capability during oxygen treatments is a rare occurrence. However, should it occur, the following actions should be taken:

If repair can be completed within 15 minutes:

- Maintain depth until repair is completed.
- After O_2 is restored, resume treatment at point of interruption.

If repair can be completed after 15 minutes but before 2 hours:

- Maintain depth until repair is completed.
- After O_2 is restored: If original table was Table 5, 6, or 6A, complete treatment with maximum number of O_2 extensions.

20-7.12.1 **Compensation.** If Table 4, 7, or 8 is being used, no compensation in decompression is needed if oxygen is lost. If decompression must be stopped because of worsening symptoms in the affected diver, then stop decompression. When oxygen is restored, continue treatment from where it was stopped.

20-7.12.2 **Switching to Air Treatment Table.** If O_2 breathing cannot be restored in 2 hours switch to the comparable air treatment table at current depth for decompression if 60 fsw or shallower. Rate of ascent must not exceed 1 fpm between stops. If symptoms worsen and an increase in treatment depth deeper than 60 feet is needed, use Treatment Table 4.

20-7.13 **Treatment at Altitude.** Before starting recompression therapy, zero the chamber depth gauges to adjust for altitude. Then use the depths as specified in the treatment table. There is no need to "Cross Correct" the treatment table depths. Divers serving as inside tenders during hyperbaric treatments at altitude are performing a dive at altitude and therefore require more decompression than at sea level. Tenders locking into the chamber for brief periods should be managed according to the Diving At Altitude procedures (Chapter 9, paragraph 9-13). Tenders remaining in the chamber for the full treatment table must breathe oxygen during the terminal portion of the treatment to satisfy their decompression requirement.

The additional oxygen breathing required at altitude on Treatment Table 5, Treatment Table 6, and Treatment Table 6A is given in Table 20-6. The requirement pertains both to tenders equilibrated at altitude and to tenders flown directly from sea level to the chamber location. Contact NEDU for guidance on tender oxygen requirements for other treatment tables.

20-8 POST-TREATMENT CONSIDERATIONS

Tenders on Treatment Tables 5, 6, 6A, 1A, 2A, or 3 should have a minimum of a 18-hour surface interval before no-decompression diving and a minimum of a 24-hour surface interval before dives requiring decompression stops. Tenders on Treatment Tables 4, 7, and 8 should have a minimum of a 48-hour surface interval prior to diving.

20-8.1 **Post-Treatment Observation Period.** After a treatment, patients treated on a Treatment Table 5 should remain at the recompression chamber facility for 2 hours. Patients who have been treated for Type II decompression sickness or who required a Treatment Table 6 for Type I symptoms and have had complete relief should remain at the recompression chamber facility for 6 hours. Patients treated on Treatment Tables 6, 6A, 4, 7, 8 or 9 are likely to require a period of hospitalization, and the Diving Medical Officer will need to determine a post-treatment observation period and location appropriate to their response to recompression treatment. These times may be shortened upon the recommendation of a Diving Medical Officer, provided the patient will be with personnel who are experienced at recognizing recurrence of symptoms and can return to the recompression facility within 30 minutes. All patients should remain within 60 minutes travel time of a recompression facility for 24 hours and should be accompanied throughout that period. No patient shall be released until authorized by a DMO.

Treatment table profiles place the inside tender(s) at risk for decompression sickness. After completing treatments, inside tenders should remain in the vicinity of the recompression chamber for 1 hour. If they were tending for Treatment Table

Table 20-6. Tender Oxygen Breathing Requirements. (Note 1)

Treatment Table (TT)		Altitude: Surface to 2499 ft	2500 ft. - 7499 ft.	7500 ft. - 10,000 ft.
TT5 Note (2)	without extension	:00	:00	:00
	with extension @ 30 fsw	:00	:00	:20
TT6 Note (2)	up to one extension @ 60 fsw or 30 fsw	:30	:60	:90
	more than one extension	:60	:90	:120
TT6A Note (2)	up to one extension @ 60 fsw or 30 fsw	:60	:120	:150 Note (3)
	more than one extension	:90	:150 Note (3)	:180 Note (3)

Note 1: All tender O_2 breathing times in table are conducted at 30 fsw. In addition, tenders will breathe O_2 on ascent from 30 fsw to the surface.

Note 2: If the tender had a previous hyperbaric exposure within 18 hours, use the following guidance for administering O_2:
For TT5, add an additional 20 minute O_2 breathing period to the times in the table.
For TT6 or TT6A, add an additional 60 minute O_2 breathing period to the times in the table.
For other Treatment tables contact NEDU for guidance.

Note 3: In some instances, tender's oxygen breathing obligation exceeds the table stay time at 30 fsw. Extend the time at 30 fsw to meet these obligations if patient's condition permits. Otherwise, administer O_2 to the tender to the limit allowed by the treatment table and observe the tender on the surface for 1 hour for symptoms of DCS.

4, 7, or 8, inside tenders should also remain within 60 minutes travel time of a recompression facility for 24 hours.

20-8.2 **Post-Treatment Transfer.** Patients with residual symptoms should be transferred to appropriate medical facilities as directed by qualified medical personnel. If ambulatory patients are sent home, they should always be accompanied by someone familiar with their condition who can return them to the recompression facility should the need arise. Patients completing treatment do not have to remain in the vicinity of the chamber if the Diving Medical Officer feels that transferring them to a medical facility immediately is in their best interest.

20-8.3 **Flying After Treatments.** Patients with residual symptoms should fly only with the concurrence of a Diving Medical Officer. Patients who have been treated for decompression sickness or arterial gas embolism and have complete relief should not fly for 72 hours after treatment, at a minimum.

Tenders on Treatment Tables 5, 6, 6A, 1A, 2A, or 3 should have a 24-hour surface interval before flying. Tenders on Treatment Tables 4, 7, and 8 should not fly for 72 hours.

20-8.3.1 **Emergency Air Evacuation.** Some patients will require air evacuation to another treatment or medical facility immediately after surfacing from a treatment. They will not meet surface interval requirements as described above. Such evacuation is done only on the recommendation of a Diving Medical Officer. Aircraft pressurized to one ata should be used if possible, or unpressurized aircraft flown as low as safely possible (no more than 1,000 feet is preferable). Have the patient breathe 100 percent oxygen during transport, if available. If available, an Emergency Evacuation Hyperbaric Stretcher to maintain the patient at 1ata may be used.

20-8.4 **Treatment of Residual Symptoms.** After completion of the initial recompression treatment and after a surface interval sufficient to allow complete medical evaluation, additional recompression treatments may be instituted. If additional recompression treatments are indicated a Diving Medical Officer must be consulted. Residual symptoms may remain unchanged during the first one or two treatments. In these cases, the Diving Medical Officer is the best judge as to the number of recompression treatments. Consultation with NEDU or NDSTC may be appropriate. As the delay time between completion of initial treatment and the beginning of follow-up hyperbaric treatments increases, the probability of benefit from additional treatments decreases. However, improvement has been noted in patients who have had delay times of up to 1 week. Therefore, a long delay is not necessarily a reason to preclude follow-up treatments. Once residual symptoms respond to additional recompression treatments, such treatments should be continued until no further benefit is noted. In general, treatment may be discontinued if there is no further sustained improvement on two consecutive treatments.

For persistent Type II symptoms, daily treatment on Table 6 may be used, but twice-daily treatments on Treatment Tables 5 or 9 may also be used. The treatment table chosen for re-treatments must be based upon the patient's medical condition and the potential for pulmonary oxygen toxicity. Patients surfacing from Treatment Table 6A with extensions, 4, 7, or 8 may have severe pulmonary oxygen toxicity and may find breathing 100 percent oxygen at 45 or 60 feet to be uncomfortable. In these cases, daily treatments at 30 feet may also be used. As many oxygen breathing periods (25 minutes on oxygen followed by 5 minutes on air) should be administered as can be tolerated by the patient. Ascent to the surface is at 20 feet per minute. A minimum oxygen breathing time is 90 minutes. A practical maximum bottom time is 3 to 4 hours at 30 feet. Treatments should not be administered on a daily basis for more than 5 days without a break of at least 1 day. These guidelines may have to be modified by the Diving Medical Officer to suit individual patient circumstances and tolerance to oxygen as measured by decrements in the patient's vital capacity.

20-8.5 **Returning to Diving after Recompression Treatment.** Divers diagnosed with AGE or Type II DCS may be medically cleared to return to diving duty 30 days after initial diagnosis and treatment by a DMO, if initial hyperbaric treatment is successful and no neurologic deficits persist. A BUMED waiver for return to diving is required if symptoms persist beyond initial treatment of AGE or Type II DCS. Refer to Bureau of Medicine and Surgery Manual (MANMED) P117 Article 15-102 for guidance.

20-9 NON-STANDARD TREATMENTS

The treatment recommendations presented in this chapter should be followed as closely as possible unless it becomes evident that they are not working. Only a Diving Medical Officer may then recommend changes to treatment protocols or use treatment techniques other than those described in this chapter. Any modifications to treatment tables shall be approved by the Commanding Officer. The standard treatment procedures in this chapter should be considered minimum treatments. Treatment procedures should never be shortened unless emergency situations arise that require chamber occupants to leave the chamber early, or the patient's medical condition precludes the use of standard U.S. Navy treatment tables.

20-10 RECOMPRESSION TREATMENT ABORT PROCEDURES

Once recompression therapy is started, it should be completed according to the procedures in this chapter unless the diver being treated dies or unless continuing the treatment would place the chamber occupants in mortal danger or in order to treat another more serious medical condition.

20-10.1 **Death During Treatment.** If it appears that the diver being treated has died, a Diving Medical Officer shall be consulted before the treatment is aborted. Once the decision to abort is made, there are a number of options for decompressing the tenders depending on the depth at which the death occurred and the preceding treatment profile.

- If death occurs following initial recompression to 60, 165, or 225 on Treatment Tables 6, 6A, 4 or 8, decompress the tenders on the Air/Oxygen schedule in the Air Decompression Table having a depth exactly equal to or deeper than the maximum depth attained during the treatment and a bottom time equal to or longer than the total elapsed time since treatment began. The Air/Oxygen schedule can be used even if gases other than air (i.e., nitrogen-oxygen or helium-oxygen mixtures) were breathed at depth.

- If death occurs after leaving the initial treatment depth on Treatment Tables 6 or 6A, decompress the tenders at 30 fsw/min to 30 fsw and have them breathe oxygen at 30 fsw for the times indicated in Table 20-6. Following completion of the oxygen breathing time at 30 fsw, decompress the tenders on oxygen from 30 fsw to the surface at 1 fsw/min.

- If death occurs after leaving the initial treatment depth on Treatment Tables 4 or 8, or after beginning treatment on Treatment Table 7 at 60 fsw, have the tenders decompress by continuing on the treatment table as written, or consult NEDU for a decompression schedule customized for the situation at hand. If neither option is possible, follow the original treatment table to 60 fsw. At 60 fsw, have the tenders breathe oxygen for 90 min in three 30-min periods separated by a 5-min air break. Continue decompression at 50, 40 and 30 fsw by breathing oxygen for 60 min at each depth. Ascend between stops at 30 fsw/min. At 50 fsw, breathe oxygen in two 30-min periods separated by a 5-min air break. At 40 and 30 fsw, breathe oxygen for the full 60-min period followed by a 15-min air break. Ascend to 20 fsw at 30 fsw/min and breathe oxygen for 120 min. Divide the oxygen time at 20 fsw into two 60-min periods separated by a 15 min air break. When oxygen breathing time is complete at 20 fsw, ascend to the surface at 30 fsw/min. Upon surfacing, observe the tenders carefully for the occurrence of decompression sickness.

20-10.2 **Impending Natural Disasters or Mechanical Failures.** Impending natural disasters or mechanical failures may force the treatment to be aborted. For instance, the ship where the chamber is located may be in imminent danger of sinking or a fire or explosion may have severely damaged the chamber system to such an extent that completing the treatment is impossible. In these cases, the abort procedure described in paragraph 20-10.1 could be used for all chamber occupants (including the stricken diver) if time is available. If time is not available, the following may be done:

1. If deeper than 60 feet, go immediately to 60 feet.

2. Once the chamber is 60 feet or shallower, put all chamber occupants on continuous 100 percent oxygen. Select the Air/Oxygen schedule in the Air Decompression Table corresponding to the maximum depth attained during treatment and the total elapsed time since treatment began.

3. If at 60 fsw, breathe oxygen for period of time equal to the sum of all the decompression stops 60 fsw and deeper in the Air/Oxygen schedule, then continue decompression on the Air/Oxygen schedule, breathing oxygen continuously. If shallower than 60 fsw, breathe oxygen for a period of time equal to the sum of all the decompression stops deeper than the divers current depth, then continue decompression on the Air/Oxygen schedule, breathing oxygen continuously. Complete as much of the Air/Oxygen schedule as possible.

4. When no more time is available, bring all chamber occupants to the surface (try not to exceed 10 feet per minute) and keep them on 100 percent oxygen during evacuation, if possible.

5. Immediately evacuate all chamber occupants to the nearest recompression facility and treat according to Figure 20-1. If no symptoms occurred after the treatment was aborted, follow Treatment Table 6.

20-11 ANCILLARY CARE AND ADJUNCTIVE TREATMENTS

WARNING **Drug therapy shall be administered only after consultation with a Diving Medical Officer by qualified inside tenders adequately trained and capable of administering prescribed medications.**

Most U.S. military diving operations have the unique advantage over most other diving operations with the ability to provide rapid recompression for the victims of decompression sickness (DCS) and arterial gas embolism (AGE). When stricken divers are treated without delay, the success rate of standard recompression therapy is extremely good.

Some U.S. military divers, such as Special Operations forces, however, may not have the benefit of a chamber nearby. Diving missions in Special Operations are often conducted in remote areas and may entail a lengthy delay to recompression therapy in the event of a diving accident. Delays to treatment for DCS and AGE significantly increase the probability of severe or refractory disease. In these divers, the use of adjunctive therapy (treatments other than recompression on a treatment table) can be provided while the diver is being transported to a chamber. Adjunctive therapies may also be useful for divers with severe symptoms or who have an incomplete response to recompression and hyperbaric oxygen.

Note that the adjunctive therapy guidelines are separated by accident type, with DCS and AGE covered separately. Although there is some overlap between the guidelines for these two disorders (as with the recompression phase of therapy), the best adjunctive therapy for one disorder is not necessarily the best therapy for the other. Although both DCS and AGE have in common the presence of gas bubbles in the body and a generally good response to recompression and hyperbaric oxygen, the underlying pathophysiology is somewhat different.

20-11.1 Decompression Sickness.

20-11.1.1 **Surface Oxygen.** Surface oxygen should be used for all cases of DCS until the diver can be recompressed. Use of either a high-flow (15 liters/minute) oxygen source with a reservoir mask or a demand valve can achieve high inspired fractions of oxygen. One consideration in administering surface oxygen is pulmonary oxygen toxicity. 100% oxygen can generally be tolerated for up to 12 hours. The patient may be given air breaks as necessary. If oxygen is being administered beyond this time, the decision to continue must weigh the perceived benefits against the risk of pulmonary oxygen toxicity. This risk evaluation must consider the dose of oxygen anticipated with subsequent recompression therapy as well.

20-11.1.2 **Fluids.** Fluids should be administered to all individuals suffering from DCS unless suffering from the chokes (pulmonary DCS). Oral fluids (half-strength glucose and electrolyte solutions) are acceptable if the diver is able to tolerate them. There is no data available that demonstrates a superiority of crystalloids (normal saline or Lactated Ringers solution) over colloids (such as Hetastarch compounds (Hespan or Hextend)) or vice versa, but D5W (dextrose in water without electrolytes) should not be used. Since colloids are far more expensive than Lactated Ringers or normal saline, the latter two agents are the most reasonable choice at this time. The optimal amount of crystalloids/colloids is likewise not well-established but treatment should be directed towards reversing any dehydration that may have been induced by the dive (immersion diuresis causes divers to lose 250-500 cc of fluids per hour) or fluid shifts resulting from the DCS. Fluid overloading should be avoided. Urinary output, in the range of 0.5cc/kg/hour is evidence of adequate intravascular volume.

Chokes (pulmonary DCS) causes abnormal pulmonary function and leakage of fluids into the alveolar spaces. Aggressive fluid therapy may make this condition worse. Consult a DMO (or NEDU) for guidance.

20-11.1.3 **Anticoagulants.** Since some types of DCS may increase the likelihood of hemorrhage into the tissues, anticoagulants should not be used routinely in the treatment of DCS. One exception to this rule is the case of lower extremity weakness. Low molecular weight heparin (LMWH) should be used for all patients with inability to walk due to any degree of lower extremity paralysis caused by neurological DCS or AGE. Enoxaparin 30 mg, or its equivalent, administered subcutaneously every 12 hours, should be started as soon as possible after injury to reduce the risk of deep venous thrombosis (DVT) and pulmonary embolism in paraplegic patients. Plastic stockings or intermittent pneumatic compression are alternatives, although they are less effective at preventing DVT than LMWH.

20-11.1.4 **Aspirin and Other Non-Steroidal Anti-Inflammatory Drugs.** Routine use of anti-platelet agents in patients with neurological DCS is not recommended, due to concern about worsening hemorrhage in spinal cord or inner ear decompression illness. Use of these agents may also be risky in combat divers who may be required to return to action after treatment of an episode of DCS.

20-11.1.5 **Steroids.** Steroids are no longer recommended for the treatment of DCS. No significant reduction in neurological residuals has been found in clinical studies for DCS adjunctively treated with steroids and elevated blood glucose levels associated with steroid administration may actually worsen the outcome of CNS injury.

20-11.1.6 **Lidocaine.** Lidocaine is not currently recommended for the treatment of any type of DCS.

20-11.1.7 **Environmental Temperature.** For patients with evidence of brain or spinal cord damage, the current evidence recommends aggressive treatment of elevated body temperature. When treating victims of neurological DCS, whenever practical, hot environments that may cause elevation of body temperature above normal should be avoided. The patient's body temperature and vital signs should be monitored regularly.

20-11.2 Arterial Gas Embolism.

20-11.2.1 **Surface Oxygen.** Surface oxygen should be used for all cases of AGE as it is for DCS.

20-11.2.2 **Lidocaine.** Lidocaine has been shown to be useful in the treatment of AGE. If it is to be used clinically, evidence suggests that an appropriate end-point is attainment of a serum concentration suitable for an anti-arrhythmic effect. An intravenous initial dose of 1 mg/kg followed by a continuous infusion of 2-4 mg/minute, will typically produce therapeutic serum concentrations. If an intravenous infusion is not established, intramuscular administration of 4-5 mg/kg will typically produce a therapeutic plasma concentration 15 minutes after dosing, lasting for around 90 minutes. Doses greater than those noted above may be associated with major side effects, including paresthesias, ataxia, and seizures.

20-11.2.3 **Fluids.** The fluid replacement recommendations for the treatment of AGE differ from those of DCS. The pathophysiology of the lesion (pulmonary barotrauma vs. tissue supersaturation with in-situ gas formation) is not the issue. The major difference in the recommendations for fluid therapy in AGE vs. DCS are because divers who suffer AGE may be less dehydrated than divers with DCS, either because they have had a shorter period of immersion or because they have had less bubble-induced endothelial damage. In addition, the CNS injury in AGE may be complicated by cerebral edema and an increased fluid load may worsen this cerebral edema and cause further injury to the diver. If fluids are used, crystalloids are probably the best choice for the reasons previously noted in the section on adjunctive therapy of DCS. Particular care should be taken not to overload the diver with fluids by adjusting IV rates to maintain just an adequate urine output of 0.5cc/kg/hour. A urinary catheter should be inserted in the unconscious patient and urinary output measured.

20-11.2.4 **Anticoagulants.** Anticoagulants should not be used routinely in the treatment of AGE. As noted previously in paragraph 20-11.1.3 on anticoagulants in DCS, Enoxaparin 30 mg, or its equivalent, should be administered subcutaneously every 12 hours, after initial recompression therapy in patients suffering from paralysis to prevent deep venous thrombosis (DVT) and pulmonary embolism.

20-11.2.5 **Aspirin and Other Non-Steroidal Anti-Inflammatory Drugs.** Routine use of anti-platelet agents in patients with AGE is not recommended.

20-11.2.6 **Steroids.** Steroids are no longer recommended for the treatment of AGE. No significant reduction in neurologic residual has been shown with adjunctive treatment with steroids for AGE and elevated blood glucose levels associated with administration of steroids may worsen the outcome of CNS injury.

20-11.3 **Sleeping and Eating.** The only time the patient should be kept awake during recompression treatments is during oxygen breathing periods at depths greater than 30 feet. Travel between decompression stops on Treatment Table 4, 7, and 8 is not a contra-indication to sleeping. While asleep, vital signs (pulse, respiratory rate, blood pressure) should be monitored as the patient's condition dictates. Any significant change would be reason to arouse the patient and ascertain the cause. Food may be taken by chamber occupants at any time. Adequate fluid intake should be maintained as discussed in paragraph 20-7.5.1.

20-12 EMERGENCY MEDICAL EQUIPMENT

Every diving activity shall maintain emergency medical equipment that will be available immediately for use in the event of a diving accident. This equipment is to be in addition to any medical supplies maintained in a medical treatment facility and shall be kept in a kit small enough to carry into the chamber, or in a locker in the immediate vicinity of the chamber.

20-12.1 **Primary and Secondary Emergency Kits.** Because some sterile items may become contaminated as a result of a hyperbaric exposure, it is desirable to have a primary kit for immediate use inside the chamber and a secondary kit from which items that may become contaminated can be locked into the chamber only as needed. The primary emergency kit contains diagnostic and therapeutic equipment that is available immediately when required. This kit shall be inside the chamber during all treatments. The secondary emergency kit contains equipment and medicine that does not need to be available immediately, but can be locked-in when required. This kit shall be stored in the vicinity of the chamber.

The contents of the emergency kits presented here are not meant to be restrictive but are considered the minimum requirement. Additional items may be added to suit local medical preferences.

The Primary Emergency Kit is described in Table 20-7. The Secondary Emergency Kit is described in Table 20-8.

Table 20-7. *Primary Emergency Kit.*

Diagnostic Equipment

- Flashlight and batteries
- Stethoscope
- Otoscope (Ophthalmoscope optional) and batteries
- Sphygmomanometer (aneroid type only, case vented for hyperbaric use)
- Reflex Hammer
- Tuning Fork (512 cps)
- Swab sticks which can be broken for sensory testing
- Tongue depressors
- Thermometer/temperature measurement capability (TempaDOT™ or non-mercury type, high and low reading core temperature thermometer)
- Pulse Oximeter
- Disposable exam gloves (sizes M–L)

Emergency Treatment Equipment and Medications

- Oropharyngeal airways (#4 and #5 Geudel-type)
- Nasal airways (#32F and #34F latex rubber)
- Lidocaine ointment (2% or 5%)
- Self-Inflating Bag-Mask ventilator with medium adult mask
- Suction apparatus with appropriate suction tips (includes whistle tip and Yankauer-type or tonsil suction) Also includes disposable hand operated suction units.
- Large-bore catheter on a needle (12 or 14 gauge) for cricothyrotomy or relief of tension pneumothorax (or alternatively, pre-packaged tension pneumothorax kit or cricothyrotomy kit such as QuickTrach™)
- BD Bard Parker Heimlich Chest Drain Valve (or other device to provide one-way flow of gas out of the chest
- Adhesive tape (2 inch waterproof)
- Elastic-Wrap bandage for a pressure bandage (2- and 4-inch)
- Appropriate Combat Tourniquet
- Bandage Scissors
- #11 knife blade and handle
- Sterile gloves (size 6 – 8)
- Surgical masks (4)
- Sterile 4X4s
- 10% povidone-iodine swabs or wipes
- 1% lidocaine solution
- #21 ga. 11/2-inch needles on 5 cc syringes
- Cravets
- 20 cc syringe

NOTE: One Primary Emergency Kit is required per chamber system, e.g. TRCS requires one.Additional Medical Equipment Authorized for Navy Use (ANU) in a chamber can be found in the Medical Equipment section of the ANU on the NAVSEA website. Contact the Senior Medical Officer at the Navy Experimental Diving Unit for any questions regarding specific pieces of medical equipment for use in the chamber.

Table 20-8. *Secondary Emergency Kit.*

Emergency Airway Equipment

Cuffed endotracheal tubes with adapters (7-9.5mm)
Syringe and sterile water for cuff inflation (10 cc)
Malleable stylet (approx. 12" in length)
Laryngoscope blades (McIntosh #3 and #4, Miller #2 and #3)
Sterile lubricant
Soft-rubber suction catheters
Intubating laryngeal mask airway (disposable LMA Fastrach™ size 4 – 5)
Qualitative end tidal CO_2 detector (colormetric indicator). Additional mechanical verification devices are also authorized (Tomey-type or 50cc catheter tip syringe or equivalent)
Chest tube (or equivalent device)
Cricothyrotomy kit (pre-packaged or equivalent device)
Christmas tree adapter (to connect one-way valve to chest tube)
Curved Kelly forceps

Intravenous Infusion Therapy

Catheter and needle unit, intravenous (16- and 18-gauge - 4 ea)
Adult interosseous infusion device (IO) for rapid vascular access
Intravenous infusion sets (2 standard drip and 2 micro-drip)
Intravenous infusion extension sets with injection ports (2)
Syringes (2, 5, 10 and 30 cc)
Sterile needles (18, 20, and 22 gauge)
3-way stopcocks
Normal saline (1 liter bag (4))
Gauze pads (sterile 2X2s)
Band aids
Arm boards
Venous tourniquet

Miscellaneous

Nasogastric tube
Urinary catheterization set with collection bag (appropriate size (12F–14F) Foley-type sterile catheters)
Disposable Minor Surgical Tray can substitute for items listed below:
- Straight and curved hemostats (2 ea)
- Blunt straight surgical scissors
- Needle driver
- Surgical soap
- Sterile towels
- Sterile gauze pads
- 10% povidone-iodine swabs or wipes
- Cotton Balls
- Assorted scalpel blades and handle
- Assorted suture material (0-silk with and without curved needles)

Sharps disposable box

NOTE 1: Whenever possible, preloaded syringe injection sets should be obtained to avoid the need to vent multi-dose vials or prevent implosion of ampules. Sufficient quantities should be maintained to treat one injured diver.

NOTE 2: One Secondary Emergency Kit is required per chamber system (i.e., TRCS requires one).

NOTE 3: A portable oxygen supply with an E cylinder (approximately 669 liters of oxygen) with a regulator capable of delivering 12 liters of oxygen per minute by mask/reservoir or 2 liters by nasal canula is recommended whenever possible in the event the patient needs to be transported to another facility.

20-12.2 **Portable Monitor-Defibrillator.** All diving activities/commands shall maintain an automated external defibrillator (AED), preferably with heart rhythm visualization capability, from an approved Authorized Medical Allowance List (AMAL). Diving activities with assigned Diving Medical Officer are recommended to augment with a fully capable monitor defibrillator.

CAUTION **AED's are not currently approved for use under pressure (hyperbaric environment) due to electrical safety concerns.**

20-12.3 **Advanced Cardiac Life Support Drugs.** All commands with chambers that participate in area bends watch shall maintain those drugs recommended by the American Heart Association for ACLS. These drugs need to be in sufficient quantities to support an event requiring Advanced Cardiac Life Support. These drugs/equipment are not required to be in every dive kit when multiple chambers/ kits are present in a single command.

In addition, medications for the treatment of anaphylaxis, which can occur related to marine life envenomation, including Epinephrine 1:1000 solution, Diphenhydramine IM or PO and Hydrocortisone Sodium Succinate IV will be maintained in adequate quantities to treat one patient.

NOTE **Some vendors supply pre-packed ACLS kits with automated replenishment programs (examples of which can be found on the Naval Expeditionary Combat Command (NECC) AMAL).**

20-12.4 **Use of Emergency Kits.** Unless adequately sealed against increased atmospheric pressure (i.e., vacuum packed), sterile supplies should be re-sterilized after each pressure exposure; or, if not exposed, at package expiration date. Drugs shall be replaced when their expiration date is reached. Not all drug ampules will withstand pressure.

NOTE **Stoppered multi-dose vials with large air volumes may need to be vented with a needle during pressurization and depressurization and then discarded.**

Both kits should be taken to the recompression chamber or scene of the accident. Each kit is to contain a list of contents and have a tamper evident seal. Each time the kit is opened, it shall be inventoried and each item checked for proper working order and then re-sterilized or replaced as necessary. Unopened kits are inventoried quarterly. Concise instructions for administrating each drug are to be provided in the kit along with current American Heart Association Advanced Cardiac Life-Support Protocols. In untrained hands, many of the items can be dangerous. Remember that as in all treatments **YOUR FIRST DUTY IS TO DO NO HARM.**

20-12.4.1 **Modification of Emergency Kits.** Because the available facilities may differ on board ship, at land-based diving installations, and at diver training or experimental units, the responsible Diving Medical Officer or Diving Medical Technician are authorized to augment the emergency kits to suit the local needs.

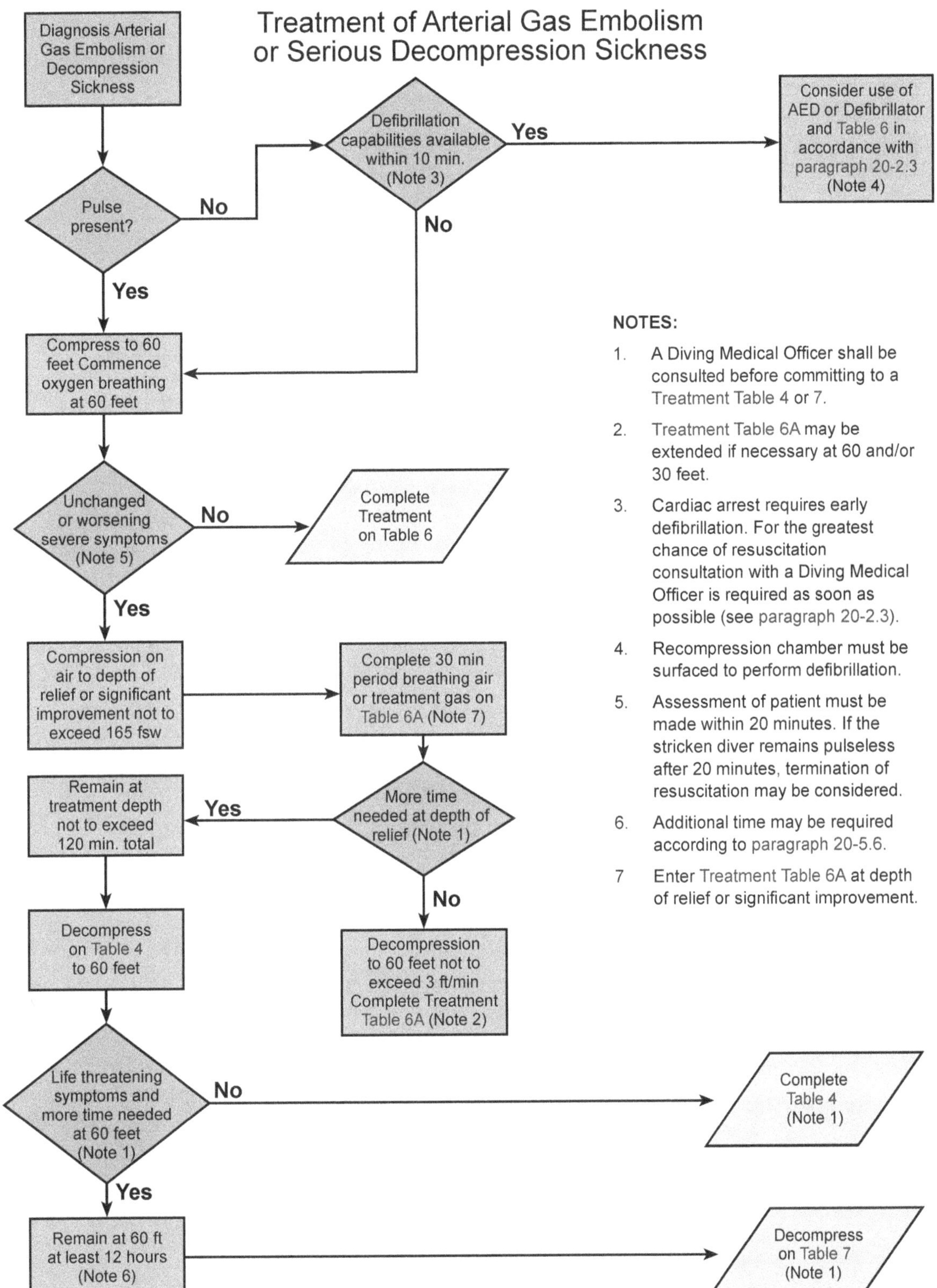

NOTES:

1. A Diving Medical Officer shall be consulted before committing to a Treatment Table 4 or 7.
2. Treatment Table 6A may be extended if necessary at 60 and/or 30 feet.
3. Cardiac arrest requires early defibrillation. For the greatest chance of resuscitation consultation with a Diving Medical Officer is required as soon as possible (see paragraph 20-2.3).
4. Recompression chamber must be surfaced to perform defibrillation.
5. Assessment of patient must be made within 20 minutes. If the stricken diver remains pulseless after 20 minutes, termination of resuscitation may be considered.
6. Additional time may be required according to paragraph 20-5.6.
7. Enter Treatment Table 6A at depth of relief or significant improvement.

Figure 20-1. Treatment of Arterial Gas Embolism or Serious Decompression Sickness.

Treatment of Type I Decompression Sickness

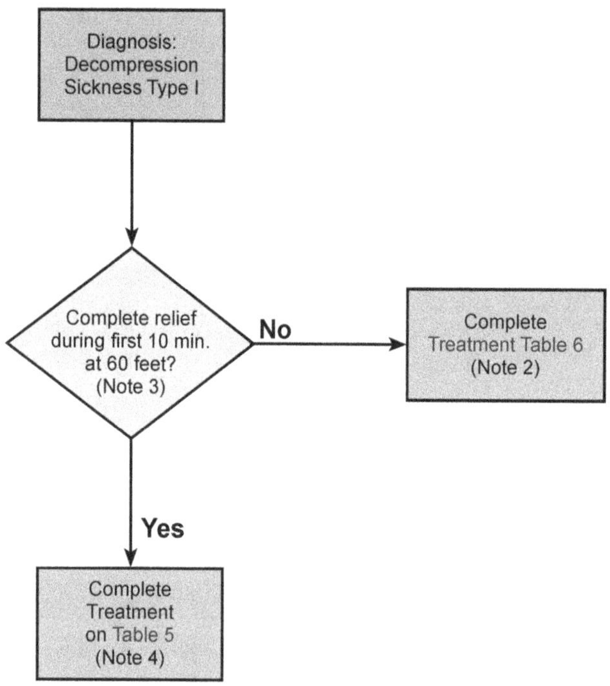

NOTES:

1. If a complete neurological exam was not completed before recompression, treat as a Type II symptom.
2. Treatment Table 6 may be extended up to four additional oxygen-breathing periods, two at 30 feet and/or two at 60 feet.
3. Diving Supervisor may elect to treat on Treatment Table 6.
4. Treatment Table 5 may be extended two oxygen-breathing periods at 30 fsw.

Figure 20-2. Treatment of Type I Decompression Sickness.

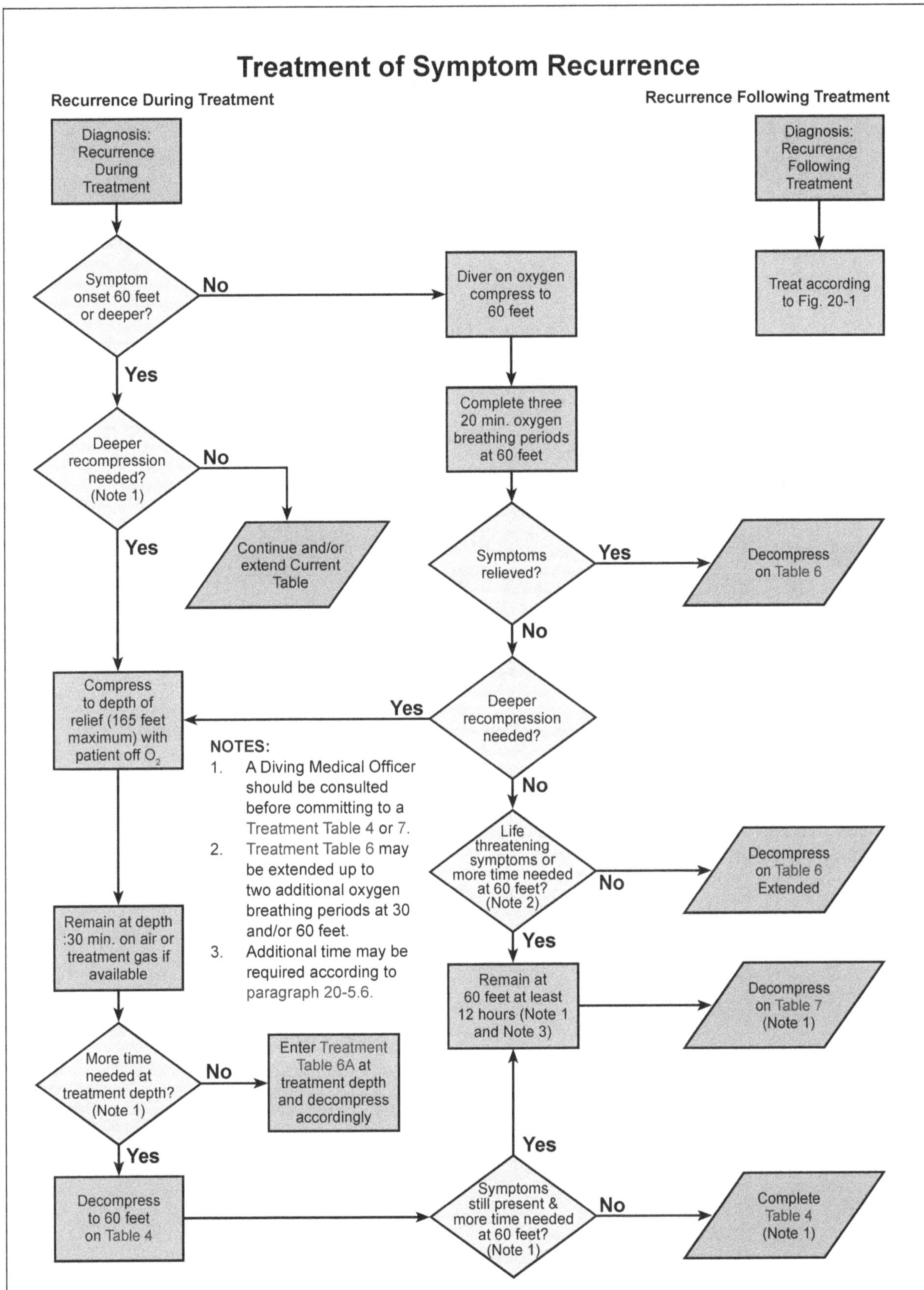

Figure 20-3. Treatment of Symptom Recurrence.

Treatment Table 5

1. Descent rate - 20 ft/min.
2. Ascent rate - Not to exceed 1 ft/min. Do not compensate for slower ascent rates. Compensate for faster rates by halting the ascent.
3. Time on oxygen begins on arrival at 60 feet.
4. If oxygen breathing must be interrupted because of CNS Oxygen Toxicity, allow 15 minutes after the reaction has entirely subsided and resume schedule at point of interruption (see paragraph 20-7.11.1.1)
5. Treatment Table may be extended two oxygen-breathing periods at the 30-foot stop. No air break required between oxygen-breathing periods or prior to ascent.
6. Tender breathes 100 percent O_2 during ascent from the 30-foot stop to the surface. If the tender had a previous hyperbaric exposure in the previous 18 hours, an additional 20 minutes of oxygen breathing is required prior to ascent.

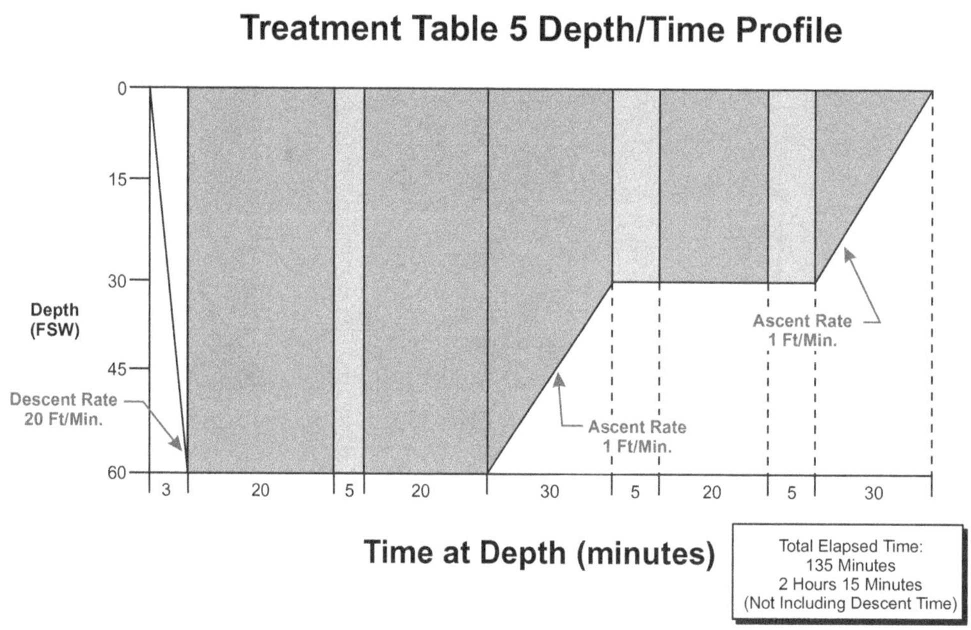

Figure 20-4. Treatment Table 5.

Treatment Table 6

1. Descent rate - 20 ft/min.
2. Ascent rate - Not to exceed 1 ft/min. Do not compensate for slower ascent rates. Compensate for faster rates by halting the ascent.
3. Time on oxygen begins on arrival at 60 feet.
4. If oxygen breathing must be interrupted because of CNS Oxygen Toxicity, allow 15 minutes after the reaction has entirely subsided and resume schedule at point of interruption (see paragraph 20-7.11.1.1).
5. Table 6 can be lengthened up to 2 additional 25-minute periods at 60 feet (20 minutes on oxygen and 5 minutes on air), or up to 2 additional 75-minute periods at 30 feet (15 minutes on air and 60 minutes on oxygen), or both.
6. Tender breathes 100 percent O_2 during the last 30 min. at 30 fsw and during ascent to the surface for an unmodified table or where there has been only a single extension at 30 or 60 feet. If there has been more than one extension, the O_2 breathing at 30 feet is increased to 60 minutes. If the tender had a hyperbaric exposure within the past 18 hours an additional 60-minute O_2 period is taken at 30 feet.

Treatment Table 6 Depth/Time Profile

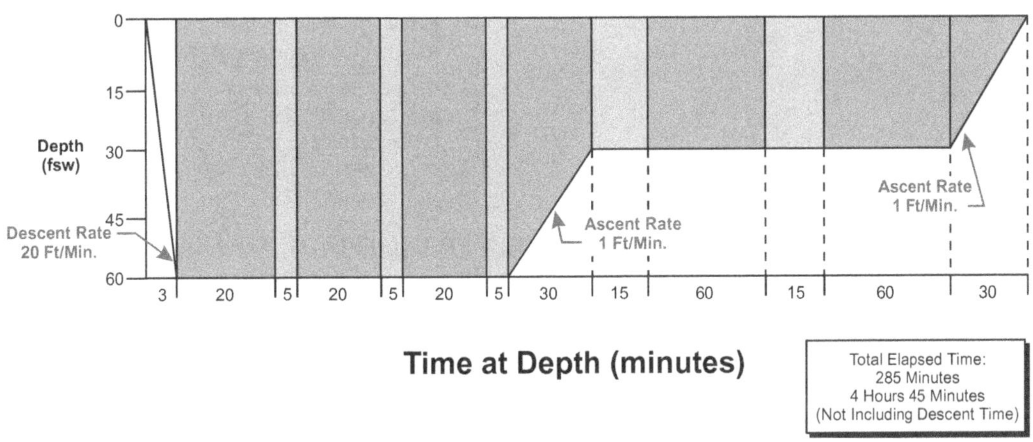

Figure 20-5. Treatment Table 6.

Treatment Table 6A

1. Descent rate - 20 ft/min.
2. Ascent rate - 165 fsw to 60 fsw not to exceed 3 ft/min, 60 fsw and shallower, not to exceed 1 ft/min. Do not compensate for slower ascent rates. Compensate for faster rates by halting the ascent.
3. Time at treatment depth does not include compression time.
4. Table begins with initial compression to depth of 60 fsw. If initial treatment was at 60 feet, up to 20 minutes may be spent at 60 feet before compression to 165 fsw. Contact a Diving Medical Officer.
5. If a chamber is equipped with a high-O_2 treatment gas, it may be administered at 165 fsw and shallower, not to exceed 3.0 ata O_2 in accordance with paragraph 20-7.10. Treatment gas is administered for 25 minutes interrupted by 5 minutes of air. Treatment gas is breathed during ascent from the treatment depth to 60 fsw.
6. Deeper than 60 feet, if treatment gas must be interrupted because of CNS oxygen toxicity, allow 15 minutes after the reaction has entirely subsided before resuming treatment gas. The time off treatment gas is counted as part of the time at treatment depth. If at 60 feet or shallower and oxygen breathing must be interrupted because of CNS oxygen toxicity, allow 15 minutes after the reaction has entirely subsided and resume schedule at point of interruption (see paragraph 20-7.11.1.1).
7. Table 6A can be lengthened up to 2 additional 25-minute periods at 60 feet (20 minutes on oxygen and 5 minutes on air), or up to 2 additional 75-minute periods at 30 feet (60 minutes on oxygen and 15 minutes on air), or both.
8. Tender breathes 100 percent O_2 during the last 60 minutes at 30 fsw and during ascent to the surface for an unmodified table or where there has been only a single extension at 30 or 60 fsw. If there has been more than one extension, the O_2 breathing at 30 fsw is increased to 90 minutes. If the tender had a hyperbaric exposure within the past 18 hours, an additional 60 minute O_2 breathing period is taken at 30 fsw.
9. If significant improvement is not obtained within 30 minutes at 165 feet, consult with a Diving Medical Officer before switching to Treatment Table 4.

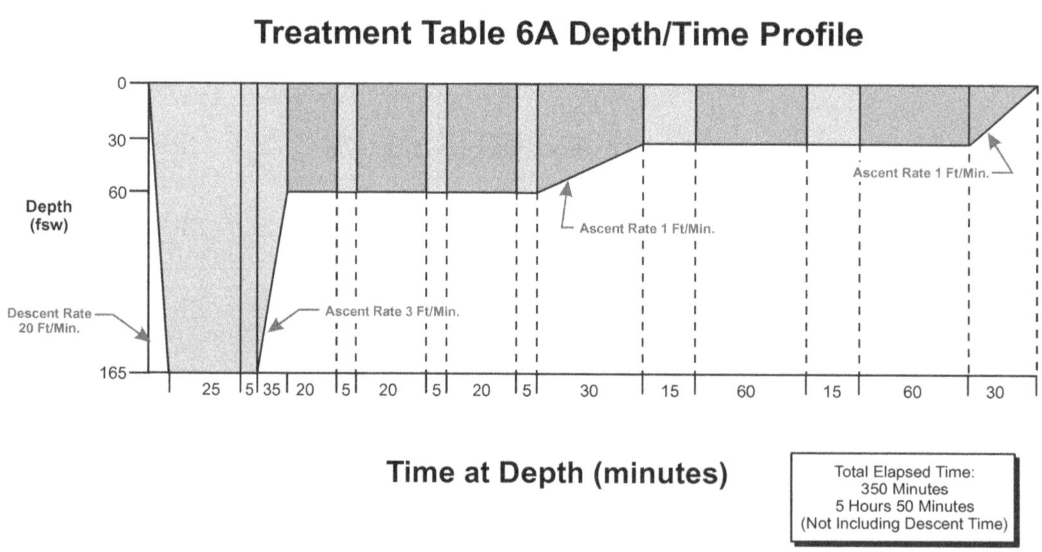

Figure 20-6. Treatment Table 6A.

Treatment Table 4

1. Descent rate - 20 ft/min.
2. Ascent rate - 1 ft/min.
3. Time at 165 feet includes compression.
4. If only air is available, decompress on air. If oxygen is available, patient begins oxygen breathing upon arrival at 60 feet with appropriate air breaks. Both tender and patient breathe oxygen beginning 2 hours before leaving 30 feet. (see paragraph 20-5.5).
5. Ensure life-support considerations can be met before committing to a Table 4. (see paragraph 20-7.5) Internal chamber temperature should be below 85° F.
6. If oxygen breathing is interrupted, no compensatory lengthening of the table is required.
7. If switching from Treatment Table 6A or 3 at 165 feet, stay a maximum of 2 hours at 165 feet before decompressing.
8. If the chamber is equipped with a high-O_2 treatment gas, it may be administered at 165 fsw, not to exceed 3.0 ata O_2. Treatment gas is administered for 25 minutes interrupted by 5 minutes of air.

Treatment Table 4 Depth/Time Profile

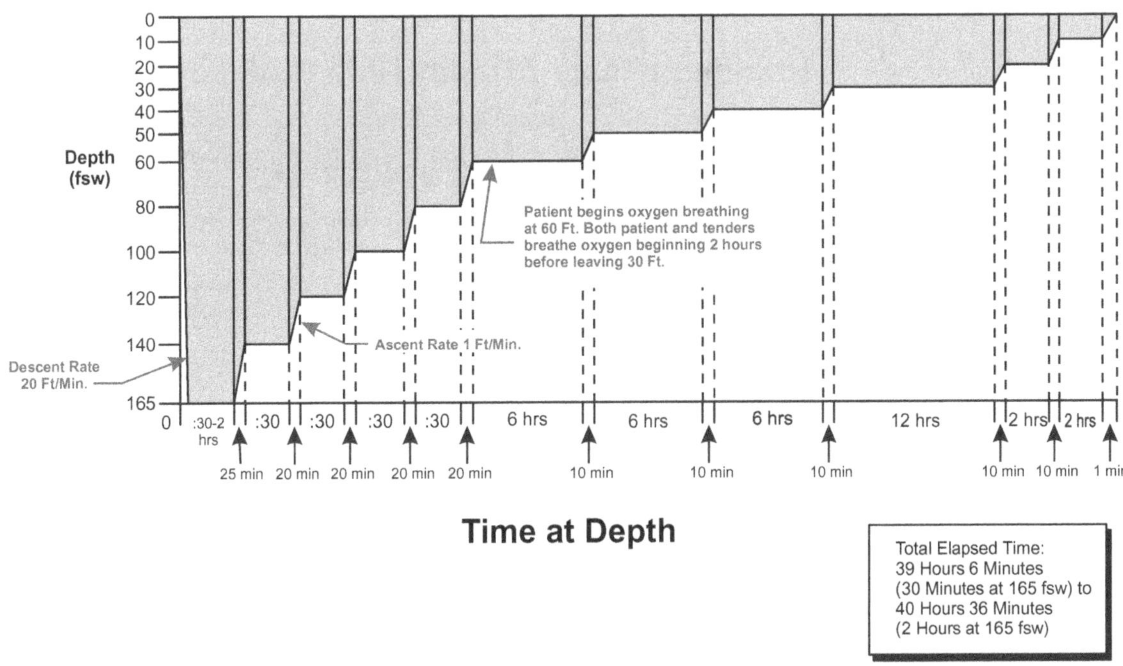

Figure 20-7. Treatment Table 4.

Treatment Table 7

1. Table begins upon arrival at 60 feet. Arrival at 60 feet is accomplished by initial treatment on Table 6, 6A or 4. If initial treatment has progressed to a depth shallower than 60 feet, compress to 60 feet at 20 ft/min to begin Table 7.
2. Maximum duration at 60 feet is unlimited. Remain at 60 feet a minimum of 12 hours unless overriding circumstances dictate earlier decompression.
3. Patient begins oxygen breathing periods at 60 feet. Tender need breathe only chamber atmosphere throughout. If oxygen breathing is interrupted, no lengthening of the table is required.
4. Minimum chamber O_2 concentration is 19 percent. Maximum CO_2 concentration is 1.5 percent SEV (11.4 mmHg). Maximum chamber internal temperature is 85°F (paragraph 20-7.5).
5. Decompression starts with a 2-foot upward excursion from 60 to 58 feet. Decompress with stops every 2 feet for times shown in profile below. Ascent time between stops is approximately 30 seconds. Stop time begins with ascent from deeper to next shallower step. Stop at 4 feet for 4 hours and then ascend to the surface at 1 ft/min.
6. Ensure chamber life-support requirements can be met before committing to a Treatment Table 7.
7. A Diving Medical Officer should be consulted before committing to this treatment table.

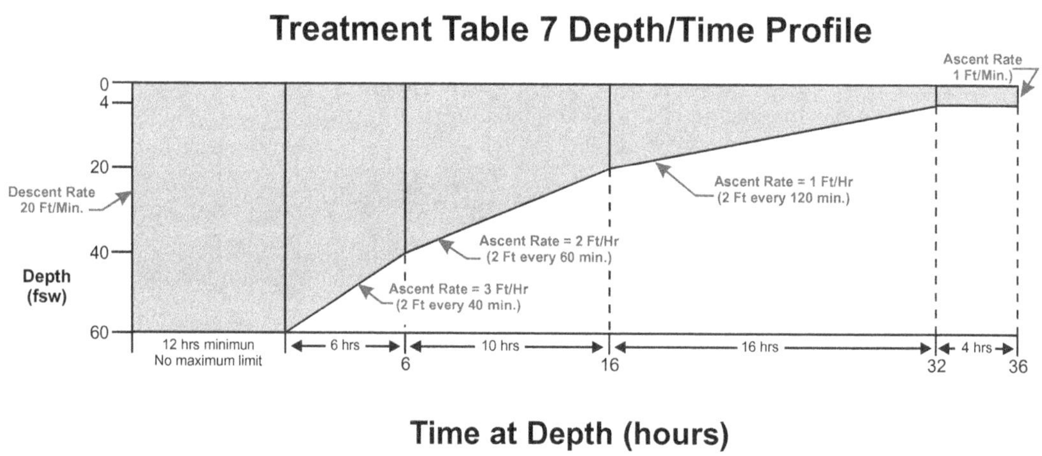

Figure 20-8. Treatment Table 7.

Treatment Table 8

1. Enter the table at the depth which is exactly equal to or next greater than the deepest depth attained in the recompression. The descent rate is as fast as tolerable.
2. The maximum time that can be spent at the deepest depth is shown in the second column. The maximum time for 225 fsw is 30 minutes; for 165 fsw, 3 hours. For an asymptomatic diver, the maximum time at depth is 30 minutes for depths exceeding 165 fsw and 2 hours for depths equal to or shallower than 165 fsw.
3. Decompression is begun with a 2-fsw reduction in pressure if the depth is an even number. Decompression is begun with a 3-fsw reduction in pressure if the depth is an odd number. Subsequent stops are carried out every 2 fsw. Stop times are given in column three. The stop time begins when leaving the previous depth. Ascend to the next stop in approximately 30 seconds.
4. Stop times apply to all stops within the band up to the next quoted depth. For example, for ascent from 165 fsw, stops for 12 minutes are made at 162 fsw and at every two-foot interval to 140 fsw. At 140 fsw, the stop time becomes 15 minutes. When traveling from 225 fsw, the 166-foot stop is 5 minutes; the 164-foot stop is 12 minutes. Once begun, decompression is continuous. For example, when decompressing from 225 feet, ascent is not halted at 165 fsw for 3 hours. However, ascent may be halted at 60 fsw and shallower for any desired period of time.
5. While deeper than 165 fsw, a helium-oxygen mixture with 16-36 percent oxygen may be breathed by mask to reduce narcosis. A 64/36 helium-oxygen mixture is the preferred treatment gas. At 165 fsw and shallower, a HeO_2 or N_2O_2 mix with a ppO_2 not to exceed 3.0 ata may be given to the diver as a treatment gas. At 60 fsw and shallower, pure oxygen may be given to the divers as a treatment gas. For all treatment gases (HeO_2, N_2O_2, and O_2), a schedule of 25 minutes on gas and 5 minutes on chamber air should be followed for a total of four cycles. Additional oxygen may be given at 60 fsw after a 2-hour interval of chamber air. See Treatment Table 7 for guidance. If high O_2 breathing is interrupted, no lengthening of the table is required.
6. To avoid loss of the chamber seal, ascent may be halted at 4 fsw and the total remaining stop time of 240 minutes taken at this depth. Ascend directly to the surface upon completion of the required time.
7. Total ascent time from 225 fsw is 56 hours, 29 minutes. For a 165-fsw recompression, total ascent time is 53 hours, 52 minutes, and for a 60-fsw recompression, 36 hours, 0 minutes.

Depth (fsw)	Max Time at Initial Treatment Depth (hours)	2-fsw Stop Times (minutes)
225	0.5	5
165	3	12
140	5	15
120	8	20
100	11	25
80	15	30
60	Unlimited	40
40	Unlimited	60
20	Unlimited	120

Figure 20-9. Treatment Table 8.

Treatment Table 9

1. Descent rate - 20 ft/min.
2. Ascent rate - 20 ft/min. Rate may be slowed to 1 ft/min depending upon the patient's medical condition.
3. Time at 45 feet begins on arrival at 45 feet.
4. If oxygen breathing must be interrupted because of CNS Oxygen Toxicity, oxygen breathing may be restarted 15 minutes after all symptoms have subsided. Resume schedule at point of interruption (see paragraph 20-7.11.1.1).
5. Tender breathes 100 percent O_2 during last 15 minutes at 45 feet and during ascent to the surface regardless of ascent rate used.
6. Patient may breathe air or oxygen during ascent.
7. If patient cannot tolerate oxygen at 45 feet, this table can be modified to allow a treatment depth of 30 feet. The oxygen breathing time can be extended to a maximum of 3 to 4 hours.

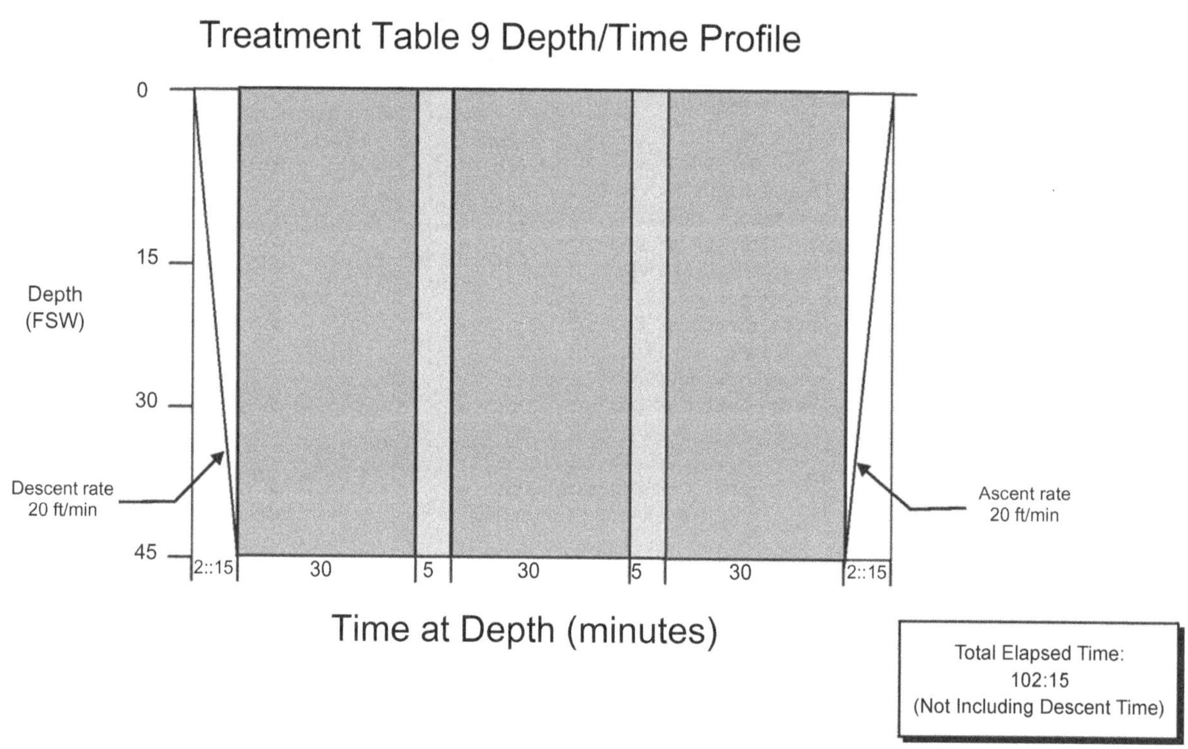

Figure 20-10. Treatment Table 9.

Air Treatment Table 1A

1. Descent rate - 20 ft/min.
2. Ascent rate - 1 ft/min.
3. Time at 100 feet includes time from the surface.

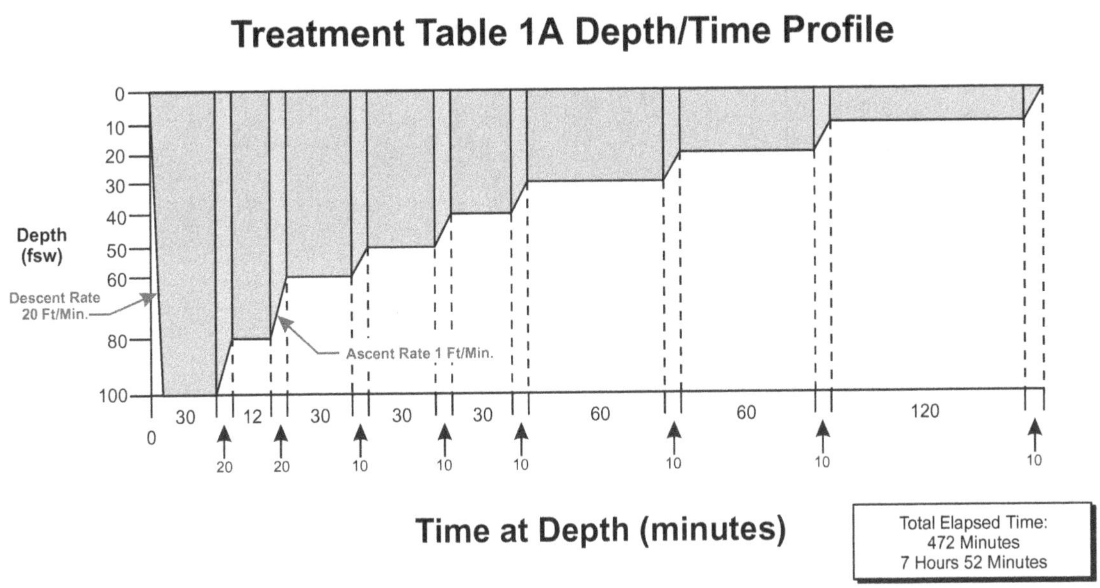

Figure 20-11. Air Treatment Table 1A.

Air Treatment Table 2A

1. Descent rate - 20 ft/min.
2. Ascent rate - 1 ft/min.
3. Time at 165 feet includes time from the surface.

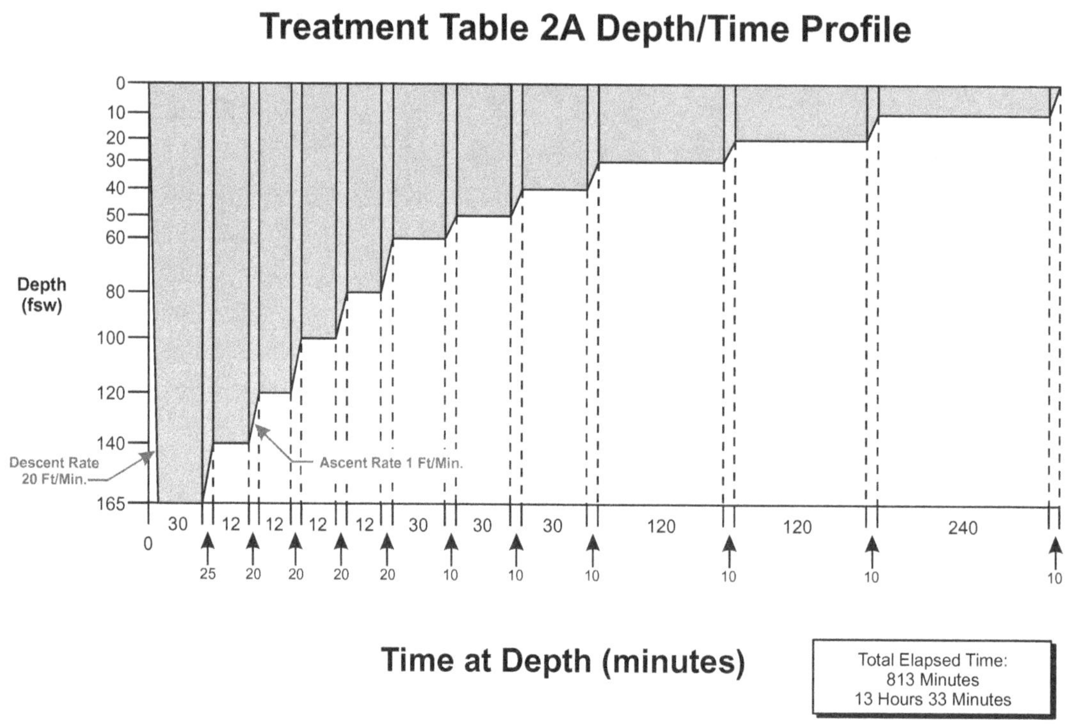

Figure 20-12. Air Treatment Table 2A.

Air Treatment Table 3

1. Descent rate - 20 ft/min.
2. Ascent rate - 1 ft/min.
3. Time at 165 feet-includes time from the surface.

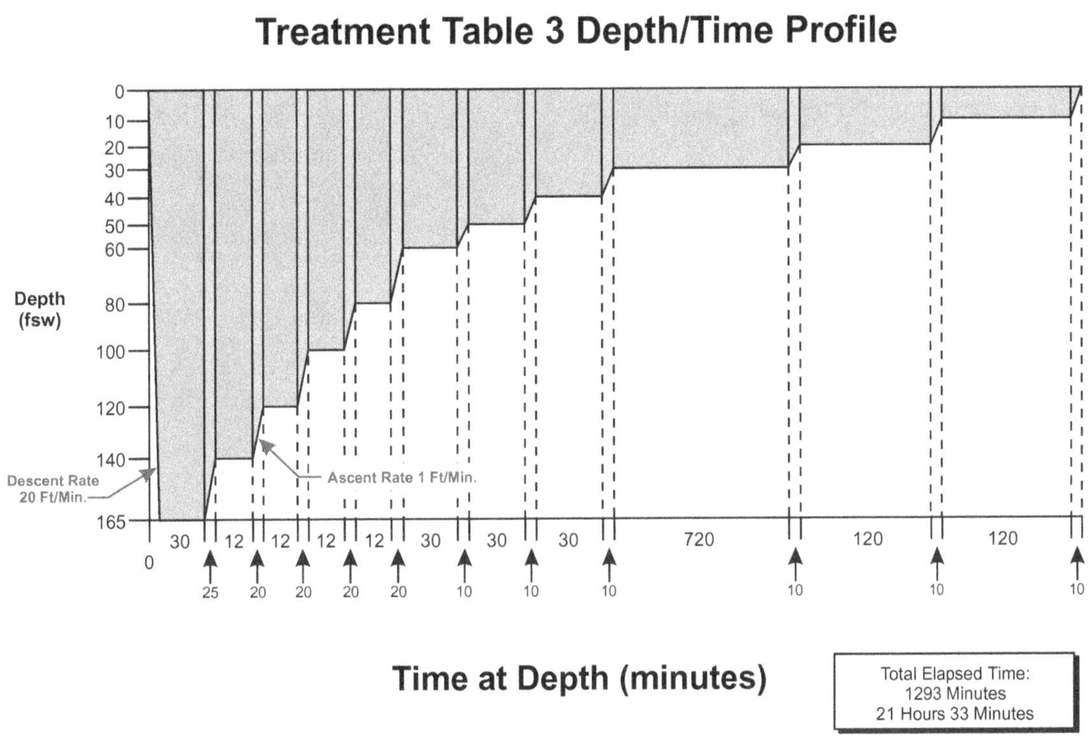

Figure 20-13. Air Treatment Table 3.

PAGE LEFT BLANK INTENTIONALLY

CHAPTER 21

Recompression Chamber Operation

21-1 INTRODUCTION

21-1.1 Purpose. This chapter will familiarize personnel with the maintenance and operational requirements for recompression chambers.

21-1.2 Scope. Recompression chambers are used for the treatment of decompression sickness and arterial gas embolism, for surface decompression, and for administering pressure tests to prospective divers. Recompression chambers equipped for hyperbaric administration of oxygen are also used in medical facilities for hyperbaric treatment of carbon monoxide poisoning, gas gangrene, and other diseases. A recompression chamber is required on site for surface-supplied air decompression dives deeper than 130 fsw and for all surface-supplied decompression helium-oxygen dives.

21-1.3 Chamber Definitions. Double-lock chambers are used because they permit tending personnel and supplies to enter and leave the chamber during treatment. Where stated:

- **On-station chamber** is defined as a certified and ready chamber at the dive site.
- **On-site chamber** is defined as a certified and ready chamber accessible within 30 minutes of the dive site by available transportation.
- **Emergency chamber** is defined as the closest recompression chamber available when a chamber is not required on station or on site. A non-certified chamber may be used if the Diving Supervisor is of the opinion that it is safe to use.

21-2 DESCRIPTION

Most chamber-equipped U.S. Navy units will have one of seven commonly provided chambers. They are:

1. Double-lock, 200-psig, 425-cubic-foot steel chamber (Figure 21-1).
2. Recompression Chamber Facility: RCF 6500 (Figure 21-2).
3. Recompression Chamber Facility: RCF 5000 (Figure 21-3).
4. Double-lock, 100-psig, 202-cubic-foot steel chamber (ARS 50 class and Modernized) (Figure 21-4 and Figure 21-5).
5. Standard Navy Double Lock Recompression Chamber System (SNDLRCS) (Figure 21-6).

6. Transportable Recompression Chamber System (TRCS) (Figure 21-7, Figure 21-8, Figure 21-9).

7. Fly-Away Recompression Chamber (FARCC) (Figure 21-10, Figure 21-11, Figure 21-12).

Select U.S. Navy units have a unique treatment option called the Emergency Evacuation Hyperbaric Stretcher (EEHS). The EEHS has a single lock and allows a patient to be administered oxygen at 60 feet while in transport to a recompression chamber. However, it does not provide hands-on access to the patient and therefore does not qualify as an on-site or on-station recompression chamber.

21-2.1 **Basic Chamber Components.** The basic components of a recompression chamber are much the same from one model to another. The basic components consist of the pressure vessel itself, an air supply and exhaust system, a pressure gauge, and a built-in breathing system (BIBS) to supply oxygen to the patient. Additional components may include oxygen, carbon dioxide, temperature and humidity monitors, carbon dioxide scrubbers, additional BIBS systems for air and treatment gases other than oxygen, a BIBS overboard dump system, and a heating/cooling system. Collectively these systems must be able to impose and maintain a pressure equivalent to a depth of 165 fsw (6 atmospheres absolute) on the diver. Double-lock chambers are used because they permit tending personnel and supplies to enter and leave the chamber during treatment.

The piping and valving on some chambers is arranged to permit control of the air supply and the exhaust from either the inside or the outside of the chamber. Controls on the outside must be able to override the inside controls in the event of a problem inside the chamber. The usual method for providing this dual-control capability is through the use of two separate systems. The first, consisting of a supply line and an exhaust line, can only be controlled by valves that are outside of the chamber. The second air supply/exhaust system has a double set of valves, one inside and one outside the chamber. This arrangement permits the tender to regulate descent or ascent from within the chamber, but always subject to final control by outside personnel.

21-2.2 **Fleet Modernized Double-Lock Recompression Chamber.** Modernized chambers (Figure 21-5) have carbon dioxide and oxygen monitors, a CO_2 scrubber system, a Built-In Breathing System (BIBS), and an oxygen dump system which together reduce the ventilation requirements. These chambers also include a chamber environment control system that regulates humidity and temperature.

21-2.3 **Recompression Chamber Facility (RCF).** The RCF series 6500 and 5000 (Figures 21-2 and 21-3) consists of two sizes of standard double lock steel chambers, each with a medical lock and easy occupant access. The RCF 6500 is capable of treating up to 12 occupants while the RCF 5000 is capable of treating 7 occupants. The systems are installed in a facility to support training, surface decompression, recompression treatment, and medical treatment operations. Each RCF includes primary and secondary air supplies comprised of compressors, purification, and storage for chamber pressurization and ventilation along with oxygen, mix treatment gas, and emergency air supply to the BIBS system. Each RCF has an atmospheric conditioning system that provides internal atmospheric scrubbing and monitoring along with temperature and humidity controls for long term treatment, gas management, and patient comfort. The RCF includes gas supply monitoring, a fire extinguishing system, ground fault interruption and emergency power. The RCF 6500 is equipped with a NATO mating flange. Both series have extra penetrations for auxiliary equipment such as patient treatment monitoring and hoods.

21-2.4 **Standard Navy Double Lock Recompression Chamber System (SNDLRCS).** The SNDLRCS (Figure 21-6) consists of a Standard Navy Double Lock (SNDL) recompression chamber and a gas supply system housed within an International Organization for Standards (ISO) container. The system is capable of supporting surface decompression, medical treatment, and training operations. Air is supplied to the system using a Air Flask Rack Assembly (AFRA) which is almost identical to the Air Supply Rack Assembly (ASRA) used in supporting a FADS 3 DLSS. Oxygen is provided by four (4) cylinders that are secured to the interior bulkhead of the ISO container. If an external supply of mixed gas is available it can also be supplied to the chamber BIBS supply.

The SNDL is a 54" diameter, double lock recompression chamber. It is outfitted with a stretcher, BIBS, gas monitoring systems, lights, and an environmental conditioning system. The chamber can comfortably accommodate 4 divers in the inner lock and 3 divers in the outer lock.

The ISO container houses the gas supply systems and the chamber. It also provides a shelter from environmental elements for the Outside Tenders and Diving Supervisor to conduct treatments. The container is both heated and air conditioned as required and also includes a fold-down desktop, a cabinet, lighting, and a vestibule.

21-2.5 **Transportable Recompression Chamber System (TRCS).** The TRCS (Figure 21-7) consists of two pressure chambers. One is a conical-shaped chamber (Figure 21-8) called the Transportable Recompression Chamber, and the other is a cylindrical shaped vessel (Figure 21-9) called the Transfer Lock (TL). The two chambers are capable of being connected by means of a freely rotating NATO female flange coupling.

The TRCS is supplied with a Compressed Air and Oxygen System (CAOS) consisting of lightweight air and oxygen racks of high pressure flasks, as well as a means of reducing the oxygen supply pressure. The chamber is capable of administering oxygen and mixed gas via BIBS.

When a recompression chamber is required on site per Figure 6-14, or surface decompression dives are planned, the full TRCS system (including both TRC and TL) shall be on site.

When a recompression chamber is not required on site per Figure 6-14, the inner lock (TRC) may be used for emergency recompression treatment.

21-2.6 **Fly Away Recompression Chamber (FARCC).** This chamber system consists of a 60-inch double lock modernized chamber in a 20' x 8' x 8' milvan (Figure 21-10 and Figure 21-11). The Fly Away Recompression Chamber (FARCC) also includes a life support skid (Figure 21-12). In addition, a stand-alone generator is provided for remote site power requirements.

21-2.7 **Emergency Evacuation Hyperbaric Stretcher (EEHS).** The Emergency Evacuation Hyperbaric Stretcher (EEHS) is a manually-portable single patient hyperbaric tube to be used to transport a diving or disabled submarine casualty from an accident site to a treatment facility while under pressure. The EEHS does not replace a recompression chamber, but is used in conjunction with a chamber. The EEHS is small enough to allow transfer of a patient, under pressure, into or out of many shore based recompression chambers owned by both the DOD, and civilian medical organizations.

21-2.8 **Standard Features.** Recompression chambers must be equipped with a means for delivering breathing oxygen to the personnel in the chamber. The inner lock should be provided with connections for demand-type oxygen inhalators. Oxygen can be furnished through a pressure reducing manifold connected with supply cylinders outside the chamber.

21-2.8.1 **Labeling.** All lines should be identified and labeled to indicate function, content and direction of flow. The color coding in Table 21-1 should be used.

21-2.8.2 **Inlet and Exhaust Ports.** Optimum chamber ventilation requires separation of the inlet and exhaust ports within the chamber. Exhaust ports must be provided with a guard device to prevent accidental injury when they are open.

21-2.8.3 **Pressure Gauges.** Chambers must be fitted with appropriate pressure gauges. These gauges, marked to read in feet of seawater (fsw), must be calibrated or compared as described in the applicable Planned Maintenance System (PMS) to ensure accuracy in accordance with the instructions in Chapter 4.

Table 21-1. *Recompression Chamber Line Guide.*

Function	Designation	Color Code
Helium	HE	Buff
Oxygen	OX	Green
Helium-Oxygen Mix	HE-OX	Buff & Green
Nitrogen	N	Light Gray
Nitrogen Oxygen Mix	N-OX	Light Gray & Green
Exhaust	E	Silver
Air (Low Pressure)	ALP	Black
Air (High Pressure)	AHP	Black
Chilled Water	CW	Blue & White
Hot Water	HW	Red & White
Potable Water	PW	Blue
Fire Fighting Material	FP	Red

21-2.8.4 **Relief Valves.** Recompression chambers should be equipped with pressure relief valves in each manned lock. Chambers that do not have latches (dogs) on the doors are not required to have a relief valve on the outer lock. The relief valves shall be set in accordance with PMS. In addition, all chambers shall be equipped with a gag valve, located between the chamber pressure hull and each relief valve. This gag valve shall be a quick acting, ball-type valve, sized to be compatible with the relief valve and its supply piping. The gag valve shall be safety wired in the open position.

21-2.8.5 **Communications System.** Chamber communications are provided through a diver's intercommunication system, with the dual microphone/speaker unit in the chamber and the surface unit outside. The communication system should be arranged so that personnel inside the chamber need not interrupt their activities to operate the system. The backup communications system may be provided by a set of standard sound-powered telephones. The press-to-talk button on the set inside the chamber can be taped down, thus keeping the circuit open.

21-2.8.6 **Lighting Fixtures.** Consideration should be given to installation of a low-level lighting fixture (on a separate circuit), which can be used to relieve the patient of the heat and glare of the main lights. Emergency lights for both locks and an external control station are mandatory. No electrical equipment, other than that authorized within the scope of certification or as listed in the NAVSEA Authorized for Navy Use (ANU) List, is allowed inside the chamber. Because of the possibility of fire or explosion when working in an oxygen or compressed air atmosphere, all electrical wiring and equipment used in a chamber shall meet required specifications.

Double-Lock Steel Recompression Chamber

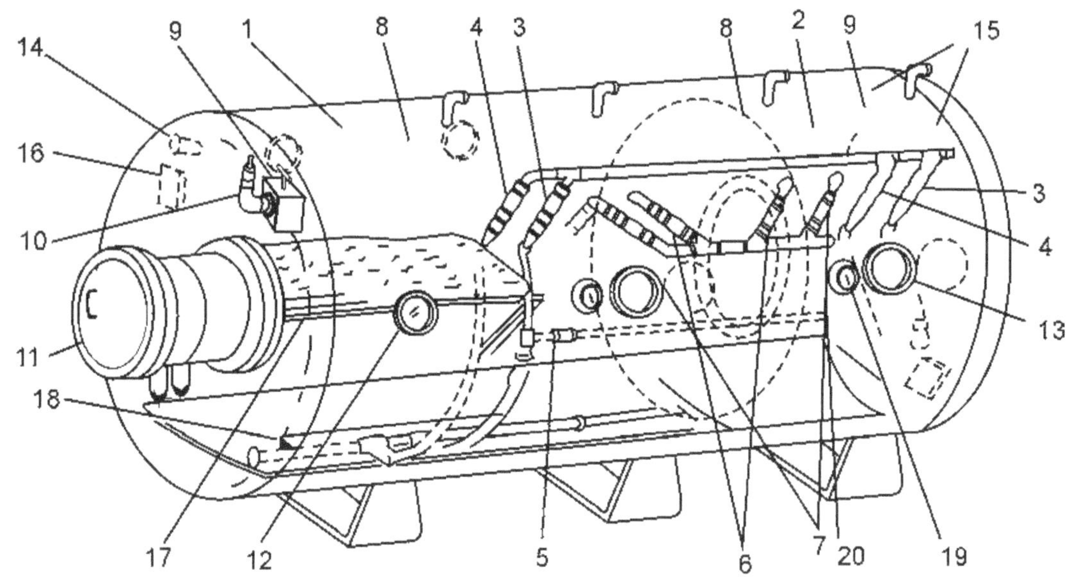

1. Inner Lock
2. Outer Lock
3. Air Supply – Two-Valve
4. Air Supply – One-Valve
5. Main Lock Pressure Equalizing Valve
6. Exhaust – Two-Valve
7. Exhaust – One-Valve
8. Oxygen Manifold
9. Relief Gag Valve (1 each lock)
10. Relief Valve – 110 psig
11. Medical Lock 18-Inch Diameter
12. View Port – Inner Lock (4)
13. View Port – Outer Lock (2)
14. Lights – Inner Lock 40 Watt (4)
15. Lights – Outer Lock 40 Watt
16. Transmitter/Receiver
17. Berth – 2′6″ × 6′6″
18. Bench
19. Pressure Gauge – Outside (2 each lock)
20. Pressure Gauge – Inside (1 each lock)

Original Design Pressure – 200 psig
Original Hydrostatic Test Pressure – 400 psig
Maximum Operating Pressure – 100 psig

Figure 21-1. Double-Lock Steel Recompression Chamber.

Recompression Chamber Facility: RCF 6500

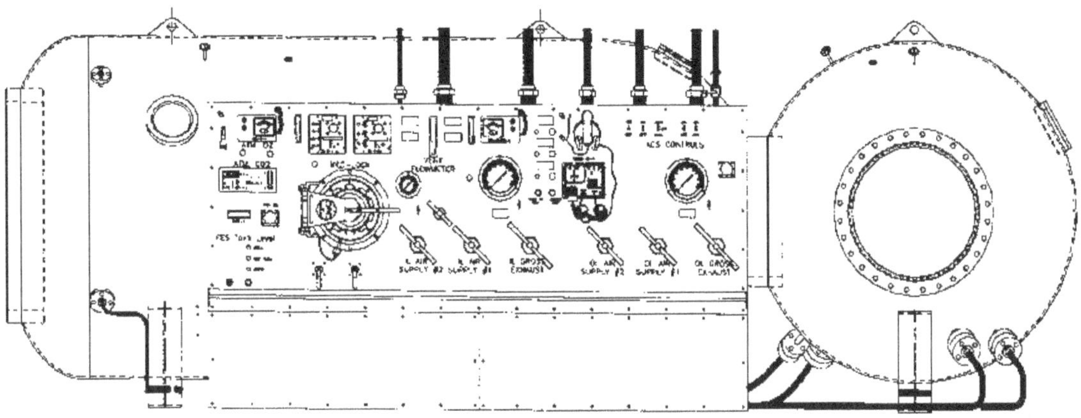

Design Pressure: 110 psig
Length: 21' 3"
Height: 7' 6"
Internal Volume (OL): 144 ft^3
Door Opening (OL): 30"

Design Temperature: 0-125°F
Diameter: 6' 6"
Height: 7' 6"
Internal Volume (IL): 440 ft^3
Door Opening (IL): 48"

Viewports: 6 @ 8" diameter Clear Opening (including 1 video port)
Medlock: 18" diameter X 20" long mounted in console with ASME Quick Actuating Enclosure
Mating Flange: NATO per STANAG 1079
Atmospheric Monitoring: Oxygen, Carbon Dioxide, Temperature
Temperature Monitoring: External Heater/Chiller with internal Blower
Scrubber: Magnetically driven, replaceable canister
BIBS: 8 masks in IL, 4 masks in OL, automatic switching with block & bleed for Oxygen/Nitrox or Heliox/Air, overboard dump, and Oxygen analysis of supply gas
Principal Communications: AC Powered Speaker/Headset w/battery backup
Secondary Communications: Sound Powered Phone
Furnishing: Two 7' Bunks, One 5' 6" Bench, One 18" X 18" Bench
Lighting: 4 Lights in IL, 2 Lights in OL
Gas Pressurization Controls: Primary and secondary air
Air Ventilation Controls: Gross vent and fine vent (with flow meter)
Fire Extinguishing System: 2 Hand Held Hoses in IL, 1 in OL

Figure 21-2. Recompression Chamber Facility: RCF 6500.

Recompression Chamber Facility: RCF 5000

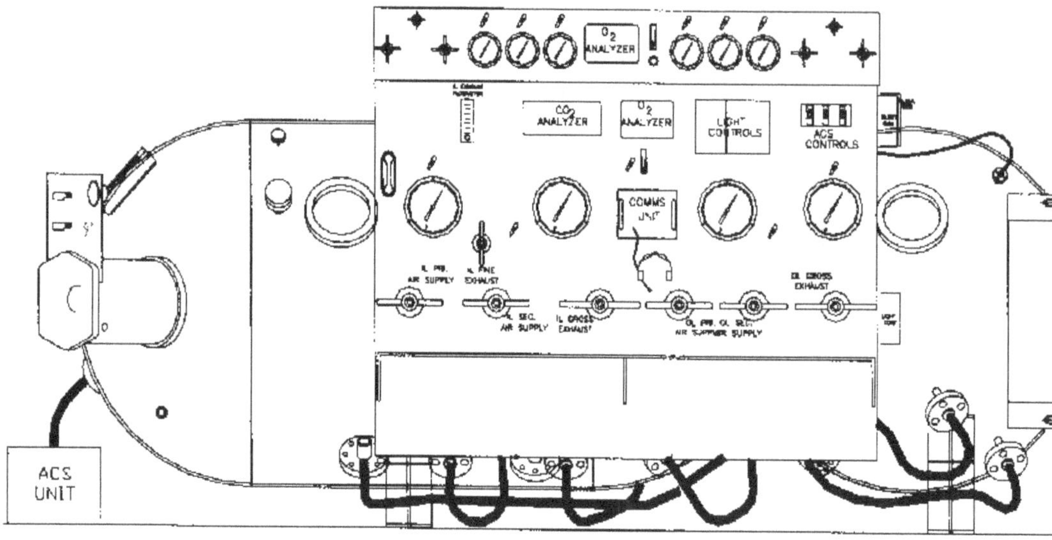

Design Pressure: 110 psig
Length: 14' 8"
Height: 5' 7"
Internal Volume (OL): 61 ft^3
Door Opening (OL): 30"

Design Temperature: 0-125°F
Diameter: 5'
Weight: 9,300 lbs.
Internal Volume (IL): 162 ft^3

Viewports: 6 @ 8" diameter Clear Opening (including 1 video port)
Medlock: 18" diameter X 20" long mounted in console with ASME Quick Actuating Enclosure
Mating Flange: NATO per STANAG 1079
Atmospheric Monitoring: Oxygen, Carbon Dioxide, Temperature
Temperature Monitoring: External Heater/Chiller with internal Blower
Scrubber: Magnetically driven, replaceable canister
BIBS: 4 masks in IL, 3 masks in OL, overboard dump, & Oxygen analysis of supply gas
Principal Communications: AC Powered Speaker/Headset w/battery backup
Secondary Communications: Sound Powered Phone
Furnishing: One Bunks, One Bench
Lighting: 2 Lights in IL, 1 Lights in OL
Gas Pressurization Controls: Primary and secondary air
Air Ventilation Controls: Gross vent and fine vent (with flow meter)
Fire Extinguishing System: Hyperbaric extinguisher

Figure 21-3. Recompression Chamber Facility: RCF 5000.

ARS 50 Class Double-Lock Recompression Chamber

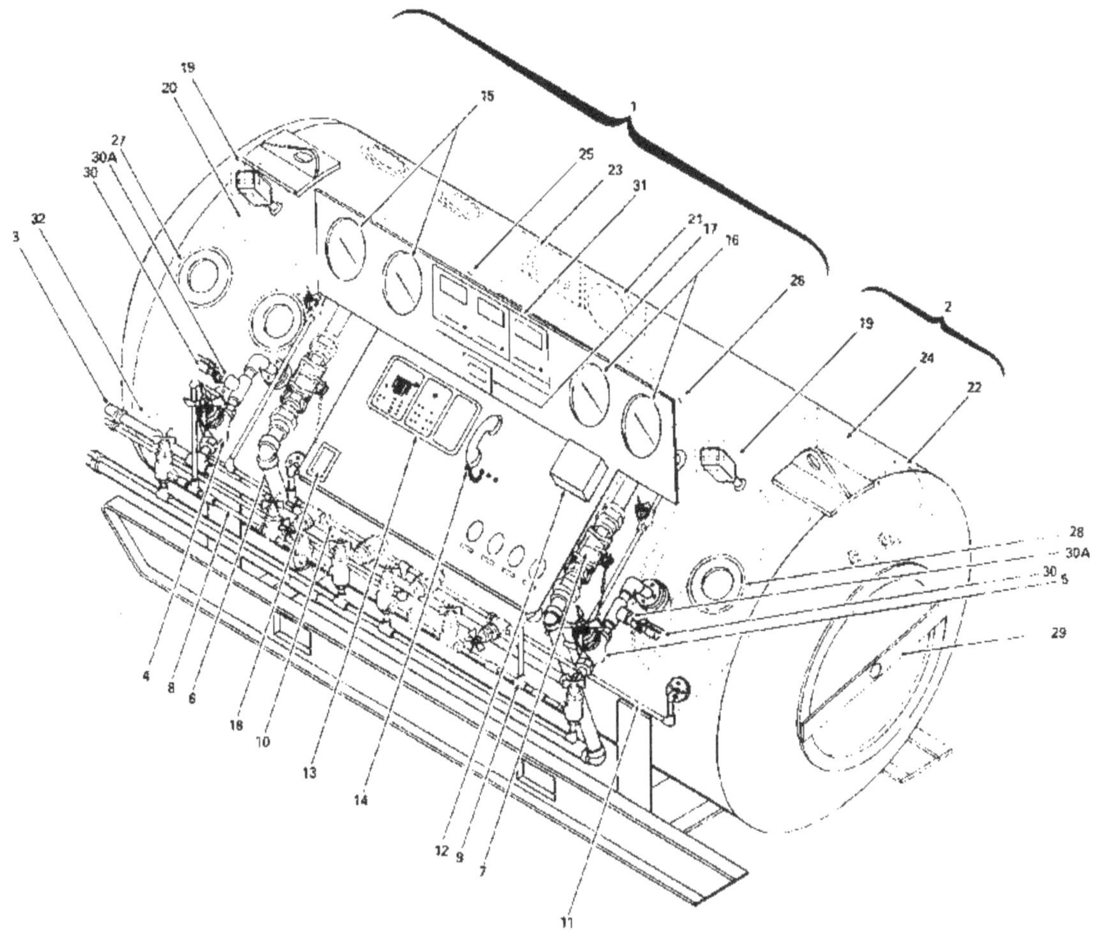

1. Inner Lock
2. Outer Lock
3. Air Supply Connection
4. Air Supply – Inner Lock
5. Air Supply – Outer Lock
6. Exhaust – Inner Lock
7. Exhaust – Outer Lock
8. BIBS Supply – Inner Lock
9. BIBS Supply – Outer Lock
10. BIBS Exhaust – Inner Lock
11. BIBS Exhaust – Outer Lock
12. Oxygen Analyzer
13. Communications
14. Sound-Powered Phones
15. External Depth Gauges – Inner Lock (2)
16. External Depth Gauges – Outer Lock (2)
17. Telethermometer
18. Ground Fault Interrupter
19. Pipe Light Assembly
20. Chiller and Scrubber Panel
23. Inner Lock Comm Panel
24. Outer Lock Comm Panel
25. Bunk Main
26. Bunk Extension
27. View Ports – Inner Lock (4)
28. View Ports – Outer Lock (2)
29. Strongback
30. Relief Valve – 100 psig

30A. Gag Valve

31. Pipe Light Controls
32. Chiller/Scrubber Penetrator

Design Pressure – 100 psig
Original Hydrostatic Pressure – 150 psig
Principal Locations – ARS-50 Class Salvage Ships

Volume – Inner Lock = 134 cubic feet
– Outer Lock = 68 cubic feet
– Total = 202 cubic feet

Figure 21-4. Double-Lock Steel Recompression Chamber.

Fleet Modernized Double-Lock Recompression Chamber

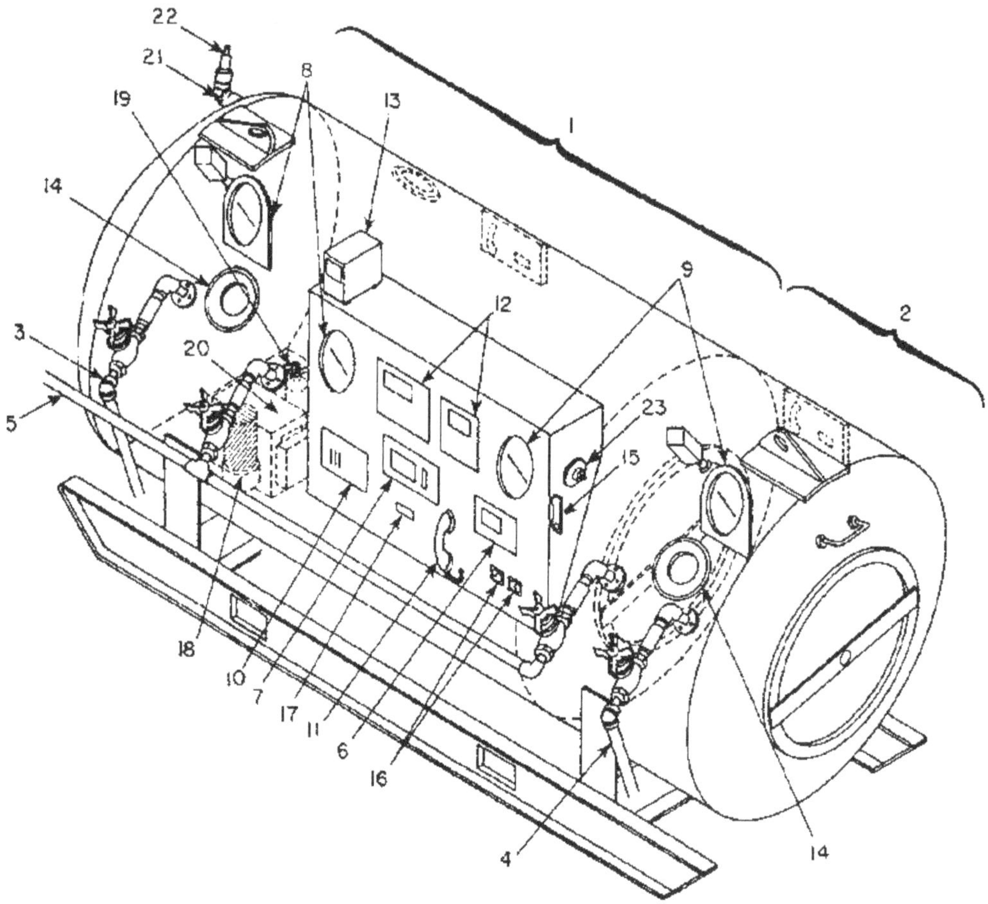

1. Inner Lock
2. Outer Lock
3. Gas Supply – Inner Lock
4. Gas Supply – Outer Lock
5. Gas Exhaust
6. O_2 Analyzer
7. CO_2 Analyzer
8. Inner-Lock Depth Gauges (2)
9. Outer-Lock Depth Gauges (2)
10. Communications Panel
11. Sound-Powered Phone
12. Pipe Light Control Panel
13. Ground Fault Interrupter
14. View Ports (5)
15. Flowmeter
16. Stopwatch/Timer
17. Telethermometer
18. CO_2 Scrubber
19. Fire Extinguisher
20. Chiller/Conditioner Unit
21. Gag Valve
22. Relief Valve – 110 psig
23. BIBS Overboard Dump Regulator – Outer Lock

Figure 21-5. Fleet Modernized Double-Lock Recompression Chamber.

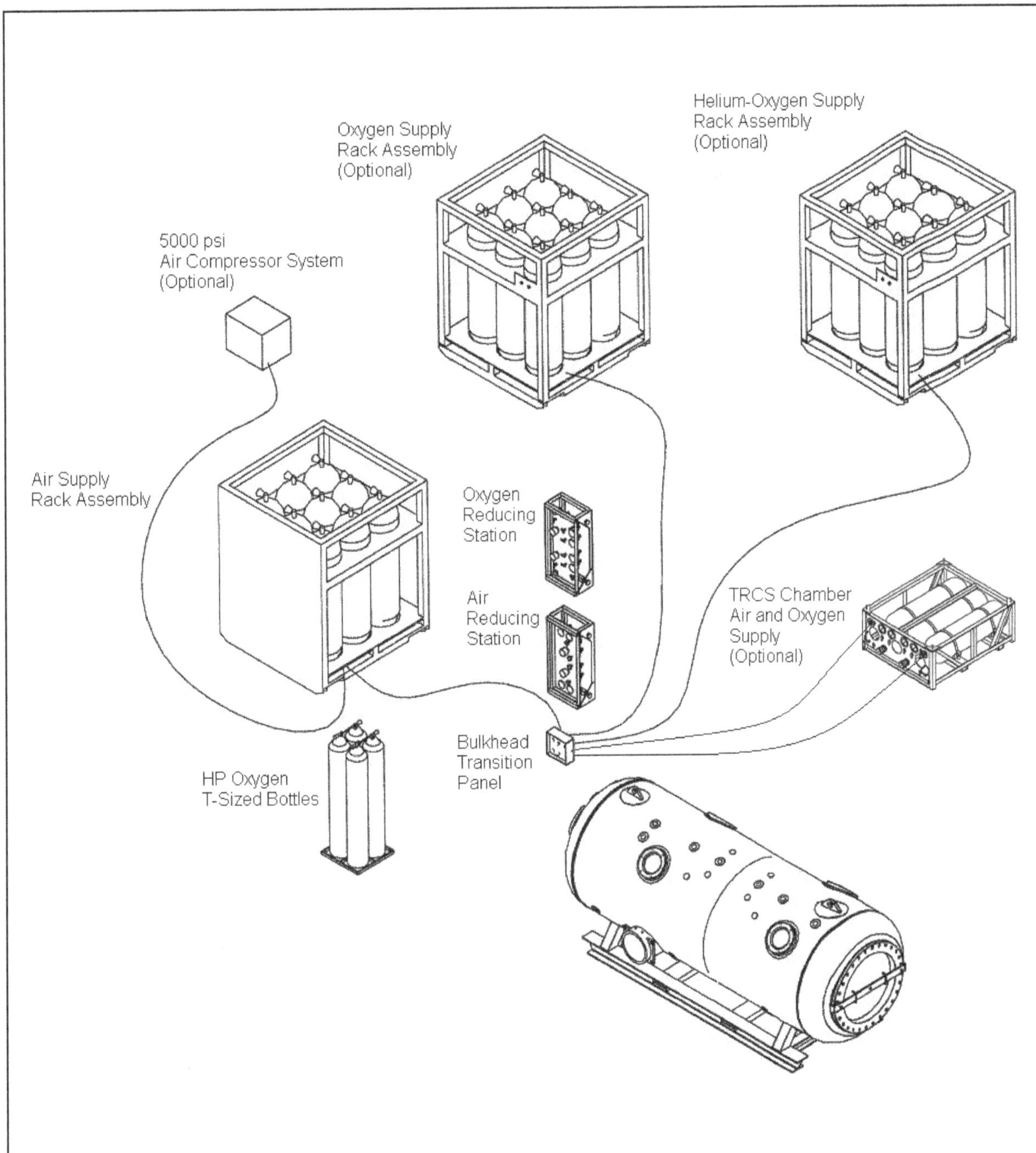

Figure 21-6. Standard Navy Double-Lock Recompression Chamber System.

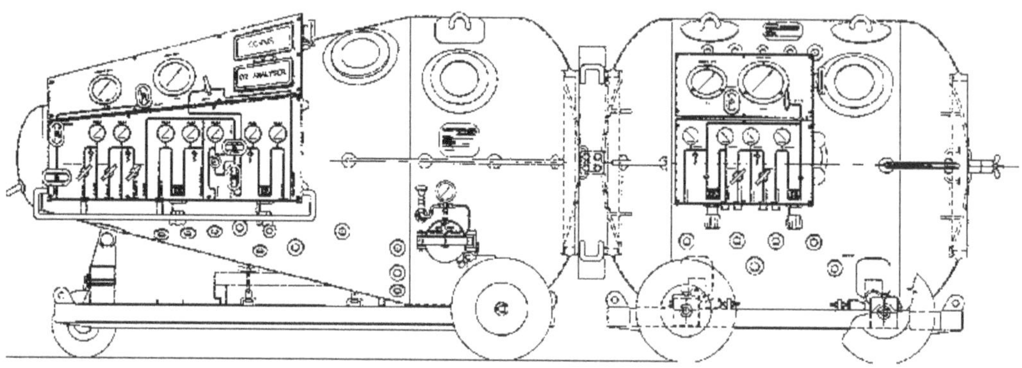

Figure 21-7. Transportable Recompression Chamber System (TRCS).

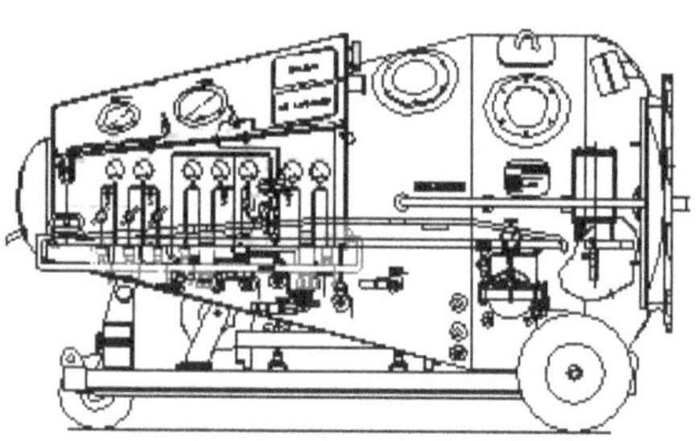

Design Pressure	110 psig
Design Temperature	0-125°F
Length	95.7″

Height	52″ with wheels, 48″ without wheels
Width	50.7″
Weight	1,268 lbs.
Internal Volume	45 cu. ft.
Door Opening	26″
View Ports	3 @ 6″ dia. Clear Opening
Medical Lock	5.75″ dia. x 11.8″ long
Mating Flange	Male per NATO STANAG 1079
Life Support Scrubber	Air driven, replaceable scrubber, canister fits in Med Lock
BIBS	2 masks – oxygen and air supply (with capability for N_2O_2 or HeO_2) – overboard dump
Atmospheric Monitoring	Oxygen and Carbon Dioxide Analyzer
Gas Supply	Primary and secondary air and O_2
Communications	Battery-powered speaker/headset phone
Furnishing	Patient litter, attendants seat

Figure 21-8. Transportable Recompression Chamber (TRC).

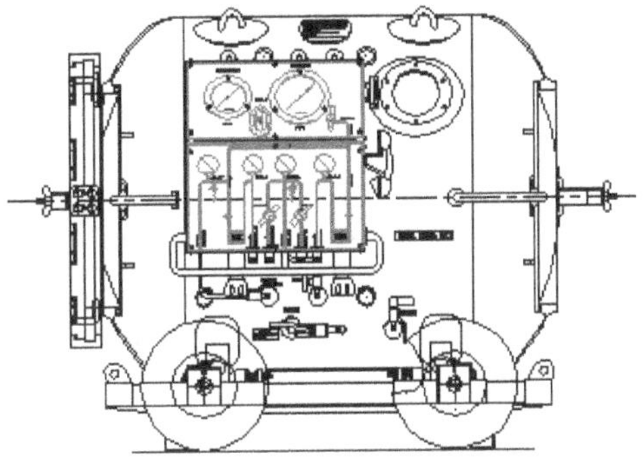

Height	52.9″
Width	54.8″
Weight	1,367 lbs.
Internal Volume	45.5 cu. ft.
Door Opening	2 doors – 26″
View Ports	2 @ 6″ dia. Clear Opening
Mating Flange	Rotating Female per NATO STANAG 1079
Life Support Scrubber	Air-driven, replaceable scrubber, canister fits in TRC Med Lock
BIBS	2 masks – oxygen and air supply – overboard dump
Atmospheric Monitoring	Oxygen and Carbon Dioxide Analyzer
Gas Supply	Primary and secondary air and O_2
Communications	Sound-powered phone

Design Pressure	110 psig
Design Temperature	0-125°F
Length	69.9″

Figure 21-9. Transfer Lock (TL).

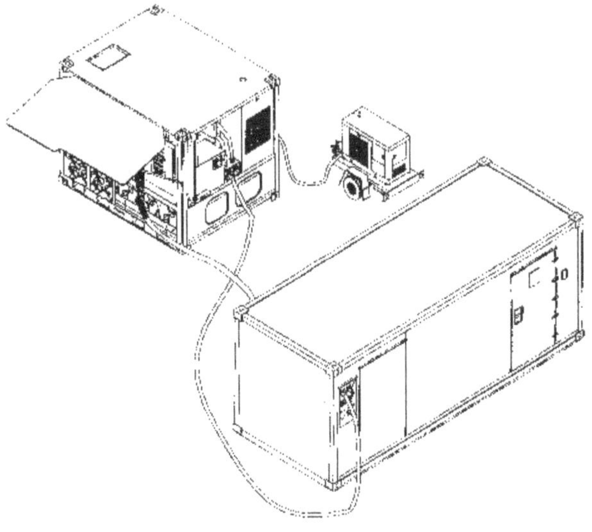

Figure 21-10. Fly Away Recompression Chamber (FARCC).

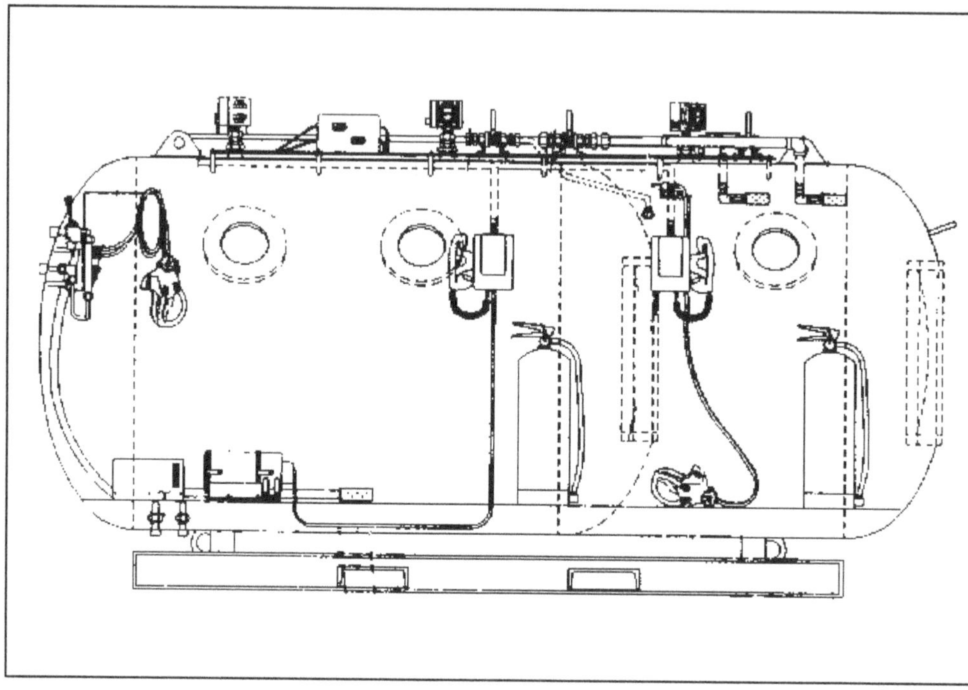

Figure 21-11. Fly Away Recompression Chamber.

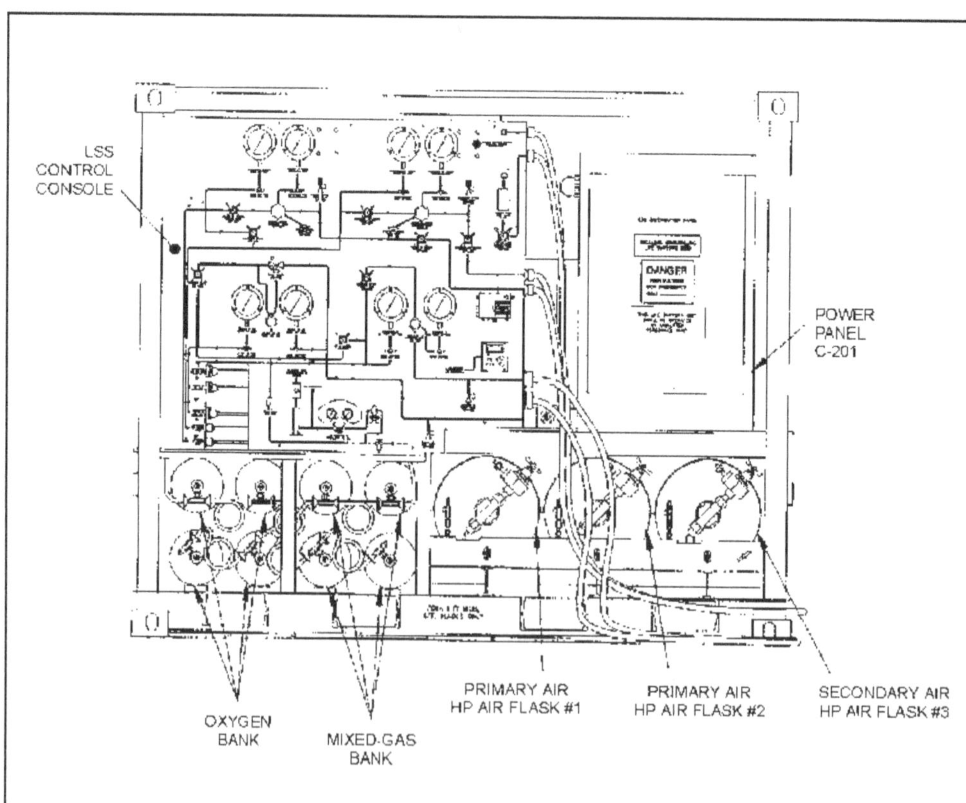

Figure 21-12. Fly Away Recompression Chamber Life Support Skid.

21-3 STATE OF READINESS

Since a recompression chamber is emergency equipment, it must be kept in a state of readiness. The chamber shall be well maintained and equipped with all necessary accessory equipment. A chamber is not to be used as a storage compartment.

The chamber and the air and oxygen supply systems shall be checked prior to each use with the Predive Checklist and in accordance with PMS instructions. All diving personnel shall be trained in the operation of the recompression chamber equipment and should be able to perform any task required during treatment.

21-4 GAS SUPPLY

A recompression chamber system must have a primary and a secondary air supply system that satisfies Table 21-2. The purpose of this requirement is to ensure the recompression chamber system, at a minimum, is capable of conducting a Treatment Table 6A (TT6A).

21-4.1 **Capacity.**

Either system may consist of air banks and/or a suitable compressor. The primary air supply system must have sufficient air to pressurize the inner lock once to 165 fsw and the outer lock twice to 165 fsw and ventilate the chamber as specified in Table 21-2.

- Primary System Capacity:

$C_p = (5 \times V_{il}) + (10 \times V_{ol}) + RV$

Where:

C_p = minimum capacity of primary system in SCF
V_{il} = volume of inner lock
V_{ol} = volume of outer lock
5 = atmospheres equivalent to 165 fsw
10 = twice the atmospheres equivalent to 165 fsw
RV = required ventilation. See paragraph 21-5.4 for Category A and B ventilation requirements. Not used for Category C, D, and E.

The secondary air supply system must have sufficient air to pressurize the inner and outer locks once to 165 fsw plus ventilate the chamber as specified in Table 21-2.

- Secondary System Requirement:

$C_s = (5 \times V_{il}) + (5 \times V_{ol}) + RV$

Where:

C_s = minimum capacity of secondary system in SCF
V_{il} = volume of inner lock
V_{ol} = volume of outer lock
5 = atmospheres equivalent to 165 fsw
RV = required ventilation. For Category A, B, and C, use 4,224 for ventilation rate of 70.4 scfm for one hour. For Category D and E, calculate air or NITROX required for two patients and one tender to breathe BIBS (when not on O_2) during one TT6A with maximum extensions.

Table 21-2. *Recompression Chamber Air Supply Requirements.*

Recompression Chamber Configuration	Primary Air Requirement	Secondary Air Requirement
CATEGORY A: No BIBS overboard dump No CO_2 scrubber No air BIBS No O_2 and CO_2 monitor	Sufficient air to press the IL once and the OL twice to 165 fsw and vent during one TT6A for one tender and two patients with maximum extensions.	Sufficient air to press the IL and OL once to 165 fsw and vent for one hour at 70.4 scfm.
CATEGORY B: BIBS overboard dump No CO_2 scrubber No air BIBS O_2 and CO_2 monitors	Sufficient air to press the IL once and the OL twice to 165 fsw and vent for CO_2 during one TT6A for one tender and two patients with maximum extensions.	Sufficient air to press the IL and OL once to 165 fsw and vent for one hour at 70.4 scfm.
CATEGORY C: BIBS overboard dump CO_2 scrubber No air BIBS O_2 and CO_2 monitors	Sufficient air to press the IL once and the OL twice to 165 fsw.	Sufficient air to press the IL and OL once to 165 fsw and vent for one hour at 70.4 scfm.
CATEGORY D: BIBS overboard dump CO_2 scrubber Air BIBS O_2 and CO_2 monitor	Sufficient air to press the IL once and the OL twice to 165 fsw. (For TRCS, sufficient air to power CO_2 scrubbers must be included)	Sufficient air to press the IL and OL once to 165 fsw and enough air for one tender and two patients (when not on O_2) to breathe air BIBS during one TT6A with maximum extensions.
CATEGORY E: BIBS overboard dump CO_2 scrubber O_2 and CO_2 monitor Spare CO_2 scrubber Secondary power supply NITROX BIBS No Air BIBS	Sufficient air to press the IL once and the OL twice to 165 fsw.	Sufficient air to press the IL and OL once to 165 fsw and enough air/NITROX for one tender and two patients (when not on O_2) to breathe air/NITROX BIBS during one TT6A with maximum extensions.
Notes: 1) Additional air source per PSOB will be required for TT4, 7 or 8. 2) For chambers used to conduct Sur "D" sufficient air is required to conduct a TT6A in addition to any planned Sur "D." 3) The requirement for BIBS overboard dump can also be satisfied with closed circuit BIBS with CO_2 scrubbers.		

21-5 OPERATION

21-5.1 **Predive Checklist.** To ensure each item is operational and ready for use, perform the equipment checks listed in the Recompression Chamber Predive Checklist, Figure 21-13.

21-5.2 **Safety Precautions.**

- Do not use oil on any oxygen fitting, air fitting, or piece of equipment.
- Do not allow oxygen supply tanks to be depleted below 100 psig.
- Ensure dogs are in good operating condition and seals are tight.
- Do not leave doors dogged (if applicable) after pressurization.
- Do not allow open flames, smoking materials, or any flammables to be carried into the chamber.
- Do not permit electrical appliances to be used in the chamber unless listed in the Authorized for Navy Use (ANU).
- Do not perform unauthorized repairs or modifications on the chamber support systems.
- Do not permit products in the chamber that may contaminate or off-gas into the chamber atmosphere.

21-5.3 **General Operating Procedures.**

1. Ensure completion of Predive Checklist.
2. Diver and tender enter the chamber together.
3. Diver sits in an uncramped position.
4. Tender closes and dogs (if so equipped) the inner lock door.
5. Pressurize the chamber, at the rate and to the depth specified in the appropriate decompression or recompression table.
6. As soon as a seal is obtained or upon reaching depth, tender releases the dogs (if so equipped).
7. Ventilate chamber according to specified rates and energize CO_2 scrubber and chamber conditioning system.
8. Ensure proper decompression of all personnel.
9. Ensure completion of Postdive Checklist.

RECOMPRESSION CHAMBER PREDIVE CHECKLIST	
Equipment	**Initials**
Chamber	
System certified	
Cleared of all extraneous equipment	
Clear of noxious odors	
Doors and seals undamaged, seals lubricated	
Pressure gauges calibrated/compared	
Air Supply System	
Primary and secondary air supply adequate	
One-valve supply: Valve closed	
Two-valve supply: Outside valve open, inside valve closed, if applicable	
Equalization valve closed, if applicable	
Supply regulator set at 250 psig or other appropriate pressure	
Fittings tight, filters clean, compressors fueled	
Exhaust System	
One-valve exhaust: Valve closed and calibrated for ventilation	
Two-valve exhaust: Outside valve open, inside valve closed, if applicable	
Oxygen Supply System	
Cylinders full, marked as BREATHING OXYGEN, cylinder valves open	
Replacement cylinders on hand	
Built in breathing system (BIBS) masks installed and tested	
Supply regulator set in accordance with OPs	
Fittings tight, gauges calibrated	
Oxygen manifold valves closed	
BIBS dump functioning	

Figure 21-13. Recompression Chamber Predive Checklist (sheet 1 of 2).

RECOMPRESSION CHAMBER PREDIVE CHECKLIST		
Equipment		**Initials**
Electrical System		
Lights		
Carbon dioxide analyzer calibrated		
Oxygen analyzer calibrated		
Temperature indicator calibrated		
Carbon dioxide scrubber operational		
Chamber conditioning unit operational		
Direct Current (DC) power supply		
Ground Fault Interrupter (GFI)		
Communication System		
Primary system tested		
Secondary system tested		
Fire Prevention System		
Tank pressurized for chambers with installed fire suppression systems		
Combustible material in metal enclosure		
Fire-retardant clothing worn by all chamber occupants		
Fire-resistant mattresses and blankets in chamber		
Means of extinguishing a fire		
Miscellaneous		
Inside Chamber:	CO_2-absorbent canister with fresh absorbent installed	
	Urinal	
	Primary medical kit	
	Ear protection sound attenuators/ear protectors (1 set per person) Must have a 1/16" hole drilled to allow for equalization.	
Outside Chamber:	Heater/chiller unit	
	Stopwatches for recompression treatment time, decompression time, personnel leaving chamber time, and cumulative time	
	Fresh CO_2 scrubber canister	
	U.S. Navy Diving Manual, Volume 5	
	Ventilation bill	
	Chamber log	
	Operating Procedures (OPs) and Emergency Procedures (EPs)	
	Secondary medical kit	
	Bedpan (to be locked in as required)	

Figure 21-13. Recompression Chamber Predive Checklist (sheet 2 of 2).

21-5.3.1 **Tender Change-Out.** During extensive treatments, medical personnel may prefer to lock-in to examine the patient and then lock-out, rather than remain inside throughout the treatment. Inside tenders may tire and need relief.

21-5.3.2 **Lock-In Operations.** Personnel entering the chamber go into the outer lock and close and dog the door (if applicable). The outer lock should be pressurized at a rate controlled by their ability to equalize, but not to exceed 75 feet per minute. The outside tender shall record the time pressurization begins to determine the decompression schedule for the occupants when they are ready to leave the chamber. When the pressure levels in the outer and inner locks are equal, the inside door (which was undogged at the beginning of the treatment) should open.

21-5.3.3 **Lock-Out Operations.** To exit the chamber, the personnel again enter the outer lock and the inside tender closes and dogs the inner door (if so equipped). When ready to ascend, the Diving Supervisor is notified and the required decompression schedule is selected and executed. Constant communications are maintained with the inside tender to ensure that a seal has been made on the inner door. Outer lock depth is controlled throughout decompression by the outside tender.

21-5.3.4 **Gag Valves.** The actuating lever of the chamber gag valves shall be maintained in the open position at all times, during both normal chamber operations and when the chamber is secured. The gag valves must be closed only in the event of relief valve failure during chamber operation. Valves are to be lock-wired in the open position with light wire that can be easily broken when required. A WARNING plate, bearing the inscription shown below, shall be affixed to the chamber in the vicinity of each gag valve and shall be readily viewable by operating personnel. The WARNING plates shall measure approximately 4 inches by 6 inches and read as follows:

> **WARNING**
> **The gag valve must remain open at all times.**
> **Close only if relief valve fails.**

21-5.4 **Ventilation.** The basic rules for ventilation are presented below. These rules permit rapid computation of the cubic feet of air per minute (acfm) required under different conditions as measured at chamber pressure (the rules are designed to ensure that the effective concentration of carbon dioxide will not exceed 1.5 percent (11.4 mmHg) and that when oxygen is being used, the percentage of oxygen in the chamber will not exceed 25 percent).

1. When air is breathed, provide 2 cubic feet per minute (acfm) for each diver at rest and 4 cubic feet per minute (acfm) for each diver who is not at rest (i.e., a tender actively taking care of a patient).

2. When oxygen is breathed from the built-in breathing system (BIBS), provide 12.5 acfm for a diver at rest and 25 acfm for a diver who is not at rest. When these ventilation rates are used, no additional ventilation is required for personnel breathing air. These ventilation rates apply only to the number of

people breathing oxygen and are used only when no BIBS dump system is installed.

3. If ventilation must be interrupted for any reason, the time should not exceed 5 minutes in any 30-minute period. When ventilation is resumed, twice the volume of ventilation should be used for the time of interruption and then the basic ventilation rate should be used again.

4. If a BIBS dump system or a closed circuit BIBS is used for oxygen breathing, the ventilation rate for air breathing may be used.

5. If portable or installed oxygen and carbon dioxide monitoring systems are available, ventilation may be adjusted to maintain the oxygen level below 25 percent by volume and the carbon dioxide level below 1.5 percent surface equivalent (sev).

21-5.4.1 **Chamber Ventilation Bill.** Knowing how much air must be used does not solve the ventilation problem unless there is some way to determine the volume of air actually being used for ventilation. The standard procedure is to open the exhaust valve a given number of turns (or fraction of a turn), which will provide a certain number of cubic feet of ventilation per minute at a specific chamber depth, and to use the supply valve to maintain a constant chamber depth during the ventilation period. Determination of valve settings required for different amounts of ventilation at different depths is accomplished as follows.

WARNING **This procedure is to be performed with an unmanned chamber to avoid exposing occupants to unnecessary risks.**

1. Mark the valve handle position so that it is possible to determine accurately the number of turns and fractions of turns.

2. Check the basic ventilation rules above against probable situations to determine the rates of ventilation at various depths (chamber pressure) that may be needed. If the air supply is ample, determination of ventilation rates for a few depths (30, 60, 100, and 165 feet) may be sufficient. It will be convenient to know the valve settings for rates such as 6, 12.5, 25, or 37.5 cubic feet per minute (acfm).

3. Determine the necessary valve settings for the selected flows and depths by using a stopwatch and the chamber as a measuring vessel.

   a. Calculate how long it will take to change the chamber pressure by 10 feet if the exhaust valve lets air escape at the desired rate close to the depth in question. Use the following formula.

$$T = \frac{V \times 60 \times \Delta P}{R \times (D + 33)}$$

Where:

T = time in seconds for chamber pressure to change 10 feet
V = internal volume of chamber (or of lock being used for test) in cubic feet (cf)
R = rate of ventilation desired, in cubic feet per minute as measured at chamber pressure (acfm)
ΔP = Change in chamber pressure in fsw
D = depth in fsw (gauge)

Example: Determine how long it will take the pressure to drop from 170 to 160 feet in a 425-cubic-foot chamber if the exhaust valve is releasing 6 cubic feet of air per minute (measured at chamber pressure of 165 feet).

1. List values from example:

T = unknown
V = 425 cf
R = 6 acfm
ΔP = 10 fsw
D = 165 fsw

2. Substitute values and solve to find how long it will take for the pressure to drop:

$$T = \frac{425 \times 60 \times 10}{6(165 + 33)}$$
$$= 215 \text{ seconds}$$

$$T = \frac{215 \text{ seconds}}{60 \text{ seconds / minute}}$$
$$= 3.6 \text{ minutes}$$

b. Increase the empty chamber pressure to 5 feet beyond the depth in question. Open the exhaust valve and determine how long it takes to come up 10 feet (for example, if checking for a depth of 165 fsw, take chamber pressure to 170 feet and clock the time needed to reach 160 feet). Open the valve to different settings until you can determine what setting will approximate the desired time. Record the setting. Calculate the times for other rates and depths and determine the settings for these times in the same way. Make a chart or table of valve setting versus ventilation rate and prepare a ventilation bill, using this information and the ventilation rules.

21-5.4.2 Notes on Chamber Ventilation.

- The basic ventilation rules are not intended to limit ventilation. Generally, if air is reasonably plentiful, more air than specified should be used for comfort. This increase is desirable because it also further lowers the concentrations of carbon dioxide and oxygen.

- There is seldom any danger of having too little oxygen in the chamber. Even with no ventilation and a high carbon dioxide level, the oxygen present would be ample for long periods of time.

- These rules assume that there is good circulation of air in the chamber during ventilation. If circulation is poor, the rules may be inadequate. Locating the inlet near one end of the chamber and the outlet near the other end improves ventilation.

- Coming up to the next stop reduces the standard cubic feet of gas in the chamber and proportionally reduces the quantity (scfm) of air required for ventilation.

- Continuous ventilation is the most efficient method of ventilation in terms of the amount of air required. However, it has the disadvantage of exposing the divers in the chamber to continuous noise. At the very high ventilation rates required for oxygen breathing, this noise can reach the level at which hearing loss becomes a hazard to the divers in the chamber. If high sound levels do occur, especially during exceptionally high ventilation rates, the chamber occupants must wear ear protectors (available as a stock item). A small hole should be drilled into the central cavity of the protector so that they do not produce a seal which can cause ear squeeze.

- The size of the chamber does not influence the rate (acfm) of air required for ventilation.

- Increasing depth increases the actual mass of air required for ventilation; but when the amount of air is expressed in volumes as measured at chamber pressure, increasing depth does not change the number of actual cubic feet (acfm) required.

- If high-pressure air banks are being used for the chamber supply, pressure changes in the cylinders can be used to check the amount of ventilation being provided.

21-6 CHAMBER MAINTENANCE

21-6.1 **Postdive Checklist.** To ensure equipment receives proper postdive maintenance and is returned to operational readiness, perform the equipment checks listed in the Recompression Chamber Postdive Checklist, Figure 21-14.

21-6.2 **Scheduled Maintenance.** Every USN recompression chamber shall adhere to PMS requirements and shall be pressure tested when initially installed, at 2-year intervals thereafter, and after a major overhaul or repair. This test shall adhere to PMS requirements and shall be conducted in accordance with Figure 21-15. The completed test form shall be retained until retest is conducted. For a permanently installed chamber, removing and reinstalling constitutes a major overhaul and requires a pressure test. For portable chambers such as the TRCS, SNDLRCS, and FARCC, follow operating procedures after moving the chamber prior to

RECOMPRESSION CHAMBER POSTDIVE CHECKLIST	
Equipment	**Initials**
Air Supply	
All valves closed	
Air banks recharged, gauged, and pressure recorded	
Compressors fueled and maintained per technical manual/PMS requirements	
View Ports and Doors	
View-ports checked for damage; replaced as necessary	
Door seals checked, replaced as necessary	
Door seals lightly lubricated with approved lubricant	
Door dogs and dogging mechanism checked for proper operation and shaft seals for tightness	
Chamber	
Inside wiped clean with Nonionic Detergent (NID) and warm fresh water	
All unnecessary support items removed from chamber	
Blankets cleaned and replaced	
All flammable material in chamber encased in fire-resistant containers	
Primary medical kit restocked as required	
Chamber aired out	
Outer door closed	
CO_2 canister packed	
Deckplates lifted, area below deckplates cleaned, deckplates reinstalled	
Support Items	
Stopwatches checked and reset	
U.S. Navy Diving Manual, Operating Procedures (OPs), Emergency Procedures (EPs), ventilation bill and pencil available at control desk	
Secondary medical kit restocked as required and stowed	
Clothing cleaned and stowed	
All entries made in chamber log book	
Chamber log book stowed	

Figure 21-14. Recompression Chamber Postdive Checklist (sheet 1 of 2).

RECOMPRESSION CHAMBER POSTDIVE CHECKLIST	
Equipment	**Initials**
Oxygen Supply	
BIBS mask removed, cleaned per current PMS procedures, reinstalled	
All valves closed	
System bled	
Breathing oxygen cylinders fully pressurized	
Spare cylinders available	
System free of contamination	
Exhaust System	
One-valve exhaust: valves closed	
Two-valve exhaust: inside valves closed	
Two-valve exhaust: outside valves opened	
Electrical	
All circuits checked	
Light bulbs replaced as necessary	
Pressure-proof housing of lights checked	
All power OFF	
Wiring checked for fraying	

Figure 21-14. Recompression Chamber Postdive Checklist (sheet 2 of 2).

manned use. Chamber relief valves shall be tested in accordance with the Planned Maintenance System to verify setting. Each tested relief valve shall be tagged to indicate the valve set pressure, date of test, and testing activity. After every use or once a month, whichever comes first, the chamber shall receive routine maintenance in accordance with the Postdive Checklist. At this time, minor repairs shall be made and used supplies shall be restocked.

21-6.2.1 **Inspections.** At the discretion of the activity, but at least once a year, the chamber shall be inspected, both inside and outside. Any deposits of grease, dust, or other dirt shall be removed and, on steel chambers, the affected areas repainted.

21-6.2.2 **Corrosion.** Corrosion is removed best by hand or by using a scraper, being careful not to gouge or otherwise damage the base metal. The corroded area and a small area around it should then be cleaned to remove any remaining paint and/or corrosion.

21-6.2.3 **Painting Steel Chambers.** Steel Chambers shall be painted utilizing original paint specifications and in accordance with approved NAVSEA or NAVFAC procedures. The following paints shall be utilized on NAVSEA carbon steel chambers:

PRESSURE TEST FOR USN RECOMPRESSION CHAMBERS

NOTE

All U.S. Navy Standard recompression chambers are restricted to a maximum operating pressure of 100 psig, regardless of design pressure rating.

A pressure test shall be conducted on every USN recompression chamber:

- When initially installed
- After repairs/overhaul
- At two-year intervals at a given location

Performance of the test and the test results are recorded on a standard U.S. Navy Recompression Chamber Air Pressure and Leak Test form (Figure 21-15).

The test is conducted as follows:

1. Pressurize the innermost lock to 100 fsw (45 psig). Using soapy water or an equivalent solution, leak test all shell penetration fittings, view-ports, dog seals, door dogs (where applicable), valve connections, pipe joints, and shell weldments.
2. Mark all leaks. Depressurize the lock and adjust, repair, or replace components as necessary to eliminate leaks.
   a. View-Port Leaks. Remove the view-port gasket (replace if necessary), wipe clean.

   CAUTION

   Acrylic view-ports should not be lubricated or come in contact with any lubricant. Acrylic view-ports should not come in contact with any volatile detergent or leak detector (non-ionic detergent is to be used for leak test). When reinstalling view-port, take up retaining ring bolts until the gasket just compresses evenly about the view-port. Do not overcompress the gasket.

   b. Weldment Leaks. Contact appropriate NAVSEA technical authority for guidance on corrective action.
3. Repeat steps 1 and 2 until all the leaks have been eliminated.
4. Pressurize lock to 225 fsw (100 psig) and hold for 5 minutes.

   WARNING

   Do not exceed maximum pressure rating for the pressure vessel.

5. Depressurize the lock to 165 fsw (73.4 psig). Hold for 1 hour. If pressure drops below 145 fsw (65 psig), locate and mark leaks. Depressurize chamber and repair leaks in accordance with Step 2 above and repeat this procedure until final pressure is at least 145 fsw (65 psig).
6. Repeat Steps 1 through 5 leaving the inner door open and outer door closed. Leak test only those portions of the chamber not previously tested.

Figure 21-15. Pressure Test for USN Recompression Chambers (sheet 1 of 3).

STANDARD U.S. NAVY RECOMPRESSION CHAMBER
AIR PRESSURE AND LEAK TEST
(Sheet 2 of 3)

Ship/Platform/Facility ______________________________

Type of Chamber:

Recompression Chamber Facility - RCF5000	Double-Lock Steel
Recompression Chamber Facility - RCF6500	Standard Navy Double Lock Recompression
Transportable Recompression Chamber (TRC)	Chamber System (SNDLRCS)
Fly-Away Recompression Chamber (FARCC)	Other*______________

NAME PLATE DATA

Manufacturer ______________________________

Date of Manufacture ______________________________

Contract/Drawing No. ______________________________

Maximum Working Pressure ______________________________

Date of Last Pressure Test ______________________________

Test Conducted by ______________________________
(Name/Rank)

1. Conduct visual inspection of chamber to determine if ready for test

   Chamber Satisfactory ____________ Initials of Test Conductor ______________

   Discrepancies from fully inoperative chamber equipment:

   ____

   ____

2 Close inner door lock. With outer lock door open pressure inner lock to 100 fsw (45 psig) and verify that the following components do not leak:
(Note: If chamber has medical lock, open inner door and close and secure outer door.)

Inner lock leak checks	Initials of Test Conductor
A. Shell penetrations and fittings	____________ Satisfactory
B. View Ports	____________ Satisfactory
C. Door Seals	____________ Satisfactory
D. Door Dog Shaft Seals	____________ Satisfactory
E. Valve Connections and Stems	____________ Satisfactory
F. Pipe Joints	____________ Satisfactory
G. Shell Welds	____________ Satisfactory

3. Increase inner lock pressure to 225 fsw (100 psig) and hold for 5 minutes.

   Record Test Pressure ______________ Satisfactory______________

   (Note: Disregard small leaks at this pressure).

Figure 21-15. Pressure Test for USN Recompression Chambers (sheet 2 of 3).

STANDARD U.S. NAVY RECOMPRESSION CHAMBER
AIR PRESSURE AND LEAK TEST
(Sheet 3 of 3)

4. Depressurize lock slowly to 165 fsw (73.4 psig). Secure all supply and exhaust valves and hold for one hour.

   Start Time ______________________ Pressure 165 fsw

   End Time ______________________ Pressure ____________ fsw

   If pressure drops below 145 fsw (65 psig) locate and mark leaks. Depressurize, repair, and retest inner lock.

   Inner Lock Pressure drop test passed ____________ Satisfactory Initials of Test Conductor.

5. Depressurize inner lock and open inner lock door. Secure in open position. Close outer door and secure.

   (Note: If chamber has medical lock, close and secure inner door and open outer door.)

6. Repeat tests of sections 2, 3, and 4 above when set up in accordance with section 5. Leak test only those portions of the chamber not tested in sections 2, 3, and 4.

7. Outer Lock Checks Initials of Test Conductor

   A. Shell penetrations and fittings ________________ Satisfactory

   B. View Ports ________________ Satisfactory

   C. Door Seals ________________ Satisfactory

   D. Door Dog Shaft Seals ________________ Satisfactory

   E. Valve Connections and Stems ________________ Satisfactory

   F. Pipe Joints ________________ Satisfactory

   G. Shell Welds ________________ Satisfactory

8. Maximum Chamber Operating Pressure (100 psig) Test (5 minute hold)

   Satisfactory ____________________ Initials of Test Conductor

9. Inner and Outer Lock Chamber Drop Test

   Start Time ______________________ Pressure 165 fsw

   End Time ______________________ Pressure ____________ fsw

   Inner and outer lock pressure drop test passed satisfactorily ______ Initials of Test Conductor

10. All above tests have been satisfactorily completed.

____
Test Director Date

____
Diving Officer Date

____
Commanding Officer Date

Figure 21-15. Pressure Test for USN Recompression Chambers (sheet 3 of 3).

- Inside:
  - — Prime coat NSN 8010-01-302-3608.
  - — Finish coat white NSN 8010-01-302-3606.
- Outside:
  - — Prime coat NSN 8010-01-302-3608.
  - — Exterior coats gray NSN 8010-01-302-6838 or white NSN 8010-01-302-3606.

For original paint specification on NAVFAC steel chambers refer to the Operation and Maintenance Support Information (OMSI) documentation delivered with the system.

21-6.2.4 **Recompression Chamber Paint Process Instruction.** Painting shall be kept to an absolute minimum. Only the coats prescribed above are to be applied. Naval Sea Systems Command will issue a Recompression Chamber Paint Process Instruction (NAVSEA-00C3-PI-001) on request.

21-6.2.5 **Stainless Steel Chambers.** Stainless steel chamber such as the TRCS and SNDLRCS do not require surfaces painted for corrosion resistance, only for cosmetic purposes. Naval Sea Systems Command will provide a Stainless Steel Recompression Chamber Paint Process Instruction on request.

21-6.2.6 **Fire Hazard Prevention.** The greatest single hazard in the use of a recompression chamber is from explosive fire. Fire may spread two to six times faster in a pressurized chamber than at atmospheric conditions because of the high partial pressure of oxygen in the chamber atmosphere. The following precautions shall be taken to minimize fire hazard:

- Maintain the chamber oxygen percentage as close to 21 percent as possible and never allow oxygen percentage to exceed 25 percent.
- Remove any fittings or equipment that do not conform with the standard requirements for the electrical system or that are made of flammable materials. Permit no wooden deck gratings, benches, or shelving in the chamber.
- Use only mattresses designed for hyperbaric chambers. Use Durett Product or submarine mattress (NSN 7210-00-275-5878 or 5874). Other mattresses may cause atmospheric contamination. Mattresses should be enclosed in flame-proof covers. Use 100% cotton sheets and pillow cases. Put no more bedding in a chamber than is necessary for the comfort of the patient. Never use blankets of wool or synthetic fibers because of the possibility of sparks from static electricity.

- Clothing worn by chamber occupants shall be made of 100% cotton, or a flame resistant blend of cotton and polyester for chambers equipped with a fire extinguisher or fixed hand-held or fire suppression system. Diver swim trunks made of 65% polyester 35% cotton material are acceptable.

- Keep oil and volatile materials out of the chamber. If any have been used, ensure that the chamber is thoroughly ventilated before pressurization. Do not put oil on or in any fittings or high-pressure line. If oil is spilled in the chamber or soaked into any chamber surface or equipment, it must be completely removed. If lubricants are required, use only those approved and listed in *Naval Ships Technical Manual* (NSTM) NAVSEA S9086-H7-STM-000, Chapter 262. Regularly inspect and clean air filters and accumulators in the air supply lines to protect against the introduction of oil or other vapors into the chamber. Permit no one to wear oily clothing into the chamber.

- Permit no one to carry smoking materials, matches, lighters or any flammable materials into a chamber. A WARNING sign should be posted outside the chamber. Example:

> **WARNING**
> **Fire/Explosion Hazard. No matches, lighters, electrical appliances, or flammable materials permitted in chamber.**

21-6.2.6.1 ***Fire Extinguishing.*** All recompression chambers must have a means of extinguishing a fire in the interior. Examples of fire protection include wetted towels, a bucket of water, fire extinguisher, hand-held hose system, or suppression/ deluge system. Refer to U.S. Navy General Specification for the Design, Construction, and Repair of Diving and Hyperbaric Equipment (TS500-AU-SPN-010) for specific requirements of fire protection systems. Only fire extinguishers listed on the NAVSEA Authorized for Navy Use (ANU) are to be used.

21-7 DIVER CANDIDATE PRESSURE TEST

All U.S. Navy diver candidates shall be physically qualified in accordance with the *Manual of the Medical Department*, Art. 15-102. Candidates shall also pass a pressure test before they are eligible for diver training. This test may be conducted at any Navy certified recompression chamber, provided it is administered by qualified chamber personnel.

21-7.1 **Candidate Requirements.** The candidate must demonstrate the ability to equalize pressure in both ears to a depth of 60 fsw. The candidate shall have also passed the screening physical readiness test in accordance with MILPERSMAN 1220-100, Exhibit 1.

21-7.2 **Procedure.**

1. Candidates shall undergo a diving physical examination by a Navy Medical Officer in accordance with the *Manual of the Medical Department,* Art. 15-102, and be qualified to undergo the test.

2. The candidates and the tender enter the recompression chamber and are pressurized to 60 fsw on air, at a rate of 75 fpm or less as tolerated by the occupants.

3. If a candidate cannot complete the descent, the chamber is stopped and the candidate is placed in the outer lock for return to the surface.

4. Stay at 60 fsw for at least 10 minutes.

5. Ascend to the surface following standard air decompression procedures.

6. All candidates shall remain at the immediate chamber site for a minimum of 15 minutes and at the test facility for 1 hour. Candidates or tenders who must return to their command via air travel must proceed in accordance with Chapter 9, paragraph 9-13.

21-7.2.1 **References.**

- *Navy Military Personnel Manual*, Art. 1220-100
- *Manual of the Medical Department*, Art. 15-102

PAGE LEFT BLANK INTENTIONALLY

APPENDIX 5A

Neurological Examination

5A-1 INTRODUCTION

This appendix provides guidance on evaluating diving accidents prior to treatment. Figure 5A-1a is a guide aimed at non-medical personnel for recording essential details and conducting a neurological examination. Copies of this form should be readily available. While its use is not mandatory, it provides a useful aid for gathering information.

5A-2 INITIAL ASSESSMENT OF DIVING INJURIES

When using the form in Figure 5A-1a, the initial assessment must gather the necessary information for proper evaluation of the accident.

When a diver reports with a medical complaint, a history of the case shall be compiled. This history should include facts ranging from the dive profile to progression of the medical problem. If available, review the diver's Health Record and completed Diving Chart or Diving Log to aid in the examination. A few key questions can help determine a preliminary diagnosis and any immediate treatment needed. If the preliminary diagnosis shows the need for immediate recompression, proceed with recompression. Complete the examination when the patient stabilizes at treatment depth. Typical questions should include the following:

1. What is the problem/symptom? If the only symptom is pain:
   a. Describe the pain:
      - Sharp
      - Dull
      - Throbbing
   b. Is the pain localized, or hard to pinpoint?
2. Has the patient made a dive recently?
3. What was the dive profile?
   a. What was the depth of the dive?
   b. What was the bottom time?
   c. What dive rig was used?
   d. What type of work was performed?
   e. Did anything unusual occur during the dive?

4. How many dives has the patient made in the last 24 hours?
    a. Chart profile(s) of any other dive(s).
5. Were the symptoms first noted before, during, or after the dive? If after the dive, how long after surfacing?
6. If during the dive, did the patient notice the symptom while descending, on the bottom, or during ascent?
7. Has the symptom either increased or decreased in intensity since first noticed?
8. Have any additional symptoms developed since the first one?
9. Has the patient ever had a similar symptom?
10. Has the patient ever suffered from decompression sickness or gas embolism in the past?
    a. Describe this symptom in relation to the prior incident if applicable.
11. Does the patient have any concurrent medical conditions that might explain the symptoms?

To aid in the evaluation, review the diver's Health Record, including a baseline neurological examination, if available, and completed Diving Chart or Diving Log, if they are readily available.

5A-3 NEUROLOGICAL ASSESSMENT

There are various ways to perform a neurological examination. The quickest information pertinent to the diving injury is obtained by directing the initial examination toward the symptomatic areas of the body. These concentrate on the motor, sensory, and coordination functions. If this examination is normal, the most productive information is obtained by performing a complete examination of the following:

1. Mental status
2. Coordination
3. Motor
4. Cranial nerves
5. Sensory
6. Deep tendon reflexes

The following procedures are adequate for preliminary examination. Figure 5A-1a can be used to record the results of the examination.

NEUROLOGICAL EXAMINATION CHECKLIST

(Sheet 1 of 2)

(See text of Appendix 5A for examination procedures and definitions of terms.)

Patient's Name: ______________________________ Date/Time: ______________________

Describe pain/numbness: ____

____

HISTORY

Type of dive last performed: ______________ Depth: __________ How long: ______________

Number of dives in last 24 hours: ____

Was symptom noticed before, during or after the dive? ______________________________

If during, was it while descending, on the bottom or ascending? _______________________

Has symptom increased or decreased since it was first noticed? _______________________

Have any other symptoms occurred since the first one was noticed? ____________________

Describe: ____

Has patient ever had a similar symptom before? _______________ When: _______________

____

MENTAL STATUS/STATE OF CONSCIOUSNESS

____

____

COORDINATION

Walk: _________

Heel-to Toe: _________

Romberg: _________

Finger-to-Nose: _________

Heel Shin Slide: _________

Rapid Movement: _________

CRANIAL NERVES

Sense of Smell (I): ________

Vision/Visual Fld (II): _________

Eye Movements, Pupils (III, IV, VI): _________

Facial Sensation, Chewing (V): _________

Facial Expression Muscles (VII): _________

Hearing (VIII): _________

Upper Mouth, Throat Sensation (IX): _________

Gag & Voice (X): _________

Shoulder Shrug (XI): _________

Tongue (XII): _________

STRENGTH (Grade 0 to 5)

UPPER BODY

Deltoids L ____ R ____

Latissimus L ____ R ____

Biceps L ____ R ____

Triceps L ____ R ____

Forearms L ____ R ____

Hands L ____ R ____

LOWER BODY

Hips

Flexion L ___ R ____

Extension L ___ R ____

Abduction L ___ R ____

Adduction L ___ R ____

Knees

Flexion L ___ R ____

Extension L ____ R ____

Figure 5A-1a. Neurological Examination Checklist (sheet 1 of 2).

NEUROLOGICAL EXAMINATION CHECKLIST

(Sheet 2 of 2)

REFLEXES

(Grade: Normal, Hypoactive, Hyperactive, Absent

Biceps	L ____	R ____	**Ankles**		
Triceps	L ____	R ____	Dorsiflexion	L ____	R ____
Knees	L ____	R ____	Plantarflexion	L ____	R ____
Ankles	L ____	R ____	**Toes**	L ____	R ____

Sensory Examination for Skin Sensation

(Use diagram to record location of sensory abnormalities – numbness, tingling, etc.)

LOCATION

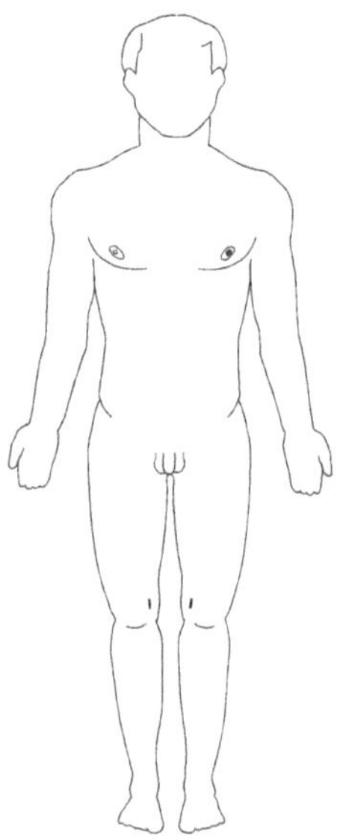

Indicate results as follows:

|||| Painful Area

≡ Decreased Sensation

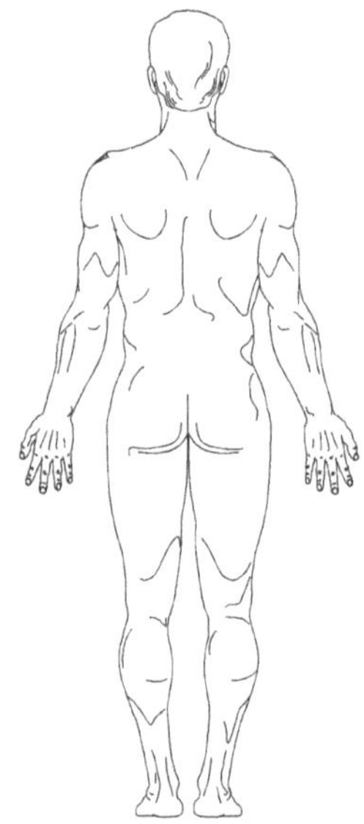

COMMENTS

____

____

____

____

____

____

____

____

Examination Performed by: ____

Figure 5A-1b. Neurological Examination Checklist (sheet 2 of 2).

5A-3.1 **Mental Status.** This is best determined when you first see the patient and is characterized by his alertness, orientation, and thought process. Obtain a good history, including the dive profile, present symptoms, and how these symptoms have changed since onset. The patient's response to this questioning and that during the neurological examination will give you a great deal of information about his mental status. It is important to determine if the patient knows the time and place, and can recognize familiar people and understands what is happening. Is the patient's mood appropriate?

Next the examiner may determine if the patient's memory is intact by questioning the patient. The questions asked should be reasonable, and you must know the answer to the questions you ask. Questions such as the following may be helpful:

- What is your commanding officer's name?
- What did you have for lunch?

Finally, if a problem does arise in the mental status evaluation, the examiner may choose to assess the patient's cognitive function more fully. Cognitive function is an intellectual process by which one becomes aware of, perceives, or comprehends ideas and involves all aspects of perception, thinking, reasoning, and remembering. Some suggested methods of assessing this function are:

- The patient should be asked to remember something. An example would be "red ball, green tree, and couch." Inform him that later in the examination you will ask him to repeat this information.
- The patient should be asked to spell a word, such as "world," backwards.
- The patient should be asked to count backwards from 100 by sevens.
- The patient should be asked to recall the information he was asked to remember at the end of the examination.

5A-3.2 **Coordination (Cerebellar/Inner Ear Function).** A good indicator of muscle strength and general coordination is to observe how the patient walks. A normal gait indicates that many muscle groups and general brain functions are normal. More thorough examination involves testing that concentrates on the brain and inner ear. In conducting these tests, both sides of the body shall be tested and the results shall be compared. These tests include:

1. **Heel-to-Toe Test.** The tandem walk is the standard "drunk driver" test. While looking straight ahead, the patient must walk a straight line, placing the heel of one foot directly in front of the toes of the opposite foot. Signs to look for and consider deficits include:

    a. Does the patient limp?

    b. Does the patient stagger or fall to one side?

2. **Romberg Test.** With eyes closed, the patient stands with feet together and arms extended to the front, palms up. Note whether the patient can maintain his balance or if he immediately falls to one side. Some examiners recommend giving the patient a small shove from either side with the fingertips.

3. **Finger-to-Nose Test.** The patient stands with eyes closed and head back, arms extended to the side. Bending the arm at the elbow, the patient touches his nose with an extended forefinger, alternating arms. An extension of this test is to have the patient, with eyes open, alternately touch his nose with his fingertip and then touch the fingertip of the examiner. The examiner will change the position of his fingertip each time the patient touches his nose. In this version, speed is not important, but accuracy is.

4. **Heel-Shin Slide Test.** While standing, the patient touches the heel of one foot to the knee of the opposite leg, foot pointing forward. While maintaining this contact, he runs his heel down the shin to the ankle. Each leg should be tested.

5. **Rapid Alternating Movement Test.** The patient slaps one hand on the palm of the other, alternating palm up and then palm down. Any exercise requiring rapidly changing movement, however, will suffice. Again, both sides should be tested.

5A-3.3 **Cranial Nerves.** The cranial nerves are the 12 pairs of nerves emerging from the cranial cavity through various openings in the skull. Beginning with the most anterior (front) on the brain stem, they are appointed Roman numerals. An isolated cranial nerve lesion is an unusual finding in decompression sickness or gas embolism, but deficits occasionally occur and you should test for abnormalities. The cranial nerves must be quickly assessed as follows:

I. **Olfactory.** The olfactory nerve, which provides our sense of smell, is usually not tested.

II. **Optic.** The optic nerve is for vision. It functions in the recognition of light and shade and in the perception of objects. This test should be completed one eye at a time to determine whether the patient can read. Ask the patient if he has any blurring of vision, loss of vision, spots in the visual field, or peripheral vision loss (tunnel vision). More detailed testing can be done by standing in front of the patient and asking him to cover one eye and look straight at you. In a plane midway between yourself and the patient, slowly bring your fingertip in turn from above, below, to the right, and to the left of the direction of gaze until the patient can see it. Compare this with the earliest that you can see it with the equivalent eye. If a deficit is present, roughly map out the positions of the blind spots by passing the finger tip across the visual field.

III. **Oculomotor, (IV.) Trochlear, (VI.) Abducens.** These three nerves control eye movements. All three nerves can be tested by having the patient's eyes follow the examiner's finger in all four directions (quadrants) and then in towards the tip of the nose (giving a "crossed-eyed" look). The oculomotor nerve can be further tested by shining a light into one eye at a time. In a normal response, the pupils of both eyes will constrict.

V. **Trigeminal.** The Trigeminal Nerve governs sensation of the forehead and face and the clenching of the jaw. It also supplies the muscle of the ear (tensor tympani) necessary for normal hearing. Sensation is tested by lightly stroking the forehead, face, and jaw on each side with a finger or wisp of cotton wool.

VII. **Facial.** The Facial Nerve controls the face muscles. It stimulates the scalp, forehead, eyelids, muscles of facial expression, cheeks, and jaw. It is tested by having the patient smile, show his teeth, whistle, wrinkle his forehead, and close his eyes tightly. The two sides should perform symmetrically. Symmetry of the nasolabial folds (lines from nose to outside corners of the mouth) should be observed.

VIII. **Acoustic.** The Acoustic Nerve controls hearing and balance. Test this nerve by whispering to the patient, rubbing your fingers together next to the patient's ears, or putting a tuning fork near the patient's ears. Compare this against the other ear.

IX. **Glossopharyngeal.** The Glossopharyngeal Nerves transmit sensation from the upper mouth and throat area. It supplies the sensory component of the gag reflex and constriction of the pharyngeal wall when saying "aah." Test this nerve by touching the back of the patient's throat with a tongue depressor. This should cause a gagging response. This nerve is normally not tested.

X. **Vagus.** The Vagus Nerve has many functions, including control of the roof of the mouth and vocal cords. The examiner can test this nerve by having the patient say "aah" while watching for the palate to rise. Note the tone of the voice; hoarseness may also indicate vagus nerve involvement.

XI. **Spinal Accessory.** The Spinal Accessory Nerve controls the turning of the head from side to side and shoulder shrug against resistance. Test this nerve by having the patient turn his head from side to side. Resistance is provided by placing one hand against the side of the patient's head. The examiner should note that an injury to the nerve on one side will cause an inability to turn the head to the opposite side or weakness/absence of the shoulder shrug on the affected side.

XII. **Hypoglossal.** The Hypoglossal Nerve governs the muscle activity of the tongue. An injury to one of the hypoglossal nerves causes the tongue to twist to that side when stuck out of the mouth.

5A-3.4 **Motor.** A diver with decompression sickness may experience disturbances in the muscle system. The range of symptoms can be from a mild twitching of a muscle to weakness and paralysis. No matter how slight the abnormality, symptoms involving the motor system shall be treated.

5A-3.4.1 **Extremity Strength.** It is common for a diver with decompression illness to experience muscle weakness. Extremity strength testing is divided into two parts: upper body and lower body. All muscle groups should be tested and compared with the corresponding group on the other side, as well as with the examiner. Table 5A-1 describes the extremity strength tests in more detail. Muscle strength is graded (0-5) as follows:

- **(0) Paralysis.** No motion possible.
- **(1) Profound Weakness.** Flicker or trace of muscle contraction.
- **(2) Severe Weakness.** Able to contract muscle but cannot move joint against gravity.
- **(3) Moderate Weakness.** Able to overcome the force of gravity but not the resistance of the examiner.
- **(4) Mild Weakness.** Able to resist slight force of examiner.
- **(5) Normal.** Equal strength bilaterally (both sides) and able to resist examiner.

5A-3.4.1.1 ***Upper Extremities.*** These muscles are tested with resistance provided by the examiner. The patient should overcome force applied by the examiner that is tailored to the patient's strength. Table 5A-1 describes the extremity strength tests. The six muscle groups tested in the upper extremity are:

1. Deltoids.
2. Latissimus.
3. Biceps.
4. Triceps.
5. Forearm muscles.
6. Hand muscles.

5A-3.4.1.2 ***Lower Extremities.*** The lower extremity strength is assessed by watching the patient walk on his heels for a short distance and then on his toes. The patient should then walk while squatting ("duck walk"). These tests adequately assess lower extremity strength, as well as balance and coordination. If a more detailed examination of the lower extremity strength is desired, testing should be accomplished at each joint as in the upper arm.

5A-3.4.2 **Muscle Size.** Muscles are visually inspected and felt, while at rest, for size and consistency. Look for symmetry of posture and of muscle contours and outlines. Examine for fine muscle twitching.

5A-3.4.3 **Muscle Tone.** Feel the muscles at rest and the resistance to passive movement. Look and feel for abnormalities in tone such as spasticity, rigidity, or no tone.

5A-3.4.4 **Involuntary Movements.** Inspection may reveal slow, irregular, and jerky movements, rapid contractions, tics, or tremors.

5A-3.5 **Sensory Function.** Common presentations of decompression sickness in a diver that may indicate spinal cord dysfunction are:

Table 5A-1. *Extremity Strength Tests.*

Test	Procedure
Deltoid Muscles	The patient raises his arm to the side at the shoulder joint. The examiner places a hand on the patient's wrist and exerts a downward force that the patient resists.
Latissimus Group	The patient raises his arm to the side. The examiner places a hand on the underside of the patient's wrist and resists the patient's attempt to lower his arm.
Biceps	The patient bends his arm at the elbow, toward his chest. The examiner then grasps the patient's wrist and exerts a force to straighten the patient's arm.
Triceps	The patient bends his arm at the elbow, toward his chest. The examiner then places his hand on the patient's forearm and the patient tries to straighten his arm.
Forearm Muscles	The patient makes a fist. The examiner grips the patient's fist and resists while the patient tries to bend his wrist upward and downward.
Hand Muscles	• The patient strongly grips the examiner's extended fingers. • The patient extends his hand with the fingers widespread. The examiner grips two of the extended fingers with two of his own fingers and tries to squeeze the patient's two fingers together, noting the patient's strength of resistance.
Lower Extremity Strength	• The patient walks on his heels for a short distance. The patient then turns around and walks back on his toes. • The patient walks while squatting (duck walk). These tests adequately assesses lower extremity strength as well as balance and coordination. If a more detailed examination of lower extremity strength is desired, testing should be accomplished at each joint as in the upper arm.
In the following tests, the patient sits on a solid surface such as a desk, with feet off the floor.	
Hip Flexion	The examiner places his hand on the patient's thigh to resist as the patient tries to raise his thigh.
Hip Extension	The examiner places his hand on the underside of the patient's thigh to resist as the patient tries to lower his thigh.
Hip Abduction	The patients sits as above, with knees together. The examiner places a hand on the outside of each of the patient's knees to provide resistance. The patient tries to open his knees.
Hip Adduction	The patient sits as above, with knees apart. The examiner places a hand on the inside of each of the patient's knees to provide resistance. The patient tries to bring his knees together.
Knee Extension	The examiner places a hand on the patient's shin to resist as the patient tries to straighten his leg.
Knee Flexion	The examiner places a hand on the back of the patient's lower leg to resist as the patient tries to pull his lower leg to the rear by flexing his knee.
Ankle Dorsiflexion (ability to flex the foot toward the rear)	The examiner places a hand on top of the patient's foot to resist as the patient tries to raise his foot by flexing it at the ankle.
Ankle Plantarflexion (ability to flex the foot downward)	The examiner places a hand on the bottom of the patient's foot to resist as the patient tries to lower his foot by flexing it at the ankle.
Toes	• The patient stands on tiptoes for 15 seconds • The patient flexes his toes with resistance provided by the examiner.

- Pain
- Numbness
- Tingling ("pins-and-needles" feeling; also called paresthesia)

5A-3.5.1 **Sensory Examination.** An examination of the patient's sensory faculties should be performed. Figure 5A-2a shows the dermatomal (sensory) areas of skin sensations that correlate with each spinal cord segment. Note that the dermatomal areas of the trunk run in a circular pattern around the trunk. The dermatomal areas in the arms and legs run in a more lengthwise pattern. In a complete examination, each spinal segment should be checked for loss of sensation.

5A-3.5.2 **Sensations.** Sensations easily recognized by most normal people are sharp/dull discrimination (to perceive as separate) and light touch. It is possible to test pressure, temperature, and vibration in special cases. The likelihood of DCS affecting only one sense, however, is very small.

5A-3.5.3 **Instruments.** An ideal instrument for testing changes in sensation is a sharp object, such as the Wartenberg pinwheel or a common safety pin. Either of these objects must applied at intervals. Avoid scratching or penetrating the skin. It is not the intent of this test to cause pain.

5A-3.5.4 **Testing the Trunk.** Move the pinwheel or other sharp object from the top of the shoulder slowly down the front of the torso to the groin area. Another method is to run it down the rear of the torso to just below the buttocks. The patient should be asked if he feels a sharp point and if he felt it all the time. Test each dermatome by going down the trunk on each side of the body. Test the neck area in similar fashion.

5A-3.5.5 **Testing Limbs.** In testing the limbs, a circular pattern of testing is best. Test each limb in at least three locations, and note any difference in sensation on each side of the body. On the arms, circle the arm at the deltoid, just below the elbow, and at the wrist. In testing the legs, circle the upper thigh, just below the knee, and the ankle.

5A-3.5.6 **Testing the Hands.** The hand is tested by running the sharp object across the back and palm of the hand and then across the fingertips.

5A-3.5.7 **Marking Abnormalities.** If an area of abnormality is found, mark the area as a reference point in assessment. Some examiners use a marking pen to trace the area of decreased or increased sensation on the patient's body. During treatment, these areas are rechecked to determine whether the area is improving. An example of improvement is an area of numbness getting smaller.

5A-3.6 **Deep Tendon Reflexes.** The purpose of the deep tendon reflexes is to determine if the patient's response is normal, nonexistent, hypoactive (deficient), or hyperactive (excessive). The patient's response should be compared to responses the examiner has observed before. Notation should be made of whether the responses are equal bilaterally (both sides) and if the upper and lower reflexes are similar. If any difference in the reflexes is noticed, the patient should be asked if there is a prior

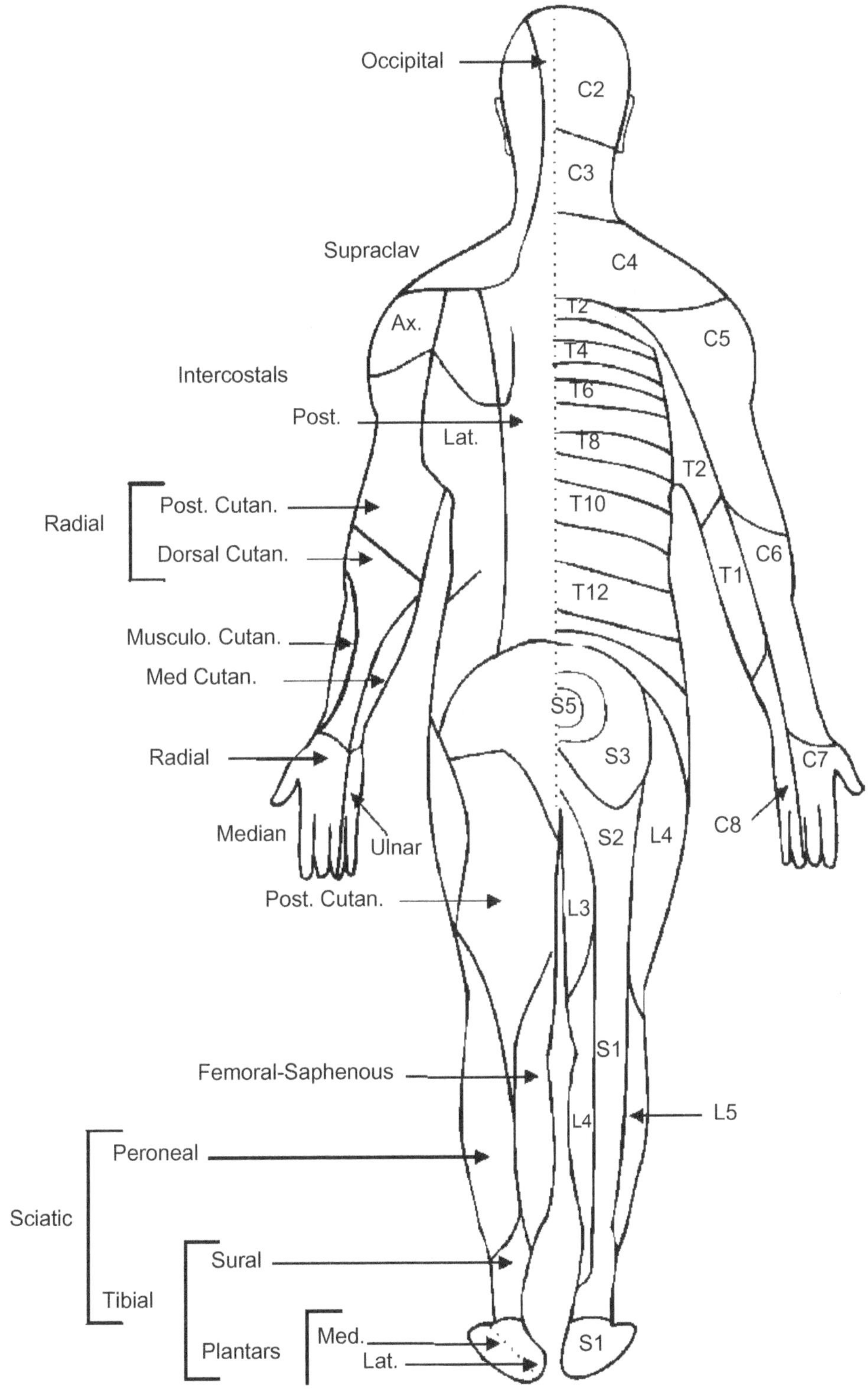

Figure 5A-2a. Dermatomal Areas Correlated to Spinal Cord Segment (sheet 1 of 2).

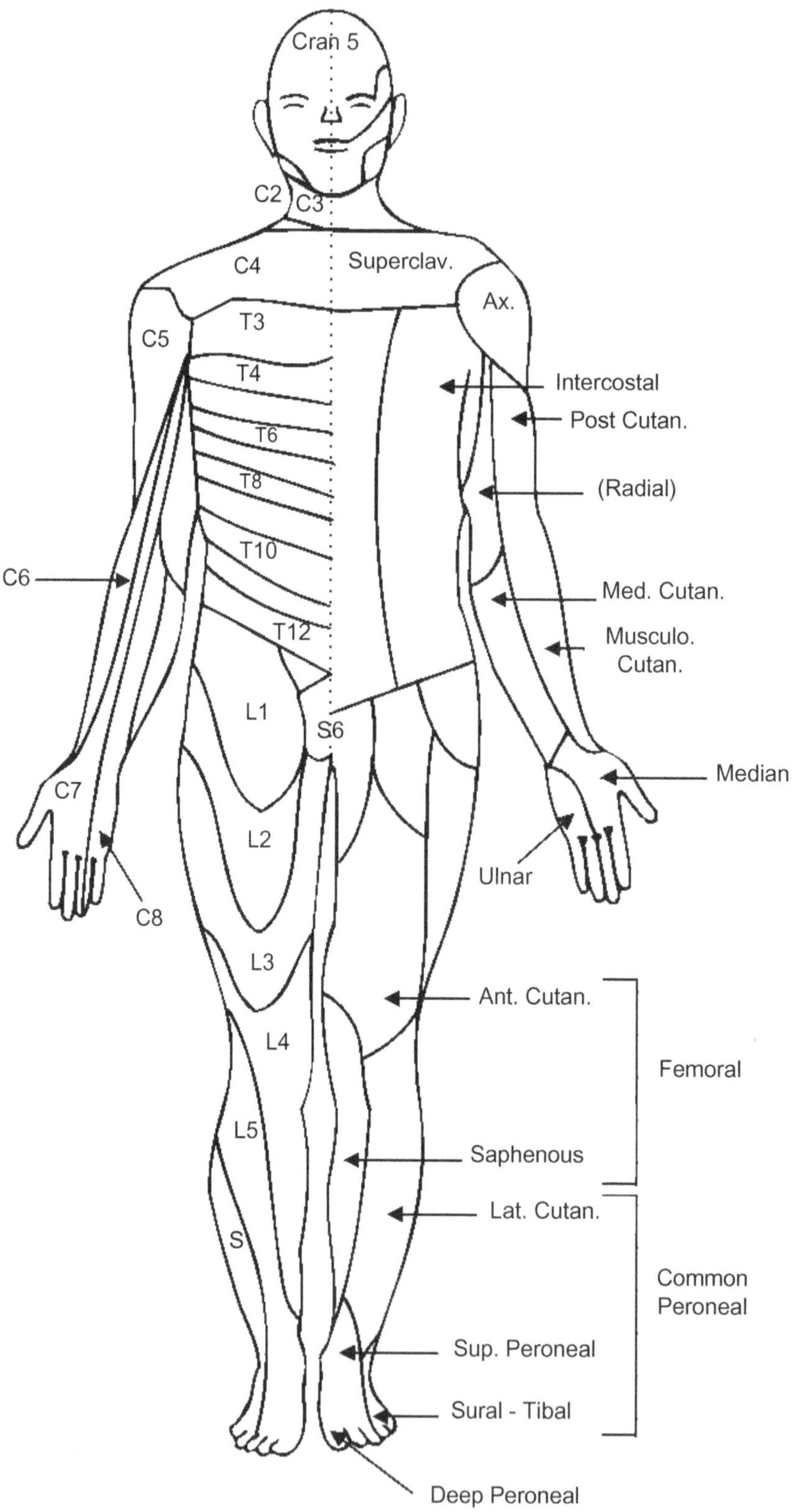

Figure 5A-2b. Dermatomal Areas Correlated to Spinal Cord Segment (sheet 2 of 2).

medical condition or injury that would cause the difference. Isolated differences should not be treated, because it is extremely difficult to get symmetrical responses bilaterally. To get the best response, strike each tendon with an equal, light force, and with sharp, quick taps. Usually, if a deep tendon reflex is abnormal due to decompression sickness, there will be other abnormal signs present. Test the biceps, triceps, knee, and ankle reflexes by striking the tendon as described in Table 5A-2.

Table 5A-2. *Reflexes.*

Test	Procedure
Biceps	The examiner holds the patient's elbow with the patient's hand resting on the examiner's forearm. The patient's elbow should be slightly bent and his arm relaxed. The examiner places his thumb on the patient's biceps tendon, located in the bend of the patient's elbow. The examiner taps his thumb with the percussion hammer, feeling for the patient's muscle to contract.
Triceps	The examiner supports the patient's arm at the biceps. The patient's arm hangs with the elbow bent. The examiner taps the back of the patient's arm just above the elbow with the percussion hammer, feeling for the muscle to contract.
Knee	The patient sits on a table or bench with his feet off the deck. The examiner taps the patient's knee just below the kneecap, on the tendon. The examiner looks for the contraction of the quadriceps (thigh muscle) and movement of the lower leg.
Ankle	The patient sits as above. The examiner places slight pressure on the patient's toes to stretch the Achilles' tendon, feeling for the toes to contract as the Achilles' tendon shortens (contracts).

PAGE LEFT BLANK INTENTIONALLY

APPENDIX 5B

First Aid

5B-1 INTRODUCTION

This appendix, covering one-man cardiopulmonary resuscitation, control of bleeding and shock treatment is intended as a quick reference for individuals trained in first aid and basic life support. Complete descriptions of all basic life support techniques are available through your local branch of the American Heart Association. Further information on the control of bleeding and treatment for shock is in the *Hospital Corpsman 3 & 2 Manual*, NAVEDTRA 10669-C.

5B-2 CARDIOPULMONARY RESUSCITATION

All divers must be qualified in cardiopulmonary resuscitation (CPR) in accordance with the procedures of the American Heart Association. Periodic recertification according to current guidelines in basic life support is mandatory for all Navy divers. Training can be requested through your local medical command or directly through your local branch of the American Heart Association.

5B-3 CONTROL OF MASSIVE BLEEDING

Massive bleeding must be controlled immediately. If the victim also requires resuscitation, the two problems must be handled simultaneously. Bleeding may involve veins or arteries; the urgency and method of treatment will be determined in part by the type and extent of the bleeding.

5B-3.1 **External Arterial Hemorrhage.** Arterial bleeding can usually be identified by bright red blood, gushing forth in jets or spurts that are synchronous with the pulse. The first measure used to control external arterial hemorrhage is direct pressure on the wound.

5B-3.2 **Direct Pressure.** Pressure is best applied with sterile compresses, placed directly and firmly over the wound. In a crisis, however, almost any material can be used. If the material used to apply direct pressure soaks through with blood, apply additional material on top; do not remove the original pressure bandage. Elevating the extremity also helps to control bleeding. If direct pressure cannot control bleeding, it should be used in combination with pressure points.

5B-3.3 **Pressure Points.** Bleeding can often be temporarily controlled by applying hand pressure to the appropriate pressure point. A pressure point is a place where the main artery to the injured part lies near the skin surface and over a bone. Apply pressure at this point with the fingers (digital pressure) or with the heel of the hand; no first aid materials are required. The object of the pressure is to compress the artery against the bone, thus shutting off the flow of blood from the heart to the wound.

5B-3.3.1 **Pressure Point Location on Face.** There are 11 principal points on each side of the body where hand or finger pressure can be used to stop hemorrhage. These points are shown in Figure 5B-1. If bleeding occurs on the face below the level of the eyes, apply pressure to the point on the mandible. This is shown in Figure 5B-1(A). To find this pressure point, start at the angle of the jaw and run your finger forward along the lower edge of the mandible until you feel a small notch. The pressure point is in this notch.

5B-3.3.2 **Pressure Point Location for Shoulder or Upper Arm.** If bleeding is in the shoulder or in the upper part of the arm, apply pressure with the fingers behind the clavicle. You can press down against the first rib or forward against the clavicle—either kind of pressure will stop the bleeding. This pressure point is shown in Figure 5B-1(B).

5B-3.3.3 **Pressure Point Location for Middle Arm and Hand.** Bleeding between the middle of the upper arm and the elbow should be controlled by applying digital pressure in the inner (body) side of the arm, about halfway between the shoulder and the elbow. This compresses the artery against the bone of the arm. The application of pressure at this point is shown in Figure 5B-1(C). Bleeding from the hand can be controlled by pressure at the wrist, as shown in Figure 5B-1(D). If it is possible to hold the arm up in the air, the bleeding will be relatively easy to stop.

5B-3.3.4 **Pressure Point Location for Thigh.** Figure 5B-1(E) shows how to apply digital pressure in the middle of the groin to control bleeding from the thigh. The artery at this point lies over a bone and quite close to the surface, so pressure with your fingers may be sufficient to stop the bleeding.

5B-3.3.5 **Pressure Point Location for Foot.** Figure 5B-1(F) shows the proper position for controlling bleeding from the foot. As in the case of bleeding from the hand, elevation is helpful in controlling the bleeding.

5B-3.3.6 **Pressure Point Location for Temple or Scalp.** If bleeding is in the region of the temple or the scalp, use your finger to compress the main artery to the temple against the skull bone at the pressure point just in front of the ear. Figure 5B-1(G) shows the proper position.

5B-3.3.7 **Pressure Point Location for Neck.** If the neck is bleeding, apply pressure below the wound, just in front of the prominent neck muscle. Press inward and slightly backward, compressing the main artery of that side of the neck against the bones of the spinal column. The application of pressure at this point is shown in Figure 5B-1(H). Do not apply pressure at this point unless it is absolutely essential, since there is a great danger of pressing on the windpipe and thus choking the victim.

5B-3.3.8 **Pressure Point Location for Lower Arm.** Bleeding from the lower arm can be controlled by applying pressure at the elbow, as shown in Figure 5B-1(I).

5B-3.3.9 **Pressure Point Location of the Upper Thigh.** As mentioned before, bleeding in the upper part of the thigh can sometimes be controlled by applying digital pressure in the middle of the groin, as shown in Figure 5B-1(E). Sometimes, however, it

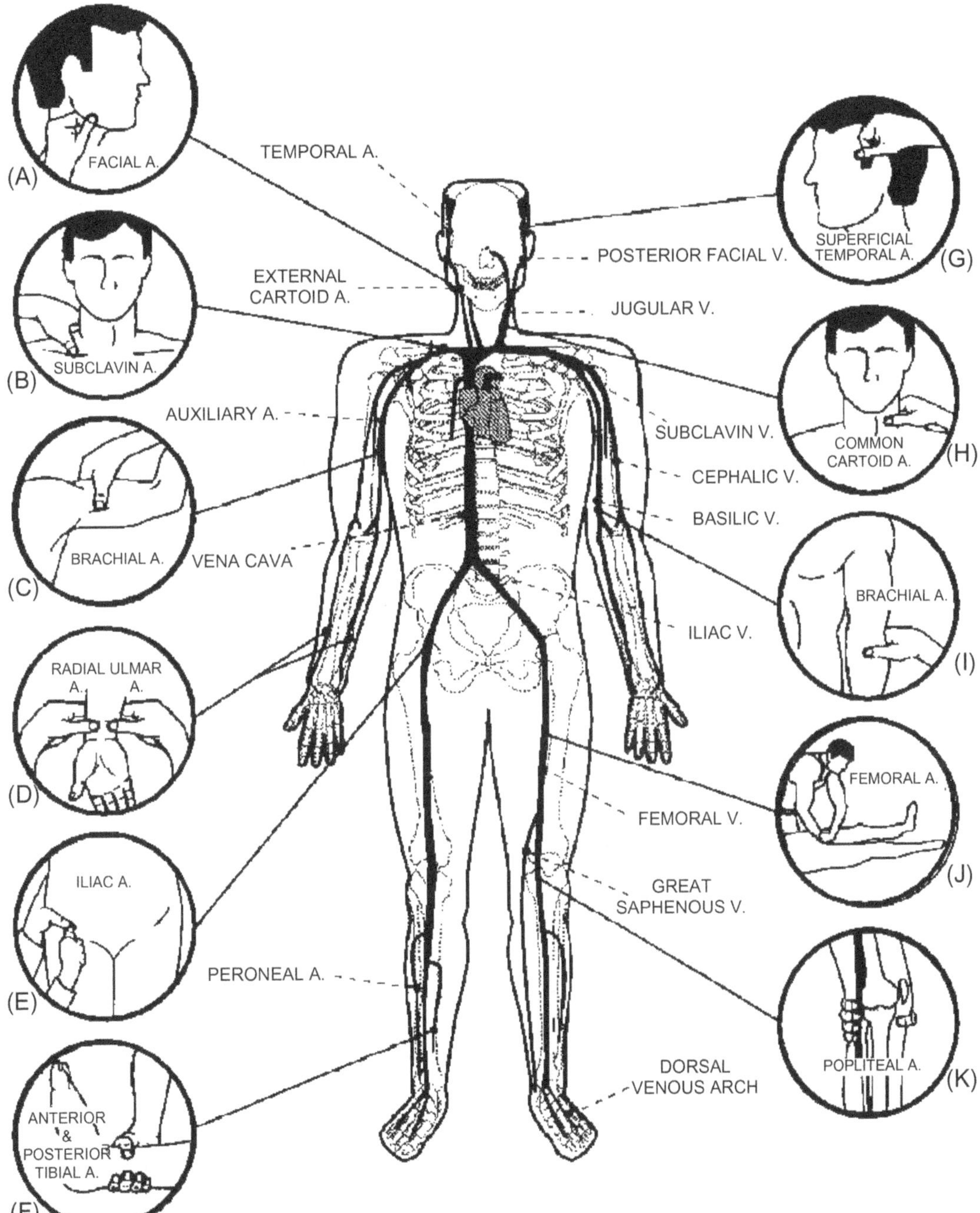

Figure 5B-1. Pressure Points.

is more effective to use the pressure point of the upper thigh as shown in Figure 5B-1(J). If you use this point, apply pressure with the closed fist of one hand and use the other hand to give additional pressure. The artery at this point is deeply buried in some of the heaviest muscle of the body, so a great deal of pressure must be exerted to compress the artery against the bone.

5B-3.3.10 **Pressure Point Location Between Knee and Foot.** Bleeding between the knee and the foot may be controlled by firm pressure at the knee. If pressure at the side of the knee does not stop the bleeding, hold the front of the knee with one hand and thrust your fist hard against the artery behind the knee, as shown in Figure 5B-1(K). If necessary, you can place a folded compress or bandage behind the knee, bend the leg back and hold it in place by a firm bandage. This is a most effective way of controlling bleeding, but it is so uncomfortable for the victim that it should be used only as a last resort.

5B-3.3.11 **Determining Correct Pressure Point.** You should memorize these pressure points so that you will know immediately which point to use for controlling hemorrhage from a particular part of the body. Remember, the correct pressure point is that which is (1) NEAREST THE WOUND and (2) BETWEEN THE WOUND AND THE MAIN PART OF THE BODY.

5B-3.3.12 **When to Use Pressure Points.** It is very tiring to apply digital pressure and it can seldom be maintained for more than 15 minutes. Pressure points are recommended for use while direct pressure is being applied to a serious wound by a second rescuer, or after a compress, bandage, or dressing has been applied to the wound, since it will slow the flow of blood to the area, thus giving the direct pressure technique a better chance to stop the hemorrhage. It is also recommended as a stopgap measure until a pressure dressing or a tourniquet can be applied.

5B-3.4 **Tourniquet.** A tourniquet is a constricting band that is used to cut off the supply of blood to an injured limb. Use a tourniquet only if the control of hemorrhage by other means proves to be difficult or impossible. A tourniquet must always be applied ABOVE the wound, i.e., towards the trunk, and it must be applied as close to the wound as practical.

5B-3.4.1 **How to Make a Tourniquet.** Basically, a tourniquet consists of a pad, a band and a device for tightening the band so that the blood vessels will be compressed. It is best to use a pad, compress or similar pressure object, if one is available. It goes under the band. It must be placed directly over the artery or it will actually decrease the pressure on the artery and thus allow a greater flow of blood. If a tourniquet placed over a pressure object does not stop the bleeding, there is a good chance that the pressure object is in the wrong place. If this occurs, shift the object around until the tourniquet, when tightened, will control the bleeding. Any long flat material may be used as the band. It is important that the band be flat: belts, stockings, flat strips of rubber or neckerchiefs may be used; but rope, wire, string or very narrow pieces of cloth should not be used because they cut into the flesh. A short stick may be used to twist the band tightening the tourniquet. Figure 5B-2 shows how to apply a tourniquet.

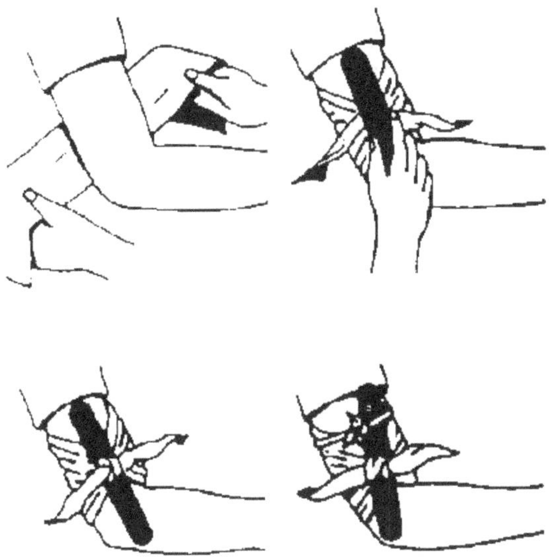

Figure 5B-2. Applying a Tourniquet.

5B-3.4.2 **Tightness of Tourniquet.** To be effective, a tourniquet must be tight enough to stop the arterial blood flow to the limb, so be sure to draw the tourniquet tight enough to stop the bleeding. However, do not make it any tighter than necessary.

5B-3.4.3 **After Bleeding is Under Control.** After you have brought the bleeding under control with the tourniquet, apply a sterile compress or dressing to the wound and fasten it in position with a bandage.

5B-3.4.4 **Points to Remember.** Here are the points to remember about using a tourniquet:

1. Don't use a tourniquet unless you can't control the bleeding by any other means.

2. Don't use a tourniquet for bleeding from the head, face, neck or trunk. Use it only on the limbs.

3. Always apply a tourniquet ABOVE THE WOUND and as close to the wound as possible. As a general rule, do not place a tourniquet below the knee or elbow except for complete amputations. In certain distal areas of the extremities, nerves lie close to the skin and may be damaged by the compression. Furthermore, rarely does one encounter bleeding distal to the knee or elbow that requires a tourniquet.

4. Be sure you draw the tourniquet tight enough to stop the bleeding, but don't make it any tighter than necessary. The pulse beyond the tourniquet should disappear.

5. Don't loosen a tourniquet after it has been applied. Transport the victim to a medical facility that can offer proper care.

6. Don't cover a tourniquet with a dressing. If it is necessary to cover the injured person in some way, MAKE SURE that all the other people concerned with the case know about the tourniquet. Using crayon, skin pencil or blood, mark a large "T" on the victim's forehead or on a medical tag attached to the wrist.

5B-3.5 **External Venous Hemorrhage.** Venous hemorrhage is not as dramatic as severe arterial bleeding, but if left unchecked, it can be equally serious. Venous bleeding is usually controlled by applying direct pressure on the wound.

5B-3.6 **Internal Bleeding.** The signs of external bleeding are obvious, but the first aid team must be alert for the possibility of internal hemorrhage. Victims subjected to crushing injuries, heavy blows or deep puncture wounds should be observed carefully for signs of internal bleeding. Signs usually present include:

- Moist, clammy, pale skin
- Feeble and very rapid pulse rate
- Lowered blood pressure
- Faintness or actual fainting
- Blood in stool, urine, or vomitus

5B-3.6.1 **Treatment of Internal Bleeding.** Internal bleeding can be controlled only by trained medical personnel and often only under hospital conditions. Efforts in the field are generally limited to replacing lost blood volume through intravenous infusion of saline, Ringer's Lactate, or other fluids, and the administration of oxygen. Rapid evacuation to a medical facility is essential.

5B-4 SHOCK

Shock may occur with any injury and will certainly be present to some extent with serious injuries. Shock is caused by a loss of blood flow, resulting in a drop of blood pressure and decreased circulation. If not treated, this drop in the quantity of blood flowing to the tissues can have serious permanent effects, including death.

5B-4.1 **Signs and Symptoms of Shock.** Shock can be recognized from the following signs and symptoms.

- Respiration shallow, irregular, labored
- Eyes vacant (staring), lackluster, tired-looking
- Pupils dilated
- Cyanosis (blue lips/fingernails)
- Skin pale or ashen gray; wet, clammy, cold
- Pulse weak and rapid, or may be normal
- Blood pressure drop
- Possible retching, vomiting, nausea, hiccups
- Thirst

5B-4.2 **Treatment.** Shock must be treated before any other injuries or conditions except breathing and circulation obstructions and profuse bleeding. Proper treatment involves caring for the whole patient, not limiting attention to only a few of the disorders. The following steps must be taken to treat a patient in shock.

1. Ensure adequate breathing. If the patient is breathing, maintain an adequate airway by tilting the head back properly. If the patient is not breathing, establish an airway and restore breathing through some method of pulmonary resuscitation. If both respiration and circulation have stopped, institute cardiopulmonary resuscitation measures (refer to paragraph 5B-2).

2. Control bleeding. If the patient has bleeding injuries, use direct pressure points or a tourniquet, as required (refer to paragraph 5B-3).

3. Administer oxygen. Remember that an oxygen deficiency will be caused by the reduced circulation. Administer 100 percent oxygen.

4. Elevate the lower extremities. Since blood flow to the heart and brain may have been diminished, circulation can be improved by raising the legs slightly. It is not recommended that the entire body be tilted, since the abdominal organs pressing against the diaphragm may interfere with respiration. Exceptions to the rule of raising the feet are cases of head and chest injuries, when it is desirable to lower the pressure in the injured parts; in these cases, the upper part of the body should be elevated slightly. Whenever there is any doubt as to the best position, lay the patient flat.

5. Avoid rough handling. Handle the patient as little and as gently as possible. Body motion has a tendency to aggravate shock conditions.

6. Prevent loss of body heat. Keep the patient warm but guard against overheating, which can aggravate shock. Remember to place a blanket under as well as on top of the patient, to prevent loss of heat into the ground, boat or ship deck.

7. Keep the patient lying down. A prone position avoids taxing the circulatory system. However, some patients, such as those with heart disorders, will have to be transported in a semi-sitting position.

8. Give nothing by mouth.

PAGE LEFT BLANK INTENTIONALLY

APPENDIX 5C

Dangerous Marine Animals

5C-1 INTRODUCTION

5C-1.1 **Purpose.** This appendix provides general information on dangerous marine life that may be encountered in diving operations.

5C-1.2 **Scope.** It is beyond the scope of this manual to catalog all types of marine encounters and potential injury. Planners should consult the recommended references listed at the end of this appendix for more definite information. Medical personnel are also a good source of information and should be consulted prior to operating in unfamiliar waters. A good working knowledge of the marine environment should preclude lost time and severe injury.

5C-2 PREDATORY MARINE ANIMALS

5C-2.1 **Sharks.** Shark attacks on humans are infrequent. Since 1965, the annual recorded number of shark attacks is only 40 to 100 worldwide. These attacks are unpredictable and injuries may result not only from bites, but also by coming in contact with the shark's skin. Shark skin is covered with very sharp dentine appendages, called denticles, which are reinforced with tooth-like centers. Contact with shark skin can lead to wide abrasions and heavy bleeding.

5C-2.1.1 **Shark Pre-Attack Behavior.** Pre-attack behavior by most sharks is somewhat predictable. A shark preparing to attack swims with an exaggerated motion, its pectoral fins pointing down in contrast to the usual flared out position, and it swims in circles of decreasing radius around the prey. An attack may be heralded by unexpected acceleration or other marked change in behavior, posture, or swim patterns. Should surrounding schools of fish become unexplainably agitated, sharks may be in the area. Sharks are much faster and more powerful than any swimmer. All sharks must be treated with extreme respect and caution (see Figure 5C-1).

5C-2.1.2 **First Aid and Treatment.**

1. Bites may result in a large amount of bleeding and tissue loss. Take immediate action to control bleeding using large gauze pressure bandages. Cover wounds with layers of compressive dressings preferably made with gauze, but easily made from shirts or towels, and held in place by wrapping the wound tightly with gauze, torn clothing, towels, or sheets. Direct pressure with elevation or extreme compression on pressure points will control all but the most serious bleeding. The major pressure points are: the radial artery pulse point for the hand; above the elbow under the biceps muscle for the forearm (brachial artery); and the groin area with deep finger-tip or heel-of-the-hand pressure for bleeding from the leg (femoral artery). When bleeding cannot be controlled by direct pressure and elevation or pressure points, a tourniquet or ligature may

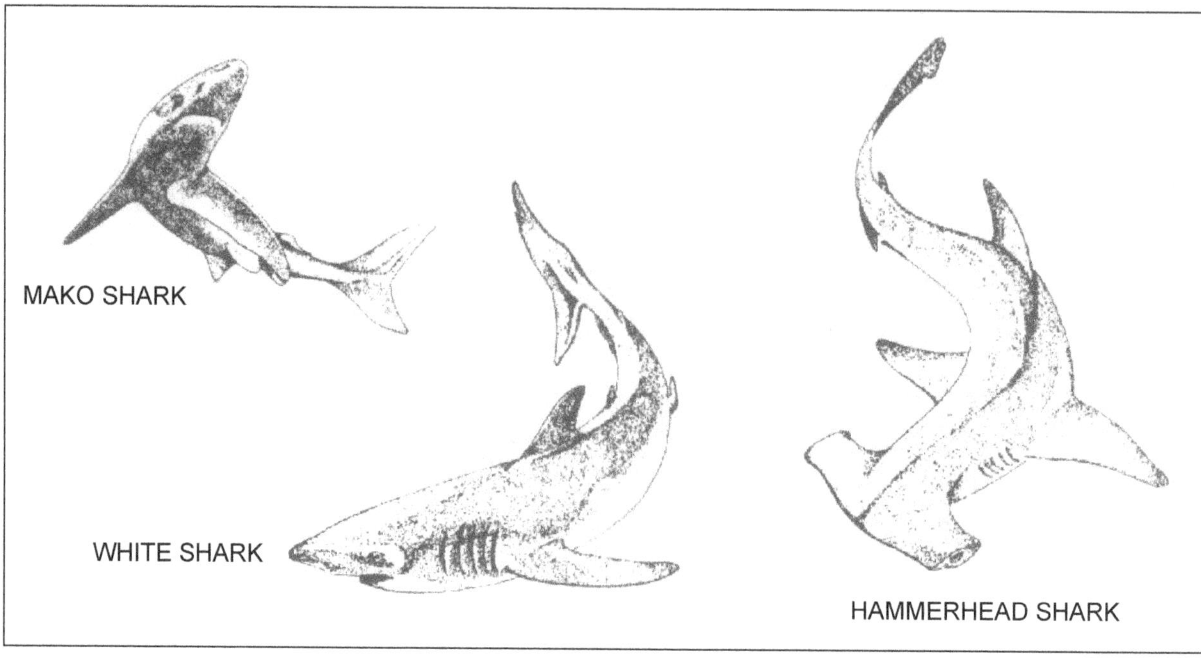

Figure 5C-1. Types of Sharks.

be needed to save the victim's life even though there is the possibility of loss of the limb. Tourniquets are applied only as a last resort and with only enough pressure to control bleeding. Do not remove the tourniquet. The tourniquet should be removed only by a physician in a hospital setting. Loosening of a tourniquet may cause further shock by releasing toxins into the circulatory system from the injured limb as well as continued blood loss.

2. Treat for shock by laying the patient down and elevating his feet.

3. If medical personnel are available, begin intravenous (IV) Ringer's lactate or normal saline with a large-bore cannula (16 or 18 ga). If blood loss has been extensive, several liters should be infused rapidly. The patient's color, pulse, and blood pressure should be used as a guide to the volume of fluid required. Maintain an airway and administer oxygen. Do not give fluids by mouth. If the patient's cardiovascular state is stable, narcotics may be administered in small doses for pain relief. Observe closely for evidence of depressed respirations due to the use of narcotics.

4. Initial stabilization procedures should include attention to the airway, breathing, and circulation, followed by a complete evaluation for multiple trauma.

5. Transport the victim to a medical facility as soon as possible. Reassure the patient.

6. Should a severed limb be retrieved, wrap it in bandages, moisten with saline, place in a plastic bag and chill, but not in direct contact with ice. Transport the severed limb with the patient.

7. Clean and debride wounds as soon as possible in a hospital or controlled environment. Since shark teeth are cartilage, not bone, and may not appear on an X-ray, operative exploration should be performed to remove dislodged teeth.

8. Consider X-ray evaluation for potential bone damage due to crush injury. Severe crush injury may result in acute renal failure due to myoglobin released from injured muscle, causing the urine to be a smoky brown color. Monitor closely for kidney function and adjust IV fluid therapy appropriately.

9. Administer tetanus prophylaxis: Tetanus toxoid, 0.5 ml intramuscular (IM) and tetanus immune globulin, 250 to 400 units IM.

10. Culture infected wounds for both aerobes and anaerobes before instituting broad spectrum antibiotic coverage; secondary infections with Clostridium and Vibrio species have been reported frequently.

11. Acute surgical repair, reconstructive surgery, and hyperbaric oxygen (HBO) adjuvant therapy improving tissue oxygenation may all be needed.

12. In cases of unexplained decrease in mental status or other neurological signs and symptoms following shark attack while diving, consider arterial gas embolism or decompression sickness as a possible cause.

5C-2.2 **Killer Whales.** Killer whales live in all oceans, both tropical and polar. This whale is a large mammal with a blunt, rounded snout and high black dorsal fin (Figure 5C-2). The jet black head and back contrast sharply with the snowy-white underbelly. Usually, a white patch can be seen behind and above the eye. The killer whale is usually observed in packs of 3 to 40 whales. It has powerful jaws, great weight, speed, and interlocking teeth. Because of its speed and carnivorous habits, this animal should be treated with great respect. There have been no recorded attacks on humans.

Figure 5C-2. Killer Whale.

5C-2.2.1 **Prevention.** When killer whales are spotted, all personnel should immediately leave the water. Extreme care should be taken on shore areas, piers, barges, ice floes, etc., when killer whales are in the area.

5C-2.2.2 **First Aid and Treatment.** First aid and treatment would follow the same general principles as those used for a shark bite (paragraph 5C-2.1.2).

5C-2.3 **Barracuda.** Approximately 20 species of barracuda inhabit the oceans of the West Indies, the tropical waters from Brazil to Florida and the Indo-Pacific oceans from the Red Sea to the Hawaiian Islands. The barracuda is a long, thin fish with prominent jaws and teeth, silver to blue in color, with a large head and a V-shaped tail (Figure 5C-3). It may grow up to 10 feet long and is a fast swimmer, capable of striking rapidly and fiercely. It will follow swimmers but seldom attacks an underwater swimmer. It is known to attack surface swimmers and limbs dangling in the water. Barracuda wounds can be distinguished from those of a shark by the tooth pattern. A barracuda leaves straight or V-shaped wounds while those of a shark are curved like the shape of its jaws. Life threatening attacks by barracuda are rare.

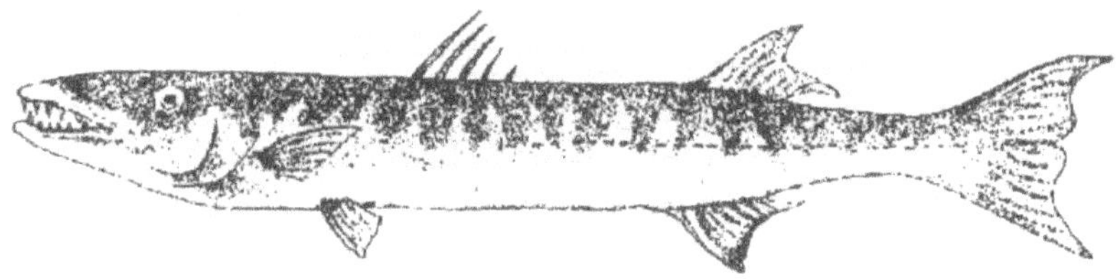

Figure 5C-3. Barracuda.

5C-2.3.1 **Prevention.** Barracuda are attracted by any bright object. Avoid wearing shiny equipment or jewelry in waters when barracudas are likely to be present. Avoid carrying speared fish, as barracuda will strike them. Avoid splashing or dangling limbs in barracuda-infested waters.

5C-2.3.2 **First Aid and Treatment.** First aid and treatment follow the same general principles as those used for shark bites (paragraph 5C-2.1.2). Injuries are likely to be less severe than shark bite injuries.

5C-2.4 **Moray Eels.** While some temperate zone species of the moray eel are known, it primarily inhabits tropical and subtropical waters. It is a bottom dweller and is commonly found in holes and crevices or under rocks and coral. It is snake-like in both appearance and movement and has tough, leathery skin (Figure 5C-4). It can grow to a length of 10 feet and has prominent teeth. A moray eel is extremely territorial and attacks frequently result from reaching into a crevice or hole occupied by the eel. It is a powerful and vicious biter and may be difficult to dislodge after a bite is initiated. Bites from moray eels may vary from multiple small puncture wounds to the tearing, jagged type with profuse bleeding if there has been a struggle. Injuries are usually inflicted on hands or forearms.

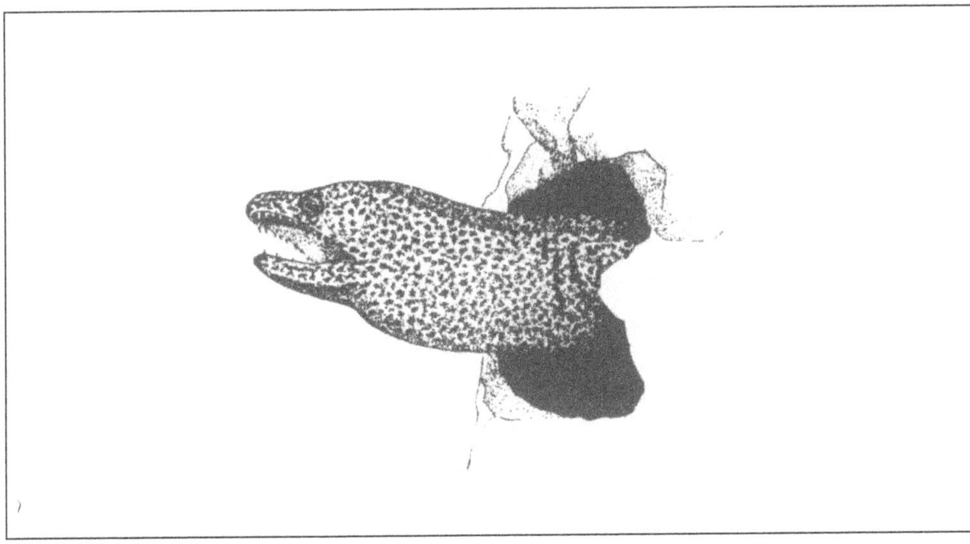

Figure 5C-4. Moray Eel.

5C-2.4.1 **Prevention.** Extreme care should be used when reaching into holes or crevices. Avoid provoking or attempting to dislodge an eel from its hole.

5C-2.4.2 **First Aid and Treatment.** Primary first aid must stop the bleeding. Direct pressure and raising the injured extremity almost always controls bleeding. Arrange for medical follow-up. Severe hand injuries should be evaluated immediately by a physician. Mild envenomation may occur from a toxin that is released from the palatine mucosa in the mouth of certain moray eels. The nature of this toxin is not known. Treatment is supportive. Follow principles of wound management and tetanus prophylaxis as in caring for shark bites. Antibiotic therapy should be instituted early. Immediate specialized care by a hand surgeon may be necessary for tendon and nerve repair of the hand to prevent permanent damage and loss of function of the hand.

5C-2.5 **Sea Lions.** The sea lion inhabits the Pacific Ocean and is numerous on the West Coast of the United States. It resembles a large seal. Sea lions are normally harmless; however, during the breeding season (October through December) large bull sea lions can become irritated and will nip at divers. Attempts by divers to handle these animals may result in bites. These bites appear similar to dog bites and are rarely severe.

5C-2.5.1 **Prevention.** Divers should avoid these mammals when in the water.

5C-2.5.2 **First Aid and Treatment.**

1. Control local bleeding.
2. Clean and debride wound.

3. Administer tetanus prophylaxis as appropriate.
4. Wound infections are common and prophylactic antibiotic therapy is advised.

5C-3 VENOMOUS MARINE ANIMALS

5C-3.1 **Venomous Fish (Excluding Stonefish, Zebrafish, Scorpionfish).** Identification of a fish following a sting is not always possible; however, symptoms and effects of venom do not vary greatly. Venomous fish are rarely aggressive and usually contact is made by accidentally stepping on or handling the fish. Dead fish spines remain toxic (see Figure 5C-5). Venom is generally heat-labile and may be decomposed by hot water. Local symptoms following a sting may first include severe pain later combined with numbness or even hypersensitivity around the wound. The wound site may become cyanotic with surrounding tissue becoming pale and swollen. General symptoms may include nausea, vomiting, sweating, mild fever, respiratory distress and collapse. The pain induced may seem disproportionately high to apparent severity of the injury. Medical personnel should be prepared for serious anaphylactic reactions from apparently minor stings or envenomation.

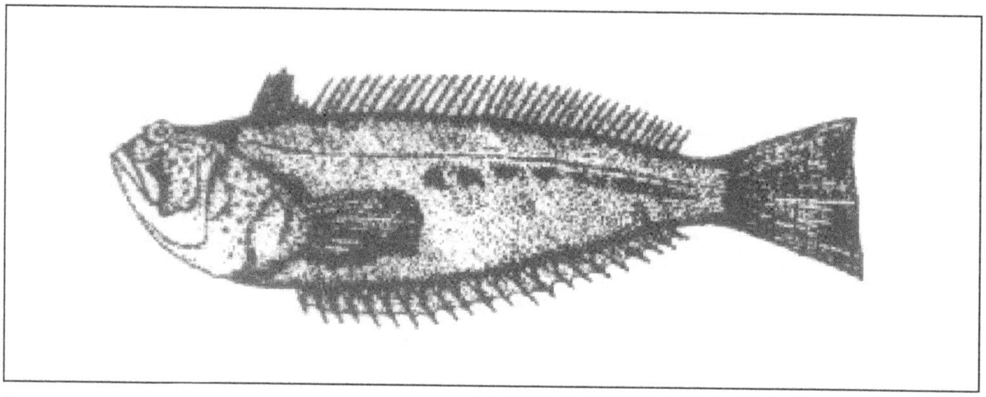

Figure 5C-5. Venomous Fish. Shown is the weeverfish.

5C-3.1.1 **Prevention.** Avoid handling suspected venomous fish. Venomous fish are often found in holes or crevices or lying well camouflaged on rocky bottoms. Divers should be alert for their presence and should take care to avoid them.

5C-3.1.2 **First Aid and Treatment.**

1. Get victim out of water; watch for fainting.

2. Lay patient down and reassure.

3. Observe for signs of shock.

4. Wash wound with cold, salt water or sterile saline solution. Surgery may be required to open up the puncture wound. Suction is not effective to remove this toxin.

5. Soak wound in hot water for 30 to 90 minutes. Heat may break down the venom. The water should be as hot as the victim can tolerate but not hotter than 122°F (50°C). Immersion in water above 122°F (50°C) for longer than a brief period may lead to scalding. Immersion in water up to 122°F (50°C) should therefore be brief and repeated as necessary. Use hot compresses if the wound is on the face. Adding magnesium sulfate (epsom salts) to the water offers no benefit.

6. Calcium gluconate injections, diazepam, or methocarbamol may help to reduce muscle spasms. Infiltration of the wound with 0.5 percent to 2.0 percent xylocaine with no epinephrine is helpful in reducing pain. If xylocaine with epinephrine is mistakenly used, local necrosis may result from both the toxin and epinephrine present in the wound. Narcotics may also be needed to manage severe pain.

7. Clean and debride wound. Spines and sheath frequently remain. Be sure to remove all of the sheath as it may continue to release venom.

8. Tourniquets or ligatures are no longer advised. Use an antiseptic or antibiotic ointment and sterile dressing. Restrict movement of the extremity with immobilizing splints and cravats.

9. Administer tetanus prophylaxis as appropriate.

10. Treat prophylactically with topical antibiotic ointment. If delay in treatment has occurred, it is recommended that the wound be cultured prior to administering systemic antibiotics.

5C-3.2 **Highly Toxic Fish (Stonefish, Zebrafish, Scorpionfish).** Stings by stonefish, zebrafish, and scorpionfish have been known to cause fatalities. While many similarities exist between these fish and the venomous fish of the previous section, a separate section has been included because of the greater toxicity of their venom and the availability of an antivenin. The antivenin is specific for the stonefish but may have some beneficial effects against the scorpionfish and zebrafish. Local symptoms are similar to other fish envenomation except that pain is more severe and may persist for many days. Generalized symptoms are often present and may include respiratory failure and cardiovascular collapse. These fish are widely distributed in temperate and tropical seas and in some arctic waters. They are shallow-water bottom dwellers. Stonefish and scorpionfish are flattened vertically, dark and mottled. Zebrafish are ornate and feathery in appearance with alternating patches of dark and light color (see Figure 5C-6).

5C-3.2.1 **Prevention.** Prevention is the same as for venomous fish (paragraph 5C-3.1.1).

5C-3.2.2 **First Aid and Treatment.**

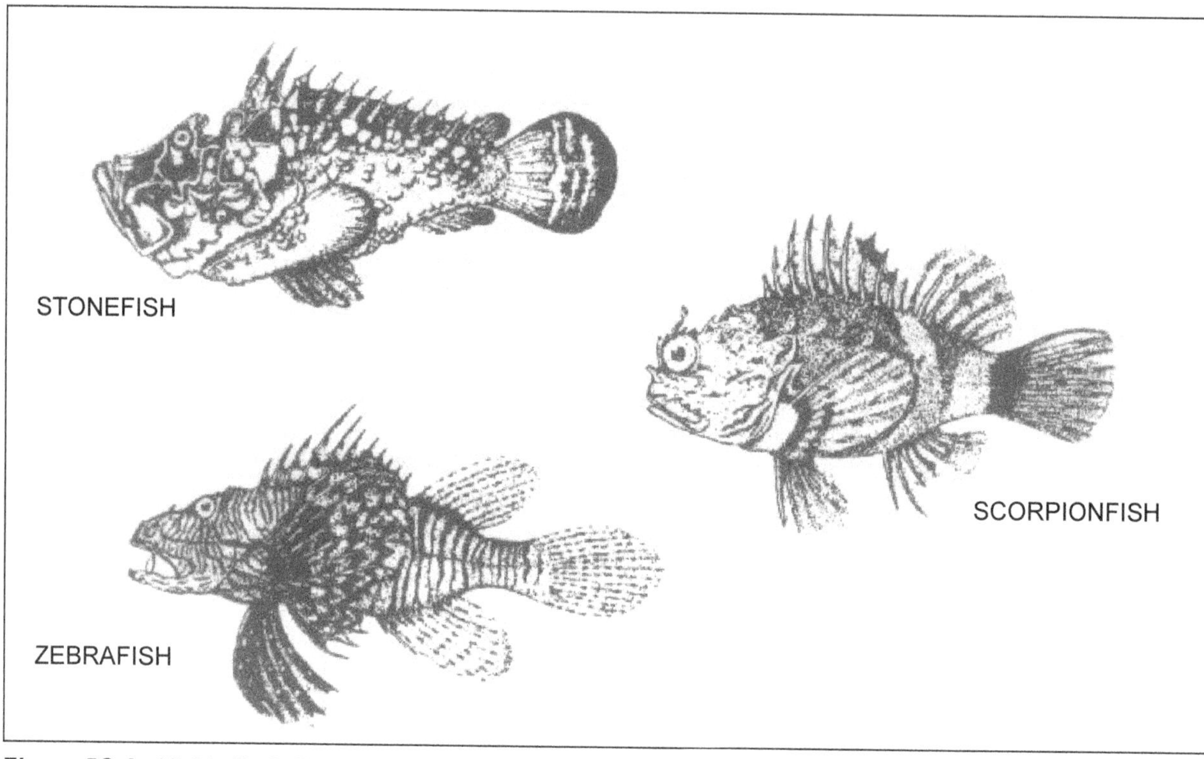

Figure 5C-6. Highly Toxic Fish.

1. Give the same first aid as that given for venomous fish (paragraph 5C-3.1.2).

2. Observe the patient carefully for the possible development of life-threatening complications. The venom is an unstable protein which acts as a myotoxin on skeletal, involuntary, and cardiac muscle. This may result in muscular paralysis, respiratory depression, peripheral vasodilation, shock, cardiac dysrhythmias, or cardiac arrest.

3. Clean and debride wound.

4. Antivenin is available from Commonwealth Serum Lab, Melbourne, Australia (see Reference 4 at end of this appendix for address and phone number). If antivenin is used, the directions regarding dosage and sensitivity testing on the accompanying package insert should be followed and the physician must be ready to treat for anaphylactic shock (severe allergic reaction). In brief, one or two punctures require 2,000 units (one ampule); three to four punctures, 4,000 units (two ampules); and five to six punctures, 6,000 units (three ampules). Antivenin must be delivered by slow IV injection and the victim closely monitored for anaphylactic shock.

5. Institute tetanus prophylaxis, analgesic therapy and antibiotics as described for other fish stings.

5C-3.3 **Stingrays.** The stingray is common in all tropical, subtropical, warm, and temperate regions. It usually favors sheltered water and will burrow into sand with only eyes and tail exposed. It has a bat-like shape and a long tail (Figure 5C-7). Approximately 1,800 stingray attacks are reported annually in the U.S. Most attacks occur when waders inadvertently step on a ray, causing it to lash out defensively with its tail. The spine is located near the base of the tail. Wounds are either of the laceration or puncture type and are extremely painful. The wound appears swollen and pale with a blue rim. Secondary wound infections are common. Systemic symptoms may be present and can include fainting, nausea, vomiting, sweating, respiratory difficulty, and cardiovascular collapse.

Figure 5C-7. Stingray.

5C-3.3.1 **Prevention.** In shallow waters which favor stingray habitation, shuffle feet on the bottom and probe with a stick to alert the rays and chase them away.

5C-3.3.2 **First Aid and Treatment.**

1. Give the same first aid as that given for venomous fish (paragraph 5C-3.1.2). No antivenin is available.

2. Institute hot water therapy as described under fish envenomation.

3. Clean and debride wound. Removal of the spine may additionally lacerate tissues due to retropointed barbs. Be sure to remove integumental sheath as it will continue to release toxin.

4. Observe patient carefully for the possible development of life-threatening complications. Symptoms can include cardiac dysrhythmias, hypotension, vomiting, diarrhea, sweating, muscle paralysis, respiratory depression, and cardiac arrest. Fatalities have been reported occasionally.

5. Institute tetanus prophylaxis, analgesic therapy, and broad-spectrum antibiotics as described for fish envenomation.

5C-3.4 **Coelenterates.** Hazardous types of coelenterates include: Portuguese man-of-war, sea wasp or box jellyfish, sea nettle, sea blubber, sea anemone, and rosy anemone (Figure 5C-8). Jellyfish vary widely in color (blue, green, pink, red, brown) or may be transparent. They appear to be balloon-like floats with tentacles dangling down into the water. The most common stinging injury is the jellyfish sting. Jellyfish can come into direct contact with a diver in virtually any oceanic region, worldwide. When this happens, the diver is exposed to literally thousands of minute stinging organs in the tentacles called nematocysts. Most jellyfish stings result only in painful local skin irritation.

The sea wasp or box jellyfish and Portuguese man-of-war are the most dangerous types. The sea wasp or box jellyfish (found in the Indo-Pacific) can induce death within 10 minutes by cardiovascular collapse, respiratory failure, and muscular paralysis. Deaths from Portuguese man-of-war stings have also been reported. Even though intoxication from ingesting poisonous sea anemones is rare, sea anemones must not be eaten.

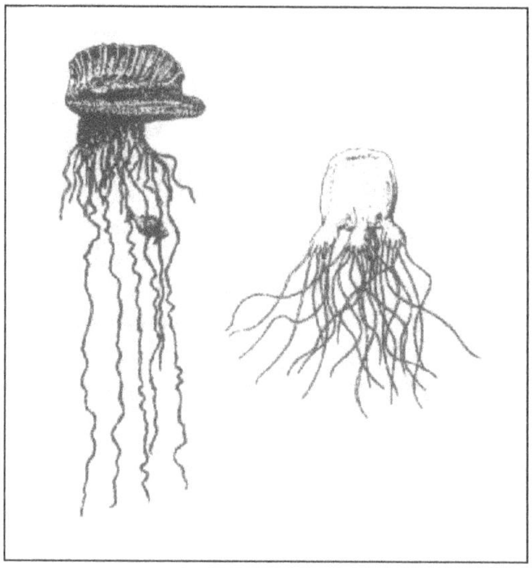

Figure 5C-8. Coelenterates. Hazardous coelenterates include the Portuguese Man-of-War (left) and the sea wasp (right).

5C-3.4.1 **Prevention.** Do not handle jellyfish. Beached or apparently dead specimens may still be able to sting. Even towels or clothing contaminated with the stinging nematocysts may cause stinging months later.

5C-3.4.2 **Avoidance of Tentacles.** In some species of jellyfish, tentacles may trail for great distances horizontally or vertically in the water and are not easily seen by the diver. Swimmers and divers should avoid close proximity to jellyfish to avoid contacting their tentacles, especially when near the surface.

5C-3.4.3 **Protection Against Jellyfish.** Wet suits, body shells, or protective clothing should be worn when diving in waters where jellyfish are abundant. Petroleum jelly applied to exposed skin (e.g., around the mouth) helps to prevent stinging, but caution should be used since petroleum jelly can deteriorate rubber products.

5C-3.4.4 **First Aid and Treatment.** Without rubbing, gently remove any remaining tentacles using a towel or clothing. For preventing any further discharge of the stinging nematocysts, use vinegar (dilute acetic acid) or a 3- to 10-percent solution of acetic acid. An aqueous solution of 20 percent aluminum sulfate and 11 percent surfactant (detergent) is moderately effective but vinegar works better. Do not use alcohol or preparations containing alcohol. Methylated spirits or methanol, 100 percent alcohol and alcohol plus seawater mixtures have all been demonstrated to cause a massive discharge of the nematocysts. In addition, these compounds may also worsen the skin inflammatory reaction. Picric acid, human urine, and fresh water also have been found to either be ineffective or even to discharge nematocysts and should not be used. Rubbing sand or applying papain-containing meat tenderizer is ineffective and may lead to further nematocysts discharge and should not be used. It has been suggested that isopropyl (rubbing) alcohol may be effective. It should only be tried if vinegar or dilute acetic acid is not available.

5C-3.4.5 **Symptomatic Treatment.** Symptomatic treatment can include topical steroid therapy, anesthetic ointment (xylocaine, 2 percent) antihistamine lotion, systemic antihistamines or analgesics. Benzocaine topical anesthetic preparations should not be used as they may cause sensitization and later skin reactions.

5C-3.4.6 **Anaphylaxis.** Anaphylaxis (severe allergic reaction) may result from jellyfish stings.

5C-3.4.7 **Antivenin.** Antivenin is available to neutralize the effects of the sea wasp or box jellyfish (Chironex fleckeri). The antivenin should be administered slowly through an IV, with an infusion technique if possible. IM injection should be administered only if the IV method is not feasible. One container (vial) of sea wasp antivenin should be used by the IV route and three containers if injected by the IM route. Each container of sea wasp antivenin is 20,000 units and is to be kept refrigerated, not frozen, at 36–50°F (2–10°C). Sensitivity reaction to the antivenin should be treated with a subcutaneous injection of epinephrine (0.3 cc of 1:1,000 dilution), corticosteroids, and antihistamines. Treat any hypotension (severely low blood pressure) with IV volume expanders and pressor medication as necessary. The antivenin may be obtained from the Commonwealth Serum Laboratories, Melbourne, Australia (see Reference 4 for address and phone number).

5C-3.5 **Coral.** Coral, a porous, rock-like formation, is found in tropical and subtropical waters. Coral is extremely sharp and the most delicate coral is often the most dangerous because of their razor-sharp edges. Coral cuts, while usually fairly superficial, take a long time to heal and can cause temporary disability. The smallest cut, if left untreated, can develop into a skin ulcer. Secondary infections often occur and may be recognized by the presence of a red and tender area surrounding the wound. All coral cuts should receive medical attention. Some varieties of coral can actually sting a diver since coral is a coelenterate like jellyfish. Some of the soft coral of the genus Palythoa have been found recently to contain the deadliest poison known to man. This poison is found within the body of the organism and not in the stinging nematocysts. The slime of this coral may cause a serious skin reaction (dermatitis) or even be fatal if exposed to an open wound. No antidote is known.

5C-3.5.1 **Prevention.** Extreme care should be used when working near coral. Often coral is located in a reef formation subjected to heavy surface water action, surface current, and bottom current. Surge also develops in reef areas. For this reason, it is easy for the unknowing diver to be swept or tumbled across coral with serious consequences. Be prepared.

5C-3.5.2 **Protection Against Coral.** Coral should not be handled with bare hands. Feet should be protected with booties, coral shoes or tennis shoes. Wet suits and protective clothing, especially gloves (neoprene or heavy work gloves), should be worn when near coral.

5C-3.5.3 **First Aid and Treatment.**

1. Control local bleeding.

2. Promptly clean with hydrogen peroxide or 10-percent povidone-iodine solution and debride the wound, removing all foreign particles.

3. Cover with a clean dressing.

4. Administer tetanus prophylaxis as appropriate.

5. Topical antibiotic ointment has been proven very effective in preventing secondary infection. Stinging coral wounds may require symptomatic management such as topical steroid therapy, systemic antihistamines, and analgesics. In severe cases, restrict the patient to bed rest with elevation of the extremity, wet-to-dry dressings, and systemic antibiotics. Systemic steroids may be needed to manage the inflammatory reaction resulting from a combination of trauma and dermatitis.

5C-3.6 **Octopuses.** The octopus inhabits tropical and temperate oceans. Species vary depending on region. It has a large sac surrounded by 8 to 10 tentacles (Figure 5C-9). The head sac is large with well-developed eyes and horny jaws on the mouth. Movement is made by jet action produced by expelling water from the mantle cavity through the siphon. The octopus will hide in caves, crevices and shells. It possesses a well-developed venom apparatus in its salivary glands and stings by biting. Most species of octopus found in the U.S. are harmless. The blue-ringed octopus common in Australian and Indo-Pacific waters may inflict fatal bites. The venom of the blue-ringed octopus is a neuromuscular blocker called tetrodotoxin and is also found in Puffer (Fugu) fish. Envenomation from the bite of a blue-ringed octopus may lead to muscular paralysis, vomiting, respiratory difficulty, visual disturbances, and cardiovascular collapse. Octopus bites consist of two small punctures. A burning or tingling sensation results and may soon spread. Swelling, redness, and inflammation are common. Bleeding may be severe and the clotting ability of the blood is often retarded by the action of an anticoagulant in the venom.

Figure 5C-9. Octopus.

5C-3.6.1 **Prevention.** Extreme care should be used when reaching into caves and crevices. Regardless of size, an octopus should be handled carefully with gloves. One should not spear an octopus, especially the large ones found off the coast of the Northwestern United States, because of the risk of being entangled by its tentacles. If killing an octopus becomes necessary, stabbing it between the eyes is recommended.

5C-3.6.2 **First Aid and Treatment.**

1. Control local bleeding.
2. Clean and debride the wound and cover with a clean dressing.
3. For suspected blue-ringed octopus bites, do not apply a loose constrictive band. Apply direct pressure with a pressure bandage and immobilize the extremity in a position that is lower than the heart using splints and elastic bandages.
4. Be prepared to administer mouth-to-mouth resuscitation and cardiopulmonary resuscitation if necessary.
5. Blue-ringed octopus venom is heat stable and acts as a neurotoxin and neuromuscular blocking agent. Venom is not affected by hot water therapy. No antivenin is available.
6. Medical therapy for blue-ringed octopus bites is directed toward management of paralytic, cardiovascular, and respiratory complications. Respiratory arrest is common and intubation with mechanical ventilation may be required. Duration of paralysis is between 4 and 12 hours. Reassure the patient.
7. Administer tetanus prophylaxis as appropriate.

5C-3.7 **Segmented Worms (Annelida) (Examples: Bloodworm, Bristleworm).** This invertebrate type varies according to region and is found in warm, tropical or temperate zones. It is usually found under rocks or coral and is especially common in the tropical Pacific, Bahamas, Florida Keys, and Gulf of Mexico. Annelida have long, segmented bodies with stinging bristle-like structures on each segment. Some species have jaws and will also inflict a very painful bite. Venom causes swelling and pain.

5C-3.7.1 **Prevention.** Wear lightweight, cotton gloves to protect against bloodworms, but wear rubber or heavy leather gloves for protection against bristleworms.

5C-3.7.2 **First Aid and Treatment.**

1. Remove bristles with a very sticky tape such as adhesive tape or duct tape. Topical application of vinegar will lessen pain.

2. Treatment is directed toward relief of symptoms and may include topical steroid therapy, systemic antihistamines, and analgesics.

3. Wound infection can occur but can be easily prevented by cleaning the skin using an antiseptic solution of 10 percent povidone-iodine and topical antibiotic ointment. Systemic antibiotics may be needed for established secondary infections that first need culturing, aerobically and anaerobically.

5C-3.8 **Sea Urchins.** There are various species of sea urchins with widespread distribution. Each species has a radial shape and long spines. Penetration of the sea urchin spine can cause intense local pain due to a venom in the spine or from another type of stinging organ called the globiferous pedicellariae. Numbness, generalized weakness, paresthesias, nausea, vomiting, and cardiac dysrhythmias have been reported.

5C-3.8.1 **Prevention.** Avoid contact with sea urchins. Even the short-spined sea urchin can inflict its venom via the pedicellariae stinging organs. Protective footwear and gloves are recommended. Spines can penetrate wet suits, booties, and tennis shoes.

5C-3.8.2 **First Aid and Treatment.**

1. Remove large spine fragments gently, being very careful not to break them into small fragments that remain in the wound.

2. Bathe the wound in vinegar or isopropyl alcohol. Soaking the injured extremity in hot water up to 122°F (50°C) may help. Caution should be used to prevent scalding the skin which can easily occur after a brief period in water above 122°F (50°C).

3. Clean and debride the wound. Topical antibiotic ointment should be used to prevent infection. Culture both aerobically and anaerobically before administering systemic antibiotics for established secondary infections.

4. Remove as much of the spine as possible. Some small fragments may be absorbed by the body. Surgical removal, preferably with a dissecting microscope, may be required when spines are near nerves and joints. X-rays may be required to locate these spines. Spines can form granulomas months later and may even migrate to other sites.

5. Allergic reaction and bronchospasm can be controlled with subcutaneous epinephrine (0.3 cc of 1:1,000 dilution) and by using systemic antihistamines. There are no specific antivenins available.

6. Administer tetanus prophylaxis as appropriate.

7. Get medical attention for deep wounds.

5C-3.9 **Cone Shells.** The cone shell is widely distributed in all regions and is usually found under rocks and coral or crawling along sand. The shell is most often symmetrical in a spiral coil, colorful, with a distinct head, one to two pairs of tentacles, two eyes, and a large flattened foot on the body (Figure 5C-10). A cone shell sting should be considered as severe as a poisonous snake bite. It has a highly developed venom apparatus: venom is contained in darts inside the proboscis which extrudes from the narrow end but is able to reach most of the shell. Cone shell stings are followed by a stinging or burning sensation at the site of the wound. Numbness and tingling begin at the site of the wound and may spread to the rest of the body; involvement of the mouth and lips is severe. Other symptoms may include muscular paralysis, difficulty with swallowing and speech, visual disturbances, and respiratory distress.

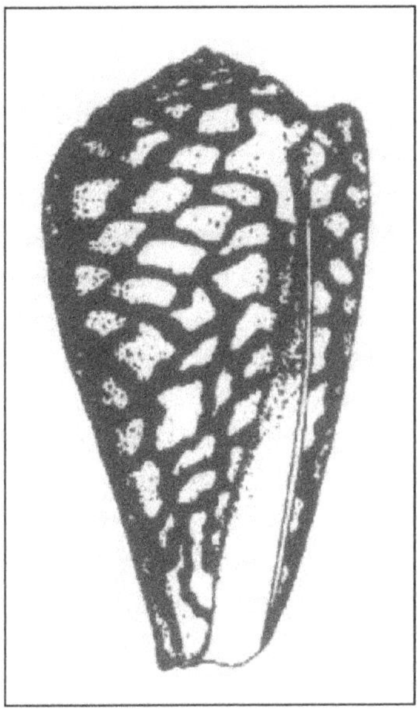

Figure 5C-10. Cone Shell.

5C-3.9.1 **Prevention.** Avoid handling cone shells. Venom can be injected through clothing and gloves.

5C-3.9.2 **First Aid and Treatment.**

1. Lay the patient down.

2. Do not apply a loose constricting band or ligature. Direct pressure with a pressure bandage and immobilization in a position lower than the level of the heart using splints and elastic bandages is recommended.

3. Some authorities recommend incision of the wound and removal of the venom by suction, although this is controversial. However, general agreement is that if an incision is to be made, the cuts should be small (one centimeter), linear and penetrate no deeper than the subcutaneous tissue. The incision and suction should only be performed if it is possible to do so within two minutes of the sting. Otherwise, the procedure may be ineffective. Incision and suction by inexperienced personnel has resulted in inadvertent disruption of nerves, tendons, and blood vessels.

4. Transport the patient to a medical facility while ensuring that the patient is breathing adequately. Be prepared to administer mouth-to-mouth resuscitation if necessary.

5. Cone shell venom results in paralysis or paresis of skeletal muscle, with or without myalgia. Symptoms develop within minutes of the sting and effects can last up to 24 hours.

6. No antivenin is available.

7. Respiratory distress may occur due to neuromuscular block. Patient should be admitted to a medical facility and monitored closely for respiratory or cardiovascular complications. Treat as symptoms develop.

8. Local anesthetic with no epinephrine may be injected into the site of the wound if pain is severe. Analgesics which produce respiratory depression should be used with caution.

9. Management of severe stings is supportive. Respiration may need to be supported with intubation and mechanical ventilation.

10. Administer tetanus prophylaxis as appropriate.

5C-3.10 **Sea Snakes.** The sea snake is an air-breathing reptile which has adapted to its aquatic environment by developing a paddle tail. Sea snakes inhabit the Indo-Pacific area and the Red Sea and have been seen 150 miles from land. The most dangerous areas in which to swim are river mouths, where sea snakes are more numerous and the water more turbid. The sea snake is a true snake, usually 3 to 4 feet in length, but it may reach 9 feet. It is generally banded (Figure 5C-11). The sea snake is curious and is often attracted by divers and usually is not aggressive except during its mating season.

Figure 5C-11. Sea Snake.

5C-3.10.1 **Sea Snake Bite Effects.** The sea snake injects a poison that has 2 to 10 times the toxicity of cobra venom. The bites usually appear as four puncture marks but may range from one to 20 punctures. Teeth may remain in the wound. The neurotoxin poison is a heat-stable nonenzymatic protein; hence, sea snake bites should not be immersed in hot water as with venomous fish stings. Due to its small jaws, bites often do not result in envenomation. Sea snake bites characteristically produce little pain and there is usually a latent period of 10 minutes to as long as several hours before the development of generalized symptoms: muscle aching and stiffness, thick tongue sensation, progressive paralysis, nausea, vomiting, difficulty with speech and swallowing, respiratory distress and failure, plus smoky-colored urine from myoglobinuria, which may go on to kidney failure.

5C-3.10.2 **Prevention.** Wet suits or protective clothing, especially gloves, may provide substantial protection against bites and should be worn when diving in waters where sea snakes are abundant. Also, shoes should be worn when walking where sea snakes are known to exist, including in the vicinity of fishing operations. Do not handle sea snakes. Bites often occur on the hands of fishermen attempting to remove snakes from nets.

5C-3.10.3 **First Aid and Treatment.**

1. Keep victim still.

2. Do not apply a loose constricting band or tourniquet. Apply direct pressure using a compression bandage and immobilize the extremity in the dependent position with splints and elastic bandages. This prevents spreading of the neurotoxin through the lymphatic circulation.

3. Incise and apply suction (see cone shell stings, paragraph 5C-3.9).

4. Transport all sea snake-bite victims to a medical facility as soon as possible, regardless of their current symptoms.

5. Watch to ensure that the patient is breathing adequately. Be prepared to administer mouth-to-mouth resuscitation or cardiopulmonary resuscitation if required.

6. The venom is a heat-stable protein which blocks neuromuscular transmission. Myonecrosis with resultant myoglobinuria and renal damage are often seen. Hypotension may develop.

7. Respiratory arrest may result from generalized muscular paralysis; intubation and mechanical ventilation may be required.

8. Renal function should be closely monitored and peritoneal or hemodialysis may be needed. Alkalinization of urine with sufficient IV fluids will promote myoglobin excretion. Monitor renal function and fluid balance anticipating acute renal failure.

9. Vital signs should be monitored closely. Cardiovascular support plus oxygen and IV fluids may be required.

10. Because of the possibility of delayed symptoms, all sea snake-bite victims should be observed for at least 12 hours.

11. If symptoms of envenomation occur within one hour, antivenin should be administered as soon as possible. In a seriously envenomated patient, antivenin therapy may be helpful even after a significant delay. Antivenin is available

from the Commonwealth Serum Lab in Melbourne, Australia (see Reference D of this appendix for address and phone number). If specific antivenin is not available, polyvalent land snake antivenin (with a tiger snake or krait Elapidae component) may be substituted. If antivenin is used, the directions regarding dosage and sensitivity testing on the accompanying package insert should be followed and the physician must be ready to treat for anaphylaxis (severe allergic reaction). Infusion by the IV method or closely monitored drip over a period of one hour is recommended.

12. Administer tetanus prophylaxis as appropriate.

5C-3.11 **Sponges.** Sponges are composed of minute multicellular animals with spicules of silica or calcium carbonate embedded in a fibrous skeleton. Exposure of skin to the chemical irritants on the surface of certain sponges or exposure to the minute sharp spicules can cause a painful skin condition called dermatitis.

5C-3.11.1 **Prevention.** Avoid contact with sponges and wear gloves when handling live sponges.

5C-3.11.2 **First Aid and Treatment.**

1. Adhesive or duct tape can effectively remove the sponge spicules.

2. Vinegar or 3- to 10-percent acetic acid should be applied with saturated compresses as sponges may be secondarily inhabited by stinging coelenterates.

3. Antihistamine lotion (diphenhydramine) and later a topical steroid (hydrocortisone), may be applied to reduce the early inflammatory reaction.

4. Antibiotic ointment is effective in reducing the chance of a secondary infection.

5C-4 POISONOUS MARINE ANIMALS

5C-4.1 **Ciguatera Fish Poisoning.** Ciguatera poisoning is fish poisoning caused by eating the flesh of a fish that has eaten a toxin-producing microorganism, the dinoflagellate, Gambierdiscus toxicus. The poisoning is common in reef fish between latitudes 35°N and 35°S around tropical islands or tropical and semitropical shorelines in Southern Florida, the Caribbean, the West Indies, and the Pacific and Indian Oceans. Fish and marine animals affected include barracuda, red snapper, grouper, sea bass, amberjack, parrot fish, and the moray eel. Incidence is unpredictable and dependent on environmental changes that affect the level of dinoflagellates. The toxin is heat-stable, tasteless, and odorless, and is not destroyed by cooking or gastric acid. Symptoms may begin immediately or within several hours of ingestion and may include nausea, vomiting, diarrhea, itching and muscle weakness, aches and spasms. Neurological symptoms may include pain, ataxia (stumbling gait), paresthesias (tingling), and circumoral parasthesias (numbness around the mouth). Sensory reversal of hot and cold sensation when touching or eating objects of extreme temperatures may occur. In severe cases, respiratory failure and cardiovascular collapse may occur. Pruritus (itching) is characteristically made worse by alcohol ingestion. Gastrointestinal symptoms usually disappear within 24 to 72 hours. Although complete recovery will occur in the majority of cases, neurological symptoms may persist for months or years. Signs and symptoms of ciguatera fish poisoning may be misdiagnosed as decompression sickness or contact dermatitis from unseen fire coral or jellyfish. Because of rapid modern travel and refrigeration, ciguatera poisoning may occur far from endemic areas with international travelers or unsuspecting restaurant patrons.

5C-4.1.1 **Prevention.** Never eat the liver, viscera, or roe (eggs) of tropical fish. Unusually large fish of a species should be suspected. When traveling, consult natives concerning fish poisoning from local fish, although such information may not always be reliable. A radioimmunoassay has been developed to test fish flesh for the presence of the toxin and soon may be generally available.

5C-4.1.2 **First Aid and Treatment.**

1. Treatment is largely supportive and symptomatic. If the time since suspected ingestion of the fish is brief and the victim is fully conscious, induce vomiting (syrup of Ipecac) and administer purgatives (cathartics, laxatives) to speed the elimination of undigested fish.

2. In addition to the symptoms described above, other complications which may require treatment include hypotension and cardiac dysrhythmias.

3. Antiemetics and antidiarrheal agents may be required if gastrointestinal symptoms are severe. Atropine may be needed to control bradycardia. IV fluids may be needed to control hypotension. Calcium gluconate, diazepam, and methocarbamol can be given for muscle spasm.

4. Amytriptyline has been used successfully to resolve neurological symptoms such as depression.

5. Cool showers may induce pruritus (itching).

5C-4.2 **Scombroid Fish Poisoning.** Unlike ciguatera fish poisoning (see paragraph 5C-4.1), where actual toxin is already concentrated in the flesh of the fish, scombroid fish poisoning occurs from different types of fish that have not been promptly cooled or prepared for immediate consumption. Typical fish causing scombroid poisoning include tuna, skipjack, mackerel, bonito, dolphin fish, mahi mahi (Pacific dolphin), and bluefish. Fish that cause scombroid poisoning are found in both tropical and temperate waters. A rapid bacterial production of histamine and saurine (a histamine-like compound) produce the symptoms of a histamine reaction: nausea, abdominal pain, vomiting, facial flushing, urticaria (hives), headache, pruritus (itching), bronchospasm, and a burning or itching sensation in the mouth. Symptoms may begin one hour after ingestion and last 8 to 12 hours. Death is rare.

5C-4.2.1 **Prevention.** Immediately clean the fish and preserve by rapid chilling. Do not eat any fish that has been left in the sun or in the heat longer than two hours.

5C-4.2.2 **First Aid and Treatment.** Oral antihistamine, (e.g., diphenhydramine, cimetidine), epinephrine (given subcutaneously), and steroids are to be given as needed.

5C-4.3 Puffer (Fugu) Fish Poisoning. An extremely potent neurotoxin called tetrodotoxin is found in the viscera, gonads, liver, and skin of a variety of fish, including the puffer fish, porcupine fish, and ocean sunfish. Puffer fish—also called blow fish, toad fish, and balloon fish, and called Fugu in Japanese—are found primarily in the tropics but also in temperate waters of the coastal U.S., Africa, South America, Asia, and the Mediterranean. Puffer fish is considered a delicacy in Japan, where it is thinly sliced and eaten as sashimi. Licensed chefs are trained to select those puffer fish least likely to be poisonous and also to avoid contact with the visceral organs known to concentrate the poison. The first sign of poisoning is usually tingling around the mouth, which spreads to the extremities and may lead to a bodywide numbness. Neurological findings may progress to stumbling gait (ataxia), generalized weakness, and paralysis. The victim, though paralyzed, remains conscious until death occurs by respiratory arrest.

5C-4.3.1 **Prevention.** Avoid eating puffer fish. Cooking the poisonous flesh will not destroy the toxin.

5C-4.3.2 **First Aid and Treatment.**

1. Provide supportive care with airway management and monitor breathing and circulation.

2. Monitor anal function.

3. Monitor and treat cardiac dysrhythmias.

5C-4.4 Paralytic Shellfish Poisoning (PSP) (Red Tide). Paralytic shellfish poisoning (PSP) is due to mollusks (bivalves) such as clams, oysters, and mussels ingesting dinoflagellates that produce a neurotoxin which then affects man. Proliferation of these dinoflagellates during the warmest months of the year produce a characteristic red tide. However, some dinoflagellate blooms are colorless, so that poisonous mollusks may be unknowingly consumed. Local public health authorities must monitor both seawater and shellfish samples to detect the toxin. Poisonous shellfish cannot be detected by appearance, smell, or discoloration of either a silver object or a garlic placed in the cooking water. Also, poisonous shellfish can be found in either low or high tidal zones. The toxic varieties of dinoflagellates are common in the following areas: Northwestern U.S. and Canada, Alaska, part of western South America, Northeastern U.S., the North Sea European countries, and in the Gulf Coast area of the U.S. One other type of dinoflagellate, though not toxic if ingested, may lead to eye and respiratory tract irritation from shoreline exposure to a dinoflagellate bloom that becomes aerosolized by wave action and wind.

5C-4.4.1 **Symptoms.** Symptoms of bodywide PSP include circumoral paresthesias (tingling around the mouth) which spreads to the extremities and may progress to muscle weakness, ataxia, salivation, intense thirst, and difficulty in swallowing. Gastrointestinal symptoms are not common. Death, although uncommon, can result from respiratory arrest. Symptoms begin 30 minutes after ingestion and may last for many weeks. Gastrointestinal illness occurring several hours after ingestion is most likely due to a bacterial contamination of the shellfish (see paragraph 5C-4.5). Allergic reactions such as urticaria (hives), pruritus (itching), dryness or scratching sensation in the throat, swollen tongue and bronchospasm may also be an individual hypersensitivity to a specific shellfish and not PSP.

5C-4.4.2 **Prevention.** Since this dinoflagellate is heat stable, cooking does not prevent poisoning. The broth or bouillon in which the shellfish is boiled is especially dangerous since the poison is water-soluble and will be found concentrated in the broth.

5C-4.4.3 **First Aid and Treatment.**

1. No antidote is known. If the victim is fully conscious, induce vomiting with 30 cc (two tablespoons) of syrup of Ipecac. Lavaging the stomach with alkaline fluids (solution of baking soda) may be helpful since the poison is acid-stable.

2. Provide supportive treatment with close observation and advanced life support if needed until the illness resolves. The poisoning is also related to the quantity of poisonous shellfish consumed and the concentration of the dinoflagellate contamination.

5C-4.5 **Bacterial and Viral Diseases from Shellfish.** Large outbreaks of typhoid fever and other diarrheal diseases caused by the genus Vibrio have been traced to consuming contaminated raw oysters and inadequately cooked crabs and shrimp. Diarrheal stool samples from patients suspected of having bacterial and viral diseases from shellfish should be placed on a special growth medium (thiosulfate-citrate-bile salts-sucrose agar) to specifically grow Vibrio species, with isolates being sent to reference laboratories for confirmation.

5C-4.5.1 **Prevention.** To avoid bacterial or viral disease (e.g., Hepatitus A or Norwalk viral gastroenteritis) associated with oysters, clams, and other shellfish, an individual should eat only thoroughly cooked shellfish. It has been proven that eating raw shellfish (mollusks) presents a definite risk of contracting disease.

5C-4.5.2 **First Aid and Treatment.**

1. Provide supportive care with attention to maintaining fluid intake by mouth or IV if necessary.

2. Consult medical personnel for treatment of the various Vibrio species that may be suspected.

5C-4.6 **Sea Cucumbers.** The sea cucumber is frequently eaten in some parts of the world where it is sold as Trepang or Beche-de-mer. It is boiled and then dried in the sun or smoked. Contact with the liquid ejected from the visceral cavity of some sea cucumber species may result in a severe skin reaction (dermatitis) or even blindness. Intoxication from sea cucumber ingestion is rare.

5C-4.6.1 **Prevention.** Local inhabitants can advise about the edibility of sea cucumbers in that region. However, this information may not be reliable. Avoid contact with visceral juices.

5C-4.6.2 **First Aid and Treatment.** Because no antidote is known, treatment is only symptomatic. Skin irritation may be treated like jellyfish stings (paragraph 5C-3.4.4).

5C-4.7 **Parasitic Infestation.** Parasitic infestations can be of two types: superficial and flesh. Superficial parasites burrow in the flesh of the fish and are easily seen and removed. These may include fish lice, anchor worms, and leeches. Flesh parasites can be either encysted or free in the muscle, entrails, and gills of the fish. These parasites may include roundworms, tapeworms, and flukes. If the fish is inadequately cooked, these parasites can be passed on to humans.

5C-4.7.1 **Prevention.** Avoid eating raw fish. Prepare all fish by thorough cooking or hot-smoking. When cleaning fish, look for mealy or encysted areas in the flesh; cut out and discard any cyst or suspicious areas. Remove all superficial parasites. Never eat the entrails or viscera of any fish.

5C-5 REFERENCES FOR ADDITIONAL INFORMATION

1. *Prevention and Treatment of Dangerous Marine Animal Injuries*, a publication by International Bio-toxicological Centre, World Life Research Institute, Colton, CA; November 1982; P.S. Auerbach and B.W. Halstead.

2. *Management of Wilderness and Environmental Emergencies*, Macmillan Publishing Co., New York, N. Y., 1983. Eds. P.S. Auerbach and E.C. Greehr.

3. *The Life of Sharks*, Columbia University Press, New York 1971. P. Budkur.

4. Commonwealth Serum Laboratories, 45 Poplar Road, Parkville, Melbourne, Victoria, Australia; Telephone Number: 011-61-3-389-1911, Telex AA-32789.

5. *Sharks*. Doubleday, Garden City, N.Y., 1970. J. Y. Cousteau.

6. *Fish and Shellfish Acquired Diseases*. American Family Physician. Vol 24: pp. 103-108, 1981. M. L. Dembert, K. Strosahl and R. L. Bumgarner.

7. *Consumption of Raw Shellfish - Is the Risk Now Unacceptable?* New England Journal of Medicine. Vol 314: pp.707-708, 1986. H. L. DuPont.

8. *Diving and Subaquatic Medicine*, Diving Medical Centre, Masman N.S.W., Australia; 1981, Second edition; C. Edmonds, C. Lowry and C. Pennefather.

9. *Poisonous and Venomous Marine Animals of the World*, Darwin Press Inc., Princeton, NJ; 1978; B. W. Halstead.

10. *Principles and Practice of Emergency Medicine*, W. B. Saunders Co., Philadelphia, PA; 1978, pp. 812-815; G. Schwartz, P. Sofar, J. Stone, P. Starey and D. Wagner.

11. *Dangerous Marine Creatures*, Reed Book Ptg., Ltd., 2 Aquatic Drive, French's Forest, NSW 20806 Australia. C. Edmonds.

12. *A Medical Guide to Hazardous Marine Life*, Second Edition, Mosby Yearbook, 1991, P.S. Auerbach.

PAGE LEFT BLANK INTENTIONALLY

Index

A

B

C

D

E

F

G

H

I

K

L

M

N

O

S

T

U

V

W

www.ingramcontent.com/pod-product-compliance
Lightning Source LLC
LaVergne TN
LVHW062132200926
848484LV00007B/1801
* 9 7 8 1 9 0 7 5 2 1 2 0 1 *